B&T

11/18/92

72.00

PRENTICE HALL SERIES IN PROCESS POLLUTION
AND CONTROL EQUIPMENT

by Nicholas P. Cheremisinoff and Paul N. Cheremisinoff

Pumps and Pumping Operations

Compressors and Fans

Filtration Equipment for Wastewater Treatment Recovery

Pumps and Pumping Operations

Nicholas P. Cheremisinoff
Paul N. Cheremisinoff

Prentice Hall

Englewood Cliffs, New Jersey 07632

Library of Congress Cataloging-in-Publication Data

Cheremisinoff, Nicholas P.
    Pumps and pumping operations / Nicholas P. Cheremisinoff, Paul N.
Cheremisinoff.
        p.   cm. -- (Process and pollution control equipment ; v. 1)
    Includes bibliographical references and index.
    ISBN 0-13-739319-9
    1. Pumping machinery.   I. Cheremisinoff, Paul N.   II. Title.
III. Series: Cheremisinoff, Nicholas P.   Process and pollution
control equipment ; v. 1.
    TJ900.C52   1993
    621.6'9--dc20   91-44761
                                                            CIP

Editorial/production supervision
and interior design: *Brendan M. Stewart*
Prepress buyer: *Mary McCartney*
Manufacturing buyer: *Susan Brunke*
Acquisitions editor: *Michael Hays*

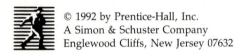

© 1992 by Prentice-Hall, Inc.
A Simon & Schuster Company
Englewood Cliffs, New Jersey 07632

The publisher offers discounts on this book when ordered in bulk quantities. For more information, write: Special Sales/Professional Marketing, Prentice Hall, Professional & Technical Reference Division, Englewood Cliffs, NJ 07632.

Printed in the United States of America
10 9 8 7 6 5 4 3 2 1

ISBN 0-13-739319-9

Prentice-Hall International (UK) Limited, *London*
Prentice-Hall of Australia Pty. Limited, *Sydney*
Prentice-Hall Canada Inc., *Toronto*
Prentice-Hall Hispanoamericana, S.A., *Mexico*
Prentice-Hall of India Private Limited, *New Delhi*
Prentice-Hall of Japan, Inc., *Tokyo*
Simon & Schuster Asia Pte. Ltd., *Singapore*
Editora Prentice-Hall do Brasil, Ltda., *Rio de Janeiro*

Contents

v

3   *Industrial Pumps and Their Applications*   58

4   Oscillating Displacement Pumps   179

Preface

This book is the first of a multivolume series, the intent of which is to provide practical guidance in the selection and use of industrial hardware for process operations and pollution control. The intent is not to recreate a textbook presentation of the theory and design principles of equipment, but rather to provide a detailed reference set for the practitioner which will assist in the proper choice, implementation, use, and troubleshooting of equipment in common industrial operations, and in pollution control and abatement.

This first volume covers machines used in the transport and handling of fluids, namely pumps and their applications. For the purposes of this volume, fluid handling operations considered are only those involving liquids, i.e., incompressible fluids. These products or materials may be single or multicomponent, as in the case of solid slurries which are often encountered in wastewater treatment operations. The volume is organized into six chapters.

Chapter 1 provides a brief overview of pumps, pumping operations, support equipment, and classes of pumps. Its purpose is to provide a quick introduction to the subject and orientation for information presented in subsequent sections. Chapter 2 covers pumps and pumping services in greater detail. A working knowledge description of major pump styles and classifications is given. Typical operating ranges and selection criteria are provided. Chapter 3 covers industrial pumps and their applications. This chapter provides extensive engineering and manufacturer's data on serviceability, operating ranges, and corrosion resistance performance for major pump styles. Detailed descriptions of typical operations are provided in this section to assist the user in selecting machinery for an intended service.

Chapter 4 provides a separate treatment of oscillating displacement pumps. Due to the complexity of design for this style machine, an extensive description of operating ranges is required. Chapter 5 addresses system analysis for pumping equipment. It provides description of the principles behind sizing pumps and in choosing equipment from manufacturers' literature. Chapter 6 covers mechanical seals. Since this is often an important component of pump maintenance, a working description of seal configurations is warranted.

Extensive appendices on engineering data and information important to pump selection and use are included in this volume. Appendix A contains extensive data tables on material selectivity for pumps, pumping components, and piping distribution systems. The data and information have been compiled directly from the service files of pump manufacturers. Appendix B provides extensive tables of engineering data pertinent to fluid (liquid) handling. Provided are physical properties data along with extensive correlations on viscosity conversions. Appendix C provides a general glossary of engineering terms which is pertinent not only to this present volume but to subsequent monographs in the series as well. Finally, Appendix D provides an extensive summary of definitions, calculation methods, and engineering guidelines in selecting and sizing pumps.

The information presented in this volume not only represents the practical experience and know-how of the authors, but that of a large number of pump manufacturers who contributed materials and effort to this work. We greatly acknowledge their contributions and advice on technical accuracy of the information provided. It must, however, be emphasized that engineering information and data provided represent typical conditions and/or serviceability. Each fluid-handling operation must be viewed as a unique problem and it is the responsibility of the practitioner to apply safe and proven procedures.

—Nicholas P. Cheremisinoff
—Paul N. Cheremisinoff

Flow Dynamics and Pumping Principles

INTRODUCTION

Routine practical problems in fluid flow often have the irritating property of being so novel that they are not treated directly by a handbook formula. The engineer must build solutions on fundamental ideas and interpret existing mathematical expressions with more than a superficial understanding of underlying principles. This chapter reviews basic relationships of fluid flow, particularly those forms which describe the flow of liquids and slurries.

The dynamics of fluids in motion can be described by a set of mathematical equations covering both the physical nature and mechanics of the phenomena. The equation as a whole represents some physical law. Individual terms have physical meaning. Satisfactory formulas will retain these meanings and will lose them only in algebraic manipulations carried out after the problem has been fully described in a mathematical language.

Innumerable forms and simplifications of the same basic relationship give the impression that one is dealing with a host of independent equations when actually there is only one. A questioning of the underlying physical law of each formula should reveal its importance and proper place in any quantitative description of a flow problem.

Every physical equation is more than one equation since it is accompanied by "ghost" equations representing physical quantities or dimensions. The ghost equations are the basis of dimensional analysis and the clue to physical meaning.

There are three basic laws of fluid flow:

1. conservation of mass
2. conservation of energy
3. Newton's second law of motion

The first two laws are essential for the solution of ordinary engineering problems. Each appears in many forms depending on the way symbols are defined, the relative importance of various terms, and the mathematical language in which the equations are written. Discussions are confined to only two forms, that for the familiar one-dimensional flow and that for the differential volume. The requirement of conservation of mass is of prime importance. For steady flow of a liquid or slurry,

$$\Delta W = O \qquad (1.1)$$

for one-dimensional flow, or

$$\frac{\partial(\rho u)}{\partial x} + \frac{\partial(\rho v)}{\partial y} + \frac{\partial(\rho \omega)}{\partial z} = O \qquad (1.2)$$

for a differential volume where ΔW is the mass rate of flow leaving the second position of a control volume drawn around a one-dimensional flow minus the mass rate of flow entering a first position. The units of terms are *lb. mass/sec.* in Equation 1.1 and *lb. mass. sec. ft.*^{3} in Equation 1.2., ρ is the constant density of the fluid which can be canceled from Equation 1.2, and u, v, ω, are the velocity components in the x-, y-, z-coordinate directions. The equations represent a universal law: At steady state what goes in must come out.

Rates of one-dimensional flow are given in three equivalent forms: the mass rate of flow W, the volume rate of flow Q, or the average volumetric velocity U. For cylindrical pipe flow these are related by the identities

$$W = \rho Q = \rho U \pi D^2 / 4 \qquad (1.3)$$

where D is the pipe diameter.

For one-dimensional flow we can consider the energy balance on a lb. mass of fluid passing through any system to be:

outgoing kinetic energy of lb. mass
+ outgoing potential energy of lb. mass
+ work done by lb. mass on fluid in front of it as it leaves the system
+ energy losses in the system per lb. mass
= incoming kinetic energy of lb. mass
+ incoming potential energy of lb. mass

+ work done on lb. mass by fluid behind it as it enters the system

+ mechanical work from an external source on lb. mass while in system

Using standard symbols and the same order of terms

$$\frac{U_2^2}{2g_c} + \frac{g_L}{g_c}Z_2 + \frac{p_2}{\rho} + \Sigma F = \frac{U_1^2}{g_c} + \frac{g_L}{g_c}Z_1 + \frac{p_1}{\rho} + W_0$$

or   (1.4)

$$(1/2g_c)\Delta U^2 + (g_L/g_c)\Delta Z + (1/\rho)\Delta p = W_0 - \Sigma F$$

where Δ refers to the quantity at the outgoing position minus the same quantity at the incoming position. Equation 1.4, the mechanical energy balance, is probably the most useful of all the forms of the energy balance. The units are ft. lb. force/lb. mass and mean exactly that.

Equation 1.4 has many uses beyond the analyses of a practical flow system. For a streamline in ideal flow, with ΣF and W_o zero, the remaining terms constitute Bernoulli's equation. For head meters Section 2 becomes the minimum flow cross section, and the equation is manipulated into a form such that Section 1 becomes of infinite diameter and zero velocity. Corrections for finite values at Section 1 and a measurable constriction are swallowed by various "coefficients" so defined that they are near unity.

The energy balance for the mass of a Newtonian fluid in a differential volume can be treated by application of the equation of motion, known as the Navier-Stokes equation. For the x-, y-, z-coordinate directions

$$-\frac{\partial p}{\partial x} + \rho\frac{g_{Lx}}{g_c} + \frac{\mu}{g_c}\left\{\frac{\partial^2 u}{\partial x^2} + \frac{\partial^2 u}{\partial y^2} + \frac{\partial^2 u}{\partial z^2}\right\} = \frac{\rho}{g_c}\left\{u\frac{\partial u}{\partial x} + v\frac{\partial u}{\partial y} + w\frac{\partial u}{\partial z}\right\}$$

$$-\frac{\partial p}{\partial y} + \rho\frac{g_{Ly}}{g_c} + \frac{\mu}{g_c}\left\{\frac{\partial^2 v}{\partial x^2} + \frac{\partial^2 v}{\partial y^2} + \frac{\partial^2 v}{\partial z^2}\right\} = \frac{\rho}{g_c}\left\{u\frac{\partial v}{\partial x} + v\frac{\partial v}{\partial y} + w\frac{\partial v}{\partial z}\right\} \quad (1.5)$$

$$-\frac{\partial p}{\partial z} + \rho\frac{g_{Lz}}{g_c} + \frac{\mu}{g_c}\left\{\frac{\partial^2 w}{\partial x^2} + \frac{\partial^2 w}{\partial y^2} + \frac{\partial^2 w}{\partial z^2}\right\} = \frac{\rho}{g_c}\left\{u\frac{\partial w}{\partial x} + v\frac{\partial w}{\partial y} + w\frac{\partial w}{\partial z}\right\}$$

Terms on the left-hand side of the equation represent first the pressure force acting, then the body forces (gravity), and finally the viscous shear forces. These forces result in a change in motion given in terms of the rate of change of momentum shown on the right-hand side of the equation. Equation 1.5 is an elegant statement that force equals mass times acceleration. Integration in one dimension with no viscous effects gives Bernoulli's equation. If the viscous term is taken to result in some unknown quantity F, the mechanical energy balance without the W_o term results. Shorn of the viscous term and with the proper boundary conditions, Equation 1.5 along with Equation 1.2 represent a mathematical statement of ideal flow. With appropriate assumptions, Equation 1.5 also describes motion in boundary layers, another real and practical phenomenon with which every engineer should be familiar.

Practical applications of the equation of motion to one-dimensional, steady flow can be achieved by drawing a control surface around a section of flow and treating the rate of flow of momentum, in terms of velocity U and flow cross section A, into and out of the section as a hypothetical reaction force,

$$F = AfU^2/g_c \qquad (1.6)$$

F is in the direction of flow for fluid entering the section and opposite the direction of flow for fluid leaving. If the reactions to which the changes of momentum give rise are substituted for the flow, a balanced system of forces and couples can be considered for the control surface fixed in space. Equation 1.6 is useful in estimating reaction forces on flow structures and analyzing momentum transfer devices such as ejectors.

Use of the mechanical energy balance requires considerable knowledge of the various Fs along the flow path. The important part of many flow problems consists of trying to pick the points in the system where the most significant energy losses occur and to establish their magnitude. Fortunately, many data exist for Fs under different flow situations. If some configuration has not been studied, an analogous system may have been studied and judicious reasoning can supply an educated guess of the right answer.

Most Fs are correlated as functions of a Reynolds number in dimensionless ratios with the kinetic energy of flow at a known or characteristic velocity.

$$N_f = \frac{F}{U_0^2/2g_c} = N_f(l\rho U_0/\mu) \qquad (1.7)$$

where U is a characteristic velocity and l a characteristic dimension. For friction loss per unit length of pipe, F/L, in a pipe of diameter D, the dimensionless correlation is

$$4f = \frac{F/L}{(U^2/2g_c)(1/D)} = 4f(D\rho U/\mu) \qquad (1.8)$$

f over a wide range of turbulent flows is nearly constant at an average 5.5×10^{-3}. Thus,

$$F \approx 2.2 \times 10^{-2}(L/D)(U^2/2g_c) \qquad (1.9)$$

amounting to an energy loss equal to the kinetic energy at a given point approximately every 45 pipe diameters of pipe length. Pressure energy must replenish the kinetic energy at a rate amounting to the total kinetic energy every 45 pipe diameters, the energy being converted successively and continuously from pressure energy to kinetic energy of mean flow, to turbulent kinetic energy, and finally to heat through viscous dissipation.

The fact that N_f tends to be constant for fully developed turbulent flow is no accident. If one is astute in the selection of U_0 selecting the highest value in a particular change of flow the constant N_f is unity or less. This is merely saying that flow brought to a maximum velocity of U_o can become completely or partially confused in its direction and all or some of its kinetic energy of mean motion disappears into turbulence.

The listing of turbulent Fs in terms of a constant number of $U_o^2/2g_c$ is called the method of *velocity heads,* the words actually referring to the dimensional variation where both terms are multiplied by g_c/g_L. The method becomes only a means of tabulating experimental data when dimensions are in heads and when the U_o is not characteristic of the flow where the F occurs but some other flow, usually that in the approach pipe.

Considering the real meaning of the concept and with experience one can guess Fs for unknown situations. For example, if U_o is taken as the largest apparent velocity in the smallest cross section, we have these approximations for abrupt changes in flow:

45° bend	$F = (\tfrac{1}{2})(U_o^2/2g_c)$
90° bend	$= 1(U_o^2/2g_c)$
large expansion	$= 1(U_o^2/2g_c)$
large contraction	$= \tfrac{1}{2}(U_o^2/2g_c)$

which can be added together for various complex flows.

At this point we should probably mention systems of equations which are as fundamental as the physical laws and very useful in organizing a complicated problem. Simply stated the rule is that we can solve only for as many unknown quantities as we can write independent equations for the physical situation. If we have more unknowns than equations, we have arbitrary choice over as many unknowns as the difference between the two. Some of the "independent" equations may be in the form of charted experimental data; some may be as vague as the fact that we have on hand a big supply of a certain size valve. Experience or simple-minded calculations can erase certain insignificant variables or set them at some logical value, but nature will still insist on satisfying the algebraic requirement before we can calculate unknowns or have a realizable flow system. Let us illustrate with an analysis of pipe networks. For each junction we can write from Equation 1.1,

$$\Sigma W = 0 \tag{1.10}$$

Around each circuit we can write

$$\Delta p = 0 \tag{1.11}$$

using the mechanical energy balance. We shall have as many independent Equations 1.10 as one less than the number of junctions and as many independent Equations 1.11 as one less than the number of circuits which can be traced with at least one new branch in it. The total will be the number of unknown quantities in the mathematical problem and in the actual flow system.

Static Pressure and Viscosity Relationships

Two special quantities characterizing fluid flow are pressure and viscosity. Static pressure p is defined as the compressive stress at a point in a fluid. A stress is force per unit area. The word *compression* indicates that the force is always acting inward at the point and normal to any surface on which it acts. In a moving fluid the static pressure is the pressure on a surface parallel to the direction of flow or the pressure on a surface moving with the flow, provided these surfaces do not disturb the flow.

The simplest definition of viscosity depends on the shear stress set up in a fluid by a velocity gradient across the shear area or, conversely, on the velocity gradient resulting from such a shear stress. Thus,

$$\mu = \tau g_c/(du/dy) \tag{1.12}$$

The force acts in the direction of flow on an area parallel to the flow but normal to the velocity gradient. For Newtonian fluids the viscosity is constant. For non-Newtonian fluids (and most slurries are non-Newtonian) the viscosity is an *apparent viscosity*, varying with the shear stress or velocity gradient and sometimes varying with time.

In laminar flows where τ is known as a function of position and μ is constant or a known variable, Equation 1.12 constitutes a differential equation with boundary conditions which can be solved for velocity. Turbulent flows are a subject for definitive experiment. However, since turbulent energy losses are relatively insensitive to viscosity, a satisfactory engineering calculation may be made of slurry flows in the manner of Newtonian fluids if ρ is taken to be the average density of the slurry mixture. Except in the cases where there is a question of the existence of turbulence or refined calculations of energy loss are desired, there is no need for the variously defined viscosity equivalents or for elaborate correlations of non-Newtonian turbulent flows.

PIPING SYSTEM CONCEPTS

Piping systems are the arteries and veins of any process-chemical plant. As such, they are the lifeblood of the unit, to pumps, vessels, and other equipment in which the desired processes are performed. Their importance in this capacity cannot be overestimated.

It is the intent of this section to survey the problems connected with the design of piping systems and possible solutions or the principles which may be used in arriving at solutions to these problems. In general, the most important features may be divided into classifications such as safety, operability and maintenance, and economy. These features are discussed in terms of plant layout, line sizing, application of valves, and details of piping layout.

Piping systems, as referred to here, include valves, fittings, and all other material making up the entire system.

The piping designer would like to see equipment located close together to keep piping as short as possible, thus resulting in an economical design. However, economy is only one of the factors which must be considered and is by no means the prime consideration. Although equipment layout greatly affects piping material cost, the following items cannot be overlooked:

1. The unit must be constructed easily and economically. While this period is a relatively short part of the life of the plant, it is a time of extremely high cost and one during which there is no income. Even minor delays or inefficiencies can materially increase the capital investment burden which subsequent production must regain.

2. The unit's design must provide adequate space for maintenance. Minor repairs and adjustments may be made while operation continues, but major maintenance must also be accommodated during complete unit shutdown periods.

3. Safety for personnel must be provided. The flammable nature of some products (e.g., petroleum) demands that utmost attention be paid to preventing fires. However, equipment must be arranged so that fire-fighting crews can quickly confine and put out a blaze without peril.

For the purpose of reviewing economical methods of line sizing, piping may be divided into two general classifications: (1) systems which do not contain pumping equipment, and (2) systems which involve fluid moving machines.

The first group must be sized on the basis of the available pressure drop for the circuit. In order to realize the desired products, process units are designed to operate at definite pressures and temperatures which are determined by the particular process and by the constituents in the feedstock. Therefore, it is the duty of the piping designer to insure these conditions. The size of each line should be chosen so that the total pressure loss for the circuit is absorbed in pipe friction and control valve pressure drop. Good control practice involves the use of a minimum of 25 percent of the total pressure loss for control, leaving approximately 75 percent for piping friction. For gravity draining systems, such as sewers, the slope of piping and the limiting velocity will determine the size of lines. This is important when piping is underground because large slopes mean greater depth of piping, which might increase the overall cost.

On the other hand, the systems involving pumps may be sized on the basis of any desired pressure loss, providing the pumping equipment can deliver the required quantity and overcome the total system friction. Here we have a choice of small, relatively inexpensive piping and high-head, large-driver pumping equipment; or large, more expensive piping and relatively low-head, small-driver pumping equipment. The decision as to which of these two choices will be made must necessarily depend on many factors. There is the problem of deciding the break-even point, that is, the point at which decreasing pipe size increases the cost of pumping equipment more than it decreases the initial piping costs. Beyond this point no benefit is obtained. Other considerations must be given to the difference between suction and discharge circuits. This is especially true for pumps which handle materials at or near their equilibrium temperatures, because of the tendency of these materials to vaporize at the pump suction, thus causing cavitation and subsequent damage to the pump. Generally much lower pressure losses are used in pump-suction circuits.

Higher values necessitate the elevation of equipment to suppress vaporization which may offset or override any savings in pipe by expenditure in foundations. Extensive investigation and much experience has shown that unit pressure losses of $\frac{1}{2}$ to 1 lb./sq. in. per 100 feet in suction circuits, and 3 to 5 lb./sq. in. per 100 feet in discharge circuits yield an economical balance between the initial cost of piping and foundations, and initial and operating cost of pumping equipment.

Piping systems which contain compressors must be handled with more care than those containing pumps. Compression equipment is generally large and costly. Because of the large volume of material handled, small increases in head requirements represent large increases in cost of equipment and operation. Therefore, much lower loses must be considered for these circuits. Often individual detailed economic studies are required for optimum results.

The greatest single factor affecting piping cost is the number of valves used. In general, these relatively high-priced items must be used with discretion to balance their cost against their necessity for operating and maintaining a unit. Valves which are required in the normal course of operation cannot be eliminated, regardless of their cost. However, valves for maintenance are often installed to provide ease and safety for maintenance personnel and to keep down time to a minimum. Much of the equipment in process units is indispensable to the operation of the unit. This means that the operation of the entire unit must be stopped to enable repairs or adjustments to be made to such pieces of equipment. In this case, isolating valves would not be required, but they may be installed to permit repairs to one piece of equipment while the rest is kept full of feedstock, ready to resume operation. The supply of standby equipment normally dictates the use of a sufficient number of valves to allow removal of any one of these items while the unit is in operation.

Many operations apply the practice of installing control valves without a block valve on each side and a bypass valve. When this is done, each control valve is furnished with a hand wheel to permit operation should the automatic control be out of order. The saving thus indicated is quite large, since the price of valves and the required manifold piping is far in excess of the price of the hand wheel. There is a certain amount of risk involved, in that this procedure will only allow manual operation of the control valve when the damage or inoperability is in the motor or spring of the valve. The stem and plug must necessarily be in good working order to permit operation with a hand wheel. This is one example of the balance between initial cost and maintenance convenience which results in a much lower initial investment and entails no extreme maintenance problem.

The mere installation of valves is not sufficient to insure easy operation or maintenance. Their location is extremely important. From the operational standpoint, it is best to locate as many valves as possible in a manner which would make them accessible from grade. However, this is not always practicable and many installations require ladders, platforms, or chain-wheel operation. The valve location chosen will determine the required method of operation. In some cases, piping may be diverted to permit selection of a location which would be accessible from grade or existing working levels. This additional piping cost must be balanced against any additional items which would otherwise be required to permit operation of such valves.

Despite the often-quoted phrase to the effect that the piping designer need only run lines from one piece of equipment to another, the problems are many and varied in accomplishing this task. Foremost among these problems is space. Lines must be run in the spaces provided without interfering with structural steel, platforms, and other piping. Sufficient room must be provided for valves, flanges, and other pieces of equipment which are larger in diameter than the piping itself. Insulation must be taken into consideration, since it extends beyond the diameter of the line. The latter is by no means a minor problem because certain low-temperature process lines require three to four inches of insulation, thus making a six-inch line perhaps twelve to fourteen inches in diameter. The piping designer must not only be aware of other piping but must also be concerned with all steel, concrete, vessel, pump, and many other equipment drawings for the entire job.

Among the other factors which must be considered during piping design are support and appearance. Support of piping is necessary to prevent overstressing the pipe or connecting equipment; appearance, to provide a good impression, is desirable though not mandatory. The finished plant is an advertisement for its owner. It is on permanent display to regular personnel, neighbors, and dignitaries; it will be visited and even lived in by a great many people. Much is accomplished toward these ends by the grouping of piping. Common supports may then be provided simultaneously, yielding a neat appearance. In the long run, common supports for piping systems save money and engineering, even though some of the piping must be diverted from the shortest possible course to make it fall in line with this system.

Another major consideration in piping layout is that of thermal stress caused by the difference in temperature which exists between erection and operation. During erection the piping is essentially at atmospheric temperature, whereas during operation the piping reaches the flowing medium temperature. The total stress involved is comprised of the stress due to internal pressure, that due to physical weight, and that due to thermal expansion. The stress due to internal pressure is set by the process; that due to weight is determined by the unsupported length of pipe and character of the flowing medium; and that due to thermal expansion is determined by the physical layout and operating temperature of the piping. The thermal stress is especially important when the reaction on equipment is being considered. For example, if a perfectly straight line were to be run between two towers, the thermal expansion of this line would tend to push the towers apart, thus giving a relatively large end reaction. This same line designed with one right-angle turn in it would produce much less end reaction. The lack of considering thermal expansion may cause much trouble. In a number of cases, it has been known to make pumps move from their foundations despite anchor bolts.

Many of the details of piping design have a large effect on overall piping costs. Changes in direction, for example, may be made by means of fittings, bends, or miters (welded changes in direction). Each of these types has its advantages and disadvantages. The miter is relatively inexpensive but causes concentrated stress and high-pressure drop due to its shape. The elbow requires two welds. The bend, depending on the pipe size involved, requires relatively large equipment to manufacture and is most difficult to transport from one place to another because of its shape. A choice will be made which must depend on the availability of material and equipment, ease of welding, and the distance and means of transportation. Another such detail of construction is the method of joining pipes at intersections. A tee fitting is relatively expensive and requires three welds, whereas a nozzle-type weld, *i.e.,* (welding one line directly into another) requires no fitting and only one weld. However, the latter type of intersection may require reinforcing and presents a difficult welding problem when the two lines are the same size. It is a fairly common practice to use tee fittings for lines of the same size and nozzle welding for lines of different size.

The third such detail which deserves mention is changes in size. Reducers are commonly used to accomplish this along the run of a line. However, reducing flanges at flanged equipment represents a more inexpensive arrangement. The decision to use any one of these methods for changes of direction, intersections, and changes in size must be tempered by consideration of the flowing medium and the availability of the piping materials as well as the cost of construction. For instance, abrasive flowing mediums would require bends rather than sharper changes in direction, regardless of the cost; alloy materials would normally dictate the use of as few fittings as possible because of availability and cost. There are many more difficulties involved and each must be considered in its proper place.

Not inconsequential is the location of instruments in piping circuits. The proper operation of both the instrument and the process depends on good selection of instrument location. For example, the accuracy of flow-measuring elements is greatly impaired by insufficient straight run of piping upstream and downstream of the element. These items cause considerable concern for the piping designer and in many cases it may be necessary to extend piping to provide a correct location for the installation of instruments.

In the final analysis, the piping designer must be aware of and must be held accountable for all these items. The piping designer's outlook must be as broad as the unit itself and the effects of this work extend to the length and width of process operation, the height of the tallest piece of equipment, and the depth of the largest foundation.

THE FUNCTION OF VALVES

The function of a valve is to start, stop, or regulate the flow of material in a closed or contained system. The nature of a valve is that of a mechanism with some moving parts that open or obstruct a passage to the flow of material. This material is usually either a liquid or a gas and sometimes a combination of the two. The material also may be a combination of liquid and solid matter, but it is possible also to have a gas and a solid mixture.

The choice of a valve for a given application depends on quite a large number of factors and conditions. The factors to be considered come under two general headings: (1) the conditions under which the valve is to operate, and (2) the consideration of commercially available equipment. This can be defined clearly, since the various kinds and makes of valves are limited in number.

It is always possible to consider a valve made to order by a valve manufacturer for a specific purpose. In general, valve manufacturers do not recommend a valve for a specific installation unless they have previously supplied equipment for the identical service because they do not know nor have control over the conditions under which the valve is to operate.

There are a number of factors which will affect valve operation. It is difficult to establish these conditions fully. The valve is usually located at an intermediate point in a process system. In the planning stages many conditions that should be considered are indefinite. Some of the major ones are:

1. The physical properties of the fluid.
2. The function of the valve in the process cycle from start-up through normal operation to shutdown; also, emergency operation.
3. The piping standards, safety standards, and other regulations or codes concerning the area in which the valve is to be installed.
4. The flow diagram of the process with reference to the pressures and temperatures at the valve for the variety of operating conditions.
5. The valve material, size, and construction.
6. Additional features required by the specific installation.

There are three general types of valves. First are block or shutoff valves for off-on service, either operated manually or positioned remotely by power from a separate compressed air, hydraulic fluid, or electric service supply. Then there are self-operating valves, such as check valves, spring-loaded valves, and float valves. There are also control valves for throttling service, positioned by an instrument controller using a separate compressed air or hydraulic service supply. There are other special valves that do not fit into this classification but to which, nevertheless, the application considerations also apply.

One of the basic differences in application of block valves from all others is that the pressure drop across the valve in the wide-open position should be a minimum. The nominal block valve size is usually the same as the line size. This is not usually true of the other types of valves, which must be chosen on the basis of required pressures and flow rates; that is, the nominal valve size must be determined for the service required. The valve must not be too small or too large, since performance and life are affected by improper size.

The function of the valve in the process cycle from start-up through normal operation and also emergency operation must be considered. Usually normal operation is the simplest, while start-up and shutdown of the process with variations of process conditions are more intricate. Emergency shutdown, due to failure of various components, presents an almost endless combination of conditions to consider. The combination of the three sets of conditions usually calls for a compromise.

The piping standards, safety standards, and other regulations or codes concerning the area in which the valve is to be installed usually are well known. It is important to consider the safety standards in conjunction with operation and maintenance.

With reference to the pressures and temperatures at the valve for the variety of operating conditions, the flow diagram of the process is primarily to be considered for throttling-control valves and self-operating valves. A serious problem arises when a hydrostatic test pressure is applied to a plant before start-up of the process. There are valves that do not conform to the A.S.A. pressure ratings, although the body end fittings may be constructed to A.S.A. standards. Examples of this are pressure-reducing valves with diaphragms that have the valve downstream pressure on the diaphragm, or valves with bellows stem seals. Another consideration is that of block valves that are operated at two different temperatures. For example, a steam gate valve that is seated when hot may be impossible to unseat when cold, due to thermal contraction of the metals.

So far, the discussion has been about the circumstances under which a valve is to operate. It is obvious that these factors must be known directly or by implication in order to choose the most suitable valve with respect to service performance, life, maintenance, original cost, and installation cost. A consideration of valve materials, size, and construction can be made after the physical properties of the fluid in the line have been reviewed.

It is desirable to know the basic physical properties of the fluid in the system. It is important to record the original conditions under which the valve was to function. Failures in operation must be reviewed in light of the original considerations. The difference in planned operating conditions and actual operating conditions will indicate what changes should be made to obtain satisfactory performance.

The physical properties of the fluid which impact on valve selection and design are as follows:

1. *Specific gravity at standard conditions and at operating conditions:* The specific gravity falls rapidly as the fluid approaches the boiling point. Flow at high velocity through a valve may cause the liquid to vaporize. This condition is known as flashing. The vapor condenses downstream, causing an effect similar to water hammer. Pressure let-down valves, specifically designed for this service, are available.

2. *Viscosity of the fluid under all operating conditions:* When the viscosity is such that the Reynolds number is less than about 5,000, the effect must be thoroughly considered. In general, viscous flows are either drawn off from

a pressurized vessel or moved by a pump. It can be assumed that if it is possible to pump the fluid, valves can be used. The major difficulties arise from the fluid packing in the recesses and around the valve stem or moving parts. It may be possible to move the stem, but so slowly that control is not possible.

3. *Nature of entrained solids:* The size of the solids must be considered relative to port openings. Solids flow similarly to an incompressible fluid until a flow restriction is reached, such as a partly open throttling-control valve; then the solids pack around the inlet port, restricting fluid flow. The valve then opens wide to pass the accumulation. The major problem presented by solids like fine granules, pipe scale, and welding beads is the damage caused to ground and lapped valve seats. The solids may be pressed into the seating surface, preventing valve closure. The damage is aggravated by wire drawing erosion by the leakage flow. It is prudent to install filters or strainers in lines to remove installation debris and pipe scale. These strainers may be omitted after lines have been cleaned. Plastic soft seats are to be recommended for low-pressure applications. Solids may build up in layers on moving parts. When these problems arise, the various types of commercially available equipment must be considered individually.

4. *pH of the liquid:* The effect of this factor is not very well known. In general, material failures are attributed to corrosion and erosion. It may be said that the pH effect is of a corrosive nature. As example of common interest today is in the handling of deionized water. The water does attack carbon steel vigorously, yet one would not say that deionized water is corrosive. Another example is given by the use of stainless steel valves on distillation columns, where carbon steel is used throughout, on the reflux lines from condensers.

5. *Steam flow—degrees of superheat or percentage saturation:* It is necessary to know the steam temperatures for the choice of valve stem packing and lubricants. Steam-reducing valves are usually used on saturated steam. Occasionally difficulties are encountered in using reducing valves on superheated steam. When it is superheated, the steam provides no lubrication for the internal mechanism of the valve.

If the saturation of the steam is less than 90 percent, slugs of condensate can wear valve trim rapidly. It is desirable to hardface control valve plugs and seats for longer wear.

The next three major considerations deal with the service which the valve is to perform. The principal reason for unsatisfactory valve performance is improper selection of a valve. It is important to consider what a given valve can do and what it cannot do. A valve may have nine desirable features for a given application and one undesirable feature, but this does not mean the valve is 90 percent satisfactory.

Having fully reviewed the nature of the fluid and the service required of the valve, it is then possible to begin consideration of the valve materials of construction, the nominal size, and the nature of the construction. The materials of construction of a valve are usually the same as the remainder of those used in the plant. When this is impossible, different materials must be used, and the economics of the situation becomes important. For example, in a glass or porcelain plant it is not possible to obtain reducing, relief, or control valves of the plant material. The economic problem arises when it is considered necessary to use a valve of more expensive material because the service requirements indicate a relatively short service life. The question to be answered is: "Is the original higher cost justified, or will it be less expensive to replace the less expensive valve periodically?" In light of the tax laws of today, the higher cost of skilled labor, and the cost of shutdown time of the process, the question cannot be answered definitely without restrictions.

The nominal size of the valve depends on the service. A block valve is usually the same as the line. The line size is usually based on a standard, setting the limit for fluid velocity and pressure drop per some unit of length. For example, the liquid velocity should be less than 15 ft./sec. in lines where fluid is being pumped from one unit to another, and should be less than 10 ft./sec. when the rate of flow is being controlled. Gas and steam lines may have a pressure drop of no more than 2 lb./sq. in. per hundred feet of pipe, or velocity of gas should be less than 10,000 ft./min. These values can be varied depending on particular installation conditions. The size of a control valve, reducing valve, or relief valve must be determined by the method given by the manufacturer of the particular piece of equipment.

The two primary service requirements of a block valve are that it have a minimum of pressure drop in the open position and tight shutoff in the closed position. The first requirement is a function of the valve design: The designer proportions the dimensions to obtain minimum restriction to flow. The design is based on previous experience and laboratory testing. The other requirement is not only a function of the design but also of the conditions under which a valve is used.

When shutoff is achieved by a metal-to-metal contact, many conditions arise to defeat successful closure. To begin with, precision manufacture and fitting are required. The effect of variations in temperature is a distortion in very small amounts. A similar distortion is caused by strains in the piping system. A third source of distortion is the fluid pressure. These distortions are considered by the valve designers as, for instance, in the use of split double wedges in gate valves.

The problem of entrained solids like pipe scale, welding beads, and scale deposits presents a major difficulty to the valve designer. The best that can be done is to hardface the seating surfaces or to use a material that is much harder than the entrained solids. This prevents scoring of the seating surfaces. It is interesting to note that when a piece of foreign matter is wedged between the seats, preventing closure, the operator usually applies all available force to screw the valve down in its seat. The result is a general warping and distortion of many of the valve parts. When the foreign matter is flushed away, the distortion remains in a permanent set—the valve cannot be shut off. It is reasonable to apply large amounts of force to operate a valve when the load is due to pressure unbalance and stuffing-box friction, but not to distort the valve in trying to make a tight shutoff.

There are many valves available today that overcome the metal-to-metal closure distortion problem. The principal method is the use of soft seats made of a plastic material. In recent years more plastic materials have become available for such use. The general result achieved has been that valves are available for some specific services that are more successful than general-purpose valves. The special-purpose valves are limited in their use generally by temperature and pressure or pressure-drop considerations. The trend in industry is to require general-purpose valves that are not restricted in use. A special-purpose valve usually fills only a very narrow band in the broad spectrum of valve application.

The other major difficulty in the use of block valves is the leakage at the stuffing box or packing gland. There is apparently no way in particular that leakage can be overcome. The designer can produce valves that are limited in use that have no stem seal. The development of plastic materials for use as valve packing is progressing. For many years the only materials used were graphite-impregnated asbestos and shredded lead. Teflon, for example, is extensively used in many valves.

It seems that the most difficult thing to do today is to achieve simple maintenance of equipment. Not only does the user expect to do very little, it is nearly impossible to obtain maintenance personnel. Proper maintenance requires judgment on the part of the individual doing the work and pride of workmanship. Without these two ingredients, maintenance becomes expensive and the results are mediocre.

The considerations to be made in selecting self-operated or instrument-operated control valves include all that has been highlighted concerning block valves. In addition, there are many facts to be considered relative to the control of the process under consideration.

In automatic control, output signals from the controller are transmitted to the final control elements which, for many process systems, consist of valves and their driving motors. Proper selection and specification of valves and valve actuators are an integral part of a process system design. Actuators are categorized on the basis of their power source with the principal types being pneumatically, hydraulically, and electrically operated.

Pneumatically-operated valve motors or actuators represent the most commonly used type employed in the process industries. Figure 1.1 illustrates a pneumatic valve motor whose operating principle is based on a diaphragm motor. The diaphragm is spring loaded in opposition to the driving air pressure such that the valve-stem position is proportional to the air pressure. The diaphragm is often of the limp type, fabricated of rubber fabric or some resistant material, and is normally supported by a back-up plate. Diaphragm motors typically have a maximum allowable valve-stem stroke (or travel) of 2 to 3 in.

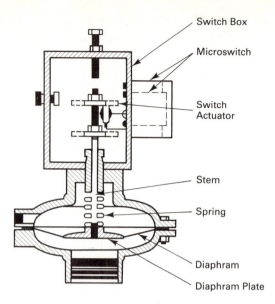

Switch Box

Microswitch

Switch
Actuator

Stem

Spring

Diaphram

Diaphram Plate

Figure 1.1 A pneumatic actuator. Static pressure from the fluid height changes the pressure supplied to the diaphragm. The pressure is opposed by spring tension. As pressure increases it causes the diaphragm to move. Diaphragm movement raises the activator stem. When the lower switch actuator contracts a microswitch lever, it closes one set of contacts. These remain closed until downward movement due to decreasing level causes the upper switch actuator to move the microswitch level in the opposite position.

Valve actuators with longer strokes are normally of the double-acting piston type; otherwise a rotary pneumatic motor, driving a rack or worm gear, can be used. Such actuators are capable of driving very large valves; for rotary motors, valve strokes can be as much as 5 ft.

It should be noted that application of tank or system air pressure to the valve motor diaphragm may not always result in the desired valve-stem position. For example, the valve stem could become frozen in its seal or stuffing box and consequently may not move at all or move only sluggishly. This problem is overcome by the incorporation of a servo mechanism which supplies sufficient pressure via a pilot amplifier to the diaphragm to ensure that the valve stem achieves the proper position. An actuator of this type is referred to as a *valve positioner*. These devices normally use a higher source of air pressure than the normal instrument air (typically 100 psig).

The forcing signal transmitted to hydraulic actuators can be either a mechanical displacement or a pneumatic or electric signal which is transformed to a mechanical displacement. The displacement in piston-type hydraulic actuators positions the cylinder valves and admits oil to either side of the piston. This action positions the piston shaft, which then drives the valve stem.

Another hydraulic-type actuator is the variable-delivery pump, in which the input displacement fixes the stroke of a number of piston pumps connected in parallel arrangement. The output-oil delivery of the pumps is proportional to the forcing displacement and is fed to a positive-displacement motor to produce an output displacement. This output displacement is also proportional to the original forcing displacement. In some cases the variable output of the piston pumps is fed to the positive-displacement motor in such a fashion that the rate of travel of the output displacement is proportional to the input displacement.

This results in the actuator generating an integrating action. Hydraulic actuators are normally employed in situations where high speeds and large forces are needed.

Electric actuators can be described as reversing motors operated by a circuit appropriate to the forcing signal. In general, these systems are slower than pneumatic and hydraulic actuators and entirely electric control systems arc often expensive.

Control valves provide the normal mechanism for adjusting inputs of process control systems. Proper valve selection is essential for a well-designed level control scheme. For systems requiring relatively short time lags and having large capacities, fairly simple control systems are adequate, in that they normally permit the use of high-sensitivity controllers with quick-opening valves. For this type of service, flow characteristics become secondary considerations and valve characteristics, such as size and material, become primary.

As time lag increases and capacity decreases, required control schemes become more complex. In general, the larger the time lag, the smaller and more precise must be the flow rate variations of the controlling medium for given changes in process conditions. Careful consideration must therefore be given to specifying control valves.

Several common control valve types are illustrated in Figures 1.2 through 1.6. Table 1.1 provides brief descriptions of various control valves.

The following definitions are used to describe the operating characteristics of control valves:

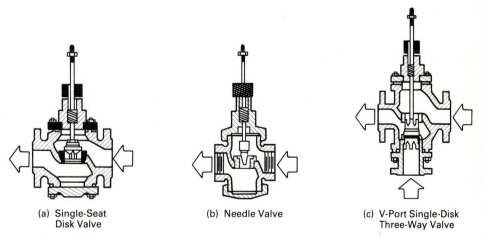

| (a) Single-Seat Disk Valve | (b) Needle Valve | (c) V-Port Single-Disk Three-Way Valve |

Figure 1.2 Various valve bodies. (A) Single-seat disk valve is used mainly on simple, high-sensitivity, small time-lag applications, and for open-and-shut service or stable load conditions and constant line pressures. (B) Needle valve is used for control of very small flows; it is employed generally where requirements are for less than $\frac{1}{2}$ in. valve size. (C) V-port single-disk three-way valve is suitable for mixing service.

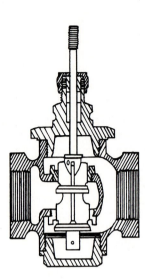

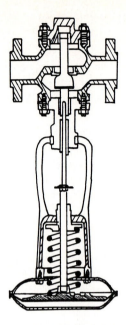

Figure 1.3 Double-seated quick-opening inner valve with guided disks. These are used where close throttling flow is needed. Often used on low-sensitivity applications.

Figure 1.4 Single-seated globe-type body with solid-plug inner valve. The throttle plug controls flow rate by means of variable port openings as determined by the curvature of the plug proper.

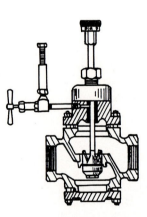

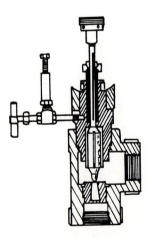

Figure 1.5 A single-seat V-port inner valve.

Figure 1.6 An angle valve.

TABLE 1.1   DESCRIPTIONS OF COMMON CONTROL VALVE TYPES

Valve type	Description	Applications	Limitations/disadvantages
Single-seated	Quick-opening disk type.	Used on simple, high-sensitivity, small time-lag applications/open-and-shut service.	Impractical for throttling control if line pressures fluctuate widely, unless controller sensitivity is very high. Not recommended for throttling control above 2-in. size.
Double-seated	Upstream pressure enters body between the seats, tending to create equal upward and downward forces. Balanced valve eliminates line pressure fluctuations.	Used for close throttling control, low-sensitivity applications.	Seating surfaces subject to wear. Cannot supply tight shutoff.
Throttle-plug	Double-seated, quick-opening inner valve with guided disks.	For large load changes. Handles large pressure drops well.	Limited to larger sizes.
V-port	See Figure 1.5.	Used where widely varying flow rates are encountered and where full throttling control over pressure fluctuations.	
Throttle-plug	Double-seated, quick-opening inner valve with guided disks.	For large load changes. Handles large pressure drops well.	Limited to larger sizes.
V-port	See Figure 1.5.	Used where widely varying flow rates are encountered and where full throttling control over entire flow range is needed. Has greatest controllable range.	
Single-seat throttling	In sizes smaller than 1 in., generally preferred to the double-seat type. Provides tight shutoff of controlling medium.	Same as single-seated.	Impractical for throttling control for lines with large pressure fluctuations.
Needle type	See Figure 1.2. Practically any required flow characteristic is possible.	For control of very small flows.	Limited to small sizes.

(*continued*)

TABLE 1.1   *Continued*

Valve type	Description	Applications	Limitations/disadvantages
Three-way type	See Figure 1.2. Provides highly stable flows.	Service requirements include: to mix fluids (e.g., mixing hot and cold water) to feed two fluids to a vessel or pipeline; used as a throttling controller to divert fluid from one vessel or pipeline to two load demands.	
Angle valves	Basically single-port valves.	Used to facilitate piping or where a self-draining piping system is needed. Used for handling fluids containing solids, slurries, and flashing fluids.	Subject to shamming.
Butterfly valves	Can be fitted with either manual or automatic valve positions. Main body may be diaphragm, cylinder, float, electric motor, or solenoid operated.	For control of low-pressure low-velocity fluids.	

Rangeability

Rangeability is the ratio of the maximum controllable flow through the valve to the minimum controllable flow. In practice, control valves do not always provide complete fluid medium shutoff. Valve seats become damaged by erosion or excessive use, or stick. Flow in the normally closed position ranges typically from 2 percent to 4 percent of the maximum flow (depending on valve type and size). These percentages are equivalent to rangeabilities of 50 to 25. Sliding-stem control valves have typical rangeabilities between 20 and 70.

$$Q = CA\sqrt{\frac{\Delta P}{\rho}} \qquad (1.13)$$

Flow control is achieved by moving the valve stem to vary the area of flow. The gain of a valve (i.e., the change of flow for a given change in stem position) depends on the change in area with stem position and also on the change in pressure drop with flow.

Control valve manufacturers have standardized on flow coefficient (C_v in Equation 1.13) to provide the flow capacities of valves. The flow coefficient is generally defined as the flow of water in gallons per minute for a pressure drop of 1 lb/in. across the fully open valve. Flows for various valve-stem positions are obtained from values of C_v and the valve characteristics. These can be plotted as percentage of maximum flow versus percentage lift or can be presented as flow-chart curves as illustrated in Figure 1.7 for one manufacturer's two-way motor valve unit. An illustrative example on the use of this chart is also given in the figure.

The inherent valve characteristic is the relationship between the stem position and the flow at constant pressure drop across a valve. Linear and equal percentage are the two common characteristics encountered. The characteristic of a parabolic plug valve is referred to as an equal-percentage characteristic. This is the kind of behavior that is desired in all characterized valves. The term *equal percentage* refers to the fact that the flow increments resulting from a given change in lift are the same percentage as the actual flow, regardless of whether the valve is nearly open or nearly closed.

Full flow is achieved with a relatively short stem movement and, consequently, the terms *low-lift valve* and *quick-opening valve* are applied.

The manufacturer should always be consulted for values of C_v and for calculation procedures, particularly when sizing valves for service in viscous liquids, flashing liquids, or gases at high pressure drops. In establishing the proper valve size, all flow conditions must be known and C_v must be determined. Based on a computed C_v value, the size for the type of valve under consideration can be selected on the manufacturer's values of C_v rating versus valve size. For liquid service, apply Equation 1.13 to estimate C_v and valve capacity.

INTRODUCTION TO PUMPS AND PUMPING

A pump is defined as a device which raises or transfers fluids. For each application, a pump is selected not only to raise or transfer liquids, but to satisfy some other requirement as well. For example, it is desired to pump a constant quantity of liquid, knowing that finite pressure differences will occur as the result of process variables. A centrifugal pump's characteristic is such that a small change in pressure differential will cause a relatively large change in flow (Figure 1.8). A positive-displacement pump, on the other hand, delivers an almost constant quantity regardless of pressure fluctuations. In order to satisfy the original requirement, therefore, a positive-displacement type of pump is clearly indicated.

In a water works, on the other hand, it is necessary to maintain a constant pressure on the mains despite fluctuations in demand. A centrifugal pump's characteristic most closely satisfies this criterion. Both pumps raise or transfer the fluid. The selection is made on the basis of additional requirements. We

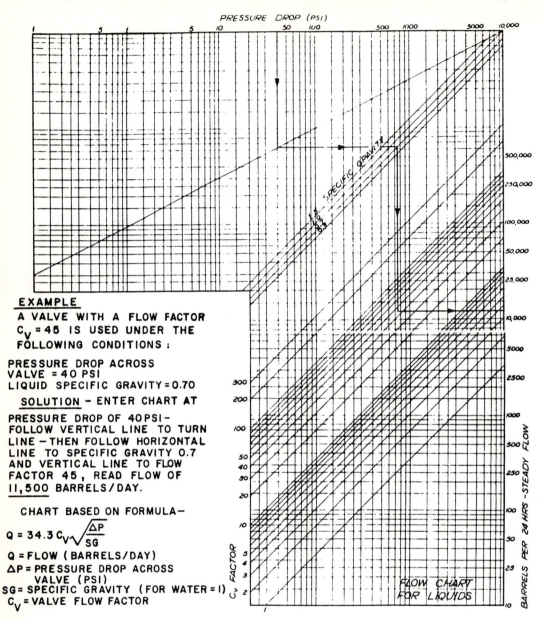

Figure 1.7   Liquid flow chart for a two-way motor valve system.

might divide all the available pump types into two main categories: positive-displacement pumps, and all other types. These classifications can be further divided as in Figure 1.9.

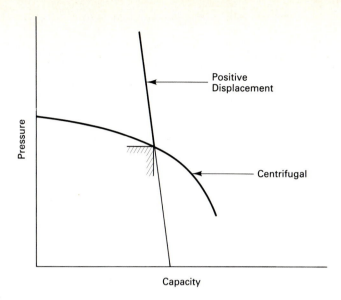

Figure 1.8 Pressure capacity characteristic at constant speed.

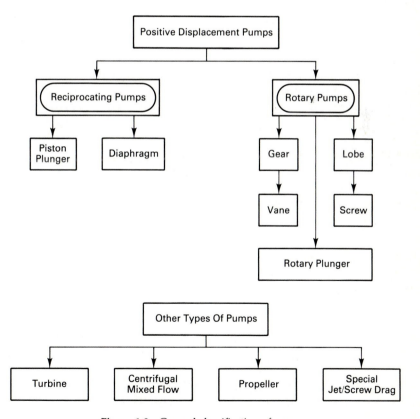

Figure 1.9 General classification of pumps.

23

In Figure 1.10 we have tried to separate one type from the other by listing advantages and disadvantages of each. Depending on the requirements of the system. It is always necessary to weigh these factors against each other. A good starting point is this:

Always use a centrifugal pump unless:

1. The viscosity is greater than 1,000 cp. (see Figure 1.11).
2. High heads and low capacities are required.
3. A constant capacity, less than 100 gpm, of a clean lubricant (such as lubricating oil) at a temperature less than 100°C is to be handled.
4. The volume percentage of undissolved gas is greater than 5 percent.

Some of the factors that influence the selection of a pump follow. Many of these are suggested in the listing of the advantages and disadvantages of the several types (see Figure 1.10).

1. Liquid to be handled
    a. Quantity and pressure
    b. Viscosity of the liquid at the pumping temperature
    c. Relation of the suction pressure at the pumping temperature to the vapor pressure
    d. Is the liquid corrosive?
    e. Is the liquid abrasive (does it contain solids)?
    f. Are undissolved gases present?
2. Requirements
    a. Must the pump meter as well as raise or transfer?
    b. Must the pump supply a variable quantity in response to process needs?
    c. Will there always be liquid for the pump to handle?
    d. Is filterability a problem?
3. Sealing
    a. Solvents (leaching of lubricant)
    b. Solids (abrasion of packing or seals)
    c. Toxicity
    d. Flammable vapors
    e. Incompatibility with ambient conditions
    f. Loss of expensive fluids
4. Safety
    a. Materials of construction to avoid fire hazard
    b. Pressure protection (particularly solids)
5. Size and position
    a. How much space is available?
    b. Where must the pump be located?

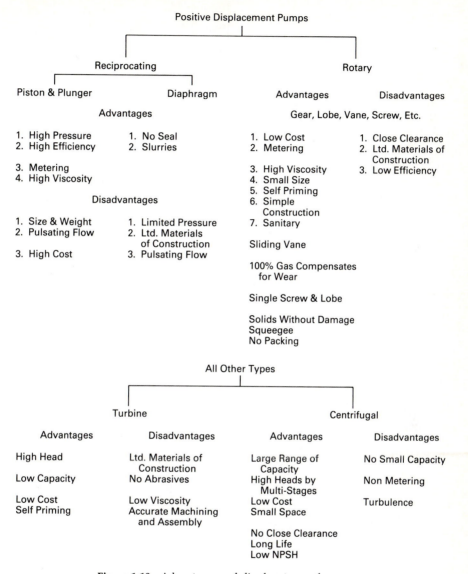

Figure 1.10 Advantages and disadvantages of pumps.

6. Scale-up problems

7. Standardization (with other types and makes already installed)

In any discussion of the problems encountered with pumps in the chemical industry today, first on the list is that of high maintenance cost. Many factors are involved in these costs, such as design, selection, maintenance and operat-

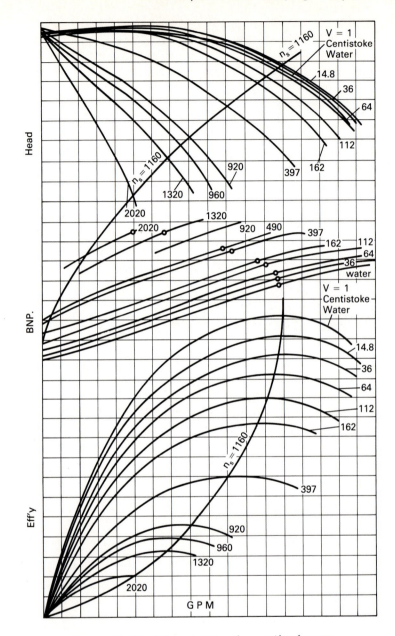

Figure 1.11 Head characteristics of a centrifugal pump.

ing procedures, installation, materials of construction, piping, and so on. In addition to playing their part in maintenance costs, these same factors must be considered from other viewpoints. For example, a sealing arrangement designed to leak 30 to 60 drops per minute for lubrication and cooling is adequate when the pump is handling water, but entirely inadequate when it is handling a flammable, toxic, or expensive liquid. Many maintenance dollars are literally poured down the drain in an effort to obtain performance not designed into the equipment. Maintenance, however, is a subject in itself. There are other problems which often stand in the way of uninterrupted production.

An often-discussed subject is *net positive suction head* (NPSH). A centrifugal pump has a required NPSH which varies with capacity (Figure 1.12). In every system in which a pump is installed, there is an available NPSH. In both instances the NPSH is referred to a plane at the center line of the impeller for a horizontal pump or to the impeller inlet for a vertical pump. The amount of NPSH required by the pump at the suction flange is that amount necessary to prevent the formation of vapor at any point in the pump suction including the inlet vane tips (Figure 1.13). In all cases, the available NPSH must exceed the required NPSH at the desired capacity.

When designing for a new installation, we can generally keep the available NPSH greater than that required by an economical pump. But in a given installation, with an available NPSH, what can we do to get a pump to operate at the desired head and capacity under these conditions? The literature will disclose that for any given head and capacity rpm varies as NPSH. This can be applied as follows. If for a given head and capacity, a 1,750-rpm pump requires 8-ft. NPSH, and the existing installation provides only 6-ft. NPSH, then a similar pump must be selected which will give the same hydraulic conditions at $(\frac{6}{8})^{3/4} \times$ 1,750 rpm, or 1,450 rpm maximum.

The selection of a special pump should be made only after a thorough investigation of the possibilities of increasing the available NPSH, such as raising the liquid level, decreasing suction-line friction losses, lowering the liquid temperature, and so on.

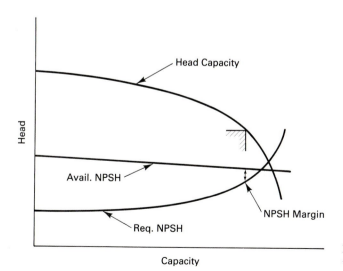

Figure 1.12 Shows important relations between head and capacity.

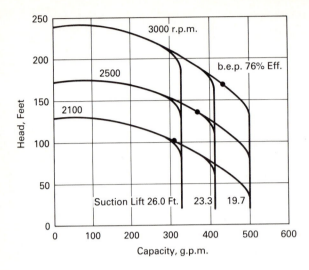

Figure 1.13 Shows effect of speed and suction lift on cavitation; 3-in-single-suction pump.

Another problem often encountered is that of pumping slurries without crystal degradation and consequent loss of filterability. Here the type of pump is involved. Centrifugal pumps have not demonstrated desirable characteristics from this standpoint. Furthermore, specific speed does not seem to be related in any way to crystal degradation; that is, for the same head and capacity, the speed of the pump seems to have little or no effect. Some improvement in filterability is gained by the use of a piston pump (Figure 1.14).

In the handling of slurries, one of the greatest problems is that of shaft seals. The greatest success in overcoming this problem has been achieved with a single outside mechanical seal with a throttle bushing in the throat of the stuffing box (Figure 1.15). A clear flushing liquid is then injected directly behind the sealing faces. This sealing liquid flows through the annulus between the shaft and the throttle bushing, preventing the solids from coming in contact with the sealing faces.

As a solution to many sealing problems, the *canned-motor* design is useful. No seal is required with this pump; the bearings supporting the motor rotor and

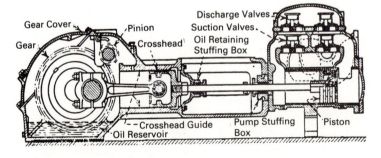

Figure 1.14 A piston-type power pump.

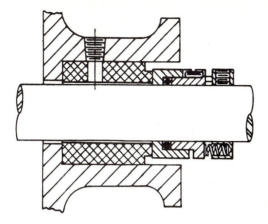

Figure 1.15 A stuffing box.

the pump impeller operate in the process fluid. This fact poses a problem in itself. Materials of construction for bearings and journals must be selected not only for their bearing properties, but for their compatibility with the process liquid. Bearing materials which have been successfully applied include graphite, the carbides, and some thin-walled plastics. Other materials are stainless steel, both 300 series and 400 series, carbides, and the harder metals in general.

Materials of construction must be carefully selected in the chemical industry. For example, materials which exhibit satisfactory corrosion rates under static conditions of many corrosive services are much less satisfactory where high fluid velocities prevent the formation of the necessary protective films.

One of the material problems to which attention is turned in the chemical industry today involves safety. In those areas where flammable liquids are handled, the use of brittle or fragile materials is discouraged. In the event of fire, the fracture of brittle materials housing flammable liquids may turn a minor fire into a major conflagration. Cast metals low in ductility are particularly susceptible to thermal shock and resulting fracture under conditions, as previously described. Nodular iron is sometimes considered as an alternate material for cast iron for its increased ductility. For standard equipment such as chemical pumps, nodular iron may replace cast iron in many services.

This chapter has provided a brief overview of some of the considerations important in selecting pumps and pumping service. The chapters that follow provide detailed information on pump types, operating characteristics, limitations and service factors, and selection bases.

2

Pumps and Pumping Services —————

This chapter presents general information and procedures for specifying pumps and process plant pumping services. Details on pump application technology are presented in subsequent discussions in this chapter. This chapter only provides supplemental information to acquaint the plant engineer with pumping technology and nomenclature. Emphasis is given to centrifugal pumps, as these are most widely encountered in the chemical process industries. For more detailed discussions on pump types, designs, and performance, the reader is referred to the suggested reading section. Note also that data presented here represent typical conditions and should therefore only be applied for approximations. The manufacturer's actual data should always be used when specifying a specific pump style for an intended service.

PUMP CLASSIFICATIONS

The pump types most often employed in process plant applications fall into the following classes: centrifugal, axial, regenerative turbine type, reciprocating, metering, and rotary. Two categories of pumps under which these classes are grouped are dynamic pumps and positive-displacement pumps.

Dynamic pumps include the centrifugal and axial classes. These pumps operate by developing a high liquid velocity and converting the velocity to pressure in a diffusing flow passage. In general, they tend to have lower efficiency than positive-displacement pumps. However, they do operate at relatively high speeds to permit high flow rate in relation to the physical size of the pump. Also, dynamic pumps usually have significantly lower maintenance requirements than positive-displacement pumps.

Positive-displacement pumps operate by forcing a fixed volume of fluid from the inlet pressure section of the pump into the discharge zone of the pump. With reciprocating pumps this is performed intermittently. In the case of rotary screw and gear pumps it is done continuously. This category of pumps operates at lower rotative speeds than dynamic pumps. Positive-displacement pumps also tend to be physically larger than equal-capacity dynamic pumps.

Table 2.1 gives a breakdown of pump types, construction styles, and descriptive information. Table 2.2 reports various performance parameters. Note that data presented in Table 2.2 are typical and descriptive in nature. Therefore, these data should be applied cautiously to specific problems.

GENERAL FEATURES OF CENTRIFUGAL PUMPS

There are a variety of design features applied to the different construction styles of centrifugal pumps. A summary of the main features follows.

Flange and Mount Bolts

The American Voluntary Standard (AVS) for centrifugal pumps for chemical industry use is applied by many manufacturers of small- to medium-sized pumps of light- to medium-duty ratings for flange, mounting bolt, and shaft center locations. AVS recommendations cover dimensional aspects to allow replacement of one manufacturer's pump in an existing installation site with that of another manufacturer. Roughly half of the centrifugal pumps manufactured in the United States follow AVS standards.

Volutes and Diffusers

Many commercial pumps have diverging channels called volutes, which are cast into the discharge zone of the casing. Flow through the volute causes the velocity to decrease from near the tip speed (around 200 fps) to the discharge line velocity (16 fps). This, in turn, causes the pressure to increase.

Single-volute passages are utilized in most pumps because of simplicity in design. However, this type of design causes an unbalanced load on the impeller because of the variation in pressure around the periphery. Double-volute configurations are employed when the unbalanced force level threatens to cause significant shaft deflection. This is typically above 500 feet per stage. Twin-discharge channels tend to balance the radial forces, thus reducing cycle stress in the shaft.

TABLE 2.1 GENERAL CHARACTERISTICS OF DIFFERENT PUMPS

Pump type	Construction style	Construction characteristics
		Dynamic-
Centrifugal (horizontal)	Single-stage overhung, process type	Impeller cantilevered beyond bearings.
	Two stage overhung	Two impellers cantilevered beyond bearings.
	Single-stage impeller between bearings	Impeller between bearings; casing radially or axially split.
	Chemical	Casting patterns designed with thin sections for high-cost alloys.
	Slurry	Designed with large flow passages.
	Canned	No stuffing box; pump and motor enclosed in a pressure shell.
	Multistage, horizontally split casing	Nozzles located in bottom half of casing.
	Multistage, barrel type	Outer casing contains inner stack of diaphragms.
Centrifugal (vertical)	Single-stage, process type	Vertical orientation.
	Multistage	Many stages with low head per stage.
	Inline	Inline installation, similar to a valve.
	High speed	Speeds to 380 rps, heads to 5800 ft (1770 m).
	Sump	Casing immersed in sump for easy priming and installation.
	Multistage, deep well	Long shafts.
Axial	Propeller	Propeller-shaped impeller.
Turbine	Regenerative	Fluted impeller. Flow path resembles screw around periphery.
		Positive
Reciprocating	Piston, plunger	Slow speeds.
	Metering	Consists of small units with precision flow control system.
	Diaphragm	No stuffing box.
Rotary	Screw	1, 2, or 3 screw rotors.
	Gear	Intermeshing gear wheels.

[a] L = low; M = medium; H = high.

Normal number of stages	Maintenance[a]	Solids tolerance	Notes
type pumps			
1	L	M-H	Capacity varies with head.
2	L	M-H	Used for heads above single-stage capability.
1	L	M-H	Used for high flows to 1083 ft (330 m) head.
1	M	M-H	Have low pressure and temperature ratings.
1	H	H	Low speed and adjustable axial clearance. Has erosion control features.
1	L-M	L	Low head capacity limits when used in chemical services.
Multi	L	M	Have moderate temperature-pressure ranges.
Multi	L	M	Used for high temperature-pressure ratings.
1	L	M	Used to exploit low net positive section head (NPSH) requirements.
1	L	M	Low-cost installation.
1	L	M	Low-cost installation.
1	M	L	High head/low flow. Moderate costs.
1	L	M-H	Low cost.
Multi	M-H	M	Used for water well service.
1	L	H	Vertical orientation.
1,2	H	M	Capacity independent of head. Low flow/high head performance.
displacement pumps			
1	H	M	Driven by steam engine cylinders or motors through crankcases.
1	M-H	L	Diaphragm and packed plunger types.
1	H	L	Used for chemical slurries. Can be pneumatically or hydraulically actuated.
1	M	M	For high-viscosity, high-flow-high-pressure services.
1	M	M	For high-viscosity, moderate-pressure/moderate-flow services.

TABLE 2.2   TYPICAL PERFORMANCE PARAMETERS OF DIFFERENT PUMPS

Pump type	Construction style	Capacity		Maximum head	
		(dm³/s)	(gpm)	(m)	(ft)
					Dynamic-type
Centrifugal (horizontal)	Single-stage over-hung	1–320	15–5,000	150	492
	Two-stage overhung	1–75	15–1,200	425	1394
	Single-stage impeller-between bearings	1–2500	15–40,000	335	1099
	Chemical	65	1000	73	239
	Slurry	65	1000	120	394
	Canned	0.1–1250	1–20,000	1500	4921
	Multistage horizontal split	1–700	20–11,000	1675	5495
	Multistage-barrel type	1–550	20–9,000	1675	5495
Centrifugal (vertical)	Single stage	1–650	20–10,000	245	804
	Multistage	1–5000	20–80,000	1830	6004
	Inline	1–750	20–12,000	215	705
	High speed	0.3–25	5–400	1770	5807
	Sump	1–45	10–700	60	197
	Multistage deep well	0.3–25	5–400	1830	6004
Axial	Propeller	1–6500	20–100,000	12	39
Turbine	Regenerative	0.1–125	1–2000	760	2493
					Positive
Reciprocating	Piston, plunger	1–650	10–10,000	345,000	1.13×10^6
	Metering	0–1	0–10	51,700	1.70×10^5
	Diaphragm	0.1–6	4–100	34,500	1.13×10^5
Rotary	Screw	0.1–125	1–2,000	20,700	6.79×10^4
	Gear	0.1–320	1–5,000	3,400	11,155

[a] Depends on strength of materials. Can be in excess of 50,000 psi.

Some process pumps employ vaned diffusers for pressure conversion rather than volutes. Included are some in-line pump styles, some axial pumps, and a few pumps designed for high-head, low-flow performance. Vaned diffusers offer the advantages of balanced radial force, compact size, and peak efficiency at high head and low flow. Disadvantages include that they are more difficult to fabricate and repair than volute pumps.

Maximum discharge pressure		Typical NPSH/requirements		Efficiency	Maximum pumping temperature	
(kPa)	(psi)	(m)	(ft)	(%)	(°C)	(°F)
pumps						
4,100	600	2–6	6.56–19.7	20–80	455	851
4,100	600	2–6.7	6.56–22.0	20–75	455	851
6,800	980	2–7.6	6.56–24.9	30–90	205–455	401–851
1,400	200	1.2–6	3.94–19.7	20–75	205	401
4,100	600	1.5–7.6	4.92–24.9	20–80	455	851
68,900	10,000	2–6	6.56–19.7	20–70	540	1004
20,100	3,000	2–6	6.56–19.7	65–90	205–260	401–500
41,400	6,000	2–6	6.56–19.7	40–75	455	851
4,100	600	0.3–6	0.98–19.7	20–85	345	653
4,800	700	0.3–6	0.98–19.7	25–90	260	500
3,400	500	2–6	6.56–19.7	20–80	260	500
13,800	2,000	2.4–12	7.87–39.4	10–50	260	500
1,380	200	0.3–6.7	0.98–22.0	45–75		
13,800	2,000	0.3–6	0.98–19.7	30–75	205	401
1,030	150	2	6.56	65–85	65	149
10,300	1,500	2–2.5	6.56–8.20	55–85	120	248
displacement pumps						
a	a	3.7	12.1	65–85	290	554
345,000	50,000	4.6	15.1	20	300	572
24,100	3,500	3.7	12.1	20	260	500
20,700	3,000	3	9.84	50–80	260	500
3,400	500	3	9.84	50–80	345	653

Impeller Configurations

Common impellers are enclosed with full discs and shrouds (these are termed *closed*). Semi open impellers are sometimes employed. These have a full back disc but no shroud. Fully open impellers have vanes but little or no disc material. These are employed occasionally in low-head, solids-handling service.

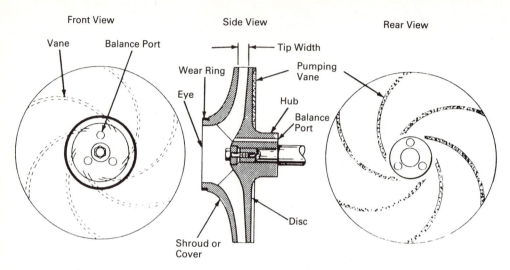

Figure 2.1 Major components of centrifugal pump impellers.

Many impellers employed are arranged from one side only (called single-suction designs). High-flow pumps employ impellers that accept suction from both sides (called double-suction designs). Very high flow rate units employ mixed-flow impellers, which involve an axial component in the direction of fluid flow. Many vertical, multistaged pumps (particularly deep well pumps) employ mixed-flow impellers to maintain small diameters for installation convenience. Low-diameter and high specific-speed combinations limit the head per stage to relatively low values (typically 115 ft to 150 ft) and thus tend to require many stages. Figure 2.1 summarizes nomenclature for centrifugal pump impellers.

Wearing Rings

Many pumps are designed with close running clearance at the suction side to separate the discharge pressure zone of the pump from the suction pressure areas and to minimize back leakage. Wearing rings are generally fitted at the close-clearance position in the pump casing and at an opposing position on the impeller, which allows easy restoration of the design clearance.

Many high-pressure centrifugal pumps are fitted with wearing rings at the back side (disc) of the impeller. This reduces the thrust force on the impeller, at the same time minimizing the pressure to which the stuffing box is exposed.

Nozzles

Many horizontal pumps are provided with suction and discharge nozzles on the top of the casing. In some horizontal pump styles, suction nozzles are located at the end of the pump, coaxial with the shaft center line. End suction pumps are most often used for services with moderate temperatures and flow rates.

Side suction connections are often used on horizontally split horizontal pumps. This arrangement permits positioning of the nozzle in the bottom half of the casing so that the bottom half does not have to be disturbed during maintenance.

Casing Construction and Orientation

Many centrifugal pumps have single casings (i.e., a single wall between the liquid under discharge pressure and the atmosphere). Double casings are employed in horizontal, multistage, high-pressure pumps and in vertical pumps. In the former, a heavy barrel-shaped casing surrounds the stack of stage diaphragms. The stack of diaphragms comprises the inner casing, while the barrel is the outer casing. Boiler-feed pumps normally employ this arrangement. In the case of vertical pumps, a vertical cylinder buried in the ground houses the pumping element. Suction liquid will enter the outer cylinder, flow to the bottom and then up through the pumping element stages. Again, the diaphragms of the stages in the pumping elements comprise the inner casing.

Casings can be joined in the same plane as the shaft axis (axially split) or perpendicular to the shaft (radially split). Axially split horizontal pumps are often called *horizontally split,* and radially split horizontal pumps are called *vertically split.* Radial jointing is employed on horizontal overhung pumps to allow easy removal of the rotor and bearing bracket assembly for maintenance purposes. It is also employed on high-pressure multistage pumps because of structural problems associated with bolting together the halves of axially split casings exposed to high internal pressure.

With horizontal pumps, their casings are supported by the base plate and the bearing brackets are, in turn, supported from the casing. Many casing-supported pumps have extensions along the sides of the casings at the center line, which rests on pedestals. These are referred to as *center-line supported.* This arrangement minimizes movement of the center line as casing temperature increases. Casings for ambient and moderate temperature service are often supported at the bottom of the casing (these are said to be *foot supported*).

Stuffing Boxes

Conventional stuffing boxes are filled with either packing material, such as braided rope or metallic foil rings, or with a mechanical seal. Pump casings designed for mounting only a mechanical shaft seal are said to have built-in seals (i.e., they cannot be converted to packing). This arrangement reduces shaft overhang and is generally less expensive than the conventional stuffing box. It is normally employed in clean, moderate suction pressure and temperature services. Disadvantages of this style include high sensitivity to dirt in service fluid, and fewer corrective measures can be applied to remedy chronic mechanical seal service problems.

Shaft Couplings

Pump shafts can be coupled to the drive shaft by a separate, removable, flexible-drive coupling, or they can be integral with the drive shaft. Most process plant pumps are coupled separately. The integral shaft style is referred to as *close coupled.*

Cooling Arrangements

Centrifugal pumps normally have several cooling features that are essential to avoid bearing overheating, to maintain alignment, and to assure proper mechanical seal performance. There are three specific areas in which cooling is necessary in high-temperature service, namely, bearing oil cooling, stuffing-box jacket cooling, and seal flushing liquid cooling. Cooling water is circulated through a bearing housing jacket, cooling coil, or external oil cooler to maintain oil temperatures below 180°F (82°C). (This is the maximum allowable temperature for proper bearing lubrication.)

Cooling water is also circulated through the jacket to remove friction heat generated by the mechanical seal. It also keeps the seal flushing liquid from flashing. Stuffing-box cooling helps to isolate the hot pump from the bearing housings.

For services between 400°F and 600°F (200°C to 315°C), the seal flushing liquid is cooled by an external cooler. This maintains temperature within the recommended limitations of the mechanical seal materials.

In addition to the previous cooling-water applications, low-pressure steam is used occasionally to cool seal plates and stuffing boxes in very high-temperature services (above 600°F).

CENTRIFUGAL PUMPS

Centrifugal pumps accomplish the generation of pressure by the conversion of velocity head into static head. The rotary motion of impellers adds energy to the service fluid in the form of a velocity increase. This velocity increase is converted into static head in the diffusing section of the casing. Centrifugal-type pumps have no valves. The flow is uniform and free of low-frequency pulsations. A pump operating at a fixed speed will develop the same theoretical head in feet of flowing fluid, regardless of density. However, the pressure corresponding to the developed head (in psi) depends on the fluid density.

The parameters that establish the maximum head (in feet of fluid) that a centrifugal pump can develop are the pump speed (rps), impeller diameter, and the number of impellers in series combination. Impeller design and blade angle mainly affect the slope and shape of the head-capacity curve and normally have no or little effects on the developed head. Conventional centrifugal pump impellers have a maximum tip speed around 200 fps (60 meter/s). In slurry pumps, impellers are limited to about half this tip speed to limit erosion problems.

Normal and maximum viscosity ranges are a major consideration in pump selection because of possible deterioration in performance with increasing viscosity. Deterioration can be both continuous and gradual. Table 2.3 can serve as a guide in selecting the proper centrifugal pump type for an application.

Most centrifugal pump types are not self-priming, meaning that they are not capable of evacuating vapor from the suction line so that liquid can flow into the line and into the pump casing without external assistance. The impellers on centrifugal pumps are designed specifically for efficient liquid pumping and are not operated at high enough tip speeds to convert them into vapor compressors. The differential head that the pump impeller can deliver is the same on vapor as on liquid. However, the equivalent differential pressure rise capability is much lower with vapor. Thus, centrifugal pump impellers are not capable of generating a significant reduction in the pressure exerted by the vapor in the suction line to permit liquid flow.

To prime a centrifugal pump before starting, the suction line and pump casing must be filled with liquid. When the suction source is at positive pressure or is positioned above the pump, priming is done by opening the suction valve and releasing (venting) the trapped vapor from a valved connection on the pump casing or discharge line (located inside the discharge block valve). Liquid then will flow into the suction line and pump casing to displace the venting vapor.

TABLE 2.3   VISCOSITY SENSITIVITY OF DIFFERENT PUMP TYPES

Type pump	Kinematic viscosity $\nu = \mu/\rho$ (mm²/sec)	Guidelines
Rotary	7	Nominal minimum viscosity for rotary pumps. Efficiency begins to decrease as viscosity increases. Viscosity should be specified for services when it exceeds this level.
Centrifugal	<30	Centrifugal pumps preferred over rotary when conditions allow application of either type.
Centrifugal	30	The head-capacity capability of a centrifugal pump begins to deteriorate.
Centrifugal	30–110	Centrifugal pump preferred over rotary despite some efficiency drop.
Positive Displacement	110–220	These are almost always used if the expected viscosity exceeds this level.
Rotary	110–650	Rotary preferred if viscosity falls in this range.

Specially designed centrifugal pumps are self-priming. One type accomplishes evacuation of the suction-line vapor by entrainment of vapor bubbles from the suction side of the impeller in a charge of liquid held in the pump casing (or sometimes held in a holdup separation bottle attached to it).

The liquid charge is recirculated to the suction side after separating the entrained vapor. There are other designs available; however, in general, self-priming centrifugal pumps have been applied on a limited basis in continuous process service.

Most centrifugal pumps have a self-venting feature. Small amounts of vapor trapped within the casing at start-up (after suction priming is complete) are swept out into the discharge line when the pump is started. Horizontally split casings are not arranged to be self-venting, however, and are equipped with specially designed valved vent connections requiring manual operation. Single-stage centrifugal pumps with top discharge connections have good self-venting performance, even though the casing shape places a small high-point vapor pocket in the top of the discharge volute.

A brief description of the major types of centrifugal pumps follows.

Single-Stage Overhung

This design has a single-stage overhung impeller. Its casing is supported at the center line. Two shaft bearings are mounted close together in the same bearing bracket, with the impeller cantilevered or overhung beyond them. This design is illustrated in Figure 2.2. This type design usually has top suction and discharge flanges, wearing rings both on the front and back of the impeller and

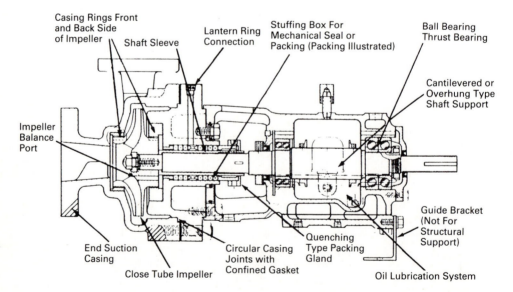

Figure 2.2 A single-stage end suction-overhung centrifugal pump.

casing, a single suction, closed impeller, and a single stuffing box fitted with a mechanical seal. There are also water-cooling options on the pedestal, stuffing box, and bearings. This pump type is well suited for high-temperature operation and can be used for handling flammable liquids.

Two-Stage Overhung

This is simply a modified version of the single-stage process pump (Figure 2.3). It is capable of higher head than its single-stage counterpart. In this style, the

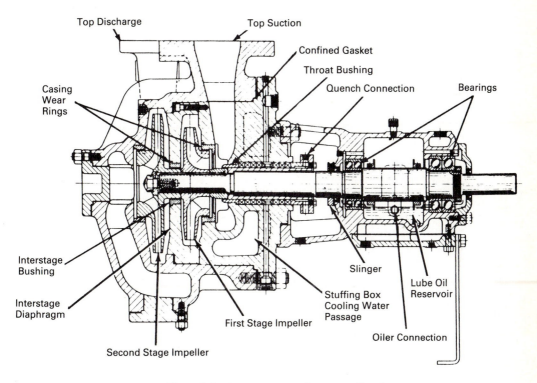

Figure 2.3 A two-stage overhung centrifugal pump.

stuffing-box pressure is roughly halfway between suction and discharge pressures.

Single-Stage (Impeller Between Bearings)

These designs have their impellers mounted between the bearings and thus have two stuffing boxes. Single-stage versions are capable of developing heads up to 1,080 feet (330 meters). Casings can be axially split for temperatures up to 500°F (260°C) and radially split for temperatures to 850°F (445°C).

In-line Pumps

These are vertical pumps with casings designed to be bolted directly to piping (similar to a valve). There are two basic configurations: coupled and close coupled. Service life and maintenance requirements for both styles are about the same. Generally, this type of pump is preferred if the cost of piping associated with installation can be reduced over the conventional horizontal styles.

High-Speed Centrifugal Pumps

These pumps are single-impeller models designed for speeds typically in the range of 170 to 280 rps and as high as 400 rps. They are capable of heads up to 5,300 feet (1,600 meters). Pumping temperatures are limited to about 500°F (260°C). High-speed pumps tend to have high NPSH requirement due to the sudden velocity increase as the liquid enters the impeller. Maintenance requirements for these pumps tend to be higher than for conventional speed, single-stage pumps; however, it is about the same as multistage models, with which they compete for high-head services.

Chemical Pumps

This class of pumps is designed with casing shapes that are cast in high-cost alloys. Casings are more frequently foot supported or bearing-bracket supported than center-line supported. These pumps are limited to relatively low-temperature, pressures and flow rates.

Slurry Pumps

Slurry pumps are used in services that have severe conditions of slurry pumping. They have a number of special design features that make them well suited for this type service:

1. Wide flow passages to avoid clogging
2. Open/semiopen impellers, which are less sensitive than closed impellers to clogging
3. Arrangements that break up large particles
4. Low-fluid velocities, as low rotating and peripheral speed are possible
5. Adjustable rotor position to restore axial clearance without dismantling the pump
6. Replaceable wearing plates and pumping plates on back of impeller, instead of wearing rings, which are subject to erosion

Canned Pumps

These are motor pump units with the rotating rotor and impeller housed entirely within a pressure casing. This type design eliminates the need for a stuffing box. The pumped fluid serves both as a lubricant for bearings and as a coolant for the motor. Designs are limited to low-flow, low-pressure, and low-temperature service.

Horizontal and Vertical Multistage Pumps

Horizontal multistage pumps are limited to about 12 stages due to the difficulty in limiting deflection over the long span between bearings. They have NPSH requirements that match those of single-stage pumps of the same capacity. These pumps are well suited for corrosive fluids handling.

Vertical multistage pumps may have as many as 24 (sometimes more) stages. Some high-differential pressure models utilize opposed thrust arrangements. Below roughly 1,200 feet (370 meters) head, pumps for as low as 1 foot (0.3 meter) NPSH at the suction flange are available. High specific-speed impellers are frequently used. The first stage is normally at the bottom of the assembly, below grade. Vertical multistage pumps need a large number of close-running clearances. Hence, these pumps are sensitive to damage by solids ingestion and by dry or two-phase generating conditions. In general, maintenance requirements for vertical multistage pumps are greater than for their horizontal counterparts.

Regenerative Turbine Pumps

Some manufacturers prefer to place regenerative turbine pumps in a separate class. In general, this type pump greatly resembles a conventional centrifugal pump but has the distinction of a much steeper head-capacity curve. The impeller consists of a solid disc with fluted vanes on each side of the perimeter, which impart energy to the liquid by multiple excursions from the impeller to the stator and back to the impeller (tracing dual screw-shaped paths along the stator annulus).

Turbine pumps have shutoff pressures typically two or three times the design level. The steepness of the head curve causes the power requirement curve to rise as flow decreases, thus peaking at shutoff. Drivers for turbine pumps must be sized for minimum flow, rather than normal flow. Usually, a safety valve is needed inside the discharge block valve. This type pump is extremely sensitive to dirt, temperature shocks, and to piping forces and moments on the pump flanges.

POSITIVE-DISPLACEMENT PUMPS

Positive-displacement pumps operate by forcing a fixed volume of liquid from the inlet pressure zone of the pump into the discharge zone of the pump. Brief descriptions of each of the major types of positive-displacement pumps follow.

Reciprocating Pumps

Reciprocating pumps produce pulsating flow, develop high shutoff or stalling pressure, display constant capacity when motor driven, and are subject to vapor binding at low NPSH conditions. There are several construction styles. One type, the direct-acting steam pump, consists of a steam cylinder end in line with a liquid cylinder end, with a straight rod connection between the steam piston and the pump piston or plunger.

Direct-acting steam pumps are available as simplex (one steam and liquid cylinder) and duplex (dual side-by-side) units. Duplex units are employed in larger-capacity services and to reduce the flow pulsations below that of the simplex. Dual pumps are designed with an interconnecting steam valve linkage arrangement so that one side pumps when the other side reaches the end of its stroke. Steam pumps consist of rod and piston design and are double acting (that is, each side pumps on every stroke). Consequently, a duplex pump will have four pumping strokes per cycle.

Another construction style is the power pump. Power pumps convert rotary motion to low-speed reciprocating motion via speed reduction gearing, a crankshaft, connecting rods, and crossheads. Plungers or pistons are driven by the crosshead drives. Rod and piston construction, similar to duplex double-acting steam pumps, are used by the liquid ends of the low-pressure, higher-capacity units. The higher-pressure units are normally single-acting plungers. This latter style generally employs three (triplex) plungers. Three or more plungers substantially reduce flow pulsation relative to simplex and even duplex pumps.

In general, power pumps have high efficiency and are capable of developing very high pressures. They can be driven either by electric motors or turbines. They are relatively expensive pumps and rarely can be justified on the basis of efficiency over centrifugal pumps. However, they are frequently justified over steam-reciprocating pumps where continuous-duty service is needed due to the high steam requirements of direct-acting steam pumps.

In general, the effective flow rate of reciprocating pumps decreases as viscosity increases because the speed must be reduced. High viscosity also leads to a reduction in pump efficiency. In contrast to centrifugal pumps, the differential pressure generated by reciprocating pumps is independent of fluid density. It is entirely dependent on the amount of force exerted on the piston.

Reciprocating pumps are most often used for sludge and slurry services, particularly where other types are inoperable or troublesome. Maintenance in such services tends to be high because of valve, cylinder, rod, and packing wear.

Metering Pumps

Metering pumps are positive-displacement pumps that provide precision control of very low flow rates. Flow rates can range from 1.6×10^{-3} to 0.16 gpm; however, there are some higher-capacity models that provide flows up to 0.66 gpm. Flow accuracy is typically within ±1 percent. Control schemes for metering pumps are used for controlling the proportioning of additives injected into the main flow stream. Other names for these type pumps are *proportioning pumps* and *controlled-volume pumps*.

Metering pumps are available in two construction styles: diaphragm and packed plunger. The diaphragm design employs a hydraulic oil barrier between the reciprocating plunger and an impervious diaphragm, which, in turn, contacts the pumped liquid. The stuffing box works in lube oil in this design, so there is no process liquid leakage. With the packed plunger arrangement, the design resembles a small version of a conventional larger plunger pump, in which the stuffing box is exposed to the service liquid.

Usually the driver consists of an electric motor. Basically, the same design criteria applied to larger motor-driven reciprocating pumps can be applied to proportioning pumps.

Capacity variations normally are handled by manual resetting of the stroke adjustment. Controls are available for automatic stroke resetting and for remote, manual stroke resetting. The typical efficiency of this type pump is around 20 percent. Generally, viscosity effects on power requirements are negligible.

Finally, it should be noted that these pumps are designed for clean service. Nozzle connections and valves of metering pumps are small and, thus, are subject to plugging and/or valve sticking when handling dirty liquids. Figure 2.4 shows a cutaway view of a metering pump.

Diaphragm Pumps

This type of positive-displacement pump operates by the periodic movement of a flexible diaphragm. It has the advantages of no stuffing boxes and high tolerance to abrasive slurries.

The diaphragm is flexed by pulsating fluid pressure on the drive side. Compressed air normally is used; however, steam and hydraulic oil systems are also available. Drive pressures normally pulsate between 0 psi and 15 psi (0–105 kPA) above the average discharge pressure level in the process stream.

Rotary Pumps

There is a wide variety of rotary pumps on the market; however, liquid services are limited primarily to external gear pumps and screw pumps. Sliding vane and internal gear pumps find limited application to process plant services.

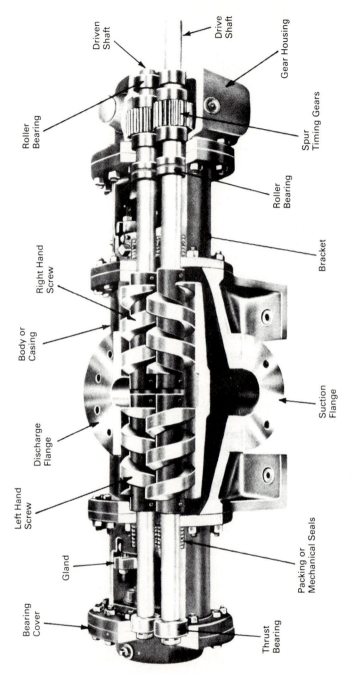

Figure 2.4 Cutaway view of a controlled-volume pump.

Driven Shaft

Drive Shaft

Gear Housing

Roller Bearing

Spur Timing Gears

Roller Bearing

Bracket

Right Hand Screw

Body or Casing

Suction Flange

Discharge Flange

Left Hand Screw

Gland

Packing or Mechanical Seals

Bearing Cover

Thrust Bearing

The single-screw pump is a special type of screw pump for handling slurries with relatively large particles. This design allows little fracturing of particles and very little abrasion damage to the pump. Single-screw pumps are employed extensively in the food processing and chemical industries for handling solid/liquid mixtures that are either abrasive or require gentle handling of the solid particles.

The working principle of the screw pump is that of Archimede's screw, invented by the Greek mathematician about 200 B.C. Screw pump applications include fuel, lube, and crude oil service; navy and marine cargo; oil burners; slurry handling; and a variety of high-viscosity materials such as polymers, copolymers and elastomers, cellulosics, syrups, fats and greases, soaps, and solvents.

Today, both single-rotor and multiple-rotor screw pumps are commercially available. Pumping action is accomplished by progressing cavities, which advance along the rotating screw from inlet to outlet. This axial flow pattern minimizes vibration, producing smooth flow.

In the twin-screw pump (Figure 2.5), two sets of screws rotate and mesh in an accurately bored casing, having rather tight operating clearances maintained between them. The mechanical displacement of fluid from inlet to outlet is generated by trapping a slug of fluid in the helical cavity (referred to as a *positive lock*), created by the meshing of the screws.

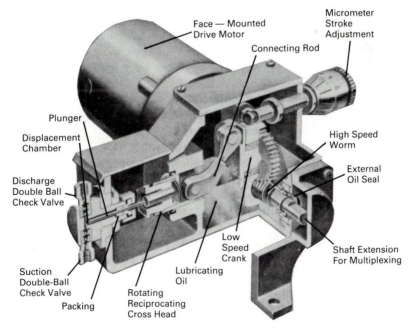

Figure 2.5 Features of the twin-screw pump. (*Courtesy of Dresser Industries, Harrison, NJ*).

In most twin-screw designs, the clearance between screws is maintained by a pair of timing gears mounted on the shafts. These gears also transmit power from the drive shaft to the driven shaft. The body consists of a casing with two precision-machined bores, which house the rotating screws. Fluid passes from the inlet chamber into the pumping chamber and then to the discharge chamber. Screw pump bodies are commonly made of cast iron, ductile iron, cast steel, or 316 stainless, depending on such factors as pressure requirement, need for corrosion or galling resistance, pumping temperature, and so on. Body bores sometimes are chrome plated to improve surface finish, antigalling characteristics, and hardness.

The screws are designed to withstand hydraulic pressure within the screw channels, which exerts both a radial and an axial force on the screws. Axial forces are balanced by using the matched pair of screws on each shaft. Radial forces tend to cause shaft deflection, and are resisted by the fluid film trapped between the screws and by the body bore diameter.

Screws and shafts may be of *pinned* or *integral* design. In the pinned screw design, the screws are machined as separate pieces and mounted on the shaft with pins or keys. This design is used for relatively low pressures, up to about 500 psi. The integral screw arrangement, with screws and shaft machined from a single forging or piece of barstock, provides much higher pressure, viscosity, and shaft torque capabilities.

Timing gears are used to transmit power from one rotor shaft to the other and to maintain the proper clearance between the pumping screws. Most screw pumps are designed with relatively small internal clearances between the pumping screws. This mandates high-precision timing gears if metal-to-metal contact is to be avoided. The positioning of timing gears relative to the pumping screws is called *the timing of the screws*. Timing gears may be the double-helical type, which maintain the proper angular and axial relationship between the pumping screws. Single-helical or spur designs also are used. The latter maintain only the angular relationship, relying on the thrust bearing to maintain the proper axial relationships between pumping screws.

Some screw pump designs work without timing gears. In these designs, there is a drive screw that is used to drive the driven screws. The design of twin-screw pump bodies and the arrangement of screw assemblies are interrelated. Figure 2.6 illustrates typical body designs.

There is a variety of novel rotary pump types commercially available. The more conventional styles are compared in Table 2.4.

The main reason for selecting rotary pumps over centrifugals is to take advantage of their high-viscosity capability. In addition, rotaries are simple in design and efficient in handling flow conditions that are generally considered too low for economic application of centrifugals. The importance of viscosity in rotary pump design is summarized in Table 2.5. Rotary pumps designed to handle high-viscosity liquids must be operated at reduced speeds (and, thus, at reduced flow rates).

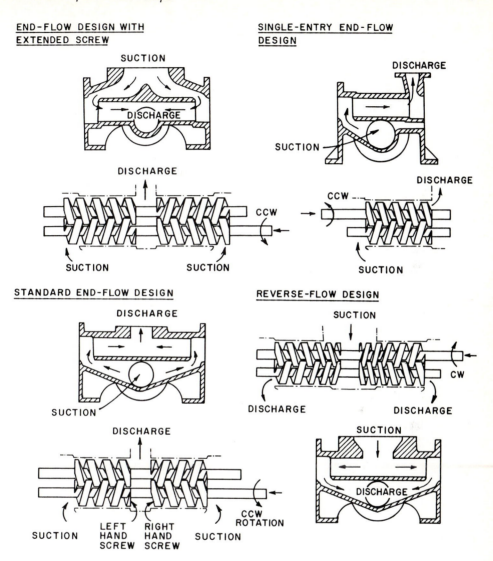

Figure 2.6 Twin-screw pump bodies and flow patterns.

Note that proper pump selection is not only dependent on the fluid viscosity level, but also on the rheological properties of liquid. For non-Newtonian fluid behavior, pump design and selection will depend also on how viscosity changes with shear rate. Shear rate/shear stress data should be included when preparing design specifications.

TABLE 2.4 COMPARISON OF ROTARY PUMP STYLES

Rotor arrangement	Description		Number of rotors	Number of stuffing boxes
	Screw type	External gear type		
Single Rotor	For solids handling services. Avoids fracturing solids and minimizes abrasion damage.	No major commercial designs.	1	1
Contact Drive of Idle Rotor (no timing gears)	Uses three rotors. Lower cost than timing gear models. Has higher NPSH requirement than timing gear styles. Pressures up to 20,700 kPa.	Uses two rotors. Differential pressure limited to 2400/3450 kPa.	2–3	1
Internal Timing Gears and Bearings	Less sensitive to solids in liquid than contact drive or gear type. Limited to liquids with sufficient lubricity for bearing lubrication. Temperature limited to 150–175°C.	Same temperature limit as screw type.	2	1
External Timing Gears and Bearings	Most versatile construction for low-viscosity high pressure (low lubricity, solids, etc.). Screws can be replaceable or integral with shaft. Timing gear may be at outboard or coupling end. Temperature limited to 370–400°C.	Differential pressure limit for standard construction, 1400 kPa; can be extended to 2800 kPa with special construction. Temperature limit same as screw type.	2	4

TABLE 2.5 DIFFERENTIAL PRESSURE LIMITATIONS FOR ROTARY PUMPS DUE TO VISCOSITY LEVELS

Kinematic viscosity (mm²/sec)	Differential pressure limitations
1.4–7.4	Standard designs with external bearings limited to 500–700 kPa.
7.4–32⁺	Novel designs extend differential pressure to 2800 kPa.
32–44	Standard designs limited to 1030 kPa.
44–75	Standard designs limited to 2400 kPa.
75–110	Standard designs limited to 3450 kPa.
>110	Novel designs extend range to 24,000 kPa.
130–640	Screw pumps reach maximum efficiency in this viscosity range.

BASIC HYDRAULICS AND DEFINITIONS IN PUMPING SERVICE

The principle of conservation of energy states that the total energy input to a closed system (Figure 2.7) is equal to the total energy output from that system. In addition to the energy in the fluid at point 1 (Figure 2.7), there is energy added by pumps. The output energy in the fluid plus the friction loss in the system is at point 2. Bernoulli's equation is restated here in its more general form (total mechanical energy balance):

$$\frac{P_1}{\rho} + Z_1 + \frac{U_1^2}{2g} + E_p = \frac{P_2}{\rho} + Z_2 + \frac{U_2^2}{2g} + H_f \tag{2.1}$$

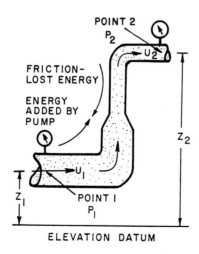

Figure 2.7 System defining the law of conservation of energy.

The left-hand side of the equation refers to point 1. The first energy term is pressure energy, P_1/ρ. The second term is another energy form, derived through the fluid's static elevation (i.e., potential energy).

The third term, $U_1^2/2g$, is an expression for kinetic energy or velocity head.

The fourth term in the equation, E_p, represents pump energy. This is the amount of energy that is added to the fluid between points 1 and 2.

The final term on the right-hand side of Equation 2.1, H_f, is the friction loss between points 1 and 2.

The continuity equation is based on the principle of conservation of mass, which states that the total mass of fluid flowing into any closed conduit is equal to the total mass of fluid flowing out of the system. More simply, this means that the flow capacity at point 1 is the same as at point 2.

$$Q = U_1A_1 = U_2A_2 \qquad (2.2)$$

where A is the cross-sectional area of the pipe.

The application of these two equations represents a generalized first step to specifying pumping requirements. Their use is best illustrated by a typical pumping problem.

Example

Figure 2.8 gives conditions for a simple pumping situation in which the total head required by the pump must be determined. The pump must be sized to pump 100 gpm and overcome the friction of the piping (which includes entrance losses, elbows, and so on) of 16 ft at that capacity. The pump takes suction from a closed vessel, which is under a vacuum of 28 in. Tg. The height of the water in the tank is 3 ft above the tank bottom. The fluid (water) flows through a 3-in. (nominal diameter) pipe to the pump and then through a 2-in. pipe, and is finally discharged at an elevation of 10 ft above the floor.

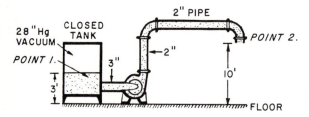

Figure 2.8 Pumping example to determine total head requirements.

Solution

To solve this problem, the Bernoulli equation must be restated in a more usable form:

$$E_p = \frac{P_2 - P_1}{\rho} + (Z_2 - Z_1) + \frac{U_2^2 - U_1^2}{2g} + H_f \qquad (2.3)$$

In the equation, P_2/p is the atmospheric pressure (i.e., zero in gauge pressure). The static elevation $Z_2 = 10$ ft and $Z_1 = 3$ ft; hence $Z = 7$ ft.

The velocity head is determined by first evaluating the fluid velocity in the 2-in. pipe. This can be obtained from the continuity equation, in which the velocity of 100 gpm in a 2-in. pipe is 9.5 fps (the details are left to the reader). The term $U_1^2/2g$ is very small since the fluid in the suction chamber is moving with such a low velocity and, hence, for practical purposes may be ignored. The friction loss in the system between the suction chamber and the discharge side is 16 ft. Adding these values, an answer of 56 ft is obtained. Hence, a pump must be selected on the basis of a total head developed of 56 ft.

In the preceding example problem, information on the friction loss of the system is given. However, in many situations the determination of the friction loss in a pipe system is an important part of the problem. A second problem will help illustrate friction loss calculations for pumping problems.

Example

Figure 2.9 defines the system for the problem. Water is being pumped at a rate of 200 gpm through a 4-in. steel pipe. The water is introduced to the pipe system through a foot valve and discharged upward by the pump through an elbow, a swing check valve, a gate valve, another elbow, and finally a sudden enlargement to an open vessel. The total length of pipe above the pump is 1,300 ft. Determine the total friction loss for the system.

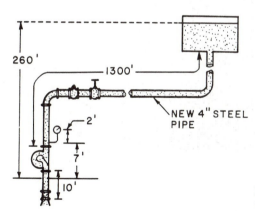

Figure 2.9 Pumping problem example to determine friction losses.

Solution

First determine the friction loss in the total length of pipe:

Total Pipe Length = 1,300 ft pipe above the pump
+ 10 ft pipe below the pump = 1,310 ft

Using standard methods for estimating friction losses based on the friction factor plot in Figure 2.10, for 200 gpm through 4-in. steel pipe, the pipe friction loss is 29.74 ft of flowing fluid.

Other losses in the problem can be determined by the friction loss formula, $H_f = KU^2/2g$; $U^2/2g = 0.395$ ft for 200 gpm and 4-in. steel pipe.

Values of coefficient K for various fittings are given in Table 2.6. Substituting into the friction-loss formula and adding up all losses, the total friction loss through all pipe and fittings is 31.02 ft.

Next the pump energy needed to pump 200 gpm through the pipe system must be determined through the Bernoulli equation:

$$E_p = \frac{P_2}{\rho} - \frac{P_1}{\rho} + \Delta Z + \frac{U_2^2}{2g} - \frac{U_1^2}{2g} + H_f$$

$P_2/\rho = P_1/\rho = 0$, since the gauge pressure on the water at the inlet and the outlet is 0 (i.e., both the sump and tank are open to the atmosphere).

The static elevation of the upper vessel is 260 feet compared with a static level of 0 if the datum line is chosen to be the lower surface of the water (Figure 2.9). Substituting values into the Bernoulli equation,

$$E_p = 0 - 0 + 260 - 0 + 0 - 0 + 31.02$$
$$E_p = 291 \text{ ft} = \text{required total pump head}$$

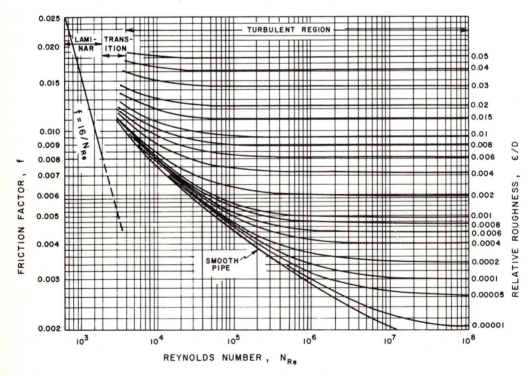

Figure 2.10 Moody friction factor chart.

TABLE 2.6 VALUES OF COEFFICIENT *K* FOR VARIOUS FITTINGS

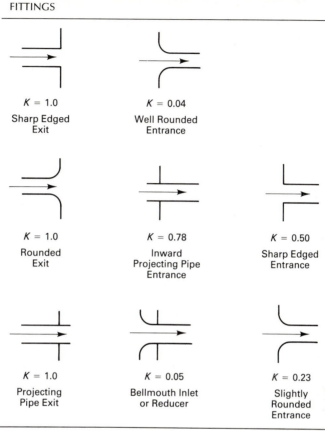

K = 1.0

Sharp Edged
Exit

K = 0.04

Well Rounded
Entrance

K = 1.0

Rounded
Exit

K = 0.78

Inward
Projecting Pipe
Entrance

K = 0.50

Sharp Edged
Entrance

K = 1.0

Projecting
Pipe Exit

K = 0.05

Bellmouth Inlet
or Reducer

K = 0.23

Slightly
Rounded
Entrance

Resistance Coefficients for Various Entrance and Exit Configurations

To carry the problem a step farther it would be useful to know the pump heads required for a range of capacities. Rather than recalculating losses for different flow rates, a short-cut method can be used by realizing that friction loss varies as the square of the velocity in the pipe, or as the square of the volumetric flow rate ($H_f = fU^2/2g$). For example, for half the flow rate, or 100 gpm, the friction loss is the square of $\frac{1}{2}$, or $\frac{1}{4} \times 31.09 = 7.8$ ft.

Similarly, 300 gpm is 1.5 times the original 200 gpm. Hence, the new friction loss is $(1.5)^2 \times 31.02$, or 70 ft.

We now have three points and can prepare a curve of friction loss versus flow, as shown in Figure 2.11. Note the change in static elevation of 260 feet is a constant regardless of flow, and is thus just added to the three values. Figure 2.11 is referred to as the system head curve. It depicts the pump energy or total head required for any capacity.

Not all head curves are alike. Various types of systems present different system head curves. Also note that while a system head curve can be prepared

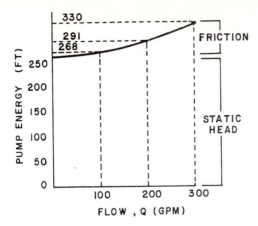

Figure 2.11 Friction loss curve for example problem.

for most systems, its accuracy may not always be precise. Distortions affect the real system head curve in many ways. Undetermined obstructions in the system, such as a clogged valve, clogged strainers in the line, or unexpected pipe roughness, can contribute to higher pipe losses. Also, the real system head curve may be lower than calculated due to more conservative estimates. For example, losses might be computed based on old pipe with scale buildup, when in fact the system uses new pipe.

SUGGESTED READINGS

AZBEL, D. S., and N. P. CHEREMISINOFF, *Fluid Mechanics and Unit Operations*. Ann Arbor, MI: Ann Arbor Science, 1983.

BERGELIN, O. P., "Flow of Gas-Liquid Mixtures," *Chem. Eng.*, 56, no. 5 (1949), 104.

BONNINGTON, S. T., and A. L. KING, "Jet Pumps and Ejectors." Cranfield, England: British Hydraulic Research Association: Fluid Engineering, 1972.

BROWN, G. G., *Unit Operations*. New York: John Wiley & Sons, Inc., 1950.

CHEREMISINOFF, N. P., *Fluid Flow: Pumps, Pipes and Channels*. Ann Arbor, MI: Ann Arbor Science Publishers, 1981.

CHURCH, A. H., *Centrifugal Pumps and Blowers*. New York: John Wiley & Sons, Inc., 1944.

COULSON, J. M., and J. F. RICHARDSON, *Chemical Engineering*. Elmsford, NY: Pergamon Press, Inc., 1961.

DATTA, R. L., "Studies for the Design of Gas Lift Pumps," *J. Imp. Coll. Chem. Eng. Soc.*, 4, no. 157 (1948).

DAUGHERTY, R. L., *Centrifugal Pumps*. New York: McGraw-Hill Book Co., 1915.

FOUST, A. S., ET AL., *Principles of Unit Operations, 2nd ed.* New York: John Wiley & Sons, Inc., 1980.

HICKS, T. G., *Pump Selection and Application*. New York: McGraw-Hill Book Co., 1957.

HICKS, T. G., and T. W. EDWARDS, *Pump Application*. New York: McGraw-Hill Book Co., 1971.

KARASSIK, I. J., *Centrifugal Pump Clinic.* New York: Marcel Dekker Inc., 1981.

KARASSIK, I. J., and E. CARTER, *Centrifugal Pump Design and Selection.* Harrison, NJ: R.P. Worthington Corp., 1981.

KARASSIK, I. J., W. C. KRUTZSCH, W. H. FRASER, and J. P. MESSINA, *Pump Handbook.* New York: John Wiley & Sons, Inc., 1976.

KIRK, R. E., and D. OTHMER, *Encyclopedia of Chemical Technology, 2nd ed.* New York: John Wiley & Sons, Inc., 1963.

METZNER, A. G., *Handbook of Fluid Dynamics.* New York: McGraw-Hill Book Co., 1961.

PERRY, R. H., and C. H. CHILTON, EDS., *Chemical Engineers' Handbook, 5th ed.* New York: McGraw-Hill Book Co., 1973.

REAVELL, E. A., "Some Aspects of Chemical Works Pumping in Acid Handling," *Proc. Chem. Eng.,* 18, no. 25 (1936).

SIMONIN, R. F., "Working of an Air Lift Water Pump," *Comp. Rend. Acad. Sci.,* 233, no. 465 (1957).

STEPANOFF, A. J., *Pumps and Blowers.* New York: John Wiley & Sons, Inc., 1965.

SWINDIN, N., *The Modern Theory and Practice of Pumping.* London: Ernest Benn Ltd., 1924.

TAYLOR, I., "The Most Persistent Problems for the Chemical Plant Designer," *Chem. Eng. Prog.,* 46, no. 637 (1950).

TETLOW, N., "Survey of Present-Day Pumping Practice in the Chemical Industry," *Trans. Inst. Chem. Eng.,* 28, no. 63 (1950).

3

Industrial Pumps and Their Applications

This chapter provides additional descriptive information on design and operating features of pumps. Specific applications are highlighted along with typical service factors. Examples of pump styles and manufacturer specifications and operating ranges are given to familiarize the reader with technical marketing literature.

EXAMPLES OF CENTRIFUGAL PUMPS IN SERVICE

As noted in Chapter 2, the centrifugal pump is among the most widely used for a multitude of industrial services. An example of a self-priming centrifugal pump powered by a gasoline engine is shown in Figure 3.1. This particular design is available from 3.5 to 5 hp. Figure 3.2 shows typical performance data based on maximum impeller diameters. Actual performance obtained in field service will vary depending on engine type, impeller spacing and diameter, and the size, length, and type of discharge hoses/pipes used. As a rule of thumb, best performance is obtained when the largest diameter and shortest possible hoses/pipes are used. By way of example, the manufacturer's specifications are given in Table 3.1.

Figures 3.3 through 3.5 show several examples of different configurations and intended services for centrifugal type pumps. Figure 3.3A shows a self-

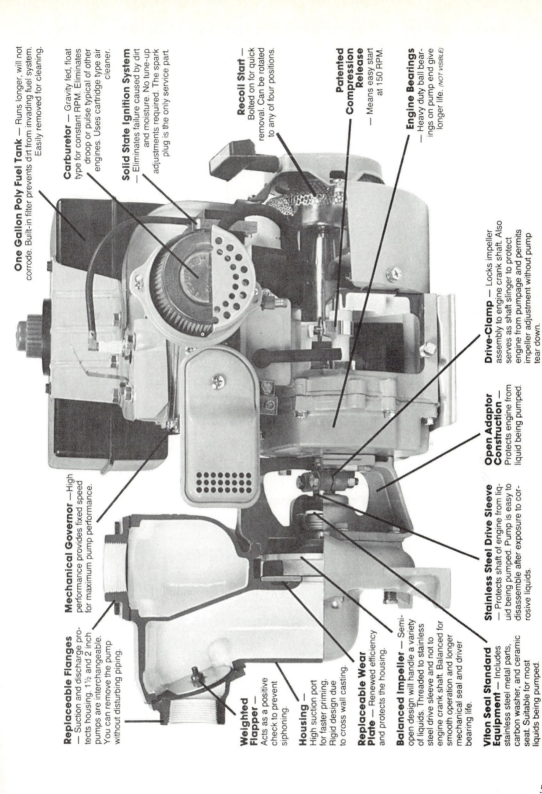

One Gallon Poly Fuel Tank — Runs longer, will not corrode. Built-in filter prevents dirt from invading fuel system. Easily removed for cleaning.

Carburetor — Gravity fed, float type for constant RPM. Eliminates droop or pulse typical of other engines. Uses cartridge type air cleaner.

Solid State Ignition System — Eliminates failure caused by dirt and moisture. No tune-up adjustments required. The spark plug is the only service part.

Recoil Start — Bolted on for quick removal. Can be rotated to any of four positions.

Patented Compression Release — Means easy start at 150 RPM.

Engine Bearings — Heavy duty ball bearings on pump end give longer life. *(NOT VISIBLE)*

Drive-Clamp — Locks impeller assembly to engine crank shaft. Also serves as shaft slinger to protect engine from pumpage and permits impeller adjustment without pump tear down.

Open Adaptor Construction — Protects engine from liquid being pumped.

Stainless Steel Drive Sleeve — Protects shaft of engine from liquid being pumped. Pump is easy to disassemble after exposure to corrosive liquids.

Viton Seal Standard Equipment — Includes stainless steel metal parts, carbon washer, and ceramic seat. Suitable for most liquids being pumped.

Balanced Impeller — Semi-open design will handle a variety of liquids. Threaded to stainless steel drive sleeve and not to engine crank shaft. Balanced for smooth operation and longer mechanical seal and driver bearing life.

Replaceable Wear Plate — Renewed efficiency and protects the housing.

Housing — High suction port for faster priming. Rigid design due to cross wall casting.

Weighted Flapper — Acts as a positive check to prevent siphoning.

Replaceable Flanges — Suction and discharge protects housing. 1½ and 2 inch pumps are interchangeable. You can remove the pump without disturbing piping.

Mechanical Governor — High performance provides fixed speed for maximum pump performance.

Figure 3.1 Example of an engine drive centrifugal pump. *(Courtesy of MP Pumps, Inc., Tecumseh Products Co., 34800 Bennet Drive, Fraser, MI 48026-1686)*

59

FLOMAX 5 - 1½" x 1½"   FLOMAX 8 - 2" x 2"

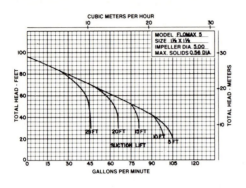
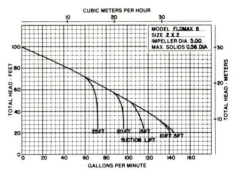

TOTAL HEAD FEET	PUMP FLOW IN U.S. GALLONS PER MINUTE			
30	97	81		
40	93	80	65	
50	78	77	63	45
60	62	62	58	43
70	45	45	45	38
80	29	29	29	27
90	12	12	12	12
SUCTION LIFT	10 FT.	15 FT.	20 FT.	25 FT.

TOTAL HEAD FEET	PUMP FLOW IN U.S. GALLONS PER MINUTE			
30	135	115		
40	119	112	96	
50	100	100	94	72
60	82	82	82	70
70	64	64	64	63
80	46	46	46	46
90	27	27	27	27
SUCTION LIFT	10 FT.	15 FT.	20 FT.	25 FT.

Figure 3.2 Example of manufacturer's performance data for pump shown in Figure 3.1. (*Courtesy of MP Pumps, Inc., Tecumseh Products Co., 34800 Bennet Drive, Fraser, MI 48026-1686*)

TABLE 3.1   TYPICAL MANUFACTURER'S SPECIFICATIONS

PUMP

Suction & Discharge	1½", 2" pipe size
Housing...............	Cast iron, bronze, aluminum or stainless steel
Impeller	Cast iron, bronze, aluminum or stainless steel
Wear Plate	Steel, bronze, aluminum or stainless steel
Shaft Sleeve...........	Stainless steel
Seals	Standard is carbon, ceramic, stainless steel & VITON, with other seal materials available. (Consult factory)
Fasteners	Stainless steel

ENGINE

		H35	H50
Horsepower		3½ H.P. 2.61 kW	5 H.P. 3.73 kW
Bore & Stroke	Inches	2½" x 1¹⁵/₁₆"	2⅝" x 2¼"
Displacement	Cubic Inches	9.51	12.18
Governor		Mechanical	Mechanical
Main Bearing PTO End		Ball Bearing	Ball Bearing
Oil Capacity	Ounces	21	19
Fuel Tank	Gallons	One	One

WARNING	DO NOT USE IN EXPLOSIVE ATMOSPHERES OR FOR PUMPING VOLATILE LIQUIDS.

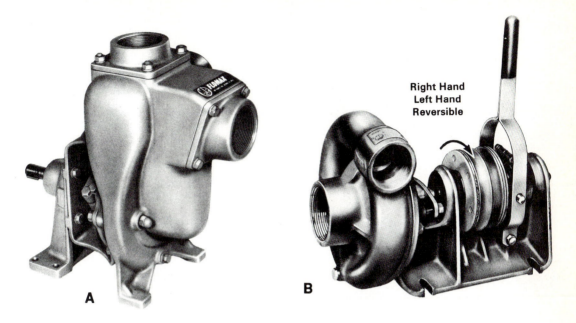

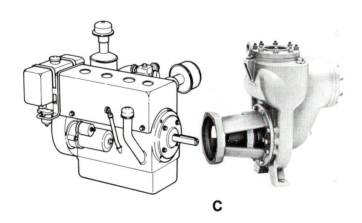

Figure 3.3 Various centrifugal pump configurations. (*Courtesy of MP Pumps, Inc., Tecumseh Products Co., 34800 Bennet Drive, Fraser, MI 48026-1686*)

primary centrifugal pump that can be used for either belt or direct drive. Construction is typically cast iron, bronze, or aluminum. It can be used with a gasoline engine, an electric motor, diesel engine, PTO, or other drive. Some models of this pump are capable of heads up to 145 feet.

Figure 3.3B shows a clutch-driven self-priming unit that is desirable for applications where intermittent operation is required, or where available power

LEFT HAND   REVERSIBLE   RIGHT HAND

Foot Mounted with Mechanical Seal STRAIGHT CENTRIFUGAL PUMPS

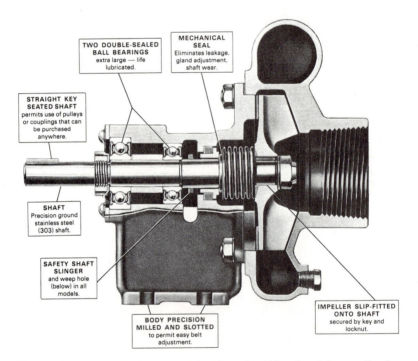

Figure 3.4   Example of a foot mounted with mechanical seal straight centrifugal pump. (*Courtesy of MP Pumps, Inc., Tecumseh Products Co., 34800 Bennet Drive, Fraser, MI 48026-1686*)

is limited. This style pump is well suited to marine applications for bilge, fire protection, wash down, fuel transfer, and as a general ballast pump. In selecting a pump of this type, one should look for the capability of quick removal of the pump from the clutch for maintenance purposes.

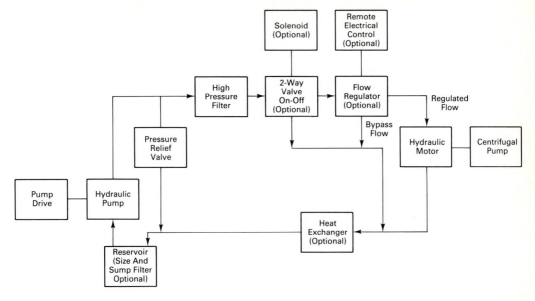

Figure 3.5 A self-priming hydraulic motor drive sealless pump. (*Courtesy of MP Pumps, Inc., Tecumseh Products Co., 34800 Bennet Drive, Fraser, MI 48026-1686*)

Figure 3.3C shows a self-priming centrifugal pump designed for mounting to standard gasoline or diesel engines rated 18 to 35 continuous bhp. This design can handle up to 750 gpm to heads up to 170 feet.

Figure 3.4 illustrates a high-capacity foot-mounted unit. These designs are typically constructed from cast iron or bronze. Housings can be rotated to meet piping requirements. These are considered general-purpose pumps.

Figure 3.5 shows a self-priming hydraulic motor drive sealless pump. A typical hydraulic system for the pumping package is shown also in this figure. The system requires the use of a good grade of hydraulic fluid (e.g., SAE #10 or #20).

One design approach to achieving quick-action priming is the use of a combination foot valve and strainer (Figure 3.6). This design is particularly applicable to all bilge piping installations but can also be used wherever a foot valve is required. This can be used with centrifugal pumps which require positive priming where the water level is below the suction on the pump. A tapped hole is also provided in the strainer housing so that a priming line may be connected directly at the strainer if so desired. If this foot valve and strainer are used in such applications, the time required to prime the pump is reduced considerably.

This unit is extremely simple in design, compact, and easy to install. Since both the foot valve and strainer are combined in one integral unit, installation is greatly simplified and piping is reduced to a minimum. The strainer is semi-spherical in shape and is provided with cast fins. This prevents foreign matter from entering the pump. The valve is a poppet-type valve and is opened by

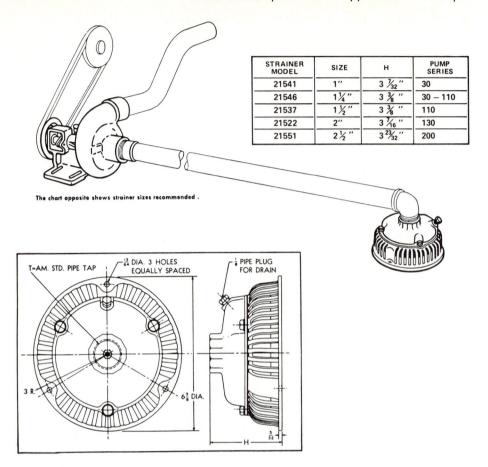

STRAINER MODEL	SIZE	H	PUMP SERIES
21541	1″	3 $\frac{7}{32}$ ″	30
21546	1 $\frac{1}{4}$ ″	3 $\frac{3}{8}$ ″	30 – 110
21537	1 $\frac{1}{2}$ ″	3 $\frac{3}{8}$ ″	110
21522	2″	3 $\frac{7}{16}$ ″	130
21551	2 $\frac{1}{2}$ ″	3 $\frac{23}{32}$ ″	200

The chart opposite shows strainer sizes recommended .

Figure 3.6 Design for a quick-action priming arrangement. (*Courtesy of MP Pumps, Inc., Tecumseh Products Co., 34800 Bennet Drive, Fraser, MI 48026-1686*)

a vacuum in the suction line. The valve is extremely light and lifts with only a slight vacuum. This combined with streamlined flow in the valve housing results in a minimum restriction to flow. The valve seal is made of a soft neoprene which insures good seating. All other parts are of bronze construction and are corrosion resistant. This unit should always be installed with the mounting base in a horizontal position.

Examples of self-priming pumps suitable for environmental control applications are shown in Figure 3.7. Shown are cast iron dewatering pumps, cast iron trash pumps, and high-pressure pumps. The self-priming centrifugal pump shown is manufactured in line sizes ranging from 1.5 to 10 inches and is available with gasoline, electric, or diesel power. Manufacturer specifications are given in Table 3.2A.

The trash pump is also self-priming and is designed to pass large solid materials in heavy concentrations. This manufacturer's unit is available in sizes ranging from 2 inches to 6 inches. Its specification sheet is shown in Table 3.2B.

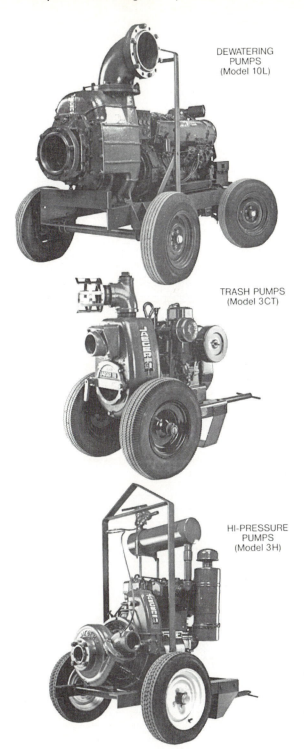

DEWATERING
PUMPS
(Model 10L)

TRASH PUMPS
(Model 3CT)

HI-PRESSURE
PUMPS
(Model 3H)

Figure 3.7 Examples of self-priming centrifugal pumps used in dewatering and trash pumping operations. (*Courtesy of MP Pumps, Inc., Tecumseh Products Co., 34800 Bennet Drive, Fraser, MI 48026-1686*)

TABLE 3.2   MANUFACTURER'S SPECIFICATIONS

Table 2A.   A range of self-priming centrifugal pumps from 1½ to 10 inches. J and M Series light weight aluminum for intermittent duties. Gasoline engine powered FM, C and L Series Heavy Duty cast iron. Available with gasoline, electric or diesel power.

MODEL	CONSTRUCTION	SIZE INCH	MAX. RATING	MAX. SOLID INCH	MAX. FLOW US/GPM	MAX. HEAD FT.
1½J5	ALUMINUM	1½	—	9/16	100	95
2J8	ALUMINUM	2	—	9/16	140	95
1½FM5	CAST IRON	1½	5M	9/16	100	95
2FM8	CAST IRON	2	8M	9/16	140	95
2FM10	CAST IRON	2	10M	3/4	210	125
3FM15	CAST IRON	3	15M	3/4	300	125
3C	CAST IRON	3	17M	15/16	374	118
3C1	CAST IRON	3	20M	13/16	460	120
4L	CAST IRON	4	40M	1¼	990	160
6L	CAST IRON	6	90M	2⁷⁄₁₆	1900	120
6PH	CAST IRON	6	—	1⅛	1550	220
10L	CAST IRON	10	—	2	3700	140

Table 2B.   Ranging from 2 to 6 inches, these self-priming pumps are designed to pass large solid materials in heavy concentrations, where conditions are tough and durability is important. Available with gasoline, electric or diesel power.

MODEL	CONSTRUCTION	SIZE INCH	MAX. RATING	MAX. SOLID INCH	MAX. FLOW US/GPM	MAX. HEAD FT.
2MT5	ALUMINUM	2	—	1	180	90
3MT8	ALUMINUM	3	—	1½	360	88
3DYN DIAPHRAGM	ALUMINUM	3	—	2	90	25
4MT10	ALUMINUM	4	—	2	540	120
4MT16	ALUMINUM	4	—	2	640	120
2CT	CAST IRON	2	10MT	1½	172	110
3CT	CAST IRON	3	18MT	1¾	360	113
4CT	CAST IRON	4	35MT	3	790	130
4LT	CAST IRON	4	35MT	3	790	130
4PT	CAST IRON	4	35MT/40M	2	825	135
6LT	CAST IRON	6	70MT	3	1400	120
6PT	CAST IRON	6	70MT/90M	3	1650	130

Table 2C.   Cast iron volutes and single stage impellers, these pumps are designed for durability and high performance for all types of jetting, irrigation and long distance pumping. Available with gasoline or diesel power.

MODEL	CONSTRUCTION	SIZE S/D INCH	MAX. FLOW US/GPM	MAX. HEAD FT.
3H	CAST IRON	3 x 3	500	380
5H	CAST IRON	5 x 4	950	475
6H	CAST IRON	6 x 6	1600	550

TABLE 3.3 SPECIFIC REQUIREMENTS FOR CENTRIFUGAL PUMPS

- Centrifugal pumps require a screen ahead of the pump to remove debris and fibrous material.

  Screw pumps require no pre-screening and pass any debris as large as the gap between screw flights. Larger objects do not clog pump; they are rejected.

- Centrifugal pumps are subject to abrasion damage caused by solids passing the screen. Repairs become necessary because of the pump's light structural elements and high turning speeds.

  Screw pumps consist of sturdily built screw flights that revolve in a trough and operate at slow speeds of approximately 13 to 110 rpm. Bearings are sealed. Screw pumps require only minimum repair and upkeep.

- Centrifugal pumps reach capacity only with full loading and are efficient only within a limited working range.

  Screw pumps are highly efficient over 70% of the operating capacity and pump from 100% to 0 flow.

- Centrifugal pumps require expensive excavations and concrete work. A deep wet well and pump house are needed, and suction and discharge piping, fittings, valves. A storage well is normally required in which, under certain conditions, influent may become septic and cause odor problems.

  Screw pumps do not require a wet well, pump house, or piping. No loss of head caused by suction and discharge piping. Sump in influent chamber need be only slightly lower than influent line. No need for a storage well to avoid on and off pump operation.

- Centrifugal pumps can disintegrate activated sludge floc.

  Screw pumps lift formed sludge floc gently, with minimum floc break-up.

The high-pressure pump has cast iron volutes and single-stage impellers. These pumps are designed for durability and high performance for all types of jetting, irrigation, and long-distance pumping.

EXAMPLES OF PUMPS IN POLLUTION CONTROL SERVICE

This section provides descriptive information on typical pumping services in pollution control application. It is not intended to be all inclusive, but rather to illustrate by way of example the diversity of services and versatility of common pump styles.

The first example is that of land disposal of liquid waste. One technology in use today involves the use of a direct subsurface injection system, which is designed to continuously deliver large volumes of liquid wastes containing up to 8 percent solids through pipe and hose line directly into the soil. The system can deliver at flow rates of up to 1,200 gpm (72,000 gph) with up to 95 percent operational efficiency. The system can operate at distances of up to several miles. The advantages of such a system are:

- Higher volume delivery and application rate than any other equal investment.
- Rapid capital recovery with low operational and maintenance costs.
- Elimination of pollutant runoff as liquid is injected 6 inches to 10 inches into the soil.
- Subsurface injection minimizes aesthetic and odor problems.
- Maximized retention of nitrogen with conditioning and micronutrient value.

- Low compaction compared to floater trucks or tanks.
- Lower potential health risks by minimizing crop and/or animal contact.
- Soil mixing during injection allows aerobic microbiological activity to occur.
- Delivers at a constant depth regardless of speed or terrain.
- Slope (grade) is less of a restricting factor for direct injection.

In general, the waste is pumped via pipeline by a high-pressure, nonplug, or chopper-type centrifugal pump to the center of a 40-acre field where it is directly injected into the soil. The waste is placed directly at the optimum depth for maximum nutrient uptake by the crop while conditioning and aerating the soil. The uniformity of distribution is controlled by using different configurations of teeth and injector points. Application rates for injecting (up to 40,000 gallons per acre) far exceed those of surface applications, with no need for postapplication cultivation.

When the unit is at the land site, and once pressurized, the waste is pumped through the pipeline to the center of a 40-acre field, and flows through two hoses into an injector manifold system. The waste flows through this distributor to the teeth and mixes directly with the soil at a controlled depth.

As the pull vehicle travels away from the pipeline (parallel to the pipeline), the feed hose (connected to the pipeline) is pulled ahead at a special "pull" elbow, at a right angle to the pipeline, by a second tractor. (Refer to Figure 3.8.)

The system allows for continuous nonstop fence-to-fence application over the entire field while minimizing hose wear and preventing rolling, as the drag hose in the system is *always* being pulled in a straight line.

A swivel arm manifold and hose control allows the sufficiently powered tractor to turn the corners at the end of the field and return *without backing up* while leaving the teeth on the soil. This eliminates the need to shut off the flow while turning corners, assuring efficient continuous flow and smoother operations.

One manufacturer of this type system is L.W.T., Inc. (Box 250, 422 Mill Street, Somerset, Wisconsin 54025). LWT's system is built in their plant, not by an outside job shop. They manufacture the injector tool bar, hose couples, hose reel, portable aluminum pipeline, flow meters, and valves, and assemble pumping units and custom-build the pipe and pump trailers, power primers, and accessories for optimum operation.

LWT uses computer programs to calculate flow in response to different parameters and equipment. This allows quick and accurate evaluation costs. The estimated maximum gallonage at distance for this system is given in Figure 3.9.

Another common pumping application is the use of hydraulic chopper pumps to move heavy solids, trash, scum, and debris. These powerful submersibles cost effectively remove more solids with less water. Hydraulics offer controls over the slurry flow and delivered horsepower. The slurry gate allows

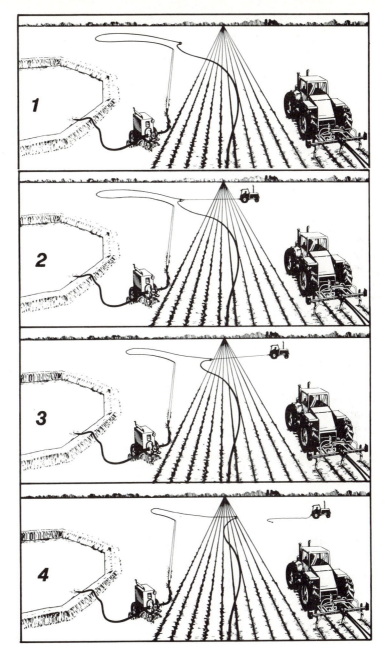

Figure 3.8 Sequential operation of land disposal technology used for liquid waste.

ESTIMATED MAXIMUM GALLONAGE AT DISTANCE
PIPE SIZE SPECIFIED

GALLONS PER MINUTE VS. DISTANCE PUMPED (Ft.)

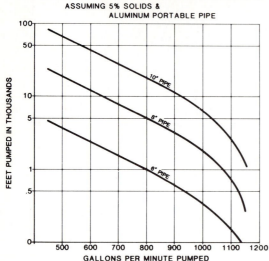

ASSUMING 5% SOLIDS &
ALUMINUM PORTABLE PIPE

10" PIPE

8" PIPE

6" PIPE

FEET PUMPED IN THOUSANDS

GALLONS PER MINUTE PUMPED

Figure 3.9 Estimated maximum gallonage at distances for LWT system. (*Courtesy of L.W.T., Inc., PO Box 250, 422 Mill Street, Somerset, WI 54025*)

heavier sludges to be mixed and then removed. For greater agitation, use a digester propeller agitator. Collapsed, this unit will enter any manhole of 24-inch minimum opening and travels on its mast. A 26-inch minimum diameter collapsible prop is hydraulically driven with up to 90 horsepower for powerful circulating and mixing. Figure 3.10 shows a digester cleaning operation.

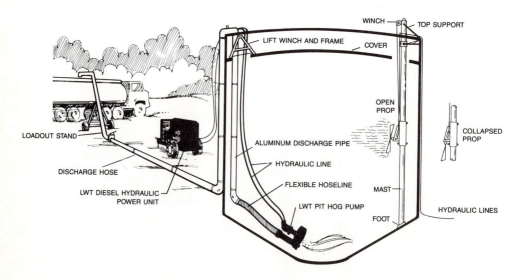

Figure 3.10 A digester cleaning operation.

The features of this type of portable pump are:

• heavy-duty construction
• moves heavy solids concentrations
• positive chopping capability
• handles minimum 3.5-inch solids
• slurry gate for agitation
• submersible for maximum efficiency
• variable flow control
• hydraulic power allows soft starts
• self-priming

Typical applications include:

• digester cleaning
• silt removal
• pond pumping
• sewer bypass
• reception pit pumping
• dewatering
• agitation
• emergency pumping
• hazardous waste

Screw Pumps

Screw pumps are an efficient means of lifting wastewater and storm water. Lifts in the 30-foot range are practical with larger standard pumps. *High-lift* pumps can extend this up to 40 feet or more. Screw pumps provide greater freedom from clogging and height operating efficiencies than other pumps. In addition, they offset variable capacity without elaborate electrical controls.

The screw pump is by far the oldest type of pump, having been in continuous use for irrigation and land drainage since the third century B.C. After the piston pump was developed in the nineteenth century, and later the centrifugal pump, the screw pump was considered obsolete because of its lower head capacities. Few screw pumps had been installed in sewage plants and few were installed for other applications, except for land drainage (refer to Figure 3.11).

Operating experience, however, proved that the centrifugal pump, although excellent for pumping water, had serious shortcomings when handling heavily contaminated wastewaters. In 1955 the Dutch, aware of the difficulties with centrifugal pumps and aware also of the fact that many wastewater pump-

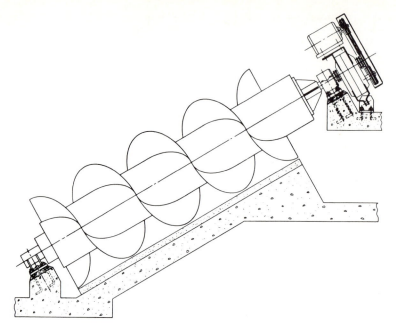

Figure 3.11 Conceptual operation of a screw pump.

ing applications do not require a high head, investigated the inherent efficiencies of the spiral-lift screw.

A prototype double-screw unit was developed employing modern refinements. The first of these screw pumps were installed for flood control in the Netherlands and then in municipal sewage installations in Europe. Based on excellent results, more installations were made in Europe and in a few years this type pump had obtained acceptance throughout the United States and Canada.

Screw pumps provide advantages over centrifugal pumps for wastewater. A prime advantage is the pump's built-in variable capacity. Pumping rate automatically adjusts to the depth of liquid in the inlet chamber. Operation is continuous; speed is constant. Screw pumps provide efficient pumping over a wide range and operate economically to 30 percent of designed capacity. Screw pumps are useful for pumping return sludge from the clarifier back to the aeration basin in activated sludge treatment. The spiral lift is a gentle lift that reduces break-up of activated sludge floc. Peripheral speeds of screw spirals are usually under 754 fpm, as compared with centrifugal pump impeller speeds of 3,500 to 4,000 fpm. Discharge from screw pumps is not pressurized. Activated sludge floc is not broken up in the effluent channel.

Screw pumps operate with a minimum of noise or vibration. Their quiet operation makes them well suited for use near residences. Pump structures can be designed to eliminate dead areas in the influent chamber and the effluent channel. This reduces the possibilities of septic conditions. Screw pump structures can be blended into a landscape and can be installed partially underground.

Typical screw pump applications in the environmental control field include:

Raw sewage lift stations

Screw pumps are nonclogging. They can handle virtually anything that is brought up in raw sewage influent—abrasive solids, rags, cans, wood. Running dry does not harm a screw pump. Motor overloads can't happen. Because screw pumps require little attention they are excellently suited for remote locations.

Sewage plant lift stations

For sewage lifts up to 40 feet. Can be installed in series (two-stage) for higher heads. Screw pumps have self-regulating lift capacity. Output is reduced with lowering level in influent chamber. Pumping rate varies with the submergence of the lower portion of the screw. Efficiency is over 70 percent.

Return activated sludge

Little floc disintegration with screw pumps. Discharge from screws into effluent channel is nonturbulent. Horsepower requirements are less than those required for centrifugal pumps. The use of screw pumps for return sludge pumping can result in improved activated sludge treatment.

Storm water pumping

Screw pumps are ideally suited for storm flows because of the large capacity at low heads. They can be installed to take discharge from storm drains or as standby units to bypass storm flow. Screw pumps can handle storm influent without prescreening.

Land drainage

Screw pumps have been used for hundreds of years for flood control. They are particularly useful for pumping large volumes of water over levees. An actual operation is shown by the photograph in Figure 3.12.

The capacity of a screw pump varies with the following engineering factors:

- diameter of the screw
- speed of the screw

Figure 3.12 Memphis, TN installation of seven 96-inch diameter, three-flight screw pumps. (*Courtesy of Lakeside Equipment Corp., 1022 E. Devon Avenue, PO Box 8448, Bartlett, Illinois 60103*)

- number of flights mounted on the screw shaft (single, double, or triple helix)
- angle of inclination of the screw
- level of influent in influent chamber
- ratio of the diameter of the screw shaft to outside diameter of the screw flights
- clearance between screw flights and trough

To some degree, increases in capacity can be obtained by making adjustments in all of these performance factors. If substantial increases in capacity are required, however, better efficiency can sometimes be obtained with multiple screw pumps.

To design a screw pump of the proper size, two things must be known: (1) the maximum flow to be pumped in gpm, and (2) the maximum lift required in feet. To get the best efficiency, other engineering considerations should be taken into account. These are: (1) inclination of the screw, and (2) screw speed.

Increasing the angle of inclination lowers the output of the screw approximately 3 percent for each degree of increase over the 22° inclination. Therefore, a screw at 30° will have a greater maximum capacity than one inclined at 38°.

There are certain design adjustments that can be made to obtain the same volume with a steeper angle. The speed of the screw can be increased, within limitations imposed by the screw's outside diameter. The screw's outside diameter can be increased.

Speed should be the highest rpm at which overflowing of the liquid over the screw shaft into the next lower chamber is avoided. If screw speed is increased beyond maximum speed, the pump will try to pump liquids over its capacity. Lower efficiency and wasted energy will result. Loss of efficiency will also result if screw speed is reduced by more than 30 percent of calculated maximum speed. Figure 3.13 shows a manufacturer's screw pump lift and capacity curve. Standard screw pump selection factors are given in Table 3.4. In this Table, *H* represents the maximum lift based on standard wall thickness of center pipes. Its definition is given by the sketch shown in Figure 3.14. In this figure, the filling point is the intake water level at which the screw pump reaches its full capacity, best efficiency, and high-power consumption. If the level rises above this point, capacity remains unchanged but power consumption and efficiency will decrease. If the level falls below the filling point, capacity, efficiency, and power consumption will be reduced. If the level falls below the touchpoint, pumping will cease.

Screw pumps can be made in single, double, or triple helix design. Output capacity decreases about 20 percent for each helix omitted. For example, the capacity of a given 2-flight screw would be approximately 80 percent of the same screw with flights. Similarly, a 1-flight screw would have 80 percent of the 2-flight screw capacity and 64 percent of the 3-flight screw capacity. The 3-flight screw will provide the most capacity in the lease space (refer to Figure 3.15).

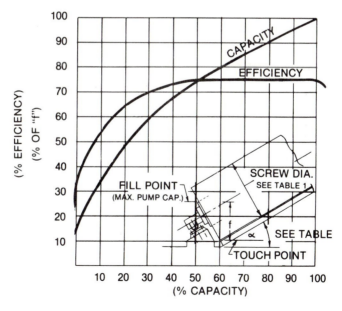

Figure 3.13 Screw pump lift and capacity curve. (*Courtesy of Lakeside Equipment Corp., 1022 E. Devon Avenue, PO Box 8448, Bartlett, Illinois 60103*)

TABLE 3.4 STANDARD SCREW-PUMP SELECTION FACTORS

Screw Dia.	Center Pipe Dia.	Max. RPM	Max Capacity @30° Slope			Max H* @ 30° Slope (for Std. Pipe only)			Max Capacity @ 38° Slope			Max H* @ 38° Slope (for Std. Pipe only)		
			1-Flight	2-Flight	3-Flight	1-Flight	2-Flight	3-Flight	1-Flight	2-Flight	3-Flight	1-Flight	2-Flight	3-Flight
12"	6⅝"	110	150	190	230	7'-3"	6'-6"	6'-0"	NOT RECOMMENDED	140	170	NOT RECOMMENDED	9'-0"	8'-3"
16"	8⅝"	91	300	370	460	8'-9"	8'-0"	7'-6"		260	330		11'-0"	10'-3"
20"	10¾"	79	500	620	770	10'-3"	9'-6"	8'-9"		450	560		13'-3"	12'-3"
24"	14"	70	740	930	1,160	12'-9"	11'-9"	11'-0"		670	830		16'-3"	15'-3"
30"	16"	60	1,275	1,600	2,000	12'-9"	11'-9"	11'-0"		1,150	1,425		16'-6"	15'-6"
36"	20"	53	1,925	2,400	3,000	15'-0"	13'-9"	12'-9"		1,725	2,150		19'-3"	18'-0"
42"	24"	48	2,750	3,425	4,275	16'-9"	15'-3"	14'-3"		2,450	3,075		21'-9"	20'-3"
48"	24"	44	3,875	4,850	6,075	15'-3"	14'-0"	12'-9"		3,450	4,300		20'-0"	18'-6"
54"	30"	41	5,000	6,250	7,800	18'-3"	16'-9"	15'-6"		4,475	5,600		24'-0"	22'-3"
60"	30"	38	6,550	8,175	10,225	16'-9"	15'-6"	14'-0"		5,800	7,250		22'-6"	20'-9"
66"	36"	36	8,050	10,075	12,575	19'-6"	17'-9"	16'-6"		7,250	9,050		26'-0"	24'-0"
72"	36"	34	10,125	12,650	15,800	18'-3"	16'-6"	15'-3"		8,975	11,225		24'-6"	22'-6"
80"	36"	31	12,300	15,350	19,200	16'-9"	15'-3"	13'-9"		10,925	13,650		22'-9"	20'-9"
84"	42"	30	14,175	17,725	22,150	19'-3"	17'-6"	15'-9"		12,575	15,725		26'-0"	23'-9"
90"	48"	29	16,575	20,700	25,900	21'-3"	19'-6"	17'-9"		14,850	18,550		28'-9"	26'-6"
96"	48"	28	19,750	24,700	30,850	20'-3"	18'-3"	16'-6"		17,525	21,900		27'-6"	25'-3"
102"	48"	27	22,450	28,075	35,100	19'-3"	17'-6"	15'-9"		19,925	24,900		26'-6"	24'-3"
108"	54"	26	26,100	32,625	40,800	21'-0"	19'-0"	17'-3"		23,175	28,950		28'-9"	26'-6"
114"	60"	25	29,100	36,375	45,475	22'-9"	20'-9"	18'-9"		26,025	32,525		31'-3"	28'-9"
120"	60"	24	33,050	41,325	51,650	21'-9"	19'-9"	17'-9"		29,325	36,675		30'-0"	27'-6"
126"	66"	23	36,325	45,400	56,750	23'-3"	21'-0"	19'-0"		32,425	40,550		32'-0"	29'-3"
132"	66"	22	40,325	50,425	63,025	22'-3"	20'-0"	18'-3"		35,800	44,725		31'-0"	28'-3"
138"	72"	22	45,650	57,050	71,325	23'-9"	21'-6"	19'-6"		40,750	50,950		33'-0"	30'-3"
144"	72"	21	49,975	62,475	78,100	22'-9"	20'-6"	18'-6"		44,350	55,450		32'-0"	29'-0"

*Maximum lift (H) based on standard wall thickness of center pipes.

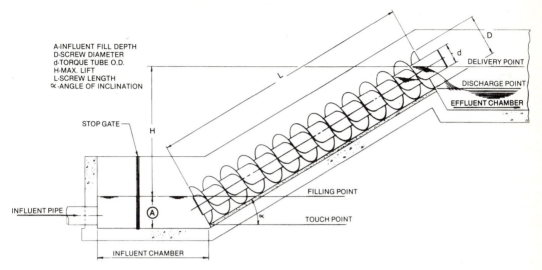

Figure 3.14 Screw pump operation and definition of terms. (*Courtesy of Lakeside Equipment Corp., 1022 E. Devon Avenue, PO Box 8448, Bartlett, Illinois 60103*)

Screw pump power requirements are based on a calculation of capacity and lift. The required brake horsepower is determined first, with the formula

$$bhp = \frac{gpm \times lift\ (H)}{2,970}$$

The motor horsepower is then derived from the bhp, taking into account limitations of the motor and gear reducer, with the formula

$$Motor\ hp = \frac{bhp}{.95}$$

Pump capacity varies directly with the pump speed in the upper speed range. Lift is the vertical distance from the filling point to the discharge point. Screw pump efficiencies in the 70 percent to 75 percent range are considered normal. The following example illustrates a design calculation.

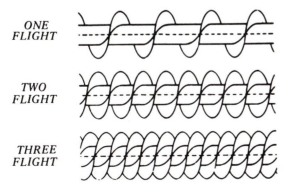

ONE FLIGHT

TWO FLIGHT

THREE FLIGHT

Figure 3.15 Screw configuration as a function of number of flights.

Demonstration of screw pump selection:

KNOWN INFORMATION

Required capacity   3,500 gpm
Required lift   15 ft.

- Determine the angle of inclination. In this case we assume a 38° angle.
- From Table 3.4 and the required pump size based on required capacity and lift, 4,300 gpm is closest to 3,500 gpm without going below it. This capacity is that of a 48-inch diameter screw pump with 3 flights. The maximum lift from the table is 18′–6″.

    NOTE: The 48-inch diameter 2-flight screw is only 50 gpm short of the 3,500-gpm design and is more economical and should be considered where this difference in capacity is not critical.

- From Table 3.4 find the speed required to pump the maximum capacity of 4,300 gpm. It is 44 rpm.
- The capacity of the screw pump is directly proportional to speed. Therefore, to find the required speed multiply the ratio of required capacity to maximum capacity. In this case

$$44 \times \frac{3,500}{4,300} = 44 \times .81 = 35.81 \text{ rpm}$$

Round off to next higher rpm = 36 rpm.

- Determine the required motor hp from the preceding formula.

Good design in screw pump installations is dependent on good design of the influent and effluent basins. Influent and effluent levels are calculated for anticipated flows.

The influent chamber must be designed to achieve full screw pump capacity. Influent should feed to the screw at the highest inflow level. Required capacity, amount of inflow, and the amount of required influent storage in the chamber must be known in order to calculate the length, width, and depth of a chamber in which the level will be maintained high enough to reach the filling point of the screw. Screw pump output efficiency and capacity fall off when the level in the influent chamber sinks below the filling point of the screw.

The influent basin should be designed so each pump can be isolated and enough room provided around the lower bearing for maintenance personnel should it be required. The basin should be large enough to prevent the water level from rising when a pump shuts off and activating a float switch to restart the pump.

If the liquid level is too high in the effluent channel, the discharge from the screws will try to flow back. Effluent level should be the highest point at which runoff will not cause backflow. If parallel screw pumps empty into a common

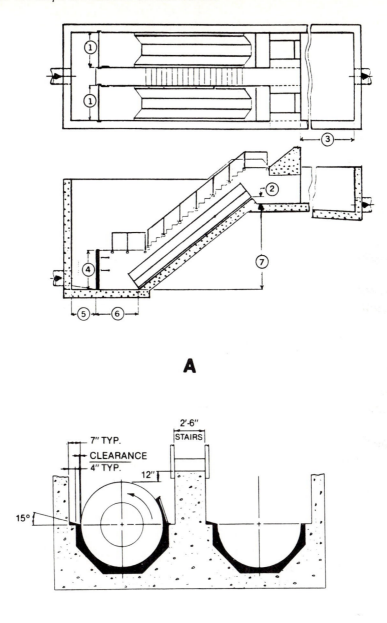

A

B

Figure 3.16 (A) Screw pump concrete layout (refer to Table 3.5 for actual dimensions). (B) Concrete framing detail. Note that concrete area shaded should be framed and poured in a second operation after the other concrete has set and dried.

TABLE 3.5   TYPICAL CONCRETE DIMENSIONS FOR STANDARD SCREW PUMPS

TABLE — TYPICAL CONCRETE DIMENSIONS FOR STANDARD SCREW PUMPS @38°

Screw ∅	(1)	(2) 2-Flgt.	(2) 3-Flgt.	(3)	(4)	(5)	(6)	(7) 2-Flgt.	(7) 3-Flgt.
12	2'-2"	6"	6"	4'-0"	2'-0"	3'-0"	5'-0"	9'-4"	8'-7"
16	2'-6"	6"	6"	4'-0"	2'-3"	3'-0"	5'-0"	11'-6"	10'-9"
20	2'-10"	6"	6"	4'-0"	2'-6"	3'-0"	5'-0"	13'-11"	12'-11"
24	3'-2"	6"	6"	4'-0"	2'-9"	3'-0"	5'-0"	17'-1"	16'-1"
30	3'-8"	6"	9"	4'-0"	3'-0"	3'-0"	5'-6"	17'-6"	16'-3"
36	4'-2"	9"	9"	5'-0"	3'-0"	4'-0"	5'-6"	20'-3"	19'-0"
42	4'-8"	9"	1'-0"	5'-0"	3'-6"	4'-0"	5'-6"	23'-0"	21'-3"
48	5'-2"	1'-0"	1'-3"	5'-0"	4'-0"	4'-0"	6'-6"	21'-2"	19'-5"
54	5'-8"	1'-3"	1'-6"	5'-0"	4'-6"	4'-0"	6'-6"	25'-2"	23'-2"
60	6'-2"	1'-6"	1'-9"	6'-0"	4'-9"	5'-0"	6'-6"	23'-7"	21'-7"
66	6'-8"	1'-6"	2'-0"	6'-0"	5'-3"	5'-0"	7'-6"	27'-5"	24'-11"
72	7'-2"	2'-0"	2'-3"	6'-0"	5'-6"	5'-0"	7'-6"	25'-6"	23'-3"
80	7'-10"	2'-0"	2'-6"	6'-0"	5'-9"	5'-0"	7'-6"	23'-11"	21'-5"
84	8'-2"	2'-3"	2'-9"	7'-0"	6'-0"	5'-0"	7'-6"	27'-3"	24'-6"
90	8'-8"	2'-6"	3'-3"	7'-0"	6'-6"	5'-0"	8'-6"	30'-0"	27'-0"
96	9'-2"	2'-9"	3'-6"	7'-0"	6'-9"	5'-0"	8'-6"	28'-8"	25'-8"
102	9'-8"	3'-0"	3'-9"	8'-0"	7'-0"	5'-0"	10'-0"	27'-6"	24'-6"
108	10'-2"	3'-6"	4'-3"	8'-0"	7'-6"	5'-0"	10'-0"	29'-7"	26'-7"
114	10'-8"	3'-6"	4'-6"	8'-0"	7'-9"	5'-0"	10'-0"	32'-5"	28'-11"
120	11'-2"	4'-0"	5'-0"	8'-0"	8'-0"	5'-0"	10'-0"	30'-9"	27'-3"
126	11'-8"	4'-3"	5'-3"	8'-0"	8'-0"	5'-0"	10'-0"	32'-10"	29'-1"
132	12'-2"	4'-6"	5'-6"	8'-0"	8'-0"	5'-0"	10'-0"	31'-9"	28'-0"
138	12'-8"	4'-9"	6'-0"	8'-0"	8'-0"	5'-0"	10'-0"	33'-9"	29'-9"
144	13'-2"	5'-0"	6'-3"	8'-0"	8'-0"	5'-0"	10'-0"	32'-8"	28'-5"

Dimension (2) should be designed so downstream conditions will not cause water to stand higher than the discharge point minus 3" to prevent spill-over into adjacent pumps.

Dimension (7) is calculated with the screw at 38° slope and the maximum lift. This dimension can be decreased but not increased.

effluent chamber, the discharge point of all screw troughs must be as high, or higher, than the backflow in the channel under the least favorable operating conditions. Poor design of the effluent channel can result in low efficiency and uneven operation of the screw. Figure 3.16 shows a recommended screw pump concrete layout (Figure 3.16A), along with details of the concrete framing (Figure 3.16B). Typical concrete dimensions for sketch A are given in Table 3.5.

Submersible and Solids-Handling Pumps

There are a variety of hydraulic, electric, and mechanical pumps which are applied to such applications as dredging operations, digesters, sewer bypass, construction sites, lagoons, wastewater treatment plans, lift stations, flood water control areas, and irrigation. Examples of a digester emptying system and a

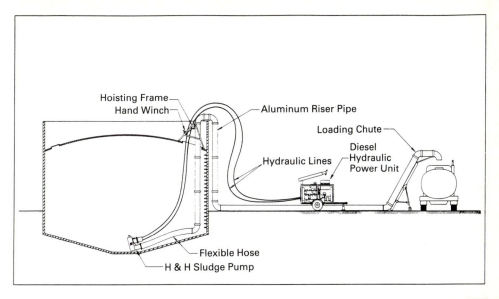

Figure 3.17 Example of a digester emptying system. (*Courtesy of H & H Pump and Dredge Company, PO Box 486, Clarksdale, MS 38614*)

sewer bypass system are shown in Figures 3.17 and 3.18, respectively. Such systems can be supplied as a turnkey operation.

Examples of pumps designed to handle high-solids loadings are shown in Figures 3.19 through 3.25. Figure 3.19 shows a diesel power unit which powers

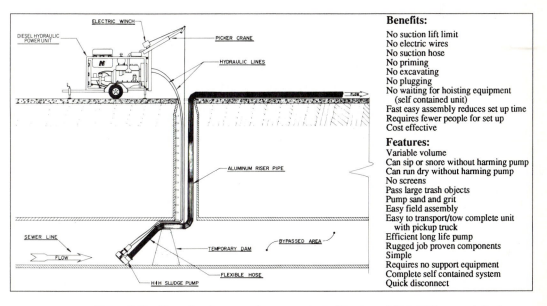

Benefits:
No suction lift limit
No electric wires
No suction hose
No priming
No excavating
No plugging
No waiting for hoisting equipment
  (self contained unit)
Fast easy assembly reduces set up time
Requires fewer people for set up
Cost effective

Features:
Variable volume
Can sip or snore without harming pump
Can run dry without harming pump
No screens
Pass large trash objects
Pump sand and grit
Easy field assembly
Easy to transport/tow complete unit
  with pickup truck
Efficient long life pump
Rugged job proven components
Simple
Requires no support equipment
Complete self contained system
Quick disconnect

Figure 3.18 Example of a sewer bypass system. (*Courtesy of H & H Pump and Dredge Company, PO Box 486, Clarksdale, MS 38614*)

Figure 3.19   Floating hydraulic submersible pump.   (*Courtesy of H & H Pump and Dredge Company, PO Box 486, Clarksdale, MS 38614*)

a high-volume, medium-head, submersible steel pump capable of passing solids and moving the most viscous materials like pulp and paper-mill waste without suction hose. Self-priming pumps are available in sizes up to 16-inch diameter, flows to 9,000 gpm. This system floating pump can be configured as a simple pump on floats or as a remote dredging unit complete with cutterhead and

Figure 3.20 An electric hydraulic power unit. (*Courtesy of H & H Pump and Dredge Company, PO Box 486 Clarksdale, MS 38614*)

windlass winch travel system. Power units range 40 to 177+ hp, feature open center hydraulic systems mounted on a highway towable trailer with an electric winch mounted on an extendable picker crane. Versatile and designed for varied and continuous contractor use, these power units can be used for powering any hydraulic pumping needs from 15 gpm to 120 gpm at 2,000 psi. Figure 3.20 shows an electric hydraulic power unit. This design combines the clean efficiency of electric power with the versatility of hydraulic pumping equipment. Electric hydraulic power units feature variable flow, highly energy-efficient operation, and are mounted on a reservoir. Ranging from 20–300+ hp, these units power low maintenance, highly durable steel sludge pumps capable of moving the most viscous materials without suction hose or auger. Typical flow rates are from 200–9,000 gpm. The maintenance benefits of electric hydraulic submersible pumps over shaft-driven, permanently mounted electric pumps are tremendous. Downtime is measured in minutes, not days. In addition, these electric sludge pump systems feature variable flow rates and can pump liquids up to 100°C.

A sectional view of a hydraulic submersible pump is shown in Figure 3.21. The hydraulic motor (1) is of the oil hydraulic gear type. The motor bolts to the top of the bearing housing (2) which contains grease lubricated bearings (3) and a grease seal (4). The motor output shaft (5) connects to the pump shaft (6) which drives an impeller hub (7) and the impeller (8).

Oil flows from the pressure port (9) to the return port (10) of the hydraulic motor, turning gears within the motor. These gears turn the motor shaft which, through direct drive, turns the pump shaft and impeller. Direct drive means the motor rpm and impeller rpm are the same.

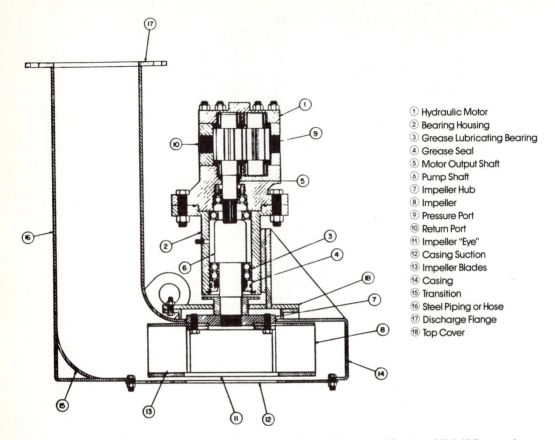

Figure 3.21 Sectional view of hydraulic submersible pump. (*Courtesy of H & H Pump and Dredge Company, PO Box 486, Clarksdale, MS 38614*)

① Hydraulic Motor
② Bearing Housing
③ Grease Lubricating Bearing
④ Grease Seal
⑤ Motor Output Shaft
⑥ Pump Shaft
⑦ Impeller Hub
⑧ Impeller
⑨ Pressure Port
⑩ Return Port
⑪ Impeller "Eye"
⑫ Casing Suction
⑬ Impeller Blades
⑭ Casing
⑮ Transition
⑯ Steel Piping or Hose
⑰ Discharge Flange
⑱ Top Cover

The impeller is of the radial flow centrifugal type. As the impeller turns, the centrifugal action creates a vacuum at the impeller "eye" (11) and the casing suction (12) which causes material to flow into the impeller blades (13).

As the material flows through the blades, increasing centrifugal force impels the fluid outward with increasing velocity. The high velocity at the blade tips (outer impeller diameter) is converted to pressure in the casing (14). A higher speed on the impeller means higher blade velocities and therefore higher pressures.

The casing discharges the flow through a transition (15) into the steel pipe (16). Other piping or hose is connected to the discharge flange (17).

These pumps are submersible, which means the casing has to be submerged in the liquid in order to pump. If the top cover (18) is above the liquid level, it will cause air to enter the casing and thus prevent liquid from being discharged. As soon as the casing is again submerged, it will self-prime and start pumping. Figure 3.22 shows a hydraulic submersible pump, diesel engine power unit, and discharge hose in operational position.

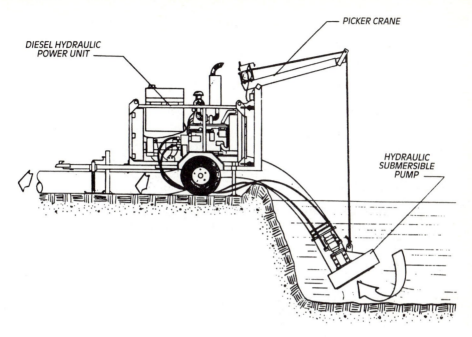

Figure 3.22 Operational configuration of a hydraulic submersible pump.

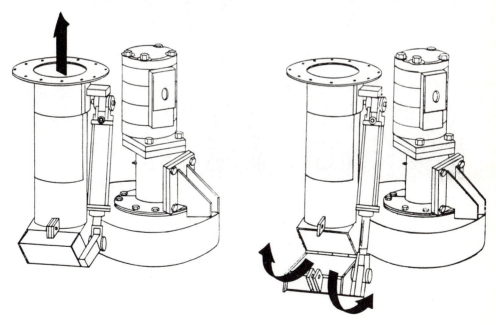

Figure 3.23 A slurry gate pump which combines mixing and pumping into a single operation.

A somewhat unique design for transporting sludge is a slurry gate pump. Illustrated in Figures 3.23 and 3.24, the design combines both a pump machine with a mixer. A hydraulic cylinder mounted on the pump head operates a door on the discharge that can be opened and closed remotely from a diesel hydraulic power unit (with slurry gate circuit). Digester sludges, thixotropic sludges, or any material with solids in suspension may be rendered more easily pumpable by the agitation and homogenizing mixing action provided by a slurry gate pump. Figure 3.24 shows a photograph of this type of pump. The operation can flip a lever at the power unit, and the submersible unit changes from a pump to an efficient slurry and mixer. The unit shown in Figure 3.24 is available in discharge sizes from 4 to 12 inches, at flows from 775 gpm to 6,000 gpm at 25 ft tdh.

Another common solids handling application is dredging. These systems can employ slurry gate pumps. Figure 3.25 shows the operating features of a turnkey system. The system shown is a fully manned 6-inch dredging process

Figure 3.24   Photograph of a slurry gate pump. (*Courtesy of H & H Pump and Dredge Company, PO Box 486, Clarksdale, MS 38614*)

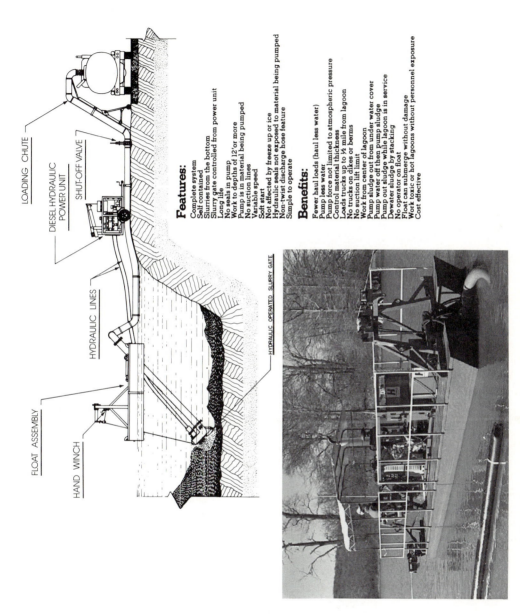

Features:

LOADING CHUTE

DIESEL HYDRAULIC POWER UNIT

SHUT-OFF VALVE

HYDRAULIC LINES

FLOAT ASSEMBLY

HAND WINCH

HYDRAULIC OPERATED SLURRY GATE

Complete system
Self contained
Slurries from the bottom
Slurry gate controlled from power unit
Long life
No seals in pump
Work to depths of 12' or more
Pump is in material being pumped
No suction lines
Variable speed
Soft start
Not affected by freeze up or ice
Hydraulic seals not exposed to material being pumped
Non-twist discharge hose feature
Simple to operate

Benefits:

Fewer haul loads (haul less water)
Pump less water
Pump force not limited to atmospheric pressure
Control material thickness
Loads trucks up to 1½ mile from lagoon
No trucks on dikes or berms
No suction lift limit
Work from center of lagoon
Pump sludge out from under water cover
Pump water off then pump sludge
Pump out sludge while lagoon is in service
Dewater sludge by stacking
No operator on float
Float can submerge without damage
Work toxic or hot lagoons without personnel exposure
Cost effective

Figure 3.25 A compact dredging system. (*Courtesy of H & H Pump and Dredge Company, PO Box 486, Clarksdale, MS 38614*)

87

useful for small dredging operations. The components of the system include a 100-hp diesel engine, the dredge pump, windlass or three-winch travel system, an 8-foot wide auger/tyne cutterhead, and 250 feet of flexible discharge tubing.

A solids handling pump designed for intermittent service is the sump pump. This pump usually comes complete with motor, discharge piping, support plate, and sump-level control. It is designed for pumping wastewater, liquids with abrasives, soft stringy solids, or cannery waste. The pump is suspended in liquid while the motor remains at floor level, away from moisture and contamination. Weather-protected motor enclosure is usually standard. Figure 3.26 shows some typical configurations. Sketch A (Figure 3.26) shows an open-shaft style used for storm drains, sewage, and industrial wastes. Motor bearings carry the thrust load through adjustable couplings. Intermediate bearings are usually bronze for grease lubrication. A renewable sleeve protects the shaft at the pump end bearing. The pump end bearing is usually rubber for waste lubrication or bronze for grease lubrication.

Sketch B in Figure 3.26 is again an open-shaft design, for low capacities and heads. This style is used when pumpage does not contain large solids. Thrust bearing is in the motor support. Intermediate bearings are grease lubricated. The lower bearing is water-flushed rubber or grease-lubed bronze.

Sketch C shows an enclosed-shaft design. This style can be either a water-flush system or equipped for oil lubrication. The design is used for conditions where pumpage contains light abrasives.

A final sump pump illustration is given in Sketch D (Figure 3.26). This

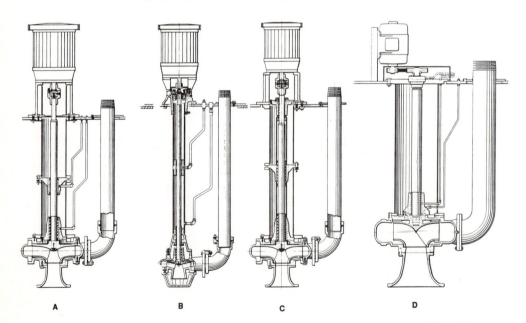

A   B   C   D

Figure 3.26   Example of sump pump styles. (*Courtesy of Cornell Pump Co., 2323 SE Harvester Drive, Portland, Oregon 97222*)

open-shaft design is used for the transfer of fruits, vegetables, and various food products in water.

The final pump style we discuss for solids/slurry handling operations is the plunger pump. A plunger pump is a positive-displacement pump. Positive-displacement pumps are recommended for handling concentrated sludge and scum, mainly because they can pump viscous liquids containing entrained gas without losing prime. Also, because it has large unobstructed passages throughout and no close tolerance parts in the path of fluid flow, a plunger pump is not susceptible to clogging, making it well suited to applications with large solids and heavy sludge.

A plunger pump consists of a plunger, a pump body, and ball check valves on both sides of the pump body. The pump uses the reciprocating motion of the plunger to move liquid through the pump. Pumping is accomplished as follows:

1. *Suction stroke (upward motion)*: The discharge ball valve closes and a low-pressure area is created in the pump body. This opens the suction ball valve and pulls liquid into the pump.

2. *Discharge stroke (downward motion)*: The suction ball valve closes and the compression of the fluid in the pump body opens the discharge ball valve and pushes the fluid out of the pump.

This operation is repeated for every shaft revolution, moving the fluid through the pump.

As a positive-displacement pump, the maximum allowable flow is determined by the pump speed (i.e., number of strokes/minute) and the stroke length. (The pump displaces a measured volume for each cycle.) The maximum pump speed has been traditionally set at 50 rpm; maximum stroke length on many pumps is 7 inches. Therefore, the maximum capacity from a single plunger is approximately 45 gpm, 90 gpm, and 135 gpm for 7.5-inch, 9-inch, and 11-inch plungers, respectively. If more capacity is required, multiple plungers driven from a common shaft are used resulting in duplex, triplex, and quadruplex pumps. One of the largest pumps manufactured is an 11-inch quadruplex with a maximum capacity of 540 gpm. Refer to Figure 3.27.

Different pump models are available for continuous operation at different pressure ranges. Some model pumps operate up to 80 psi (180 feet); others operate to 100 psi (230 feet); and some pumps operate to 130 psi (300 feet). In these different models the capacity ranges are the same. Use of heavier components (castings, shafts, bearings, and so on) allow the respective pump models to operate at the higher pressures.

Plunger pumps are primarily manufactured in cast iron. Some components can be made of other materials for corrosion or abrasion resistance. Materials of the main components are:

1. *Base plate*: Fabricated from ASTM A36 carbon steel. Substantial members are used to resist buckling.

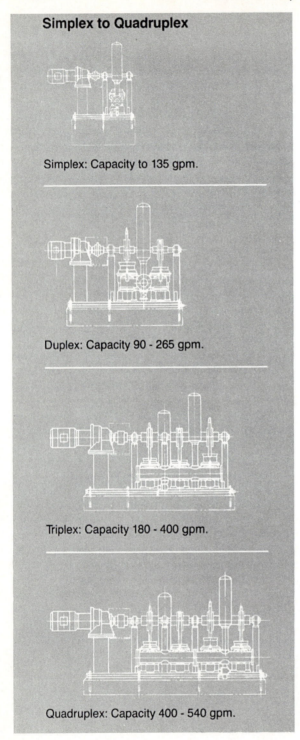

Simplex to Quadruplex

Simplex: Capacity to 135 gpm.

Duplex: Capacity 90 - 265 gpm.

Triplex: Capacity 180 - 400 gpm.

Quadruplex: Capacity 400 - 540 gpm.

Figure 3.27 Plunger pump configurations.

2. *Pump shaft*: turned ground and polished, stress-proof steel, 25,000 psi minimum.

3. *Bearing liner*: Marine-grade Babbitt, replaceable-type insert. Phenolic is available as an option.

4. *Bearings*: Self-aligning, grease-lubricated, tapered, double-row roller bearings. Self-aligning, grease-lubricated ball bearings are also available.

5. *Pump body, valve chambers, connecting rod*: ASTM A48, Class 30 minimum cast iron.

6. *Valve seats*: Cast iron standard on many pumps; 89 Durometer neoprene rubber is standard. Stainless steel is optional on all pumps for corrosion resistance and for high-pressure applications.

7. *Plungers*: Polished cast iron is standard with the following options:
   a. Gas nitride (hardened) plunger
   b. Chrome oxide (flame-sprayed ceramic)
   c. Urethane coated

In general plunger pumps can be used in sludges up to 12 percent solids (by volume). With the proper application of a service factor and possibly a larger drive, plunger pumps can be used in installations with higher than 12 percent solids.

When sizing a pump, use the following steps:

1. Determine required capacity. Then from coverage chart, select the plunger size and number of plungers.
2. Determine the operating pressure and again from a coverage chart, select the pump model.
3. From a performance curve, select the required motor horsepower, always selecting the motor horsepower to the right of the point of conditions.

Plunger pumps are typically installed in municipal and industrial wastewater applications. Some examples are:

1. *Primary sludges*: The pump's large clearances and solids capability allow it to handle sticks, stringy solids, stones, and so on, settled out of plant influent.
2. *Thickened sludges*: Plunger pumps can handle heavy sludges up to 12 percent.
3. *Septage*: Large solids, rags, and trash will not clog plunger pumps.
4. *Scum*: Plunger pump's self-priming and dry-run capability allow scum wet well to be pumped completely dry, thereby removing all the scum, including the floating portion.

5. *Process feed*: As a positive-displacement pump, it can meter sludge to belt-filter presses, rotary-drum vacuum filters, or other processes requiring accurate measured flow.

The advantages and features of a plunger pump are:

1. *Self-priming*: Plunger pumps can self-prime and pull high-suction lifts. This allows the pump to be placed above an in-ground tank eliminating the need for a deep dry well or pump room below the tank. (Excellent for septage-receiving facility).

2. *Runs dry*: The plunger pump can run dry without damage. Should the tank be pumped dry or suction flow be temporarily interrupted, the pump will continue to run without damage until suction flow is restored (scum applications).

3. *Large openings*: The plunger pump has large openings that can handle 3-inch solids, sticks, stringy solids, and rags without clogging. Dual-ball check valves can be added to further decrease the possibility of clogs (primary sludge applications).

4. *Flexibility of design flow*: Each plunger pump has adjustable stroke lengths. By changing the stroke, the flow rate can be changed. The same pump can be used for several different design conditions, eliminating the cost of multiple pumps or a variable-speed drive.

5. *Positive, controlled output*: Flow from a plunger pump is independent of the pressure it must pump against, allowing it to be used as a metering (feed) pump.

6. *Slow operating speed*: Plunger pumps operate at 50 rpm so bearing failures are virtually eliminated. Wear to pump bodies and valve chambers is minimal due to low fluid velocities through the pump.

7. *Easy and inexpensive to maintain*: Normal wear and replacement parts are few, inexpensive, and easily accessed. Maintenance time required is low and highly trained technicians are not required.

8. *Rugged and dependable*: Plunger pumps have been used in wastewater applications for over 50 years. They are heavy duty and can operate in the most difficult installations. Heavy construction and liberal service factors result in many years of trouble-free installation. In addition, because of its mode of operation and design, should considerable wear occur, the plunger pump will continue to operate with only minimal loss of efficiency.

9. *High efficient and trouble-free drive*: These pumps use a totally enclosed, oil-lubricated gear reducer to reduce an 1,800-rpm motor to the 50-rpm pump speed required. The selected reducer has a 35:1 reduction ratio in single reduction. Since the motor is direct coupled to the reducer, the life of motor bearings is very high. The single-reduction feature provides a

highly efficient drive system and the totally enclosed unit requires low maintenance.

Portable Pumps for Drums and Containers

Since 1950, portable drum pumps have provided the safest and most efficient means of transferring chemicals from drums and carboys. Drums and carboys are emptied safely and efficiently by the pump that is specifically designed for the task—a portable drum pump. The inherent design of drum pumps allows chemicals to be completely transferred in the shortest possible time with complete safety.

A quality drum pump fulfills four criteria:

Mobility

Pumps range in weight from 7.5 pounds to 26 pounds depending on tube construction and type of motor used. This low weight allows for a one-man operation in most cases. Semipermanent piping or large hoses are avoided with a drum pump. A drum pump can be moved from one drum to another with a minimum of labor.

Versatility

A sealless design prevents the damage by inadvertent dry running. Flow rates up to 50 gpm and heads up to 38 feet are available for high-flow applications or head conditions. Viscosities up to 750 cs and specific gravities up to 1.8 can be pumped with a drum pump. Some pumps provide both a high-flow rotor and a high-head impeller with each pump. A tool is provided for instantaneous change over if conditions of service change.

To be truly versatile a drum pump is built on a modular system in which all motors and pump tubes are interchangeable. This feature allows an end user to purchase numerous tubes for various chemicals and use one motor as the power source. A simple hand wheel couples the self-aligning motor and tube in seconds, thus eliminating the need to purchase a single pump for each application.

In instances where materials are above the centrifugal range, a progressive cavity design is available with an integral-speed reducer for viscosities up to 100,000 ssu. The positive-displacement pump is also interchangeable with the motors from the centrifugal line.

Compatibility

A wide range of materials, polypropylene, stainless steel 316, PVDF, aluminum and Hastelloy C, are available for guaranteed compatibility of fluids being handled. The low price of pump tubes enables a user to purchase the proper material for construction without fear of marginal compatibility.

Safety

A portable drum pump avoids the safety hazards associated with drum unloading by forklift. Manual unloading via tilt cart is not cost effective or safe. The dangers of environmental contamination are eliminated when a drum pump is used versus forklift unloading or tilt-cart unloading.

Portable electric drum and container pumps have been on the market since about 1950. During these early days there was understandably only one motor of specified power available and the materials were also limited at that time to two grades which, in contrast to today, allowed only restricted applications.

Very soon, however, the drive motors of the drum pumps were brought into line with the service requirements both in terms of power and degree of protection. And the wide range of liquids which required pumping also necessitated the use of ever better and more resistant material for the pumping units.

Today, the portable drum and container pump has become firmly established among the wide array of pumps and has become indispensable in factories where it is necessary to pump out or transfer the most varied liquids rationally and safely.

The drum pump is a high-power, reliable, portable unit. Its application is in most cases, although not exclusively, determined by its mobility.

As long as liquids are transported and stored in drums and tanks, it will not be possible to do without portable drum pumps. Particularly in the highly industrialized areas, these pumps have become a significant means of rationalization in all branches of industry.

The fluid being pumped must be handled with care. Water-like liquids as well as highly viscous fluids can be pumped equally well. The pumps are of simple construction, easy to dismantle and clean.

The absolutely latest state of the art in the field of portable drum and container pumps is the multipurpose drum pump, a new generation of pumping unit which meets the four criteria for safe efficient chemical transfer outlined earlier.

A portable multipurpose pump tube system allows:

1. *A choice of pump tube material*: polypropylene, stainless steel, and aluminum are available as suitable and reliable pump tube materials for almost any liquid. Additionally we offer PVDF and hastelloy C for special cases.

2. *A choice of sealing arrangement*: Two reliable types of sealing are available within the pump tube system—with mechanical seal or sealless. The pump tubes with mechanical seal are recommended for applications involving crystallizing, curing, and setting agents. Dry running must be avoided. In all other cases, preference should be given to the sealless design which will not fail if subjected to limited periods of dry running.

3. *A choice of impeller system*: Experience shows that two functions are most commonly required—overcoming higher pressure differentials (hose length, viscosity, difference of level) and hence the need for a higher deliv-

ery head, and free outflow and-low pressure losses, but with a requirement for higher delivery. With a multipurpose pump tube, each pump comes complete with both a high-head impeller and a high-volume rotor. In a matter of minutes, the impeller/rotor can be changed to meet the performance required.

Double-diaphragm pumps are also applicable to this service. The diaphragm pump can run dry without damage and handles solids up to 1/8 inch. It can be designed with a convertible manifold which allows each chamber to handle different liquids simultaneously, which is excellent for ink transfer. These pumps are well suited for sludges, slurries, and highly viscous materials both organic and nonorganic.

The following constitute safety procedures when transferring chemicals from drums with portable pumps:

1. Make sure motor enclosure corresponds to the material being transferred. For example, *do not pump an explosive or hazardous material with an open motor or plastic pump tube.*

2. Flammable or explosive liquids should be pumped with a hazardous-duty electric motor or pneumatic motor in conjunction with a metal pump tube marked zone "O".

3. Be sure the motor nameplate corresponds to the power source used to operate the pump.

4. Before placing the pump tube in material, make sure it is compatible with the material being transferred.

5. The motor should be in the *off* position before connecting the motor to an operational outlet.

6. All connections should be secured. The banding style of clamps on all hose connections is recommended.

7. Discharge end of hose should be in the discharge container securely before turning pump unit on.

8. Do not submerge any motor in liquid.

9. When solids are present in the material being transferred, use a strainer to prevent damage to the pump tube.

10. After using the pump unit, flush with a neutral solution and store the pump on a wall hanger.

11. Do not run the mechanical seal pump tube dry.

12. Operating unit should never be left unattended.

13. Prior to use read operating instructions completely.

Figure 3.28 illustrates proper bonding and grounding arrangements.

It's important to note that the high-velocity flow of material through a pump may cause static electricity. All pumps and containers must be properly

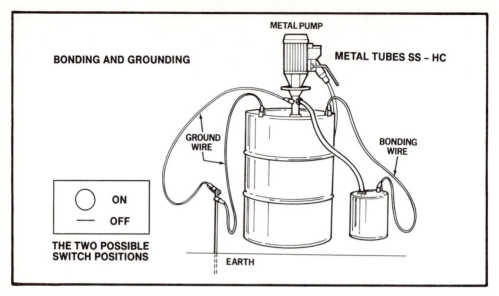

Figure 3.28   Bonding and grounding.

grounded and bonded to prevent static discharge and sparking which could cause electric shock, fire, or explosion.

One should use pneumatic motors in conjunction with metal pump tubes when pumping flammable or explosive liquids or in a hazardous environment.

Normal safety procedures must be used at all times when operating any piece of machinery. Hose clamps, hose connections, and all fittings should be secure before operating the pump. Proper grounding and bonding should be employed at all times. Protective clothing and safety goggles must be worn at all times when operating a pump.

Table 3.6 provides corrosion resistance data for different materials to various chemical media. This can be used as a guide to tubing and pump material selection. A more extensive listing is given in Appendix A.

Use of Sampling Pumps and Composite Samplers

The assemblies consist of a motor, gear train, sealed rechargeable batteries, external detachable cables, and the internal circuitry required to perform both operating and recharging functions. Some units operate on 115V AC 50/60 Hz while others require 230V AC 50/60 Hz.

Many commercial units can be operated with internal batteries for approximately 2.5 hours. Larger pump heads consume more power and operate for a shorter time. A test light and push-button switch are provided to give a visual indication of the internal battery's condition. For extended field use, an external 12V battery can be used to operate the pump. An external cable is provided with an adapter which will fit a standard cigarette lighter socket. This permits convenient access to a car or boat battery.

TABLE 3.6   CORROSION RESISTANCE DATA

The data in the following table gives the corrosion resistance of the five materials to various chemical media. This information may be considered only as a basis for recommendation, but not as a guarantee. Where compatibility is in question, material should be tested under actual field conditions to determine the proper choice. All test data listed is at room temperature (72°F) unless otherwise stated.

R = Recommended
F = Fair, should be tested under field conditions
D = Not Recommended
— = Unknown

B = Minor to moderate effect
* = Use only Air, or EX-UL Motors in conjunction with Stainless Steel and Hastelloy C pump tubes marked zone "O" when pumping these flammable chemicals.

	POLYPROPYLENE MAX F 130	KYNAR MAX F 180	ALUMINUM MAX F 190	STAINLESS STEEL 316 MAX F 200	HASTELLOY C MAX F 200
*Acetaldehyde	D	—	B	R	R
Acetamide	—	—	R	R	—
*Acetate Solvents	D	D	B	R	R
Acetic Acid, 10-80%	R	R	—	F	R
Acetic Acid, Glacial	R	R	B	F	R
Acetic, 80%	—	R	—	F	R
Acetic Anhydride	B	B	B	F	R
*Acetone	D	D	R	R	R
*Acetyle Chloride	D	D	D	B	—
*Acetylene	D	—	R	R	R
Alcohols					
*Allyl	D	D	—	R	R
*Amyl	D	D	B	R	R
*Butyl	D	D	B	R	R
*Ethyl	D	D	B	R	R
*Methyl	D	D	B	R	R
*Propyl	D	D	R	R	R
Aluminum Chloride	R	R	B	D	R
Aluminum Fluoride	R	R	—	D	R
Aluminum Hydroxide	R	R	R	R	—
Aluminum Nitrate	R	R	B	R	—
Aluminum Oxychloride	F	—	—	—	—
Aluminum Sulfate	R	R	R	R	R
Aluminum Sulfate Acid	—	—	—	—	F
Aluminum (All types)	R	—	—	—	
*Ammonia, Aqua, 10%	D	D	—	R	R
*Ammonia, Aqueous	D	—	—	R	R
*Ammonia (Concentrated)	D	—	—	R	—
Ammonium Bifluoride	R	R	D	R	B
Ammonium Carbonate	R	R	D	R	B
Ammonium Chloride	R	R	—	F	R
Ammonium Fluoride, 10%	—	—	—	—	R
Ammonium Fluoride, 25%	R	—	—	—	R
Ammonium Hydroxide	R	R	D	R	R
Ammonium Nitrate	R	R	B	R	R
Ammonium Nitrite	—	—	—	—	—
Ammonium Oxalate	R	—	—	R	R
Ammonium Persulfate	R	—	D	R	R
Ammonium Phosphate, Dibasic	R	R	F	R	R
Ammonium Phosphate, Monobasic	R	R	F	R	R
Ammonium Phosphate, Tribasic	R	R	F	R	R
Ammonium Sulfate	R	R	F	R	F
Ammonium Sulfide	—	—	—	—	R
Ammonium Thio-Sulfate	—	—	—	R	—
Ammonium Thicyanate	—	—	—	—	—
*Amyl Acetate	D	D	F	R	F
*Amyl Chloride	D	R	D	R	R
*Amyl Chloronapthalene	—	—	—	—	—
*Amyl Napthalene	—	—	—	—	—
Aniline (Kynar, R to 70°F)	F	R	F	R	F
Aniline Dyes	—	—	F	F	—
Aniline Hydrochloride	—	—	D	D	—
Animal Fats	—	—	R	R	—
Anisole	F	—	—	R	R
Antimony Chloride	R	—	—	—	R
Anti-Freeze	R	—	R	R	R
Antimony Trichloride	R	—	—	U	—
Aqua Regia (80%), (Kynar, R, 70°)	D	R	D	D	D
Arsenic Acid	R	R	D	F	R
Barium Carbonate	R	R	F	F	F
Barium Chloride	R	R	D	F	R
Barium Hydroxide	R	R	D	R	R

(continued)

TABLE 3.6   *Continued*

	POLYPROPYLENE MAX F 130	KYNAR MAX F 180	ALUMINUM MAX F 190	STAINLESS STEEL 316 MAX F 200	HASTELLOY C MAX F 200
Barium Sulfate	—	R	D	R	—
Barium Sulfide	R	R	D	R	—
Beer	D	—	R	R	—
Beet Sugar Liquors	R	—	R	R	R
Benzaldehyde (Kynar, R, 70°F)	D	R	F	R	—
Benzene, Benzol	D	D	F	R	F
Benzene Sulfonic Acid	—	D	D	F	—
Benzoic Acid	D	R	F	R	R
*Benzyl Alcohol	D	—	—	—	—
Benzyl Chloride	—	—	—	—	—
Bismuth Carbonate	R	—	—	—	R
Black Liquors	R	—	—	—	F
Bleach, 12.5% Active	R	—	—	R	R
Borax (Sodium)	R	R	F	R	R
Boric Acid	R	R	F	R	R
Brine Acid	—	—	—	—	R
Bromic Acid	R	—	—	D	R
Bromine Liquid	—	—	—	D	R
Bromine Water	—	R	D	F	—
*Butane	D	—	R	R	—
*Butyl Acetate	D	D	R	F	F
Butyl Phthalate	R	—	—	—	—
*Butylene	—	D	R	R	—
Butyl Phenol	—	—	—	—	R
Butyric Acid	R	R	F	R	R
Calcium Bisulfite	R	R	F	F	—
Calcium Carbonate	R	R	F	R	R
Calcium Chlorate	R	R	F	R	R
Calcium Chloride	R	R	F	R	R
Calcium Hydroxide	R	R	F	R	R
Calcium Hyprochlorite	R	R	F	R	R
Calcium Nitrate	R	R	F	F	—
Calcium Sulfate	R	R	F	R	F
Calcium Sulfite	R	—	R	F	—
Cane Sugar Liquors	R	R	R	R	R
Carbon Dioxide	R	R	R	R	—
*Carbon Disulfide	D	D	F	R	—
Carbonic Acid	R	R	R	R	R
Carbon Tetrachloride	D	R	F	R	R
Castor Oil	R	—	—	R	R
Cellosolve	R	R	F	F	—
*Cetyl Alcohol	D	—	—	R	—

	POLYPROPYLENE MAX F 130	KYNAR MAX F 180	ALUMINUM MAX F 190	STAINLESS STEEL 316 MAX F 200	HASTELLOY C MAX F 200
Chlorine (Dry)	—	R	—	—	—
Chlorine (Wet)	—	R	D	D	R
Chlorine Liquid	D	R	D	D	R
Chloroacetic Acid	—	—	—	—	F
*Chlorobenzene	D	D	F	R	R
Chlorobenzyl Chloride	—	—	—	—	—
Chloroform	D	R	D	R	F
Chlorosulfonic Acid, 100%	D	D	D	D	F
Chromic Sulfuric Acid	D	—	—	—	—
Chromic Acid, 10% (Kynar, R, 120°F)	R	R	F	R	R
Chromic Acid, 50% (Kynar, R, 120°F)	F	R	F	F	R
Cider	F	R	F	R	R
Citric Acid	F	R	F	R	R
Copper Chloride	R	R	D	D	—
Copper Cyanide	R	R	D	R	R
Copper Nitrate	R	R	D	R	R
Copper Sulfate	R	R	D	R	R
Cottonseed Oil	R	—	—	R	R
Cresylic Acid (Kynar, R, 150°F)	—	R	F	R	F
*Cyclohexane	D	D	R	R	—
*Cyclohexanol	—	D	F	F	—
*Cyclohexanone	D	D	F	F	—
Decane	D	—	—	R	—
Detergents	R	—	R	R	—
Dextrose	—	—	—	R	R
*Diacetone Alcohol	—	—	R	R	—
*Dichloroethylene	D	—	—	—	R
Diesel Fuels	D	R	R	R	R
*Diethyl Ether	D	D	F	B	—
*Diisobutylene	—	D	D	F	—
Dimethyl Formamide	—	—	R	R	—
Dioctyl Phthalate	—	—	R	R	—
Distilled Water	R	R	—	R	R
*Epichlorohydrine	—	D	R	R	—
*Ethanolamine	—	D	—	R	—
*Ether	—	D	R	R	F
*Ethyl Acetate	D	D	F	R	F
*Ethyl Chloride	D	D	F	R	F
*Ethyl Ether	—	—	F	R	—

TABLE 3.6 *Continued*

	POLYPROPYLENE MAX F° 130	KYNAR® MAX F° 180	ALUMINUM MAX F° 190	STAINLESS STEEL 316 MAX F° 200	HASTELLOY C® MAX F° 200
*Ethylene Chloride	D	D	F	R	F
*Ethylene Dichloride	D	D	D	R	F
Ethylene Glycol	R	R	R	R	—
*Ethylene Oxide	—	D	R	—	—
Fatty Acids	—	R	F	R	R
Ferric Chloride	R	R	D	F	F
Ferric Nitrate	R	R	D	R	R
Ferric Sulfate	R	R	D	R	R
Ferrous Chloride	R	R	D	D	F
Ferrous Sulfate	R	R	D	F	F
Fluoboric Acid	R	R	—	F	R
Fluosilicic Acid	—	F	D	—	—
Formaldehyde					
(Kynar, R, 120°F)	R	R	R	R	F
Formic Acid	R	R	D	R	R
Freon (Seal-less Pump					
Tubes Only)	—	—	—	R	—
Fruit Juices	R	—	F	R	—
Furfural	D	F	R	R	F
Gallic Acid (Kynar, R, 70°F)	F	R	R	F	F
*Gasoline	D	D	R	R	R
Gelatine	D	D	R	R	R
Glucose	R	R	R	R	—
Glycerine	R	R	R	R	R
Glycolic Acid (Kynar, R, 70°F)	R	R	—	—	R
Glycols	—	R	F	F	—
*Heptane	D	D	R	R	—
*Hexane	D	D	R	R	R
Hydrobromic Acid	F	R	D	D	R
Hydrochloric Acid (20%)	R	R	D	D	R
Hydrochloric Acid (37%) (Hot)	—	R	D	D	D
Hydrochloric Acid (37%) (Cold)	R	R	D	D	R
Hydrofluoric Acid (20%)	R	R	D	F	F
Hydrofluoric Acid (50%)	D	R	D	D	F
Hydrofluoric Acid (75%)	D	R	D	D	—
Hydrofluoric Acid					
(Conc.) (Hot)	—	R	D	D	—
Hydrofluoric Acid					
(Conc.) (Cold)	—	R	—	—	—
Hydrofluosilicic Acid (20%)	R	R	D	D	—

	POLYPROPYLENE MAX F° 130	KYNAR® MAX F° 180	ALUMINUM MAX F° 190	STAINLESS STEEL 316 MAX F° 200	HASTELLOY C® MAX F° 200
Hydrogen Fluoride	R	—	—	R	R
*Hydrogen Peroxide	D	D	D	R	R
*Hydrogen Sulfide (Cold)	D	D	F	R	—
*Hydrogen Sulfide (Hot)	D	D	F	R	—
Hypochlorous Acid	—	R	D	D	—
Iodine (Kynar, R, 150°F)	D	R	D	D	F
*Isopropyl Ether	D	D	R	R	—
				D	
*Jet Fuel (JP3, JP4, JP5)	D	D	D	R	—
Kerosene	D	D	R	R	—
*Lacquers	D	D	R	R	—
*Lacquer Solvents	—	—	R	R	—
Lactic Acid (Kynar, R, 70°F)	R	R	F	R	—
Lard Oil	R	R	R	R	—
Latex	—	—	R	R	—
Lead Acetate	R	R	D	F	—
Lubricating Oils	—	R	R	R	—
Magnesium Carbonate	R	R	—	R	F
Magnesium Chloride	R	R	D	R	R
Magnesium Hydroxide	R	R	D	R	—
Magnesium Sulfate	R	R	F	R	F
Maleic Acid	F	R	F	R	—
Mercuric Chloride					
(Dilute Solution)	R	R	D	D	F
Mercuric Cyanide	R	R	D	R	—
*Methyl Acetone	—	D	R	R	—
Methyl Chloride	D	D	D	R	—
*Methyl Ethyl Ketone	D	D	R	R	—
*Methyl Isobutyl Ketone	D	D	—	R	—
Methylene Chloride	D	D	F	R	R
Milk	R	R	R	R	—
*Monoethanolamine	—	D	F	R	—
Muriatic Acid—See Hydrochloric					
*Naptha	D	D	—	R	F
*Napthalene	D	D	F	F	—
Nickel Chloride	R	R	D	R	—
Nickel Sulfate	R	R	D	R	F
Nitric Acid (5-10%)	R	R	D	R	R

(continued)

TABLE 3.6   *Continued*

	POLYPROPYLENE MAX F° 130	KYNAR® MAX F° 180	ALUMINUM MAX F° 190	STAINLESS STEEL 316 MAX F° 200	HASTELLOY C® MAX F° 200
Nitric Acid (20%)	R	R	D	R	R
Nitric Acid (50%)	D	R	D	R	R
Nitric Acid, Concentrated (Kynar, R, 120°F)	—	R	D	R	F
Nitric Acid, Red Fuming	—	F	R	R	—
Nitrobenzene	D	D	F	F	F
Oleic Acid	R	R	F	R	—
Oleum	D	D	F	R	—
Oxalic Acid, (Cold)	R	—	F	R	F
Palmitic Acid	—	R	F	R	—
Perchloric Acid (Kynar, R, 125°F)	—	R	D	D	—
Perchloroethylene	—	—	—	R	—
Petrolatum	—	—	F	R	—
Petroleum Oils, Refined	R	—	—	R	R
Phenol (Carbolic Acid) (Kynar, R, 150°)	F	R	F	R	R
Phosphoric Acid, 20%	R	R	D	F	R
Phosphoric Acid, 20-40%	R	R	D	R	F
Phosphoric Acid, 45%	—	R	D	F	—
Phosphorus, Yellow	—	—	—	R	R
Phosphorus, Red	—	—	—	R	R
Photographic Solutions	R	—	F	R	R
Plating Solutions:					
Chrome 40	R	R	F	R	—
Copper	R	R	F	R	—
Gold	R	—	F	R	—
Iron	R	—	F	R	—
Lead	R	—	F	—	—
Nickel	R	R	F	—	—
Silver	R	R	F	R	—
Tin	R	—	F	R	—
Zinc	R	R	F	R	—
Potassium Bicarbonate	—	R	F	F	R
Potassium Bromide	R	R	F	R	R
Potassium Carbonate	R	R	F	R	F
Potassium Chlorate	R	R	F	R	—
Potassium Chloride	R	R	F	F	R
Potassium Chromate	—	R	R	F	R
Potassium Dichromate	R	R	R	R	F

	POLYPROPYLENE MAX F° 130	KYNAR® MAX F° 180	ALUMINUM MAX F° 190	STAINLESS STEEL 316 MAX F° 200	HASTELLOY C® MAX F° 200
Potassium Hydroxide (Kynar, R, 150°F)	R	R	D	R	F
Potassium Nitrate	—	R	F	R	F
Potassium Permanganate	F	R	F	F	R
Potassium Sulfate	R	R	R	F	F
*Pyridine	D	D	F	F	R
Propionic Acid	—	R	—	B	R
Rust Inhibitors	R	—	—	R	—
Sea Water	R	R	F	R	R
Silicone Oil	R	—	F	R	—
Silver Nitrate	R	R	D	R	—
Soap Solutions	R	—	F	R	—
Sodium Acetate	D	D	B	R	R
Sodium Bicarbonate	R	R	R	R	F
Sodium Bisulfate	R	R	D	R	R
Sodium Bisulfite	R	R	R	R	R
Sodium Borate	—	R	B	F	—
Sodium Bromide	R	—	—	R	R
Sodium Carbonate	R	R	B	R	R
Sodium Chlorate, 50%	R	R	B	R	R
Sodium Chloride	R	R	B	R	R
Sodium Cyanide	R	R	D	R	R
Sodium Hydroxide, 20%	R	R	D	R	R
Sodium Hydroxide, 50%	R	D	D	B	R
Sodium Hydroxide, 80%	R	U	D	D	F
Sodium Hyprochlorite to 20%	D	R	D	D	R
Sodium Metaphosphate	D	—	R	R	—
Sodium Nitrate	R	R	R	R	—
Sodium Perborate	R	—	B	D	—
Sodium Phosphate	—	R	D	B	—
Sodium Silicate	R	R	D	R	—
Sodium Sulfphate	R	R	B	R	B
Sodium Sulfide	R	R	D	R	—
Sodium Thiosulphate	R	R	B	R	—
Stannic Chloride	—	R	D	D	—
Stearic Acid	D	R	B	R	R
Sulfate Liquors	R	—	B	D	R
Sulfur	R	—	R	R	—
Sulfur Chloride (Kynar, R, 70°F)	D	R	D	D	—

TABLE 3.6 *Continued*

	POLYPROPYLENE MAX F. 130	KYNAR® MAX F. 180	ALUMINUM MAX F. 190	STAINLESS STEEL 316 MAX F. 200	HASTELLOY C® MAX F. 200		POLYPROPYLENE MAX F. 130	KYNAR® MAX F. 180	ALUMINUM MAX F. 190	STAINLESS STEEL 316 MAX F. 200	HASTELLOY C® MAX F. 200
Sulfur Dioxide	D	R	R	R	B	1,1,1 Trichloroethylene	—	—	—	D	—
Sulfuric Acid, 10%						Trichloroethylene	D	R	B	R	R
(Kynar, R, 230°F)	R	R	D	F	R	Triethylamine	—	D	—	—	—
Sulfuric Acid, 10%-75%						*Turpentine	D	D	R	R	R
(Kynar, R, 175°F)	R	R	D	B	R						
Sulfuric Acid, 66° Baumé						Vegetable Oil	R	R	R	R	R
(Kynar, 120°F)	D	R	D	F	R	Vinegar	R	R	D	R	R
Sulfurous Acid						*Vinyl Chloride	—	—	—	—	R
(Kynar, R, 212°F)	R	R	D	B	F						
						Water, Acid	R	—	D	R	R
Tannic Acid (Kynar, R, 230°F)	R	R	D	R	F	Water, Distilled	R	R	B	R	R
Tanning Liquors	—	—	D	R	R	Water, Fresh	R	R	R	R	R
Tar Bituminous	—	—	—	B	—	Water, Salt	R	R	B	R	R
Tartaric Acid (Kynar, R, 250°F)	R	R	D	R	R						
*Tetrahydrofuran	D	D	—	R	—	*Whiskey and Wines	D	D	D	R	R
Tetralin	—	—	R	R	—	White Liquor	R	—	—	R	R
Titanium Tetrachloride	—	—	D	F	R						
*Toluene, Toluol	D	D	R	R	R	*Xylene or Xylol	D	D	R	R	R
Tomato Juice	R	—	R	R	R						
Transformer Oil	R	—	R	R	R	Zinc Chloride	R	R	D	R	F
Trichloroacetic Acid						Zinc Sulfate	R	R	D	R	R
(Kynar, R, 70°F)	—	R	D	D	F						

Using a second cable with a standard AC plug, the unit can be plugged into a standard AC wall outlet. This permits operating the pump on AC power, where available, and conserving the internal batteries. The internal batteries can be recharged with either 12V DC or an AC source.

Units are designed for pumping liquid samples in locations where complete portability is required, such as along rivers, sewer systems, lakes, and shallow wells. The pump will operate at a relatively constant speed, although some reduction in speed will exist as the internal batteries discharge during operation.

Several pump head sizes are available. Even though the motor runs at only one speed, various flow rates are possible by choosing the head size which offers the most satisfactory flow. With the sampler placed on the shore or on a boat, the unit is actually raising the water sample only the short distance from the surface of the lake up to the sampler, while hydrostatic water pressure at the lower depths will cause the water to raise inside the tubing to the surface of the lake. This combination permits the sampler to draw liquid from depths greater than 30 feet.

Silicone tubing is recommended for use inside the pump head. For economy, the tubing outside the pump head can be a heavy-walled tygon tubing. A thin-wall tubing may collapse when lifting samples over great heights. The tubing in the pump head should be changed periodically (silicone tubing life equals approximately 110 hours of pump operating time).

Most systems are designed for intermittent-duty applications: continuous, 24 hour a day operation will require frequent replacement of the motor brushes and periodic recharging of the internal batteries when in that mode.

Batteries will partially discharge if the unit is not used for long periods of time. It is recommended that the sampler be given at least one complete cycle (operate motor for 2 hours, then recharge overnight with AC) once every 6 months to maintain batteries at their fully charged capacity.

When the tubing is lowered into the water to be sampled, the tubing will have a tendency to float near the surface due to the trapped air in the tubing. When long lengths are used, the tubing may also curl and be difficult to handle. A stainless steel cylindrical weight can be used to eliminate these problems. A clamp holds the weight onto the tubing. The weight will cause the tubing to hang straight in the water minimizing curling and floating.

Bearings within the pump head are permanently lubricated and do not require maintenance. The pump head should be periodically disassembled and cleaned, particularly if the tubing within the head has ruptured during pump operation. Use only soap and water for cleaning. Do not submerge in water. While the head is disassembled, inspect all parts for obvious wear or damage and replace if necessary. The drive gears generally do not require maintenance. The control circuit has solid-state components which do not need servicing. For most commercial units, the motor brush life should exceed 1,000 hours under normal operation. They can be replaced if necessary. An example of a variable-speed sampling pump is shown in Figure 3.29.

Composite samples provide a means for on-site sampling of virtually any type of water or liquid waste. It can lift liquid up to 25 feet. The sample size and frequency of sampling can be selected to meet the desired operating conditions. Left unattended, the system will automatically purge the line, and then deliver the sample to a container. A prepurge cycle minimizes cross contamination between samples, and it also unclogs the inlet from floating debris. The pump drive, rechargeable batteries, container, and control circuitry are usually contained within a weather-resistant ABS plastic case. The case can be locked to prevent unauthorized persons from changing the control settings. Sample contamination is minimized since the liquid only contacts the tubing and several stainless steel fittings. One set of fittings connects the tubing to the sample bottle. The other set is mounted in the wall of the carrying case, so that the sampler can be operated with the case closed and locked. Figure 3.30 shows a photograph of this system.

Several optional pieces of equipment can be purchased to increase the versatility of the unit. Where 115V AC 60 Hz power is available, an AC adapter can be installed inside the unit. The batteries will be kept recharged while the system is sampling. This eliminates the need for periodically removing and

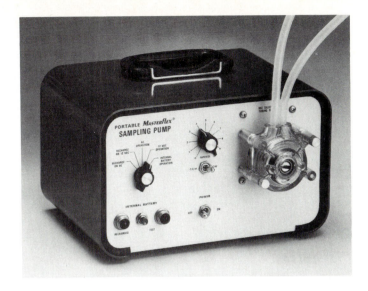

Figure 3.29 Photograph of a variable-speed sampling pump. (*Courtesy of Barnant Co., 28 W092 Commercial Avenue, Barrington, Illinois 60010*)

recharging the batteries. A flow-meter monitoring system is available so that the composite sampler can be cycled on a flow proportional basis rather than on a time cycle. Two removable plugs are provided in the case wall so that the AC input line for the AC adapter and a signal input line from the externally located flow-meter system can pass into the locked case. A brief description of the controls follows:

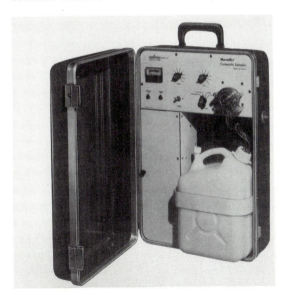

Figure 3.30 Photograph of self-contained portable composite sampler. (*Courtesy of Barnant Co., 28 W092 Commercial Avenue, Barrington, Illinois 60010*)

Power switch

Aside from energizing the sampler control circuits, the switch must be on to enable the AC adapter to recharge the batteries. The battery test switch will function with the power switch either on or off.

Function switch

1. *Continuous*: The pump will rotate clockwise, delivering liquid to the sample bottle. This permits the operator to draw a sample without using the time cycle or the purge cycle. To operate, set the function switch in this position and turn the pump on and off with the power switch.
2. *EXT/Test*: In this position, the internal clock is reset to zero and does not operate. The system will go through a purge-fill cycle each time the sample test switch is depressed and released. This allows the operator to measure the sample before setting the system into its automatic time cycle mode. The system can be made to cycle by closing an external set of contacts. The switch position also permits the AC adapter to recharge the batteries without operating the sampler.
3. *15 Minute*: If the function switch is advanced to this position from the Ext/ Test position, the internal clock circuit will be energized. The sampler will cycle every 15 minutes, first going into a purge, and then a fill mode.
4. *30 Minute*: This is similar to the preceding position, except that the time cycle is once every 30 minutes.

Battery test

When depressed, the meter will indicate the condition of the batteries. They should be recharged if a weak condition is indicated. The switch will operate with the function switch in any position. It will operate with the power switch either off or on. It should only be used when the motor is *not* running. Using it while the motor is operating provides an inaccurate reading.

Sample test

With the function switch in Ext/Test, and the power switch on, depress and release the sample test switch. The system will go through one complete purge-fill cycle and then return to a standby condition. The external signal line connected to the input jack must be open, not closed. The length of purge time (pump rotates counterclockwise) and fill time (pump rotates clockwise) will depend on the control knob settings.

Purge

To minimize cross contamination of samples, the system will go through a purge or flushing cycle before drawing a sample of liquid. The purge control knob allows the operator to adjust the length of the purge cycle. To reduce battery power consumption, the purge time should be adjusted to the shortest time period which will empty the tubing. This time period is related to the tubing length and lift height. The tubing should be cut as short as reasonably possible. Extra length only increases the current drain from the batteries by requiring a greater running time.

The composite sampler is normally shipped and stored with the power switch off and the function switch in the Ext/Test position. In preparation for sampling, if not so located, reset the switches to these positions. With the power switch off, depress the battery test switch to check the condition of the batteries. If weak, either recharge or replace with other fully recharged batteries. Install the tubing in the pump head following the manufacturer's directions. At the site, check that one end of the external sampling tubing is firmly attached to the case fittings and that the other end is inserted into the liquid to be sampled. Use a tubing weight if necessary. Select the control knob settings for the lift height and sample size desired. Unscrew the cap from the container and hold over a graduated beaker or other container. Turn on the power switch and depress the sample test switch. The sampler will go through one purge and fill cycle. Measure and record the sample delivered. Again depress the sample test switch, taking a second sample. If the second sample is more than 10 percent larger than the first, the system may not be completely purging the line. Inspect the tubing during the purge cycle and/or adjust the purge control knob to a higher dial setting. Once the system is purging completely, the fill control can be reset (if necessary) to provide the desired sample size. When the purge and fill are satisfactory, attach the cap back onto the container. Reset the function control to either the 15-minute or 30-minute position (whichever is desired). The system will go through one purge-fill cycle and then continue to cycle once every time period until the power switch is turned off. One should select a sample size and time cycle which will not cause the container to overflow. Manufacturer's specifications for the unit shown in Figure 3.30 are given in Table 3.7.

As noted earlier, it is recommended that silicon tubing be used in the pump head. This tubing has good chemical resistance, dimensional stability, and greater life than most other types of tubing. Some types will show a tendency to take a "set" with use (so that a considerable change in flow will occur during operation). Other types will require more torque to operate, and hence reduce the battery life between rechargings.

Tubing is normally supplied with the pump. Tubing dimensions are carefully chosen so that the tubing will perform satisfactorily. If some other source for the tubing is used, dimensions should be chosen to match the manufactur-

TABLE 3.7   SPECIFICATIONS FOR SAMPLER SHOWN IN
FIGURE 3.30

Sampler Specifications	
Operating Temperature Range:	+10°C (+50°F) to +50°C (+122°F)
Clock Accuracy:	±12 minutes in 24 hours
Sample Accuracy:	±10%
Pump Speed:	64 rpm
AC Operation:	lift height versus total volume collected in 24 hours
Up to 20 feet:	2 gallons (15-minute or 30-minute cycle)
20 to 25 feet:	1 gallon (15-minute or 30-minute cycle)
Battery Operation:	lift height versus total volume collected in 24 hours
Up to 15 feet:	2 gallons (15-minute or 30-minute cycle)
15 to 20 feet:	1 gallon (30-minute cycle only)
20 to 25 feet:	0.5 gallon (30-minute cycle only)

er's specifications. Oversized tubing will create overcompression, resulting in early tubing failure. Undersized tubing will travel in the pump head, perhaps jamming at some point inside the pump and preventing the pump from rotating. Improper bore dimensions may reduce the liquid flow capacity or the lifting capability of the pump by improper occlusion.

While the tubing inside the case should be silicon, the tubing outside of the case can be any material which is chemically and physically satisfactory. Thin-walled tubing is not recommended since it may collapse when the pump is in its lifting or suction mode. Tygon tubing due to its low cost and high strength should be acceptable in most sampling operations. If a tubing weight or long lengths of tubing are used, it is suggested that a hose clamp be used to secure the tubing to the fitting in the carrying case wall.

When sampling bodies of water such as lakes or rivers, the tubing will have a tendency to float near the surface due to the trapped air in the tubing (before it is completely filled with water during cycling).

When long lengths of tubing are lowered into shallow wells or large-diameter well pipe, the tubing may curl and be difficult to handle. The stainless steel cylindrical weight described earlier is available to minimize these problems, by causing the tubing to hang relatively straight downward. A clamp holds the weight onto the tubing. The weight is drilled so that even if it rests on the bottom of the riverbed or well, the liquid can still flow into the tubing through holes cross drilled in opposing sides of the weight.

Using two pump heads on the sampler will overload the system, creating excessive wear on the drive train and internal circuitry. Operate the sampler with only one pump at a time. A length of tubing connects the fittings in the side wall of the carrying case to the container. If the sampler is left operating so that the bottle accidentally overfills, the drain line permits the excess liquid to drain outside of the case instead of spilling inside, and perhaps shorting out the electronic circuitry.

Using optional equipment, a composite sampler can be adjusted to complete a purge-fill cycle each time a given volume of liquid passes, instead of sampling on a time period. This equipment consists of an electronic flow meter and a batch controller. The flow meter develops an electrical signal as the liquid passes through it. This signal is sent to the batch controller which can be adjusted to a given volume of liquid. Each time the desired volume passes the flow meter, the batch controller actuates a set of contacts. This in turn causes the composite sampler to go through one purge and fill cycle. The length of purge and the size of the sample delivered will depend on the composite sampler dial settings, while the frequency at which the sampler cycles will depend on the settings of the batch controller.

To operate the system, the signal output of the batch controller must be connected to the sampler through a phone jack located inside the battery compartment, directly below the meter. Normally, any two-conductor cable can be used, but if excessive electrical interference is present, a shielded cable is recommended.

The following example is given to clarify the arrangement. Assume a 50-ml sample is desired for each 100 gallons of liquid. Set the controls on the batch controller for 100 gallons; power off. Set the composite sampler purge and fill controls. Move the function switch to the Ext/Test position and turn on the power switch. Unscrew the cap from the bottle, hold over a graduated container and depress the sample test switch. During this trial run, determine if the purge and fill settings provide satisfactory results.

If not, reset the sampler controls and take another trial run. When satisfactory, turn off the sampler power switch, and attach the cap back onto the bottle. Insert the signal output cable plug into the phone jack in the sampler, and energize the flow meter/batch controller. The dials will monitor the passage of the liquid. Now turn the sampler power switch on. After the 100-gallon volume has been indicated, the composite sampler will go through one purge-fill cycle, delivering the 50-ml sample into the container. The system will continue to monitor the flow, periodically cycling the sampler until the units are turned off manually.

To assist in the selection of tubing for a particular pumping application, Table 3.8 can be used as a general guide. One should always pretest tubing before proceeding with its use, especially when handling hazardous materials. An immersion test is advisable to check compatibility. Immerse a sample length of tubing in a closed vessel containing the fluid to be pumped for at least 48 hours. Examine the tubing for signs of swelling, embrittlement, or other evidence of deterioration.

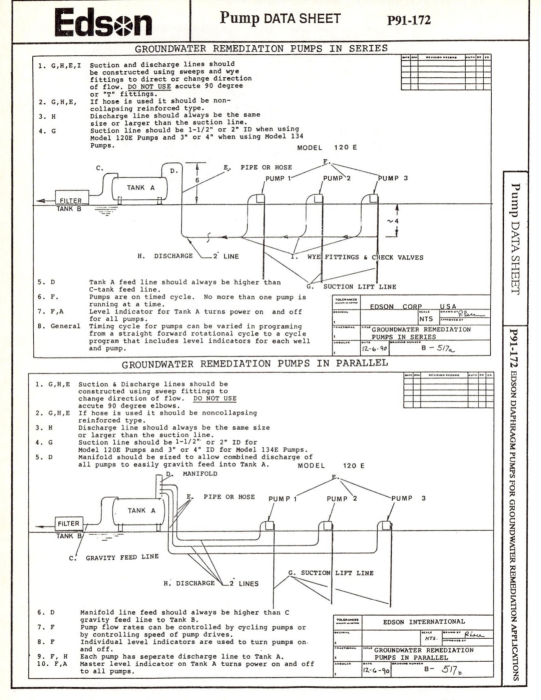

Figure 3.35 Manufacturer's pump data sheet illustrating series and parallel pump arrangements. (*Courtesy of Edson International, Industrial Products Div., Industrial Park Road, New Bedford, MA 02745*)

Figure 3.36 Photograph of a portable pump kit. *(Courtesy of Edson International Industrial Products Div., Industrial Park Road, New Bedford, MA 02745)*

ing board, 32-inch pump handle, quick clamp hose fittings, 15 feet of lightweight noncollapsing hose, and either Hypalon or Viton diaphragms and valve.

Finally, Figure 3.37 shows a heavy-duty diaphragm pump used for transferring industrial waste and water from sump to filtration systems. Because the pump works like a shovel, transferring up to 2.25 gallons per stroke, the model

Figure 3.37 Heavy-duty diaphragm pump for transferring industrial waste. *(Courtesy of Edson International, Industrial Products Div., Industrial Park Road, New Bedford, MA 02745)*

reduces the emulsification of already naturally separated wastes. This effectively improves filter performance by preventing the eggbeater characteristic of other pumps. At up to 60 strokes per minute this model can transfer up to 8,000 gallons per hour and is self-priming from vertical lifts up to 20 feet. The unit is a self-contained electric motor, air cylinder, and gasoline engine diaphragm pump that is designed to be nonclogging. Built with 3-inch or 4-inch NPT inlet and outlet waterways, the unit will pump liquids made up of 50 percent solid matter and will generally pump any solid that will flow through the hose. The performance ability of the unit coupled with the ability to run dry which virtually allows unmonitored operation make it ideal for transfer applications and secondary surface water containment applications. The pump and rocker arm are constructed of close-gain cast iron. These are powered through a series of spur gears and an adjustable manganese crank arm assembly. The standard electric motor is 2 hp, three-phase, 208-230/460 Volt. The air drive is an 8-inch cylinder bore, 3-1/2-inch stroke air motor. The gasoline engine is a 9 hp cast iron block. All surfaces are coated with a high-quality weather, chemical, and abrasive-resistant marine enamel paint. When applications require it, other coatings are available upon request.

PUMPS FOR METERING AND HANDLING CHEMICALS

Metering pumps are high-precision machines used extensively in pilot, plant, and numerous process operations. A metering plunger-type pump is designed to deliver a controlled volume of liquid at a specific adjustable rate of flow determined by stroke length, plunger diameter, and plunger speed (5 pm). These designs usually are equipped with a stop-adjust screw mounted in a heavy-duty eccentric which permits flow-rate adjustment while at rest from 0 to 100 percent of pump capacity. There are a vast number of metering pump applications which do not require frequent adjustment of capacity.

Diaphragm-type metering pumps are similar to their plunger counterpart, except that the plunger is isolated from the pumped liquid by a flexible diaphragm. Suction-discharge pulses are transmitted through a hydraulic fluid to the diaphragm, which in turn displaces the pumped liquid. Some designs incorporate a gas-venting system that prevents the accumulation of vapors which can seriously affect pumping performance.

Diagrams highlighting the features of these two types of pumps are shown in Figures 3.38 and 3.39. Plunger-type metering pumps can be purchased to operate at pressures ranging from 1.0 psi (0.068 bar) to typically 10,000 psi (690.5 bar) and flows from 0.1 gph (0.378 lph) to in excess of 2,000 gph (7,570 lph). The unit shown in Figure 3.32 has a single plunger. It operates from a single motor and gear reducer. The single plunger, because of its inherent reciprocating action, produces an intermittent pulsating flow and utilizes full power only 50 percent of the time. The other half of the cycle, the suction stroke, requires only a small amount of power.

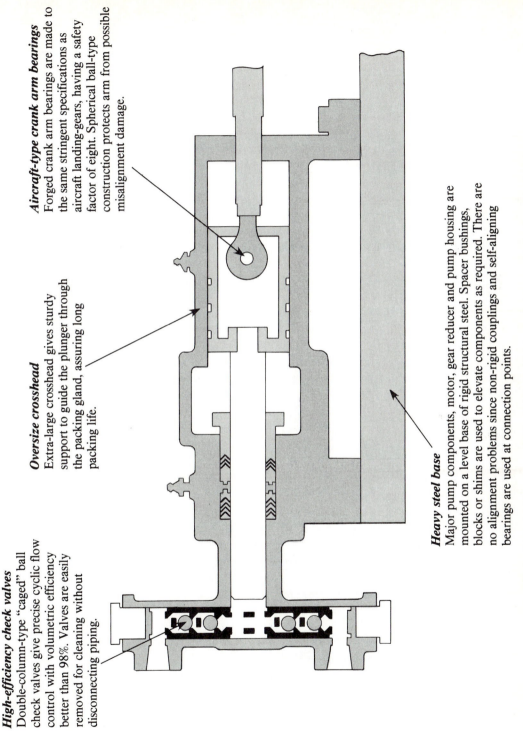

High-efficiency check valves
Double-column-type "caged" ball check valves give precise cyclic flow control with volumetric efficiency better than 98%. Valves are easily removed for cleaning without disconnecting piping.

Oversize crosshead
Extra-large crosshead gives sturdy support to guide the plunger through the packing gland, assuring long packing life.

Aircraft-type crank arm bearings
Forged crank arm bearings are made to the same stringent specifications as aircraft landing-gears, having a safety factor of eight. Spherical ball-type construction protects arm from possible misalignment damage.

Heavy steel base
Major pump components, motor, gear reducer and pump housing are mounted on a level base of rigid structural steel. Spacer bushings, blocks or shims are used to elevate components as required. There are no alignment problems since non-rigid couplings and self-aligning bearings are used at connection points.

Figure 3.38 Sketch of a piston-type metering pump. (*Courtesy of Clark-Cooper Corp., 464 N. Randolph Avenue, Cinnaminson, NJ 08077*)

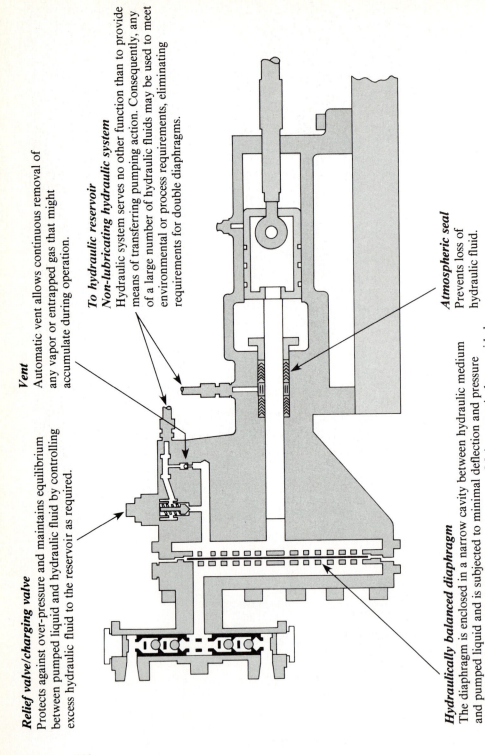

Relief valve/charging valve
Protects against over-pressure and maintains equilibrium between pumped liquid and hydraulic fluid by controlling excess hydraulic fluid to the reservoir as required.

Vent
Automatic vent allows continuous removal of any vapor or entrapped gas that might accumulate during operation.

To hydraulic reservoir
Non-lubricating hydraulic system
Hydraulic system serves no other function than to provide means of transferring pumping action. Consequently, any of a large number of hydraulic fluids may be used to meet environmental or process requirements, eliminating requirements for double diaphragms.

Atmospheric seal
Prevents loss of hydraulic fluid.

Hydraulically balanced diaphragm
The diaphragm is enclosed in a narrow cavity between hydraulic medium and pumped liquid and is subjected to minimal deflection and pressure differential, even under high discharge pressures. Major stress is thus avoided and long diaphragm life is assured.

Figure 3.39 Sketch of a diaphragm-type metering pump. *(Courtesy of Clark-Cooper Corp., 464 N. Randolph Avenue, Cinnaminson, NJ 08077)*

By having the same motor and reducer drive and second plunger, all of the available power can be utilized. With the two plungers operating 180° out of phase, one drawing liquid while the other discharges, liquid flow is continuous with minimal pulsation and output is doubled.

One dual-plunger pump costs less to operate than two single-plunger pumps for the same output. Often, dual-plunger pumps can simplify a process. Two different liquids can be metered in a rationing action. By adjusting the stroke length of the two plungers, virtually any ratio can be achieved. Three, four, and more multiplexed plunger and diaphragm-type pumps are available for complex ratio applications, either batch or on-stream continuous blending.

Many metering pumps for high-accuracy repeatable metering of chemicals, pharmaceuticals and other liquids, slurries and suspensions feature caged ball check valves. A caged design assures highest volumetric accuracy and efficiency. Made to precise tolerances, the caged check valve assemblies provide relatively unrestricted flow, assuring longer pump life. Automatic hydraulic rotation of the balls provides self-cleaning, cutting maintenance and assuring

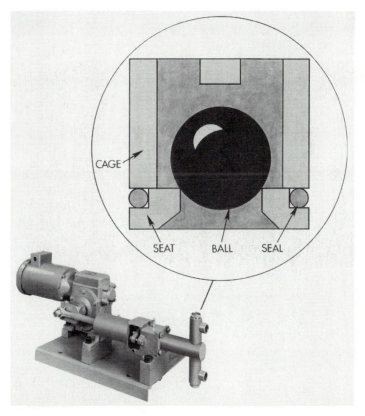

Figure 3.40 Photograph of high-precision metering pump with caged ball check valve.

even wear of all components for firm, effective valve closures. Pumps equipped with the exclusive caged check valves provide high-efficiency metering with better than ±1 percent repeated accuracy. Metering pumps are available in a wide variety of corrosion-resistant materials for flow rates from 0.1 gph (0.378 lph) to over 2,000 gph (7,570 lph). Figure 3.40 shows an actual pump along with the details of the caged ball check valve.

Modular check valve assemblies facilitate pump maintenance, making it easier, faster, and less costly. Stacked one over the other in a straight-column reagent end, the modular assemblies can be removed quickly and easily, simply by unscrewing the two threaded end plugs (refer to Figure 3.41). Reassembly is just as easy because the modular valve assemblies are completely interchangeable and self-aligning. Each modular assembly (i.e., ball, seat, and cage) has its own sealing ring which prevents bypass leakage, thereby eliminating corroded-in valve seats. Another sealing ring prevents corrosive liquids from reaching the threads of the two end plugs, thereby eliminating corroded-in end plugs.

Table 3.9 provides pressure/capacity ranges for different size plunger pumps provided by one manufacturer.

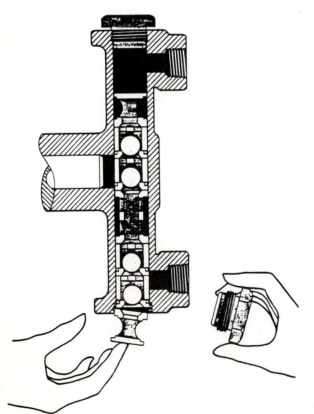

Figure 3.41 Illustrates maintenance on a modular "drop-out" ball check valve assembly. (*Courtesy of Clark-Cooper Corp., 464 N. Randolph Avenue, Cinnaminson, NJ 08077*)

TABLE 3.9 PRESSURE/CAPACITY RANGES FOR PLUNGER PUMPS

To determine capacities for multiple-plunger pumps,
double, triple or quadruple capacities shown below.

Plunger Diameter	Strokes Per Minute	Maximum gph (lph)	MAXIMUM PUMPING PRESSURE—psi (bar)							
			1/4 HP		1/3 HP		1/2 HP		3/4 HP	
			psi	(bar)	psi	(bar)	psi	(bar)	psi	(bar)
Table 1: **Model CP1A and** **Model DP1A, Single-Plunger** 1/4 in. (6.35 mm)	60 72 90	2.06 (7.79) 2.48 (9.38) 3.08 (11.65)	2180 1985 1580	(150.3) (136.8) (108.9)	3140 2630 1900	(216.4) (181.3) (131.0)	4750 3960 3170	(327.5) (273.0) (218.5)	7150 5940 4760	(492.9) (409.5) (328.1)
3/8 in. (9.53 mm)	60 72 90	4.88 (18.47) 5.76 (21.80) 7.32 (27.70)	974 880 706	(67.1) (60.6) (48.6)	1400 1170 845	(96.5) (80.6) (58.2)	2120 1760 1410	(146.1) (121.3) (97.2)	3180 2640 2120	(219.2) (182.0) (146.1)
Maximum Stroke Length 3.0 in. (76.20 mm) Connections 1/4 in. (6.35 mm) 1/2 in. (12.70 mm)	60 72 90	8.24 (31.18) 9.89 (37.43) 12.37 (46.82)	545 495 396	(37.5) (34.1) (27.3)	785 660 475	(54.1) (45.5) (32.7)	1190 990 790	(82.0) (68.2) (54.4)	1960 1480 1190	(135.1) (102.0) (82.0)
Table 2: **Model CP1B and** **Model DP1B, Single-Plunger** 5/8 in. (15.87 mm)	60 72 90	12.92 (48.90) 15.50 (58.66) 19.33 (73.16)	350 316 252	(24.1) (21.7) (17.3)	500 420 304	(34.4) (28.9) (20.9)	760 630 506	(52.4) (43.4) (34.8)	1140 945 758	(78.6) (65.1) (52.2)
Maximum Stroke Length 3.0 in. (76.20 mm) Connections 3/8 in. (9.53 mm) 7/8 in. (22.22 mm)	60 72 90	25.29 (95.72) 30.34 (114.83) 37.93 (143.56)	178 161 129	(12.2) (11.1) (8.8)	256 214 155	(17.6) (14.7) (10.6)	385 322 257	(26.5) (22.2) (17.7)	580 480 386	(39.9) (33.0) (26.6)

Plunger Diameter	Strokes Per Minute	Maximum gph (lph)	MAXIMUM PUMPING PRESSURE—psi (bar)									
			1/2 HP		3/4 HP		1.0 HP		1-1/2 HP		2.0 HP	
			psi	(bar)	psi	(bar)	psi	(bar)	psi	(bar)	psi	(bar)
Table 3: **Model CP2A and** **Model DP2A, Single-Plunger** 5/8 in. (15.87 mm)	60 72 90	17.22 (65.17) 20.70 (78.34) 25.83 (97.76)	598 500 398	(41.2) (34.4) (27.4)	915 748 598	(63.0) (51.5) (41.2)	1195 995 798	(82.3) (68.6) (55.0)	1710 1480 1195	(117.9) (102.0) (82.3)	2390 1995 1593	(164.7) (137.5) (109.8)
7/8 in. (22.22 mm)	60 72 90	33.71 (127.59) 39.52 (149.58) 50.58 (191.44)	340 254 203	(23.4) (17.5) (13.9)	425 380 307	(29.3) (26.2) (21.1)	593 510 407	(40.8) (35.1) (28.0)	875 465 525	(60.3) (32.0) (36.1)	1220 1000 813	(84.1) (68.9) (56.0)
Maximum Stroke Length 4.0 in. (101.60 mm) Connections 1/2 in. (12.70 mm) 1-1/8 in. (28.57 mm)	60 72 90	55.76 (211.05) 66.91 (253.25) 83.65 (316.61)	184 153 123	(12.6) (10.5) (8.4)	257 231 185	(17.7) (15.9) (12.7)	360 310 247	(24.8) (21.3) (17.0)	528 460 319	(36.4) (31.7) (21.9)	740 616 493	(51.0) (42.4) (33.9)
Table 4: **Model CP2B and** **Model DP2B, Single-Plunger** 1-1/2 in. (38.10 mm)	60 72 90	99.10 (375.09) 119.00 (450.41) 148.70 (562.32)	104 86 69	(7.1) (5.9) (4.7)	144 130 104	(9.9) (8.9) (7.1)	202 173 139	(13.9) (11.9) (9.5)	297 260 179	(20.4) (17.9) (12.3)	416 346 278	(28.6) (23.8) (19.1)
Maximum Stroke Length 4.0 in. (101.60 mm) Connections 3/4 in. (19.05 mm) 2.0 in. (50.80 mm)	60 72 90	176.30 (667.29) 211.50 (800.52) 264.40 (1000.75)	59 48 39	(4.0) (3.3) (2.6)	81 73 58	(5.5) (5.0) (3.9)	113 97 78	(7.7) (6.6) (5.3)	167 146 100	(11.5) (10.0) (6.8)	233 195 156	(16.0) (13.4) (10.7)

Plunger Diameter	Strokes Per Minute	Maximum gph (lph)	MAXIMUM PUMPING PRESSURE—psi (bar)									
			1-1/2 HP		2.0 HP		3.0 HP		5.0 HP		7-1/2 HP	
			psi	(bar)	psi	(bar)	psi	(bar)	psi	(bar)	psi	(bar)
Table 5: **Model CP3A and** **Model DP3A, Single-Plunger** 1-1/2 in. (38.10 mm)	60 72 90	148.70 (562.82) 178.50 (675.62) 223.90 (847.46)	199 172 139	(13.7) (11.8) (9.5)	278 232 184	(19.1) (15.9) (12.6)	420 334 280	(28.9) (23.0) (19.3)	690 575 460	(47.5) (39.6) (31.7)	1000 864 690	(68.9) (59.5) (47.5)
Maximum Stroke Length 6.0 in. (152.40 mm) Connections 1-1/2 in. (38.10 mm) 2.0 in. (50.80 mm)	60 72 90	264.40 (1000.75) 317.30 (1200.98) 396.60 (1501.13)	111 97 78	(7.6) (6.6) (5.3)	156 130 104	(10.7) (8.9) (7.1)	234 187 157	(16.1) (12.8) (10.8)	390 355 260	(26.8) (24.4) (17.9)	583 485 390	(40.1) (33.4) (26.8)
Table 6: **Model CP3B and** **Model DP3B, Single-Plunger** 2-1/2 in. (63.50 mm)	60 72 90	413.00 (1563.20) 513.60 (1943.97) 619.60 (2345.18)	71 62 50	(4.8) (4.2) (3.4)	100 84 66	(6.8) (5.7) (4.5)	149 120 100	(10.2) (8.2) (6.8)	250 208 166	(17.2) (14.3) (11.4)	375 312 230	(25.8) (21.5) (15.8)
Maximum Stroke Length 6.0 in. (152.40 mm) Connections 2.0 in. (50.80 mm) 3.0 in. (76.20 mm)	60 72 90	594.80 (2251.31) 711.50 (2693.02) 892.30 (3377.35)	49 43 34	(3.3) (2.9) (2.3)	69 58 46	(4.7) (3.9) (3.1)	102 83 70	(7.0) (5.7) (4.8)	172 143 115	(11.8) (9.8) (7.9)	258 216 173	(17.7) (14.8) (11.9)

Many processes ideally lend themselves to digital batch blending using metering pumps as the basic controlling medium. There are several important reasons for this statement and the design engineer who is familiar with them will have a successful installation at lower possible cost, lowest operating cost, and with maximum accuracy and system reliability. Digital batch blending using metering pumps is a relatively new application It is one that offers many possibilities to the design engineer.

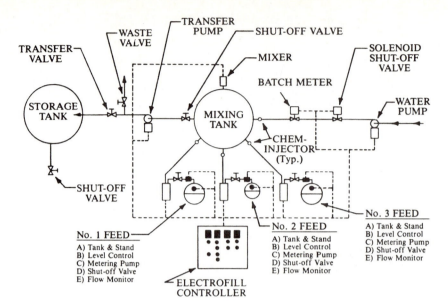

Figure 3.42   A digital batch blending system. (*Courtesy of Clark-Cooper Corp.,*
464 N. Randolph Avenue, Cinnaminson, NJ 08077)

Figure 3.42 illustrates a typical digital batch blending system. The advantages of this type of processing can be applied equally to many industrial applications. For example:

1. *Low Volume.* Batch blending is ideally used where lower production rates in the order of 10 gph to 10,000 gph of product are required, for example, specialty products or feedstock preparation for a continuous process.

2. *Dry Chemicals.* Some processes require the addition of dry chemicals. Unless these can be put into solution for pumping, the best approach is to use digital batch blending where dry chemicals are added at the mixing stage. Automatic feeding of dry chemicals can easily be included in the digital batch blending control system.

3. *Viscous or Slow Soluble Chemicals.* Chemicals that are highly viscous and require slow pumping or take time to become assimilated are also best mixed in a digital batch blending system.

4. *Equipment Cost.* Since there is no expensive analog instrumentation including indicating and/or recording instruments, the cost of the system is very low. Yet the system is fully automatic and therefore affords the same labor savings.

5. *Maintenance.* Digital batch blending is a microprocessor-based system, using electrical pulses from metering pumps and occasionally from flow meters. This type of equipment is easily maintained by ordinary electrical/mechanical repairmen and requires a minimum knowledge of electricity or electronics.

Digital batch blending is a repeat cycle of operations with built-in safety and fail-safe circuits. In fact, the system digital batch controller is programmed to monitor the input signals and unless the microprocessor receives the total number of required pulse inputs for each component of the batch, the system will automatically alarm, notify the operator that the batch is incomplete, and indicate which component is out of specification and by how much. The operator then simply adds the missing amounts without losing the batch. The system, therefore, is self-monitoring and specially trained operators are not required.

Since the variety of products, plant conditions, and economics is infinite, we cannot begin to cover every situation that may be encountered. We have tried, however, to list some of the conditions under which batch blending can be justified. Depending on the application, a wide variety of equipment may be used in any one system. All systems, however, have two components in common: metering pumps and the microprocessor-based digital batch controller. These two components are the nucleus of the system. Additional components may include transfer pumps, high-capacity feed pumps, high-capacity meters with pulse transmitter, mixers, tanks, valves, and other piping and/or electrical accessories.

Metering pumps are unique in their ability to both pump and meter a product. Accuracy of ±1.0 percent over a wide flow range of 001 gph to 1,000 gph nominal is capable with high-precision pumps. They have a linear, predictable output and they can be fabricated in a variety of corrosion-resistant materials, including plastics and all alloys.

Metering pumps as applied to batch process blending completely eliminate the use of either costly analog instrumentation or low-productivity mechanical weigh beams and scales, which require that each component of the batch be added individually, which increases the mixing time. With metering pumps, however, all of the components are fed to the process simultaneously and mixing is continuous. At the end of the feed cycle, the final product is ready for use.

Because of the unique reciprocating plunger action of the metering pump, it is possible for extremely small quantities of additives to be blended into larger systems. Examples are dyes, deodorants, vitamins, and so on which may only represent one part in 1,000 or less of the total batch.

The metering pump with its inherent accuracy will inject this small quantity in an exact ratio. Flow rates as low as 0.0001 gallons per stroke are obtained.

Metering pumps when used in digital batch blending must be manufactured to include certain components that make them suitable to this type of processing. For example, the start-stop position of a metering pump on digital batch blending is extremely important. Figure 3.43 shows a typical metering pump equipped with a pulse transmitter and clutch brake that assures the pump's starting and stopping on each cycle to within ±1.0 degree of its setting. This produces a repeatable accuracy of better than 0.5 percent.

In a typical digital batch blending process, all of the feed streams are broken down mathematically into percentages, then into true capacity, that is, 5 gal/batch, 20 gal/batch, 100 gal/batch, and so on. A batch cycle time is then

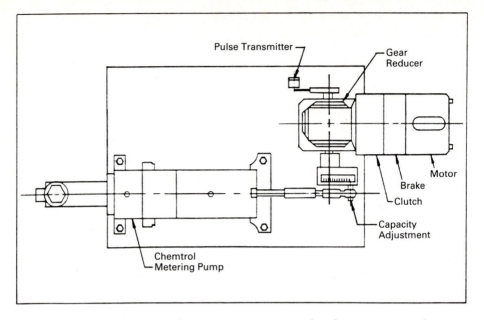

Figure 3.43 Top view of series metering pump with pulse transmitter and clutch brake. (*Courtesy of Clark-Cooper Corp., 464 N. Randolph Avenue, Cinnaminson, NJ 08077*)

established. This may be 15 minutes, 30 minutes, or 1 hour. Once the cycle time has been established, the pump size is selected that has the capacity to inject the given volume within the allotted time. The number of strokes of the pump is then computed and this number becomes the pulse level that is preset on the digital batch controller. As an example, consider 60 gallons of additive were required to be injected into a batch with a cycle time of 30 minutes. A pump would be selected with a capacity of 120 gallons per hour (60 gal/30 min). Assume this pump produced this capacity at a speed of 60 strokes per minute. Then 1,800 strokes or pulses would be equal to 30 gallons of additive. The pump pulse transmitter transmits the pulses to the digital batch controller. When 1,800 counts are reached, the digital batch controller will automatically activate the metering pump brake/clutch, precisely stopping the pumping action. The digital batch controller can be made to accommodate any number of inputs so that extremely complex formulations are practical and any number of variations, such as formula variation, are easily incorporated into the system. The digital batch controller can be set to any variation within the capacity of the metering pump volume. Multipumping installations can be arranged so that the number of metering pumps feeding the batch can be varied. This versatility allows the use of one basic system for the manufacture of a wide variety of products. Any number of other automatic functions can be incorporated in the system, such as temperature control, automatic mixing, and automatic transfer at the end of the mixing cycle.

Selection, Installation, and Sizing Criteria

There are many types of metering pumps available for processes having precise volumetric pumping requirements. Metering pumps are used in municipal and industrial applications ranging from water and waste treatment to chemical and petrochemical processes.

The major difference between metering (or controlled-volume) pumps and the less precise rotary-centrifugal pumps is in the flow characteristics. The rotary pump produces a relatively uniform flow. The metering pump, a reciprocating device, produces a characteristic pulsating flow.

This pulsating flow requires an approach to system design entirely different from steady-state flow. Systems using small-diameter pipe or extreme pipe length on the discharge may be subject to high-acceleration pressures. The resulting hydraulic shock could damage the piping or the pump. If the suction piping is improperly designed, the result could be cavitation causing loss of volumetric efficiency or worse, loss of prime.

Metering pump accuracy stems from positive displacement of the liquid being pumped. Reciprocating pumps all employ some form of driven plunger and/or diaphragm. A pulsing flow is created between a pair of check valves by alternately increasing and decreasing pressure on a confined volume of liquid.

The piston-diaphragm pump is a metering pump with a hydraulically driven diaphragm. The drive unit can be an AC motor or an SCR-controlled DC motor. The drive and capacity-adjusting mechanisms are usually integrated. They include some means of regulating the stroke length of the piston. Thus, for a specific speed and piston diameter, capacity is determined by the stroke length. This is the final variable in figuring volumetric displacement. Between the liquid-end assembly and the piston is an intermediate-fluid area. It allows the piston to be hydraulically coupled to the diaphragm. The intermediate fluid, a carefully selected oil, is contained in this area. A system of valves is used to maintain the hydraulic coupling for reliable performance. Finally, the liquid end consists of the diaphragm, ports, and inlet and outlet check valves. The diaphragm provides an interface between the oil and the liquid being pumped. As the diaphragm flexes, the inlet and outlet check valves alternately open and close to allow a positive pumping action to occur.

The pump's drive mechanism is a variable-eccentric type: The piston is driven through its stroke by an eccentric cam attached to a connecting rod. The radius of eccentricity is infinitely variable through its range, thus producing a wide range of stroke adjustment. At a fixed pumping speed and specified piston size, this variable-eccentric cam is the means of controlling pump capacity. It varies the length of the displacement stroke.

Known as *amplitude modulation of stroke*, this rotating cam produces a displacement that resembles a sine wave when plotted against cycle or time (see Figure 3.44). The acceleration and deceleration of the liquid being pumped are gradual and continuous. This allows efficient valve response and minimizes shock stresses on the drive parts. When the cam's radius of eccentricity is re-

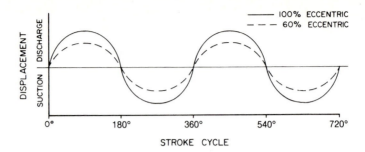

Figure 3.44   Illustrates amplitude modulation of stroke.

duced, the stroke length of the piston is shortened. This results in lower capacity and appears as a lower-amplitude sine wave. Similarly, when the radius of eccentricity of the cam is increased the stroke length of the piston is lengthened; this results in increased capacity and a higher-amplitude sine wave. The shift in amplitude in response to changing stroke length provides the name for this type of drive mechanism (*variable-eccentric*).

The piston reciprocates within a cylinder in the pump's crosshead. At the beginning of the suction stroke, the piston draws the diaphragm back by means of a hydraulic coupling. This creates a partial vacuum within the chamber. The static pressure in the discharge line seats the discharge valve, preventing any backflow of liquid from the discharge piping. At the same time, the suction check valve is lifted from its seat by the incoming flow of liquid. This liquid is being forced into the pump by the static suction pressure and the partial vacuum created by the piston. The result is net positive inlet pressure.

At the completion of the suction stroke, the pumping chamber is full of liquid. As the piston starts forward on the discharge stroke, it builds up sufficient pressure to overcome the positive inlet pressure and seat the suction valve. As the chamber pressure rises to and above the pressure in the discharge line, the discharge valve opens allowing the liquid to flow from the pump.

As mentioned earlier, the piston reciprocates within an accurately sized and precisely bored cylinder in the pump's crosshead. The piston moves through an established stroke length displacing an exact volume of oil. The oil serves as an intermediate fluid between the piston and the diaphragm. As the piston stroke displaces oil, the diaphragm flexes causing the process fluid to enter or leave the pump. As there is no significant pressure differential across the diaphragm, it is said to be hydraulically balanced. This ensures that no accuracy or efficiency will be lost due to ballooning of the diaphragm or through the inability of the diaphragm to move through the entire displacement. Both defects are associated with high pressure and long stroke conditions in mechanical diaphragm pumps. A system of four valves located throughout the liquid end maintains the balanced hydraulic coupling.

The loss of a small amount of oil is inherent in a piston-pump design. If provision is not made for refilling this oil, the diaphragm will eventually flatten against the backing plate or the oil will vaporize. Either will impair pump operation. The oil refill valve performs this refill function.

Two criteria limit the setting of this valve. It must be set for a pressure slightly higher than the vapor pressure of the oil. This will allow maximum motion of the diaphragm without oil vaporization. Also, the net suction pressure must be at least 2 psi higher than the oil refill valve setting. This is to avoid vacuum relieving and loss of pumping efficiency.

The moving parts in the drive are partially submerged in the oil used for the hydraulic coupling. This causes air to become entrained in the oil. Also, most oil has some air dissolved in it as supplied. In order to provide accurate metering, air must be purged from the hydraulic system.

The air-purge valve does the purging. On each stroke of the pump, the valve opens and allows any trapped air to be vented. When no air is present, a slight amount of oil is vented. The oil refill valve senses the loss of oil volume and refills the oil on each stroke.

A pressure-relief valve protects the diaphragm and thrust-carrying parts of the drive from excess pressure by relieving excess oil. Excess pressure can occur when the actual suction lift exceeds the level designed into the pump (by closing a suction-line valve) or if the discharge line is inadvertently dead-ended (by closing a discharge line valve). Both will cause the pump's design pressure to be exceeded. The valve, which is adjustable, should be set to relieve at 10 percent to 15 percent above the process pressure, not only to protect the pump but also for the process piping.

In the case of excessive suction lift, the setting of the oil refill valve will be exceeded. This will allow too much oil into the intermediate area between the piston and the diaphragm. The diaphragm will eventually become pinned against the front plate. The ensuing excess pressure will cause the valve to relieve.

In the case of a closed discharge line, the pump will continually try to oppose the static-discharge pressure, thus building up excessive pressures. When the setting of the pressure-relief valve is reached, the valve will open. This relieves oil to the pump's reservoir, thus venting excess pressure. If the pump is the only pressure-producing component in the system, the pressure-relief valve will serve as protection for the entire system.

Piston-diaphragm pumps are available with one, two, or three pistons linked to a common drive unit. Each piston is hydraulically coupled to its own liquid end. The liquid ends can have manifolded sections, manifolded discharges, or manifolded sections and discharges. According to the number of commonly driven pistons, the pumps are named simplex, duplex, or triplex.

There are two major reasons for the multiplexing of a piston-diaphragm pump.

1. More than one pump head is used when greater capacity is required. The flow that can be expected from a duplex or triplex pump is double or triple

the mean flow obtainable from a simplex. In order to achieve the higher flow, the suction and discharge lines must be manifolded.

2. Multiple liquid ends, again with suction and discharge lines manifolded, tend to smooth the output flow. At the end of every discharge stroke, there is a period of zero flow. As a result, the output of a simplex pump has a flow pattern characterized by distinct pulses (see Figure 3.45). The pistons of a duplex pump are coupled in such a manner that the peak-discharge flows are 180° apart. This means that when one head is in its discharge mode the other head is in it suction mode. The result is a reduction of zero-flow period and some smoothing of flow pulsations. The peaks of the discharge flows of a triplex pump are only 120° apart. This results in a flow output that is as close to uniform as possible with a reciprocating pump. When a triplex pump is used, pulsations in flow rate

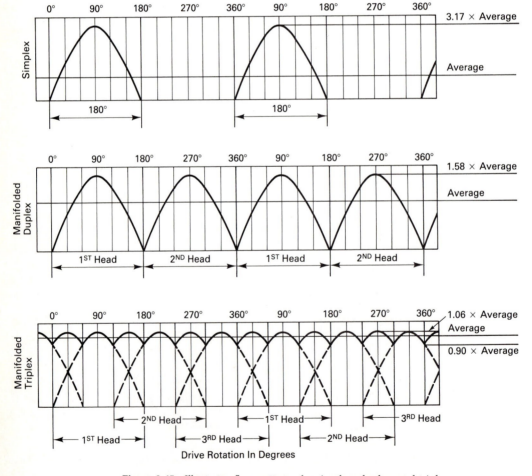

Figure 3.45 Illustrates flow patterns for simplex, duplex, and triplex.

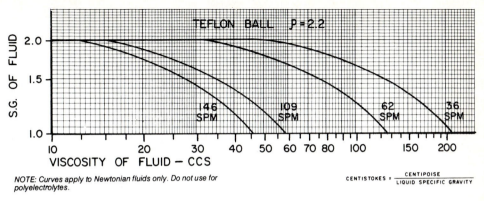

NOTE: *Curves apply to Newtonian fluids only. Do not use for polyelectrolytes.*

$$CENTISTOKES = \frac{CENTIPOISE}{LIQUID\ SPECIFIC\ GRAVITY}$$

Figure 3.46   Stroking speed limits for valves.

are not detectable 50 pipe diameters downstream from the discharge manifold. For this reason, a standard flow meter may be successfully employed in the pump discharge (downstream from the manifold) for a direct indication of the pumping rate.

As described, the flow capacity of the pump depends on stroking speed, stroke length, and piston diameter. The maximum capacity of the pump is set by selection of a stroking speed and piston diameter. This capacity is then adjustable through a 10:1 range by adjusting the stroke length.

Fluid-specific gravity, viscosity, and stroking speed can all affect the closing efficiency of the suction and discharge valve balls. Proper performance can be achieved by calculating the maximum stroking speed for the viscosity and specific gravity of the solids to be pumped. In order to do this, refer to Figure 3.46. To avoid loss of pumping efficiency, select the stroking speed nearest to the right and above the intersection of the specific gravity and viscosity coordinates. Note that all stroking speeds are expressed in strokes per minute (spm).

Example

What is the maximum stroking speed for a fluid with a viscosity of 99 centipoise and a specific gravity of 1.35?

1. Convert centipoise to centistoke by dividing by the specific gravity. 99 ÷ 1.35 = 73 centistokes (see Figure 3.47).
2. Plot the intersection of 73 centistokes viscosity and specific gravity of 1.35 on the graph.
3. Select the stroking speed nearest to this point that is to the right of and above this point. Maximum stroking speed = 62 spm.

When considering the sizing of the piston-diaphragm pump, four selections must be made:

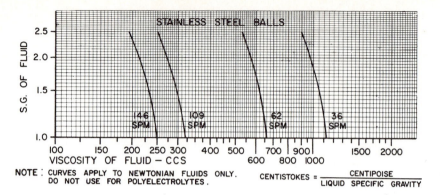

Figure 3.47   Chart relating viscosity to fluid specific gravity. (*Courtesy of Wallace & Tiernan, Inc., Belleville, NJ*)

1. Stroking speed, e.g., 36 spm, 62 spm, 109 spm, or 146 spm
2. Piston diameter, e.g., $\frac{15}{16}$ inch, $1\frac{1}{4}$ inch, $1\frac{11}{16}$ inch, $2\frac{1}{4}$ inch, or 3 inch
3. Number of heads, simplex, duplex, or triplex
4. Horsepower of the powered drive unit, $\frac{1}{3}$ to 3 in AC motors or $\frac{1}{2}$ to 5 in SCR-controlled DC motors

Selection of the pump arrangement that will deliver the required maximum flow capacity is a function of stroking speed, piston diameter, and the number of pump heads. As discussed earlier, the maximum allowable stroking speed will be set by the viscosity and specific gravity of the fluid to be pumped. This factor puts an upper limit on pumping capacity of that particular fluid for any given arrangement of the pump.

Although fluid characteristics determine the maximum stroking speed, the pump can be run at this level or at any stroking speed below the maximum. In point of fact, the optimum stroking speed is the slowest speed which will provide the maximum required pumpage while not exceeding the stroke-speed-limit criterion. That is, if the maximum allowable stroking speed for a particular liquid is 109 spm, but 62 spm will provide more than enough flow, use 62 spm as the optimum stroking speed. Selecting the lower speed will minimize wear on thrust-carrying parts of the pump and permit the use of smaller-diameter piping in the suction and discharge systems.

Piston diameter must be selected on pressure requirements first and on capacity requirements second. The maximum allowable pressures for the various diameter material combinations are listed in Table 3.10.

For a pressure requirement below 125 psi, the head configuration that will provide the required capacity at the optimum stroking speed should be selected (either $1\frac{1}{4}$-inch or $2\frac{1}{4}$-inch piston). If the maximum speed limits the capacity so that neither piston diameter will provide adequate flow using the simplex group, multiplexing of pump heads will have to be considered. For a pressure

TABLE 3.10   ALLOWABLE PRESSURES FOR MATERIALS COMBINATIONS

PISTON DIAMETER	HEAD MATERIAL	MAXIMUM DISCHARGE PRESSURE RATING
15/16"	PVC	150 psi
15/16"	SS 316	700 psi
1-1/4"	PVC	150 psi
1-1/4"	SS 316	400 psi
1-11/16"	PVC	125 psi
1-11/16"	SS 316	225 psi
2-1/4"	PVC	125 psi
2-1/4"	SS 316	125 psi
3"	PVC	70 psi
3"	SS 316	70 psi

requirement in excess of 125 psi, the appropriate combination of maximum speed and multiple heads for the $1\frac{1}{4}$-inch piston must be selected.

Table 3.11 compares one manufacturer's arrangements of piston-diaphragm pumps. Mean flow is the actual value to be used for capacity considerations. Peak flow values represent the magnitude of the instantaneous pulsations of the pump. These peak flows will be considered later.

The final consideration in pump sizing is the horsepower rating of the drive unit for the various pump arrangements, speeds, and pressures. Referring back to the example of optimum versus maximum stroking speed, one can see that 109 spm, the maximum speed in the example, will allow only a maximum discharge pressure of 100 psi for a $1\frac{1}{4}$-inch piston with a PVC head using a $\frac{1}{3}$

TABLE 3.11   ARRANGEMENTS OF PISTON–DIAPHRAGM PUMPS

15/16" Dia. Piston	Simplex		Duplex (manifolded)		Triplex (manifolded)	
	Mean	Peak	Mean	Peak	Mean	Peak
36 SPM	8.0	25	16	25	24	26
62 SPM	14.0	44	28	44	42	45
109 SPM	24.5	77	49	77	73.5	79
146 SPM	33.0	104	66	104	99	106
1-1/4" Dia. Piston						
36 SPM	15.0	47	30	47	45	48
62 SPM	26.0	82	52	82	78	83
109 SPM	45.5	143	91	143	136.5	146
146 SPM	60.5	190	121	190	181.5	194
1-11/16" Dia. Piston						
36 SPM	28.5	89	57	89	85.5	91
62 SPM	49.5	155	99	155	148.5	159
109 SPM	87.0	273	174	273	261	279
146 SPM	116.5	366	233	366	349.5	374
2-1/4" Dia. Piston						
36 SPM	53.0	166	106	166	159	170
62 SPM	91.0	286	182	286	273	292
109 SPM	160.0	502	320	502	480	514
146 SPM	214.0	672	428	672	642	687
3" Dia. Piston						
36 SPM	95.0	298	190	298	285	305
62 SPM	163.0	512	326	512	489	523
109 SPM	286.0	898	572	898	858	918
146 SPM	383.0	1203	766	1203	1149	1229
	Nominal	×3.14	Nominal	×1.57	Nominal	×1.07

TABLE 3.12 MATERIALS AND CAPACITY CHARACTERISTICS

PISTON DIAMETER (INCHES)	VALVE MATERIAL	STROKING SPEED (SPM)	CAPACITY (GPH)			MAXIMUM DISCHARGE PRESSURE (PSIG) MOTOR HORSEPOWER REQUIRED INDUCTION (VARIABLE SPEED)				
			SIMPLEX	DUPLEX	TRIPLEX	1/3(1/2)*	1/2(3/4)*	3/4(1)*	1(1 1/2)*	1 1/2(2)*
15/16	316 S.S CARP 20 HAST C	36	8.0	16.0	24.0	620	690	700		
		62	14.0	28.0	42.0	360	545	700		
		109	24.5	49.0	73.5	215	320	480	645	700
		146	33.0	66.0	99.0	155	235	355	480	700
	PVC	36	8.5	17.0	25.5	150				
		62	15.0	30.0	45.0	150				
		109	27.0	54.0	81.0	150				
		146	36.0	72.0	108.0	110	150			
	PTFE	36	9.0	18.0	27.0	100				
		62	15.5	31.0	46.5	100				
		109	27.0	54.0	81.0	100				
		146	36.5	73.0	109.5	100				
1 1/4	316 S.S. CARP 20 HAST C	36	15.0	30.0	45.0	310	400			
		62	26.0	52.0	78.0	180	280	400		
		109	45.5	91.0	136.5	100	150	230	310	400
		146	60.5	121.0	181.5	80	125	190	275	400
	PVC	36	15.5	31.0	46.5	150				
		62	27.0	54.0	81.0	150				
		109	48.0	96.0	144.0	100	150			
		146	64.0	128.0	192.0	80	125	150		
	PTFE	36	16.0	32.0	48.0	100				
		62	27.5	55.0	82.5	100				
		109	48.5	97.0	145.5	100				
		146	65.0	130.0	195.0	80	100			
1 11/16	316 S.S CARP 20 HAST C	36	28.5	57.0	85.5	200	225			
		62	49.5	99.0	148.5	110	170	225		
		109	87.0	174.0	261.0	65	95	145	195	225
		146	116.5	233.0	349.5	45	70	110	145	225
	PVC	36	29.0	58.0	87.0	125				
		62	50.0	100.0	150.0	100	125			
		109	88.0	176.0	264.0	55	85	125		
		146	118.0	236.0	354.0	40	60	90	125	
	PTFE	36	29.0	58.0	87.0	100				
		62	50.5	101.0	151.5	100				
		109	88.5	177.0	265.5	55	80	100		
		146	118.5	237.0	355.5	40	60	90	100	
2 1/4	316 S.S. CARP 20 HAST C PVC	36	53.0	106.0	159.0	100	125			
		62	91.0	182.0	273.0	55	85	125		
		109	160.0	320.0	480.0	30	45	70	95	125
		146	214.0	428.0	642.0	25	35	55	80	125
	PTFE	36	53.0	106.0	159.0	100				
		62	91.0	182.0	273.0	55	85	100		
		109	160.0	320.0	480.0	30	45	70	95	100
		146	214.0	428.0	642.0	25	35	55	80	100
3	PVC 316 S.S. PTFE CARP 20	36	95.0	190.0	285.0	60	70			
		62	163.0	326.0	489.0	35	50	70		
		109	286.0	572.0	858.0	20	30	45	60	70
		146	383.0	766.0	1149.0	15	20	35	45	70

*Horsepower requirements: first figure is for induction motors, figure in brackets is for variable speed motors.
NOTE: Capacities listed are at maximum discharge pressure and a base motor speed of approxiamtely 1700 rpm.
¾ HP minimum for DC drive triplex arrangements.

hp AC motor. Reducing speed to the more desirable 62 spm allows a maximum discharge pressure of 150 psi, the limit of the head, with no increase on motor size. In this case, the lower stroking speed is desirable. It reduces wear on the pump and permits use of a smaller motor. If a $1\frac{1}{4}$-inch piston driven at 62 spm will provide adequate flow, the combination is ideal from the standpoint of pump sizing.

There are pressure-temperature limitations that must be considered in pump material selection. Pump materials include elastomers, plastics, and alloys selected to provide reliable performance with many corrosive fluids pumped at various pressures and temperatures. The PVC liquid ends have temperature and chemical resistances corresponding to those of Schedule 80 threaded PVC pipe. The stainless steel liquid ends will, of course, provide higher resistance to pressure and temperature.

When selecting PVC, consider its loss of strength at high temperatures and reduce the maximum pressure accordingly for reliable operation. Table 3.13 illustrates the pressure limitations on PVC liquid ends at different temperatures. As PVC becomes brittle at low temperatures, the PVC liquid end should not be subjected to temperatures below 35°F.

As mentioned earlier, the 316 stainless steel liquid end can tolerate substantially higher temperatures. Temperatures for process fluids run to 180°F with pressures to 400 psi for the $1\frac{1}{4}$-inch piston and 125 psi for the $2\frac{1}{4}$-inch piston.

Its important to note that a piston-diaphragm pump is designed to provide a high degree of repeatability through a wide range of pressures. The capacity that can be expected at atmospheric pressure will be nearly that of the theoretical displacement as indicated in Table 3.11. As the discharge pressure increases, a small decrease in capacity can be expected. The rate of capacity fall-off at higher pressures is approximately 1 percent to 1.5 percent per 100 psi.

The ultimate check of actual pump output is on-site calibration. This may be done easily using a graduate and two valves in the suction line.

TABLE 3.13 PRESSURE LIMITS OF PVC LIQUID ENDS

	Pressure Limits of PVC Liquid Ends				
	Maximum Discharge Pressure for Piston Diameters				
Temperature (°F)	$\frac{15}{16}''$	$1\frac{1}{4}''$	$1\frac{11}{16}''$	$2\frac{1}{4}''$	$3''$
75	150	150	125	125	70
100	150	150	125	125	70
110	150	150	120	120	70
115	144	144	108	108	70
120	120	120	96	96	70

(Courtesy of Wallace & Tiernan, Inc., Belleville, NJ)

To obtain optimum performance, a pump must be applied correctly. It must be selected and sized correctly, and it must be installed in a carefully planned and designed pumping system. In order to approach these topics logically, this discussion is broken down into three parts: typical installation schematics; piping installation considerations; and pump installation considerations.

Installation Schematics

The pump's versatility permits a wide variety of system schematics to be used. Most important, however, are the two basic systems of flooded suction and suction lift. In suction lift, the center line of the liquid end is located at an elevation higher than the supply tank. In flooded suction, the supply is located at a higher elevation than the center line of the liquid end. Figures 3.48 and 3.49

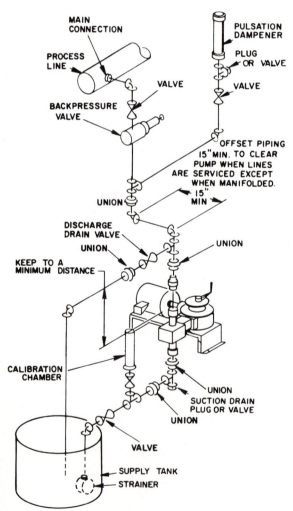

Figure 3.48 Suction-lift installation with optional accessories. (*Courtesy of Wallace & Tiernan, Inc., 25 Main Street, Belleville, NJ 07109-3057*)

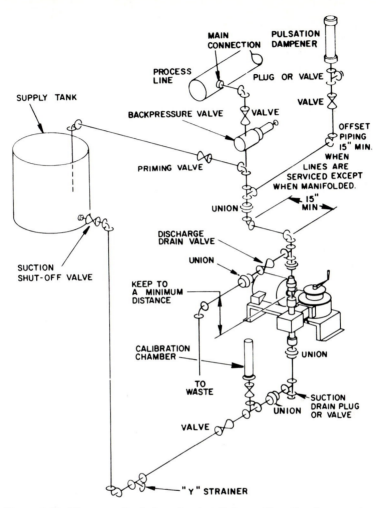

Figure 3.49 Illustrates flooded suction installation with optional accessories. (*Courtesy of Wallace & Tiernan, Inc., 25 Main Street, Belleville, NJ 07109-3057*)

depict these two basic systems. They show necessary components as well as optional accessories.

Two manual ball valves and a graduate in the suction line for on-site pump calibration can be useful additions to any pump installation. A pulsation dampener to reduce pressure peaks installed in the discharge line minimizes vibration and wear due to long lines and/or high stroke speeds (not needed with triplex-pump arrangement). A back-pressure valve in the discharge line will augment the operation of the discharge valve on the pump; increase static discharge pressure to eliminate free-wheeling and resultant overfeeding; and eliminate static siphoning due to high-static-suction pressures or low-static-discharge pressures.

Suction

1. To keep the suction piping as short as possible, locate the pump close to the supply tank.
2. Install a drain valve in the lowest point of the piping.
3. In a suction-lift situation, always run the suction piping vertically from the supply to the pump. If this is not practical, the piping must slope upward in the direction of flow. This helps remove air at start-up and prevents accumulation of entrained air or gas released in the suction line from the fluid being pumped.
4. Suction piping should be thoroughly tightened at the joints and fittings. This prevents air, which will reduce volumetric efficiency, from being drawn into the system. A continuous length of tubing is preferred over pipe and fittings.
5. Where possible, always install a strainer on the entrance to the suction line.

Discharge

1. The pipe-pressure rating should be well in excess of the pump's pressure-relief-valve setting.

Suction and Discharge

1. Piping must be sized to handle the peak instantaneous flow from the pump. On a simplex arrangement, this is over three times the mean flow.
2. Install unions near the pump suction and discharge ports in order to facilitate removal of the liquid end for routine maintenance.
3. Do not route PVC pipe through cold areas where the liquid could freeze or the pipe could become brittle.
4. When high-temperature fluids are being pumped, pay close attention to the pressure derating factors (mentioned earlier) for PVC pipe.
5. When the stainless steel liquid end is used, select pipe material that will not cause galvanic corrosion at the pump connections.
6. Be sure that there are no burrs, sharp edges, or machine oil in the piping. Lines should be blow or flushed clean before final installation to the pump.
7. Piping should be adequately supported so as not to place a strain on the pump connections. Do not "spring" the piping when connecting to the pump.
8. When supporting the pipe with hangers, allowance should be made for expansion and contraction due to extreme temperatures. Lay out suction and discharge lines with 90-degree traverses between the fixed ends. Do not clamp the pipes to the supports. The pipe must be supported vertically without being restricted horizontally.

9. Both suction and discharge piping should be straight, short, and uncomplicated with as few fittings as possible. When fittings must be used, employ 45-degree or long-radius 90-degree elbows. All piping should be sloped up to eliminate the possibility of vapor pockets.

10. Avoid radical restrictions in the line at any point, including the final point of discharge. This is particularly important if scaling of the pipe is likely to occur.

11. When shutoff valves are installed, they should open to the full inside diameter of the line.

Installation Considerations

1. Wherever possible, locate the pump below or at the same elevation as the supply. This is desirable because high-suction lifts tend to cause cavitation, resulting in difficulty with priming and loss of pump efficiency.

2. The pump should be bolted to a concrete foundation above floor level. This will protect the base of the pump from washdowns. The floor should be reasonably level. Be sure the foundation is high enough to accommodate system piping.

3. Space around the pump should be sufficient to allow access for various adjustments and routine maintenance. Be sure to leave enough room around the liquid ends for removal and servicing.

4. The pump should be installed away from elevated tanks and other equipment where overflow could cause damage.

Pump/System/Piping Design Guidelines

Proper piping system design is critical to reliable pump operation and to accurate metering of the chemical. The suction and discharge systems must be considered separately when calculating pipe sizes and pressure heads and when selecting components for inclusion in the system. The effect of each on the other must also be considered in order to optimize pump efficiency and accuracy.

The main concern in design of the suction system is adequate inlet pressure, or *suction pressure*. The suction pressure forces the product into the pumping chamber. If the suction pressure is adequate, the liquid will not only accelerate fast enough, but will also reach the required peak velocity (velocity necessary to satisfy the suction requirements of the pump). If the suction pressure is insufficient, cavitation (product vaporization) could occur. This causes reduction in pumpage or, in extreme cases, loss of pump delivery. Another possible result of inadequate suction pressure is activation of the oil refill valve and possible vaporization of the hydraulic fluid. If, at the other extreme, the available suction pressure is too high, the discharge head may be exceeded. This will result in freewheeling (overfeeding or dynamic syphoning).

The four pressure elements that contribute to suction pressure are:

1. atmospheric pressure (Pa)
2. static pressure, the pressure acting on the suction system under shutdown or no-pumping conditions (Ps)
3. suction-line-friction losses (Pf)
4. suction-line-inertia pressure (Pl)

The absolute static suction pressure is equal to the algebraic sum of the first two pressure heads cited previously. Expressed in engineering symbols:

$$\text{Static-Suction Pressure} = Pa + Ps$$

Note that static pressure (Ps) can be either a head (positive) or a lift (negative).

The fourth element of inertia pressure (Pl) requires some explanation. Force is required to accelerate the mass of liquid in the suction line toward the pump. This force is provided by a decrease in suction pressure at the pump inlet during the first half of the suction stroke. The mass must then decelerate during the second half of the suction stroke. The force required to accomplish this results in an *increase* in pressure at the pump inlet. The pressure pulsation that results from this decreasing and increasing force on the liquid is referred to as the *suction-line-inertia pressure*. The maximum value of the pressure pulsation which occurs at the beginning and end of every *suction* stroke will be termed *suction Pl*.

The *dynamic suction pressure* is the static suction pressure, as defined previously, less friction loss Pf plus or minus the suction Pl. Expressed in engineering symbols:

$$\text{Dynamic-Suction Pressure} = Pa \pm Ps - Pf \pm Pl$$

The adequacy of the suction pressure requires calculation of the *minimum* static and dynamic suction pressures.

$$\text{Minimum Static-Suction Pressure} = Pa \pm Ps$$
$$\text{Minimum Dynamic-Suction Pressure} = Pa \pm Ps - Pf - \text{Suction Pl}$$

(Note: To calculate the preceding minimum suction pressures, the minimum Ps should be used as referred to the center line of the liquid end of the pump.)

The consideration of the possibility of freewheeling during the suction stroke requires calculation of maximum suction pressures.

$$\text{Maximum Static-Suction Pressure} = Pa \pm Ps$$
$$\text{Maximum Dynamic-Suction Pressure} = Pa \pm Ps - Pf + \text{Suction Pl}$$

(Note: For these determinations, the maximum Ps should be used.)

There are two major concerns when considering the design of the discharge system: The maximum discharge pressure must not exceed the pressure rating of the pump head; the minimum discharge pressure must be greater than the maximum suction pressure in order to prevent freewheeling during the discharge stroke.

The four pressure elements that contribute to discharge pressure are:

1. atmospheric pressure (Pa)
2. static pressure, the pressure acting on the discharge system under shut-down conditions (Ps)
3. discharge-line-friction losses (Pf)
4. discharge-line-inertia pressure (Pl)

Minimum and Maximum Static Discharge Pressure = Pa ± Ps.

The minimum static discharge pressure will occur when Ps is at its lowest and the maximum static discharge pressure will occur when Ps is at its highest. Furthermore, the value of Ps will be additive when the point of application is higher than the center line of the liquid end. It will be subtractive when the point of application is lower than the center line of the pump.

The discharge Pl acts on the discharge system in a manner similar to that of the suction system. Force is required to accelerate the mass of liquid in the discharge line away from the pump toward the point of application. This force is provided by an increase in pressure at the pump outlet during the first half of the discharge stroke. The mass must then decelerate during the second half of the discharge stroke. The force required to accomplish this results in a decrease in pressure at the pump outlet. The pressure pulsation that results from this increasing and decreasing force on the pump's outlet is referred to as the *discharge-line-inertia pressure*. The maximum value of the pressure pulsation which occurs at the beginning and end of every discharge stroke will be termed *discharge Pl*.

The dynamic discharge pressure can be figured, knowing the minimum and maximum static discharge pressures and the discharge Pl.

Minimum Dynamic Discharge Pressure = Pa ± Ps + Pf − Discharge Pl
Maximum Dynamic Discharge Pressure = Pa ± Ps + Pf + Discharge Pl

As indicated earlier, flow friction loss does act in reciprocating pump systems. Experience has indicated, however, that if the piping system is designed to operate satisfactorily from the inertia pressure standpoint, the calculated pipe diameter will be of a size that flow friction losses will be negligible and may be excluded from the calculation. Only in cases where the fluid to be pumped is of a high viscosity will flow friction loss need to be considered. A method of inclusion of Pf into the calculation is discussed at the end of this section. (Note: For triplex arrangements, Pf consideration will be more critical.)

The following symbols are used in the following calculation procedures:

ρ: Specific gravity of process fluid.

M-factor: Composite pump factor relating stroking speed and piston diameter. It is used in sizing of suction and discharge lines. M-factor values are tabulated on the nomograph in Figure 3.50, later in this chapter.

Pl: Inertia pressure, the magnitude of the pressure pulsation above and below static pressure. The value of Pl is a function of the length and diameter of the suction or discharge line, specific gravity, and the M-factor. Pl is expressed in pounds per square inch (psi). Separate values of Pl are calculated for the suction system and for the discharge system.

Ls or Ld: Length of suction or discharge lines (feet).

Pa: Atmospheric pressure (psia).

Pvp: Vapor pressure of the liquid at pumping temperature (psia).

Pf: Flow friction loss (psi).

Ps: Static pressure acting on the system under shutdown or no-pumping conditions (psi).

As noted earlier, there are certain minimum design criteria that must be met for proper pump operation and accurate metering. Using these criteria, the values of suction Pl, discharge Pl, and corresponding pipe sizes may be established.

I. Simplex and Triplex Pump
  A. Suction System Criteria
    1. Minimum dynamic suction pressure should exceed Pvp by at least 5 psi (to prevent cavitation). And dynamic suction pressure must be greater than 10 psia.
    2. Maximum dynamic suction pressure must be at least 2 psi less than minimum static discharge pressure (to prevent freewheeling during the suction stroke).
    3. Suction Pl must not exceed 15 psi. (15 psi is the established limit above which pulsations become audible and produce vibrations which can damage plastic piping. If noise is not objectionable, Pl values higher than 15 psi can be used for well-anchored metallic piping systems. But the maximum dynamic pressure must not exceed the rating of the pipe or pump and all other criteria must be satisfied.)
  B. Discharge System Criteria
    1. Maximum dynamic discharge pressure must not exceed the maximum pressure rating of the pump head (see Table 3.10).
    2. Minimum dynamic discharge pressure must be at least 2 psi greater than maximum static suction pressure (to prevent freewheeling during the discharge stroke).
    3. Discharge Pl must not exceed 15 psi. (Again, 15 psi is the established limit above which pulsations become audible and produce vibrations which can damage plastic piping.)

II. Duplex Pump
  A. Suction System Criteria
    1. Same as for the simplex and triplex pumps, outlined earlier (I.A.1).
    2. Maximum dynamic suction pressure must be at least 2 psi *less* than the minimum dynamic discharge pressure (to prevent freewheeling during the suction stroke). Differs from the simplex and triplex pumps (I.A.2) because, in the case of duplex pumps, maximum dynamic suction pressure and minimum dynamic discharge pressure occur simultaneously.
    3. Same as simplex and triplex pumps (I.A.3).
  B. Discharge System Criteria
    1. Same as simplex and triplex pumps (I.B.1).
    2. Minimum dynamic discharge pressure must be at least 2 psi greater than maximum dynamic suction pressure (to prevent freewheeling during the discharge stroke).
    3. Same as simplex and triplex pumps (I.B.3).

The triplex arrangement employs the same basic design criteria as that of the simplex. The major difference, as stated earlier, is the handling of flow-friction loss. The inertia pressures for the triplex will be only one half that of the simplex. As a result, the pipe sizes used will be substantially smaller, for the triplex and friction losses will be more significant when considering minimum pressure limits. The handling of flow-friction loss is covered later on in this chapter.

Example 3.1: Application of Design Criteria for a Simplex Pump

Maximum pressure rating for PVC head: 150 psi (164.7 psia). Refer to Figure 3.51.

Suction Pressures

$$\text{Minimum Static Suction Pressure} = 14.7 - (.433 \times 10 \times 1.2) = 14.2 \text{ psia}$$
$$\text{Maximum Static Suction Pressure} = 14.7 + (.433 \times 8.0 \times 1.2) = 18.9 \text{ psia}$$
$$\text{Minimum Dynamic Suction Pressure} = 14.2 - 4.0 = 10.2 \text{ psia}$$
$$\text{Maximum Dynamic Suction Pressure} = 18.9 + 4.0 = 22.9 \text{ psia}$$

Discharge Pressures

$$\text{Minimum Static Discharge Pressure} = 14.7 + (.433 \times 20 \times 1.2) + 20$$
$$= 45.1 \text{ psia}$$

$$\text{Maximum Static Discharge Pressure} = 14.7 + (.433 \times 20 \times 1.2) + 30$$
$$= 55.1 \text{ psia}$$

$$\text{Minimum Dynamic Discharge Pressure} = 45.1 - 6.0 = 39.1 \text{ psia}$$

$$\text{Maximum Dynamic Discharge Pressure} = 55.1 + 6.0 = 61.1 \text{ psia}$$

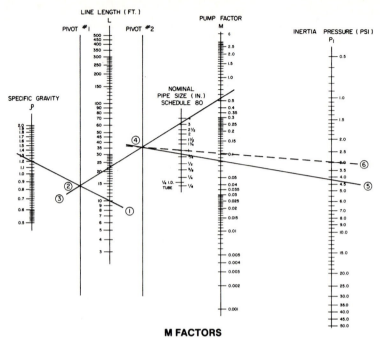

M FACTORS

	STROKING SPEED (S=Simplex, D=Duplex, T=Triplex)							
PISTON	36 SPM		62 SPM		109 SPM		146 SPM	
DIAMETER	S&D	T	S&D	T	S&D	T	S&D	T
15/16"	.0299	.0149	.0914	.0457	.2988	.1494	.5247	.2624
1-1/4"	.0531	.0266	.1625	.0813	.5313	.2656	.9328	.4664
1-11/16"	.0968	.0484	.2962	.1481	.9682	.4841	1.7000	.8500
2-1/4"	.1721	.0861	.5265	.2633	1.7213	.8606	3.0223	1.5112
3"	.306	.153	.936	.468	3.06	1.53	5.373	2.6865

METHOD

1. CONNECT ℓ AND L VALUES WITH STRAIGHT EDGE. ①
2. MARK PIVOT LINE #1 WHERE STRAIGHT EDGE CROSSES. ②
3. CONNECT MARK ON PIVOT LINE #1 AND "M" FACTOR WITH STRAIGHT EDGE. ③
4. MARK PIVOT LINE #2 WHERE STRAIGHT EDGE CROSSES. ④
5. CONNECT MARK ON PIVOT LINE #2 WITH DESIRED OR KNOWN VALUE OF P₁ ⑤ AND READ PIPE SIZE OR CONNECT THE MARK ON PIVOT LINE #2 WITH PIPE SIZE ⑥ AND READ RESULTING P₁.

EXAMPLE

CALCULATE THE SIZE OF THE SUCTION PIPING FOR: A 43 SERIES SIMPLEX PUMPING 1200 GPD OF A FLUID WITH A SPECIFIC GRAVITY OF 1.2. THE PUMP HAS A 1¼INCH-DIAMETER PISTON OPERATING AT 109 STROKES PER MINUTE. THE SUCTION AND DISCHARGE LINES ARE 10 FEET AND 30 FEET.
SUCTION P₁=4.2 PSI (PER EXAMPLE 2) AND DISCHARGE P₁=15 PSI (PER EXAMPLE 3).

Figure 3.50   Suction and discharge piping sizing nomograph for a manufacturer's pump series. (*Courtesy of Wallace & Tiernan, Inc., 25 Main Street, Belleville, NJ 07109-3057*)

Note: Static pressure (in psi) = .433 psi/ft × feet of head × ρ. Examining these numbers versus the criteria for a simplex pump, we find:

- I.A.1 The minimum dynamic suction pressure exceeds PVD by more than 5 psi (10.2 − 1 = 9.2 psi), and the minimum dynamic suction pressure exceeds 10 psia (10.2 psia).

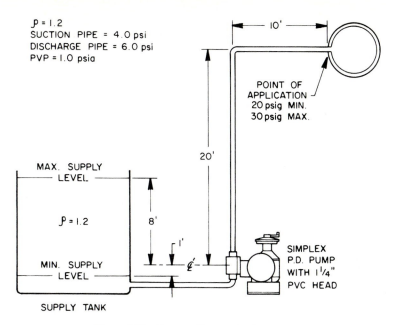

$\rho = 1.2$
SUCTION PIPE = 4.0 psi
DISCHARGE PIPE = 6.0 psi
PVP = 1.0 psia

|← 10' →|

POINT OF
APPLICATION
20 psig MIN.
30 psig MAX.

20'

MAX. SUPPLY
LEVEL

$\rho = 1.2$ 8'

MIN. SUPPLY
LEVEL

SIMPLEX
P.D. PUMP
WITH 1¼"
PVC HEAD

SUPPLY TANK

Figure 3.51 System diagram for Example 3.1.

- I.A.2 The maximum dynamic suction pressure is substantially less than the minimum static discharge pressure (45.1 − 22.9 = 22.2; 22.2 psi is a greater difference than the required 2 psi).
- I.A.3 Suction Pl, 4.0 psi, is less than 15 psi.
- I.B.1 The maximum dynamic discharge pressure of 61.1 psia is well under the 164.7 psia pressure rating of the PVC head.
- I.B.2 The minimum dynamic discharge pressure is substantially higher than the maximum static suction pressure (39.1 − 18.9 = 20.2 psi; 20.2 psi is a greater difference than the 2 psi required).
- I.B.3 The discharge Pl, 6.0 psi, is less than 15 psi.

Conclusion: The system in Example 3.1 satisfies all the criteria of a well-designed simplex pump system.

The criteria outlined previously limit the inertia pressure that a pumping system can tolerate and still function in a precise manner. Using these criteria, and the definitions of the various suction and discharge pressures (8 in total), the theoretical maximum allowable design values of suction Pl and discharge Pl can be calculated. The value of Pl that satisfies *all* criteria for a particular system shall be used in designing that system. This value will turn out to be the least value calculated from the criteria, but not greater than 15 psi or lower than 0 psi (to be explained later).

DESIGN VALUE OF SUCTION Pl, SIMPLEX OR TRIPLEX PUMP:

1. Calculate the design value of suction Pl per the first criterion (I.A.1). Suction Pl = Minimum Static Suction Pressure − (Pvp + 5) and Suction Pl = Minimum Static Suction Pressure − 10.
2. Calculate the design value of suction Pl per the second criterion (I.A.2). Suction Pl = Minimum Static Discharge Pressure − (Maximum Static Suction Pressure + 2).
3. Suction Pl shall not exceed 15 psi (I.A.3).

Example 3.2: Figuring Suction Pl

Refer to the pumping system of Example 3.1. Note the values for the various suction pressures and the equations resulting from the criteria for calculating the maximum allowable suction Pl.

1. Pl = 14.2 − (1.0 + 5) = 8.2 and Pl = 14.2 − 10 = 4.2
2. Pl = 45.1 − (18.9 + 2) = 24.2
3. Pl shall not exceed 15 psi.

Selecting a suction Pl of 4.2 psi meets all criteria for the suction system of the simplex pump and is the maximum allowable suction Pl of the suction system.

DESIGN VALUE OF DISCHARGE Pl, SIMPLEX PUMP:

1. Calculate the design value of discharge Pl per the first criterion (I.B.1). Discharge Pl = Pump head rating (psia) − Maximum Static Discharge Pressure.
2. Calculate the design value of discharge Pl per the second criterion (I.B.2). Discharge Pl = Minimum Static Discharge Pressure − (Maximum Static Suction Pressure + 2).
3. Discharge Pl shall not exceed 15 psi.

Example 3.3: Figuring Discharge Pl

Refer to the pumping system of Example 3.1. Note the values for the various discharge pressures and the equations resulting from the criteria for calculating the maximum allowable discharge Pl.

1. Pl = 164.7 − 55.1 = 109.6 psi
2. Pl = 15.1 − (18.9 + 2) = 24.2 psi
3. Pl shall not exceed 15 psi.

Selecting a discharge Pl of 15 psi meets all the criteria for the discharge system of the simplex pump and is the maximum allowable discharge Pl of the discharge system. (Note: In the original example, a discharge Pl of 6.0 psi was assumed. As explained later, a larger discharge pipe size than was required to accommodate the allowable discharge Pl was used to bring the Pl down from 15 psi to 5 psi.)

The procedure used to find the maximum allowable suction and discharge Pl for duplex pumps is similar to that of simplex or triplex pumps. The major difference is the calculation of the maximum allowable Pl to satisfy the free-wheeling criteria. In this situation the minimum-design margin of 2 psi exists between two dynamic pressures rather than between one dynamic and one static (already known) pressure. Since both dynamic pressures are a function of suction- and discharge-line sizes (as yet unknown), the determination of the maximum allowable suction Pl and discharge Pl must be done by trial and error. Two factors help simplify calculation of Pl values which meet both suction and discharge freewheeling criteria:

1. *Minimum static discharge pressure.* Maximum Static Suction Pressure = Suction Pl + Discharge Pl + 2. Because neither suction Pl nor discharge Pl can exceed 15 psi, it follows from the freewheeling criteria that suction Pl and discharge Pl must be less than 15 + 15 + 2 or 32 psi. If 32 psi is exceeded, one of the other two criteria will apply as the least Pl satisfying all three criteria.

2. We do know that the sum of suction Pl and discharge Pl equals Suction Pl + Discharge Pl = Minimum Static Discharge Pressure (Maximum Static Suction Pressure + 2).

To calculate the suction Pl and discharge Pl (freewheeling), use the following formula to distribute the sum of the total allowable Pl into separate values of suction Pl and discharge Pl after this sum has been found:

$$\frac{\text{Suction Pl}}{\text{Discharge Pl}} = \frac{\text{Ls}}{\text{Ld}}$$

Example 3.4: Calculation of Suction Pl and Discharge Pl for a Duplex Pump

Assume that the pump of Example 3.1 is a duplex pump with a suction-line length of 10 feet and a discharge-line length of 30 feet.

$$\text{Minimum Static Discharge Pressure} = 45.1 \text{ psia}$$

$$\text{Maximum Static Suction Pressure} = 18.9 \text{ psia}$$

$$\text{Ls} = 10 \text{ feet}$$

$$\text{Ld} = 30 \text{ feet}$$

1. The sum of the allowable Pl values = $45.1 - (18.9 + 2) = 24.2$ psi. This is less than 32 psi, so the calculation may proceed.
2. Suction Pl + Discharge Pl = 24.2 psi
3. Discharge Pl = 24.2 − Suction Pl
4. $\dfrac{\text{Suction Pl}}{\text{Discharge Pl}} = \dfrac{10}{30} = .333$
5. $\dfrac{\text{Suction Pl}}{24.2 - \text{Suction Pl}} = .333$
6. Suction Pl = .333 (24.2) − .333 Suction Pl
7. 1.33 Suction Pl = 8.1
8. Suction Pl = 6.0 psi (for freewheeling criterion)
9. Discharge Pl = 24.2 − 6.0 = 18.2 *psi* (for freewheeling criterion)

Reviewing the maximum allowable suction Pl and discharge Pl for the preceding duplex pump, we find:

SUCTION CRITERIA:

- II.A.1 Pl = 8.2 psi (vapor pressure criterion); Pl = 4.2 psi (minimum dynamic suction pressure)
- II.A.2 Pl = 6.0 psi (freewheeling criterion)
- II.A.3 Pl shall not exceed 15 psi

A suction Pl of 4.2 psi will be the maximum allowable design value for the suction system.

DISCHARGE CRITERIA:

- II.B.1 Pl = 109.6 psi (head rating criterion)
- II.B.2 Pl = 18.2 psi (freewheeling criterion)
- II.B.3 Pl shall not exceed 15 psi.

A discharge Pl of 15.0 psi will be the maximum allowable design value for the discharge system.

Having calculated the maximum allowable suction Pl and discharge Pl that will satisfy all the design criteria of an efficient and accurate piston-diaphragm pump, the appropriate pipe sizes to accommodate these pressure pulsations and peak flows can be determined. To facilitate selection of these pipe sizes, a nomograph can be used to find both suction and discharge pipe sizes. Suction line length (Ls) and suction Pl is used for suction pipe sizing. Discharge line length (Ld) and discharge Pl is used for discharge pipe sizing. The example in Figure 3.51 illustrates how to apply this nomograph for sizing the suction piping. The same procedure applies to sizing discharge piping.

Following is a summary of the steps involved in designing the suction and discharge piping system:

1. Make a sketch of typical installation schematics.
2. Calculate the minimum and maximum static suction and discharge pressures.
3. Determine the maximum allowable design value of suction Pl per the three suction criteria and resultant formulae for the given head arrangement. (Remember that the maximum allowable Pl will be the least of the three answers in order to satisfy all three criteria.)
4. Refer to the pipe sizing nomograph and find the correct suction-pipe size.
5. Having found the correct pipe size, find the actual suction Pl for that selected pipe size.
6. Determine the maximum allowable design value of discharge Pl per the three discharge criteria and resultant formulae for the given head arrangement. (Remember that the maximum allowable Pl will be the *least* of the three answers in order to satisfy all three criteria.)
7. Refer to the pipe-sizing nomograph and find the correct discharge-pipe size.
8. Having found the correct pipe size, find the *actual* discharge Pl for that selected pipe size.

A pump designed in accordance with these steps will operate at optimum efficiency due to proper suction conditions. It will not waste electrical energy or be damaged by internal pressure surges; and it will not deliver in excess of its catalog capacity as a result of freewheeling.

As noted earlier, flow-friction loss is applicable to pump suction and discharge piping. Flow-friction loss (Pf) depends on pipe diameter and length-specific gravity and viscosity of the fluid, and flow rate. As illustrated in Figure 3.52, maximum Pf occurs at midstroke of the reciprocating piston which coincides with the maximum instantaneous flow of the pump.

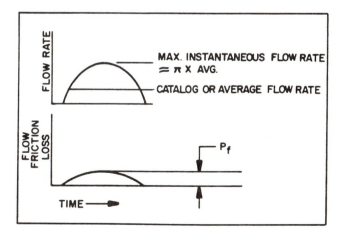

Figure 3.52 Flow friction loss versus time.

The resultant pressure variation, due to inertia pressure and flow-friction loss, will reach a maximum value during the first half of the suction stroke and again during the first half of the discharge stroke. The point at which (Pl + Pf) maximum occurs is a function of the ratio of the values of Pl and Pf.

As can be seen from Figure 3.53, (Pl + Pf) maximum occurs during the first half of the suction and discharge strokes. Note also that the sum of Pl + Pf, during the second half of the suction and discharge strokes, reaches a maximum value at *the end* of each stroke. This value is equal to Pl. Thus, a piping system designed on the basis of Pl alone, such that freewheeling cannot occur during the second half of the suction and discharge strokes, can never have freewheeling occur as the result of adding flow-friction loss. That is, Criterion #2 (freewheeling) cannot be violated by the effects of flow-friction loss.

Flow-friction loss can affect both suction and discharge Criterion #1. That is, if a pump is working near its maximum pressure rating (Pl + Pf), maximum can cause the maximum dynamic discharge pressure to be in excess of this rating. Also, when (Pl + Pf) maximum is used in place of Pl in Suction Criterion #1, the minimum dynamic suction pressure may fall below 10 psia. This causes oil refill valve trouble or, if below Pvp + 5, cavitation.

Small-pipe diameters increase the possibility that flow-friction loss will affect a design based on Pl values alone. The following situations can result in small pipe sizes and indicate a check of flow-friction loss. Minimum pipe diameters result from the use of the Pl concept when:

1. Required line lengths are short.
2. The pump operates at low speed.
3. The system layout and static pressure conditions permit the utilization of high Pl design values.

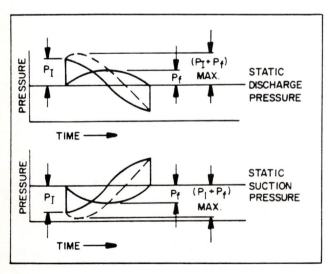

Figure 3.53 The combined effects of Pl and Pf.

Minimum pipe diameters as higher flows also result with a triplex. The pipe-sizing nomograph (Figure 3.51) shows M-factors for triplex pumps are *one-half* those of the simplex or duplex. The result is a pipe size smaller than one would expect for the flow and Pl. Thus, flow-friction loss can be a significant factor in designing the suction and discharge piping system for a triplex pump.

There are no hard-and-fast rules which can indicate whether or not to check for flow-friction-loss effects. Small-diameter suction lines may be affected by friction losses when the liquid viscosity is as low as 20 centistokes. Large-diameter lines, especially discharge lines, may not be affected by viscosity when it is in the hundreds of centistokes.

You should, however, check for flow-friction-loss effects when:

1. Fluid viscosity exceeds 20 centistokes.
2. Triplex pump is used at any viscosity level.
3. Pipe diameters are adequate for the design flow, but small enough to cause concern.

Although flow-friction-loss formulae can be found in any hydraulic book, a nomograph, "Flow-Friction-Loss Nomograph" (Figure 3.54) is included to simplify these calculations. Although instructions for its use are included on the nomograph, the following points should be understood when figuring flow-friction loss:

1. Virtually all piping systems designed on the Pl concept will operate in the laminar- or streamline-flow condition where flow-friction loss is independent of pipe roughness. This laminar-flow condition almost always occurs when viscosity is greater than 5 centistokes. The nomograph checks flow-friction loss under conditions of laminar flow, therefore, $\frac{q}{dk}$ must be less than .664, where $q = \pi X$ average peak flow in gpm; d = inside pipe diameter in inches; and K = kinematic viscosity in centistokes.
2. The pipe length to be used (Le) is equivalent pipe length. Le is equal to the actual pipe length used in the Pl design plus friction loss through elbows, valves, and so on expressed in terms of feet of pipe producing an equivalent loss. Table 3.14 provides equivalent pipe lengths for commonly used fittings and accessories. More extensive data can be found in Appendix D.
3. The value of the flow rate (q) used in the nomograph is the maximum instantaneous flow rate which occurs at midstroke. This is found by multiplying the manufacturer's catalog flow rate of a simplex, duplex, or triplex by 3.14, 1.57, or 1.07. Note that the flow is expressed in gallons per minute.

Once Pf is found, the value of (Pl + Pf) maximum must be calculated: (Pl + Pf) maximum = $\sqrt{Pl^2 + Pf^2}$.

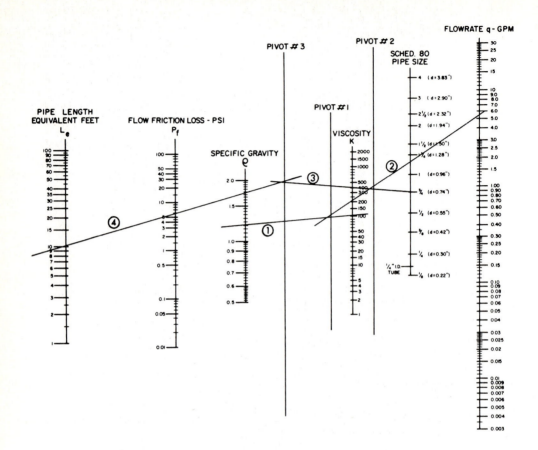

FLOW FRICTION LOSS IN SMALL DIAMETER PIPES UNDER CONDITIONS OF LAMINAR FLOW

$$P_f = \frac{\rho K q L_e}{3668\, d^4} \qquad \text{FANNING'S EQUATION}$$

P_f = FLOW FRICTION LOSS - PSI
ρ = SPECIFIC GRAVITY OF LIQUID
K = KINEMATIC VISCOSITY OF LIQUID - CENTISTOKES
q = MAXIMUM INSTANTANEOUS FLOWRATE - GPM
d = INTERNAL DIAMETER OF PIPE - INCHES
L_e = PIPE LENGTH - EQUIVALENT FEET

NOTE: THE ABOVE EQUATION AND THIS NOMOGRAPH ARE VALID ONLY WHEN LAMINAR FLOW EXISTS I.E: WHEN THE FOLLOWING CONDITION EXISTS. $\frac{q}{dK}$ IS LESS THAN 0.664

EXAMPLE   A LIQUID HAVING A SPECIFIC GRAVITY OF 1.2 IS FLOWING AT A RATE OF 5 GPM THROUGH A 3/4" SCHEDULE 80 PIPE HAVING AN EQUIVALENT LENGTH OF 10 FEET. THE VISCOSITY OF THE LIQUID IS 100 CENTISTOKES AND THE FLOW FRICTION LOSS IS 5.4 PSI.

METHOD   1. CALCULATE $\frac{q}{dk}$ TO ENSURE FLOW IS LAMINAR.

2. CONNECT ρ AND K VALUES WITH STRAIGHTEDGE ① AND MARK PIVOT #1 WHERE STRAIGHTEDGE CROSSES IT.
3. CONNECT MARK ON PIVOT #1 AND q WITH STRAIGHTEDGE ② AND MARK PIVOT #2 WHERE STRAIGHTEDGE CROSSES IT.
4. CONNECT MARK ON PIVOT #2 AND PIPE SIZE WITH STRAIGHTEDGE ③ AND MARK PIVOT #3 WHERE STRAIGHTEDGE CROSSES IT.
5. CONNECT MARK ON PIVOT #3 AND EQUIVALENT PIPE LENGTH WITH STRAIGHTEDGE ④ AND READ FLOW FRICTION LOSS WHERE STRAIGHTEDGE INTERSECTS THE P_f SCALE.

Figure 3.54 Flow-friction loss diagram. (*Courtesy of Wallace & Tiernan, Inc., 25 Main Street, Belleville, NJ 07109-3057*)

TABLE 3.14   FLOW RESISTANCE OF VALVES & FITTINGS (APPROXIMATE FEET OF EQUIVALENT PIPE)

	\frac{3}{8}	\frac{1}{2}	\frac{3}{4}	1	1\frac{1}{4}	1\frac{1}{2}	2	2\frac{1}{2}	3
				Nominal Pipe Size—Inches					
Tank/Suction Line Entrance Loss	1	1	1	2	2	2	3	4	5
Open Globe Valve	12	16	22	27	37	44	56	65	80
Open Angle Valve	6	8	11	15	18	21	28	33	43
Open Gate Valve	—	—	1	1	1	1	1	1	2
Standard Elbow	1	2	2	3	4	4	5	6	8
45° Elbow	1	1	1	1	2	2	3	3	4

SUMMARY

The basic steps for checking the effects of flow-friction loss are as follows:

Step 1. Calculate the value of q/dk. This must be less than .664 to allow use of the "Flow-Friction Loss Nomograph." In virtually all cases where the viscosity exceeds about 5 centistokes, this criterion will be met. Where it is not met, standard procedures for finding flow-friction loss with turbulent flow must be used.

Step 2. Determine the Pf from the nomograph (for laminar flow) or from a hydraulics handbook (for turbulent flow), or the procedures given in Appendix D.

Step 3. Determine (Pl + Pf) maximum from the nomograph.

Step 4a. If the pipe in question is a suction line, check that the minimum static suction pressure (Pl + Pf) maximum is greater than Pvp + 5, and that the minimum static suction pressure (Pl + Pf) maximum is greater than 10 psia.

Step 4b. If the pipe in question is a discharge line, check that maximum static discharge pressure + (Pl + Pf) maximum does not exceed the pressure rating of the pump head.

Step 5. If the criteria described in steps 4a and/or 4b are violated, a larger pipe size must be used. This will decrease the effects of flow-friction loss to a tolerable level. A new Pf and Pl must be calculated for the larger pipe size. This new (Pl + Pf) maximum must be used to check that the preceding requirements are met.

The examples discussed so far contain suction and discharge pressures which allow all design criteria to be satisfied. Situations may be encountered where suction pressure is high relative to discharge pressure (typical when pumping into open tanks). Such installations must be checked for the possibility of freewheeling, both when the pump is operating and when it is stopped. There are two indicators of possible problems due to freewheeling.

1. High inertia pressures caused by long suction and/or discharge lines increase the possibility of dynamic freewheeling. A pump may freewheel during the second half of the discharge stroke or during the second half of both strokes. As depicted in Figure 3.55, a pump will freewheel whenever the instantaneous discharge pressure is less than the instantaneous suction pressure.

2. Whenever the maximum allowable design value of suction Pl and/or discharge Pl from Criterion #2 is close to or equal to 0, there is the possibility of freewheeling. This is an indication that the minimum static discharge pressure is close to the maximum static suction pressure. Obviously, if the design value of Pl works out to less than 0 (then the maximum static

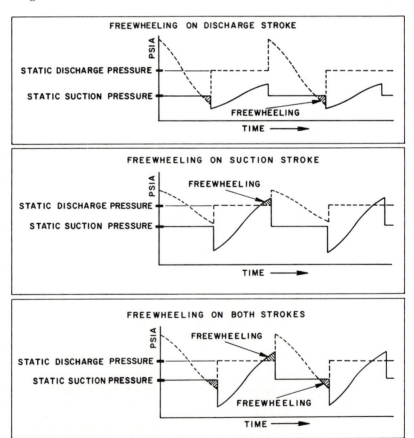

Figure 3.55 Freewheeling curves. (*Courtesy of Wallace & Tiernan, Inc., 25 Main Street, Belleville, NJ 07109-3057*)

suction pressure is more than the minimum static discharge pressure less 2 psi), freewheeling will definitely occur. Although an allowable Pl of slightly more than zero may mean that the pump will not freewheel, the pipe sizes required to accommodate this small Pl may be very large. Designing for a higher allowable Pl (increased in static discharge pressure or decrease in static suction pressure) will allow smaller pipe sizes to be used.

The three curves in Figure 3.55 show that free-wheeling can be prevented or eliminated by:

1. Decreasing discharge Pl
2. Decreasing suction Pl
3. Increasing static discharge pressure
4. Decreasing static suction pressure

TO REDUCE DISCHARGE Pl:

1. Decrease stroking speed and use a larger head and/or multiple heads to obtain the required flow capacity.
2. Increase the diameter of the discharge line. The pipe-sizing nomograph (Figure 3.51) will allow determination of the decreased discharge Pl for given pipe sizes.
3. If possible, decrease the length of the discharge line by relocating the pump or the point of application.
4. Decrease the "apparent" length of the discharge line by utilizing a vented riser (opened-pulse dampener) or discharge headbox in the line adjacent to the pump. Pl acts only in the short run between the pump and the riser or headbox. Steady-state-flow conditions exist from the riser or headbox to the point of application. In a sense, the pump behaves as if the vented riser or headbox is the new point of application.

The vented riser will work only when a shortening of the discharge line would improve system design.

TO REDUCE SUCTION Pl:

1. Decrease stroking speed and use a larger head or multiple heads to obtain the required flow capacity.
2. Increase suction line diameter. The pipe-sizing nomograph will allow determination of the decreased suction Pl for given pipe sizes.
3. Decrease suction line length by relocating the pump or moving the supply tank closer to the pump.
4. Decrease the "apparent" length of the suction line by using a vented riser or suction headbox.

A sample schematic of well-planned layouts are given in Figure 3.57. Careful consideration should be given to the piping layout in order to avoid settling out of the slurry. Settling during long shutdown periods can clog pump discharge valves. Coarse material from a long vertical run usually settles back into the valve when flow ceases. To avoid this, use an elbow in the discharge fitting of the pump followed by a horizontal run of at least one foot.

Use smooth plastic pipe to reduce a plug's resistance to flushing. The best solution to settling in piping is to minimize vertical runs. In vertical pipes, heavy particles move as long as flow continues at sufficient velocities. But they accumulate at low velocities, and when flow stops they settle to clog the line.

In horizontal lines, settling occurs only until velocity in the top of the pipe (cross section) is high enough to carry solids through. As velocity increases, the flow area becomes larger and more solids are picked up. Keep vertical legs short and to a minimum. If possible, avoid vertical runs over one foot. If they are necessary, keep in mind that water supply pressure of 10 psi per foot of pipe may be required to clear the lines after a stoppage. Don't use elbows for 90-degree direction changes. Use plugged tees or crosses. They permit rodding out deposits and provide temporary flushing connections.

High velocities aid in maintaining a slurry suspension. Increasing discharge line diameters decreases flow velocities and may lead to settling in piping. In cases where large-diameter PVC pipe would be required because of inertia pressure limitations (15 psi), flexible hose or iron pipe (both of which can withstand higher inertia pressures) should be considered as a first choice.

If deposits are likely (calcium carbonate scaling from lime slurries), use flexible plastic tubing or rubber hose rather than rigid pipe for the discharge. Normal flexing from pump pulsations will dislodge scale. A flexible discharge line also allows long-radius, low-friction bends and direction changes. The suction line should be rigid pipe within the supply tank to avoid tangling with the mixer.

On long vertical runs of slurry piping, install manual ball valves at the far end of the discharge piping as depicted in the installation schematics. They will relieve backpressure at the discharge to facilitate emergency flushing. Always add a relief valve as shown on the schematics to protect against dead-ending or excess pressure from severe plugs in long vertical runs. In cases where syphoning will occur, do not use a backpressure valve—it will not work well with slurries. Use a suction headbox instead. With a second-level control valve, the suction headbox becomes an alternate location to introduce flushing water. To prevent syphoning, the discharge pressure can also be increased by pumping to an elevated headbox from which the slurry flows to the application point by gravity. The margin between static suction and discharge pressure can be increased further by decreasing suction pressure with a suction headbox.

Settling during short shutdowns is unavoidable. Thus, a flush connection should be located as shown in the schematics. The flush should be applied immediately before or after the shutdown; the flushing system can be manual or automatic. If automatic, it will provide a time sequence flushing cycle.

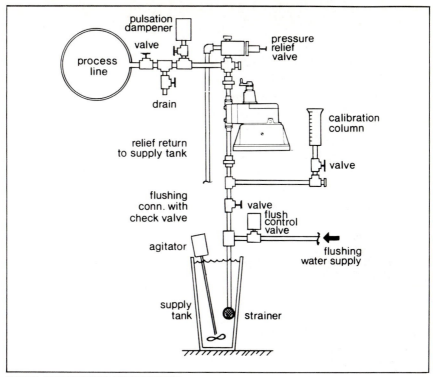

TYPICAL SLURRY PIPING ARRANGEMENTS

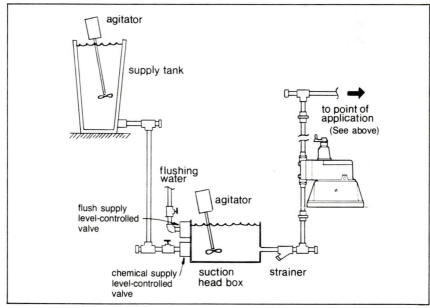

Figure 3.57 Typical slurry piping arrangements.

If a standby pump is used, keep the discharge runs from both the primary and standby pump in the horizontal plane. Do not use a vertical pipe to manifold a standby pump to an on-line slurry pump. The slurry will fall back into and eventually plug up the discharge of the standby pump. Offset the standby pump and connect it into the discharge line of the on-line pump with at least one foot of horizontal pipe. Refer to Figure 3.58.

In planning a metering pump installation, it is imperative that the factors relating to suction lift be thoroughly understood and observed.

To begin the discharge stroke, the drive eccentric forces the diaphragm into the pumping chamber (head). This forces liquid out through the discharge valve. At the end of this stroke, the drive eccentric breaks contact with the diaphragm push rod and the diaphragm is forced *out of* the pump head by the return spring. This reduces pressure in the pump head. When head pressure drops below supply pressure, liquid is forced into the head from the supply system.

One side of the diaphragm is exposed to atmosphere; the other is exposed to pressure from the suction and discharge systems alternately. During the suction stroke, the spring is working against the difference between atmosphere and pressure in the pump head. A strong spring would be required to return the diaphragm against low pump head pressure, while a relatively light spring would return the diaphragm against higher pump head pressure. Low pump head pressure is associated with high-suction lift; higher head pressure is associated with low-suction lift or flooded suction conditions. It follows then that the ability of a pump to operate on suction lifts is a function of the strength of the diaphragm return spring.

The motor supplies the energy to return the diaphragm by "storing" it in the return spring as it is compressed during the previous discharge stroke. Total energy required of the drive motor equals that required to drive the diaphragm against discharge pressure plus that required to compress the return spring. For a given motor horsepower, it is obvious that *maximum* discharge pressure capa-

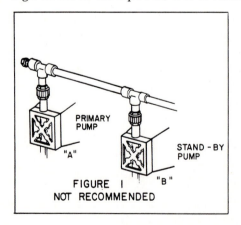
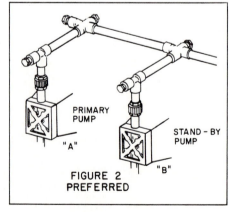

Figure 3.58   Proper layout arrangement for a standby pump.

bility can be achieved by minimizing the energy required to compress the spring. As noted previously, however, minimizing spring force reduces the pump's ability to operate on high-suction lifts. Obviously, a compromise has to be made in selecting the proper return spring force.

Note that the minimum pumping chamber pressure (minimum dynamic suction pressure) occurs at or very near the start of the suction stroke. Neglecting friction loss, it is equal to minimum static suction pressure (absolute static pressure in the pumping head when the liquid level in the supply tank is minimum) less inertia pressure P_1.

The factors relating to the ability of a pump to operate in situations of suction lift should now be apparent. The "lift" itself is only one of the factors contributing to minimum pump chamber pressure. The other is P_1 which is a function of pump stroking speed and diaphragm diameter as well as suction-line length and diameter, factors which vary from job to job.

It is important to bear in mind that the dynamic suction-lift capability of a pump assumes that the pump has been primed and is running. Although the self-priming capability of a pump varies from job to job, in general, installations with suction lifts over 10 feet of water should be avoided if a pump is to self-prime.

Pumps with high-suction lift and threaded-pipe suction lines are more prone to loss of prime because of leakage at joints. This may be especially troublesome in start-stop operation where a foot valve and a flexible suction line may be required. If priming aggressive or toxic materials, it may be impractical to use water to help the priming (as with sulfuric acid where heat will be generated). The suction lift should be limited to allow the pump to self-prime.

Pumps feeding against high discharge pressures prime with more difficulty than pumps working against low pressures. A bypass line to supply may be required in applications with high discharge pressures.

We may encounter situations where suction pressure is high relative to discharge pressure (typically when pumping into open tanks, and so on). Such installations must be checked for the possibility of syphoning, both when the pump is stopped and running. High values of Pl caused by long suction and/or discharge lines increase the possibility of dynamic syphoning, or freewheeling. A pump may freewheel during the second half of the discharge stroke, during the second half of the suction stroke, or during the second half of both strokes. A pump will freewheel when the instantaneous discharge pressure is less than the instantaneous suction pressure.

1. How can freewheeling be prevented or eliminated in the case of an existing installation? The preceding curves suggest the following (refer to Figures 3.59 and 3.60).
a. Decrease discharge system P_1.
b. Decrease suction system P_1.
c. Increase static discharge pressure.
d. Decrease static suction pressure.

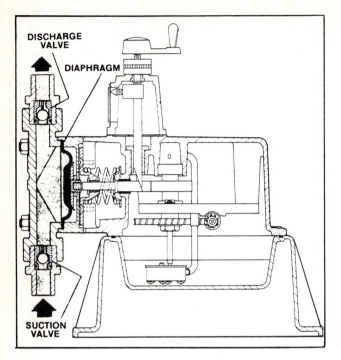

DISCHARGE
VALVE

DIAPHRAGM

SUCTION
VALVE

Figure 3.59 Diaphragm pump schematic.

   2. How can discharge system P_1 be reduced?

a. Decrease pump stroking speed and use a larger head to obtain the required capacity.

b. Increase the diameter of the discharge line.

c. Decrease the length of the discharge line by relocating the pump or the point of application.

d. Decrease the "apparent" length of the discharge line by utilizing a vented riser (open-end pulsation dampener) in the line adjacent to the pump, discharge pressure conditions permitting. P_1 then acts only in the short run between pump and riser; steady-state flow exists between riser and the point of chemical application. In a sense, the process (as viewed by the pump) becomes the riser which simulates an open tank.

A variation of this is the discharge headbox.

   3. How can suction system Pl be reduced?

a. Decrease pump stroking speed.

b. Increase diameter of suction line.

c. Decrease length of suction line by relocating pump.

d. Decrease "apparent" length of the suction line by utilizing a vented riser (open-end accumulator adjacent to the pump). P_1 acts in the length of pipe between riser and pump; steady-state flow exists between supply tank and riser. In a sense, the pump views the riser as the supply tank.

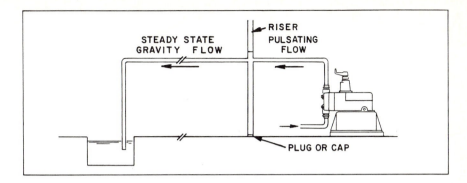

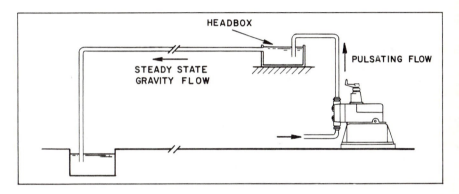

Figure 3.60 Recommended layouts to prevent freewheeling.

A variant of this is the suction headbox with a float valve for level control. This approach should be carefully considered: Flow to the pump can be interrupted by a plugged float valve or by the headbox overflowing should the float lever hang up. Refer to Figure 3.61.

4. How can static discharge pressure be increased? (Refer to Figure 3.62.)

a. The simplest way to do this is to utilize a vented riser similar to that discussed in paragraph 2d, except that the connection from riser to process is made at an elevation equivalent to the required static pressure. An additional advantage of such a system is that the process connection to the riser is above the maximum level in the supply tank. Thus, all possibility of static syphoning (pump shutdown) is eliminated.

b. In some installations where the maximum static suction pressure is relatively high (supply tank on floor above pump floor) and/or where insufficient headroom exists for the installation of a riser of the height necessary to produce the required head on the pump, a backpressure valve can be employed. A backpressure valve is essentially a spring-loaded check valve. Refer to Figure 3.63.

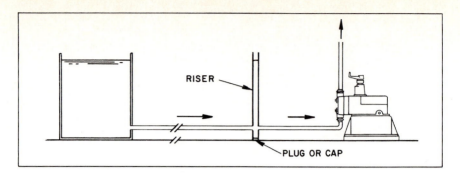

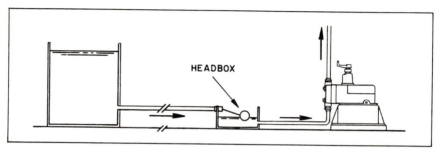

Figure 3.61 Recommended layouts to reduce suction system pressure.

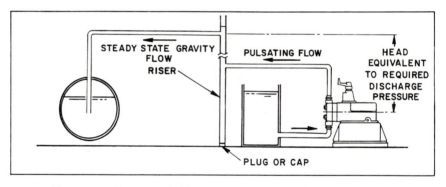

Figure 3.62 Recommended layout to increase static discharge pressure.

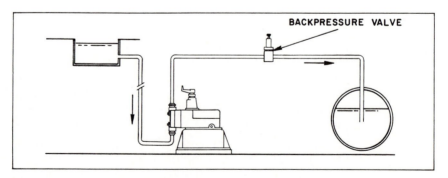

Figure 3.63 Illustrates use of a backpressure valve.

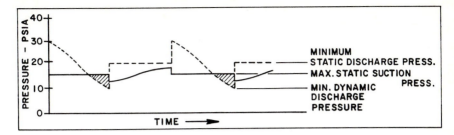

Figure 3.64 Shows system pressure conditions with and without a backpressure valve.

Note that the addition of a backpressure valve in no way affects the value of Pl in the discharge line. Its set pressure value is, however, the static pressure in the discharge system as "seen" by the pump. Figure 3.64 is a graphical illustration of system pressure conditions with and without a backpressure valve.

In the previous example, the system is freewheeling during the second half of the discharge stroke due to the high discharge system Pl. To operate correctly, minimum dynamic discharge pressure should be at least 2 psi higher than maximum static suction pressure (simplex pump situation).

From this we determine that the setting of a backpressure valve to correct the situation should be 15 + 2 + 10 = 27 psia = 12.3 psig. (A practical setting would be a nominal 15 psig.) The revised pressure characteristics are illustrated in Figure 3.65.

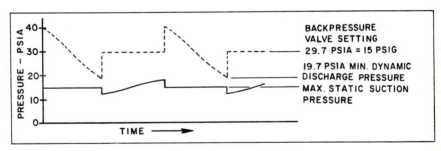

Figure 3.65 Revised pressure characteristics for example.

    c. In some instances the addition of a backpressure valve is not the sole solution to freewheeling problems. If P_1 is higher than say 15 psi, something should be done to reduce the value, especially if plastic pipe is being used. Excessive Pl in metallic piping systems can usually be tolerated by the piping but results in inordinately high backpressure valve settings to prevent freewheeling. Thus, we simply waste pump horsepower. Paragraph 2d illustrates the use of the vent riser to reduce discharge system P_1. Figure 3.66 is a sketch of a system utilizing the vented riser to reduce P_1 and a backpressure valve to increase static discharge pressure.

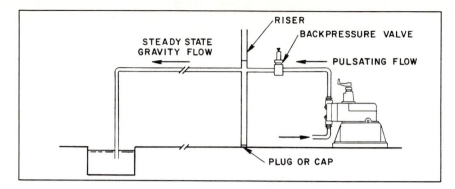

Figure 3.66 Sketch of system using a vented riser to reduce P1 and a backpressure valve to increase static discharge pressure.

In instances where static suction pressure can be higher than static discharge pressure (a static-syphoning situation), a backpressure valve can normally be expected to prevent the contents of the supply tank from flowing through the pump to the process when the pump is not running. But as the valve wears, some minor leakage must be expected. In addition, the valve can be jammed open by contaminants in the chemical which arrived with the chemical or which dropped into the supply tank. For this reason it is strongly recommended that the process takeoff from the riser be accomplished at an elevation higher than the maximum supply elevation as illustrated in paragraph 4a. The combination of a riser to prevent static syphoning and to reduce discharge system Pl coupled with a backpressure valve to increase static discharge pressure is sketched in Figure 3.67.

5. How can static suction pressure be decreased? This method of correcting freewheeling should be reserved until the previous more practical and more commonly used methods are considered. The supply tank can be relocated, but this is rarely practical on an existing installation. The second method is to utilize the suction headbox as illustrated in paragraph 3d as a variant of the suction line riser. In addition to decreasing the

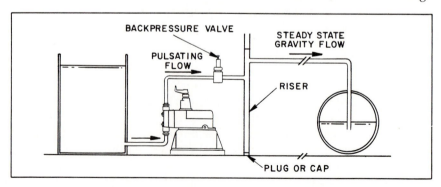

Figure 3.67 Shows combination of riser with backpressure valve.

"apparent" length of the suction line and thus reducing suction line Pl, the headbox and float valve reduce static suction pressure to a low and constant value. The headbox approach, however, may not be the best alternative for the reasons outlined in paragraph 3d.

6. As noted earlier, duplex arrangements are more prone to freewheel than simplex pumps since the minimum dynamic discharge pressure and maximum dynamic suction pressure conditions occur simultaneously. The same methods used to prevent freewheeling as outlined previously can be used in duplex situations. The determination of suction and discharge Pl values in duplex situations is, however, more complex than for simplex pump systems.

The diversity of services for metering pumps is enormous. Table 3.15 provides a listing of selected applications to illustrate their versatility.

Chemical Compatibility

Chemical compatibility or corrosion resistance is a major consideration in any pumping service. It is essential that one understands the materials that are being handled, the conditions or environment in which they are to be handled, and the limitations of the materials of pump construction. Pumps can be constructed from a wide range of materials. For illustration, Figures 3.68 and 3.69

Figure 3.68 Example of Teflon-constructed pump. (*Courtesy of Osmonics, Inc., 2022 West 11th Street, Upland, CA 91786*)

TABLE 3.15 EXAMPLES OF SELECTED APPLICATIONS

Bleach Producers - Dose perfume such as eucalyptus oil in proportion to bleach flow to containers.

Boilers - Full range of boiler chemicals for both heating and power boilers fed automatically or manually.

Bottling Plants - Use pumps for filter flocculation, water purification with hypochlorite, injection of wetting agents and alkaline cleaning solutions into bottle washers.

Boys' Clubs - Swimming pool applications for hypochlorite, alum, or soda ash feed. Slurry feed to diatomite filters.

Breweries - Slurry feed to filters. Chlorinate process water of water used for washing floors, walls and equipment.

Camps, summer - Chlorinate drinking water, remove H2S or iron: often having swimming pools. (See Swimming Pools)

Car Washes - Injection of soaps or detergents into washing water.

Cement Manufacturers - Use metering pumps for addition of air entraining chemicals and grinding aids to Portland cement.

Cheese Plants - Chlorinate process water. Use pumps for addition of citric acid or hydrogen peroxide to cottage cheese tanks.

Citrus Plants - Slime control.

Electroless Plating - Add replenishment chemicals.

Electronic Component Manufacturers - May pump coatings for parts.

Electroplaters - Feed brighteners or replenishing solutions. Use slurry feed to plating solution filters. Add chemicals for pH control, cyanide destruction in waste waters.

Exterminators - May meter insecticides into spray water.

Feed Mills - Boiler treatment chemicals, scale control chemicals in humidification systems. May add liquid vitamin and antibiotic chemicals to feeds.

Fish Processing - Chlorinate wash water and flume conveyor water. Inject detergents into mechanical conveyor cleaning sprays.

Fishing Vessels - Chlorinate wash waters for decks and holds.

Flooring Manufacturers - Sometimes meter liquid wax onto finished floor tiles.

Foundries - May meter detergent as part of magnetic test for flaws, may also meter sand conditioning chemicals.

Fruit Packers - Wash water chlorination. May inject chemical treatment to enhance or preserve color and storage life of fruit.

166

Coal Washing Plants - Dosing of flocculant aids ahead of centrifugation for dewatering coal slurry.

Concrete Block Manufacturers - Use pumps for addition of chemicals to boilers and to steam lines for the curing kiln.

Condensers - Add chemicals for prevention of scale, algae growth and pH control.

Cooling Towers - Inject a number of different chemicals to control corrosion, adjust pH, prevent algae growth or delignification.

Dairies - Often chlorinate general water supply and water used in washing butter. May super chlorinate wash down water. May feed iodine or potassium permanganate solutions for sanitizing milking equipment and udder wash water.

De-aerating Heaters - Sodium sulfite or hydrazine may follow DA heater as an oxygen scavenger.

Defense Department Bases - Chlorinate drinking water systems, often fluoridate or feed corrosion control chemicals and may use slurry feeders on diatomite filters.

Die Casting Machines - Add sequestering or pH control chemicals to die cooling water. Scale prevention is critical.

Dishwashing Machines - (in hospitals, restaurants, schools, etc.) chlorinate sterilizing rinse water.

Distilleries - Chlorinate general water supply and treat wastes for pH control. May use slurry feeders for diatomite filters. Boiler and condenser chemical treatment may be necessary.

Fuel Oil - Catalytic treatments sometimes containing abrasive powders may be fed for control of slag and vanadium corrosion in power boilers.

Furnaces, Industrial - May inject scale and corrosion preventative chemicals into cooling water lines.

Golf Courses - Inject liquid fertilizer into sprinkling system.

Grain Storage Elevators - May add insecticides to grain going into storage.

Greenhouses - Fertilizer injection is a common application. May also inject pH control chemicals and insecticides.

Heat Exchangers - Inject corrosion or scale prevention chemicals on the water side.

Highway Departments - Frequently chlorinate drinking water at rest stops and road side service facilities.

Homes - Iron removal, sulfur removal, acid water correction and chlorination for purification are the common applications.

Hospitals - Pump boiler treatment chemicals. May chlorinate drinking water supply, whirlpool baths and sewage.

Hotels - (see Swimming Pools) Drinking water may be chlorinated. Corrosion control chemicals may be added. Boilers and cooling towers may be treated.

Hydroponic Vegetable and Plant Growers - May feed nutrient solutions or pH control solutions to the growing beds.

(continued)

TABLE 3.15 *Continued*

Irrigation - May inject fertilizer, herbicides and rust control chemicals into irrigation lines. May also chlorinate to prevent algae formation.

Laboratories - Metering pumps may be used in experimentation and pilot plants.

Laundromats - Waste water treatment with chlorine, alum or activated carbon may be required. Boiler treatment is common.

Lumber Mills - Add chemical to dip tanks for prevention of "Blues-tain" floatation.

Meat Packers - Waste treatment including heavy chlorination is frequently required. Wash water and process water may be chlorinated. Boiler treatment is common.

Mines - Ore floatation often requires chemical additives. Drinking water chlorination, water reclaimation, flocculant feeding.

Motels - Water supply is often individual and must be chlorinated and treated for acidity, iron or sulfur removal. Swimming pools are common. (See Swimming Pools)

Municipalities - Corrosion control, iron removal, hydrogen sulfide removal are often a necessity. Chlorination is common. Fluoridation and sewage treatment.

Nurseries - Fertilizer, insecticides, pH adjustment chemicals are often injected into irrigation water.

Nursing Homes - Often feed chemicals for corrosion control, iron removal, hydrogen sulfide removal and purification. Chlorination of drinking water is common on individual supplies, and sewage effluent sanitaion may also be required.

Oil Producer and Refiners - Chlorination of water used in secondary recovery is common. Wetting agents are often injected. Chemicals to break the oil water emulsion and to control corrosion are frequently used.

Poultry Producers - Pump sterilizing chemicals, nutrients, vitamins, and antibiotics into water supplies. Scale control chemicals may be added to humidification systems.

Powder Metallurgy - Sintering dies have cooling water connections and treatment chemicals are often added.

Power Plants - Pump algicides and corrosion control chemicals into condenser cooling water. Generally pump boiler treatment chemicals including filming amines for condensate return lines.

Printers - May meter viscosity control fluids for inks or offset dampener fluids.

Pulp Mills - May feed pitch dispersants, waste treatment chemicals.

Refineries - Cooling tower treatments, corrosion control chemicals and pH adjustment chemicals including concentrated sulfuric acid are often used.

Schools - Fluoridation and chlorination of domestic water supply may be required. Iron, sulfur or acid water treatment may be needed. Swimming pool may be invloved. (See swimming Pools)

Scrubbers - Adjust pH of scrubber water by NaOH injection. In some processes also inject H_2SO_4.

Sewage Plants and Oxidation Ponds - Chlorination for BOD and odor reduction, flocculation, pH adjustment.

Ships - Frequently chlorinate potable water supplies. May also use metering pumps for boiler treatment.

Shrimp processing - Like many food processing industries, sterilization of wash and process water may be required.

Sintering Process - (See Powder Metallurgy)

Sprinkling Systems - Inject liquid fertilizer into system.

Painting Processes - May meter flocculants and other chemicals into overspray removal waters.

Paper Mills - Many pumps may be used by a single mill to feed defoaming chemicals, slimicides, pH control chemicals, dyes, pitch dispersants, wire life extenders, biocides and others. Also for feeding polymers in plant waste treatment.

Parks - Chlorination of drinking water is common. Swimming pools may be included. (See Swimming Pools) Liquid fertilizers for sprinkling system or lawns.

Photographic Labs - Wetting agents are sometimes automatically fed into film wash baths. Sometimes algae control chemicals are pumped.

Pilot Plants - Chemical process plants often set up small scale reaction processes which require chemical metering pumps.

Pipe Line Stations - Cooling tower and condenser waters, algicide and phosphate injection.

Plastic Extruders - May meter pigments into extrusion machine; may treat cooling water supplies.

Plastic Molders - Frequently treat mold cooling waters to prevent corrosion, rust or scale build up.

Plating, Metal - Organic brighteners may be metered, replenishment chemicals may be pumped and filter aid slurry may be added continuously. pH adjustment of waste flows may be required.

Polymer Flocculant Aids - Fed to increase speed and efficiency of setting systems. Often very high viscosity.

Poultry Processors - May heavily chlorinate washdown and process waters.

Sugar Mills - Often pump bacteria control chemicals and chemicals designed to increase sucrose yield.

Swimming Pools - Sterilization is always required, usually by chlorination. Alum for flocculation, soda ash for pH control and filter aid slurry may also be pumped.

Taverns - Chlorination of private drinking water supply is often required.

Textile Mills - dyes, pH adjustment chemicals and wetting agents may be used in processing. Waste treatment may require chlorination and other chemicals.

Textile Printers - May meter inks and dyes in the printing process.

Trailer Courts - Chlorination of drinking water supply and sewage effluent treatment is often required by Health Departments.

Vegetable Packers - Chlorination of process wash water is common. Sometimes either coloring, flavoring or preservative chemicals may be added.

Vibratory Finishers - Soap, rust inhibitors and polishing agents may be pumped.

Waste Water Systems - May require chemical treatment for pH control, flocculation, sanitation, BOD control, etc.

Wire Coaters - Meter varnish to coating machines.

Wool Scouring - (See Textile Mills)

YMCA, YWCA, YWCA, YMHA - Swimming Pool treatment is almost universal. Summer camps may require swimming pool treatment and sometimes drinking water treatment.

TABLE 3.16 CHEMICAL COMPATIBILITY CHARTS

Liquid	Hast. C	Carp. 20	316 S.S.	Hypalon	Viton	PVC	TFE	Nordel	Kynar-150°F	Ceramic	Kalrez	Ref. no.
Acetaldehyde	A	A	A	C	C	C	A	B	C	A	B	57
Acetate solvents	A	A	A	C	C	C	A	N	A	A	A	57
Acetic acid, crude	A	A	A	C	C	C	A	B	A	A	N	2, 57
Acetic acid, pure	A	A	A	C	C	C	A	B	A	A	N	2, 57
Acetic acid (10%)	A	A	A	B	B	C	A	B	A	A	A	2, 3
Acetic acid (80%)	A	B	B	C	C	C	A	B	A	A	B	2, 57
Acetic anhydride	A	A	A	C	C	C	A	C	A	A	A	2
Acetone	A	A	A	A	A	A	A	A	A	N	A	
Acetylene	A	A	A	B	B	A	A	A	A	N	N	
Acrylonitrile	B	A	A	C	C	A	A	N	A	A	N	58
Aluminum chloride	B	B	B	A	A	A	N	A	A	A	A	5
Aluminum hydroxide	A	A	A	A	C	A	A	A	A	A	A	6
Aluminum nitrate	A	A	A	B	B	A	A	A	A	N	A	
Aluminum sulfate	A	A	A	A	A	A	A	A	A	A	A	3, 4
Alums	A	B	B	C	C	A	A	A	A	A	A	
Amines	A	A	A	A	C	A	A	A	A	N	A	
Amines (filmine) B	A	A	A	C	C	A	A	N	N	N	N	
Ammonia anhydrous (liq.)	A	A	A	B	B	C	A	A	A	A	A	
Ammonia Solutions	A	A	A	B	A	A	A	A	C	A	A	21
Ammonium carbonate	A	A	A	A	A	A	A	A	A	N	A	
Ammonium chloride	A	A	A	A	A	A	A	A	A	N	A	2, 4, 7
Ammonium hydroxide	A	B	B	A	A	A	A	A	N	N	A	8
Ammonium monophosphate	A	A	A	A	A	A	A	A	A	A	A	9

Liquid	Hast. C	Carp. 20	316 S.S.	Hypalon	Viton	PVC	TFE	Nordel	Kynar-150°F	Ceramic	Kalrez	Ref. no.
Calcium nitrate	A	A	A	A	A	A	A	A	A	A	A	
Calcium sulfate	A	A	A	A	A	A	A	A	A	A	A	
Cane sugar liquors	A	A	A	A	A	A	A	A	A	N	N	14
Carbolic acid (phenol)	A	A	A	C	A	A	A	B	A	A	A	11, 14, 57
Carbon bisulfide	A	A	A	C	A	A	A	A	A	A	B	
Carbonic acid	A	A	A	C	A	A	A	N	A	N	A	14, 57
Carbon tetrachloride	A	A	A	A	A	A	A	A	A	A	A	13, 3
Chloracetic acid	A	C	C	C	A	C	A	C	C	A	A	54
Chlorobenzene (dry)	A	A	A	C	A	C	A	C	A	A	A	
Chloroform	A	B	B	C	A	C	A	C	A	A	A	
Chlorsulfonic acid	B	B	B	C	C	A	A	C	C	A	A	
Chromic acid	B	A	A	A	A	A	A	C	A	A	A	19, 58
Citric acid	A	A	A	A	A	A	A	A	A	A	A	20
Copper acetate	A	C	C	C	C	A	A	N	A	N	A	
Copper chloride	A	C	C	B	C	A	A	A	A	A	A	5
Copper cyanide	A	N	A	B	A	A	A	N	A	A	A	3
Copper nitrate	B	A	A	A	A	A	A	A	A	A	A	3
Copper sulfate	A	A	A	A	A	A	A	A	A	A	A	21
Creosote	A	A	A	C	C	C	A	C	A	A	A	3
Cresylic acid (50%)	A	C	A	C	C	C	A	N	C	A	A	
Cyclohexane	N	N	A	A	A	A	A	A	A	A	A	
Detergent	A	A	A	N	A	A	A	N	N	N	N	
Diethylamine	A	A	A	C	C	C	N	A	A	A	A	57
Diethylene glycol	A	A	A	C	C	C	A	N	A	A	N	
Dowtherms	N	A	A	B	B	C	N	C	B	N	N	
Ethers (ethyl)	A	A	A	C	B	C	A	C	B	A	A	
Ethyl acetate	A	A	A	C	C	C	A	A	C	A	A	

Table of chemical resistance ratings (headers appear on a preceding page). Each chemical is followed by its reference number(s) and a row of material-resistance ratings (A, B, C, N).

Chemical	Ref.	Ratings
Ammonium di-phosphate	9	A A A A A A A A A A A A A A A A
Ammonium tri-phosphate	9	A A A A A A A A A A A A A A A A
Ammonium nitrate	2, 10	B B A A A A A A A A N A A A A A
Ammonium sulfate	2	B B A B A A A B C A A A A A A A
Ammonium sulfide	58	N A A C C A A C A C A A N A A A
Amyl acetate	11, 12	A A A C A B A A A B A A A A A A
Amyl alcohol		A B A C C A A A A C A A A A A A
Amyl chloride		A A A A B A A A A A A A A A A A
Aniline	13	A A C C C A C A A C A B C B A A
Aniline dyes		A A A A B B A A A N A A B A A A
Arsenic acid	14	A B B C B C A N A A A A N A A A
Barium carbonate	15	A A C A A A A A A A A A A A A A
Barium chloride		A A B B A B A A A A N A A A A A
Barium hydroxide	14, 5	B A A A B B A A A A A A N A A A
Barium sulfate		A N A C A A A A A A A A N A A A
Barium sulfide		N A B A A A A A A A N A A A A A
Beer		A A A A A A A A A A N A A A A A
Beet sugar liquors		A A A C A A A N A A A A A A A A
Benzene or benzol	13, 14	A A B C A A A A C B A A A A A A
Benzaldehyde		A A C C A A A A B A A A A A A A
Benzoic acid		A B C A A A A A A A A A A A A A
Black sulfate liquor	57	N A A B B A N A A A A A A A A A
Borax (sodium borate)	14, 54	A A A A B A A A A A A A A A A A
Boric acid	16	A A A A B A A A A A A A A A A A
Butane		A A A A A B A A A A N A A A A A
Butadiene		A A A A B A A A A A N A A A A A
Butyl acetate		A A C N B C A A C N A A C A A A
Butyl alcohol	17	A A A A B A A A A A A A A A A A
Butyric acid	14	A A A B B A A A A A A A A A A B
Calcium bisulfite		A A A A A A A A A B A A A A A A
Calcium carbonate	15	B A A A A A A A A A A A B A A N
Calcium chlorate		A A A A A A A A A A A A A A A A
Calcium chloride	18	A A A A A B A A A A A A A A A A
Calcium hydroxide	15	A A A A A A A A A A A A A A C N
Calcium hypochlorite	15	A B C A A A A A A B A A A A B A

Chemical	Ref.	Ratings
Ethyl alcohol	12	A A A A A A A A A A A A A A A A
Ethyl chloride		A A A C A A A A A C A B N A A A
Ethyl mercaptan		A N A N N A A N N A A N N N A A
Ethylene chloride	22	N A C B C A A A B A A A A A A A
Ethylene glycol	12	A A A A A A A A A A A A A A A A
Ethylene oxide		A A A C C C A A C C A A A A A A
Fatty acids	14, 54	B C C A A B C A A C A A A A A A
Ferric chloride	23, 6	B A A A A A A C A B B A A A A A
Ferric nitrate		A A B A A A A A A A A A A A A A
Ferric sulfate	24	A A A A A A A A A A N A A A A A
Ferrous chloride		C C C C A A A A C C A A A A A A
Ferrous sulfate	14	B B B B A A A A B A A A A A A A
Filter aid	15	A A A A A A A A A A A A A A A A
Fluosilicic acid	6, 25, 26	A A B A A A A A A B A A A C A C
Formaldehyde		A A A A A C A A A A A A A A A A
Formic acid	3, 58	A A A B B B A A A A A A A B A B
Fruit juices		A A A C A A A A A N A A B A A N
Furfural	57	B B A C A C A A C B A B B A A A
Gallic acid (5%)		B B A C A A A B B A A A B B A A
Gasoline		A A A A A A A A A A A A C A A A
Glucose		A A A A A A A A A A A A A A A A
Glycerol (glycerin)	6, 11, 27	A A A A A A A A A N A A A A A A
Heptane, Hexane		A A A A A A A A C A A A A A A A
Hydrazine (35%)	28	B N A B A A A B N A B C N A B A
Hydrobromic acid	29	B C C N A A A A B C A N A A A A
Hydrochloric acid (37%)	5, 30	B A C A A A A A B C A A A B A A
Hydrocyanic acid		A A A A A A A A A A A A A A A A
Hydrofluoric acid	6, 26, 25	A C A A A A A A A C A A A A C A
Hydrofluosilicic acid	6, 25, 26, 57	A B A A B B A A A B A A A A A A
Hydrogen peroxide	4, 31, 59	A B A A B A A A A B A A B A A A
Hydrogen sulfide	11, 3	A A A A A A A A A A A A A A A A
Inks	19	A A A A A A A A A A A A N N A A
Iodine solution		A C C A C A B A A C A A A C A A

(continued)

171

TABLE 3.16 Continued

Right-hand portion of the table:

Liquid	Ref. no.	Hast. C	Carp. 20	316 S.S.	Hypalon	Viton	PVC	TFE	Nordel	Kynar-150°F	Ceramic	Kalrez
Potassium chromate		A	A	A	A	A	A	A	A	A	A	A
Potassium cyanide		A	A	A	A	A	A	A	A	A	N	A
Potassium hydroxide	42	A	A	A	A	C	A	A	A	A	N	A
Potassium nitrate		A	A	A	A	A	A	A	A	A	C	A
Potassium permanganate	5, 43	A	A	A	A	A	A	A	A	A	A	A
Potassium monophosphate		A	A	A	A	A	A	A	N	N	N	N
Potassium diphosphate		A	A	A	A	A	A	A	A	A	A	A
Potassium sulfate	41, 5	A	A	A	N	A	A	N	A	N	N	A
Potassium sulfide		N	A	A	N	B	A	A	A	A	N	A
Potassium sulfite		N	A	A	B	A	A	A	A	A	A	A
Propane (liq.)		A	A	A	A	B	A	A	C	N	A	N
Propyl alcohol	12, 58	A	A	A	A	B	B	A	N	A	N	A
Propylene glycol		A	A	A	A	A	C	A	N	A	N	A
Resins & rosins		A	A	A	N	N	N	N	N	N	N	A
Sea water		C	A	B	A	A	A	A	B	N	A	A
Silver nitrate		A	A	A	A	A	A	A	B	N	N	A
Soap solutions (stearates)	6, 57	A	A	A	A	A	A	A	A	A	A	A
Sodium acetate		A	A	A	C	A	A	A	A	A	A	A
Sodium aluminate 27Be		A	A	A	A	A	B	A	A	A	A	A
Sodium bicarbonate		A	A	A	A	A	A	A	A	A	A	A
Sodium bisulfate (to 100°F)		A	A	A	A	A	A	A	A	A	A	A
Sodium bisulfite (to 100°F)		A	A	A	A	A	A	A	A	A	A	A

Left-hand portion of the table:

Liquid	Ref. no.	Hast. C	Carp. 20	316 S.S.	Hypalon	Viton	PVC	TFE	Nordel	Kynar-150°F	Ceramic	Kalrez
Kerosene	32, 57	A	A	A	A	C	A	A	C	A	A	A
Lactic acid		A	A	A	A	A	A	A	A	A	A	A
Lead acetate	15	A	A	A	A	C	A	A	A	A	A	A
Lime slurries		A	A	A	A	A	A	A	A	A	N	A
Linseed oil		A	A	A	A	A	A	A	B	A	A	N
Magnesium carbonate		A	A	A	A	A	A	A	A	A	A	N
Magnesium chloride	6, 33, 34	A	A	C	A	A	A	A	A	A	A	N
Magnesium hydroxide	6, 15	A	A	A	A	A	A	A	A	A	N	N
Magnesium nitrate		A	A	A	A	A	A	A	A	A	A	A
Magnesium sulfate	14, 5	A	A	A	A	A	A	A	A	A	A	A
Maleic acid (dilute)	5, 14	A	A	A	A	C	B	A	C	C	A	A
Malic acid	14	A	A	A	B	A	A	A	N	A	A	A
Melamine resins		A	A	A	A	N	A	A	N	N	A	N
Mercuric chloride	5	A	A	C	A	A	A	A	A	A	N	A
Mercuric cyanide		A	A	A	A	A	A	A	N	A	A	A
Mercury		B	A	A	A	A	A	A	N	A	A	A
Methyl acetate	57, 54	N	A	A	C	C	C	A	N	N	N	A
Methyl acetone		N	A	A	A	B	C	A	N	N	N	N
Methyl alcohol	35	A	A	A	C	C	N	A	N	A	A	A
Methylamine		N	A	A	C	C	N	N	N	C	N	A
Methyl bromide		A	A	A	A	A	C	A	N	C	N	A
Methyl cellosolve		A	A	A	C	C	N	A	B	A	A	A
Methyl chloride (Liq.)		A	A	A	A	C	C	A	C	A	A	A
Methylethylketone		A	A	A	C	C	C	A	C	C	A	A
Methylene chloride	36, 14	A	A	A	B	B	A	A	N	A	N	A
Molasses		A	A	A	A	A	A	A	N	A	A	A
Monochloracetic acid	54	A	B	C	N	N	A	N	N	A	A	A

172

Left column table:

Chemical	Ref.	Ratings
Morpholine	57	A A C A A C A A A
Naphtha	13	A A C A A A C A A
Naphthalene	11	A A A A A A C A A
Nickel chloride	14	A A A A A A A A A
Nickel nitrate	14	A A A A A A A A A
Nickel sulfate	14	A N A A A A A A A
Nicotinic acid		A N A C A A A A N
Nitric acid (10%)	60	B B A A N A A A A
Nitric acid (70%) to 100°F	60	A B A C A B A A A
Nitrobenzene		B B C B A A C A A
Oils, animal		A A C C A A A A A
Oils cottonseed	11, 58	A A A A A A A A A
Oils fuel	37, 14	A A A A A A C A A
Oleic acid	3	A A C A C A A C C
Oleum (20–25%)		A A C B C A B A C
Oxalic acid	54	A A B A A A A A A
Palmitic acid	3, 54	A N C A B A B A A
Perchloroethylene (dry)	11	N A A C A A C A B
Perchloric acid (10%)		N C B N B N A N A
Phenol (carbolic acid)	11	A C A N N B A N B
Phosphoric acid	2, 6, 11, 39	A A A A A A A A A
Phosphorous trichloride		A N C A C A A A A
Picric acid	57	A A A C N B A A N
Potassium bicarbonate		A A A A A N A A A
Potassium tetraborate		A A A A A A A A A
Potassium bromate		N N N A N A N N N
Potassium bromide		N N N A N N N N N
Potassium carbonate	40	A A A A A N A A A
Potassium chlorate	3	A A A A A A A A A
Potassium chloride	5, 41	A B A A A A A A A

Right column table:

Chemical	Ref.	Ratings
Sodium borate	14	A A A A A A A A A A A A A A
Sodium carbonate	44	A A A A A A A A A A A A A A
Sodium chlorate	14	A A A A A A A A A A A A A C
Sodium chloride	4, 3	A B A B A A A A A A A A A A
Sodium chlorite (to 20%)	45	A C C C N A A A A A A A A A
Sodium chromate		A A N N A A A N A A A A N A
Sodium cyanide		A A A A N A N A A A A A N A
Sodium fluoride	25, 46	B A B A A A A A N A C A A A
Sodium hydroxide 20%	5, 3, 6	A A A A A A A A A A A A C A
Sodium hydroxide 50%		A C A C A A C B A A A C A A
Sodium hypochlorite	5, 3, 6; 30, 13, 47	A C A B A A A B A A A N A A
Sodium nitrate	48	A A A A A A A A A A A N A A
Sodium perborate		A A A B A B A A A A N N A A
Sodium peroxide	6	A A A A B B A A A A A A A A
Sodium monophosphate		A A A A A A A A A A A A A A
Sodium di- or triphosphate		A A A A A A A A A A A A A A
Sodium polyphosphate		A A A A A A A A A A A A A A
Sodium silicate	49	A A B A A A A A A A A A A A
Sodium sulfate	50	A A A A A A A A A A A A A A
Sodium sulfide	1, 48	A A A B A A A A A A A N A A
Sodium sulfite	44	A A B A A A A A A A A A A A
Sodium borate		A A A A A A A A A A A A A A
Sodium thiosulfate (hypo)	51	A A B A B A A A A A A A A A
Starch		A A A A A A A A A A N A A A
Stearic acid	37	A A B A B A A A A A A B A A
Sugar solutions	14	A A B N B A A A A A N A A A
Sulfur, molten		A C A C A A A A N C A A A N
Sulfur chloride	57	A A A A A A A A A A A A A N
Sulfuric acid (0–40%)	5	A C A A A B A B A A A A A A

(continued)

TABLE 3.16 Continued

Liquid	Ref. no.	Hast. C	Carp. 20	316 S.S.	Hypalon	Viton	PVC	TFE	Nordel	Kynar-150°F	Ceramic	Kalrez
Sulfuric acid (40–95%)	5, 55, 59	A	A	C	A	A	A	A	C	A	A	A
Sulfuric acid (95–100%)	56, 59	A	A	B	B	A	B	A	C	A	A	A
Sulfurous acid		A	A	B	A	A	A	A	C	N	A	A
Tannic acid	52	A	A	A	A	A	A	A	C	A	A	A
Tartaric acid	6, 44	A	A	A	A	A	A	A	B	A	A	A
Titanium dioxide		A	A	A	A	A	B	B	A	N	N	N
Toluol & toluene	36	A	A	A	C	A	C	A	C	B	A	A
Trichlorethylene	57	A	A	A	C	A	C	A	C	A	A	A
Turpentine	13	A	A	A	C	A	A	A	C	A	A	A
Urea formaldehyde		A	A	A	N	N	N	N	N	A	A	N
Varnish & solvents	14	A	A	A	C	A	A	A	N	A	A	A
Vinegar		A	A	A	A	N	A	A	N	N	A	A
Vinyl acetate		A	A	A	C	C	C	A	C	A	A	N
Water, deionized		A	A	A	A	A	A	A	A	A	A	A
Water, salt		C	A	B	A	A	A	A	A	N	A	A
Whiskey and wines	58	A	A	A	A	A	A	A	N	A	A	A
Xylene or xylol	13	A	A	A	C	A	C	A	C	A	A	A
Zinc chloride	6, 33, 53	A	A	C	A	A	A	A	A	A	A	A
Zinc hydrosulfite		A	A	B	B	A	A	A	N	N	N	A
Zinc sulfate		A	A	A	A	A	A	A	A	A	A	A

15. Use slurry valves
16. PVC to 105°F, SS to 180°F
17. PVC to 100°F, SS to 100°F
18. SS to 70°F dilute, PVC to 125°F
19. PVC to 100°F, 50%, SS to 70°F, 5%
20. PVC to 100°F, 25%, SS to 180°F, 50%
21. PVC to 100°F, SS to 160°F
22. Viton, Nordel to 120°F
23. Hast. C to 10%
24. PVC to 125°F, 36%, SS to 180°F 10%
25. Fluoridation requires an anti-syphon pump installation. Consult local regulations for details.
26. PVC to 30%
27. PVC to 125°F, 50%, SS to 70°F, 5%
28. May cause surface pitting to SS
29. PVC to 125°F, 48%
30. Hypalon to 130°F
31. PVC to 100°F, 50%, SS to 100°F, 50%
32. PVC to 70°F, 10%, SS to 70°F, 10%
33. Carp. 20 5%
34. SS to 70°F, 5%, PVC 125°F SAT
35. PVC to 100°F, SS to 70°F
36. Viton to 100°F
37. Hypalon to 150°F
38. SS to 70°F, 10%
39. PVC to 125°F, 80%, SS to 70°F, 80%
40. PVC to 100°F, SAT, SS to 180°F, 50%
41. SS to 180°F, 5%
42. PVC to 70°F, 50% or to 125°F, 30%, SS to 180°F, 50%
43. SS to 140°F, 10%
44. SS to 180°F, 50%
45. PVC to 105°F
46. PVC to 125°F, 4%, SS to 70°F, 5%
47. PVC to 125°F, 15%, SS to 70°F, 5%

Rating Key
A Acceptable
B Satisfactory where minor attack is acceptable
C Should not be used
N Information lacking
Unless otherwise noted, concentration of aqueous solutions are saturated.
All ratings are at room temperature unless specified.

174

REFERENCE NOTES

1. Warning: Dried residue of spilled solutions is explosive.
2. Hast. C to 180°F
3. SS to 180°F
4. Carp. 20 to 180°F
5. PVC to 125°F
6. Hypalon to 180°F
7. SS to 125°F 10%, PVC to 125°F
8. PVC to 125°F, 29%, SS to 180°F, 29%
9. SS to 70°F, 5%
10. PVC to 105°F, 40%, SS to 180°F SAT
11. Viton to 180°F
12. PVC to 100°F Pure
13. Viton to 158°F
14. SS to 140°F

48. SS to 125°F
49. PVC to 125°F, 41Be, SS to 140°F, 41 Be
50. PVC to 125°F, 30%
51. PVC to 125°F, 50%, SS to 70°F, 50%
52. PVC to 100°F, 10%, SS to 150°F
53. PVC to 100°F, SS to 180°F, 70%
54. Kalrez to 180°F
55. Kalrez to 150°F
56. Kalrez to 100°F
57. Kynar to 70°F
58. Kynar to 120°F
59. Kynar to 120°F, 30%
60. Kynar to 100°F

(*Data Courtesy of Wallace & Tiernan, Inc. Belleville, N.J. 07109-3057*)

175

highlight some designs with unique characteristics. Figure 3.68 shows a high-temperature Teflon pump featuring a leak-free seal and dual bellows that operate at pressures up to 80 psig (5.5 bars). This unit is recommended for operations up to 180°C (212°F). All wetted surfaces are manufactured from PTFE Teflon to provide corrosion resistance.

Examples of multistage centrifugal pumps and diaphragm pumps with polypropylene and Teflon construction are illustrated in Figure 3.69. The multistage centrifugal pumps feature stainless steel and Noryl™ construction. These are high-efficiency, pulse-free units. The diaphragm pumps are of Teflon construction. They have high-temperature capability (to 300°F). These are designed as nonmetallic for corrosive environments.

Manufacturers will typically supply chemical compatibility charts such as those given in Table 3.16. These can serve as a guide in material selection. Table 3.17 provides a description of the materials listed in Table 3.16. Appendix A

TABLE 3.17   MATERIAL ANALYSIS OR DESCRIPTION

The following is typical analysis and description of products named in Materials Section.

Carpenter 20	High nickel-chrome alloy with nominal analysis; C 0.7% Mn 0.75%, Si 1.0%, Cr 20%, Ni 25%, Mo 20% and Cu 3.0%.
Ceramic	99% aluminum oxide.
Hastelloy C	High nickel-chrome alloy of following analysis; Ni 54%, Cr 15.5%, Co 2.5%, Mo 16%, W 4%, Fe 5.5%, C 0.08%, others— 3%.
Hypalon*	A chlorosulphonated polyethylene.
Kalrez*	Perfluoroelastomer.
Kynar**	Polyvinylidene fluoride.
Nordel*	A terpolymer of ethylene, propylene and a nonconjugated diene.
PVC	Polyvinyl chloride.
Stainless 316	AISI 316–Cr 16–18% Ni 10–14%, C 0.08%, Mn 2%, Si 1%, P 0.045%, S 0.03%, Mo 2–3%.
TFE	Fluorocarbon resin of tetrafluoroethylene polymer.
Vitron*	Copolymer of vinylidene fluoride and perfluoropropylene or hexafluoropropylene.

* Trade names of E.I. DuPont de Nemours & Co., Inc.

** Trade name of Pennwalt Corp.

(*Data Courtesy of Wallace & Tiernan Inc. Belleville, N.J. 07109-3057*)

TABLE 3.18 GUIDE TO PUMP STYLE/TYPE AND SELECTION

	PUMP LINE	DESCRIPTION	SIZES	CAPACITIES TO	HEADS TO	APPLICATIONS
	Slurry and Solids Handling Pumps	Heavy-wall abrasion resistant designs. Enclosed or semi-open, and nonclogging impellers.	1¼" to 30"	30,000 gpm (6800 M³/hr)	700' TDH (215 M)	Slurries to 70% Wgt. Abrasives/Corrosives. Large solids.
	Screw Centrifugal Pumps	Unique screw centrifugal impeller. Gentle handling, low NPSHR.	3" to 20"	20,000 gpm (4600 M³/hr)	300' TDH (90 M)	Low shear, low turbulence, low emulsification, low crystal degradation, large and stringy solids, food handling applications.
	Vertical Cantilever Shaft Pumps	Ultimate in sump pump reliability. No submerged bearings. No packing or seals.	1" to 30"	30,000 gpm (6800 M³/hr)	300' TDH (90 M)	Critical sump or tank applications. Abrasives and Corrosives. Unattended installations. Long life and low maintenance.
	Submersible and Immersible Pumps	Portable, free-standing, and "Fast-Out". Motors feature dry rotor/stator design to 600 HP.	1" to 20"	20,000 gpm (4600 M³/hr)	300' TDH (90M)	Slurry and solids handling. Open sump applications. Agitation and cutters available.
	Heavy-Duty Chemical and Process Pumps	Engineered pumps in all alloys for demanding duty. Accepts all types of mechanical seals.	1" to 20"	20,000 gpm (4600 M³/hr)	700' TDH (215 M)	High reliability chemical process applications for corrosives, high temperature, high pressure, critical services.
	Axial Flow Pumps	Horizontal elbow pumps and vertical pump designs available for services to 800°F (425°C) and 1500 PSIG (104 BAR) system pressures.	6" to 66"	250,000 gpm (57,000 M³/hr)	5' to 170' TDH (2-52 M)	Evaporators, crystallizers reactors, circulators, and other high volume transfer.
	API-610 Slurry and Process Pumps	Centerline supported process pumps. In-line barrel multi-stage pumps. Solids handling pumps with replaceable liners.	1½" to 16"	12,000 gpm (2700 M³/hr)	2000' TDH (610 M)	Refining, petrochemical applications for hot oils, light hydrocarbon, coal/oil slurries, coke, catalyst, slurry, synthetic fuels.
	In-Line Process Pumps	TRUE-LINE vertical pumps, and LPI/REIMA® pumps specifically designed for low-flow, high head applications.	1½" to 3"	2 gpm to 350 gpm (0.5-80 M3/hr)	6900' TDH (2100 M)	Low flow, high head applications for light hydrocarbons, oils, solvents, chemicals, and water.
	Chlorine and Toxic Liquid Pumps	Designed for flange mounting on top-entry storage vessels. Several safety features and instrumentation for monitoring and control.	1" to 4"	600 gpm (140 M³/hr)	800' TDH (244 M)	Liquid chlorine, phosgene, hydrogen cyanide, bromine, hydrofluoric acid, etc.
	Molten Metals and Salts Pumps	Cantilever shaft design engineered to accommodate thermal expansion, metallurgical properties, and corrosion.	1" to 8"	3500 gpm (800 M3/hr)	300' TDH (90 M)	Molten metals: lead, tin, zinc magnesium, etc. Molten salts: Sodium/potassium chloride, carbonate salts, etc. to 1650°F (900°C)
	Cryogenic Pumps	Single and multi-stage pumps. Top suction design for rapid start. Inducers for low NPSHR.	1" to 6"	1200 gpm (275 M³/hr)	2000' TDH (610 M)	Liquid oxygen, nitrogen, light hydrocarbons, and other liquified gases to −320°F (−200°C).
	Jacketed Pumps	Fully jacketed pump components including casings, columns, discharge pipes, and mechanical seals. Welded jackets designed for steam or other heat transfer liquids.	1" to 8"	3500 gpm (800 M3/hr)	300' TDH (90 M)	Molten sulfur, inorganics, plastics, and other liquids requiring close temperature control.

Figure 3.69 Examples of multistage centrifugal and diaphragm pumps with plastic construction. (*Courtesy of Osmonics, Inc., 2022 West 11th Street, Upland, CA 91786*)

provides an extensive database on chemical compatibility and corrosion performance of materials.

In closing this chapter, Table 3.18 is included as a guide to pump style selection.

4

Oscillating Displacement Pumps

The oscillating displacement pump uses one of the oldest pump principles. Known to have been used in Greece long before the birth of Christ, piston pumps are still among the most extensively applied in a wide variety of technical services.

The present widespread application of the oscillating displacement pump is due to its unique properties. Conspicuous among these is its high efficiency—a characteristic that is particularly valuable in today's era of energy shortages. Although it has been replaced in some services by the less costly centrifugal pump, the oscillating displacement pump is finding many new applications, especially in high-pressure engineering. Here, the basic principle of oscillating displacement is employed in varied designs for such services as processing, power hydraulics, and high-pressure purification.

The following is a brief survey of applications and types of oscillating displacement pumps:

1. Drinking water supply
   - Deep-well piston pumps
   - Pressure-booster piston pumps

2. Crude oil production
   - Deep-hole reciprocating piston pumps
   - Hydraulic deep-hole pump systems
   - Drill-hole, piston-type scavenger pumps
3. Power hydraulics
   - Axial piston pumps
   - Radial piston pump
   - In-line piston pumps
4. Natural gas exploration and chemistry
   - Natural gas-driven piston pumps
   - Air-driven diaphragm pumps
5. Homogenizing by throttles split friction
   - Homogenizing pumps of in-line piston-type construction
6. Boiler feed water supply
   - Steam-driven piston pumps
   - Electric piston pumps
7. Lubrication
   - Lubricating pumps for in-line piston-type arrangement
8. Heart prostheses
   - Oscillating or pulsating blood pumps
9. Fuel injection in combustion engines
   - In-line piston pumps
10. Volumetric liquid metering
    - Mostly stroke-adjustable diaphragm, piston, bellows, and tube-type metering pumps
11. High-pressure pumping of process technology liquids
    - Multicylinder piston pumps
    - Single- or multicylinder diaphragm pumps

Because of their importance in process technology and to permit detailed treatment of their design and application, only metering pumps and high-pressure discharge pumps are discussed in this chapter.

Process technology imposes the highest requirements for proper choice of materials because of the different media involved. Toxic, abrasive, and corrosive media are to be controlled at high and low temperatures, as well as in the form of suspensions, abrasive slurries, and pastes. The pressure ratings encountered range up to several thousand bar.

The differences between metering and discharge (conveying) pumps are dictated by the job to be performed. Metering means conveying, measuring, and adjusting the metering flow. That is, compared to discharge pumps, which only convey, two additional tasks have to be completed. As will be shown, this has a strong influence on the technical design of pumps.

OPERATING PRINCIPLES

Kinematics of Displacement

The chronological progression of the stroke motion of frequently used oscillating drive units (Figure 4.1) more or less corresponds with harmonic motion. The rotary cam-type drive unit exhibits harmonic motion exactly. At a not too large rod ratio with respect to the displacement speed, the direct-thrust, crank-type drive unit hardly deviates from harmonic motion; however, it markedly deviates from it in regard to acceleration values. The pulsation of the single cylinder ($i = 1$, $i = 2$) can be smoothed out by superimposing several cylinders, as demonstrated in Figure 4.2 or a single-cylinder pump and a three-cylinder pump having an identical flow rate, Q_m. Figure 4.3 shows the flow-rate pulsation for multicylinder pumps at $\lambda = 0$ (left hand) and $\lambda = 0.225$ (right hand). The variances are caused by the deviations from the harmonic motion at $\lambda = 0$.

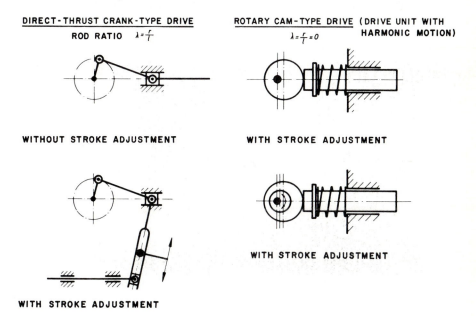

Figure 4.1 Frequently used oscillating drive units.

With regard to $i = 5$ and $i = 6$, the results are identical for the different λ. The fluctuation of Q_m is shown in the figures. Odd numbers of cylinders (triplex, quintuplex) produce the smallest pulsations.

For simplification, we can dispense at this point with analysis of the influence by volumetric efficiency; that is, the elasticities on the kinematics of displacement which, roughly, is only significant up to average pressures.

The pulsation in flow rate, as demonstrated in Figures 4.2 and 4.3, produces corresponding accelerations and delays of the liquid compounds in the

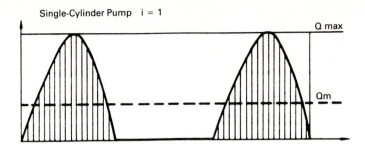

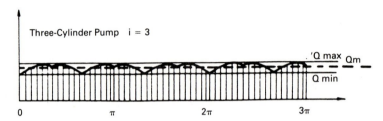

Figure 4.2 Pulsation of single- and three-cylinder pumps of equal average flow rate.

system. The acceleration or delay of the liquid results from the differential quotient of the instantaneous flow rate, $Q(t)$. Conditions as shown in Figure 4.4 ($\xi = 0.225$) are representative for the single-cylinder pump.

At commencement of suction there is a high initial acceleration and a final delay at the end, as is easily understood in terms of physics. The opposite is true during the pressure stroke.

Flow Rate Characteristics

The average rate of flow is approximated as follows:

$$Q_n = n \times h \times F_K \times \rho \times \left(1 - C\frac{\rho}{h/h_{100}} \times \eta_G\right) \tag{4.1}$$

where n = stroke frequency

h = stroke length for pumps with constant stroke length is $h/h_{100} = 1.0$

F_K = cross section of piston

ρ = density

The individual variables can be summarized into three factors:

$$Q_m = Q_{mth} \times \eta_E \times \eta_G \tag{4.2}$$

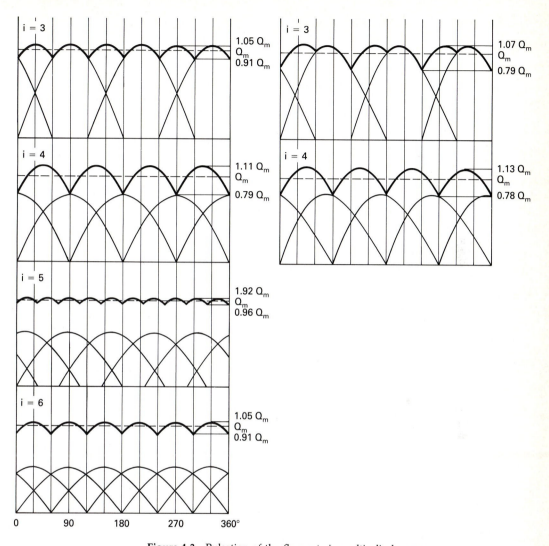

Figure 4.3 Pulsation of the flow rate in multicylinder pumps.

1. The theoretical flow rate, Q_{mth}, includes the rating variables of the pump and can be approached only under ideal conditions.

2. The extent of elasticity, η_E, takes into consideration the elasticity influence of the pump influence of the pumped medium and of the pump chamber, which depends on pressure and length of stroke. The constant, C, is dependent on the relative compressibility of the liquid, $\Delta V / \Delta 0$, and on the elasticity constant of the pump chamber.

3. The internal and external leakage losses, which are also influenced by the viscosity, are included in the quality level, η_G.

Direct-Thrust Crank Drive

$\lambda = 0,225$

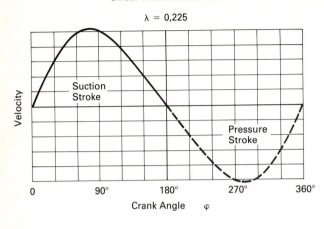

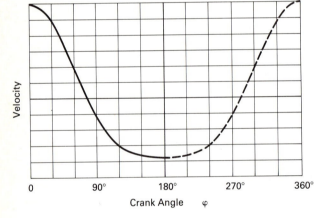

Figure 4.4 Velocity and acceleration of a single-cylinder pump.

Elasticity level η_E and quality level η_G combine to form volumetric efficiency, $\eta_F = \eta_E \times \eta_G$.

Usual values for water at average performance for η_F are as follows:

Piston pumps	Up to 100 bar	95–98%
	Up to 500 bar	93–96%
Diaphragm pumps	Up to 100 bar	85–95%
	Up to 500 bar	80–90%

The effect of elasticity on the displacement process is shown in Figures 4.5a and 4.5b. It is easy to find one's way through the indicator diagram by using the explanations (Figures 4.5b and 4.5c). Note that condition $Q_m = 0$ occurs when the parenthetical expression of Equation 4.1 becomes zero, which is the case in the limiting stroke length:

$$h_0 = C \times \rho \times h_{100}$$

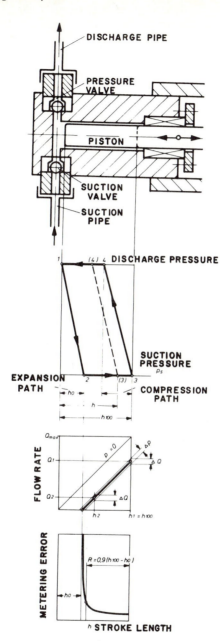

Figure 4.5 Displacement action and its impact on operating characteristics. Legend: 1 = pressure valve closes; 1-2 = release from discharge pressure to suction pressure; 2 = suction valve opens; 2-3 = suctioning of pumped liquid; 3 = suction valve closes; 3-4 = compression from suction pressure to discharge pressure; 4 = pressure valve opens; 4-1 = displacement into pressure pipe.

This is of extreme importance for stroke-adjustable metering pumps because an adjustment like $h < h_0$ no longer has any influence on the flow rate. The relation $Q_m = f(h)$ is admittedly linear, as per Equation 4.1. However, it bisects the axis of abscises offset by $h = h_0$; thus, the flow rate depends on the stroke length (Figure 4.5c) in a linear, but not proportional, mode.

Figure 4.6 provides a collection of measured results for pumps of very different sizes. As shown by this figure, there is a substantial increase of h_0, that is, the parallel shift to the right at small stroke volume and high pressure (for instance, line f). With regard to stroke-adjustable metering pumps, it is particularly necessary that the characteristic form, $h_0 1$, be very acutely observed be-

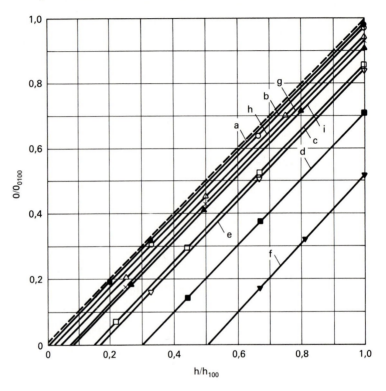

Curve	Symbol	Pump Type	Flowrate (liter/hr) $h/h_{100}=1.0$	Outlet Pressure (bar)
a	●	MH	2000	5.0
b	○	MH	500	30
c	□	MH	100	300
d	■	MH	100	600
e	▽	MH	15	500
f	▼	MH	2	2000
g	×	FM	500	3.5
h	△	MM	50	5.0
i	▲	K	70	1000

Figure 4.6 Plot of dimensionless flow rate as a function of stroke length. Pump types: MH = diaphragm pump head with hydraulic pump drive; FM = fellows-type pump head with mechanical diaphragm drive; K = piston pump head.

cause of the available range of adjustment. At lower pressures and larger stroke volumes $h_0 = 0$, and the $Q(h)$ line almost runs through the zero point.

The influence of the pumped medium on the $Q(h)$ characteristic is great because of the very different elasticity rate of the liquid (Figure 4.7). Particular attention must be paid to the liquefied gases whose elasticity rate is both large and temperature dependent. With small stroke volumes, the relative clearance volume is rather large, which increases the elasticity effect.

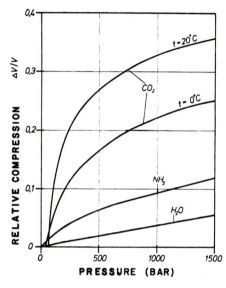

Figure 4.7 Relative compressibility of liquids as a function of pressure.

For different pressures, the $Q(h, \rho)$ performance characteristic of a small oscillating metering pump is represented in Figure 4.8. The lines of resultant pressure run in a parallel mode in a variant displacement motion, $h_0(\rho)$.

At low pressure the parallel shift (Figure 4.6) is infinitely small, so the pressure influence can almost be neglected. How pressure rigid the flow rate pressure characteristic can quantitatively be in oscillating displacement pumps in comparison to other pump types is shown in Figure 4.9. This is the reason why oscillating displacement pumps are used so successfully as metering pumps.

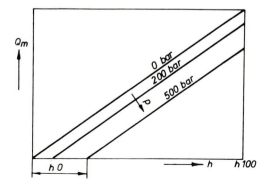

Figure 4.8 Flow rate is dependent on stroke length and pressure for a small, high-pressure metering pump.

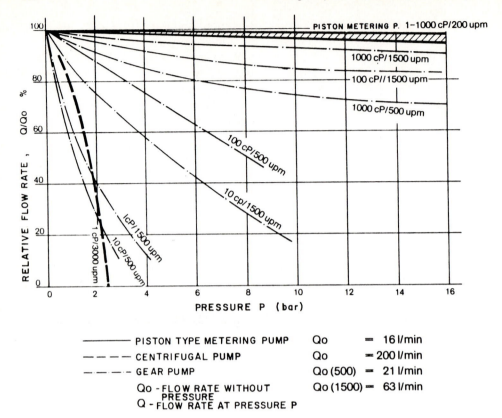

Figure 4.9   Flow rate versus pressure characteristics for different pump types.

According to Equation 4.1, the relation $Q = f(n)$ is linear and proportional, as is demonstrated in Figure 4.10 at different stroke adjustment, h/h_{100}. The range of adjustment can, therefore, be increased by adjusting stroke frequency when it is no longer possible to adjust stroke length.

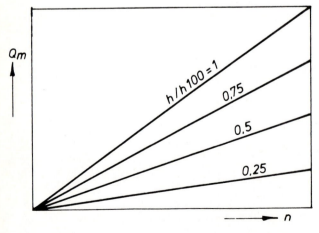

Figure 4.10   Flow rate is dependent on stroke frequency.

EFFICIENCY

Excellent efficiency rates are attained in oscillating displacement pumps because of their negligible losses by leakage, in combination with the inherent advantages of hydrostatic energy transmission. This is especially important for high-performance pumps. Development of oscillating displacement pumps to achieve much higher stroke frequencies and, thus, reduction of specific machine weights, has removed the barrier between centrifugal and piston pumps. It frequently pays to carry out a profitability comparison study. Figure 4.11 emphasizes this comparison of efficiency between piston and centrifugal pumps at partial load for a performance level of the piston pump of 200–300 kW. Conspicuous here is the enormous margin of efficiency of the piston pump at partial load, as well as at full load.

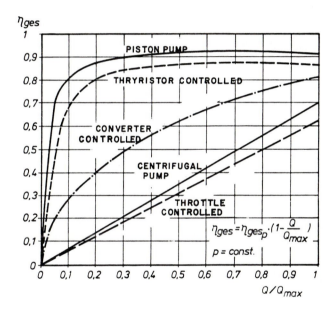

Figure 4.11 Efficiency comparison between oscillating displacement and centrifugal pumps.

Metering Error

Analyzing the variables that influence metering error is of great importance in determining the application of metering pumps. Table 4.1 shows which influences affect the individual factors of Equation 4.1 and, thus, can vary the flow rate. Once the average erratic value of all individual variables in Equation 4.1

$$\pm\Delta A_K, \ \pm\Delta\rho, \ \pm\Delta n, \ \pm\Delta h, \ \pm\Delta p, \ \pm\Delta\eta_G$$

is known, the mean mass flow fluctuation rate then can be determined in accordance with the law of error propagation.

TABLE 4.1 INFLUENCES ON FLOW-RATE FLUCTUATION

Analytical connection	Influences on the fluctuation of the individual variables	Pump plus installation	Caused by pump medium	Marginal value
$Q_m = A_K \cdot \rho \cdot n \cdot h \left(1 - C \dfrac{P}{h/h_{100}}\right) \cdot \eta_G$	Viscosity, stroke frequency, pressure, temperature	X	X	X
	Adjustment accuracy, load (delivery pressure)	X		X
	Type of drive, load (delivery pressure)	X		X
	Pressure, temperature, gas content		X	X
	Production allowances, wear and tear	X	X	

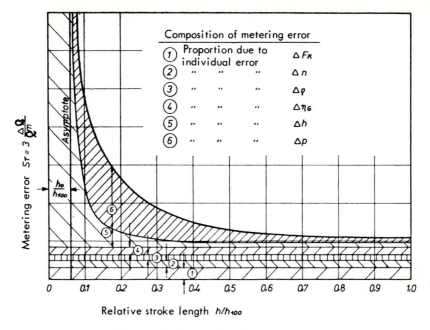

Figure 4.12 Composition of metering error.

Figure 4.12 shows the composition of the metering error. In comparison, Figure 4.13 represents the metering error measured for a piston pump. The graph shows that a stroke-adjustment ratio of $1:10$ still produces good metering accuracy.

The smaller the stroke length, h, the larger the metering error becomes. At

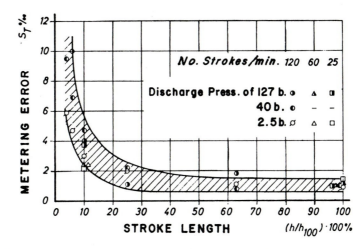

Figure 4.13 Measured metering error depends on stroke adjustment, h/h_{100}, counterpressure, and stroke frequency.

$h = h_0 = C_p h_{100}$ (see Figure 4.5), it heads toward infinity. Therefore, it is not possible to operate a metering pump at high metering accuracy over a broad stroke-adjustment range at random (perhaps $1:100$ or even $1:1,000$). The solution to the problem must be adjustment through stroke length and stroke frequency of distribution of the flow-rate range among several individual pumps. To maintain a high metering accuracy across as large a flow-rate range as possible, it is necessary to aim at the following targets:

- smallest dead spaces independent of stroke length
- working species of small elasticity
- constant marginal values

In most metering pumps and in piston pumps in particular, the elasticity value C is small. The pressure fluctuations within a broad stroke-length range have only a very weak effect on the metering error. This provides the explanation for one of the known properties of the volumetric metering pump, namely, its metering accuracy nearly independent of pressure fluctuations.

EFFECT OF INFLUENCING VARIABLES ON FLOW RATE AND METERING ERROR

Oscillating pumps operate on a volumetric basis. The flow of the material conveyed is directly influenced by its density, ρ, which is dependent on pressure and temperature (Figure 4.14). In liquified gases in particular, this dependence is considerable so that, for instance, an accurate mass metering operation requires constant levels of pressure and temperature. The metering error due to density variation of the liquid is reproducible at constant marginal values (pressure and temperature). Gas bubbles in the pumped liquid vary the average density to the degree that the gas portion changes. The only measures possible to eliminate the density influence are the constant stabilization of the density or the use of a density-dependent control system.

Pressure and Elasticities

In Figure 4.15, taking a small high-pressure diaphragm pump as an example, the performance characteristic *volumetric efficiency/discharge pressure* is explained in a nondimensional plotting (maximum discharge pressure is 600 bar). This shows that the consequence of a pressure fluctuation, ΔP_{at}, diminishing stroke adjustment, h/h_{100}, is a rate of ever increasing volumetric efficiency fluctuations which, as has been explained already, causes the metering error to become infinite at $h = h_0$. Temperature fluctuations also affect the volumetric efficiency via the elasticity of medium (see Figure 4.16).

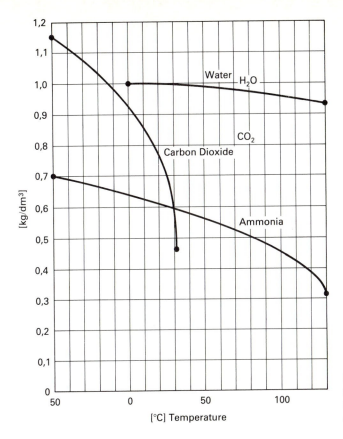

Figure 4.14 Reports specific gravity of some liquids. (Note that more extensive data tables are reported in Appendix B.)

Gas bubbles in the medium to be pumped may increase considerably its total elasticity and produce strong disturbances because of the compressibility of gases being larger by powers of ten. Frequently gas develops first in the working chamber of the pump. For instance, when pumping under vacuum, gas or air can penetrate into the pumping space from the outside around the sealed area. For such applications, seals having correspondingly lower leakage rates must be used (good seal-casing features and high seal compaction). To block up the piston gland by means of a liquid buffer would be an appropriate action. Vapor (cavitation) starts to form when the vapor pressure of the liquid is reached anywhere in the system. It goes without saying that any evidence of cavitation in the working and hydraulic areas must be suppressed. It not only causes catastrophic metering errors and drop of the flow rate, but also the generation of loud noise and damage to the pump. Cavitation can always be avoided by properly dimensioning and installing the pump. Gas dissolved in the liquid does not cause any disturbance as long as it remains in solution. Pressure drop or temperature increase, however, can shift the balance of solution so that the

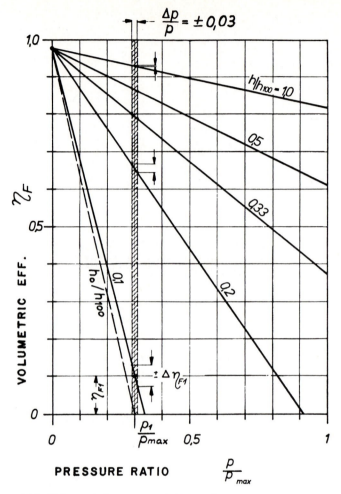

Figure 4.15 Volumetric frequency as function of pressure and stroke length for a small-diaphragm pump.

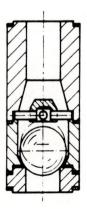

Figure 4.16 A suspension valve. (*Courtesy of American Lewa, Inc., Holliston, MA*).

gas is released. To preclude disturbances created by gas bubbles, various preventive measures can be taken:

1. Keep gas bubbles away from the pump by appropriately sized piping or by superimposing a degassing reservoir.
2. Avoid conditions that cause or promote the formation of gas in the pump housing by a correct selection of the variable parameters during layout.
3. Prevent the settling of gas bubbles by optimally designing the working chamber of the pump in a flow-technical respect. This step is particularly important for metering the smallest flow rates because the long dwell time of the liquid in the pump cylinder promotes gas evaporation.

Volumetric Efficiency

The volumetric efficiency takes leakages and reflux losses into account. In accordance with Table 4.1, this efficiency is dependent on viscosity (therefore, also on temperature), stroke frequency, and pressure.

Leakage Losses in Moving Glands

In extreme cases, leakage flow in piston glands can lead to a total stoppage of pumping (for instance, if there is no maintenance over an extended period of time). If the unit is subjected to competent maintenance and proper rating, the leakage flow can, however, be kept to a minimum so that its influence on the flow rate can be neglected. Where media are involved that are hard to seal, the problem, more often than not, is solved by the use of glandless pumps.

Reflux Losses of Valves

If liquid refluxes into the working chamber of the pump through the pressure valve or into the suction pipe through the suction valve, the conveying performance of the pump is diminished; this may result in considerable metering errors. There are two essential requirements for proper functioning of the valves:

1. The moving valve body (ball cone, disc) must close tightly during the period of time anticipated for this purpose. Valve leakages are caused by wear and tear, corrosion, or jamming solid particles between guide and seat areas. By means of suitable materials (hardness, corrosion resistance), or by using suspension valves (Figure 4.16) specially designed for polluted media, these causes of error almost always can be eliminated, or at least reduced to tolerable conditions. Due to the small flow speed in contaminated media or suspensions, there is the increased risk in smaller stroke lengths for solid particles to settle in the valve and cause disturbances. The higher the stroke frequency of the pump, the smaller the reflux losses at

the valve-seal areas. In this respect, a smaller high-speed pump is superior to a larger one of lower stroke frequency.

2. The valve must open or close at the proper timing. The influence of the valve kinematics on flow rate and metering error is hard to understand and has not yet been clarified entirely. Whereas the opening timing plays a subordinate role, it is obtained by force by the displaces via the hydraulic coupling—the closing timing and the timed progression of the valve-stroke curve have a distinct effect in the proximity of the closing point.

INFLUENCES ON VALVE KINEMATICS

Figure 4.17 shows some stroke curves for valves that close without delay; that is, they close at the timing scheduled. The lower the stroke frequency and the viscosity of the medium and the heavier the weight of the obturator (closing body) and its spring load, the flatter the curve's progression.

In essence, the closing delay of a valve with fixed geometry depends on the influence rates as shown in Table 4.2. The same tendency as in Table 4.2 is also

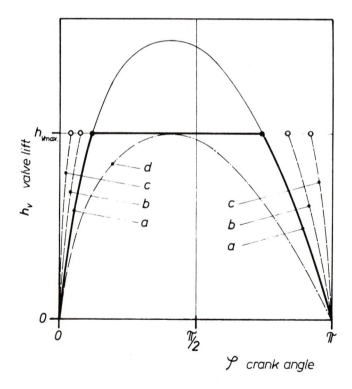

Figure 4.17 Curves of the valve strokes for varying parameters (i.e., spring loading, obturator body weight, stroke frequency, fluid viscosity).

TABLE 4.2 INFLUENCE RATES ON CLOSING DELAY

Closing delay	High	Low
Viscosity	High	Low
Stroke Frequency	High	Low
Weight of Closing Body	Low	High
Spring Load	Low	High

shown in Figure 4.18, which represents the relationship determined from measurements between closing angle, $\xi = wt$ (without delay means = 180°), viscosity and spring load at average stroke frequency.

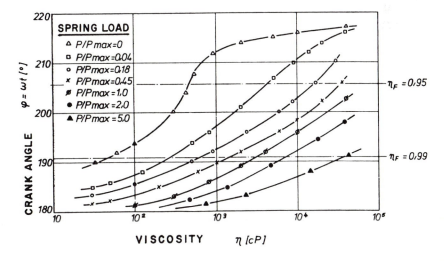

Figure 4.18 Closing delay of valve depends on viscosity and spring load.

A large closing delay leads with certainty to large reflux losses and, thus, to a smaller flow-rate and metering errors. It is a question whether the pump has been properly rated to be able to keep the closing delay time so small that it no longer has any appreciable influence. There is no problem with the design rating where low-viscosity media are involved. Highly viscous media require a careful synchronization among the freely selectable variables such as stroke frequency and spring loading, whereby the layout latitude is narrowed by the size of the pump (low stroke frequencies result in larger and, thus, more expensive pumps) and by the risk of cavitation during the suction stroke (high spring-load rates generate high entry-pressure losses). Figure 4.19 represents a stroke curve with closing delay. A heavily delayed closing action also can be seen in the indicator diagram.

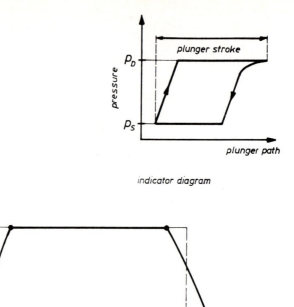

indicator diagram

Figure 4.19 Closing delay of valve (i.e., indicator diagram and stroke curve).

SUCTION ACTION, ENTRY-PRESSURE LOSS, CAVITATION

The objective in examining the suction action is to determine the pressure progression in the pump's working chamber. Figure 4.20 is a schematic of a pump head and Figure 4.21 describes this suction action with a piston pump head having a spring-loaded ball valve as an example, assuming a low operating pressure P_{DF} of the pump. Obviously, conditions for other compression ratios are quantitatively different. However, in principle, the described presentation is applicable to all cases. When considering pressure loss, it is appropriate to examine separately the opening phase, P_{E1} and the flow phase, P_{E2}. For consideration of the opening phase, the piston is in the upper dead position (O.T. in Figure 4.20). Let us assume that the working space pressure (see Figure 4.21, point a) has the value P_{DF}. Due to the elasticities of liquid and working space walls, as well as leakage of the pressure valve, the return expansion curve, b, does not progress in a vertical direction, but does in an inclined mode. In order

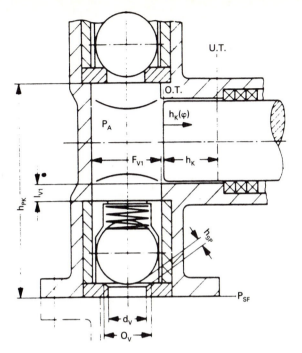

Figure 4.20 Schematic of pump head.

to open, the suction valve first requires equilibrium of the static forces at the moving part (for instance, the ball).

The pressure drop arises due to gravity, and resilience, less buoyancy and adhesive force by a finite width of the valve seat. To begin with, the pressure drops to P_{A1} in the working space, with the total elasticity of the working space permitting one finite piston motion with the suction valve still closed. Then comes the time of pulling away (see Figure 4.21, point c). By means of the finite piston speed, V_K, the liquid masses are suddenly hydraulically coupled to the piston, causing movement at the suction side. Depending on conditions such as size of mass, size of delay angle, and so on, an additional pressure drop to P_{A1} min will develop although the adhesive force collapses and becomes zero. Due to the liquid now quickly flowing into the relaxed space, the suction valve is fully opened.

The pressure drop, P_{sf}-P_{A1}, is that portion created by jerky acceleration of the valve ball and by the total liquid beyond the pressure decrease on the suction side due to the adhesive force. After the suction valve has opened and the fade-out process has terminated (see Figure 4.21), the flow phase will commence during which the pressure in the pump's working space is determined almost solely by the variations of friction and mass pressure of the quasistationary flow in the valve.

Amplitude and duration of the oscillations between opening and flow phases depend on the prevalent damping rate, natural frequency of elasticity,

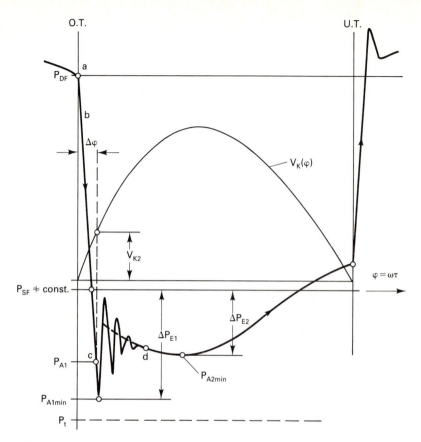

Figure 4.21 Curve progression for pressure in the working space of pump during suction.

and mass of the oscillating system. It follows from the explanations about the pressure drop at the entry to the pump head that to determine the pressure progression during the suction action, both phases will have to be observed. If a test is made for cavitation, the lowest absolute pressure can arise in the pump's working space during one or the other phase, depending on conditions, and thus cause cavitation. It is known from measurements that for determination of the cavitation limit in oscillating displacement pumps only the flow phase will have to be observed. The opening phase and the minimum pressure arising therein are dominating so briefly that possibly occurring vapor bubbles disappear again by the end of the suction stroke.

Furthermore, we know from the measurements mentioned that the entry-pressure loss, $\Delta P_E = \Delta P_{E2}$, can be determined from static pressure drop measurements. To this end, one determines the drag coefficient, (e.g., see Figure 4.22, f, ball valve) in dependence on the Reynolds number and from it one determines P_E, which describes in detail the formation of bubbles unclear various cavitation

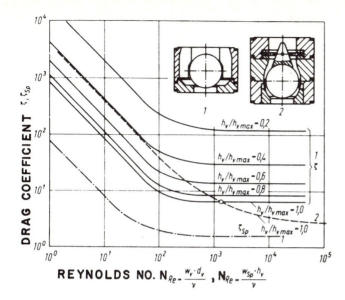

conditions. Disturbing to the pumping function is the beginning remaining cavitation for which the disturbance can be determined by means of an indicator diagram. However, material damage or overload jolts already happen at an earlier point of time when, for instance, vapor bubbles occur during the stroke action, but will have regressed, partially under heavy jolts, by the end of stroke.

DESIGN OF OSCILLATING DISPLACEMENT PUMPS

To highlight the differences in design, metering pumps and conveying pumps are treated on a separate basis, although there are, of course, flowing transitional intergrades. We can characterize the essential differences as follows.

Metering Pumps

The tasks to be performed by metering include such functions as pumping, measuring, and adjusting. Thus, a metering pump is a combination of working machine (pump), measuring device, and regulator. Primarily, metering pumps must be rated for their metering function; therefore, the stroke volume of the displaces must be as consistently accurate as possible. Exact valve operation, smallest elasticities, and minimum leakages are important. This is why stroke frequencies, as a rule, are less than 300 strokes per minute. Due to the required recipe-metering operation, the unit design is applied customarily, mostly the modular block system providing a large selection of different materials and types. Precision adjustability (stroke, stroke frequency) is the characteristic quality feature. The measuring device and the regulator functions of the metering pump make it partly dependent on control circuits and impose demands on its dynamic behavior. To compete with other metering systems, especially with centrifugal pumps having throughflow measuring provision, the capacity of oscillating displacement pumps seldom exceeds 200 kW and is usually in the 20-kW range.

Conveying Pumps

The job required of these pumps is the maximum possible economical conveyance of the pumped medium against the existent pressure. Economy, and efficiency in particular, are the dominant parameters.

The three-plunger pump (triplex pump, refer to Figure 3.27 in Chapter 3) is predominant because the rather pulsation-free flow rate and even torque allow maximum stroke frequencies, which exceed 1,500 strokes per minute (spm). The variants in materials and types are not so numerous as they are for metering pumps. By contrast, their performance range extends far beyond 100 kW and imposes extreme demands on heavy precision machinery construction.

Detailed Designs of Metering Pumps

The most economical solution to the existing problems for widely different requirements is the use of modular systems, by means of which the metering pump most suitable for the application can be put together from series modular groups on a "tailor-made" basis.

The basic modules are shown in Figure 4.23. An almost unlimited number of metering elements of quite different size and design can be mounted side by side. A horizontal arrangement allows low installation cost, better serviceability, cleaner leakage disposal, and simple disassembly. The most important modules are the drive, motor, actuator, and pump head.

Drive Unit and Motor

Successful designs for stroke-adjustable drive units include such diverse constructions as eccentric systems, crank-type swivel drive units, rocker-arm systems, and spring-action cam systems (rotary cam). Stroke-adjustable drive units are designed for an approximate performance range from 0.1 kW to 100 kW.

As an example, Figures 4.24 and 4.25 illustrate the proven rocker-arm system (crank slot). Bearance fulcrum A of rocker arm 3 can be infinitely varied by hand wheel during standstill and operation. At the same time, the upper dead position of piston rod 4 remains unchanged; this maintains constant clearance volume and very favorable characteristics. The kinematics remains quasi-harmonic for all stroke lengths. Where higher performance is required, the horizontally opposed arrangement (by removing cover, second pump head can be added to same drive unit) would permit a substantial increase in the machine's output (Figure 4.25).

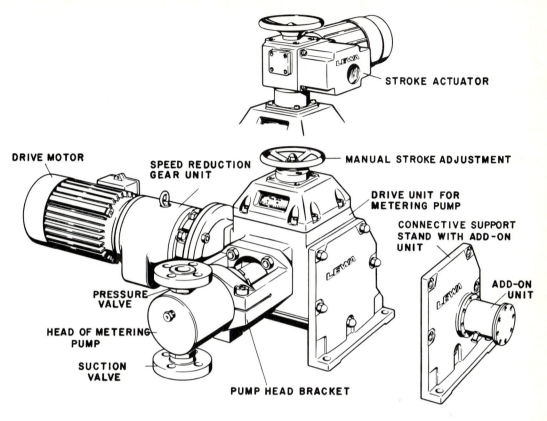

STROKE ACTUATOR

DRIVE MOTOR

SPEED REDUCTION
GEAR UNIT

MANUAL STROKE ADJUSTMENT

DRIVE UNIT FOR
METERING PUMP

CONNECTIVE SUPPORT
STAND WITH ADD-ON
UNIT

ADD-ON
UNIT

PRESSURE
VALVE

HEAD OF METERING
PUMP

SUCTION
VALVE

PUMP HEAD BRACKET

Figure 4.23 Modules of metering pumps. (*Courtesy of American Lewa, Inc., Holliston, MA*).

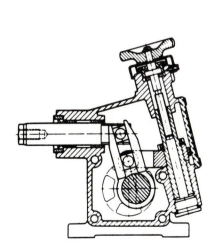

Figure 4.24 Rocker-arm system. (*Courtesy of American Lewa, Inc., Holliston, MA*).

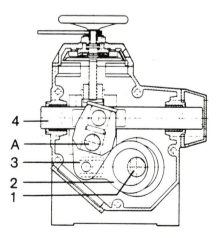

4
A
3
2
1

Figure 4.25 Rocker-arm system in horizontally opposed configuration. (*Courtesy of American Lewa, Inc., Holliston, MA*).

Quite popular in the area of application for smaller outputs is the spring-action cam-drive system shown in Figure 4.26 as a diaphragm pump head with direct diaphragm-link-rod control. Here, the kinematics is discontinuous because of the phase-angle control (Figure 4.27) at partial stroke, which causes jerking accelerations. This is why the range of application remains restricted to small flow rates. For micrometering, the jerking piston motion at partial stroke presents advantages for the valve control.

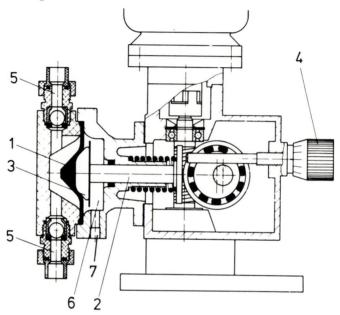

Figure 4.26   Spring-action cam-drive unit. (*Courtesy of American Lewa, Inc., Holliston, MA*).

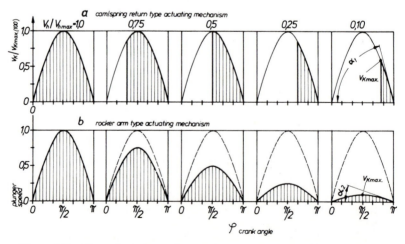

Figure 4.27   Chronological progression of piston speed at partial stroke for (A) spring-action cam-drive units, and (B) rocker-arm drive units. (*Courtesy of American Lewa, Inc., Holliston, MA*).

Besides three-phase asynchronous motors, variable-speed DC and AC motors, as well as mechanical and hydraulic variable-speed drives, are used to drive metering pumps. There are a large number of special metering pump drive units-hydraulic, pneumatic, pulsation-free. An explosion-proof solenoid drive (Figure 4.28) is a popular type used for digital proportioning of small quantities.

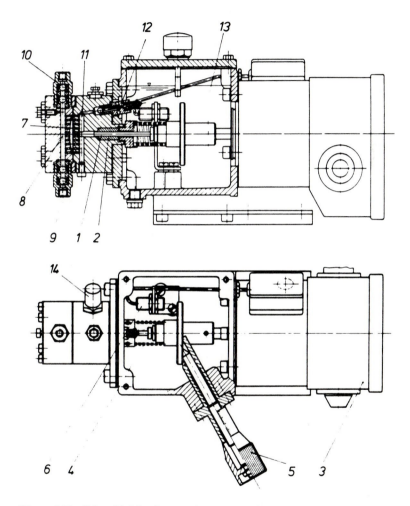

Figure 4.28 Solenoid drive for metering pump. (*Courtesy of American Lewa Inc., Natick, MA*)

The oil-flooded solenoid 3 drives directly via a semiconductor control system, piston I of a diaphragm pump head having hydraulic diaphragm actuation. The stroke-length adjustment 5 allows control of the flow rate. Solenoid-type metering pumps are the most economical solution to the digital proportioning of small flow rates. The main flow-proportional pulse sequence of the flow meter actuates the solenoid metering pump via an electrical control unit (Figure 4.29).

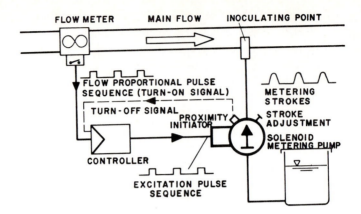

Figure 4.29 Digital proportioning with solenoid metering pump.

Heavy-duty demands are made on stroke actuators with respect to instrumentation and actuation behavior. Linearity, reproducibility, continuous actuating periods, maximum number of switch operations, and smallest actuation steps are advantages provided. Electric actuators (Figure 4.30) without exception operate on the basis of a positive-negative three-step action (right-hand operation, left-hand operation, standstill). In part, they are also directly excitable by means of a 0-0 mA signal. Explosion-proof protection and high-impermeability grades are required. However, in the area of explosion-proof protection, pneumatic instrumentation usually is preferred. Preponderantly, for this

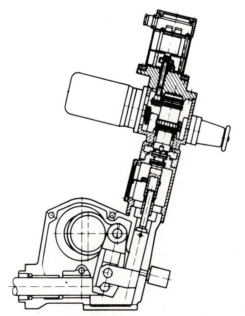

Figure 4.30 An electric actuator. (*Courtesy of American Lewa, Inc., Holliston, MA*).

purpose, double-acting hydropneumatic piston-type actuators (Figure 4.31), controlled through a position controller, are used in the field of metering pump construction. The operating air acts on the actuating piston across the oil surfaces and via reservoirs.

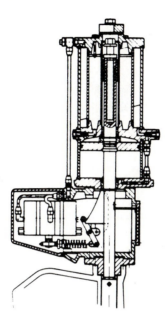

Figure 4.31   A pneumatic actuator. (*Courtesy of American Lewa, Inc., Holliston, MA*).

Pump Heads

Pump heads preferably are made from stainless and acid-resisting steels, nickel alloys, titanium, nickel, ceramics, glass, and various types of plastic materials.

Pump Valves

The majority of all metering pumps operate with self-acting valves, which, depending on conditions, are designed as multiball, cone, or disc-type valves, with or without spring loading. In particular, ball valves are used for small nominal widths because they offer optimal properties and features with regard to choice of material, manufacture, and resistance to dirt and wear (Figure 4.32). Pump valves of larger nominal width and in particular for higher stroke frequencies must have light, movable parts for positive operation and, therefore, are mostly designed to be easily spring-loaded ball or disc valves (Figure 4.33).

Spring loading is used to offset unfavorable geodetic conditions and to obtain by force undelayed valve movement where high stroke frequencies or viscous media are involved.

For the metering of suspensions and contaminated media, special flow favorable designs of valves are used in which clogging or jamming is avoided

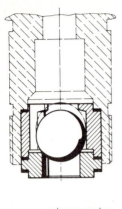
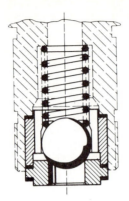

Figure 4.32 Ball valves with or without spring loading. (*Courtesy of American Lewa, Inc., Holliston, MA*).

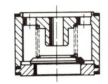

Figure 4.33 Cone and plate valves. (*Courtesy of American Lewa, Inc., Holliston, MA*).

(see Figure 4.16). The following theorem applies: metering accuracy decreases with deteriorating guiding of the ball. On the other hand, however, the operating safety increases. For very unfavorable conditions, such as high viscosity, nonhomogeneous media, and so on, nonself-acting but separately energy-controlled valves (mechanical, electrical, pneumatic) are sometimes used. Piston pumps can be used for the great majority of applications because small leakages are permissible.

Pistons and Piston Glands

Pump pistons are sealed by means of differently formed packings or rough-pressed seal rings (Figure 4.34). Compared to rings, packings have an advantage in that their lips are pressed on the piston by the pressure during the pressure stroke. In other words, they seal off without having to be tensioned from the outside. Seal rings must be pretensioned from the outside and retensioned from time to time. They require the very best casing quality features, especially at the extreme end of the stuffing box. Maintenance of plunger glands having seal rings is costlier than with packings, but they are by far not so susceptible to damage. As far as the life of the piston glands is concerned, the decisive factors are sufficient hardness and surface quality of the piston. Pistons of hardened steel with spray-on wear protection as well as pistons of aluminum oxide, ceramics, and carbide metal are in use. Because of the more favorable compression distribution, a two-step design (Figure 4.34 C) is more appropriate for higher pressures. Here the seal for the flushing area is separately put under tension. The design shown in Figure 4.34 D is used to serve the injection-type

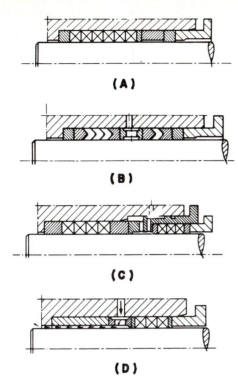

(A)

(B)

(C)

(D)

Figure 4.34 Piston glands: (A) piston gland with seal rings (has flushing capability); (B) piston gland with profiled seal rings and flushing lantern; (C) packing with separately tensionable medium and flushing liquid seal; and (D) packing with volumetric flushing possibility. (*Courtesy of American Lewa, Inc., Holliston, MA*).

flushing of the stuffing box. A suitable liquid (usually by means of a volumetric pump) is pressed into the pump's working chamber during the suction stroke so that the flushing medium, but not the pumped medium so hard to seal, makes contact with the plunger gland.

DESIGN EXAMPLES

After suitable pump heads have been chosen, it is relatively easy to mount them on, and connect the pistons to, the drive unit (or units). This type of assembly has the advantage of maintaining total separation of the pumped medium from the drive unit—a feature that is absolutely essential when corrosive materials are being handled.

Figures 4.35 through 4.38 represent piston-type pump heads of various

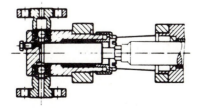

Figure 4.35 Piston head in metal design for average pressures (300 bar maximum). (*Courtesy of American Lewa, Inc., Holliston, MA*).

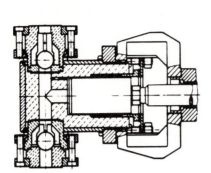

Figure 4.36   Piston head in plastic design (20 bar maximum). (*Courtesy of American Lewa, Inc., Holliston, MA*).

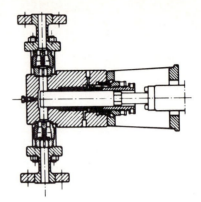

Figure 4.37   High-pressure piston head (700 bar maximum) with double-tensionable piston seal. (*Courtesy of American Lewa, Inc., Holliston, MA*).

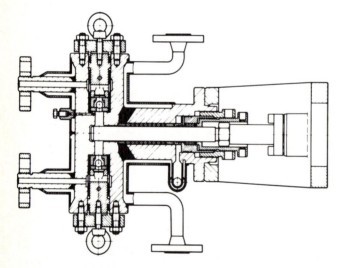

Figure 4.38   Head in cross-head design, completely jacketed for metering units. (*Courtesy of American Lewa, Inc., Holliston, MA*).

designs. Pump heads for high-pressure application (for instance, 3,500 bar) require that special strength problems be overcome.

DIAPHRAGM AND BELLOWS PUMP HEADS

Where critical media (toxic and dangerous ones plus those that are subject to wear) are involved, grandless metering pumps having a diaphragm or bellows as a displacer are used more and more because they offer the distinct advantages of nearly totally maintenance-free operation and of excellent sealing features. Two widely used, proved-in-service systems are:

1. *Diaphragms pumps with plastic diaphragms.* These allow larger diaphragm deflection and, thus, high specific displacement. As to pressure and temperature, these pumps are limited to approximately 350 bar and 120°C, respectively.

2. *Diaphragm pumps with metal diaphragms.* These allow smaller diaphragm deflection rates, but are suitable for pressures up to 4000 bar and also for higher temperatures.

Hydraulic Diaphragm Head with Plastic Diaphragm

The principle is illustrated in Figure 4.39. Pump working chamber I is separated from the hydraulic system 3a by a specially shaped plastic diaphragm 2 (usually PTFE). The diaphragm permits a considerable rate of deflection and is clamped around its entire circumference. Piston 4 sealed with gland 5 displaces the hydraulic fluid (usually hydraulic oil) and, thus, effects an elastic deformation of the diaphragm. The operation to fill the hydraulic chamber is controlled by position scanning of the diaphragm via the snifter gate 9, which connects to hydraulic chambers 3b and 3a in the appropriate position of the diaphragm. When the diaphragm reaches the rear contact area, a replenishing valve automatically refills the hydraulic chamber.

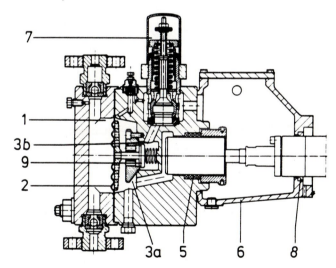

Figure 4.39 Diaphragm head with PTFE diaphragm. (*Courtesy of American Lewa, Inc., Holliston, MA*).

As soon as the diaphragm leaves the rear contact, the gate closes and keeps any negative pressure effect away from the replenishing valve.

This is why the system is also safe from excessive pressures without having a front diaphragm contact face. Safety from excessive pressure is provided by pressure relief valve 7. Automatic bleeder valves carry out a permanent venting of the hydraulic system. When the diaphragm is oscillating freely, the pump's working chamber is totally insensitive to dirt. Therefore, suspensions, slurries, and other viscous liquids can be pumped without hesitation. The system is maintenance free, automatically secured, and also sealed against outside influences. Changing the diaphragm is a simple operation. Because of their thickness and the small notch sensitivity of plastic, the diaphragms are insensitive to contamination. The excess pressure safeguard of the system on the hydraulic side automatically prevents damage in cases of accident. The advantages provided by the safety devices not in contact with the medium are self-evident.

Diaphragm Head with Metallic Diaphragm

As per Figure 4.40, the hydraulic system operates on the basis of leakage replenishment from reservoir o through replenishing valve n. Characteristic is the small deflection rate of metallic diaphragms and their amplitude limitation due to the concave or sometimes flat walls of the pump chamber. Safety valve f in the hydraulic system provides protection against excess pressure as well as compensation when the hydraulic space is overloaded. This permits the system to operate reliably and to be independent of excess-pressure or low-pressure conditions. The diaphragm, a thin, cold-rolled sheet (for instance, made from austenitic steel or titanium), is loaded, almost up to the yield point, between lapped surfaces by the use of expansion bolts. The hydraulic drive principle as per Figure 4.40 and the fact that the diaphragm at its extreme positions will

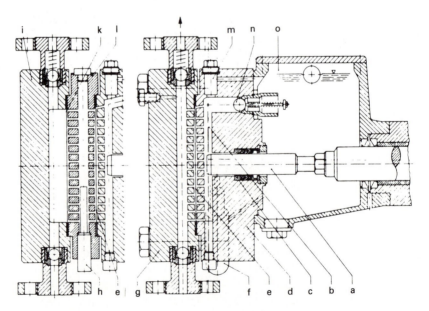

Figure 4.40 Diaphragm head with hydraulic displacer drive and metallic diaphragm. (*Courtesy of American Lewa, Inc., Holliston, MA*).

contact the contour (or backup) plates require clean liquids free from dirt and contamination on both process and hydraulic sides. The diaphragm is sensitive to notching to a degree that indentations made by particles will reduce its life. If particles cannot be avoided by the use of dirt traps or if they are, as in suspensions, even scheduled to be part of the medium, the twin-diaphragm model, as shown in the left-hand side of Figure 4.40, assures a freely oscillating unit. This reduces the system's susceptibility to dirt. Diaphragm pumps with metallic diaphragms have no limit as to pressure. The hydraulic medium sets the temperature limits. However, the application of diaphragm pumps having metallic diaphragms should remain restricted to that area for which plastic diaphragms cannot be used. Metallic diaphragms are much more sensitive than are plastic units. They require more care during installation, in particular.

Diaphragm Head with Mechanically Driven Plastic Diaphragm

Diaphragm pumps with mechanical diaphragm drive (see Figure 4.26) are widely used for lower-pressure applications, smaller flow rates and less stringent requirements with regard to metering accuracy. Their characteristics are not strictly linear and are noticeably influenced by pressure and temperature.

Diaphragm pumps are built for performance rates beyond 200 kW. They meet with great success in laboratories, pilot plants, and research operations because they can be put into use on a universal basis, do not require any maintenance, and because the liquid to be handled is sealed off hermetically.

For the purpose of metering very small flow rates, diaphragm metering pumps offer advantages because contaminations derived from the piston gland (minute particles) cannot occur. This is why small high-pressure diaphragm metering pumps are particularly applied in metering operations in the field of liquid chromatography, where requirements call for flow rates of 100–200 cm³/hr and pressure rates of up to 400 bar. Diaphragm pumps are built for pressures of up to 3,000 bar. They have the great advantage of the piston gland being located at the problem-free hydraulic side.

Bellows Head with Mechanically Driven Bellows

In their application, bellows pumps are almost exclusively of glass/PTFE design. Due to the parts made of glass (Figure 4.41), the pressure is limited to only a maximum of 5 bar. The characteristics of the metered flow are sharply linear and somewhat pressure dependent.

Life of Diaphragms and Rupture Control

The life of the diaphragms is an essential criterion for the application of diaphragm pumps. The life of the various diaphragm designs differs greatly as follows:

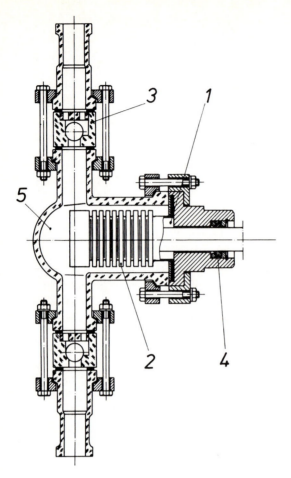

Figure 4.41 Glass-type bellows head. (*Courtesy of American Lewa, Inc., Holliston, MA*).

**HYDRAULICALLY DRIVEN
DIAPHRAGM**

PTFE   up to 15,000 hours
Metallic   up to 5,000 hours

MECHANICALLY DRIVEN DIAPHRAGM OR BELLOWS

Elastomer diaphragm   up to 3,000 hours
PTFE bellows   up to 10,000 hours

Several methods are used to signal rupture of a bellows or diaphragm. In mechanical bellows or diaphragm drives, level switches (Figure 4.42) are used to signal any ruptures.

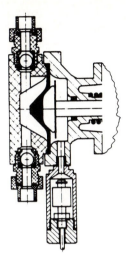

Figure 4.42 A rupture indicator switch. (*Courtesy of American Lewa, Inc., Holliston, MA*).

Where hydraulic diaphragm pumps are involved, the use of the sandwich-type diaphragm is predominant to signal ruptures. In this method, two or three hydraulically coupled diaphragms, with a connection to sensor 4 and check valve 2, are used to determine selectively the pressure rise during rupture of one of the sandwiched diaphragms (Figure 4.43).

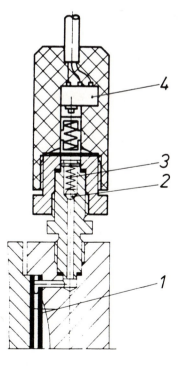

Figure 4.43 Sandwich-type diaphragm with pressure switch system. (*Courtesy of American Lewa, Inc., Holliston, MA*).

Much rarer is the application of the conductive or capacitive method of signaling diaphragm rupture by electrode through the twin-diaphragm arrangement, as per Figure 4.40. This system requires considerable experience and care in operational surveillance and is now restricted to special cases where metallic diaphragms and high temperatures are involved.

Special Types of Diaphragm Pumps

Remote-Head designs have gained importance in nuclear technology. Here, the drive is located outside of the radioactively dangerous area. It is only the liquid part that is exposed to those dangerous conditions and connected to the drive via a hydraulic actuation line. Diaphragm pumps with operable cover for cleaning purposes in hygienic operations are also worthy of notice.

CONVEYING PUMPS

In this class of pumps, the horizontal triplex pump (three-plunger pump) is predominant, but the standup, multicylinder design (for instance, quintuplex) as well as the horizontally opposed arrangement and the double-acting pumps are in use.

Drive Unit and Motor

Figure 4.44 shows a typical drive unit in triple-crank construction for external gear mounting. For larger outputs, the forced-feed lubrication method is used throughout. Stroke-frequency control is achieved by means of hydraulic torque converter, variable-speed gear unit, or DC motor. For lower performance rates, the gearing is frequently dispensed with in favor of belt drives. There are also integrated drives in the box of the direct-thrust, crank-type gear units. Where heavy-duty pumps are involved, it is preferable to use gearless coupling methods for motor and drive units, with speeds of 750–1,500 rpm prevailing. This also achieves an optimum of noise abatement. For less frequent applications, such as large, high-pressure diaphragm pumps and high-pressure circulation, the horizontally opposed arrangement of drive units (Figure 4.45) generally is used.

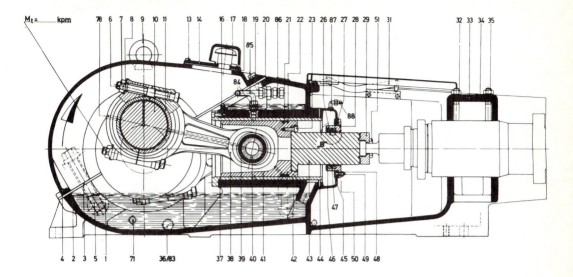

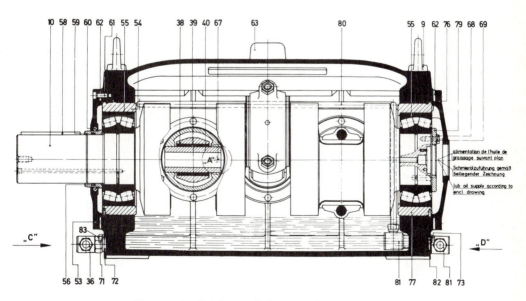

Figure 4.44   Triplex crank-drive unit for heavy-duty application.

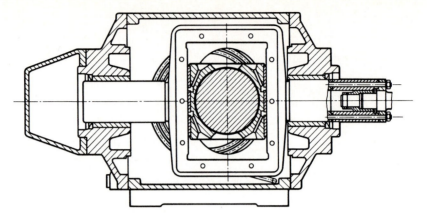

Figure 4.45   A horizontally opposed drive unit arrangement. When two pump heads are mounted, the right-hand cover is removed. (*Courtesy of American Lewa Inc., Natick, MA*)

Pump Heads

The compact cylinder block arrangement is preferred for simpler applications. Where more complex pumping tasks in chemical process technology are to be handled, separation of the individual cylinders becomes a necessity because this is the only way to be sure that each pumping element is wholly suited to the characteristics of the various media and to the processing conditions for each of them. Each of the individual cylinders (separated into valve crosshead and stuffing-box unit) is mounted by means of the drive-unit bracket and a transverse yoke. Suction and discharge piping are manifolded by the headers (Figure 4.46). Because the area around the bore holes (valve and piston axis) is subject to extreme stress especially under high, or even extremely high, pressure over an extended operating period the item 12 (Figure 4.46) may develop stress corrosion cracks. Therefore, it is considered to be a throwaway part, and the pump is designed to permit easy replacement. Pump head designs as shown in Figure 4.47 are customary for extreme high-pressure metering pumps. The part (T piece) being subjected to the highest stress will be designed for easy replaceability. For pressures of more than 2,000 bar, it is usually designed for autofrettage (radial expansion).

For specific pump head designs, the three-plunger construction otherwise customary for conveying pumps is not used in most cases because the center distance of the plungers is not sufficient to accommodate the pump heads. This is why large high-pressure, diaphragm-type process pumps arc frequently when designed for horizontally opposed arrangements.

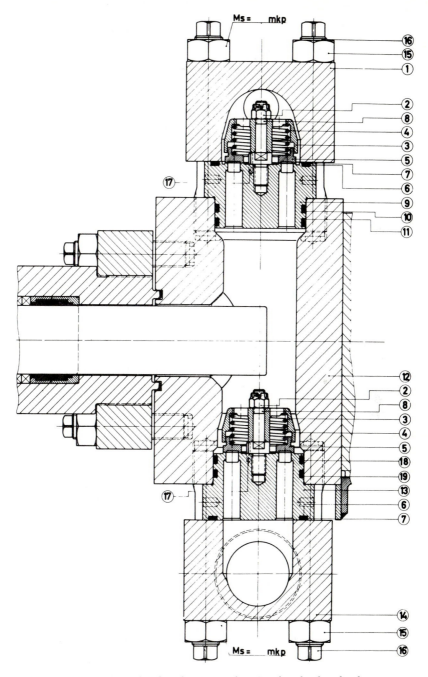

Figure 4.46 A piston head with separate housing for gland and valve cross head. (*Courtesy of American Lewa Inc., Natick, MA*)

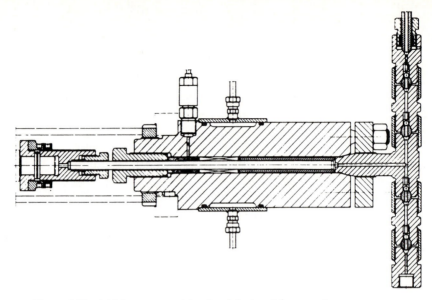

Figure 4.47   A high-pressure piston head designed for operating pressures up to 3,000 bar. (*Courtesy of American Lewa Inc., Natick, MA*)

For handling slurries, slow-speed and double-acting, low-pressure, tube-type diaphragm pumps (Figure 4.48) are often used. Their advantage is that tube is well flushed throughout so that sedimentation does not occur. However, as many applications prove, this objective can be achieved by the use of dia-

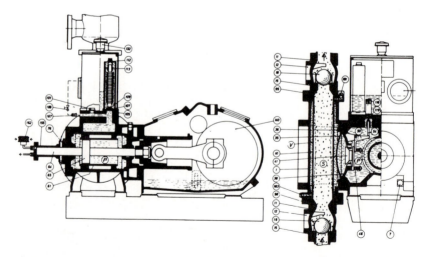

Figure 4.48   Double-acting (twin-cylinder), low-pressure diaphragm tube pump. (*Courtesy of American Lewa Inc., Natick, MA*)

phragm pumps as per Figure 4.39, provided the flow channels are generously rated.

Piston Glands and Valves

Besides the basic design features described earlier, additional features sometimes are required. For example, in larger high-pressure three-plunger pumps having stroke lengths of approximately 100–200 mm, especially for difficult media such as abrasive ones, piston glands with injection flushing (Figure 4.49, items 15/33/23/24) and lubrication or flushing (see Figure 4.34 D) are utilized. There are also recent developments such as split glands for an oscillating piston movement. Increasingly, they find application in extra high-speed pumps. Circular plate valves are used for larger outputs (Figure 4.46) to increase the throughflow cross section. The ball valve is still the predominant unit used for slow-speed slurry diaphragm pumps (Figure 4.48).

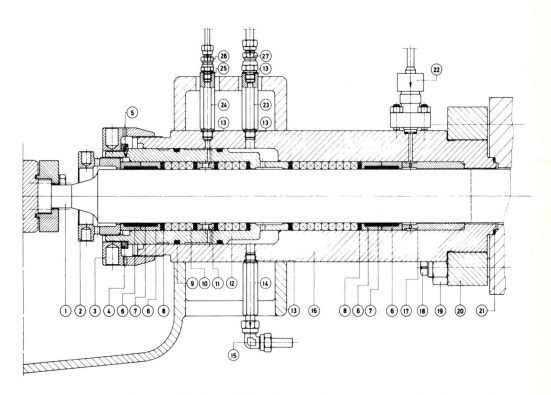

Figure 4.49 A stuffing box with injection-type flushing. (*Courtesy of American Lewa Inc., Natick, MA*)

INSTALLATION LAYOUT FOR OSCILLATING
DISPLACEMENT PUMPS

Because of the pulsating flow, the layout of the piping for oscillating displacement pumps is much more difficult than for centrifugal pumps. In addition to time-variable flow speeds and because of the oscillating displacement motion, accelerations and delays of the liquid masses emerge which must be taken into account.

The objective of all calculations for installation is the avoidance of detrimental conditions that disturb or prevent proper operation of the facility. Above all, harmful cavitation must be eliminated. The consequence of cavitation is a drop of the flow rate, erratic operation of the pump, knocking of the pump valves, pressure thrusts, and considerable wear and tear. Other types of disturbances of the pump function concern overloading and such conditions as are known under the term *overpumping*, where the pressure variance in the pump does not suffice for the planned control of the self-acting valves. Under simplifying conditions, all these problems now can be controlled on an analytical basis and solved either manually or by the use of digital computers.

5

System Analysis for Pumping Equipment

CHARACTERISTICS OF CENTRIFUGAL PUMPS

The selection of a centrifugal pump for an energy-efficient pumping system requires an understanding of the principles of mechanics and physics that can affect the pumping system and the pumped liquid. The efficiency of a centrifugal pump is also dependent on the behavior of the liquids being pumped. The principles of centrifugal pump operation that govern head and flow must be understood clearly before pump performance can be evaluated accurately.

The behavior of a fluid depends on its state, that is, liquid or gas. Liquids and gases offer little resistance to changes in form. Typically, fluids such as water and air have no permanent shape and readily flow to take the shape of the containing enclosure when even a slight shear loading is imposed. Factors affecting behavior of fluids include:

- viscosity
- specific gravity
- vapor pressure

Viscosity is the resistance of a fluid to shear motion—its internal friction. The molecules of a liquid have an attraction for each other. They resist move-

223

ment and repositioning relative to each other. This resistance to flow is expressed as the viscosity of the liquid. Dynamic viscosity also can be defined as the ratio of shearing stress to the rate of deformation. The viscosity of a liquid varies directly with temperature; therefore, viscosity is always stated at a specific temperature.

Liquid viscosity is very important in analyzing the movement of liquids through pumps, piping, and valves. A change in viscosity alters liquid-handling characteristics in a system; more or less energy then may be required to perform the same amount of work. In a centrifugal pump, an increase in viscosity reduces the pressure energy (head) produced while increasing the rate of energy input. In piping systems, a liquid with a high viscosity has a high-energy gradient against which a pump must work, and more power is required than for pumping low-viscosity liquids.

Specific gravity is the ratio of the density of one substance to that of a reference substance at a specified temperature. Water at 4°C is used as the reference for solids and liquids. Air generally is used as the reference for gases. The specific gravity of a liquid affects the input energy requirements, or brake horsepower (bhp), of centrifugal pumps. Brake horsepower varies directly with the specific gravity of the liquid pumped. For example, water at 4°C has a specific gravity of 1.0 Table 5.1 includes specific gravity values for water at selected temperatures.

TABLE 5.1   SPECIFIC GRAVITY VALUES FOR WATER
AT SELECTED TEMPERATURES

Water temperature		Water specific gravity		Effect of specific gravity on energy input constant flow
(°C)	(°F)		(kW)	(bhp)
4	39.2	1.0	74.6	100
60	140	0.983	73.3	98.3
100	212	0.958	71.5	95.8
125	257	0.939	70.0	93.9
150	302	0.917	68.4	91.7

Specific gravity affects the liquid mass but not the head developed by a centrifugal pump, as shown in Table 5.2. The specific gravity also affects the energy required to move the liquid and, therefore, must be used in determining the pump's horsepower requirement.

Vapor pressure is the pressure at which a pure liquid can exist in equilibrium with its vapor at a specified temperature. Fluids at temperatures greater than their specified (critical) temperatures will exist as single-phase liquids (vapors), with no distinction between gas and liquid phases. At less than the critical temperature, two fluid phases can coexist; the denser fluid phase exists as a liquid and the less dense phase as a vapor. At a specific temperature, the

TABLE 5.2 EFFECT OF SPECIFIC
GRAVITY ON PUMP HEAD

Specific gravity	Linear head		Gauge pressure	
	(ft)	(m)	(psi)	(bar)
0.75	246	75	79.9	5.51
1.00	246	75	106.5	7.34
1.20	246	75	127.8	8.81

liquid phase is stable at pressures exceeding the vapor pressure and the gas phase is stable at pressures less than the vapor pressure.

For a fluid to exist in a liquid state, its surface pressure must be equal to or greater than the vapor pressure at the prevailing temperature. For example, water has a vapor pressure of 0.1781 psia at 10°C and 14.69 psia at 100°C. The vapor pressure of a volatile liquid (such as ether, alcohol, or propane) is considerably higher than that of water at the same temperature; consequently, much higher pressures must be applied to maintain volatile materials in their liquid states. The surface pressure of a liquid must be greater than its vapor pressure for satisfactory operation of a centrifugal pump.

Centrifugal pump characteristics remain constant unless an outside influence causes a change in operating conditions. Three conditions can alter pump performance:

1. changes in impeller or casing geometry
2. increased internal pumping losses caused by wear;
3. variation of liquid properties

For example, if the impeller passages become impacted with debris, the head-flow relationship will be reduced. Similarly, performance will decline if mechanical wear increases the clearance between the rotating and stationary parts of the pump.

Except for specially designed pumps, most centrifugal pumps can handle liquids containing approximately 3 percent to 4 percent of gas (by volume) without an adverse effect on performance. An excess of gas will reduce the flow of liquid through the pump and, under certain conditions, flow will cease, setting up a condition that may damage the pump.

The function of a pump is to move liquids by imparting pressure energy (head) to the liquid. The ability of the pump to perform its function is based on the Bernoulli theorem, which states that energy cannot be created or destroyed, but can only be converted in form. A pump converts mechanical energy into pressure energy. Part of the converted energy is required to overcome inertia and move the liquid; most of the remaining energy is stored in the liquid as elevated pressure, which can be used to perform useful work outside the pump. A centrifugal pump is basically a velocity machine designed around its impeller. The interaction between the impeller and its casing produces the characteristics of head or pressure energy.

To review, the developed head is a function of the difference in velocity between the impeller vane diameter at entrance and the impeller vane diameter at exit. The expression of theoretical head can be related to the law of a falling body: $H = u^2/2g$ where H = height or head (ft), u = velocity of moving body (fps), and g = acceleration of gravity (32.2. ft/s^2).

When the height of fall is known (for example, $H = 100$ ft), the terminal velocity can be determined (in this case $u = 80.3$ fps). Conversely, if the direction of motion is reversed, a liquid exiting through an impeller vane tip at a velocity of 80.3 fps reaches a velocity of 0 fps at 100 ft above the impeller tip (refer to Figure 5.1).

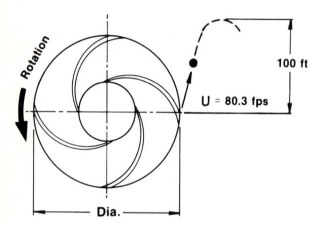

Figure 5.1 Example to determine theoretical impeller diameter.

When the developed head required is known, the theoretical impeller diameter for any pump at any rotational speed can be determined from the equation for peripheral velocity of a round rotating body:

$$D = 229.2 \ U/N \tag{5.1}$$

where D = unknown impeller diameter (in.)

u = velocity (derived from $u = \sqrt{2gH}$) (fps)

N = rotational speed of the pump (rpm)

If the pump head is 100 ft and rotational speed is 1,750 rpm, the formula indicates that a 10.52-in. theoretical diameter impeller is required.

The function of a centrifugal pump is to transport liquids by transferring and converting mechanical energy (foot-pounds of torque) from a rotating impeller into pressure energy (head). By the transfer of this energy to a liquid the liquid can perform work, move through pipes and fittings, rise to a higher elevation, or increase the pressure level. Because a centrifugal pump is a velocity machine, the amount of mechanical energy per unit of fluid weight transferred to the liquid depends on the peripheral velocity of the impeller, regardless of fluid density. This energy per unit of weight is defined as pump head and is expressed in feet of liquid. If the effect of liquid viscosity is ignored, the head developed by a given pump impeller at a given speed and flow rate remains constant for all liquids.

Head is the vertical height of a column of liquid. The pressure this liquid column exerts on the base surface depends on the specific gravity of the liquid. A 10-ft column of liquid with a specific gravity of 0.5 exerts only 2.16 psi; a 10-ft column of liquid with a specific gravity of 1.0 exerts 4.33 psi. The formula for converting feet of head to pressure is

$$H = 2.31P/S_g \tag{5.2}$$

where P = pressure (psi)

S_g = specific gravity

The proper selection of a centrifugal pump requires that pressure be converted to feet of head. If not, a pump that is incapable of imparting the required energy to a liquid may be installed before the inadequacy is discovered. The heads required to produce a pressure of 100 psi for three liquids of different specific gravities are given in Table 5.3.

TABLE 5.3   HEADS REQUIRED
TO PRODUCE 100 psi PRESSURE FOR
DIFFERENT LIQUIDS

Specific gravity	Head equivalent of 100 psi for liquids of different specific gravity
	Head (ft of liquid)
0.75	308
1.0	231
1.2	193

From this analysis a single-stage pump selected for pumping a liquid with a specific gravity of 1.2 will have the lowest peripheral velocity at the impeller tip, while the pump for a liquid with 0.75 specific gravity will have the highest peripheral velocity.

The volumetric flow rate is a function of the peripheral velocity of the impeller and the cross-sectional areas in the impeller and its casing; the larger the passage area, the greater the flow rate. Consequently, the physical size of the pump increases with higher flow requirements for a given operating speed. The liquid flow rate is directly proportional to the area of the pump passages and can be expressed as

$$Q = UA/0.321 \tag{5.3}$$

where Q = flow (gpm)

u = velocity (fps)

A = (in.²)

For example, a centrifugal pump having a 15-fps liquid velocity at the discharge flange can pump approximately 330 gpm through a 3-in. diameter opening. If the flow requirements are increased to approximately 1,300 gpm and the velocity remains at 15 fps, a 6-in. diameter opening will be required. Liquid flow is also directly proportional to the rotational speed that produces the velocity. The four preceding equations establish two relationships:

1. Head is directly proportional to the square of the liquid velocity.
2. Flow is directly proportional to the peripheral velocity of the impeller.

Cavitation in a centrifugal pump can be a serious problem. Liquid pressure is reduced as the liquid flows from the inlet of the pump to the entrance to the impeller vanes. If this pressure drop reduces the absolute pressure on the liquid to a value equal to or less than its vapor pressure, the liquid will change to a gas and form vapor bubbles. The vapor bubbles will collapse when the fluid enters the high-pressure zones of the impeller passages.

This collapse is called *cavitation* and results in a concentrated transfer of energy, which creates local forces. These high-energy forces can destroy metal surfaces; very brittle materials are subject to the greatest damage. In addition to causing severe mechanical damage, cavitation also causes a loss of head and reduces pump efficiency. Cavitation will also produce noise.

If cavitation is to be prevented, a centrifugal pump must be provided with liquid under an absolute pressure that exceeds the combined vapor pressure and friction loss of the liquid between the inlet of the pump and the entrance to the impeller vanes.

NET POSITIVE SUCTION HEAD (NPSH)

NPSHA (net positive suction head available) is the absolute pressure of the liquid at the inlet of the pump. NPSHA as a function of the elevation, temperature, and pressure of the liquid is expressed in units of absolute pressure (psia). Any variation of these three liquid characteristics will change the NPSHA. An accurate determination of NPSHA is critical for any centrifugal pump application.

The net positive suction head required (NPSHR) by a specific centrifugal pump remains unchanged for a given head, flow, rotational speed, and impeller diameter, but changes with wear and liquids.

Specific speed is a correlation of pump flow, head, and speed at optimum efficiency. It classifies pump impellers with respect to their geometric similarity. Specific speed is usually expressed as

$$N_s = N\sqrt{Q}/H^{3/4} \qquad (5.4)$$

where N_s = pump specific speed

Q = flow at optimum efficiency (gpm)

The specific speed of an impeller is defined as the revolutions per minute at which a geometrically similar impeller would run if it were of a size that would discharge 1 gpm against a head of 1 foot. Specific speed is indicative of the impeller's shape and characteristics.

Centrifugal pumps are divided into three classes:

1. radial flow
2. mixed flow
3. axial flow

There is a continuous change from the radial-flow impeller (which develops head principally by the action of centrifugal force) to the axial-flow impeller (which develops most of its head by the propelling or lifting action of the vanes on the liquid.) Typically, centrifugal pumps also can be categorized by physical characteristics relating to the specific speed range of the design (refer to Figure 5.2). Once the values for head and capacity become established for a given application, the pump's specific speed range can be determined and specified to ensure the selection of a pump with optimal operating deficiency.

GENERAL PUMP CHARACTERISTICS

Differing pump hydraulic characteristics will result in one pump being better suited for a given application than another. This section examines head-flow curves, pump performance curves, variations in curve shape as a function of specific speed, and pump affinity laws.

Head-Flow Curve

The head-flow curve of an ideal pump with ideal (frictionless) fluid is a straight line whose slope from zero flow to maximum flow varies with the impeller exit vane angle. For example, a given impeller and casing combination with an impeller vane exit angle of 25° will have a greater maximum flow than a similar impeller with a 15° vane exit angle (Figure 5.3). However, the actual head-flow characteristic of a centrifugal pump is not an ideal straight line. Its shape is altered by friction, leakage, and shock losses that occur in impeller and casing passages.

Single or multistage, and single (illustrated) and double suction, Francis-type impellers, operating in volute or diffuser type casings, produce medium to high capacity at medium to low speeds. Specific speed range is 2000 to 4000.

Single or multistage, double (illustrated) and single suction, volute and diffuser design centrifugal pumps deliver medium capacity and medium heads. Specific speed range is 1000 to 2000.

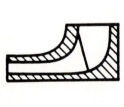

Single suction, horizontal or vertical centrifugal pumps with narrow port impellers have low capacities and deliver high heads. Specific speed range is 500 to 1000.

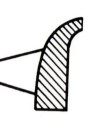

Single or multistage pumps with mixed flow and propeller-type impellers have very high capacities and deliver very low heads. Specific speed range is 6000 to 10,000.

Single or multistage single suction centrifugal pumps, in volute or diffuser-type casings with mixed flow impellers, deliver high capacity at low head. Specific speed range is 4000 to 6000.

Figure 5.2 Centrifugal pump configurations and specific speed ranges.

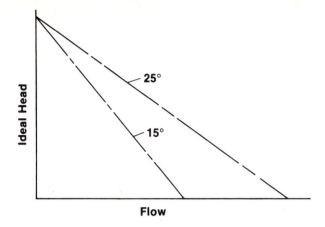

Figure 5.3 The ideal head curve.

Friction losses in a centrifugal pump are proportional to the surface roughness and the wetted areas of the impeller and casing. Leakage losses result from the flow of liquid between the clearance of rotating and stationary parts, such as impeller to case wear ring clearances. Shock losses occur as the liquid enters the impeller entrance vanes and as the liquid flows from the impeller into the casing. These internal losses characteristically reduce pump performance from the ideal to the actual total head-flow curve shown in Figure 5.4. The flow at which the sum of all these losses is the least determines the point of maximum efficiency.

Mechanical losses in bearings, packings, and mechanical seals further reduce pump efficiency. Although mechanical losses may be calculated, the results are generally not accurate; actual pump performance can be determined only by testing.

A centrifugal pump is designed around its impeller. The function of the casing is to collect the liquid leaving the impeller and convert its kinetic energy into pressure energy. A case is designed in conjunction with an impeller to

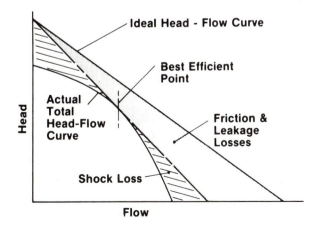

Figure 5.4 Deviation from the ideal head-flow curve due to internal losses.

achieve the most efficient match for conversion of liquid velocity to pressure energy. The interaction between the impeller and its casing for a given pump determines the pump's unique performance characteristics.

Performance Curves

Pump performance curves depict the total head developed by the pump, the brake horsepower required to drive it, the derived efficiency and the net positive suction head required to drive it, and the derived efficiency and the net positive suction head required over a range of flows at a constant speed. Pump performance can be shown as a single line curve depicting one impeller diameter (Figure 5.5) or as multiple curves for the performance of several impeller diameters in one casing (Figure 5.6).

The performance characteristics of pumps are classified by discharge specific speed and have the approximate curve shapes shown in Figure 5.7. Ac-

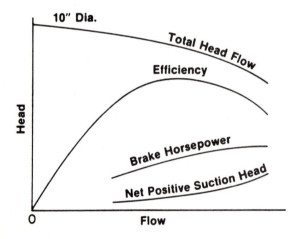

Figure 5.5 Pump performance curve for a single impeller diameter.

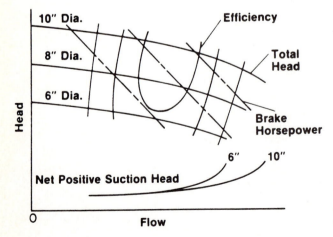

Figure 5.6 Pump performance curves for several impeller diameters.

BEP is the reference to the flow point at which the best pump efficiency occurs.

All specific speed values are in English units.

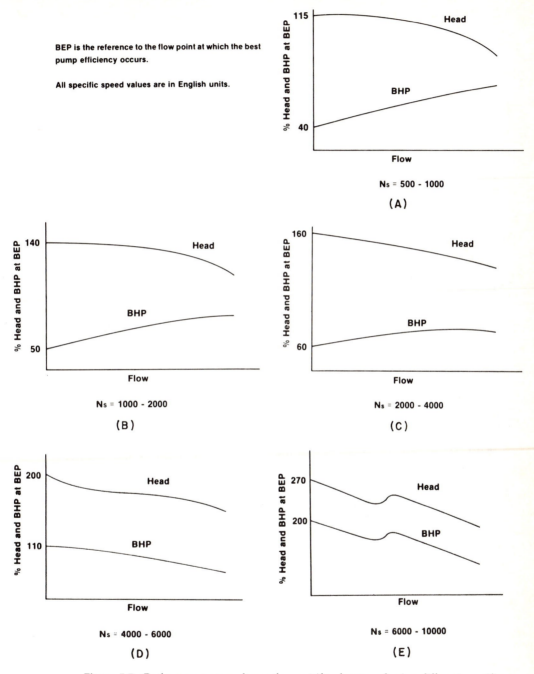

Figure 5.7 Performance curve shapes for centrifugal pumps having different specific speeds.

- flood irrigation
- high service
- low service
- municipal booster
- raw water
- river intake
- sewage ejector
- soiler feed
- sprinkler irrigation
- storm water

Items that affect the cost and operation of piping systems include noise, water hammer, pipe aging, and friction loss.

Flow noise, or *hiss*, contains all sound frequencies at an evenly distributed sound level, called *white noise*. This noise is generated by turbulent flow through system pipes and increases with velocity, or it is generated by cavitation in the system. Flow noise occurs in pumps, valves, elbows, tees, or wherever flow changes direction or velocity.

Conduction of noise proceeds through the liquid, as well as through the pipe material, and is usually controlled by flexible fittings, sound insulating pipe supports, and pipe insulation.

Pipe resonance can result in objectional sound in a piping system. Resonance is a single tone sound magnified and emitted by a length of piping. It occurs when the natural frequency of a length of piping is matched by the frequency of some regular energy source such as the vanes in an impeller passing the volute tongue (cutwater) as shown in Figure 5.17. An illustration of this type of resonance is shown in Figure 5.18. Other regular energy sources, such as the rotating elements in a bearing or the power supply frequency, also can cause pipe resonance. Pipe resonance can be corrected by increasing or decreasing the resonating length or weight of pipe under the direction of an acoustical consultant.

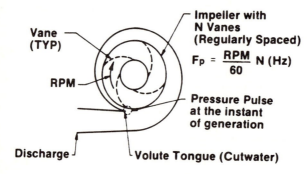

Figure 5.17 Noise problems in pumps generated by vanes in an impeller passing the volute tongue.

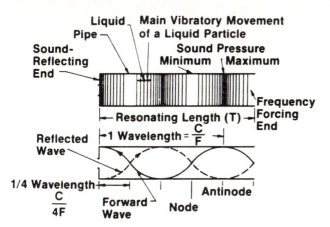

Figure 5.18 Noise generated by pipe resonance.

Water hammer is the result of a strong pressure wave in a liquid caused by an abrupt change in flow rate. As an illustration, for the maximum possible instantaneous head increase above the normal head due to a water-hammer pressure wave, the following expression is applicable:

$$H_{wh} = \frac{Cu}{g} \tag{5.11}$$

where C = velocity of sound in the liquid (ft/s), e.g., C for 15.6°C water is 4,820 ft/s

u = normal velocity in the conduit before closing valve (ft/s)

g = acceleration due to gravity (32.2 ft/s2).

Equation 5.11 is known as Joukowsky's law. Assuming u = 15 ft/s, H_{wh} for 15.6°C water is 2,245 ft, or a pressure surge of 972 psi. This is the maximum possible pressure rise by instantaneous closing of the valve and may be more than the system can withstand.

Pipe Deterioration

Some deterioration of the inside surfaces of a piping system normally occurs with age. This deterioration may result from corrosion, deposits, or a combination of both, which impede flow by increasing the relative roughness ratio, ε/D, where ε = height of protrusion on a side in feet or inches, and, D = inside diameter of pipe in feet or inches. Table 5.5 contains representative values for the height of the internal surface protrusions.

For old material, the values listed may be increased by 100 or more times due to corrosion or deposits. Past experience is generally the most accurate guide to determine the extent of pipe deterioration caused by aging.

TABLE 5.5   REPRESENTATIVE VALUES FOR THE
HEIGHT OF INTERNAL SURFACE PROTRUSIONS IN
PIPES

Material (new)	ε (ft)	ε (in.)
Copper or Brass Tubing	5×10^{-6}	4.2×10^{-7}
Steel Pipe	1.5×10^{-4}	1.3×10^{-5}
Galvanized Steel	5×10^{-4}	4.2×10^{-5}
Concrete	4×10^{-3}	3.3×10^{-4}

Pipe Friction

Pipe friction is resistance to flow and results in a loss of head, which is expressed as friction head H_f, in feet of liquid. Methods of calculating friction head can be found in any standard textbook on fluid mechanics and only supplementary notes are given here. First, proper viscosity conversion is sometimes confusing in evaluating friction losses. If absolute viscosity is expressed in centipoise (cp), the following conversions should be used:

$$2.089 \times 10^{-5} \times cp = \text{lb-s/ft}^2$$

which, in turn, can be converted into kinematic viscosity (ft^2/s) by dividing by the mass density (lb-s^2/ft^4).

If kinetic viscosity is expressed in centistokes (cs), use the following conversion:

$$1.076 \times 10^{-5} \times cs = \text{ft}^2/\text{s}$$

If viscosity is expressed in SSU, there is no distinction between the absolute and the kinetic designations. Convert SSU to kinematic viscosity as follows.
For SSU larger than 100:

$$1.076 \times 10^{-3} \times \left(0.0022 \text{ SSU} - \frac{1.35}{\text{SSU}}\right) = \text{ft}^2/\text{s}$$

For SSU equal to 100 or less:

$$1.076 \times 10^{-3} \times \left(0.00226 \text{ SSU} - \frac{1.95}{\text{SSU}}\right) = \text{ft}^2/\text{s}$$

The following examples illustrate these conversions:

Example 5.1

Given $-40°C$ water with kinematic viscosity of 1.58 cs. Convert viscosity into ft^2/s units.

Solution

$$1.076 \times 10^{-5} \times 1.58 \text{ cs} = 1.70 \times 10^{-5} \text{ ft}^2/\text{s}$$

Example 5.2

Given ethylene glycol at 21.1°C with a viscosity of 88.4 SSU. Convert viscosity into ft²/s units.

Solution

$$1.076 \times 10^{-3} \times \left(0.00226 \times 88.4 - \frac{1.95}{88.4}\right) = 1.912 \times 10^{-4} \text{ ft}^2/\text{s}$$

Example 5.3

Given ethyl alcohol at 20°C with specific weight of 49.4 lb/ft³ and absolute viscosity of 1.2 cp. Convert viscosity into ft²/s units.

Solution

$$2.089 \times 10^{-5} \times 1.2 = 2.507 \times 10^{-5} \text{ lb-s/ft}^2$$

$$\text{Mass Density} = \frac{49.4 \text{ lb/ft}^3}{32.2 \text{ ft/s}^2} = 1.5342 \frac{\text{lb-s}^2}{\text{ft}^4}$$

$$\frac{2.507 \times 10 \text{ lb-s/ft}^2}{1.5342 \text{ lb-s}^2/\text{ft}^4} = 1.63 \times 10^{-5} \text{ ft}^2/\text{s}$$

Figure 5.19 provides viscosity classification equivalents and Table 5.6 provides numerical relationships among viscosity classification systems. Once proper viscosities have been determined, friction factors can be obtained.

Appendix B provides conversion charts and numerical data for different viscosity classification systems.

Through many years, simplified variations of the Darcy-Weisbach formula have been devised for limited areas of application. The accuracy of such variations is supported by successful use and allows the user to bypass the diagram interpretation and calculation necessary with the basic formula. The use of specific formula variations outside their known areas of successful application may lead to inaccurate projections of friction head.

This simplified derivation of the Hazen-Williams formula can be used to calculate friction head for liquids having a kinematic viscosity of 1.1 cs. (Water at 60°F has a viscosity of 1.13 cs.)

$$\text{Friction Head, } H_f = 1,045 \left(\frac{GPM}{C}\right)^{1.85}\left(\frac{L}{D^{4.87}}\right) \tag{5.12}$$

Viscosity Classification Equivalents

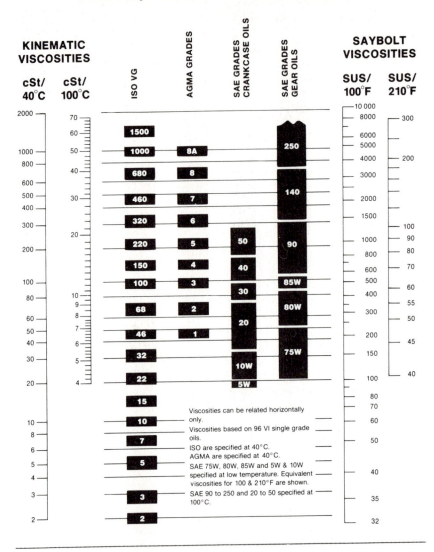

ISO VISCOSITY CLASSIFICATION SYSTEM

Many petroleum products are graded according to the ISO Viscosity Classification System, approved by the International Standards Organization (ISO). Each ISO viscosity grade number corresponds to the mid-point of a viscosity range expressed in centistokes (cSt) at 40°C. For example, a lubricant with an ISO grade of 32 has a viscosity within the range of 28.8-35.2, the midpoint of which is 32.

Rule-of-Thumb: The comparable ISO grade of a given product whose viscosity in SUS at 100°F is known can be determined by using the following conversion formula:

$$\text{SUS @ 100°F} \div 5 \cong \text{cSt @ 40°C}$$

Figure 5.19 Viscosity classification equivalents.

TABLE 5.6 NUMERICAL RELATIONSHIPS AMONG VISCOSITY CLASSIFICATION SYSTEMS

ISO Vis Grade	cSt @ 40°C	SSU @ 100°F (assume 95 VI)	ASTM/BSI Grade	SSU @ 100°F	SAE Crankcase Oils	SSU @ 100°F (assume 110 VI)	AGMA Industrial Gear Oils	SSU @ 100°F	SAE Gear Oils	SSU @ 100°F (assume 90 VI)
2	1.98–2.42	32.8–34.4	32	33–34						
3	2.88–3.52	36.0–38.2	36	36–38						
5	4.14–5.06	40.4–43.5	40	40–43						
7	6.12–7.48	47.2–52.0	50*	46–50	5W	140 Max.				
10	9.00–11.0	57.6–65.3	60	55–62						
15	13.5–16.5	75.8–89.1	75	72–83					75W	102 min.
22	19.8–24.2	105–126	105	97–116	10W	140–202				
32	28.8–35.2	149–182	150	136–165	20	168–366				
46	41.4–50.6	214–262	215	193–235	20W	202–500	1	193–235	80W	251 min.
68	61.2–74.8	317–389	315	284–347	30	366–560	2,2EP	284–347	85W	531 min
100	90.0–110	468–575	465	417–510	40	560–812	3,3EP	417–510		
150	135–165	709–871	700	625–764	50	812–1,272	4,4EP	626–765	90	747–1,844
220	198–242	1,047–1,283	1,000	917–1,121	60**	1,272–1,561	5,5EP	916–1,122		
320	288–352	1,533–1,881	1,500	1,334–1,631	70**	1,561–2,085	6,6EP	1,335–1,632	140	1,844–4,378
460	414–506	2,214–2,719	2,150	1,918–2,344			7EP, 7 Comp.	1,919–2,346		
680	612–748	3,298–4,048	3,150	2,835–3,465			8EP, 8 Comp.	2,837–3,467		
1,000	900–1,100	4,854–5,975	4,650	4,169–5,095			8A Comp.	4,171–5,089	250	4,378 min
1,500	1,350–1,650	7,365–9,080	7,000	6,253–7,643						

*Currently is not considered a standardized grade in the USA.
** Not part of SAE classification.

For Estimating Purposes Only

where GPM = gallons per minute

C = pipe roughness factor ranging from 60 to 160

D = inside diameter of pipe (in.)

L = length of pipe (ft)

The Hazen-Williams formula is generally used for cast iron pipes of 3-inch and larger diameter. As pipes deteriorate, the roughness factor, C decreases. This decrease in C depends on the pipe material, pipe linings, pipe age, and the characteristics of the liquid.

For liquids with viscosities other than 1.1 cs, methods more accurate than the Hazen-Williams formula should be used in determining friction head. For example, friction head computed with the Hazen-Williams formula will increase by 20 percent, at a viscosity of 1.8 cs and decrease by 20 percent at a viscosity of 0.29 cs. In addition, friction loss will occur in fittings such as elbows, tees, and valves. Many handbooks and manuals provide means for determining these losses. Manufacturers' data must be used to determine friction loss for mechanical equipment installed in the system.

SYSTEM HEAD CURVES AND COMPONENTS

A system head curve is the graphic representation of the head required at all flows to satisfy the system function. Regardless of the mechanical configuration, function, or means of control, the system head curve or system head band is used to define total head versus flow for any piping system. System head curve development is a combination of calculations and the designer's "best feel" for the variable conditions.

The three components that make up total system head are static head, design working head, and friction head. These components are defined as follows:

1. *Static head* is the vertical difference in height between the point of entry to the system and the highest point of discharge.
2. *Design working head* is the head that must be available in the system at a specified location to satisfy design requirements.
3. *Friction head* is the head required by the system to overcome the resistance to flow in pipes, valves, fittings, and mechanical equipment.

Total system head at a specified flow rate is the sum of static head, design working head, and friction head. The values of static head and design working head may be zero. Figure 5.20 illustrates a total system head curve made up of the friction head, design working head, and static head.

For most systems, the total system head requirements are best illustrated by a band formed by two total system head curves. This band of system require-

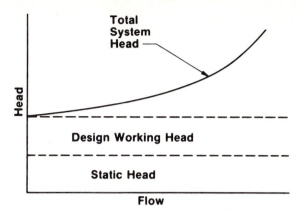

Figure 5.20 Total system head curve components.

ments is a result of variable factors that affect total system head calculations. The following list of variables should be considered in projecting system requirements.

Static head will vary as a result of change in elevation of highest point of discharge of the system.

Design working head is usually treated as a constant component of total system head.

Friction head at any specified flow will vary as a result of:

- method of calculation or source of tabulated data
- change in viscosity resulting from a change in liquid temperature
- deterioration of the piping system
- load distribution
- systems differences between design and "as-built"
- manufacturing tolerances of mechanical equipment
- accumulation of solids in the system

Most systems are not completely defined by a single-line system head curve. Those variables that are applicable determine the band of maximum and minimum system head requirements. The band between maximum and minimum head requirements at any given flow will be significant in some systems and not in others. For all systems, interaction of total head available and total head required must be evaluated to achieve satisfactory operation.

SYSTEM OPERATION AND CONTROL

For a system to operate at design flow, total head available to the system must be equal to the total system head required. For proper evaluation of system operation at flow rates other than design, a comparison of total available head and system required head must be made based on the means selected to control system flow. The reference point used to determine total available head must be the same as that used to determine total system head required. The point of entry to the system, which is also the pump discharge connection, is used in this text as the reference point. Total available head at the point of entry to the system is the sum of static head and pump head.

Static head available is the vertical difference in height between the point of entry to the system and the liquid supply, minus any friction head in the supply pipe and fittings. Static head may be relatively constant. The decrease in static head with increasing flow is the result of friction head in the supply pipe. In some applications static head will vary and is normally illustrated by a band of maximum and minimum values.

When the liquid supply is below the point of entry to the system, as shown in Figure 5.21, static head available is a negative value.

The difference in elevation between the fluid supply and the point of entry to the system must be developed by the pump. Therefore, pump head must include static head and friction head in the supply pipe and fittings.

In systems in which the liquid supply is from a pressurized main, the pressure must be expressed in feet of liquid. The equivalent height of the liquid above the system entry point is static head available. In Figure 5.22, with a 100 psig pressure at point A and a liquid with a specific gravity of 1.0, the static head

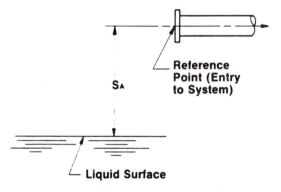

Figure 5.21   A case in which the static head available is a negative value.

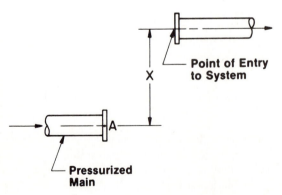

Figure 5.22   Example problem to determine static head available.

available will be 231 feet minus elevation height, X. In return piping systems, in which none of the working fluid is discharged from the system, static head available is zero.

Pump head is the total head developed by a pump. Its value at any given flow rate must be obtained from the pump manufacturer's performance curve.

A curve of total available head versus flow is developed by adding static head and pump head at several values of flow. The total available head curve is slightly steeper than the pump head curve due to the slope of the static head curve. To evaluate the operation of a system, the total system head curve or curves and the total available head curve are plotted on a common graph. The two curves will intersect at maximum flow for that system. As shown in Figure 5.23, system flow greater than 100 percent cannot occur since total system head required exceeds the total available head.

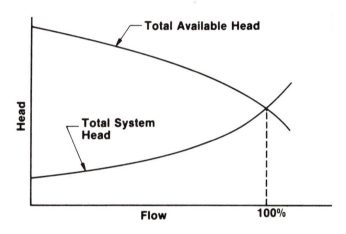

Figure 5.23 The total system head required exceeds the total available head above 100 percent flow.

Consider a system where the total available head must be represented by two curves forming a band as shown in Figure 5.24A. This is a common situation that is a result of maximum and minimum values of static head available. It is now possible for system flow to exceed the design flow of 100 percent, since the pump selection was based on the minimum value of static head available at 100 percent flow.

Figure 5.24B illustrates a system with a band of total system head and a band of total available head. At any given time, system operation can occur anywhere within the area *A-B-D-C-A*. While the pump is selected for proper operation at design flow, operation at flows other than design may result in problems such as noise, cavitation, low pump efficiency, pump or system damage, or poor system function. To avoid these problems, should they occur, some means of system control must be employed.

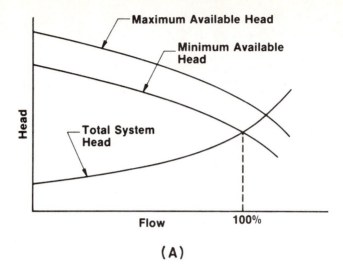

(A)

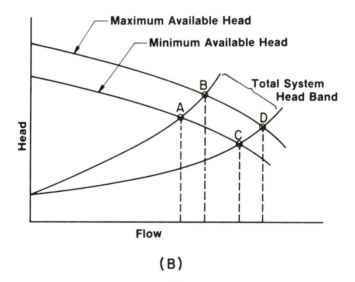

(B)

Figure 5.24 Interpretation of flow-head curves.

Control of system operation is achieved by altering total system head, total available head, or both. Each method has advantages and disadvantages depending on the system, its design function, and the characteristics of total available head. Methods of control should be selected that best satisfy the operation and cost objectives of the system.

Total System Head Alteration

Control of system operation by altering total system head is accomplished with valves that change the friction head component of total system head. Two methods of valve control are bypassing and throttling, each having a different effect on system head.

Valve bypassing usually is accomplished with valves that have three ports: one for entrance of liquid and two that determine the liquid flow path, as shown in Figure 5.25A. Movement of the valve mechanism reduces the cross-sectional area of one port and simultaneously increases the cross-sectional area of the other port. This action causes increased head through one flow path and re-

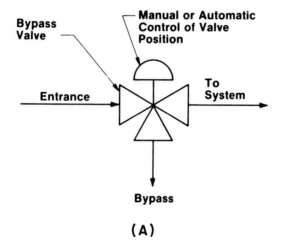

(A)

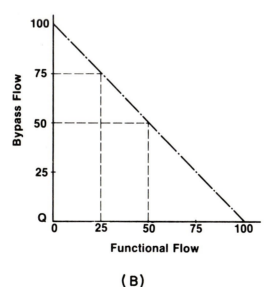

(B)

Figure 5.25 Valve bypassing. Flow remains constant at the entrance of the bypass valve as flow changes in the functional portion of the system (see Sketch B).

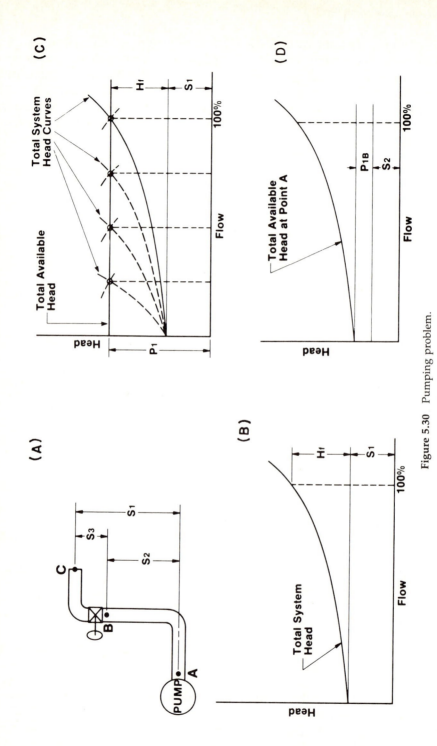

Figure 5.30 Pumping problem.

256

shown in Figure 5.30B and includes static head S_1 and friction head H_f between points A and C, with the valve in a fixed position.

A variable-speed pump with a pressure monitor at the point of entry to the system, A, will produce a constant total available head at A for any system flow. The setting for the pressure monitor is determined from the following formula:

$$P_1 = P_2 + H_f + S \tag{5.13}$$

where

P_1 = total available head, in feet, to be maintained at the pressure monitor

P_2 = design working head, in feet

H_f = full-flow friction head, in feet, between the monitor and the location in the system where P_2 is to be maintained, or from the monitor to the end of the system (A to C) when P_2 is zero

S = static head, in feet, between the monitor and the location in the system where P_2 is to be maintained, or from the monitor to the highest point of discharge; static head, S, is applicable only in systems employing pressure monitors

In Figure 5.30A, P_2 is zero and $P_{1A} = H_r + S_1$.

As system flow is throttled, a series of new system head curves will be produced, as shown in Figure 5.30. System flow will occur at those points where the system head curves intersect the available head curve. At any flow of less than 100 percent, there is more head available than required, and the excess is consumed as friction head across the throttling valve. Now consider the same system with the pressure setting monitor located in Figure 5.30 at point B. The system has not changed, and the total system head required is still as shown in Figure 5.30. The pressure to be maintained at the monitor is determined from the same formula:

$$P_1 = P_2 + H_f + S$$

Design working head, P_2, is zero, and static head is equal to S_3. H_f is the full-flow friction from B to C. The pressure setting is $P_{1B} = H_f + S_2$. The total head that must be available at point A to satisfy the monitor at point B will be equal to the pressure monitor setting, P_{1B}, plus the static head, S_2, plus the friction head from A to B. This is illustrated in Figure 5.30D.

As system flow is throttled, a series of system head curves will be produced, as shown in Figure 5.31. System flow will occur at those points where the system head curves intersect the total available head curve. By comparing Figures 5.30C and 5.31, the throttling losses are less with the pressure monitor at B than they are with the pressure monitor at A for specific values of flow. This can result in lower pump operating speed, lower horsepower required, and lower operating cost.

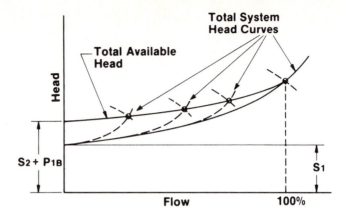

Figure 5.31   Head curves developed for example.

Sensing Multiple-Branch Systems

In parallel piping systems with two or more branches, a single pressure monitor must be set to satisfy the branch with the maximum calculated friction head. In some instances, load variation in the system can cause the pressure in nonmonitored branches to drop below acceptable limits, even though the monitored branch is satisfied. This type of problem can be resolved by analyzing the various branches based on the location and setting of a single pressure monitor. Additional pressure monitors can be added to those branches that experience the low-pressure condition to restore system operation to acceptable levels.

Evaluation of the total available head band and the total system head band over the expected range of system flow is necessary to make decisions that will affect the cost and operation of a pumping system. The decisions are based on problem statements and evaluation of the possible solutions with respect to the cost and functional objectives of the system design. The following questions outline some of the considerations that should be reviewed to assure satisfactory operation of a pumping system:

1. If the system is throttled for reduced system flow, does the maximum total available head exceed the system working pressure? If so, the alternatives include:
   a. Increase system working pressure.
   b. Select a pump with a "flatter" head capacity characteristic.
   c. Consider variable-speed pump drive.
   d. Add a bypass valve or pressure-reducing valve to the system.
2. Does pump horsepower increase with decreasing flow? If it does, the alternatives include:
   a. Add a bypass valve to system.
   b. Consider a variable-speed pump drive.
   c. Select several lower specific-speed pumps to operate in parallel.

3. Will operation at reduced flows occur for significant periods of time? If so, pump efficiency will be reduced and the service life of pump bearings, packing, mechanical seals, and close-clearance wearing rings usually will be shortened, and the alternatives include:
   a. Select smaller pump for the reduced flow operation.
   b. Add bypass valve to system.
   c. Consider variable-speed pump drive.

4. Will actual system head at design flow be considerably less than the calculated value? If so, the system flow will exceed the design flow, and the alternatives include:
   a. Reduce the pump impeller diameter.
   b. Install a valve to throttle system flow.
   c. Consider a variable-speed pump drive.

5. In systems planned for two or more pumps operating in parallel, have all the operating possibilities been considered? For example, one pump alone can operate at higher flow than its rating, as selected for parallel operation. For the higher-flow operation the pump will require more net positive suction head and may be subject to undesirable hydraulic loads, motor overload, increased noise level, and shortened life.

There are only a few of the many problems that can be encountered in the course of designing a pumping system. In many instances, several solutions can provide equally satisfactory results if operating power requirements are not considered. If design objectives include minimum power consumption, an energy evaluation of the pumping system must be made.

TOTAL SYSTEM EVALUATION

Total system evaluation involves developing information about the system and its pumping equipment necessary to satisfy a system's design requirements, concluding with a projection of total power requirements. To perform the evaluation, system head and total available head must be established. These two components are combined with pump power requirements and system load profile to project total system operating costs.

Load profile expresses the measure of work executed in a system compared to a unit of time. Work performed has a direct relationship to flow; therefore, load expressed as flow provides a common base for any system energy requirement projection. System load profile can be illustrated with a curve as shown in Figure 5.32; however, it is generally easier to use the flow/time relationship in tabular form, as illustrated in Table 5.7. The tabular format organizes total operational time at specific flow values and simplifies identifying high-flow/time concentration areas, which should be the focal point for preliminary pumping equipment selection.

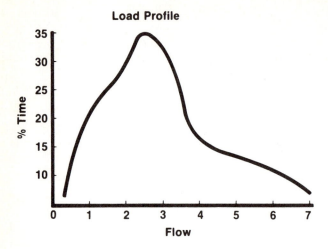

Figure 5.32 System load profile plot.

Figure 5.33 (top) illustrates how the flow/time relationship can be combined with a system head curve to aid in pump selection procedures. In this example, a significant amount of system operational time will occur in the flow ranges of 20 percent to 30 percent and 60 percent to 75 percent of the total design flow.

Pumping equipment should be selected to provide maximum operating efficiencies in these flow ranges. One possible pump equipment selection for this system would include a single pump selected for the system head and flow requirements in the 20 percent to 30 percent range, as well as two identically sized pumps to share the total system capacity equally at full system flow, but selected so that each will be in its best efficiency range when operated independently in the 60 percent to 75 percent system flow range (Figure 5.33, bottom).

The time base for computing load requirements should be of sufficient length to incorporate all variables affecting system operation. The minimum practical load profile time length is one complete cycle, from minimum to maximum flow requirements. A twelve-month time base for total system load evaluation generally encompasses all production, load, and climatic variables for a

TABLE 5.7   TABULATED FLOW/TIME DATA

Flow (gpm)	Hours	Flow (gpm)	Hours
280	87	4,480	788
840	1,138	5,600	613
1,680	1,841	6,440	350
2,520	2,629	7,000	87
3,640	1,227	---	8,760

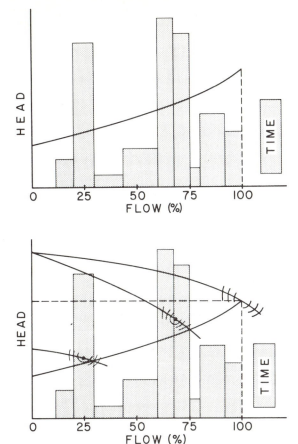

Figure 5.33 How the flow/time relationship can be combined with a system head curve.

system and can be used to project system operating requirements and costs. System load profile is the central component of any energy evaluation process.

In projecting total system energy requirements, total available head must be reviewed. If static head available is variable with respect to system flow, time, or both, a profile of these relationships should be made. This projection should represent the magnitude of change compared to the length of time the change is expected to occur, thus developing a correlation with system load profile. If the accuracy of these data is questionable, a band of minimum to maximum values should be established from which a total system power requirement band can be developed.

In projecting total power requirements for any liquid handling system, pump horsepowers at all significant flow rates between minimum and maximum conditions must be developed. Significant flow rates can be determined by reviewing the system load profile. To determine power inputs at various flow rates, it is necessary to know whether the pumping equipment shares the total system flow or head equally or unequally, and whether the pumping equipment operates at fixed speed, infinitely variable speed, or a combination of both. The system monitoring means also must be known when infinitely variable-speed pumps are employed.

When pumps are combined to share total system flow or head equally, the system power requirement at any flow rate can be computed based on individual pump horsepower. This definition is compatible with any range or quantity of pumps employed to satisfy all system flow requirements. The energy input will be governed by the number of pumps necessary to satisfy those requirements. In this case, the flows or heads of individual pumps combine equally to satisfy total system requirements and the total power required at any flow rate is the sum of the power required by the individual pumps.

When unequal-sized pumps are combined to satisfy system flow or head requirements (i.e., unequal percentage of total required), special consideration must be given to the pump operating sequence to ensure that maximum system efficiency is achieved.

Initial pump selections should be compared to the system head band and system load profile to establish a basic operating sequence for significant system flow values. Horsepower computations for the individual pumps are then made and recorded on a data form. To ensure that the selected order of pump operation results in the minimum power input necessary to satisfy the system requirements, rearrange the pump operating sequence at individual system flow rates, compute new pump horsepower requirements, and record this information.

This procedure will illustrate that for any system flow rate there is a single combination of pumps that requires minimum power input to satisfy the system requirements.

PROJECTING PUMP HORSEPOWER

Constant Speed

To compute constant-speed pump horsepower for parallel, series, equal, or unequal combinations, all pump characteristics necessary are available on the manufacturers' performance curves. For pump operation with fixed-speed change increments, the characteristics of flow, head, and brake horsepower can be calculated from the manufacturers' curves using the affinity laws. Therefore, power requirements for all constant-speed pumps and pumps of fixed-speed increments can be projected from available data.

Variable Speed

To compute variable-speed pump horsepower, it is first necessary to determine system monitor type and location. Type and location will determine head available, and head available will determine power required. Unfortunately, there is no simple mathematical relationship that can be developed for computing the interacting characteristics of systems and variable-speed pumps. The derivation of the affinity laws is helpful in computing performance for individual

$$Q_2 = Q_1 \sqrt{H_2/H_1} \tag{5.14}$$

variable-speed pumps. In this formula, $Q_1 - H_1$ represents any single value of flow and head taken from the total system head curve. $Q_2 - H_2$ represents a point on a pump total head curve of known operating speed that will satisfy the formula. The relationship of these values is shown in Figure 5.34. When the formula has been satisfied, the pump operating speed necessary to produce system conditions $Q_1 - H_1$ can be calculated.

In using the proposed formula, values known include system flow, Q_1, system head, H_1, and full pump operating speed, N_2. Values unknown are the equivalent pump flow, Q_2, equivalent pump head, H_2, and the unknown speed, N_1, at which the pump must operate to produce the required system flow and

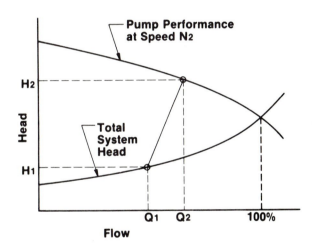

Figure 5.34 Relation of pump affinity law to total system head curve.

head. When making variable-speed pump brake horsepower calculations, it is necessary to know the equivalent point on the full-speed characteristic curve that equates to the required system flow and head, since the efficiency at the equivalent point will reoccur at the reduced flow and head condition. Variable-speed pump brake horsepower is computed from the flowing formula:

$$BHP_1 = \frac{Q_1 \times H_1 \times \text{Specific Gravity}}{3,960 \times Eff_2} \tag{5.15}$$

where Eff_2 is the efficiency from the full-speed pump curve at Q_2, H_2 expressed as a decimal value, Q_1 is U.S. gpm, and H_1 is feet of head. The reduced pump operating speed necessary to satisfy the system flow and head can be projected from the affinity law derivation:

$$N_1 = \left(\frac{Q_1}{Q_2}\right) N_2 \qquad (5.16)$$

This value is used when computing brake horsepower requirements for variable-speed equipment where drive efficiency is a function of output-to-input speed ratio. The expected range of pump operating speed should be determined and compared by the pump manufacturer to the critical speed of the pump. If operating speed occurs at or near pump critical speed, it will be necessary to modify the pump or to select a different pump for the application.

Determining the precise $Q_1 - H_2$ point on the pump curve usually will require reiterative calculation. The triangulation procedure illustrated in Figure 5.35 locates an intercept point on the pump curve that closely approximates the actual flow and head solution, $Q_2 - H_2$. The accuracy of that approximated point is affected by the relative magnitude of H_1, H_2, and the slopes of the pump and system head curves. To satisfy the formula, and thus determine the point on the pump curve of the known speed that produces the required system values $Q_1 - H_1$ at reduced speed, it is first necessary to calculate the initial $Q_2 - H_2$ condition as a reference point. This is done by entering known system conditions $Q_1 - H_1$ and an initial value of H_2 in the formula. The initial value of H_2 is taken from the known pump performance curve at system flow, Q_1. By solving the formula for the initial value of Q_2, a reference point for initial $Q_2 - H_2$ is established. By connecting the initial (reference) values of $Q_2 - H_2$ with required system condition $Q_1 - H_1$, an intersection point with the known pump curve is established. This intersection point becomes the approximate $Q_2 - H_2$ value that will, at reduced operating speed, produce system conditions $Q_1 - H_1$.

In a parallel pumping system where system flow is shared equally by all pumps, the procedure is identical to that for a single-pump system, except that system flow, Q_1, must be divided by the number of pumps, P_n operating. Figure 5.35B illustrates the triangulation process for a system incorporating two equally-sized pumps operating in parallel.

When pumps selected to share unequal portions of total system flow are operating, the individual pump solution formula must be modified to $Q_1 = Q_2 \sqrt{H_1/H_2}$. This is necessary since the flow produced by each pump for a selected system flow, Q_1, is unknown. To determine the flow and head developed by unequal-sized variable-speed pumps when they are both contributing to total system flow, it is first necessary to solve the basic triangulation process for system flow, Q_1, on the combined pump characteristic curve (Figure 5.35C). The intersection point that occurs on the combined pump curve establishes the H_2 value for the individual pumps. Use of the formula $Q_1 = Q_2 \sqrt{H_1/H_2}$ determines individual pump flow, $Q_1(P_1)$ and $Q_1(P_2)$, for the system operation at Q_1. System power requirements at flow Q_1 are determined by adding the horsepower requirements of each pump.

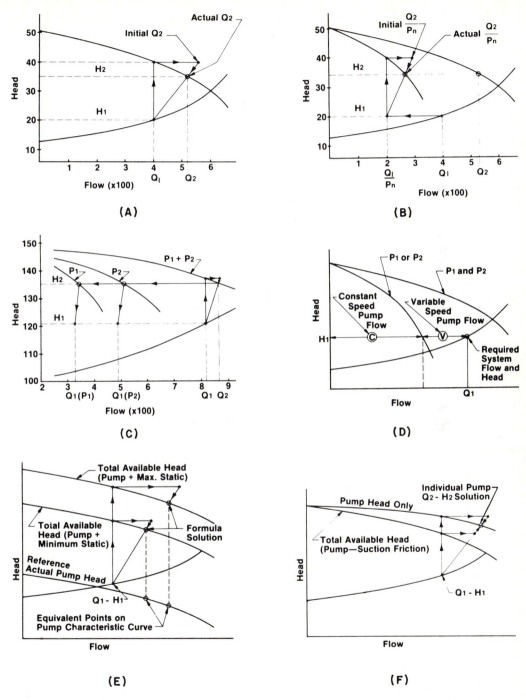

Figure 5.35 Triangulation procedure on pump curves to determine actual flow-head solution.

When a variable-speed pump is combined with a constant-speed pump, the constant-speed pump will always produce flow at the point where the developed head matches the required system head. When system flow requirements are such that the constant-speed and variable-speed pumps must be operated in parallel, flow produced by the variable-speed pump will be the difference between required system flow, Q_1, and constant-speed pump flow at system head, H_1.

Figure 5.35D illustrates a two-pump system in which both pumps are of equal size—one constant speed and one variable speed. With a required system flow of Q_1 and head of H_1, the two pumps will develop unequal percentages of the total required flow. As system flow changes, the percentage of flow contribution by each pump will change.

To determine total system power requirements under these circumstances, it is necessary to exercise the basic triangulation procedure for the variable-speed pump based on the known capacity it will develop at the required head of the system. The horsepower of the variable-speed pump is calculated and added to the horsepower required by the constant-speed pump at its developed head and flow.

Projecting variable-speed pump performance under conditions of variable static available head requires the formula solution to be equated to the actual developed head of the pump. Use of the normal projection formula is sufficient, and the equivalent point on the pump head curve will result from projecting the actual Q_2, H_2 solution and subtracting the static head. This will determine accurately the equivalent point on the pump head curve that equates to the system required head and flow. Figure 5.35E illustrates the effect of variable available static head.

When friction head in the supply pipe and fittings significantly reduces pump head, it is necessary to project the triangulation solution to the pump head curve only. This is illustrated in Figure 5.35F. The projection values can be computed from either the total available head curve or the pump curve, but the intersection solution always must be related to the equivalent point on the individual pump head curve. The accuracy of the triangulation procedure will depend to a great extent on the accuracy of pump and system curve data used.

In the employment of variable-speed drives whose efficiency is a function of output-to-input speed, system operating efficiency can be improved when major system operating conditions require pump speed changes below the nominal speeds available from standard, multiple-speed AC induction motors. Careful comparison of the operating speed range to the system load profile should be made. If the time/load relationship is of sufficient magnitude, motor speed can be reduced to improve the variable-speed drive efficiency by altering output-to-input speed ratios.

Numerous contingencies must be considered in the analysis of this type of application to ensure that all mechanical and hydraulic considerations are taken into account.

EQUIPMENT POWER COMPARISONS AND TOTAL ENERGY PROJECTION

Accurate comparisons of the various kinds of pumping equipment under consideration for a given application make it necessary to establish a common point of reference. That reference point is the power that the prime mover must supply at its output shaft:

1. *Constant speed.* To project the total system power required for a constant-speed pump, divide the pump horsepower by the prime-mover efficiency.
2. *Variable speed.* The horsepower at the output shaft of the prime mover for a pump driven by a variable-speed device is derived by dividing the pump brake horsepower by both the prime-mover efficiency and the variable-speed device efficiency. This applies whether the variable-speed device is between the pump and the prime mover or ahead of the prime mover.

Do not be misled by the lower net efficiency of the equipment, which includes the variable-speed device. Via pump operation, which is closely aligned to system needs, significantly valuable operating cost savings can be obtained, particularly in systems that have variable loads. The greater initial cost of the variable-speed equipment often can be amortized in a relatively short time, after which a net reduction in overall system operating costs can be obtained.

To project the total energy requirements for any system with any combination of pumping equipment and to ensure that all answers are projected from a common base, a structured procedure for the analysis should be followed; such a procedure is outlined in Figure 5.36. Each item in the procedure will have a significant impact on the total energy requirement of a specified pump-equipment combination and system operating procedure.

To make comparisons of dissimilar equipment or to evaluate the impact of system operation change, it is necessary to follow the evaluation procedure for each specified condition. The individual steps of the procedure are defined as follows:

Step 1. At the point of entry to the evaluation procedure, it is necessary to have established and assembled all information relative to the system design requirements. An accurate projection of system flow and head requirements, as illustrated by a system head curve or band, is necessary at this stage of evaluation.

Step 2. The load profile of the system should be reviewed to make preliminary pump equipment selections and to establish the evaluation base for the system under consideration. If future system conditions such as expansion will significantly alter the load profile, these considerations should be confirmed at this time.

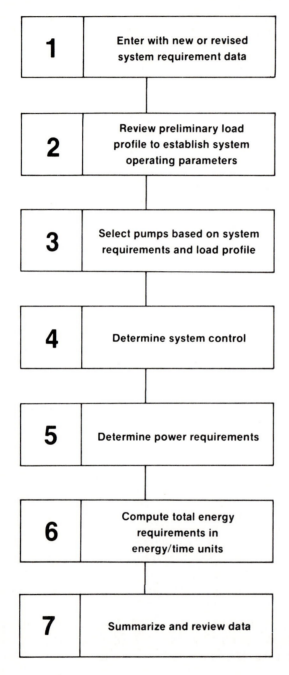

Figure 5.36 Summary of procedure to project total energy requirements.

Step 3. Pump selection evaluation. Beyond the normal mechanical and hydraulic criteria associated with pump selection for a specific application, the selection procedure should encompass the primary operating efficiency at the point of selection and the amount of efficiency deviation that will be caused by system flow variation. These values then should be compared to the operating time at specified flow rates to ensure that the selection offers the maximum potential operating efficiency through the range of maximum flow/time concentration.

Step 4. Preliminary decisions must be made relative to the control means that will be employed to satisfy system flow and head requirements. The decision to control system required head, total available head, or both, will be a significant factor in the total energy requirements projected for the system being evaluated.

Step 5. This stage of the evaluation requires a projection of the pumping equipment power requirements over the entire system operating flow range. The pump required power can be organized in tabular or graphical form, as best suits the user. In tabular form, the power requirements of the pumping system, compared to required system flow, should be of sufficient quantity to allow the user an analysis of all significant operating areas, as dictated by the flow/time profile.

Step 6. The load profile is used to compute the energy requirements of the selected pumping equipment and system operating procedures. When compared to the power projection in Step 5, the load profile will determine the total energy requirements of the system, thus concluding the evaluation in energy time units (brake horsepower hours, kilowatt hours, and so on).

Step 7. Determination of the total energy required to satisfy all projected system operating conditions is the basis from which all other considerations and comparisons can be made. This is the reviewing and decision-making phase of the total system evaluation procedure. If the energy summary is adequate, the evaluation procedures are concluded; if not, the evaluation can be repeated with new or modified data to determine what effect the change will have.

When the total system energy answer has been determined, normal decision-making processes to select that combination of pumping equipment and system operating procedures that most effectively satisfy the design goals of the system can begin. Total system evaluation ensures that all projections, comparisons, and decisions are based on as much factual information as possible.

When comparison of various combinations of pump equipment and system operating procedures are desired, total system evaluation allows the comparisons to be made on an equal basis for the requirements of the system. Total system evaluation is a structured, fundamental process for determining the

POWER COMPARISON DATA

FROM:
INDIVIDUAL _____

COMPANY _____

ADDRESS _____
           (Street)   (City)   (State)   (Zip)

TELEPHONE (   ) _____

SYSTEM FLOW	SYSTEM HEAD	PUMP SUCTION SUPPLY
Maximum _____ GPM	Constant Pressure	☐ Constant
Minimum _____ GPM	Static _____ (FT)	☐ Variable
Liquid _____	Delivery _____ (FT)	Minimum _____ (FT)
Specific Gravity _____	Friction _____ (FT)	Maximum _____ (FT)
Viscosity _____	Total _____ (FT)	

LOAD PROFILE—ESTIMATED OR MEASURED

GPM									
Hours									

(Fill in as many conditions as applicable.)

APPLICATION, SERVICE OR PROCESS BEING CONSIDERED:

_____

_____
(Please provide simple piping sketch.)

PRESENT ELECTRICAL COST:

_____ ¢ Per KWH

TYPE OF COMPARISON TO BE MADE
(check one)

☐ Peerless variable speed drives with existing pumps *to* existing constant speed pumps.

☐ Peerless variable speed drives with new Peerless pumps *to* existing constant speed pumps.

☐ Peerless variable speed drives with new Peerless pumps *to* proposed constant speed pumps.

PUMPS PROPOSED FOR/OR EXISTING IN SYSTEM

Quantity _____

Design Capacity _____ (GPM EACH)

Design Head _____ (FT EACH)

Driver _____ HP _____ RPM _____
(Please provide performance curve for each pump)

Figure 5.37 Power comparison data sheet for pump energy comparison evaluation.

pump and system combination that best suits the cost-benefit functionality requirements of the user.

Figure 5.37 is a recommended data sheet for recording the information for the energy analysis.

6

Mechanical Seals for Pumps

The most frequently encountered requirement for sealing a rotating shaft is found on a centrifugal pump. Figure 6.1 illustrates the liquid end of a typical centrifugal pump. The liquid to be pumped enters the suction inlet at the eye of the impeller. As the impeller rotates at relatively high speed, liquid occupying the space between the vanes is held captive by the close clearance that exists between the front of the impeller and the pump housing or volute. Having no other path of escape, the liquid is centrifugally forced to the outside diameter of the impeller, where it flows out through the discharge nozzle. At the point of discharge, the pressure of the liquid is many times higher than it is at the suction inlet, which accounts for the centrifugal pump's ability to push liquid to heights far above the pump. However, this same discharge pressure flows down behind the impeller to the drive shaft, which is connected to a driver outside the pump. This is the shaft that must be efficiently sealed if the pump is to be of any practical use.

To make the job of sealing the drive shaft easier, most pump manufacturers relieve some of the pressure behind the impeller. For example, on an open-impeller pump this is usually accomplished in either one of two ways, or a combination of both. Considering that the clearance between the back of the impeller and the pump back head is actually much closer than shown in Figure 6.1, small pumping vanes are positioned on the back side of the impeller. Although not as large or as efficient as the vanes on the front of the impeller,

271

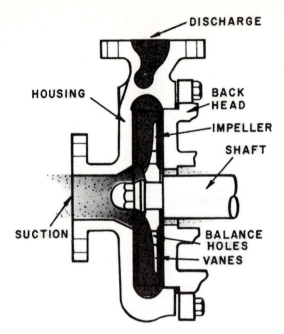

Figure 6.1 Sketch of centrifugal pump liquid end.

these smaller vanes will significantly reduce the pressure of the liquid trying to escape to the atmosphere from behind the impeller.

In addition to pumping vanes, balance holes can also be drilled through the impeller to the suction eye, thus relieving the high pressure behind the impeller to the low-pressure suction eye of the impeller. By decreasing the pressure differential between the front and rear of the impeller these methods also greatly reduce the axial thrust on the shaft, thereby prolonging thrust-bearing life. Even with the existence of pumping vanes and balance holes, suction pressure, plus usually a small percentage of discharge pressure, always surrounds the drive shaft. Therefore, shafts must be fitted with reliable seals.

Figure 6.2 shows a typical stuffing box sealed with square rings of com-

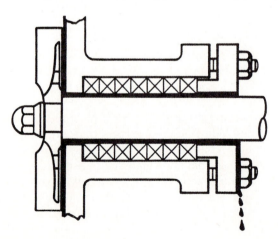

Figure 6.2 A typical stuffing box with compression packing.

pression packing. If this packing were to rub against the shaft, without lubrication present to prevent a buildup of frictional heat it would soon destroy itself. Therefore, a certain amount of the liquid being pumped, or some other liquid injected at high pressure from an outside source, must be allowed to flow between the packing and the shaft. Because of the surface irregularities of packing, eccentricity between the stuffing-box bore and the shaft, as well as normal shaft runout, a significant amount of packing must be used to compensate for irregularities. The packing requires a generous amount of lubrication. When this flow of lubrication exits from the packing, it becomes identified as leakage.

Other inconveniences associated with packing are adjustment and sleeve or shaft wear. It has been suggested already that because of certain irregularities, a particular ring of packing is periodically in contact and out of contact with the shaft. When the packing wears or loses its resiliency, leakage increases. The packing then must be adjusted to bring the leakage under control.

These same irregularities also exert uneven pressure on the shaft and restrict lubrication flow at certain locations, producing wear on the shaft or shaft sleeve. When this wear has been allowed to become acute, stuffing-box leakage increases significantly. The only solution to reducing the leakage is to incur the expense of a new or repaired shaft or shaft sleeve.

Figure 6.3 shows the parts of a simple mechanical seal. The coil spring, O-ring shaft packing and seal ring fit over the shaft and rotate with it. The spring

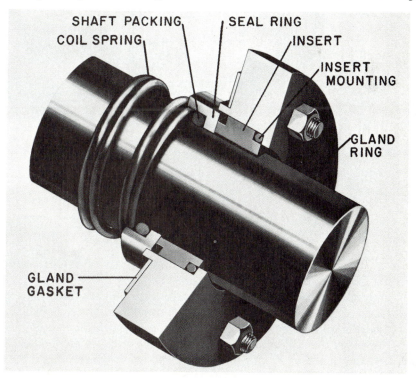

Figure 6.3   A mechanical seal. (*Courtesy of Durametallic Corp., Kalamazoo, MI*)

can be made from any one of a number of materials. The selection of the material depends on the corrosion resistance required to withstand attack from the product being pumped. Similar considerations are evaluated when selecting the O-ring shaft packing material and seal-ring material. The insert and insert O-ring mounting are fitted into a bore provided in the gland ring. This entire assembly is then attached to the pump stuffing box, remaining stationary as the shaft rotates. The insert is made from carbon graphite, which offers a good bearing surface for a variety of seal-ring materials to rotate against. It is inert to corrosive attack by most chemicals at a wide range of temperatures.

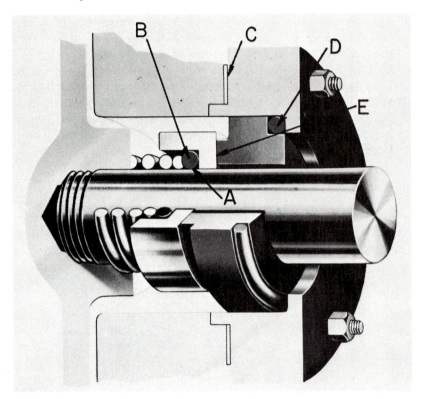

Figure 6.4 A stuffing box with mechanical seal. (*Courtesy of Durametallic Corp., Kalamazoo, MI*)

Figure 6.4 shows this simple seal completely assembled in the pump stuffing box. The coil spring rests against the back of the pump impeller, pushing the seal ring forward so that it remains in intimate contact with the stationary insert. Friction between the rotating pump shaft and shaft packing causes the seal ring to rotate with the shaft. Friction between the spring and the impeller, as well as the spring and the compressed O-ring, is responsible for the spring rotating with the seal ring and shaft packing. The insert is positioned within the gland bore. The gland itself is bolted to the face of the stuffing box.

Friction between the O-ring insert mounting and the inside diameter of the gland bore holds this part of the seal stationary as the shaft rotates within the bore of the insert.

The path of the liquid down the shaft is blocked by the O-ring shaft packing at point *A* in Figure 6.4. Liquid attempting to pass under the seal ring is blocked by the O-ring shaft packing at point *B*. Liquid attempting to pass over the seal ring is finally blocked from reaching the atmospheric side of the pump by the gland gasket, point *C*, and the O-ring insert mounting, point *D*. The path of escape remaining is the seam between the rotating seal ring and the stationary insert, point *E*.

The face of the seal ring and the face of the stationary insert are lapped to a flatness that is measured in millionths of an inch. Therefore, these faces remain in contact throughout their entire contact surface area, thereby providing a nearly positive seal. Just as in the case of packing, the faces must be lubricated. Because the area of the bearing surface is only a fraction of that encountered with packing and the contact pressure is evenly distributed throughout the interface, much less lubrication is required. Consequently, a correspondingly smaller amount of lubrication passes between the seal faces to exit as leakage. Whereas most packing depends on a measurable flow of liquid between it and the shaft for lubrication, the mechanical seal faces ride on a microscopic film of liquid that is able to migrate between them. When this film reaches the atmosphere, it can be classified as leakage; however, it is usually so slight that it will vaporize before it can be detected visually. Where liquids that do not readily evaporate, such as wax, are being sealed, minor accumulations may be detectable under the gland, but only after many hours of operation.

SEAL DESIGNS

Single-Coil Spring Seals

The seal shown in Figure 6.4 is limited to a narrow range of applications by virtue of its simplicity. Because the seal depends on friction to impart drive to the rotary unit, it is limited to use with nonlubricating-type liquids, such as water. Therefore, if it is desired to use such a seal in conjunction with liquids having lubricating properties, it becomes necessary to drive the seal mechanically.

Two methods of converting simple seals to positive drives are shown in Figure 6.5. Both methods utilize a collar, which is held in position on the shaft with a set screw. In Figure 6.5, tabs on either end of the spring engage notches in the collar and the seal ring. In this manner, rotational drive is transmitted from the collar to the seal ring by the spring. Figure 6.5 shows two pins extending over the spring from the collar to the seal ring.

There are two limitations to single-coil spring seals. Both limitations are related to the use of a coil spring installed over the shaft. Such a spring should

(A)

Figure 6.5 Positive drive for (A) single-coil spring seal and (B) multicoil spring seal. (*Courtesy of Durametallic Corp., Kalamazoo, MI*)

be used only with relatively low-shaft surface speeds because it has a tendency to distort at high-surface speeds. Such distortion will only magnify the natural tendency of a coil spring to push harder on one side of the seal ring than the other. The end result is an uneven liquid film between the faces, which can result in excessive leakage and wear. The other limitation is economics. Pumps can be purchased in a variety of shaft sizes. Consequently, a user who stores seal parts may find it necessary to carry large inventories of springs to cover the range of sizes required.

Multiple-Spring Seals

Figure 6.6 illustrates a more sophisticated seal design utilizing a number of small coil springs neatly spaced around the rotary unit. These small springs are not as susceptible to distortion at high speeds and exert an even closing pressure on the seal ring at all times. In addition, the same-sized springs can be used in

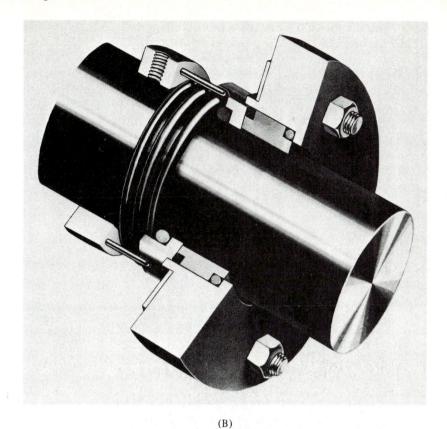

(B)

Figure 6.5 *Continued*

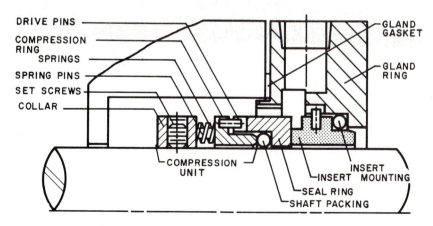

Figure 6.6 A mechanical seal with multiple seals.

conjunction with a wide variety of shaft sizes. This general design is often referred to as a conventional mechanical seal.

The seal in Figure 6.6 is a positive-drive seal. The collar is normally secured to the shaft by set screws. Pins extend from the collar to the compression ring, which presses against, or energizes, the shaft packing. Additional pins extend from the compression ring to the seal ring, completing the drive train from the collar. The collar, compression ring, pins, springs, and set screws are usually confined in an assembly that is securely put together at the factory. This assembly is called the *compression unit.* The seal points of this multiple-spring seal are identical to those outlined for the single-coil spring seal shown in Figure 6.4.

Bellows Seals

A third basic seal design offered by some manufacturers is the bellows seal, shown in Figure 6.7. Two forms of the bellows seal are the elastomer bellows (Figure 6.7A) and the metal bellows (Figure 6.7B).

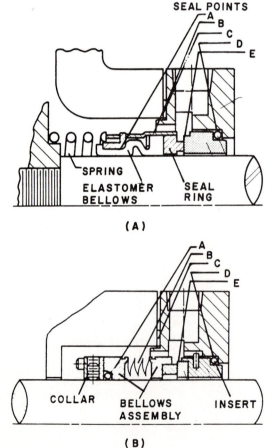

Figure 6.7 (A) Elastomers bellows seal. (B) Metal bellows seal.

The main advantage of a bellows-type seal is that the seal ring is free to move forward as the insert wears, without being retarded by the friction between the shaft-packing O ring and the shaft. In most applications this is not an important consideration; however, in some situations the pump shaft or shaft sleeve metallurgy is not compatible with the product and corrosion takes place, freezing the conventional O-ring shaft packing to the shaft. This leads to leakage. Although a bellows seal is not nearly so sensitive to shaft or shaft-sleeve corrosion, such a problem lies not with the conventional seal, but with the improper selection of shaft or shaft-sleeve material.

The elastomer bellows seal is energized by a single-coil spring, which fits over the shaft and bellows. Consequently, this seal is subject to the same speed and spare parts inventory considerations. Additional limitations are imposed by the bellows itself, which replaces the O-ring shaft packing found in the conventional designs. While a satisfactory bellows material usually can be selected to meet the environmental requirements of the stuffing box, the bellows itself is not as closely confined in a conventional seal as the shaft packing. Therefore, the bellows is much more susceptible to fatigue failure.

This same lack of confinement adds to the seriousness of a bellows failure, as opposed to a conventional shaft-packing failure. If conventional shaft packing begins to deteriorate, its confinement generally dictates that the resulting leakage is restricted and, therefore, growing progressively worse as the shaft packing continues to fail. This often provides the user with sufficient warning to plan for a shutdown to repair the seal. However, in the event of a bellows rupture, the resulting leakage is often great enough to necessitate immediate repairs.

Sealing Points and Drive of Elastomer Bellows

The sealing points of an elastomer bellows seal are primarily the same as encountered in conventional seals. The bellows sleeve grips the shaft at the rear of the rotary unit assembly, transmitting drive to the seal ring. Although the seal illustrated is a friction-drive seal, the increased friction offered by the bellows sleeve exempts this seal from the nonlubricating liquid limitation imposed on the conventional friction-drive seal. Figure 6.7A illustrates the sealing points and defines the nomenclature of an elastomer bellows seal.

Metal Bellows

The metal bellows seal shown in Figure 6.7B permits the bellows seal design to be used at higher speeds and temperatures than the elastomer bellows design. The bellows itself provides the necessary pressure to maintain face contact, thereby eliminating the necessity for a spring. Since the bellows is a one-piece unit supported around its entire circumference and along its total length, even pressure is applied to all points of the seal ring. It is this configuration of the bellows that allows the seal to be applied to higher shaft speeds than can be accommodated by a single-coil spring over the shaft.

Aside from being able to accommodate higher shaft speeds, the metal bellows also can be applied to high and cryogenic temperatures. Until recently, conventional seals had to rely on elastomers or TFE resins for secondary sealing members (shaft packing and insert mounting). These materials have a rather narrow temperature range, depending on their composition. Even though a particular material might be suitable for use in an extreme temperature, this did not necessarily mean that it has the necessary resistance to chemical attack by the liquid that has to be sealed. Although the combined problem of temperature and chemical resistance could be generally overcome by creating an artificial, compatible environment in the stuffing box, such a solution requires the use of costly environmental controls. Therefore, the metal bellows seal promises an economic solution to many high- and low-temperature sealing problems.

Figure 6.7B shows that the metal bellows is actually a series of plates called *convolutions*. Each one of these convolutions is a separate ring that has been welded to similar rings on either side. Unfortunately, from a single piece of material it is not possible to make a bellows with the necessary mechanical properties. It is the necessity for welding that currently presents the possibility of fatigue failure, resulting in ruptured bellows. Such a rupture can result in severe leakage. Although there has been considerable effort to improve its reliability, the need for a bellows of this type for high-temperature seal applications has been reduced significantly by the recent introduction of all-graphite secondary packings for conventional seals that are capable of functioning in high- or low-temperature extremes. They are also inert to the deteriorating effects of most chemicals. When considering the bellows seal, note that when the rotary face requires replacement, the entire bellows assembly also must be replaced.

Balanced Seals

One of the most frequently encountered environmental conditions requiring a slight modification in seal design is pressure. If an inside mechanical seal is called on to seal high pressures, provisions must be made to ensure that all of the pressure on the seal is not trying to push the seal ring to the atmospheric side of the stuffing box.

Figure 6.8 is a cross section of a conventional inside unbalanced seal. Almost all of the stuffing-box pressure is exerting a closing force on the seal ring. Only a very small portion of the seal-ring face is exposed over the top of the insert, allowing a proportionately small amount of pressure to work against the seal ring in the opposite direction, in addition to the opening force exerted by the liquid film between the faces. If the closing force becomes great enough, the liquid film between the faces is literally squeezed out. Deprived of lubrication and highly loaded, the faces soon destroy themselves. The solution to this problem is to balance the seal. Figure 6.9 illustrates a conventional inside seal that has been balanced. Notice that a step in the shaft sleeve moves the insert face radially inward without decreasing the width of the face itself. The seal ring, on the other hand, remains mounted on the original shaft diameter, which

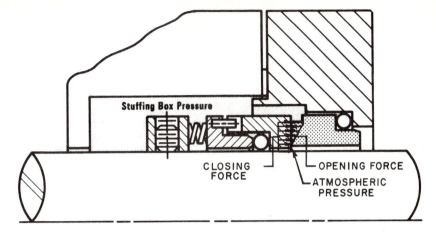

Figure 6.8 Sketch of inside unbalanced seal.

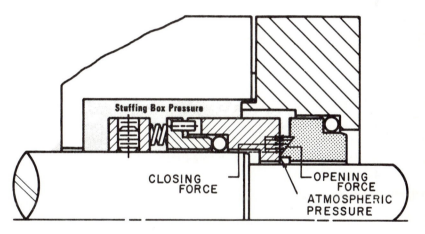

Figure 6.9 Sketch of an inside balanced seal.

means that the closing force remains unchanged. Because more of the seal-ring face is exposed to the hydraulic pressure working to open the seal, the design is balanced. When comparing the surface areas of the seal ring available for the hydraulic pressure to work against, it becomes apparent that the opening force is slightly less than the closing force. This is intentional to ensure that the faces are kept in contact at all times.

The method of achieving balance on an outside seal is the same as with an inside seal, except that the action is the reverse. Instead of counterbalancing a portion of the closing force imposed by stuffing-box pressure, a portion of the stuffing-box pressure is added to the closing force, counteracting the opening force exerted by the liquid film between the faces.

Figure 6.10 is a cross section of an outside balance seal. The shaft packing is forced against the collar, leaving an area under the seal ring exposed to

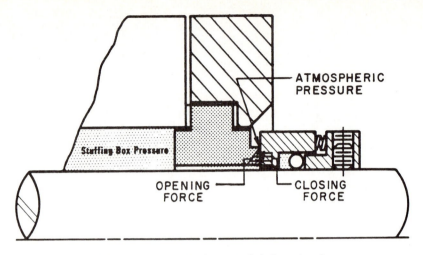

Figure 6.10   Sketch of an outside balanced seal.

stuffing-box pressure. The closing force exerted by the stuffing-box pressure, acting against the shoulder of the seal ring, is slightly greater than the opening force exerted by the liquid film between the faces, thereby keeping the faces in contact at all times.

High-Pressure Seals

Seals for extremely high pressures demand more design sophistication than standard balanced seals. Not only is the ratio between the opening and closing force more critical, but the entire seal must withstand elevated pressure and ensure that the faces do not become distorted.

Specifically, the entire rotary unit cross section is thicker. Special springs are provided to ensure proper face loading on this very closely balanced design. Drive pins are also heavier to accommodate more torque. The insert also has a larger cross section to withstand the pressure. The gland thickness is greatly increased to prevent distortion. Grinding of the bottom of the gland counterbore, against which the insert is held (insert backup), is important to ensure flatness. At high pressures, any high point on this surface would most certainly be transmitted to the insert face, resulting in leakage. Figure 6.11 illustrates a high-pressure seal design.

High-Speed Seals

To conserve space and achieve a more efficient weight-to-horsepower ratio, more attention is being given to high-speed pumps with shaft speeds in excess of 6,500 fpm. Because dynamic forces begin to exceed the limitations of a conventional rotary unit at this speed, the roles of the insert and rotary unit are

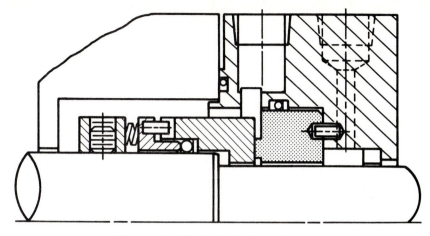

Figure 6.11 Sketch of a high-speed seal.

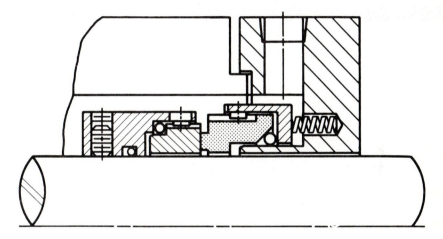

Figure 6.12 Sketch of a high-speed seal.

reversed. The springs become stationary, loading a stationary, rather than rotating, seal ring. The insert is flexibly mounted inside a collar, which is set screwed to the shaft. Rotational drive is transmitted to the insert via lugs in the collar, rather than pins. Similar lugs hold the seal ring stationary. Figure 6.12 illustrates the configuration of a typical high-speed seal.

MECHANICAL SEAL ARRANGEMENTS

Single inside mechanical seals (Figure 6.8) are the most common type of mechanical seal used. The materials of construction are selected to withstand the corrosive nature of the liquid in the stuffing box. They are easily modified to include environmental controls and can be balanced to withstand extremely high stuffing-box pressures.

Inside seals require a suitable stuffing-box housing for installation and cannot be adjusted without dismantling the equipment.

If an extremely corrosive liquid is encountered that has satisfactory face lubricating properties, an outside seal offers another alternative to expensive metallurgies necessary for an inside seal to resist corrosion. Outside seal arrangements have only the insert, seal ring, and secondary seals exposed to the product. All of these components can be nonmetallic. The metallic rotary unit parts are exposed only to the atmosphere.

Outside seals also can be an advantage when a piece of equipment is found to have a stuffing box that will not accommodate an inside seal. Although outside seals are easier to adjust and troubleshoot because they are exposed, they have the disadvantage of being vulnerable to damage from impact. Of greater importance, however, is the pressure limitation of an outside unbalanced seal. By contrast to an inside seal, the hydraulic pressure works to open, rather than close, the seal faces. Therefore, the seal design depends entirely on the springs to maintain face contact. Even though a degree of balance can be designed into an outside seal, all outside seals are limited to use in conjunction with moderate stuffing-box pressures.

Double Seals

When selecting a seal for a pump, liquids are sometimes encountered that are not compatible with a single inside mechanical seal. Often these liquids carry abrasive materials in suspension that would rapidly wear the faces; or the liquid may be so corrosive that extremely expensive materials would have to be supplied for the seal components. There are two solutions to this problem. One is the application of environmental controls. The other requires the use of seals, which fall under a different design classification such as the *double seal.*

Figure 6.13 depicts a double-seal installation. A liquid, such as water, is injected into the stuffing box at point *A* and exits at point *B*. Assuming that the water entering point *A* is available to the stuffing box at a higher pressure than the product behind the impeller, a valve is located beyond point *B* to maintain this pressure between the double seal at the desired level, and yet allow a minimal amount of water circulation through the box to carry away heat generated by friction at the seal faces. With this arrangement, the water, at a higher pressure than the product trying to enter the box, surrounds the double seal and provides lubrication to both sets of seal faces. The inboard seal prevents the water from entering the pump, while the other seal prevents the water from escaping to the atmosphere.

The differential pressure across the inner seal is the difference in pressure between the sealing-liquid pressure and the product pressure acting on the stuffing box. The differential pressure across the outer seal is the difference in pressure between the sealing liquid and the atmosphere. Either, or both, of the

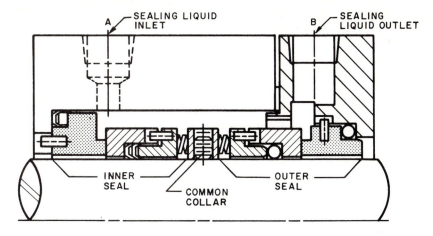

Figure 6.13 Sketch of a double inside seal.

seals may be balanced if the differential pressures exceed the limitations of unbalanced seals.

Another variation in seal arrangement is the double-tandem seal, illustrated in Figure 6.14. The purpose of this seal is not to create an artificial environment, as is the case with the double seal, but to provide a backup seal in the event the inner seal fails. The inner seal functions in a manner identical to a conventional single inside seal. The cavity between the inner and outer seal is flooded from a closed reservoir. The liquid in the reservoir provides lubrication to the outer seal. Because the space between the seals is only flooded and not under pressure, the product, not the liquid in the reservoir, lubricates the faces of the inner seal. If the inner seal fails, the resulting pressure rise in the area between the seals is sensed at the reservoir, where it either can be registered on a gauge or can activate an alarm. In any event, a failure of the inner seal can be detected, while the outer seal assumes the responsibility of sealing the shaft until repairs can be made.

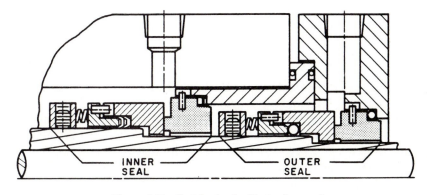

Figure 6.14 Sketch of a double tandem seal.

Since the purpose of the tandem seal does not require that the liquid between the seals be at a higher pressure than the product pressure, the inner seal can be balanced for high stuffing-box pressures without requiring a higher sealing-liquid pressure. If a conventional double seal were used in place of a tandem seal, product pressures would have to be limited to prevent the hydraulic pressure from opening the inner-seal faces. However a tandem seal is capable of being balanced to accept high stuffing-box pressures.

Requirements for tandem or double seals are sometimes encountered where the equipment stuffing box is too shallow to accommodate the more conventional varieties of these two fundamental designs. Therefore, it becomes necessary to turn to an alternate seal arrangement. In this case, the alternative is the inside-outside double seal.

The inside-outside double seal is assembled, as the name implies, with one seal inside the stuffing box and one outside the stuffing box. Both seals rotate against opposite ends of the same stationary insert. Figure 6.15 illustrates an

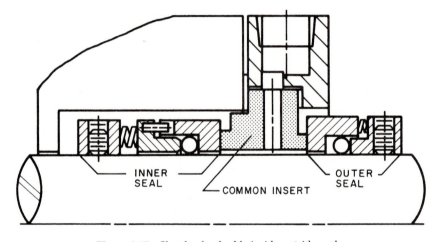

Figure 6.15   Sketch of a double inside-outside seal.

inside-outside double-seal assembly, using an unbalanced inner seal and a balanced outer seal.

Whether the inside-outside arrangement is to be considered a tandem or double seal depends on its function. If the liquid used between the seals is at a higher pressure than the product in the stuffing box, then the purpose of the design is to lubricate the inner seal with a liquid other than the product. This is the role of a true double seal, creating an artificial environment in which the mechanical seal can operate. However, if the liquid is circulated between the seals at a lower pressure than the equipment stuffing-box pressure, the role of the inner seal remains identical to that of any single seal in the event the inner seal fails. A situation such as this would identify the inside-outside assembly as a tandem seal.

MATERIALS OF CONSTRUCTION

The environment in which the seal will operate must be analyzed to determine the proper materials of construction. The corrosive properties of the liquid, temperature, and temperature variation must be considered.

Lubricity is another important factor. Liquids with lubricating qualities similar to, or less than, that of water require the use of face materials that are known to be capable of maintaining satisfactory wear rates without the benefit of good lubrication. In addition, whatever face materials are selected must have compatible wearing surfaces. This is not generally a problem since one of the seal-face materials is most always carbon. However, in rare applications in which either the temperature or corrosive properties of the liquid exceed the limitations of carbon, substitute materials must be selected.

Metallic-face materials are constructed from Kalamate/Stellite®, Duramate®, Ni-Resist®, bronze, or Hastalloy®. Kalamate is a trade name for seal rings made from solid cast stellite. Stellite is a good wearing material with a corrosion resistance greater than that of the 300 series stainless steels. Its composition is primarily chrome and cobalt, with small percentages of nickel, iron, and tungsten. Stellite is considered generally satisfactory for use in conjunction with a wide range of liquids with a similarly wide range of lubricating properties. Stellite, however, is not recommended for use in water because the water has a tendency to leach out the cobalt.

Duramate is a hardened stainless steel that is compatible with carbon as a wearing surface in the presence of liquids possessing lubricating properties equal to, or better than, light hydrocarbons. Duramate has a corrosion resistance roughly equivalent to that of 304 stainless steel.

Ni-Resist is an economical material that is sometimes used for mildly corrosive services. Ni-Resist is a 42 percent to 67 percent cast iron material containing 13 percent to 32 percent nickel, 10 percent copper, and small amounts of other metals. When used, Ni-Resist is usually applied to the sealing of petroleum products.

Bronze has proven to be an excellent face material when mated against tungsten carbide in abrasive services. It has greater abrasion resistance that carbon and has been used extensively on water flood, pipeline, and crude oil services where abrasion commonly occurs.

Bronze also has been used advantageously when the fragility of carbon limits the use of the latter material. Its higher modulus of elasticity and other improved physical properties dictates its use in high-pressure applications. Due to its relatively low-corrosion resistance, caution should be used in applying bronze in corrosive liquids.

Hastelloy is almost chemically inert. It is used primarily in compression units requiring a high degree of corrosion resistance. As a seal ring, Hastelloy displays poor wear characteristics and is only used occasionally. Before the introduction of ceramics and cost reductions for tungsten carbide, Hastelloy was used as a seal face where extreme corrosion resistance was required.

Face materials are also constructed from ceramics, which offer excellent wear characteristics due basically to their hardness. They also possess chemically inert properties that often make their use desirable when nonmetallic face-material combinations are specified for difficult liquids.

Care must be exercised when recommending ceramic seal rings since they are very susceptible to fracture from thermal or mechanical shock. In spite of this disadvantage, ceramic seal rings remain popular because of their economical cost. Ceramic seal rings fall into two major classifications: solid and ceramic faced.

Carbides are another major construction material. A seal face of tungsten carbide possesses qualities of superior hardness, a wide range of chemical inertness, and excellent antifrictional properties. Consequently, it is undoubtedly the most versatile face material available. However, cost dictates that the use of tungsten carbide be confined to applications where its performance will prove to be a distinct advantage over that of other face materials. As an example, tungsten carbide is routinely applied to high-pressure applications where, in the

TABLE 6.1 COMMONLY USED GRADES OF TUNGSTEN CARBIDE

Grade	Binder	Remarks
62-6 Tung-Car "M" Tung-Car "A"	Nickel	Standard grade. Wide range of applications. Suited for hot H_2O, petroleum products and most chemicals.
62-1 Tung-Car "M1"	Cobalt	Used in environments where nickel binder of grade 62-6 would be attacked, such as ammonia.
62-3	Tantalum	Called tantalum carbide because of binder. Very expensive, chemically inert. Only used in presence of strong oxidizing agents.

solid form, its high modulus of elasticity is a valuable asset in preventing face distortion. Tungsten carbide also exhibits outstanding wear characteristics in the presence of liquids with extremely poor lubricating qualities. Table 6.1 lists different grades of tungsten carbide offered by one manufacturer.

New materials that exhibit superior hardness and wearability are constantly being tested and evaluated for use as seal faces. One of these materials that has shown superior qualities as a seal face and the use of which is increasing rapidly is silicon carbide.

Silicon carbide's high, uniform hardness and high density make it an extremely abrasion-resistant material that outwears all other materials. Because of its high thermal conductivity combined with its high tensile strength and low thermal expansion, silicon carbide has excellent thermal shock resistance and, unlike tungsten carbide, does not heat check. These properties along with its unlimited corrosion resistance and low coefficient of friction, make silicon carbide an excellent seal-face material.

TABLE 6.2   TEMPERATURE LIMITATIONS OF SEAL FACE MATERIALS

Material	Construction	Maximum Temperature	
		(°F)	(°C)
Stellite Face	Welded Stellite face on metal ring	350	177
Kalamate	Solid-cast Stellite ring	450	232
Tung-Car	Solid tungsten carbide ring	750	400
Tung-Car "M"	Solid tungsten carbide element mounted in metal body	350	177
Tung-Car "A"	Solid tungsten carbide element mounted in Duramate body	750	400
Duramate	Solid machined stainless steel ring	600	316
Ni-Resist	Solid nickel-cast iron ring	350	177
Bronze	Solid leaded-bronze ring	350	177
Duramic	Solid ceramic ring	350[a]	177[a]
Peramic	Solid pure ceramic ring	350[a]	177[a]
55 Face	Ceramic facing on metal ring	350[a]	177[a]
No. 5 Carbon	Solid carbon-graphite ring	525	275
No. 6 Carbon	Solid carbon-graphite ring	525	275
Silicon Carbide "V"	0.030' thick conversion of silicon carbide on carbon-graphite substrate	700	371
Silicon Carbide 1	Solid silicon carbide ring	3000	1650
Glass-Filled Duraflon	Solid ring	350	177

[a]Subject to fracture from thermal shock.

TFE resin (Teflon®)[1] is another material with high chemical inertness. It has a tendency to "cold-flow," however, stability is achieved through glass impregnation. Glass-filled TFE resin is a poor wearing material and is, therefore, limited to very low-pressure velocity ratings. Typical temperature limitations for different types of face materials are given in Table 6.2.

SECONDARY SEAL MATERIALS

Secondary seals applied to mechanical seals are defined as the shaft packing and insert mounting. Because they are static and must seal imperfect surfaces, secondary seals must possess a degree of resiliency that will allow them to seat perfectly against adjacent surfaces. The most elastic or resilient of all secondary seals are elastomers, which are applied in the configuration of O rings. However, the temperature and chemical resistance of elastomeric materials are often exceeded, thus requiring the use of other less resilient materials. Because these other materials lack the elasticity of elastomers, they assume configurations that tend to compensate for their limited resiliency. Several elastomer O-ring types are described in the following paragraphs.

---

[1] Registered trademark of E.I. duPont de Nemours and Company, Inc., Wilmington, Delaware.

Buna N (Nitrile)

This material, often referred to as nitrile, is a copolymer of butadiene and acrylonitrile. Buna N has an excellent resistance to petroleum products, also finding wide acceptance in water applications.

Buna N can be exposed to temperatures ranging from −40°F to +225°F (−40°C to +107°C).

Inherently, Buna N does not possess good resistance to ozone, sunlight, or weather. Therefore, spare O rings should not be stored in direct sunlight or adjacent to electrical equipment that may generate ozone.

Neoprene

Neoprene, also called chloroprene rubber, was among the earliest of the synthetic rubbers available to seal manufacturers. It continues to be used for refrigerants, such as freon and ammonia, and other mild services. Like Bruna N, Neoprene can be subjected to temperatures ranging from −40°F to +225°F (−40°C to +107°C).

Butyl

This is an all-petroleum product produced by copolymerizing isobutylene and isoprene. Butyl rubber will resist the deteriorating effects of many mild liquids, such as MEK and acetone, but should not be used in petroleum oils. Its temperature range is −40°F to +225°F (−40°C to +107°C).

Silicone Rubber

Silicone elastomers are made from silicone, oxygen, hydrogen, and carbon. They usually display poor tensile strength, tear, and abrasion resistance. Silicone O rings are not recommended for use in most petroleum fluids or ketones. They can be applied successfully to temperatures ranging from −80°F to +400°F (−62°C to +204°C).

Fluorosilicone

Fluorosilicone combines the good high- and low-temperature properties of silicone rubber with the resistance to petroleum oils and hydrocarbon fuels. Like silicone rubber, it can be in a temperature range of −80°F to +400°F (−62°C to +204°C).

Hypalon

A chlorosulfonated polyethylene, Hypalon displays good acid resistance. Depending upon the liquid, it can be exposed to temperatures ranging from −40°F to +225°F (−40°C to +107°C). Note, however, that the compression and permanent set characteristics of Hypalon are not as desirable as found in most other elastomeric materials. Therefore, it is used only as a secondary seal when other materials will not stand up when exposed to a particularly difficult fluid.

Viton®

A versatile compound, Viton is a copolymer of vinylidene fluoride and hexafluoro propylene made by Du Pont. Also known as fluorocarbon rubber, it has excellent tolerance for temperatures ranging from 0°F to +400°F (−18°C to +204°C).

Viton is applicable to petroleum oils, diester-base lubricants, silicone fluids, halogenated hydrocarbons, water, low-temperature steam, a wide variety of acids and many other fluids too numerous to mention. Fluorocarbon rubber will not tolerate ketones, anhydrous ammonia, amines and hot hydrofluoric, or chlorosulfolnic acids.

Kel-F®

Kel-F is a fluorocarbon rubber made by the 3M Company. It is a material similar to Viton. Therefore, the two materials generally can be used interchangeably.

EPR/EPDM

EPR and EPDM are made from ethylene and propylene monomers. EPR actually stands for *ethylene propylene rubber*. EPDM is a very similar compound containing a small amount of a third monomer, making the compound *ethylene propylene diene monomer* (i.e., a terpolymer). Although the ethylene propylene compounds are relatively new, they have found wide acceptance in the sealing industry. EPRs have excellent corrosion resistance to dilute acids and alkalis, ketones, alcohols, water, steam and phosphate ester-base hydraulic fluids, but are not recommended for petroleum oils or diester-base lubricants. They are applicable for temperatures ranging from −40°F to +300°F (−40°C to 150°C).

Kalrez®

Kalrez is Du Pont's perfluoroelastomer. Kalrez combines many of the elastomeric properties of Viton and the chemical and heat resistance of TFE resin with certain advantages over both. Kalrez can be applied for temperatures ranging from 0°F to +500°F (−18°C to 260°C). Kalrez has excellent corrosion resistance to solvents, inorganic and organic acid and bases, strong oxidizing agents, metal halogen compounds, hot mercury, chlorine, fuels, and heat-transfer fluids.

Durafite®

Durafite is a relatively new secondary packing material. It is an all-graphite product containing no resin binders or organic fillers. It is used in environments where the temperature exceeds the limitations of other secondary sealing materials.

Durafite is especially inert. It is vulnerable to attack only by oxidizing liquids. In addition, it is highly resistant to all forms of nuclear radiation. Durafite can be applied in temperatures ranging from cryogenics, $-450°F$ ($-268°C$), to higher than $750°F$ ($400°C$).

Table 6.3 gives typical temperature limitations of secondary seal materials.

ENVIRONMENTAL CONTROLS

Environmental controls are necessary to control dirt and heat, or to cool the seal area, and under certain conditions to keep the product away from the seal entirely.

Temperature Control

The temperature surrounding the seal in a stuffing box is an important consideration. Certain parts of the seal, such as the secondary packings and carbon, may deteriorate under exposure to extremely high temperatures. As an example, Kalrez is the most stable elastomer material at high temperatures. Nevertheless, its use is limited to temperatures below $500°F$ ($260°C$). The binder in most carbons begins to break down at temperatures above $500°F$ ($260°C$). Although the temperature limitations of these two materials are generously high, seal faces and secondary packings cannot be selected entirely on their ability to withstand high temperature. These materials also must be able to resist chemical attack by the liquid surrounding them. Consequently, the mechanical seal specialist is not always at liberty to apply the most temperature-resistant materials. Often a compromise between temperature resistance and resistance to the corrosive properties of the liquid must be made. In making this compromise, it often becomes necessary to protect some of the materials by cooling the seal area.

Occasionally, the material being pumped solidifies at ambient temperatures. In its liquid state, however, this same product may act as an excellent clean lubricant. Such is the case with wax which, in order to be pumped, must be heated. The stuffing-box area surrounding the mechanical seal also must be heated to ensure that the seal will operate in a liquid environment.

Another important temperature consideration is the boiling point of the liquid in the stuffing box. Although the product must remain below its boiling point if it is to be pumped, stuffing boxes can be heat traps. They may not efficiently dissipate the additional heat that can be expected to be generated at the seal faces and through turbulence created by the seal spinning within the

TABLE 6.3 TEMPERATURE LIMITATIONS OF SECONDARY SEAL MATERIALS

Material	Form	Minimum Temperature Limit[a]		Maximum Temperature Limit[a]	
		(°F)	(°C)	(°F)	(°C)
Buna N (Nitrile)	"O" ring	−40	−40	+225	+107
Neoprene	"O" ring	−40	−40	+225	+107
Butyl	"O" ring	−40	−40	+225	+107
Silicone	"O" ring	−80	−62	+400	+204
Fluorosilicone	"O" ring	−80	−62	+400	+204
Hypalon	"O" ring	−40	−40	+225	+107
Kel-F	"O" ring	0	−18	+350	+177
Viton	"O" ring	0	−18	+400	+204
EPR/EPT (ethylene propylene)	"O" ring	−40	−40	+300	+150
Kalrez	"O" ring	0	−18	+500	+260
Duraflon (TFE resin)	"V" ring	−100	−73	+350	+177
Glass-Filled Duraflon	"V" ring	−175	−115	+450	+232
Durafite (pure graphite)	Square ring	−450	−268	+750	+400

[a]The minimum and maximum temperature limits listed are general limits and will vary with the liquid in contact with the elastomer.

close confines of the stuffing box. Although the amount of heat generated in this manner may not be great, it can be cumulative to a certain degree, thereby eventually raising the temperature of the hot product in the stuffing box to its boiling point. If this occurs, the seal faces do not enjoy the benefits of a liquid film between them. As a general rule, the temperature of the liquid in the stuffing box should be maintained at least 25°F (14°C) below the boiling point of the liquid at the stuffing box pressure. Vaporization also may cause precipitation of salts, minerals, or other abrasive impurities. Accelerated face wear is then imminent, ultimately culminating in a complete failure of the mechanical seal.

Cooling normally is accomplished by flushing. Flushing can be defined as the introduction of a liquid into a stuffing box at a higher pressure than stuffing-box pressure. To be effective, the liquid being injected must flow into the box at a rate that is sufficiently high to prevent the product behind the impeller from entering the stuffing box. Since most stuffing boxes are restricted at the throat, velocity is imparted to the flush liquid leaving the box, thereby preventing the pumped product from entering the stuffing box.

Two methods of flushing can be employed. The most commonly used is the bypass-flush system. When it is desired to use a bypass flush for cooling, a recirculation line is run from the discharge nozzle of the pump through a heat exchanger to a connection in the gland ring located over the seal faces. Since discharge pressure is inevitably higher than stuffing-box pressure, circulation is ensured. It is important to note that some pump designs subject the stuffing box to pressures equal to, or very near, full discharge pressure. In such cases, bypass is achieved by running a recirculation line from the gland ring back to the pump suction or other low-pressure source. Flushing in the gland ring for an inside seal, shown in Figure 6.16A, consists of a drilled and tapped hole in the gland ring, which channels the liquid directly over the seal faces.

Flushing an outside seal is shown in Figure 6.16B. It is important to note that in addition to the tapped hole in the gland ring, a hole is drilled through the stationary insert to channel the flush into the space between the insert inside diameter (i.d.) and the shaft. To seal the insert outside diameter (o.d.) to the gland bore, the insert is mounted in two O rings.

In the arrangement described, the flush flow is directed through a heat exchanger before entering the stuffing box. However, a simple bypass flush without a heat exchanger is an effective temperature stabilizing system that is also used regularly in conjunction with single mechanical-seal installations. Although this type of flush does not cool the liquid being bypassed to or from the stuffing box, it is extremely effective in maintaining the temperature within the stuffing box at the same level as the pumped product. This is accomplished by carrying away any heat generated by the seal in the stuffing box. A bypass flush of this type gives the seal engineer a distinct advantage in selecting a seal with the knowledge that the temperature inside the stuffing box will be the same as the system temperature.

A second method of cooling a stuffing box by injecting a flush over the seal

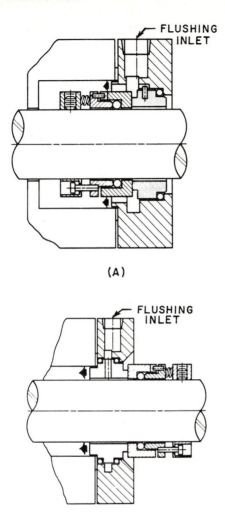

Figure 6.16 (A) Flushing-inside seal.
(B) Flushing-outside seal.

faces is accomplished by the use of a circulating or pumping ring. One type is shown in Figure 6.17. It consists of an auxiliary pump of low head and capacity that is mounted on the process pump shaft inside the stuffing box. Liquid is carried between the slots in the pumping ring as the shaft rotates. It is restricted from evacuating these ports by the close clearance that has been intentionally created between the o.d. of the pumping ring and the bore of the stuffing box. When the ports pass under the low-pressure area at the discharge hole in the stuffing box, captive liquid flows out through a heat exchanger. After passing through the heat-exchanger coils, it returns to the stuffing box via a flushing connection in the gland ring located directly over the seal faces.

The circulating ring is often preferred over a bypass-flush system when a heat exchanger must be used. Depending on pump pressures and tempera-

Figure 6.17 A circulating ring. (*Courtesy of Durametallic Corp., Kalamazoo, MI*)

tures, a circulating-ring feature may allow the use of a smaller heat exchanger because of its low flow rate. Note that most bypass-flush systems originate at the pump discharge; they depend on relatively high flow rates to create enough velocity at the throat of the stuffing box to purge the area of hot process liquid. This increase in flush line pressure, coupled with a shorter residence time in the heat exchanger, often requires the use of a heavier heat-transfer device with greater cooling coil area.

When installing a circulating-ring cooling system, caution must be exercised in locating the heat exchanger. Unduly long flush lines with unnecessary turns can easily set up frictional losses that exceed the discharge head of the circulating ring. Therefore, the heat exchanger should be mounted as close to the pump stuffing box as possible, preferably directly over the stuffing box.

The capacity of the circulating ring to develop head is limited by the peripheral speed of the shaft. A minimum peripheral speed of 800 fpm is required to deliver 0.5 gpm. Table 6.4 can be used for computing circumferential speeds in fpm from shaft size and rpm.

Cooling often is used when the liquid is dead-ended in the stuffing box and provisions for flushing are uneconomical. However, the word *cooling* can be

TABLE 6.4 CONVERSION OF SHAFT SIZE AND rpm TO PERIPHERAL SPEED

Diameter in Inches			Revolutions per Minute									
			2,400	2,600	2,800	3,000	3,200	3,400	3,600	3,800	4,000	4,400
			1,200	1,300	1,400	1,500	1,600	1,700	1,800	1,900	2,000	2,200
			600	650	700	750	800	850	900	950	1,000	1,100
1/4	1/2	1	157	170	183	196	209	223	236	249	262	288
1/2	1	2	314	340	367	393	419	445	471	497	524	576
3/4	1 1/2	3	471	510	550	589	628	668	707	746	785	863
1	2	4	628	681	733	785	838	890	942	995	1,047	1,152
1 1/4	2 1/2	5	785	851	916	982	1,047	1,113	1,178	1,244	1,309	1,440
1 1/2	3	6	942	1,021	1,100	1,178	1,257	1,335	1,414	1,492	1,571	1,728
1 3/4	3 1/2	7	1,100	1,191	1,283	1,375	1,466	1,558	1,649	1,741	1,832	2,016
2	4	8	1,257	1,361	1,466	1,571	1,675	1,780	1,885	1,990	2,094	2,304
2 1/4	4 1/2	9	1,414	1,531	1,649	1,767	1,885	2,003	2,121	2,238	2,356	2,592
2 1/2	5	10	1,571	1,702	1,833	1,964	2,094	2,225	2,356	2,487	2,618	2,880
2 3/4	5 1/2	11	1,728	1,872	2,016	2,160	2,304	2,448	2,592	2,736	2,880	3,168
3	6	12	1,885	2,042	2,199	2,356	2,513	2,670	2,827	2,984	3,143	3,456
3 1/4	6 1/2	13	2,042	2,212	2,382	2,552	2,723	2,893	3,063	3,233	3,403	3,744
3 1/2	7	14	2,199	2,382	2,566	2,749	2,932	3,115	3,299	3,482	3,665	4,032
3 3/4	7 1/2	15	2,356	2,552	2,749	2,945	3,142	3,338	3,534	3,731	3,927	4,320
4	8	16	2,513	2,723	2,932	3,142	3,351	3,560	3,770	3,979	4,189	4,608
4 1/4	8 1/2	17	2,670	2,893	3,115	3,338	3,560	3,783	4,006	4,228	4,451	4,896
4 1/2	9	18	2,827	3,063	3,299	3,534	3,770	4,006	4,241	4,477	4,712	5,184
4 3/4	9 1/2	19	2,985	3,233	3,482	3,731	3,979	4,228	4,477	4,725	4,974	5,472
5	10	20	3,142	3,403	3,665	3,927	4,189	4,451	4,712	4,974	5,236	5,760
5 1/4	10 1/2	21	3,299	3,573	3,848	4,123	4,398	4,673	4,948	5,223	5,498	6,048
5 1/2	11	22	3,456	3,744	4,032	4,320	4,608	4,896	5,184	5,472	5,760	6,336
5 3/4	11 1/2	23	3,613	3,914	4,215	4,516	4,817	5,118	5,419	5,720	6,021	6,623
6	12	24	3,770	4,084	4,398	4,712	5,027	5,341	5,655	5,969	6,283	6,912
6 1/2	13	26	4,084	4,424	4,764	5,105	5,445	5,786	6,126	6,466	6,807	7,487
7	14	28	4,398	4,764	5,131	5,498	5,864	6,231	6,597	6,963	7,330	8,063
7 1/2	15	30	4,712	5,105	5,498	5,890	6,283	6,676	7,069	7,461	7,854	8,639
8	16	32	5,027	5,445	5,864	6,283	6,702	7,121	7,540	7,959	8,378	9,215
8 1/2	17	34	5,341	5,785	6,231	6,676	7,121	7,566	8,011	8,456	8,901	9,791
9	18	36	5,655	6,126	6,597	7,069	7,540	8,011	8,482	8,954	9,425	10,367
9 1/2	19	38	5,969	6,466	6,964	7,461	7,959	8,456	8,954	9,451	9,948	10,943
10	20	40	6,283	6,807	7,330	7,854	8,378	8,901	9,425	9,948	10,472	

misleading since the same design approach can be used to transmit heat to the seal, thereby preventing solidification or crystallization of the product.

Whether this provision is used for cooling or heating, the design and method of heat transfer remain the same. The gland ring is machined to include two drilled and tapped holes leading to an annular groove in the counterbore of the gland ring. The heat-transfer fluid, usually water or steam, is circulated through this annular groove picking up from, or imparting heat to, the stationary insert for the purpose of controlling seal-face temperatures. Since carbon is a relatively poor conductor of heat, other methods of temperature control prove to be more efficient. However, cooling of this type can be particularly effective when combined with a water-jacketed stuffing box, as shown in Figure 6.18.

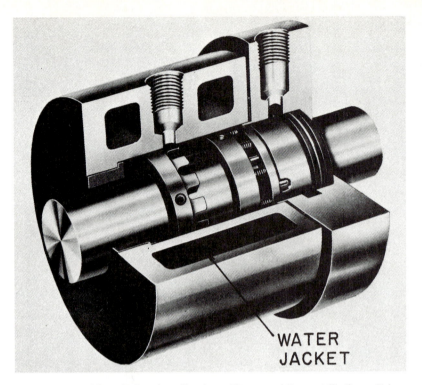

WATER
JACKET

Figure 6.18 Water-jacketed stuffing box. (*Courtesy of Durametallic Corp., Kala-mazoo, MI*)

For higher-temperature steam heating, a two-piece welded gland ring is available for steam pressures to 300 psia and temperatures to 417°F (214°C). Most installations that involve any type of steam heating are of such a nature that the heating should be left on at all times, regardless whether the equipment is operating. Careful preheating to dissolve all solids is necessary following a complete shutdown.

Quenching the outside seal is still another approach to temperature control. Quenching an outside mechanical seal involves immersing a large portion of the assembly in water, thereby approaching the combined heat-transfer efficiency of flushing and gland cooling.

Figure 6.19 shows an outside, quenched seal. Two pipe taps in the gland, 180° apart, allow water to be run over the seal ring and insert nose. The outlet connection should always be lower than the inlet connection, since the water must drain through the outlet connection. In addition to cooling, this arrangement also has been used as an effective means for washing away any leakage that may occur at the seal faces. The method can be used as a safeguard when handling dangerous liquids or to prevent the accumulation of abrasive solids when handling liquids that crystallize on contacting the atmosphere. Some designs are available with a quench collar. The unit shown in Figure 6.20 can

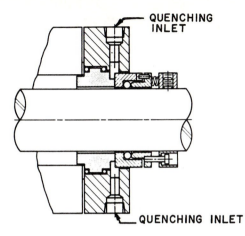

Figure 6.19 Quench outside seal. (*Courtesy of Durametallic Corp., Kalamazoo, MI*)

accommodate up to 3 gpm of flow at heads of 5 feet to 10 feet on the outlet. This means that the inlet and outlet pipe connections can be placed on a horizontal center line if so desired. The quench-collar arrangement is also ideal when quenching is used to carry away toxic or noxious vapors emitted from the mechanical seal.

Figure 6.20 Quench collar. (*Courtesy of Durametallic Corp., Kalamazoo, MI*)

Inside seals also may be quenched. Once more, unpressurized water is allowed to flow through the gland and drain out through a connection at the bottom. The quenching water is retained in the gland by means of a throttle bushing or auxiliary compression packing. However, when an inside seal is quenched, the water naturally tends to follow the path of least resistance, which is directly behind the insert. By contrast to an outside quenched seal, where water is constantly running over the seal faces, exchange of water is at a much lower rate under the insert of an inside seal. Therefore, other forms of cooling inside seals are usually employed, providing greater efficiency when cooling is required. However, this method is effective for use as a safeguard or to prevent the accumulation of solids from cystallization.

CONTAMINATED FLUIDS HANDLING

Quite often, extremely corrosive or dirty liquids are encountered that are difficult to seal even with the use of the most exotic materials. In such cases, the mechanical seal must be isolated from the destructive product.

For dirty or contaminated fluids, a bypass-flush arrangement can be used to flood the stuffing box with clean liquid. However, this can be accomplished only by inserting filters or cyclone separators into the flush line. Caution must be exercised when selecting either of these two options rather than a double-seal design. Extremely dirty fluids will rapidly clog the filters. On the other hand, cyclone separators are only 90 percent efficient in removing solids down to 2.5 microns in size under optimum pressure and flow conditions. Therefore, filters should be called on to clean flush systems that are only lightly contaminated with solid particles. Cyclone separators should be applied only to moderately dirty systems or where the abrasive material is known to be at least $9\ \mu$ in size.

External liquid flushing involves the introduction of a clean, clear liquid to the stuffing box from an external source. The first point to be considered with the external flushing arrangement is that to be effective, it must be injected into the box at a pressure greater than the stuffing-box pressure. Once more, as is the case with a bypass flush, the stuffing box must be provided with a throat bushing or some effective throat restriction so that it will be completely purged of the pumped product by the external liquid flush.

A definite measurable amount of product dilution will occur when using an external liquid flush. The amount of dilution will depend on the design of the throat restriction and the pressure differential between the flush liquid and the product behind the impeller.

The external flush arrangement has proven satisfactory and is widely accepted as standard procedure by many process industries handling abrasive liquids and slurries where some dilution of the product can be tolerated. The external flush will provide a clean liquid to the seal faces and prevent abrasive solids from entering the stuffing box.

Stuffing-Box Throat

The throat of a pump is that portion of the pump casing that forms a close clearance with the pump rotating shaft. The throat is also the bottom face of the stuffing box and, therefore, is commonly referred to as the stuffing-box throat. In addition to forming a close clearance with the shaft to reduce leakage of the pumped liquid to the stuffing box, it also acts as a shoulder to retain the stuffing-box packing. On pumps incorporating packing, the clearance between the throat and the pump shaft must be maintained relatively small to prevent extrusion of the packing.

For most commercially designed centrifugal pumps, the normal throat clearance between the shaft and the stuffing-box throat (without throat bushing) is 0.030- to 0.040-inch radial clearance. On pumps with mechanical seals it is often necessary to reduce this clearance to limit flow past the throat, to dead-end the liquid in the stuffing box or to raise the pressure of the liquid in the stuffing box.

When jackets are incorporated into the stuffing-box design to facilitate effective cooling or heating of the liquid within, it is normally recommended to dead-end the liquid in the stuffing box. The addition of a restriction device at the throat of the stuffing box dead-ends the liquid in the stuffing box and impedes mix with the pumped liquid.

Bypass lines (recirculation lines) are frequently used to ensure proper lubrication and cooling of mechanical seals when the pressures acting on the stuffing box are near the vaporization pressure of the liquid being pumped. Since most stuffing-box pressures are only slightly in excess of suction pressure, it is necessary to pressurize the stuffing box to prevent vaporization of the liquid in the stuffing box due to the small amount of frictional heat being generated at the mechanical seal faces and heat generated by turbulence. Bypass lines under discharge pressure will not raise the pressure in the stuffing box unless there is restriction at the stuffing-box throat. In no case, however, can the pressure in the stuffing box be raised to equal full discharge pressure.

The most common use of restriction devices at the stuffing-box throat is in conjunction with external flushing to the stuffing box. External flushing involves the introduction of a clean, cool liquid to the stuffing box from an external source. It is used whenever the pumped liquid will not provide a suitable environment for proper mechanical seal operation and life. When the pumped liquid is a slurry containing solids or abrasives that would not be suitable for lubrication of the mechanical seal faces, external flushing at a pressure higher than stuffing-box pressure is introduced to prevent the pumped liquid from entering the stuffing box. The mechanical seal then operates in the environment of the external flushing liquid. External flushing also is used to prevent hazardous, toxic, extremely corrosive, hot liquids (or liquids that tend to vaporize or solidify) from entering the stuffing box, and to provide a suitable environment for mechanical seal operation.

When external flushing is necessary to provide a proper mechanical seal environment, and small amounts of dilution of the pumped product can be tolerated, single mechanical seals with a restriction device at the stuffing-box throat are used.

To exclude the pumped liquid from entering the stuffing box, it is necessary to create a velocity at the stuffing-box throat. Velocities of 10 ft/s to 15 ft/s at the throat have proven satisfactory for this purpose. From Figures 6.20 and 6.21 it can be seen that with normal throat radial clearances of 0.030 in. to 0.040 in., flushing liquid flow rates in the magnitude of 10 gpm are required to obtain the desired velocity at the throat.

The primary purpose of a restriction device at the throat of the stuffing box is to lessen the radial clearance at the throat, thereby reducing the amount of

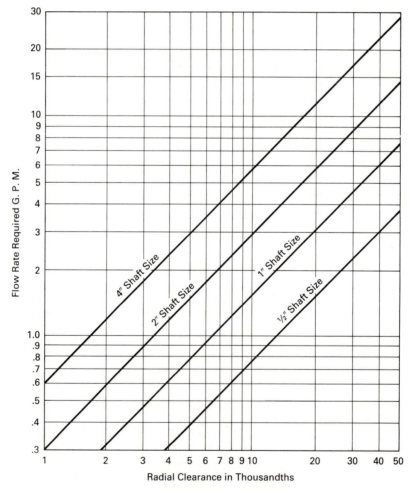

Figure 6.21   Flow rates required to create 15 fps velocity. (*Courtesy of Durametallic Corp., Kalamazoo, MI*)

flushing liquid required to create the desired velocity. In no case will the device control the flow.

The simplest type of throat-restriction device is the fixed throat bushing. This is a bushing, normally of a nonsparking, nongalling material such as carbon, which is pressed into a bore at the throat of the stuffing box.

The fixed throat bushing shown in Figure 6.22A is the arrangement used in standard chemical-type pumps. The throat of the stuffing box is remachined to accept a 0.003- to 0.005-in. interference fit with the outside diameter of the throat bushing. An alternate arrangement designed for refinery-type pumps is shown in Figure 6.22B. Here, the bushing is mounted for opposition to internal pump casing pressure in the event of flushing liquid pressure loss. For horizontally

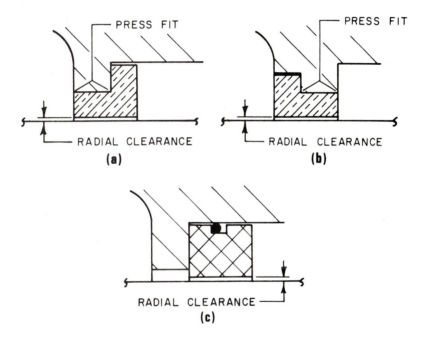

Figure 6.22 Sketch of fixed throat bushings.

split case pumps, where it is not possible to press the bushing into the throat, the bushing may be mounted in an O ring, as shown in Figure 6.22C.

The practical radial clearance for fixed throat bushings is 0.006-in. to 0.008 in. Clearances smaller than this may result in the shaft touching the bore of the bushing due to shaft deflection or shaft whip.

The following method is used to calculate the flushing liquid flow rate and flushing liquid pressure required.

1. From Figure 6.21, determine the flow rate required for the radial clearance and shaft size.

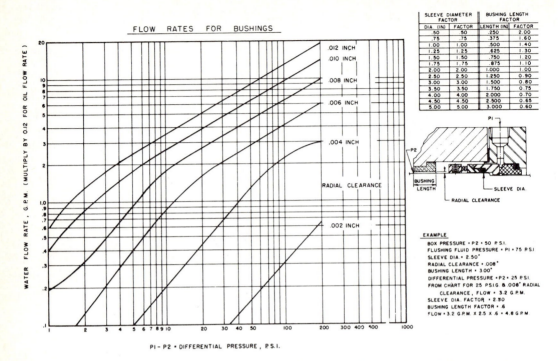

Figure 6.23   Flow rates for close clearance bushings. (*Courtesy of Durametallic Corp., Kalamazoo, MI*)

2. From Figure 6.23, correct the flow rate for shaft size and bushing length.
3. For the corrected flow rate and radial clearance, determine the differential pressure required across the bushing.
4. The flushing liquid pressure required is the stuffing-box pressure plus the differential pressure obtained from the preceding step 3.

Floating Throat Bushings

The radial clearance for fixed throat bushings as described previously must be sufficient to prevent rubbing of the shaft in the inside diameter since the fixed bushing does not compensate for shaft deflection or shaft whip. For smaller radial clearances and resulting lower flushing liquid flow requirements, the bushing must compensate for shaft deflection and/or shaft whip. The floating throat bushing as shown in Figure 6.24 is allowed to move radially, thereby centering itself around the shaft, regardless of the location of the shaft with respect to the stuffing-box bore.

    The carbon bushing is held stationary in the bottom of the stuffing box by spring pressure. There is a liberal clearance between the bushing o.d. and the stuffing-box bore so the bushing can move radially, thereby centering itself

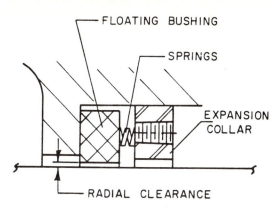

FLOATING BUSHING

SPRINGS

EXPANSION
COLLAR

RADIAL CLEARANCE

Figure 6.24 Sketch of floating throat bushing.

around the shaft. The springs are held in an expansion collar, which incorporates a tapered set screw tap which, on tightening of the set screw, expands the collar to grip in the stuffing-box bore. Due to the ability of the bushing to move radially, closer radial clearances can be utilized to reduce flushing fluid flow rates. Flushing liquid flow rates and flushing liquid pressure requirements are calculated in the same manner as for the fixed throat bushings.

Regardless of which throat restriction device is to be used, a minimum flushing flow rate is required for proper cooling and lubrication of the mechanical seal. Figure 6.25 is a guide to the minimum flushing flow rates that are required for proper operation of the mechanical seal.

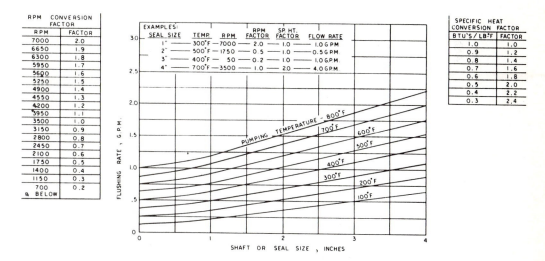

Figure 6.25 Sealing and flushing liquid requirements. (*Courtesy of Durametallic Corp., Kalamazoo, MI*)

The function of a double seal has been discussed already. It is noteworthy that double seals perform the same environmental function as single seals flushed from an outside source. Double seals are usually applied in situations where the nature of the product is such that it would rapidly deteriorate a single seal. Although the single seal theoretically could be flushed with a compatible liquid from another source, many processes cannot tolerate the resulting contamination or dilution of the product that is inherent with external liquid flushing. Therefore, the mechanical seal user looks to a double seal as a means of flooding the stuffing box with a clean, cool liquid without incurring the consequences of this liquid mixing with the product being pumped.

All double seals require a sealing liquid between the seals for proper lubrication and cooling. This sealing liquid must be maintained at a pressure of 15

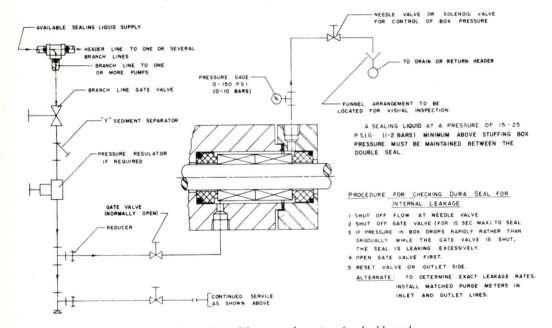

Figure 6.26   Water supply system for double seals.

psi to 25 psi above the stuffing-box pressure. In addition, the sealing liquid must be circulated. The amount of sealing liquid flow required can be estimated from Figure 6.25. Figure 6.26 is a suggested arrangement for a water supply system for double seals.

When selecting a sealing liquid system for double-tandem seals, one must consider not only the amount of cooling that is required to dispose of the heat generated by the outer or secondary seal, but also how to dispose of the primary seal weepage that occurs during normal operation. In a double-tandem seal, the primary seal weepage is contained in the sealing liquid system for the secondary seal. Unless this collection of normal primary seal weepage is properly drained or vented, a pressure buildup will occur in the secondary seal system.

SAFETY FEATURES

When a mechanical seal fails, leakage occurs. It is the responsibility of the user and the seal manufacturer to determine whether the leakage resulting from a seal failure poses a threat to the equipment, personnel, or environment. If any leakage of the process fluid is hazardous, then provisions must be made for controlling the leakage.

Two options are available to the user for controlling leakage. One is to use double or tandem seals, both of which will prevent leakage of the process fluid to the atmosphere in the event the inner or primary seal fails. The other alternative is to use a single mechanical seal and make provisions for collecting the leakage should a seal failure occur (i.e., a vent and drain).

The vent and drain can be applied to the gland rings of inside seals in addition to other features already discussed. The purpose of a vent and drain is as the term implies. As an example, when handling liquified petroleum gases, any leakage past the seal will be a highly explosive gas that can be "vented" to safe external disposal. The drain connection at the bottom of the gland, 180° from the vent, permits hazardous liquid leakage to be carried away to a point where it can be collected safely.

The vent and drain are provided in the gland ring by machining two drilled and tapped openings behind the stationary insert, as illustrated in Figure 6.27. Leakage is discouraged from traveling along the shaft to the atmosphere by a restriction, called a throttle bushing, incorporated into the rear of the gland ring. With the vent and drain connections backed up by a throttle bushing, leakage takes the path of least resistance to the point of collection or disposal.

Most throttle bushings used in conjunction with vent and drain connections are of the fixed type, being pressed into a bore at the rear of the gland ring and adequately backed up by the back of the gland ring itself. Throttle bushings are made from either carbon or bronze, both of which are nonsparking materials.

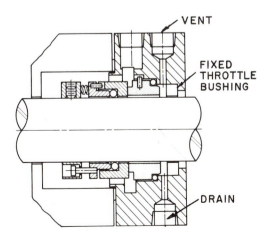

Figure 6.27 Sketch of vent and drain with fixed throttle bushing.

Figure 6.27 shows a typical fixed throttle bushing application. Since the throttle bushing is not free to float and center itself around the shaft, 0.025-in. diameter clearance is provided between the i.d. of the bushing and o.d. of the shaft. It has been found that this is the optimum clearance to adequately accommodate slight vertical misalignment of the gland ring and/or shaft runout.

An alternate gland throttle bushing arrangement is the floating variety. This type of bushing is free to center itself around the shaft, thereby allowing the application of a closer radial clearance of 0.002 in. between the bushing and the shaft. The additional restricting efficiency provided by the floating throttle bushing is especially popular for applications in which it is desirable to achieve maximum restriction by the use of a throttle bushing.

The application of an auxiliary stuffing box behind a mechanical seal is illustrated in Figure 6.28. In this case, an auxiliary gland follower flange compresses two rings of suitable packing against the shaft. This arrangement will provide a nearly positive auxiliary shaft seal in the event the primary mechanical

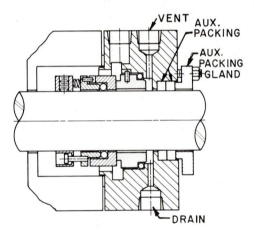

Figure 6.28 Sketch of vent and drain with auxiliary stuffing box.

seal fails. However, it is necessary that the packing be properly lubricated and cooled to avoid excessive wear on the shaft and packing. This entails circulating a liquid, usually water, through the vent and drain connections.

Although an auxiliary stuffing box provides the user with a tight backup seal, the necessity of furnishing lubrication to the packing may prove objectionable. Therefore, one may prefer to use a throttle bushing design that does not require the circulation of any liquid through the vent and drain connections.

SEAL SELECTION AND APPLICATION

A number of factors must be considered in selecting the proper mechanical seal for an application. Seal size, rpm, pressure, temperature, chemical and physical characteristics of the process liquid, and the equipment to which the seal is to be applied all come into play in seal selection. The actual selection is often a compromise among several of these factors along with space availability, customer specifications, economics, and the availability of environmental controls.

PROCESS LIQUID CHARACTERISTICS

Corrosiveness

The corrosive nature of the process liquid will determine the metallurgy, face materials, and type of secondary sealing materials that can be used. Note that while some materials of construction possess excellent resistance to corrosion, they may have poor wearing characteristics, while others that provide excellent wear properties exhibit poor corrosion resistance. Seal manufacturers do not always agree on design and material selection for a given application. This is due to differences in design, standardization of parts and materials, and economics. Also, while some material combinations may be less expensive in initial cost, the extension of seal life through the use of upgraded materials, nevertheless, may be an excellent investment. This is especially true when complicated and costly maintenance procedures are involved.

For very corrosive applications, outside seals are the preferred arrangement. The use of outside seals with proper material selection can result in a seal design in which no metallic parts are in contact with the process liquid. Small parts such as springs and pins thus would not be subject to process liquid corrosion. Very often, using an outside seal arrangement also will allow the use of a more economical material for metallic components.

Density or Specific Gravity

Liquid density or specific gravity is an indication of lubricity and can be used as a general guide in seal selection. However, it must be remembered that specific gravity is a guide only. Generally, liquids with specific gravities above 0.65 possess lubricating qualities sufficient for most seal face combinations.

For liquids with specific gravities below 0.65, balanced seals are required. In addition, tungsten carbide is preferred as one of the sealing faces because of its excellent wearing characteristics in the absence of a good lubricating film between the faces.

Certain face combinations can be used only in good lubricating liquids. For example, ceramic versus ceramic can only be used in liquids possessing lubricating qualities equal to that of oil, concentrated sulfuric acid, or concentrated nitric acid.

Vapor Pressure and Boiling Point

Vapor pressure is an important consideration when the process liquid is being handled at a pressure at or near its vapor pressure. Light hydrocarbons, for example, especially liquified petroleum gases (LPG), require a stuffing-box pressure-temperature relationship that will ensure a stuffing-box pressure of at least 25 psi to 50 psi greater than the vapor pressure of the product at the stuffing-box temperature.

One method of increasing the stuffing-box pressure using the process liquid is through the installation of a bypass recirculation line and a throat restriction device. Note, however, that there must be sufficient pressure differential between the pump discharge pressure and stuffing-box pressure to be effective.

An alternative to increasing the stuffing-box pressure for proper stuffing-box pressure-temperature relationship is cooling the stuffing box. Experience has proven that stuffing-box temperatures should be maintained 25° to 50°F (15° to 30°C) below the boiling point of the product at the corresponding stuffing-box pressure. Stuffing-box water jackets, external coolers, and so on can be installed for this purpose.

The volatility aspect of a process stream also must be considered from the safety standpoint. This may involve incorporating seal features such as vents and drains with a gland-ring bushing or auxiliary stuffing box.

Viscosity

Viscosity, like density, is an indication of a liquid's lubricating qualities. Liquids with higher viscosities usually possess better lubricating characteristics.

For liquids with viscosities less than 3,000 SSU or 700 cp, standard seal designs may be applied without restriction or modification. Special consideration must be given to the sealing of liquids with higher viscosities.

Liquids with viscosities of 3,000 SSU to 7,000 SSU (700 ccp to 1,600 cp) require that the seal be of a positive-drive design with a heavy-duty compression unit. This is to compensate for the increased shear that occurs at the sealing faces under a viscous environment.

Mechanical seals applied to liquids with viscosities higher than 7,000 SSU (1,600 cp) require both (1) positive drive with heavy-duty compression units, and (2) environmental controls for proper lubrication of the sealing faces. With these high viscosities, no liquid film between the sealing faces can possibly exist using the process liquid. Single seals with external flushing, single seals with external lubrication, or double seals with an external sealing liquid must be used.

Abrasives

If the liquid being handled is dirty, contains abrasives, or is a slurry, all of which would not properly lubricate the seal faces, an arrangement to provide a clean liquid environment for the seal must be applied.

There are numerous approaches to providing an artificial environment for the mechanical seal if the process liquid is unsuitable. Considerable thought should be given to the selection of the most economical environmental control system before the seal selection is made. The selection of the mechanical seal design and materials of construction should be based on the actual conditions in the stuffing box. If the seal is to operate under environmental controls, the selection should be based on the conditions imposed by the environmental control, rather than the operating conditions of the equipment to which the seal is applied.

Physical and Chemical Changes

Process liquids that go through physical or chemical change with slight variations from the temperature at which the liquid is being handled may require temperature controls in the sealing area to ensure that adequate lubrication is maintained at the seal faces. Likewise, liquids with high melting points or that tend to crystallize on exposure to the atmosphere will require temperature control to ensure that the liquid does not solidify at the faces.

STUFFING-BOX PRESSURE

The pressure that exists in the stuffing box varies with equipment design. For example, on most end-suction centrifugal pumps, the stuffing box is subject to the pump-suction pressure or suction pressure plus a small percentage of discharge pressure. Vertical turbine pumps, on the other hand, often subject the stuffing box to full pump-discharge pressure. Many multistage pumps have one stuffing box subjected to pump-suction pressure, while the other stuffing box is subjected to an intermediate-stage discharge pressure. All this points out that knowledge of more than pump-suction and discharge pressures may be necessary to determine the stuffing-box pressure for which the seal must be selected.

The best source of stuffing-box pressure information is the original equipment manufacturer. Most pump manufacturers have tested their equipment hundreds of hours to determine stuffing-box pressures and can readily provide this information for any given set of operating conditions.

For equipment already in the field, the easiest way to determine the stuffing-box pressure is to install a pressure gauge into a tapped hole leading directly to the stuffing box.

In discussing pressure and its influence on seal selection or application, it is necessary to include velocity. Both laboratory testing and field experience have shown that velocity is as important a factor as pressure in determining the performance of mechanical seals. It has been demonstrated that unbalanced seals can be used at relatively high pressures, if the velocities are correspondingly low. On the other hand, when velocities are high, balanced seals are sometimes required even with low pressures.

In the past, pressure limitations of unbalanced seals have been established arbitrarily, regardless of size, speed, mating face materials, or liquid handled. Some seal manufacturers have used a value of 200 psig, others 150 psig or 100 psig, for unbalanced seals. Some refineries have adopted standards that require the use of balanced seals whenever the pressure in the stuffing box exceeds 50 psig. Such pressure limitations are established largely from field experience.

Modern sealing technology recognizes the importance of such factors as shaft or seal size, peripheral speed, contact mating face materials, and the characteristics of the liquid handled. It is not to be expected, for example, that for a specific pressure and liquid an unbalanced 3-in. shaft seal operating at 3,500 rpm would have the same life expectancy as a 1-in. shaft seal operating under the same conditions at 1,750 rpm.

Recognition of these factors has led to the establishment of *pressure-velocity* limitations for inside unbalanced seals. The velocity portion of this relationship is the peripheral speed expressed in ft/min (fpm) which can be arrived at by the formula:

$$\text{Velocity (fpm)} = \frac{\text{Shaft Size (in.)} \times \text{rpm}}{3.82} \qquad (6.1)$$

Figure 6.29 shows the relationship between shaft diameter, rpm, and peripheral speed in fpm in the form of a nomogram.

Note that wear differences exist with different liquids handled, even though seal size, speed, and pressure may be identical. Therefore, two sets of curves are provided by seal manufacturers, one for water and water solutions, and one for liquids that could be considered at least equal to gasoline in lubricity. This provides conservative estimates for many acids and corrosive liquids; although they present corrosive problems, they still maintain a good liquid film between the faces.

Pressure-velocity ratings based on peripheral speed now have been established for all of the more popular combinations of mating face materials presently used. From these ratings, pressure-velocity curves have been plotted for both water and water solutions, and gasoline and equivalent liquids, at both 1,750 and 3,500 rpm. The established ratings are based on a seal life of 15,000 hours on either continuous or intermittent service.

Figure 6.30 is a pressure-velocity curve for inside unbalanced seals for water solutions when the stuffing-box temperature does not exceed 160°F (70°C). For water temperatures in excess of 160°F (70°C), special considerations must be given, as will be discussed later.

Figure 6.31 is a pressure-velocity curve for inside unbalanced seals for acids, hydrocarbons, and other liquids with similar lubricity. This curve may be used for light hydrocarbon applications where the specific gravity is about 0.65. It does not apply to light hydrocarbons or LPG applications in which the specific gravity is below 0.65. Where such a condition is encountered, balanced seals are required regardless of speed, seal size, or pressure.

To use the pressure-velocity curves, enter the chart at the correct seal size and read up to the appropriate material-speed curve. At this point, read across to obtain the maximum recommended stuffing-box pressure for an inside unbalanced seal.

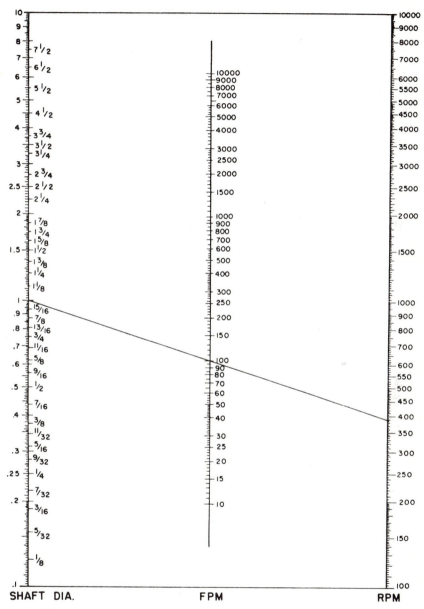

A straight edge line up with the values on any pair of the vertical lines will cross the third line at the corresponding value. For example, a straight edge set at 1″ on the shaft diameter line and 382 RPM on the R.P.M. line will cross the FPM line at 100 FPM, the corresponding surface speed.

Figure 6.29   Nomogram defining the relationship between shaft size, rpm and peripheral speed. (*Courtesy of Durametallic Corp., Kalamazoo, MI*)

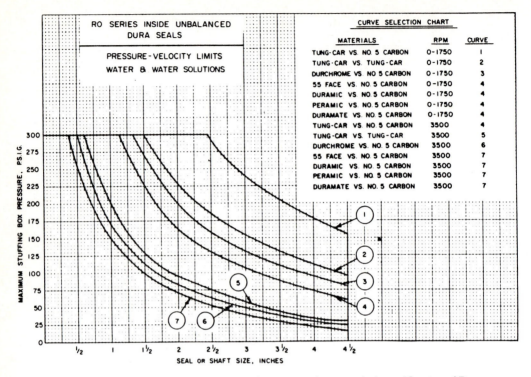

Figure 6.30 Pressure-velocity curves for water and water solutions. (*Courtesy of Durametallic Corp., Kalamazoo, MI*)

If the actual pressure involved in the application exceeds the maximum recommended stuffing-box pressure, then a change in the material selection to one with a higher pressure-velocity rating or a balanced seal is required.

Pressure-velocity ratings and curves also have been established for other seal designs and special material combinations. For example, Figure 6.32 shows ratings for another series of seals. These ratings are somewhat lower than those in Figure 6.30 since this series of seals incorporates friction drive.

Vacuum Conditions

Pressure and/or vacuum also determine whether inside- or outside-type unbalanced seals can be used for a given application. Outside unbalanced seals are recommended for use with up to approximately 35 psig of positive stuffing-box pressure. They are also recommended for use on negative pressures. In such applications, the vacuum adds to the spring load on the seal. Vacuum applications require environmental controls such as lubrication or product recirculation to maintain a liquid film between the seal faces. With satisfactory environmental controls, outside mechanical seals are capable of handling very high vacuums.

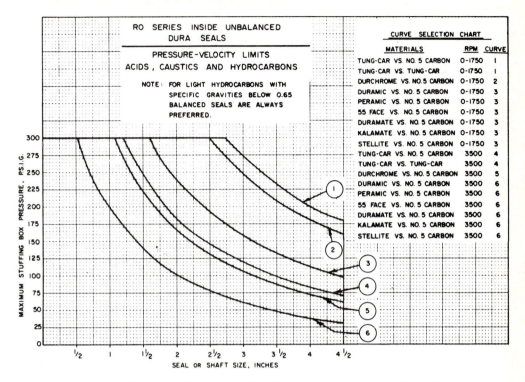

Figure 6.31 Pressure-velocity curves for acids, caustics, and hydrocarbons. (*Courtesy of Durametallic Corp., Kalamazoo, MI*)

Outside seals offer ease of assembly, ease of repair, and visual performance of the seal since the rotation portion of the seal operates in the atmosphere outside the stuffing box rather than inside the stuffing box. When space permits, outside unbalanced seals are sometimes preferred in low-pressure, corrosive service. Quench collars provide further protection from corrosive atmospheres.

Balanced seals have also pressure limitations that vary with peripheral speed and seal-face materials, although they are substantially higher than the corresponding unbalanced seal limits. In many cases, the pressure limitations of balanced seals are due to inherent design and material limitations, regardless of peripheral speed. Figure 6.33 is a pressure-velocity curve for inside balanced seals for water and water solutions.

Outside unbalanced seals are limited to 35 psig stuffing-box pressure, since one must depend solely on the spring load to keep the seal faces closed. However, outside seals have many advantages, especially in corrosive applications. Furthermore, in many cases the seal must tolerate pressures higher than 35 psig. Balanced outside seals are designed to withstand stuffing-box pressures up to 400 psig.

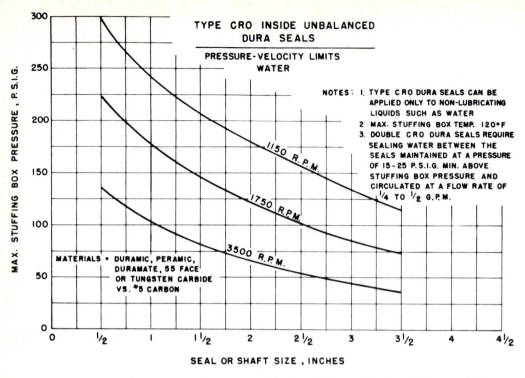

Figure 6.32 Pressure-velocity curves. (*Courtesy of Durametallic Corp., Kalamazoo, MI*)

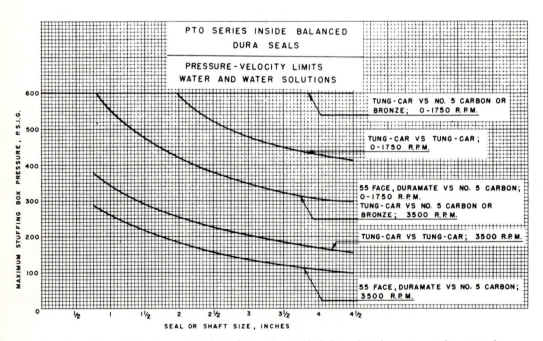

Figure 6.33 Pressure-velocity curves for inside balanced seals, water, and water solutions. (*Courtesy of Durametallic Corp., Kalamazoo, MI*)

STUFFING-BOX TEMPERATURE

Until recently, the use of elastomers for secondary sealing members in a mechanical seal limited the temperature that could be tolerated in the stuffing box. Advancements in metal bellows technology and flexible nonelastomeric materials have increased the temperature limits of mechanical seals, provided the liquid has adequate lubricating qualities at the process temperature.

Assuming that the process liquid will provide adequate lubrication of the sealing faces, once a seal design is selected for the pressures that will be incurred and the materials have been selected for the corrosive nature of the process liquid, the temperature limitations of each material should be checked. Temperature limitations of commonly used face materials and secondary sealing materials are given in Tables 6.2 and 6.3.

Process liquids that require close temperature control to maintain adequate lubricating or suitability as a sealing liquid between the faces may require environmental controls, as has been discussed throughout this chapter.

Equipment Design

Equipment design, as well as operating conditions, affects the selection of a mechanical seal. Noted earlier, the stuffing-box pressure is a function of equipment design. Other features of the equipment, including shaft or shaft-sleeve design, method of assembly, type of installation, and so on, also play important roles in seal selection.

Most small chemical pumps are not equipped with shaft sleeves due to the problems in sealing under the sleeves in highly corrosive services. This prevents the use of inside balanced seals that require a balanced step in a shaft sleeve. Balanced seals that do not require a balanced step in the shaft sleeve are available and can be accommodated into a $\frac{3}{8}$-inch packing space. Pumps equipped with hook-type shaft sleeves (Figure 6.34A) can easily be modified for inside balanced seals. However, to apply standard outside seals, the shaft sleeve must extend at least 2 in. beyond the face of the stuffing box.

Clamped-style shaft-sleeve designs (Figure 6.34B) normally cannot be modified since sealing under the sleeve is accomplished on the driver end and the sleeve locates the impeller on the liquid end. The location of the joint at the outer end of the sleeve must be known if outside seals are to be used.

The threaded-sleeve design in Figure 6.34C can easily be modified or cut off to use a cartridge-seal design. The sleeve design in Figure 6.34D, which uses a threaded nut outside of the stuffing box, presents several problems. For inside balanced sleeves requiring a step in the shaft sleeve, the sleeve thickness must be sufficient to allow proper stepdown but maintain ample thickness in the shaft-sleeve nut. Outside seals cannot be installed on the sleeve nut since runout of the outside diameter of the nut is excessive. Cartridge seals are used extensively on pumps with this type of sleeve design.

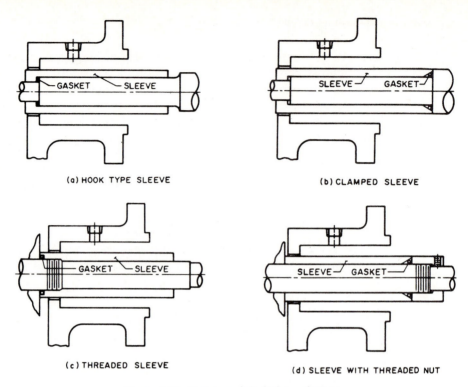

(a) HOOK TYPE SLEEVE

(b) CLAMPED SLEEVE

(c) THREADED SLEEVE

(d) SLEEVE WITH THREADED NUT

Figure 6.34   Sketches of shaft sleeve designs.

Vertically Split End-Suction Centrifugal Pumps

Enclosed-type impellers with balance holes and wearing rings generally provide suction pressure to the stuffing box, or suction pressure plus 10 percent to 15 percent of discharge pressure. Enclosed impellers without balance holes or wearing rings impose close to full discharge pressure on the stuffing box. Fully open impellers or semiopen impellers with balance holes and pump-out vanes on the back of the impeller impose suction pressure on the stuffing box.

Most pumps of this type for chemical services have solid shafts or hook-type shaft sleeves, as shown in Figure 6.34A. Pumps for general industrial services have hook-type shaft sleeves or clamped sleeves, as shown in Figure 6.34B.

Single-Stage Horizontally Split Case Pumps

These pumps normally have enclosed impellers with balance holes and wearing rings, which provide suction pressure to the stuffing boxes. Threaded sleeves such as those shown in Figures 6.34C and 6.34D are used normally.

Multistage Horizontally Split Case Pumps

In this design, stuffing-box pressures vary with impeller and staging arrangements. Usually the stuffing box on the suction end of the pump is subjected to suction pressure, while the stuffing box on the discharge end of the pump is subjected to discharge pressure of the first or other intermediate stage. Pressure-equalizing lines from the high-pressure stuffing box to the low-pressure stuffing box are often furnished to partially equalize the pressure of the stuffing box.

Threaded-shaft sleeves like those in Figures 6.34C or 6.34D are normally used. However, cartridge-seal designs are also in common use since they per-

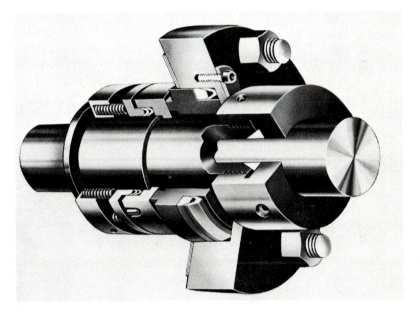

Figure 6.35 A cartridge seal design. (*Courtesy of Durametallic Corp., Kalamazoo, MI*)

mit seal installation, adjustment, and removal without removing the upper half of the casing. Any mechanical seal can be incorporated into a cartridge design, typically shown in Figure 6.35. The complete seal assembly, including the gland ring, is mounted on a sleeve that is secured to and driven by the shaft with a drive collar. The entire cartridge assembly is installed over the pump shaft and bolted to the stuffing box. The seal setting, adjustment, and drive are all accomplished outside of the stuffing box. The total length of the cartridge assembly should be taken into consideration in the design of the spacer on the coupling, so that the assembly on the driver end can be removed without disturbing the driver.

Vertical In-line Pumps

Vertical in-line pumps are vertical pumps with the suction and discharge nozzles in line. The driver is supported exclusively and the pump and the suction and discharge connections have a common center line, which intersects the shaft axis.

There are three types of vertical in-line pumps. The first is a one-piece pump/motor shaft design in which the impeller is mounted directly on an extended motor shaft. The second is a rigid coupling design in which the impeller is mounted on a stub shaft, which is coupled to the motor shaft by means of a rigid coupling. The third design incorporates a separate bearing housing, which makes this design very similar to a horizontal end-suction pump.

In most cases, the pressures acting on the stuffing box are similar to horizontal end-suction pump designs. Enclosed-type impellers with balance holes and wearing rings generally provide suction pressure to the stuffing box, or suction pressure plus 10 percent to 15 percent of discharge pressure. Enclosed impellers without balance holes or wearing rings impose close to full discharge pressure on the stuffing box. Fully open or semiopen impellers with balance holes and pump-out vanes on the back of the impeller impose suction pressure on the stuffing box.

The extended motor shaft and bearing housing designs incorporate hook-type shaft sleeves, as shown in Figure 6.34A. The rigid coupling design has a solid shaft. This stub shaft is provided with a balance step to accommodate standard inside balanced seals.

The stuffing box on all vertical pumps must provide for proper venting. Venting can be accomplished through a flush connection in the gland ring or through a connection in the stuffing box located near the seal faces.

Vertical Multistage Pumps

The stuffing-box pressure varies with pump design, specifically with impeller and staging arrangements. Some designs impose full discharge pressure on the stuffing box while others impose only suction pressure. Installations of this type require a thorough knowledge of the actual stuffing-box pressure.

This design normally requires adjustment of the impeller clearances after the pump has been completely assembled. This adjustment is accomplished by raising the pump shaft before the coupling is connected. As a result, only cartridge seals or outside seals can be applied since final setting of the seal must be made only after the coupling is connected.

Gland rings for vertical pumps and all other pumps mounted in a vertical position should provide for venting, bleedoff, or the equivalent to prevent a vapor trap in the stuffing-box area.

Rotary Pumps

Stuffing-box pressures vary considerably with this pump design, and thorough knowledge of the actual stuffing-box pressure is mandatory.

Many pumps of this type require that the pump be completely assembled before the seal can be installed since the shaft cannot be pulled through the stuffing box. Installation of inside seals requires that an access hold in the stuffing box be provided to tighten the set screws in the seal collar. Double seals are commonly used since rotary pumps are normally used on high-viscosity applications in which the pumped liquid would not lubricate the seal faces properly.

INSTALLATION AND OPERATION

A multitude of seal designs and materials has been devised to meet operating demands and conditions in the chemical and petroleum processing industries. Generally, however, all seal designs consist of two basic elements: one part secured to and *rotating* with the shaft, and the other part stationary and secured to the equipment.

The fundamental objective of mechanical seal operation is the development and maintenance of a liquid film between the seal faces. This microscopic film of liquid acts as a cushion for the seal faces to ride on, thus minimizing or preventing wear, as well as minimizing or preventing leakage.

Modern research and manufacturing methods have produced in the mechanical seal a dependable, precision-bult device that is designed to be installed in a proper and specific manner in mechanically sound equipment and maintained by recommended practices.

Equipment Check Points

Sealing performance and seal life depend to a great extent on the operating condition of the equipment to which the seal is being applied. Time spent in checking the equipment prior to seal installation has always proved to be a sound maintenance procedure to prevent excessive seal failures and to keep overall maintenance expense at a minimum.

The following conditions of the equipment should be checked:

   1. *Stuffing-box space.* To properly receive the seal, radial space and depth of the stuffing box must be the same as the dimensions shown on the assembly drawing that accompanies the seal.

   2. *Lateral or axial movement of shaft-end play.* Install a dial indicator, as shown in Figure 6.36A, so that the stem bears against the shoulder on the

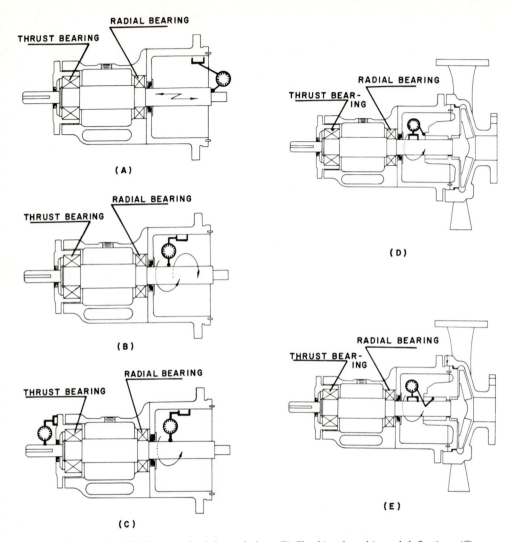

Figure 6.36 (A) Where to check for end play. (B) Checking for whip and deflection. (C) Checking for bent shaft. (D) Checking for stuffing box face squareness. (E) Checking for bore concentricity.

shaft. Use a soft hammer or mallet to tap the shaft lightly on one end and then on the other. Total indicated end play should be held between 0.001 in. and 0.004 in. A mechanical seal cannot function satisfactorily with a great amount of end play or lateral movement. If the hydraulic condition changes, which happens frequently, the shaft may "float," resulting in sealing troubles.

Minimum end play is a very desirable condition for the following reasons:

a. Excessive end play, or lateral shaft movement resulting in the shaft floating, can cause pitting, fretting, or wear at the point of contact between the shaft packing in the mechanical seal and the shaft or sleeve outer diameter. This condition may be alleviated if rubber elastomer materials can be used for shaft packing.

b. As the mechanical seal driving element is locked to the shaft or sleeve, any excessive end play or lateral movement will result in either overloading or underloading the springs, causing excessive wear and/or seal leakage.

c. A floating shaft can cause chattering, which results in chipping of the seal faces, especially the carbon element.

d. Ideal mechanical seal performance requires a uniform wear pattern and the maintenance of a liquid film between the mating contact faces. Excessive end play or lateral movement of the shaft due to defective thrust bearings can thus reduce seal life and performance by disturbing both the established wear pattern and lubricating film between the contact faces.

   3. *Radial movement of shaft (whip or deflection).* Install the dial indicator as shown in Figure 6.36B and as close to the radial bearing as possible. Lift the shaft or exert light pressure at the impeller end. If more than 0.0020 ~ 0.003 in. of radial movement occurs, investigate bearings and bearing fittings. An oversize radial bearing bore caused by wear, improper machining, or atmospheric corrosion will cause excessive radial shaft movement, resulting in shaft whip and deflection. *Minimum radial shaft* movement is important for the following reasons:

a. Excessive radial movement can cause wear, fretting, or pitting of the shaft packing or secondary sealing element at the point of contact between the shaft packing and the shaft sleeve outer diameter.

b. Excessive wear at the mating contact faces will occur when excessive shaft whip or deflection is present, due to defective radial bearings or bearing fits. The contact area of the mating faces will be increased, resulting in increased wear and the elimination, or reduction, of the lubrication film between the faces, further reducing seal life.

c. Defective radial bearings or bearing fits result in excessive vibration of the pump, resulting in reduced seal life and performance.

   4. *Shaft run out (bent shaft).* Clamp the dial indicator to the pump housing as shown in Figure 6.36C and indicate the shaft runout at two or more points on the outer diameter of the impeller end of the shaft. Also, indicate the shaft run out at the coupling end of the shaft. If the runout exceeds 0.0015 − 0.003 in., remove the shaft and straighten or replace.

   A bent shaft can lead to many seal failures and poor sealing performance. Bearing life is greatly reduced and the operating conditions of both radial and thrust bearings can be impaired, resulting in reduced seal life and performance for reasons outlined previously. Bent shafts also can cause vibration.

5. *Stuffing-box face squareness or perpendicular to the shaft axis.* With the pump stuffing-box cover bolted in place, clamp the dial indicator to the shaft as shown in Figure 6.36D, with the stem against the face of the stuffing box. The total indicator runout should not exceed 0.003 in. When the face of the stuffing box is out of square or not perpendicular to the shaft axis, the result can be serious malfunction of a mechanical seal for the following reasons:

a. The stationary gland plate that holds in position the stationary insert, or seat, is bolted to the face of the stuffing box. Misalignment at this point will cause the gland to cock, resulting in cocking of the stationary element. This results in seal wobble, that is, operating in an elliptical plane. This condition is a major factor in causing fretting, pitting, and wear of the mechanical seal shaft packing at the point of contact with the shaft or sleeve.

b. A seal that is wobbling on the shaft also can cause wear of the drive pins. Erratic wear at face contact causes poor seal performance and life.

6. *Concentricity of stuffing-box bore.* With the same dial indicator set up as previously described, place the indicator stem well into the bore of the stuffing box. The stuffing box should be concentric to the shaft axis to within a 0.005-in. total indicator reading. If the shaft is eccentric to the box bore, check the slope, or looseness, in the pump bracket fits at location *A* in Figure 6.36E. Rust, atmospheric corrosion, or corrosion due to leakage of gaskets can cause damage to these fittings, which makes it impossible to ensure a stuffing box that is concentric with the shaft.

A possible remedy for this condition is the welding of the corroded area and remachining to proper dimensions. Another possible solution is to center the stuffing box and dowel it in place. Eccentricity alters the hydraulic loading of the seal faces, reducing seal life and performance.

7. *Driver alignment and pipe strain.* Driver alignment is an extremely important consideration, and periodic checkups should be a common practice. Steps to ensure proper coupling and driver alignment are well known and too lengthy to discuss in this chapter. The effect of pipe strain is also well known as to the damage it can cause to pumps, bearings, and seals.

In most plants it is customary to blind the suction and discharge flanges of inactive pumps. These blinds should be removed before the pump driver alignment is made. Otherwise the alignment job is incomplete.

After the blinds have been removed and as the flanges on the suction and discharge are being connected to the piping, check the dial indicator reading on the outer diameter of the coupling half and observe movement of the indicator dial as the flanges are being secured. Deviation indicates that pipe strain is present. If severe strain exists, then corrective measures should be taken or damage to pump and unsatisfactory seal service can result.

Seal Check Points

The following points should be checked:

1. Ensure that all parts are kept clean, especially the running faces of the seal ring and insert.
2. Check the seal rotary unit and make sure the drive pins and/or spring pins are free in the pinholes or slots.
3. Check the set screws in the rotary unit collar to see that they are free in the threads. Note that set screws should be replaced after each use.
4. Check the thickness of all gaskets against the dimensions shown on the assembly drawing. Improper gasket thickness will affect the seal setting and spring load imposed on the seal.
5. It is always advisable to check the fit of the gland ring to the equipment. Make sure there is no interference or binding on the studs or bolts or other obstructions. Be sure the gland ring pilot, if any, enters the bore with a reasonable guiding fit for proper seal alignment.
6. Make sure all rotary unit parts of the seal fit over the shaft. Particular care should be given to V rings. They must be placed on the shaft individually and never installed on the shaft while they are seated in the seal ring or rotating assembly.
7. Check both running faces of the seal (seal ring and insert) to be sure there are no nicks or scratches. Imperfections of any kind on either of these faces will cause seal leakage.

Seal Installation

The following general guidelines are helpful for seal installations.

1. Remove all burrs and sharp edges from the shaft or shaft sleeve, including sharp edges of keyways and threads. Worn shaft or sleeves should be replaced.
2. Check the stuffing-box bore and stuffing-box face to ensure they are clean and free of burrs.
3. The assembly drawing provided by the manufacturer gives the seal setting dimensions from the face of the stuffing box to the back of the rotary unit collar. To physically set the seal to this dimension, it is first necessary to scribe a reference mark on the shaft or sleeve that will coincide with the location of the face of the stuffing box. From this reference mark on the shaft, it is a simple matter to get the proper dimension to the back of the

rotating collar. A flat scribing tool can be used to establish the reference mark on the shaft or sleeve.

4. The shaft or sleeve should be oiled lightly prior to seal assembly to allow the seal parts to move freely over it. This is especially desirable when assembling the seal collar, as the bore of the collar usually has only a few thousandths of an inch clearance, and care should be used to avoid getting the color cocked.

5. Install the rotary unit parts on the shaft or sleeve in proper sequence.

6. Set the back of the collar at the proper distance from the original reference mark on the shaft or sleeve. Tighten all set screws firmly and evenly.

7. Be extra careful when passing the seal gland ring and insert over the shaft. Do not bring the insert against the shaft, as it may chip away small pieces from the edge of the running face.

8. Wipe the seal faces clean and apply a clean oil film prior to completing the equipment assembly.

9. Complete the equipment assembly, taking care when compressing the seal into the stuffing box.

10. Seat the gland ring and gland ring gasket to the face of the stuffing box by tightening the nuts or bolts evenly and firmly. Be sure the gland ring is not cocked and tighten the nuts or bolts only enough to effect a seal at the gland ring gasket. Excessive tightening of the gland ring nut or bolt will cause distortion that will be transmitted to the running face, resulting in leakage.

If the seal assembly drawing is not available, the proper seal setting dimension for inside seals can be determined as follows:

Establish a reference mark on the shaft or sleeve flush with the face of the stuffing box (point A in Figure 6.37). Next, determine how far the face of the insert will extend into the stuffing-box bore (B). This dimension is taken from the face of the gasket.

Determine the compressed length of the rotary unit (C) by compressing the rotary unit to the proper spring gap, which is marked on the collar (D). This dimension (C), added to the distance the insert extends into the stuffing box (B), will give the seal setting dimension (E) from the reference mark on the shaft or sleeve to the back of the seal collar.

When the face of the insert is recessed in the gland ring, it will be necessary to determine the dimension from the face of the stuffing box to the face of the insert (B in Figure 6.37B). This dimension will again be measured from the face of the gasket. Determine the compressed length of the rotating unit as previously explained and subtract the dimension that the insert face is recessed in the gland (B) from the compressed rotary unit length (C), and you will have the seal setting dimension (E) from the reference mark to the back of the seal collar.

Outside seals are set with the spring gap (A in Figure 6.37C) equal to the dimension stamped on the seal collar. Cartridge seals are set at the factory and

A - STUFFING BOX FACE
B - INSERT FACE EXTENDS INTO BOX
C - ROTARY COMPRESSED LENGTH
D - SPRING GAP
E - SEAL SETTING FROM REF. MARK

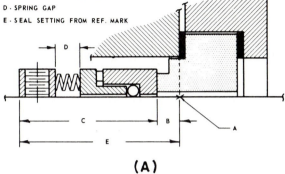

(A)

A - STUFFING BOX FACE
B - INSERT FACE RECESSED IN GLAND
C - ROTARY COMPRESSED LENGTH
D - SPRING GAP
E - SEAL SETTING FROM REF. MARK

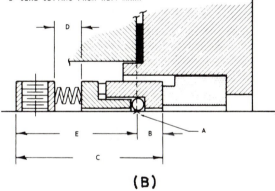

(B)

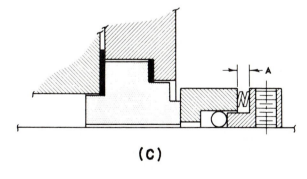

(C)

Figure 6.37 (A) Setting for inside seal. (B) Setting for inside seal with recessed insert. (C) Setting for outside seal.

are installed as complete assemblies. These assemblies contain spacers that must be removed after the seal assembly is bolted into position and the sleeve collar is positioned in place.

START-UP PROCEDURES

The first and foremost consideration in starting up equipment with mechanical seals is to ensure that the seal faces are immersed in a liquid from the very beginning so that they will not be scored or damaged from dry operation. The following recommendations in reference to seal start-up procedure will cover most types of seal installations. If followed where they are applicable, successful seal start-up and normal seal life can be expected.

1. Caution the electrician not to run the equipment dry while checking motor rotation. A slight turnover will not harm the seal, but full speed for several minutes under dry conditions will destroy or severely damage the rubber faces.

2. The stuffing box of the equipment, especially centrifugal pumps, should always be vented prior to start-up. Even though the pump has a flooded suction, it is still possible that air may be entrapped in the top portion of the stuffing box after the initial liquid purge of the pump.

3. Check installation for need of priming. Priming may be necessary in applications with a low or negative suction head.

4. Where cooling or bypass recirculation taps are incorporated in the seal gland, piping must be connected to or from these taps prior to start-up. These specific environmental control features must be used to protect the organic materials in the seal and ensure its proper performance. Cooling lines should be left open at all times or whenever possible. This is especially true when a hot product may be passing through standby equipment while it is not on the line. Many systems provide for product to pass through the standby equipment so the need for additional product volume or an equipment switch is only a matter of pushing a button.

5. On hot operational equipment that is shut down at the end of each day, it is highly advisable to leave the cooling water on at least long enough for the seal area to cool below the temperature limits of the organic materials in the seal.

6. Face lubricated-type seals must be connected from the source of lubrication to the tap openings in the seal gland prior to start-up. This is another predetermined environmental control feature that is mandatory for proper seal function. Where double seals are to be operated, it is necessary that the lube feed lines be connected to the proper ports for both circulatory or dead-end systems prior to equipment start-up. This is very important, as all types of double seals are dependent on the controlled pressure and flow of the sealing fluid to function properly. Even before the shaft is

rotated, the sealing liquid pressure must exceed the product pressure opposing the seal. Be sure a vapor trap does not prevent the lubricant from reaching the seal faces promptly.

7. Thorough warm-up procedures include a check of all steam piping arrangements to be sure that all are connected and functioning, as products that will solidify must be fully melted before start-up. It is advisable to leave all heating arrangements on during shutdown to ensure a liquid condition of the product at all times. Leaving the heat on at all times further facilitates quick start-ups and equipment switchovers, which may be necessary during a production cycle.

Thorough chilling procedures are necessary on some installations, especially LPG applications. LPG always must be kept in a liquid state in the seal area, and start-up is usually the most critical time. Even during operation it may develop that the recirculation line piped to the stuffing box may have to be run through a cooler to overcome frictional heat generated at the seal faces. LPG requires a stuffing-box pressure that is greater than the vapor pressure of the product at pumping temperature (25 psi to 50 psi differential is desired).

8. A squealing noise indicates insufficient liquid at the seal faces (that is, the faces are running dry). To avoid reduced seal life and possible seal damage, the installation should be checked immediately to determine how much liquid can be brought to, or retained in, the seal area. A recirculation line from the discharge to the stuffing box will usually overcome this condition.

9. Dry operation from loss of suction or inadequate suction is sometimes encountered and should be corrected immediately to prevent seal face damage or failure. If suction is lost at start-up, it is advisable to again vent the stuffing-box chamber before restarting the equipment. Inadequate suction piping should be replaced as soon as possible, as it is not only detrimental to the function of the seal but to the pump itself. Many pump and seal problems are due to improper piping.

10. New plant start-up should consider the presence of dirt and foreign materials in the system. Dirt frequently accumulates in the system during construction. This condition cannot be eliminated entirely, but the proper cleansing and flushing of lines and equipment, prior to actual on-the-line running, can eliminate a great many seal failures. The use of strainers, cyclone separators, and even filters on critical installations is advisable for new plant initial start-up, as a large percentage of seal failures occurs at that time.

11. If a seal leaks slightly at start-up, allow a reasonable amount of time for it to adjust itself. Liquids that contain good lubricating qualities naturally will take longer to wear in the seal than those with lesser qualities. When a seal starts out with a slight leak and gets progressively less with running, it is indicative of leakage across the seal faces and continued running will cure it. Where leakage occurs immediately and remains constant, unaffected by running, it usually indicates secondary-seal (shaft-packing) damage or the seal faces that are warped out-of-flat. Out-of-flatness condition will be recognized readily from the wear pattern established on the seal faces where high spots or intermittent contact areas will show up.

12. Unless absolutely necessary, *do not open the seal faces for inspection.* After a seal has been running, a wear pattern is established between the two faces that microscopically mates these two faces. Since it is highly improbable that the two faces can be put back together in their exact original wear pattern, disturbing the seal in any manner probably will necessitate establishing a new pattern.

Concentricity and proper alignment of the equipment are necessary to eliminate undue stresses and strains on bearings and to maintain proper running clearances. Pipe strains also can cause stresses and contribute to misalignment of the equipment to a point where they will cause a malfunction of the seal. Impeller unbalance, caused by improper machining or presence of foreign materials lodged in the vanes, will set up vibrations and a condition of dynamic unbalance that a seal will not be able to tolerate.

Malfunction of seals can further be caused by improper stuffing-box machining. The stuffing-box bore must be concentric to the shaft, and the face of the stuffing box perpendicular to the axis of the shaft within 0.003 in. total indicator reading. Eccentric boring, or out-of-square gasket surfaces, will cause the seal to be drawn into a position of misalignment. The shaft or sleeve must have a radial surface free from scratches or nicks that allow seal leakage past the secondary seal or shaft packing. Pump throat bushings with proper diameter clearances are also necessary for good seal performance.

Excessive cavitation can cause seal failure, in addition to the damage that may occur to the pump. Cavitation causes shaft vibration, shaft deflection, and bearing failures. It can also cause excessive heat buildup in the stuffing box due to lack of product and eventual loss of the liquid film between the seal faces, resulting in ultimate seal failure. (Here again, the shaft packing may be caused to oscillate against the shaft, causing grooving and accelerated corrosion). Pump cavitation is usually indicated by excessive noise in the pump casing.

Ordinarily, the greatest amount of shaft deflection is encountered on single-stage end-suction pumps when they are operated at or near shutoff. The resulting shaft deflection and overheating can, in turn, cause seal failure. The more modern pumps are designed so that this problem is practically eliminated.

The major problem when operating pumps at or near shutoff is the excessive heating of the product in the seal area. The size of the pump, the product being handled and the product pumping temperature all have a definite bearing on when this overheating condition will occur. Some liquids can boil or flash to vapor in the stuffing box, strictly from normal frictional heat at the rubbing faces, plus the heat generated from operating at or near shutoff. When these conditions prevail, ultimate seal failure results. Bypass or recirculation lines frequently can offset this condition.

Standby Equipment

Standby equipment having mechanical seals should not be allowed to stand idle for long periods of time. Rotate them at least once a week, if only for a few moments. Better still, designate a certain time for standby equipment to be placed on the line for a predetermined period of time. This practice will usually require only a minimum of time and effort, as almost all standby equipment has common suction and discharge piping and it is simply a matter of opening and closing the proper valves.

TROUBLESHOOTING

The following notes summarize troubleshooting problems with mechanical seals:

1. Seal spits and sputters in operation.
    a. Product vaporizing and flashing across the seal faces.
    b. Take steps to maintain product in a definite liquid condition.
    c. Have balance design checked with seal manufacturer.
    d. Requires accurate readings of stuffing-box pressure, temperature, and gravity of product.
2. Seal leaks and ices gland ring.
    a. Product is vaporizing and flashing across seal faces.
    b. Icing may score faces, especially on the carbon insert. Seal faces should be replaced or relapped before starting up again after vaporizing condition has been corrected.
3. Seal drips steadily.
    a. First, determine location of leakage.
    b. Check gland gasket for proper compression.
    c. Faces not flat enough.
    d. Gland bolts possibly too tight, causing warpage of gland and insert.
    e. Shaft packing nicked or pinched during installation.
    f. Carbon insert cracked, or face of insert or seal ring chipped during installation.
    g. Seal faces scored from foreign particles between them.
    h. Leakage of liquid under equipment shaft sleeve.
4. Seal squeals during operation.
    a. Inadequate amount of liquid at sealing faces. A squealing seal is a dry seal, but not all dry seals squeal.
    b. Bypass flush line may be needed; if one is already in use, it may need to be enlarged to produce more flow.

☆ – RECOMMENDED, PERFORMANCE IS NOT AFFECTED
★ – FAIR, SOME LOSS OF PROPERTIES MAY OCCUR
■ – NOT RECOMMENDED, MATERIAL IS UNSUITABLE FOR SERVICE
□ – NO DATA AVAILABLE AT THIS TIME
( ) – BRACKETS AROUND A RATING LETTER, NO DATA IS AVAILABLE, BUT THE RATINGS ARE MADE ON THE BASIS OF EXPOSURE TESTS IN SIMILAR CHEMICAL GROUPS.

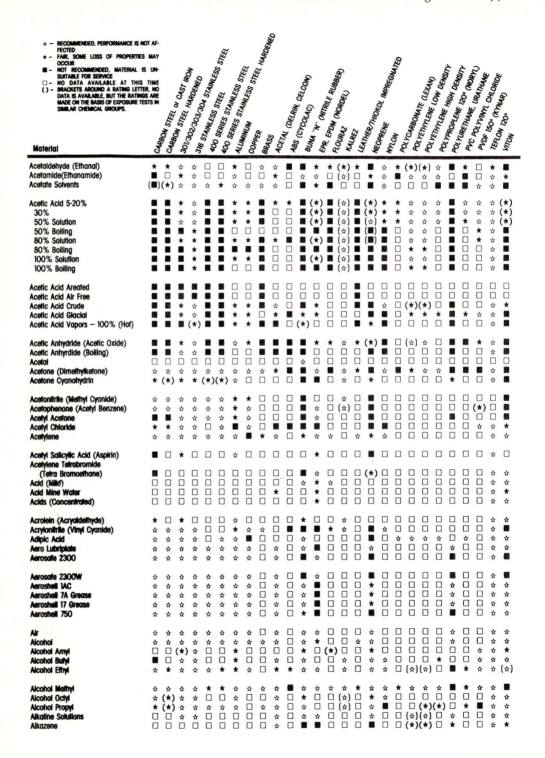

Column headers (left to right):
CARBON STEEL or CAST IRON · CARBON STEEL HARDENED · 301/302/303/304 STAINLESS STEEL · 316 STAINLESS STEEL · 400 SERIES STAINLESS STEEL · 400 SERIES STAINLESS STEEL HARDENED · ALUMINUM · COPPER · BRASS · ACETAL (DELRIN CELCON) · ABS (CYCOLAC) · BUNA "N" (NITRILE RUBBER) · EPR, EPDM (NORDEL) · FLOURAZ · KALREZ · LEATHER/THIOKOL IMPREGNATED · NEOPRENE · NYLON · POLYCARBONATE (LEXAN) · POLYETHYLENE LOW DENSITY · POLYETHYLENE HIGH DENSITY · POLYPROPLENE 120° (NORYL) · POLYURETHANE, URATHANE · PVC POLYVINYL CHLORIDE · PVDF 150° (KYNAR) · TEFLON 120° · VITON

Material	(27-column chemical-resistance rating grid)
Acetaldehyde (Ethanal)	★ ★ ☆ ☆ □ □ ☆ □ ☆ □ ■ ■ ★ ★ (★) ★ ■ ☆ ★ (★)(★) ☆ ■ ★ □ ★ ■
Acetamide (Ethanamide)	■ □ ★ ☆ □ □ ☆ ☆ ☆ ★ □ ☆ ☆ □ (☆) □ ★ ☆ ■ ☆ ☆ ☆ □ ■ □ ★ ■
Acetate Solvents	(■)(★) ☆ ☆ ☆ ★ ☆ ☆ ☆ ☆ □ ■ ★ ■ □ □ ■ ☆ □ □ □ ■ ■ ■ ☆ ☆ ■
Acetic Acid 5-20%	■ ■ ★ ☆ ■ ■ ☆ ★ ★ ★ (★) ■ (☆) ■ (★) ★ ★ ☆ ☆ ☆ ■ ☆ ☆ ☆ (★)
30%	■ ■ ★ ☆ ■ ■ ★ ★ □ □ (★) ■ (☆) ■ (★) ★ ★ ☆ ☆ ☆ ■ ☆ ☆ ☆ (★)
50% Solution	■ ■ ☆ ☆ ■ ■ ★ □ □ □ (★) ■ (☆) ■ (☆) ★ ★ ☆ ☆ ☆ ■ ☆ ☆ ☆ (★)
50% Boiling	■ ■ ★ ★ □ □ □ □ □ □ (★) ■ (☆) ■ (■) □ ☆ ☆ □ □ ■ □ ★ ☆ ■
80% Solution	■ ■ ★ ★ ■ ■ ★ ★ ★ ■ (★) ■ (☆) ■ (■) □ ☆ ☆ □ □ ■ □ ★ ☆ ■
80% Boiling	■ ■ ■ ■ □ □ □ □ □ □ (☆) ■ (☆) ■ □ □ ☆ ★ □ □ ■ ★ ★ ☆ ■
100% Solution	■ ■ ★ ☆ ■ ■ ★ ★ ■ □ (★) ■ (☆) ■ □ □ ☆ ☆ □ □ ■ □ ☆ ☆ ■
100% Boiling	■ ■ ★ ☆ □ □ □ □ □ □ (☆) ■ (☆) ■ □ □ ■ ★ □ □ ■ ★ ★ ☆ ■
Acetic Acid Areated	■ ■ ■ ■ ■ ■ □ □ □ ☆ □ □ □ □ □ □ ■ □ □ □ ☆ □ □ □ ☆ □
Acetic Acid Air Free	■ ■ ■ ■ ■ ■ □ ■ ☆ □ □ □ □ □ □ ■ □ □ □ ☆ □ □ □ ☆ □
Acetic Acid Crude	■ ■ ★ ☆ ■ ★ ★ ■ ☆ □ ☆ ★ □ □ ■ ☆ □ (★)(★) □ ■ □ □ ☆ ★
Acetic Acid Glacial	■ ■ ★ ☆ ■ ★ ★ □ ★ ■ ★ ★ □ □ ☆ ■ □ ★ ★ ★ ■ ☆ ☆ ★ ■
Acetic Acid Vapors – 100% (Hot)	■ ■ (★) ■ ■ ★ ★ ■ ■ (★) □ □ □ ■ ★ □ □ □ □ □ □ □ ☆ ☆ ■
Acetic Anhydride (Acetic Oxide)	■ ■ ★ ☆ ■ ■ ☆ ★ ■ ■ ■ ★ ★ ☆ ★ (★) ■ □ (☆) ☆ ■ ■ ★ ☆ ■
Acetic Anhyrdide (Boiling)	■ ■ ☆ ☆ ■ □ □ ■ ■ □ ☆ □ □ □ □ □ □ □ □ □ □ ☆ ☆ □
Acetal	□ □ □ □ □ □ □ □ □ □ ☆ □ ☆ □ □ □ □ □ □ □ □ □ □ □ □
Acetone (Dimethylketone)	☆ ☆ ☆ ☆ ☆ ☆ ☆ ☆ ☆ ★ ■ ☆ ■ ☆ ★ ■ ☆ ■ ★ ☆ ☆ ■ ■ ■ ☆ ■
Acetone Cyanohydrin	★ (★) ★ ★ (★)(★) ☆ □ □ □ ■ ■ ☆ □ ★ □ □ □ □ □ ★ □ □ ☆ ■
Acetonitrile (Methyl Cyanide)	☆ ☆ ☆ ☆ ☆ ★ ★ □ □ □ ■ □ □ ☆ □ ■ □ □ □ □ □ □ □ □ ■
Acetophenone (Acetyl Benzene)	☆ ☆ ☆ ☆ ☆ ★ ☆ □ □ ■ □ ☆ ☆ (☆) □ ■ □ □ □ □ □ □ (★) □ ■
Acetyl Acetone	■ ■ ☆ ☆ ☆ ☆ ☆ ☆ □ ☆ □ ☆ ☆ □ ■ □ ■ □ □ □ ■ □ □ □ ■
Acetyl Chloride	★ ★ ☆ ☆ □ ☆ ■ ☆ □ ■ ■ ■ □ ☆ □ ■ □ □ □ □ □ □ ☆ ☆ ★
Acetylene	☆ ☆ ☆ ☆ ☆ ☆ ■ ★ ☆ □ ★ ☆ ☆ □ ☆ ★ ☆ □ □ □ □ □ ☆ ☆ ☆
Acetyl Salicylic Acid (Aspirin)	■ □ ★ □ □ □ □ ☆ □ □ □ ★ □ □ □ ■ □ □ □ □ □ □ □ ☆ □
Acetylene Tetrabromide (Tetra Bromoethane)	■ □ □ □ □ □ □ □ □ ■ ☆ □ □ ■ □ (★) □ □ □ □ □ □ □ ☆ □
Acid (Mild)	□ □ □ □ □ □ □ □ □ □ ☆ ★ ☆ □ □ ■ □ □ □ □ □ □ □ ☆ ☆
Acid Mine Water	□ □ □ □ □ □ □ □ ★ □ ☆ □ ★ □ □ ■ □ □ □ □ □ □ □ ☆ ★
Acids (Concentrated)	□ □ □ □ □ □ □ □ □ ★ □ ■ □ □ ■ □ □ □ □ □ □ □ □ ☆ ☆
Acrolein (Acryaldehyde)	★ □ ★ □ □ □ ☆ □ □ □ ★ □ □ ☆ □ ■ □ □ □ □ □ □ □ ☆ ☆
Acrylonitrile (Vinyl Cyanide)	☆ ☆ ☆ ☆ ☆ □ ☆ ★ ☆ ■ ■ ■ ★ ☆ ■ □ □ □ □ □ □ □ ☆ ■
Adipic Acid	☆ ☆ ☆ ☆ □ ☆ ■ □ □ □ ☆ ☆ □ ■ □ □ ☆ ☆ ☆ □ □ ☆ ☆ ☆
Aero Lubriplate	☆ ☆ ☆ ☆ ☆ ☆ ☆ □ ☆ □ ☆ □ ■ □ □ □ ☆ □ □ □ □ ☆ ☆ ☆
Aerosafe 2300	☆ ☆ ☆ ☆ ☆ ☆ ☆ □ ☆ □ □ ■ □ □ □ □ □ □ □ □ □ ☆ ☆ ☆
Aerosafe 2300W	☆ ☆ ☆ ☆ ☆ ☆ ☆ □ ☆ □ □ ☆ □ □ □ □ □ □ □ ■ □ □ ☆ ☆ ■
Aeroshell IAC	☆ ☆ ☆ ☆ ☆ ☆ ☆ □ ☆ ☆ □ ☆ □ ■ □ □ □ ★ □ □ □ □ ☆ ☆ ☆
Aeroshell 7A Grease	☆ ☆ ☆ ☆ ☆ ☆ ☆ □ ☆ ☆ □ ■ □ □ □ ★ □ □ □ □ □ ☆ ☆ ☆
Aeroshell 17 Grease	☆ ☆ ☆ ☆ ☆ ☆ ☆ □ ☆ ☆ □ ■ □ □ □ ★ □ □ □ □ □ ☆ ☆ ☆
Aeroshell 750	☆ ☆ ☆ ☆ ☆ ☆ ☆ □ ☆ ☆ □ ■ □ □ □ ☆ □ □ □ □ □ ☆ ☆ ☆
Air	☆ ☆ ☆ ☆ ☆ ☆ ☆ ☆ ☆ □ ☆ ☆ ☆ □ □ ☆ ☆ □ ☆ ☆ ☆ □ ☆ ☆ ☆
Alcohol	☆ ☆ ☆ ☆ ☆ ☆ ☆ ☆ ☆ ☆ ☆ ☆ ☆ ☆ ★ ☆ ☆ □ ☆ ☆ ☆ □ ☆ ☆ ☆
Alcohol Amyl	□ □ (★) ☆ ☆ □ □ ★ □ □ □ ★ □ (★) □ □ ★ □ ☆ □ ☆ □ □ ★ ☆ ★
Alcohol Butyl	■ □ ☆ ☆ □ □ ★ □ □ ☆ □ ☆ □ □ □ ☆ □ ☆ ☆ ★ □ □ ★ ☆ ☆ (★)
Alcohol Ethyl	☆ ★ ☆ ☆ ☆ ★ ★ ☆ ☆ □ ★ ★ ☆ □ ☆ □ ☆ ☆ □ (☆)(☆) □ ■ ★ ☆ ☆ (☆)
Alcohol Methyl	☆ ☆ ☆ ☆ ★ ★ ☆ ☆ ☆ ☆ ■ ☆ ☆ ☆ ★ ☆ ☆ ★ ☆ ☆ ☆ ■ ★ ☆ ☆ ■
Alcohol Octyl	☆ (★) ☆ ☆ □ □ ☆ □ □ □ ★ □ ☆ □ ■ □ ☆ □ ☆ □ ☆ □ □ ☆ ☆ ☆
Alcohol Propyl	★ (★) ☆ ☆ ☆ ☆ ☆ □ ☆ ☆ □ ☆ □ □ (☆) □ ☆ ■ □ (★)(★) □ ★ ■ ☆ ☆
Alkaline Solutions	□ □ ☆ ☆ □ □ □ □ □ ☆ □ ☆ ☆ □ □ ■ □ □ (☆)(☆) □ ☆ □ ☆ ☆
Alkazene	□ □ □ □ □ □ □ □ ☆ □ ■ ■ ■ □ □ ■ □ □ (★)(★) □ ★ □ □ ☆ ★

Legend:

- ☆ — RECOMMENDED, PERFORMANCE IS NOT AFFECTED
- ★ — FAIR, SOME LOSS OF PROPERTIES MAY OCCUR
- ■ — NOT RECOMMENDED, MATERIAL IS UNSUITABLE FOR SERVICE
- □ — NO DATA AVAILABLE AT THIS TIME
- ( ) — BRACKETS AROUND A RATING LETTER, NO DATA IS AVAILABLE, BUT THE RATINGS ARE MADE ON THE BASIS OF EXPOSURE TESTS IN SIMILAR CHEMICAL GROUPS.

Material	CARBON STEEL or CAST IRON	CARBON STEEL HARDENED	301/302/303/304 STAINLESS STEEL	316 STAINLESS STEEL	400 SERIES STAINLESS STEEL	400 SERIES STAINLESS STEEL HARDENED	ALUMINUM	COPPER	BRASS	ACETAL (DELRIN, CELCON)	ABS (CYCOLAC)	BUNA "N" (NITRILE RUBBER)	EPR, EPDM (NORDEL)	FLOURAZ	KALREZ	LEATHER/THIOKOL IMPREGNATED	NEOPRENE	NYLON	POLYCARBONATE (LEXAN)	POLYETHYLENE LOW DENSITY	POLYETHYLENE HIGH DENSITY	POLYPROPLENE 120°	POLYURETHANE, URATHANE	PVC POLYVINYL CHLORIDE (NORYL)	PVDF 150° (KYNAR)	TEFLON 120°	VITON
Allyl Alcohol	☆	☆	□	□	□	□	★	★	□	□	□	□	□	□	□	□	□	■	☆	☆	☆	□	□	□	□	☆	★
Allyl Bromide	☆	(☆)	□	□	□	□	☆	☆	□	□	□	■	□	□	□	□	■	□	☆	☆	☆	□	□	□	□	☆	★
Allyl Chloride	■	(■)	★	★	(★)	(★)	■	□	□	□	□	■	■	□	□	□	■	□	☆	☆	☆	□	□	□	□	☆	★
Almond Oil (Artificial)	□	□	★	★	(★)	(★)	□	□	□	□	□	□	★	□	□	□	■	□	□	□	□	□	□	□	□	☆	■
Alum	■	■	★	★	■	■	□	□	□	☆	☆	★	☆	☆	(☆)	□	★	□	□	□	□	□	□	□	□	☆	☆
Aluminum Acetate	■	■	★	★	★	(☆)	☆	☆	□	□	□	□	□	□	(☆)	□	★	□	□	□	□	□	□	□	□	☆	☆
Aluminum Ammonium Sulfate (Alum)	□	□	□	□	□	□	□	□	□	□	□	★	□	□	□	□	★	□	□	□	□	□	□	□	□	☆	☆
Aluminum Bromide	□	□	□	□	□	□	□	□	□	□	□	☆	□	□	□	□	☆	□	□	□	□	■	□	□	□	☆	☆
Aluminum Chloride	■	■	(★)	★	(★)	(★)	■	■	■	★	□	☆	☆	☆	(☆)	□	■	□	★	★	□	□	□	□	□	☆	☆
Aluminum (Chrome 5% Sol.)	□	□	☆	☆	□	□	□	□	□	□	□	□	☆	□	□	□	☆	□	□	□	□	□	□	□	□	☆	□
Aluminum Fluoride	■	■	■	■	☆	★	■	□	□	□	□	☆	□	(☆)	□	☆	☆	☆	□	□	□	■	★	☆	☆	☆	☆
Aluminum Hydroxide	□	□	☆	☆	☆	★	☆	☆	★	□	☆	☆	☆	(☆)	☆	☆	■	☆	☆	☆	□	☆	☆	☆	☆	☆	☆
Aluminum Hydroxide – Saturated	□	□	☆	☆	□	□	□	□	□	□	□	☆	□	(☆)	□	☆	☆	☆	□	□	□	□	□	☆	☆	☆	★
Aluminum Nitride	■	■	★	★	★	★	□	□	□	□	□	☆	□	(☆)	□	☆	☆	☆	□	□	□	■	□	☆	☆	☆	☆
Aluminum Phosphate	□	□	□	□	□	□	□	□	□	□	□	☆	□	☆	□	☆	☆	☆	□	□	□	□	□	☆	☆	☆	☆
Aluminum Potassium Sulfate	■	■	★	★	■	■	☆	☆	□	□	□	☆	□	(☆)	□	☆	☆	☆	□	□	□	□	□	☆	☆	☆	☆
2% Solution	■	■	☆	☆	■	■	☆	☆	□	□	□	□	□	□	□	□	□	□	□	□	□	□	□	□	□	☆	☆
10% Solution	■	■	☆	☆	■	■	☆	☆	□	□	□	□	□	□	□	□	□	□	□	□	□	□	□	□	□	☆	☆
10% Boiling	□	□	★	★	□	□	□	□	□	□	☆	□	□	□	□	□	□	□	□	□	□	□	□	□	□	☆	☆
Saturated Boiling	□	□	■	■	★	□	□	□	□	■	□	□	□	□	□	□	■	□	□	□	□	□	□	□	□	☆	☆
Aluminum Palmitate	□	□	□	□	□	□	□	□	□	□	□	□	☆	□	□	□	☆	□	☆	☆	☆	□	■	□	☆	□	□
Aluminum Salts	□	□	□	□	□	□	□	□	□	□	□	☆	☆	□	□	□	☆	□	☆	☆	☆	□	■	☆	□	□	□
Aluminum Sulfate	■	■	(★)	(★)	■	■	★	★	□	☆	☆	☆	☆	(☆)	□	★	□	(☆)	(☆)	□	■	★	□	☆	☆	☆	☆
Amines	□	□	☆	☆	□	□	□	□	□	★	★	(☆)	(☆)	■	★	□	☆	□	□	□	□	□	□	□	☆	☆	□
Ammonia, Gas, Cold	□	□	□	□	□	□	□	□	□	□	☆	☆	(■)	(☆)	□	☆	□	□	□	□	□	□	□	□	(☆)	☆	(★)
Ammonia, Gas, Hot	□	□	□	□	□	□	□	□	□	■	★	(■)	(☆)	□	★	□	□	□	□	□	□	□	□	□	(☆)	☆	■
Ammonia, Anhydrous	☆	☆	☆	☆	☆	☆	■	■	□	□	★	☆	☆	☆	■	□	☆	□	■	★	□	■	□	■	☆	☆	■
Ammonia Aqueous	□	□	□	□	□	□	□	□	□	★	□	☆	☆	□	□	□	☆	□	□	□	□	□	□	□	☆	☆	□
Ammonia Liquids	(☆)	□	★	★	★	★	■	□	□	■	□	□	☆	□	★	□	★	■	■	★	★	☆	□	□	☆	☆	■
Ammonia Nitrate	☆	(☆)	☆	☆	☆	(☆)	■	□	□	■	□	□	☆	□	☆	□	■	☆	□	□	□	□	□	□	☆	☆	■
Ammonium Acetate	★	□	(☆)	(☆)	□	□	☆	□	□	□	□	□	□	□	□	□	☆	□	□	□	□	□	□	□	□	☆	☆
Ammonium Bicarbonate	★	★	☆	☆	☆	★	★	■	□	■	□	☆	★	■	(☆)	☆	☆	☆	□	(☆)	(☆)	□	□	□	□	☆	(★)
Ammonium Bifluoride	■	■	■	★	□	■	■	□	□	★	□	☆	★	■	□	■	□	☆	□	□	□	□	□	★	☆	☆	☆
Ammonium Carbonate	★	★	★	★	★	★	★	■	□	☆	☆	★	■	(☆)	☆	☆	□	(☆)	□	□	□	□	☆	☆	☆	☆	☆
Ammonium Chloride 1%	■	■	★	★	★	■	☆	★	★	☆	☆	☆	(☆)	□	★	□	☆	□	(☆)	□	□	□	□	(☆)	□	☆	□
10% Solution Boiling	■	■	★	★	★	■	□	□	□	■	☆	☆	(☆)	□	■	□	☆	□	(☆)	□	□	□	□	(☆)	□	☆	□
28% Solution Boiling	■	■	★	★	★	■	□	□	□	■	☆	☆	(☆)	□	■	□	☆	□	(☆)	□	□	□	□	(☆)	□	☆	□
50% Solution Boiling	■	■	★	★	★	■	□	□	□	■	☆	☆	(☆)	□	■	□	☆	□	(☆)	□	□	□	□	(☆)	□	☆	□
Ammonium Dichromate	☆	☆	☆	☆	□	□	☆	■	■	□	□	☆	☆	□	(☆)	□	☆	□	□	□	□	□	□	□	□	☆	□
Ammonium Diphosphate	■	■	★	★	★	★	■	■	□	□	☆	☆	□	(☆)	□	☆	□	□	□	□	□	□	□	□	□	☆	□
Ammonium Fluoride	★	★	★	★	(★)	(★)	★	□	□	★	□	☆	□	☆	□	★	□	□	□	□	□	□	□	□	☆	☆	★
Ammonium Hydroxide	■	■	★	★	★	■	■	■	□	☆	★	★	★	☆	☆	□	☆	☆	■	★	★	■	☆	☆	☆	☆	★
3 Molar	■	■	★	★	★	■	■	■	□	☆	□	□	□	□	☆	□	□	□	□	□	□	□	☆	☆	☆	☆	★
Concentrated	■	■	★	★	★	■	■	■	□	☆	□	□	□	□	☆	□	□	□	□	□	□	□	☆	☆	☆	☆	★
Ammonium Monophosphate	■	■	★	★	★	★	■	■	□	□	☆	☆	□	(☆)	□	☆	□	□	□	□	□	□	□	□	□	☆	□
Ammonium Nitrate	★	★	☆	☆	☆	☆	★	■	■	☆	□	☆	☆	■	☆	☆	★	□	(☆)	(☆)	□	□	★	☆	☆	☆	☆
Ammonium Nitrite	★	★	☆	☆	☆	☆	★	■	□	☆	□	☆	☆	□	□	☆	□	□	(☆)	(☆)	□	□	☆	■	☆	☆	☆
Ammonium Oxalate – 5% Sol.	■	■	★	★	(★)	■	★	★	□	□	□	□	□	□	(☆)	□	□	☆	★	☆	★	□	☆	□	□
Ammonium Persulfate Sol.	■	■	★	★	★	■	■	■	■	□	□	■	□	(☆)	□	■	□	□	□	□	■	★	□	☆	□

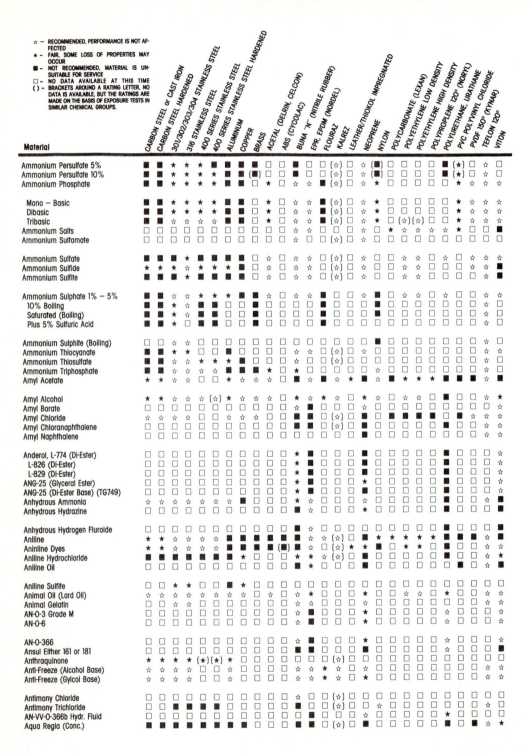

☆ – RECOMMENDED, PERFORMANCE IS NOT AFFECTED
★ – FAIR, SOME LOSS OF PROPERTIES MAY OCCUR
■ – NOT RECOMMENDED, MATERIAL IS UNSUITABLE FOR SERVICE
☐ – NO DATA AVAILABLE AT THIS TIME
( ) – BRACKETS AROUND A RATING LETTER, NO DATA IS AVAILABLE, BUT THE RATINGS ARE MADE ON THE BASIS OF EXPOSURE TESTS IN SIMILAR CHEMICAL GROUPS.

Column headings:
CARBON STEEL or CAST IRON · CARBON STEEL HARDENED · 301/302/303/304 STAINLESS STEEL · 316 STAINLESS STEEL · 400 SERIES STAINLESS STEEL · 400 SERIES STAINLESS STEEL HARDENED · ALUMINUM · COPPER · BRASS · ACETAL (DELRIN, CELCON) · ABS (CYCOLAC) · BUNA "N" (NITRILE RUBBER) · EPR, EPDM (NORDEL) · FLOURAZ · KALREZ · LEATHER/THIOKOL IMPREGNATED · NEOPRENE · NYLON · POLYCARBONATE (LEXAN) · POLYETHYLENE LOW DENSITY · POLYETHYLENE HIGH DENSITY · POLYPROPLENE 120° (NORYL) · POLYURETHANE, URATHANE · PVC POLYVINYL CHLORIDE · PVDF 150° (KYNAR) · TEFLON 120° · VITON

Material

- Ammonium Persulfate 5%
- Ammonium Persulfate 10%
- Ammonium Phosphate

- Mono – Basic
- Dibasic
- Tribasic
- Ammonium Salts
- Ammonium Sulfamate

- Ammonium Sulfate
- Ammonium Sulfide
- Ammonium Sulfite

- Ammonium Sulphate 1% – 5%
- 10% Boiling
- Saturated (Boiling)
- Plus 5% Sulfuric Acid

- Ammonium Sulphite (Boiling)
- Ammonium Thiocyanate
- Ammonium Thiosulfate
- Ammonium Triphosphate
- Amyl Acetate

- Amyl Alcohol
- Amyl Borate
- Amyl Chloride
- Amyl Chloranaphthalene
- Amyl Naphthalene

- Anderol, L-774 (Di-Ester)
- L-826 (Di-Ester)
- L-829 (Di-Ester)
- ANG-25 (Glyceral Ester)
- ANG-25 (Di-Ester Base) (TG749)
- Anhydrous Ammonia
- Anhydrous Hydrazine

- Anhydrous Hydrogen Fluoride
- Aniline
- Aniline Dyes
- Aniline Hydrochloride
- Aniline Oil

- Aniline Sulfite
- Animal Oil (Lard Oil)
- Animal Gelatin
- AN-O-3 Grade M
- AN-O-6

- AN-O-366
- Ansul Either 161 or 181
- Anthraquinone
- Anti-Freeze (Alcohol Base)
- Anti-Freeze (Gylcol Base)

- Antimony Chloride
- Antimony Trichloride
- AN-VV-O-366b Hydr. Fluid
- Aqua Regia (Conc.)

Legend:

- ☆ — RECOMMENDED, PERFORMANCE IS NOT AFFECTED
- ★ — FAIR, SOME LOSS OF PROPERTIES MAY OCCUR
- ■ — NOT RECOMMENDED, MATERIAL IS UNSUITABLE FOR SERVICE
- □ — NO DATA AVAILABLE AT THIS TIME
- ( ) — BRACKETS AROUND A RATING LETTER, NO DATA IS AVAILABLE, BUT THE RATINGS ARE MADE ON THE BASIS OF EXPOSURE TESTS IN SIMILAR CHEMICAL GROUPS.

Material	Carbon Steel or Cast Iron	Carbon Steel Hardened	301/302/303/304 Stainless Steel	316 Stainless Steel	400 Series Stainless Steel	400 Series Stainless Steel Hardened	Aluminum	Copper	Brass	Acetal (Delrin, Celcon)	ABS (Cycolac)	Buna "N" (Nitrile Rubber)	EPR, EPDM (Nordel)	Flouraz	Kalrez	Leather/Thiokol Impregnated	Neoprene	Nylon	Polycarbonate (Lexan)	Polyethylene Low Density	Polyethylene High Density	Polyproplene 120°	Polyurethane, Urathane	PVC Polyvinyl Chloride	PVDF 150° (Kynar)	Teflon 120°	Viton
Argon	☆	☆	☆	☆	☆	☆	□	☆	□	☆	■	☆	(☆)	(☆)	□	■	□	□	☆	☆	☆	☆	□	□	☆	☆	
Aristo Wax	□	□	□	□	□	□	□	□	□	□	☆	□	□	□	□	☆	□	□	□	□	□	□	□	□	□	□	□
Arochlor 1248	★	★	★	★	□	□	☆	□	□	□	■	★	■	(☆)	□	■	□	□	□	□	□	□	□	□	□	☆	☆
1254	□	□	□	□	□	□	□	□	□	□	■	★	□	□	□	☆	□	□	□	□	□	□	□	□	□	☆	☆
1260	□	□	□	□	□	□	□	□	□	□	☆	□	□	□	□	☆	□	□	□	□	□	□	□	□	□	☆	☆
Aromatic Fuel 30%, Mil.	□	□	□	□	□	□	□	□	□	□	□	□	□	□	□	■	□	□	□	□	□	□	□	□	□	☆	☆
Aromatic Fuel 50%	□	□	□	□	□	□	□	□	□	□	★	■	☆	☆	■	■	□	□	□	□	□	□	■	□	□	☆	☆
Aromatic Hydrocarbons	★	□	□	★	□	□	□	☆	□	□	■	■	□	□	□	■	□	□	□	□	□	□	□	□	□	☆	☆
Arsenic Acid	■	■	★	★	★	(★)	☆	☆	□	□	☆	☆	☆	□	(☆)	★	☆	□	□	□	☆	☆	□	☆	☆	☆	☆
Arsenic Trichloride	■	■	■	■	■	□	☆	■	□	□	☆	☆	☆	□	(☆)	□	□	□	□	□	□	☆	□	☆	☆	☆	□
Arsenious Acid	■	■	☆	☆	■	■	☆	□	□	□	☆	★	☆	□	□	☆	□	□	□	□	□	□	☆	■	☆	☆	□
Ascorbic Acid	■	■	☆	☆	□	□	□	□	□	□	☆	■	□	□	□	■	□	□	□	□	□	□	□	□	☆	☆	☆
Askarel	□	□	□	□	□	□	□	□	□	□	★	■	□	□	□	☆	□	□	□	□	□	■	□	□	□	□	☆
Asphalt	☆	☆	☆	☆	★	★	☆	☆	★	☆	□	★	■	☆	(☆)	☆	■	☆	□	☆	☆	□	★	☆	□	☆	☆
Asphalt Topping	☆	☆	☆	☆	☆	☆	☆	☆	☆	☆	□	■	□	□	□	☆	□	□	□	□	□	□	□	□	□	☆	■
ASTM Oil, No. 1	☆	☆	☆	☆	☆	☆	☆	☆	☆	☆	□	☆	■	☆	☆	☆	☆	☆	□	☆	☆	☆	□	☆	☆	☆	☆
No. 2	☆	☆	☆	☆	☆	☆	☆	☆	☆	☆	□	☆	■	□	□	☆	★	□	□	□	□	★	□	□	□	☆	☆
No. 3	☆	☆	☆	☆	☆	☆	☆	☆	☆	☆	□	☆	■	□	□	☆	■	□	□	□	□	★	□	□	□	☆	☆
No. 4	☆	☆	☆	☆	☆	☆	☆	☆	☆	☆	□	★	■	□	□	☆	■	□	□	□	□	★	□	□	□	☆	☆
ASTM Reference Fuel A	□	□	□	□	□	□	□	□	□	□	□	☆	■	□	□	□	★	□	□	■	□	□	□	☆	□	☆	☆
B	□	□	□	□	□	□	□	□	□	□	□	☆	■	□	□	□	■	□	□	■	□	□	□	★	□	☆	☆
C	□	□	□	□	□	□	□	□	□	□	□	★	■	□	□	□	■	□	□	■	□	□	□	■	□	☆	☆
ATL-857	□	□	□	□	□	□	□	□	□	□	□	★	■	□	□	□	■	□	□	■	□	□	□	■	□	□	☆
Atlantic Dominion F	□	□	□	□	□	□	□	□	□	□	□	☆	□	□	□	□	★	□	□	■	□	□	□	■	□	□	☆
Aurex 903R (Mobil)	□	□	□	□	□	□	□	□	□	□	□	■	□	□	□	□	★	□	□	■	□	□	□	☆	□	□	☆
Automatic Transmission Fluid	☆	☆	☆	☆	☆	☆	☆	□	☆	□	□	☆	■	□	□	□	★	□	□	■	□	□	□	★	□	☆	☆
Automatic Brake Fluid	☆	☆	☆	☆	☆	☆	☆	□	☆	□	□	■	☆	□	□	□	★	□	□	□	□	□	□	☆	□	☆	■
Automotive Gasoline (Standard)	☆	☆	☆	☆	☆	☆	☆	□	☆	□	□	☆	■	□	□	□	■	□	□	☆	☆	□	■	★	□	☆	☆
Aviation Gasoline, Mil.	☆	☆	☆	☆	☆	☆	☆	□	☆	□	□	★	■	□	□	□	■	□	□	☆	☆	□	□	☆	□	☆	☆
Baking Enamels, Synthetic	☆	☆	☆	☆	☆	☆	☆	□	☆	□	□	☆	□	□	□	☆	□	□	□	□	□	□	□	□	□	☆	☆
Banana Oil	□	□	□	□	□	□	□	□	□	□	□	☆	□	□	□	☆	□	□	□	□	□	□	□	□	□	☆	□
Barbeque Sauce	■	■	☆	☆	□	□	□	□	□	□	□	□	□	□	□	☆	□	□	□	□	□	□	□	□	□	☆	□
Bardol B	□	□	□	□	□	□	□	□	□	□	□	■	☆	□	□	□	■	□	□	□	□	□	■	□	□	☆	☆
Barium Carbonate	☆	☆	☆	☆	☆	■	☆	□	□	□	☆	□	(☆)	□	□	☆	☆	□	□	□	□	□	□	☆	☆	☆	☆
Barium Chloride	★	★	★	★	★	★	☆	□	□	□	☆	☆	☆	(☆)	□	☆	(■)	□	□	□	□	☆	★	☆	☆	☆	
Barium Chloride – 5%	★	★	☆	☆	☆	☆	□	□	□	□	☆	☆	(☆)	(☆)	☆	☆	■	□	□	☆	☆	□	□	☆	☆	☆	□
Aqueous Solution (Hot)	★	★	★	★	■	■	■	□	■	□	□	☆	□	□	□	■	□	□	□	□	□	□	☆	☆	☆	□
Barium Cyanide	★	★	☆	☆	★	★	■	■	■	□	□	☆	(☆)	□	□	☆	□	□	□	□	□	■	□	□	☆	☆	
Barium Hydrate	□	□	☆	☆	□	□	□	□	□	□	☆	☆	□	□	□	☆	□	□	□	□	□	□	□	□	☆	☆
Barium Hydroxide	■	■	☆	☆	□	□	■	□	☆	□	☆	☆	□	(☆)	★	☆	☆	□	□	□	□	★	☆	☆	☆	☆
Barium Nitrate	□	□	☆	☆	□	□	□	□	☆	□	☆	☆	□	(☆)	□	☆	☆	□	□	□	□	□	☆	☆	□	☆
Barium Salts	□	□	☆	☆	□	□	□	□	☆	□	☆	☆	□	□	□	□	☆	□	□	□	□	□	☆	□	□	☆
Barium Sulfate	★	★	☆	☆	☆	☆	★	★	☆	□	☆	☆	□	(☆)	☆	☆	□	□	□	□	□	☆	☆	☆	□	☆
Aqueous Solution (Hot)	★	★	☆	☆	■	■	★	★	□	□	□	☆	□	□	□	■	□	□	□	□	□	□	☆	☆	□	☆
Barium Sulfide	■	■	☆	☆	★	★	■	■	☆	□	□	☆	(☆)	□	☆	★	□	□	□	□	□	□	☆	☆	□	☆
Bayol D	□	□	□	□	□	□	□	□	□	□	☆	■	□	□	□	★	□	□	□	□	□	■	□	□	□	☆
Bayol 35	□	□	□	□	□	□	□	□	□	□	☆	■	□	□	□	★	□	□	□	□	□	★	□	□	□	☆
Beef Extract	■	■	☆	☆	□	□	□	□	□	□	□	☆	□	□	□	☆	□	□	□	□	□	□	□	□	☆	☆
Beer (Alcohol Ind.)	☆	☆	☆	☆	☆	☆	★	★	☆	□	★	☆	(☆)	□	☆	☆	☆	□	☆	□	□	□	■	□	☆	☆
Beer (Beverage Ind.)	■	■	☆	☆	☆	☆	★	■	☆	□	☆	☆	(☆)	□	☆	☆	☆	□	□	□	□	■	□	□	☆	☆

Legend:

☆ – RECOMMENDED, PERFORMANCE IS NOT AFFECTED

★ – FAIR, SOME LOSS OF PROPERTIES MAY OCCUR

■ – NOT RECOMMENDED, MATERIAL IS UNSUITABLE FOR SERVICE

□ – NO DATA AVAILABLE AT THIS TIME

( ) – BRACKETS AROUND A RATING LETTER, NO DATA IS AVAILABLE, BUT THE RATINGS ARE MADE ON THE BASIS OF EXPOSURE TESTS IN SIMILAR CHEMICAL GROUPS.

Material	CARBON STEEL or CAST IRON	CARBON STEEL HARDENED	301/302/303/304 STAINLESS STEEL	316 STAINLESS STEEL	400 SERIES STAINLESS STEEL	400 SERIES STAINLESS STEEL HARDENED	ALUMINUM	COPPER	BRASS	ACETAL (DELRIN, CELCON)	ABS (CYCOLAC)	BUNA "N" (NITRILE RUBBER)	EPR, EPDM (NORDEL)	FLOURAZ	KALREZ	LEATHER/THIOKOL IMPREGNATED	NEOPRENE	NYLON	POLYCARBONATE (LEXAN)	POLYETHYLENE LOW DENSITY	POLYETHYLENE HIGH DENSITY	POLYPROPLENE 120°	POLYURETHANE, URATHANE	PVC POLYVINYL CHLORIDE	PVDF 150° (KYNAR)	TEFLON 120°	VITON
Beet Sugar Liquids	☆	☆	☆	☆	★	★	☆	□	■	★	★	☆	☆	□	□	☆	□	★	★	□	■	□	☆	☆			
Beet Sugar Liquors	★	★	☆	☆	★	★	☆	☆	■	☆	□	☆	☆	□	□	☆	★	☆	□	☆	■	□	☆	☆			
Benzaldehyde	☆	☆	☆	☆	★	★	☆	□	■	■	☆	■	☆	(☆)	□	■	★	☆	■	■	■	■	☆	■			
Benzene (Benzol)	☆	☆	☆	☆	☆	☆	☆	☆	☆	■	■	■	■	(☆)	★	■	☆	■	★	★	★	■	■	■	□	☆	☆
Benzenesulfonic Acid	□	□	☆	☆	□	□	■	★	□	□	■	■	■	□	□	★	□	□	□	□	□	■	□	☆			
Benzine (Gasoline)	☆	☆	☆	☆	☆	☆	☆	☆	☆	☆	■	☆	★	□	□	★	■	☆	□	☆	★	□	□	☆			
Benzochloride	□	□	□	□	□	□	□	☆	□	□	■	■	☆	□	□	■	☆	□	□	□	□	□	□	☆			
Benzoic Acid	■	■	☆	☆	★	★	☆	☆	■	□	☆	☆	(☆)	□	★	★	☆	□	■	★	☆	☆					
Benzophenone	★	★	★	★	★	★	★	★	★	□	★	(☆)	□	□	☆	□	□	□	□	□	☆	☆					
Benzyl Acetate	☆	☆	☆	☆	□	□	☆	□	□	■	□	(☆)	□	□	■	★	★	★	■	□	☆	■					
Benzyl Alcohol	★	★	★	★	★	★	★	★	☆	□	★	□	(☆)	□	★	□	★	■	■	■	★	□	☆	□			
Benzyl Benzoate	★	★	★	★	★	★	☆	★	□	☆	■	□	(☆)	□	■	□	☆	☆	■	□	☆	□					
Benzyl Chloride (Chlorotoluene)	☆	☆	★	★	★	★	■	★	□	■	☆	□	(☆)	□	■	□	■	☆	☆	■	☆	□					
Benzyl Dichloride (Benzal Chloride)	★	★	☆	☆	□	□	★	□	□	☆	■	□	(☆)	□	★	■	□	■	☆	□							
Benzol	☆	☆	★	★	□	□	★	□	□	☆	■	■	■	□	★	■	□	■	□	□	☆	☆					
Bichloride of Mercury	□	□	□	□	□	□	□	□	□	☆	■	☆	(☆)	□	□	□	□	□	□	□							
Biphenyl (Diphenyl)	☆	☆	□	□	□	□	☆	□	□	■	■	■	□	■	□	□	□	□	□	□							
Bismuth Subcarbonate (Bismuth Carbonate)	□	□	★	★	□	□	□	□	□	☆	☆	(☆)	□	☆	□	□	□	□	□	□							
Black Point 77	□	□	□	□	□	□	□	□	□	□	☆	□	□	□	□	□	□	□	□	■	□	□	☆				
Black Sulphate Liquor	★	★	☆	☆	☆	☆	■	■	■	□	★	★	□	□	☆	☆	□	□	■	□	☆	☆					
Blast Furnace Gas	□	□	☆	☆	□	□	★	□	□	■	■	■	□	■	□	☆	☆	□	□	■	□	☆	☆				
Bleach Liquor	□	□	□	□	□	□	□	□	□	■	■	☆	□	★	□	■	□	□	□	■	□	☆	☆				
Bleach Solutions	□	□	☆	□	□	□	□	□	□	★	☆	☆	(☆)	□	□	☆	□	□	□	■	□	☆	☆				
Bleaching Powder (Wet)	□	□	☆	■	□	□	★	■	□	□	☆	□	□	□	□	□	□	□	□	■	□	☆	☆				
Blood (Meat Juices - Cold)	□	□	★	☆	□	□	□	☆	★	☆	■	□	□	□	□	☆	☆	□	□	■	□	☆	★				
Borax	★	★	☆	☆	☆	☆	■	☆	■	★	□	☆	(☆)	(☆)	☆	■	■	☆	☆	□	☆	☆					
Bordeaux Mixtures	■	■	☆	☆	□	□	■	□	□	☆	☆	☆	□	□	★	☆	□	□	☆	☆	☆						
Boric Acid	■	■	★	★	■	■	☆	☆	■	☆	☆	☆	☆	□	☆	☆	□	□	☆	☆	☆						
Boron Fuels (HEF)	□	□	□	□	□	□	□	□	☆	■	☆	□	□	■	□	□	□	■	□	☆	☆						
Brake Fluid (Non-Petroleum)	☆	☆	☆	☆	☆	☆	☆	☆	☆	■	□	☆	□	★	□	□	□	■	□	☆	☆						
Bray GG - 130	□	□	□	□	□	□	□	□	□	★	■	□	□	★	□	□	□	■	□	□	☆						
Brayco 719-R (VV-H-910)	□	□	□	□	□	□	□	□	□	★	■	□	□	★	□	□	□	■	□	□	☆						
885 (Mil-L-6085A)	□	□	□	□	□	□	□	□	□	★	■	□	□	★	□	□	☆	■	□	□	☆						
Brayco 910	□	□	□	□	□	□	□	□	□	★	■	□	□	★	□	□	□	■	□	□	☆						
Bret 710	□	□	□	□	□	□	□	□	★	☆	□	□	★	□	□	□	■	□	□	☆							
Brine	★	★	☆	★	□	□	□	□	☆	□	☆	(☆)	(☆)	☆	☆	□	☆	□	☆	□	(☆)	☆	★				
Brewery Slop	☆	☆	☆	☆	□	□	□	□	☆	□	☆	☆	□	☆	□	□	□	□	☆	☆							
Brom-113	□	□	□	□	□	□	□	□	□	★	■	□	□	■	□	□	□	□	□	☆	☆						
-114	□	□	□	□	□	□	□	□	□	★	■	□	□	■	□	□	□	□	□	☆	★						
Bromine (Dry)	■	■	■	■	■	■	■	□	■	□	■	■	□	☆	□	■	■	□	□	■	□	☆	☆				
Bromine (Wet)	■	■	■	■	■	■	■	□	■	□	■	★	□	☆	□	□	□	□	■	□	(☆)	☆	☆				
Bromine-Anhydrous	■	■	■	■	■	■	■	□	■	□	■	■	□	☆	□	□	□	□	□	■	□	☆	☆				
Bromine Pentafluoride	□	□	□	□	□	□	■	□	□	□	☆	□	□	□	□	□	□	□	■	□	■	☆					
Bromine Trifluoride	■	■	■	★	□	□	■	□	□	□	□	(★)	□	□	□	□	□	□	■	□	■	☆					
Bromine Water	■	■	■	■	□	□	★	■	□	□	■	□	□	■	□	☆	★	☆	□	□	□	□	☆	☆	★		
Bromobenzene	★	★	☆	☆	□	□	☆	□	□	■	■	(☆)	□	□	☆	□	□	■	□	□	☆	■					
Bromochloromethane	★	★	★	★	□	□	■	□	□	■	■	★	★	□	☆	□	□	■	□	□	☆	☆					
Bromochloro Trifluoroethane	□	□	□	□	□	□	□	□	□	□	■	□	□	□	□	□	□	■	□	□	☆	☆					
Bromotoluene	☆	☆	☆	☆	□	□	☆	□	□	■	□	□	□	□	□	□	□	□	□	□	☆	★					

Legend:

☆ – RECOMMENDED, PERFORMANCE IS NOT AFFECTED
★ – FAIR, SOME LOSS OF PROPERTIES MAY OCCUR
■ – NOT RECOMMENDED, MATERIAL IS UNSUITABLE FOR SERVICE
☐ – NO DATA AVAILABLE AT THIS TIME
( ) – BRACKETS AROUND A RATING LETTER, NO DATA IS AVAILABLE, BUT THE RATINGS ARE MADE ON THE BASIS OF EXPOSURE TESTS IN SIMILAR CHEMICAL GROUPS.

Material	CARBON STEEL or CAST IRON	CARBON STEEL HARDENED	301/302/303/304 STAINLESS STEEL	316 STAINLESS STEEL	400 SERIES STAINLESS STEEL	400 SERIES STAINLESS STEEL HARDENED	ALUMINUM	COPPER	BRASS	ACETAL (DELRIN, CELCON)	ABS (CYCOLAC)	BUNA "N" (NITRILE RUBBER)	EPR, EPDM (NORDEL)	FLOURAZ	KALREZ	LEATHER-THIOKOL IMPREGNATED	NEOPRENE	NYLON	POLYCARBONATE (LEXAN)	POLYETHYLENE LOW DENSITY	POLYETHYLENE HIGH DENSITY	POLYPROPLENE 120° (NORYL)	POLYURETHANE, URATHANE	PVC POLYVINYL CHLORIDE	PVDF 150° (KYNAR)	TEFLON 120°	VITON
Bronzing Liquid	☐	☐	☆	☆	☐	☐	☐	☐	☐	☐	■	★	☐	☐	☐	☐	■	☐	☐	☐	☐	☐	☐	☐	☐	☆	■
Bunker Oil	☆	☆	☆	☆	☐	☐	☆	☆	☐	☆	☐	☆	■	☐	☐	☆	■	☆	☐	☐	☆	☆	☐	■	☐	☆	☆
Butadiene (Monomer)	☆	☆	☆	☆	☆	☆	☆	☆	☆	☐	■	■	■	(☆)	★	■	☆	■	■	■	■	■	☐	☐	☆	☆	☆
Butane	☆	☆	☆	☆	☆	☆	☆	☆	☆	★	■	☆	■	☐	☐	★	☆	☐	■	■	■	■	☐	■	☐	☆	☆
Butane, 2, 2-Dimethyl	☆	☆	☆	☆	☆	☆	☆	☆	☆	☐	■	☐	☐	☐	☐	★	☐	☐	☐	■	■	■	☐	■	☐	☆	☆
2, 3-Dimethyl	☆	☆	☆	☆	☆	☆	☆	☆	☆	☐	■	☐	☐	☐	☐	★	☐	☐	☐	■	■	■	☐	■	☐	☆	☆
Butanol (Butyl Alcohol)	☆	☆	☆	☆	☆	☆	☆	☆	☆	★	☆	☆	☆	☐	☐	☆	★	☆	☆	☆	☆	☆	■	★	☐	☆	☆
1-Butene, 2-Ethyl	☆	☆	☆	☆	☆	☆	☆	☆	☐	☐	■	☐	☐	☐	☐	☆	☐	☐	☐	■	■	■	☐	■	☐	☆	☆
Butter	■	■	☆	☆	☆	☆	☆	☐	☐	☆	★	☆	☐	☐	☐	☆	★	☐	☐	☆	☆	☆	☐	☆	☐	☆	☆
Butter – Animal Fat	☐	☐	☆	☆	☆	☆	☆	☐	☐	☐	☐	☆	☐	☐	☐	☆	★	☐	☐	☐	☐	☆	☐	☆	☐	☆	☆
Buttermilk	■	■	☆	☆	★	★	☆	■	■	☆	★	☆	☐	☐	☐	☆	☆	☐	☐	☆	☆	☆	☐	☆	☐	☆	☆
Butyl Acetate	☆	☆	☆	☆	☆	★	■	☆	☆	■	■	★	☆	☐	■	☆	■	★	★	★	■	■	■	☆	■
Butyl Acetyl Ricinoleate	☐	☐	☐	☐	☐	☐	☐	☐	☐	☐	★	☐	☐	☐	☐	★	☐	☐	☐	☐	☐	☐	☐	☐	☐	☆	☆
Butyl Acrylate	★	★	☆	☆	☆	☆	★	☆	☆	☐	★	☆	☐	☐	■	☆	☆	☆	☆	■	■	★	☆	☆	☆
Butyl Alcohol (Butanol)	★	★	☆	☆	★	☆	★	☆	☆	☐	★	☆	☆	☐	☐	☆	★	☆	☆	☆	☆	★	■	★	☐	☆	☆
Butyl Amine (Aminobutane)	☐	☐	☆	☆	☐	☐	☐	☐	☐	■	■	■	(☆)	☐	☐	☐	☐	☐	☐	☐	☐	☐	☐	☐	☐	☐
N-Butyl Benzoate	☐	☐	☐	☐	☐	☐	☐	☐	☐	☆	☐	(☆)	☐	☐	■	☐	☐	☐	☐	☐	☐	☐	☐	☐	☐	☆
Butyl Butyrate	☐	☐	☐	☐	☐	☐	☐	☐	☐	☆	☐	(☆)	■	☐	☐	☐	☐	☐	☐	☐	☐	☐	☐	☐	☐	☆
Butyl Carbitol	☐	☐	☐	☐	☐	☐	☐	☐	☐	★	☐	☐	■	☐	☐	☐	☐	☐	☐	☐	☐	☐	☐	■	☐	☆	☆
Butyl Cellosolve	☐	☐	☆	☆	☆	☆	☐	☐	☐	★	☐	☐	■	☐	☐	☐	☐	☐	☐	☐	☐	☐	■	■	☐	☆	☆
Butyl Cellosolve Adipate	☐	☐	☐	☐	☐	☐	☐	☐	☐	★	☐	☐	■	☐	☐	☐	☐	☐	☐	☐	☐	☐	☐	☐	☐	★
Butyl Chloride (Chlorobutane)	★	★	★	★	☐	☐	■	☐	☐	■	■	☐	(☆)	☐	☐	☐	■	☐	☐	☐	☐	☐	☐	☐	☐	☆	☆
Butyl Stearate	★	★	★	★	★	☆	★	☆	★	☐	☐	★	■	☐	(☆)	☐	★	☐	☐	☆	☆	☆	☐	☆	☐	☆	☆
Butylamine	☆	☆	☆	☆	☆	☆	☐	■	☐	☐	☐	☐	(☆)	☐	☐	☐	■	☐	☐	☐	☐	☐	☐	☐	☐	☐
Butylene	☆	☆	☆	☆	☆	☆	☆	☆	☆	☐	★	■	☆	(☆)	☐	☆	■	☐	☐	☐	☐	☐	☐	■	☐	☆	☆
N-Butyl Ether	☐	☐	☐	☐	☐	☐	☐	☐	☐	■	■	☐	☐	☐	☐	☐	■	☐	☐	☐	☆	☆	☐	★	☐	☆	☐
Butyl Oleate	☐	☐	☐	☐	☐	☐	☐	☐	☐	■	★	☐	■	☐	☐	☐	■	☐	☐	☐	☐	☐	☐	☐	☐	☆
Butyraldehyde	☐	☐	☐	☐	☐	☐	☐	☐	☐	■	★	☐	■	☐	☐	☐	■	☐	☐	☐	☐	■	☐	■	☐	☆	☐
Butyric Acid	■	■	★	★	■	■	☆	★	■	☆	■	★	☆	☐	★	■	☐	■	☐	■	☆	☆	■	☆	☆	★
Butyric Anhydride	☆	☆	☆	☆	☐	☐	☆	☐	☆	☐	☐	☐	☐	☐	☐	☐	■	☐	☐	☐	☐	☐	☐	☐	☐	☆	☆
Butyronitrile	☐	☐	☐	☐	☐	☐	☐	☐	☐	☐	☐	☐	☐	☐	☐	☐	■	☐	☐	☐	☐	☐	☐	☐	☆	☐
Cadmium Sulfate (25% Concentration)	★	★	☆	☆	☐	☐	☐	★	☐	☐	☐	☆	☆	☐	(☆)	☐	☆	☐	☐	☐	☐	☐	☐	☐	☐	☐	☆
Calcine Liqurs	☐	☐	☐	☐	☐	☐	☐	☐	☐	☐	☐	☆	☆	☐	☐	☐	☐	☐	☐	☐	☐	■	☐	☐	☐	6	☆
Calcium Acetate	★	★	★	★	☆	(★)	■	★	☐	☐	☐	★	☆	☐	(☆)	☐	★	☐	☐	☆	☆	☆	☐	■	☐	☆	☆
Calcium Acid Sulphate	☐	☐	☐	☐	☐	☐	☐	☐	☐	☐	☐	☆	☐	☐	☆	☐	☐	☐	☐	☐	☐	☐	☐	☐	☐	☐	☐
Calcium Bisulfate	6	6	☐	☆	☐	☐	☐	☐	☐	☐	☐	☆	☆	☐	☐	☐	☆	☐	☐	☐	☐	☐	☐	☆	☆	☆	☆
Calcium Bisulfide	☐	☐	★	★	☐	☐	■	■	☐	☆	☐	☆	☆	(☆)	☐	☐	☆	☐	☐	☐	☐	☐	☐	☆	☆	☆	☆
Calcium Bisulfite	■	■	☐	☐	■	■	☐	☆	★	■	■	☆	■	(☆)	☆	☐	☆	☐	☐	☆	☆	☆	☐	★	☆	☆	☆
Calcium Carbonate	★	★	☆	☆	☆	☆	■	★	☐	☆	☐	☆	☆	(☆)	☆	☆	☆	☐	☐	☆	☆	☆	☐	■	☆	☆	☆
Calcium Chlorate	★	★	★	☆	★	★	☆	☆	☐	☆	☐	☆	☆	(☆)	☐	☆	☆	☐	☐	☆	☆	☆	☐	☆	☆	☆	☆
Calcium Chloride	★	★	★	★	■	■	■	★	■	■	★	☆	☆	(☆)	☆	☆	■	☐	☐	☆	☆	☆	☐	☆	☆	☆	☆
Calcium Cyanide	☆	☆	☐	☐	☐	☐	☐	☐	☐	☐	☐	☆	☆	(☆)	☐	☐	☐	☐	☐	☐	☐	☐	☐	☐	☐	☆	☆
Calcim Hydrosulfide	☐	☐	☐	☐	☐	☐	☐	☐	☐	☐	☐	☆	☐	(☆)	☐	☐	☐	☐	☐	☐	☐	☐	☐	☐	☐	☆	☆
Calcium Hydroxide – 10% (Boiling)	★	★	☆	☆	☆	(☆)	■	■	■	☐	☐	★	☆	(☆)	★	☐	■	☆	☆	☆	☆	■	☆	☆	☆	☆
20% Solution (Boiling)	☐	☐	☆	☆	☐	☐	■	■	■	☐	☐	☆	☐	☐	☐	☐	☐	☐	☐	☐	☐	☐	☐	☐	☐	☆	☆
50% Solution (Boiling)	☐	☐	■	★	☐	☐	■	■	■	☐	☐	☆	☐	☐	☐	☐	☐	☐	☐	☐	☐	☐	☐	☐	☐	☆	☆
Calcium Hypochloride	☐	☐	☐	☐	☐	☐	☐	☐	☐	☐	■	☆	☆	☐	☐	☐	☐	☐	☐	☐	☐	■	☐	☐	☐	☆	☐

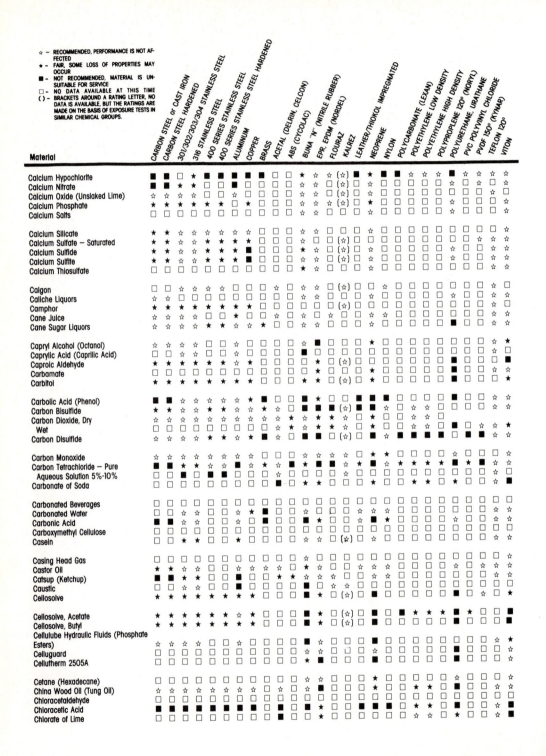

☆ – RECOMMENDED, PERFORMANCE IS NOT AFFECTED
★ – FAIR, SOME LOSS OF PROPERTIES MAY OCCUR
■ – NOT RECOMMENDED, MATERIAL IS UNSUITABLE FOR SERVICE
□ – NO DATA AVAILABLE AT THIS TIME
( ) – BRACKETS AROUND A RATING LETTER, NO DATA IS AVAILABLE, BUT THE RATINGS ARE MADE ON THE BASIS OF EXPOSURE TESTS IN SIMILAR CHEMICAL GROUPS.

Material	CARBON STEEL or CAST IRON	CARBON STEEL HARDENED	301/302/303/304 STAINLESS STEEL	316 STAINLESS STEEL	400 SERIES STAINLESS STEEL	400 SERIES STAINLESS STEEL HARDENED	ALUMINUM	COPPER	BRASS	ACETAL (DELRIN, CELCON)	ABS (CYCOLAC)	BUNA "N" (NITRILE RUBBER)	EPR, EPDM (NORDEL)	FLOURAZ	KALREZ	LEATHER/THIOKOL IMPREGNATED	NEOPRENE	NYLON	POLYCARBONATE (LEXAN)	POLYETHYLENE LOW DENSITY	POLYETHYLENE HIGH DENSITY	POLYPROPLENE 120° (NORYL)	POLYURETHANE, URATHANE	PVC POLYVINYL CHLORIDE	PVDF 150° (KYNAR)	TEFLON 120°	VITON
Calcium Hypochlorite	■	■	□	★	■	■	■	■	■	□	□	★	☆	☆	(☆)	■	★	■	■	☆	☆	☆	■	☆	☆	☆	☆
Calcium Nitrate	■	■	★	★	□	□	■	□	□	□	□	☆	☆	□	(☆)	□	☆	□	□	□	□	□	☆	□	☆	□	☆
Calcium Oxide (Unslaked Lime)	☆	☆	☆	☆	□	□	☆	□	□	□	□	☆	☆	□	(☆)	□	☆	□	□	□	□	□	□	□	□	☆	□
Calcium Phosphate	★	★	★	★	★	★	□	★	□	□	□	☆	☆	☆	(☆)	□	★	□	□	□	□	□	☆	□	□	□	☆
Calcium Salts	□	□	□	□	□	□	□	□	□	□	□	☆	☆	□	□	□	☆	□	□	□	□	□	☆	□	□	□	☆
Calcium Silicate	★	★	☆	☆	☆	☆	☆	☆	□	□	□	☆	☆	□	□	□	☆	□	□	□	□	□	□	□	□	□	☆
Calcium Sulfate – Saturated	★	★	☆	☆	★	★	★	★	□	□	□	☆	□	□	(☆)	□	□	□	□	□	□	□	□	□	☆	☆	☆
Calcium Sulfide	★	★	☆	☆	★	★	★	■	□	□	□	★	☆	□	(☆)	□	☆	□	□	□	□	□	☆	□	□	☆	☆
Calcium Sulfite	★	★	☆	☆	★	★	★	■	□	□	□	☆	☆	□	(☆)	□	☆	□	□	□	□	□	☆	□	□	☆	☆
Calcium Thiosulfate	□	□	□	□	□	□	□	□	□	□	□	★	☆	□	□	□	☆	□	□	□	□	□	□	□	□	☆	☆
Calgon	□	□	☆	☆	☆	☆	□	□	□	☆	□	☆	☆	□	(☆)	□	□	☆	□	□	□	□	□	□	□	☆	□
Caliche Liquors	☆	☆	□	□	□	□	□	□	□	□	□	☆	☆	□	□	□	☆	□	□	□	□	□	☆	□	□	☆	☆
Camphor	★	★	★	★	★	★	★	★	□	□	□	□	□	□	(☆)	□	□	□	□	□	□	□	□	□	□	☆	□
Cane Juice	☆	☆	☆	☆	□	□	★	□	□	☆	□	☆	□	☆	□	□	☆	☆	□	□	□	□	□	□	□	☆	☆
Cane Sugar Liquors	☆	☆	☆	☆	★	★	☆	☆	★	□	□	☆	☆	□	□	□	☆	□	□	□	□	□	■	□	□	☆	☆
Capryl Alcohol (Octanol)	☆	☆	☆	☆	□	□	☆	□	□	□	□	☆	■	□	□	□	★	□	□	□	□	□	□	□	□	☆	★
Caprylic Acid (Caprilic Acid)	□	□	☆	☆	□	□	☆	□	□	□	□	■	□	□	□	□	□	□	□	□	□	□	□	□	□	☆	□
Caproic Aldehyde	★	★	★	★	★	★	☆	★	□	□	□	□	★	□	(☆)	□	□	□	□	□	□	□	■	□	□	□	■
Carbamate	□	□	□	□	□	□	□	□	□	□	□	□	★	□	□	□	★	□	□	□	□	□	■	□	□	☆	☆
Carbitol	★	★	★	★	★	★	★	★	□	□	□	★	★	□	(☆)	□	★	□	□	□	□	□	■	□	□	□	★
Carbolic Acid (Phenol)	■	■	☆	☆	☆	☆	☆	★	■	□	□	■	★	□	□	■	■	■	□	□	□	□	■	□	□	☆	☆
Carbon Bisulfide	★	★	☆	☆	★	★	☆	☆	★	☆	□	■	■	■	(☆)	■	■	☆	□	☆	☆	□	□	□	□	☆	☆
Carbon Dioxide, Dry	☆	☆	☆	☆	☆	☆	☆	☆	☆	☆	★	☆	★	★	☆	□	★	□	□	☆	☆	□					
Wet	□	□	□	□	□	□	□	□	□	☆	★	☆	★	★	☆	□	★	□	□	☆	☆	□	■	□	☆	☆	★
Carbon Disulfide	☆	☆	☆	☆	★	★	☆	★	■	☆	□	■	■	□	(☆)	□	■	☆	■	■	■	■	□	■	■	☆	☆
Carbon Monoxide	☆	☆	☆	☆	☆	☆	☆	☆	□	□	□	☆	☆	☆	☆	□	★	★	□	□	□	□	☆	□	□	□	☆
Carbon Tetrachloride – Pure	■	■	★	★	☆	☆	■	☆	★	☆	■	★	■	■	☆	★	■	☆	★	★	★	★	■	★	■	☆	☆
Aqueous Solution 5%-10%	□	□	■	□	■	■	□	□	□	☆	□	□	□	□	☆	□	□	□	□	□	□	□	□	□	□	☆	□
Carbonate of Soda	□	□	□	□	□	□	□	□	□	■	□	★	★	□	□	□	★	□	□	★	★	□	★	□	□	☆	■
Carbonated Beverages	□	□	□	□	□	□	□	□	□	□	□	□	□	□	□	□	□	□	□	□	□	□	□	□	□	□	□
Carbonated Water	□	□	☆	☆	□	□	☆	★	■	□	□	☆	□	□	□	☆	☆	☆	□	□	□	□	□	□	□	☆	☆
Carbonic Acid	■	■	☆	☆	□	□	☆	□	■	□	□	■	★	□	□	☆	■	★	□	□	□	□	☆	□	□	☆	☆
Carboxymethyl Cellulose	□	□	□	□	□	□	□	□	□	□	□	□	□	□	□	□	□	□	□	□	□	□	□	□	□	□	□
Casein	□	□	★	★	□	□	★	□	□	□	□	☆	☆	□	(☆)	□	☆	□	□	□	□	□	□	□	□	☆	☆
Casing Head Gas	□	□	□	□	□	□	□	□	□	□	□	☆	□	□	□	□	□	□	□	□	□	□	□	□	□	□	☆
Castor Oil	★	★	☆	☆	□	□	☆	☆	☆	☆	□	★	☆	□	□	☆	☆	☆	□	□	□	□	☆	□	☆	☆	☆
Catsup (Ketchup)	■	■	★	★	□	□	■	□	□	★	★	☆	☆	☆	□	□	☆	☆	□	□	□	□	□	□	□	☆	☆
Caustic	□	□	☆	☆	□	□	■	□	□	□	□	■	□	☆	☆	□	□	□	□	□	□	□	□	□	□	□	□
Cellosolve	★	★	★	★	★	★	★	★	□	□	□	■	★	□	(☆)	□	■	□	□	□	□	□	■	□	☆	□	★
Cellosolve, Acetate	★	★	★	★	★	★	☆	★	□	□	□	■	★	□	(☆)	□	■	□	■	★	★	★	■	★	□	□	■
Cellosolve, Butyl	★	★	★	★	★	★	★	★	□	□	□	■	★	□	(☆)	□	■	□	□	□	□	□	■	□	□	□	■
Cellulube Hydraulic Fluids (Phosphate Esters)	☆	☆	☆	☆	□	□	☆	□	□	□	□	■	☆	□	□	□	■	□	□	□	□	□	□	□	□	☆	★
Celluguard	□	□	□	□	□	□	□	□	□	□	□	☆	☆	□	□	□	☆	□	□	□	□	□	■	□	□	□	☆
Cellutherm 2505A	□	□	□	□	□	□	□	□	□	□	□	★	■	□	□	□	■	□	□	□	□	□	■	□	□	□	☆
Cetane (Hexadecane)	□	□	□	□	□	□	□	□	□	□	□	☆	☆	□	□	□	★	□	□	□	□	□	☆	□	□	□	☆
China Wood Oil (Tung Oil)	☆	☆	☆	☆	☆	☆	☆	☆	□	☆	□	☆	■	□	□	□	★	□	□	★	★	□	■	□	□	☆	☆
Chloracetaldehyde	□	□	□	□	□	□	□	□	□	□	□	□	□	□	□	□	□	□	□	□	□	□	□	□	□	□	□
Chloracetic Acid	■	■	■	■	■	■	■	■	□	■	□	■	★	□	□	■	■	■	□	★	★	□	■	□	□	☆	■
Chlorate of Lime	□	□	□	□	□	□	□	□	□	■	□	□	★	□	□	□	□	□	□	☆	☆	□	★	□	□	☆	■

Legend:

☆ — RECOMMENDED, PERFORMANCE IS NOT AFFECTED
★ — FAIR, SOME LOSS OF PROPERTIES MAY OCCUR
■ — NOT RECOMMENDED, MATERIAL IS UNSUITABLE FOR SERVICE
□ — NO DATA AVAILABLE AT THIS TIME
( ) — BRACKETS AROUND A RATING LETTER, NO DATA IS AVAILABLE, BUT THE RATINGS ARE MADE ON THE BASIS OF EXPOSURE TESTS IN SIMILAR CHEMICAL GROUPS.

Column headers (left to right):

1. CARBON STEEL or CAST IRON
2. CARBON STEEL HARDENED
3. 301/302/303/304 STAINLESS STEEL
4. 316 STAINLESS STEEL
5. 400 SERIES STAINLESS STEEL
6. 400 SERIES STAINLESS STEEL HARDENED
7. ALUMINUM
8. COPPER
9. BRASS
10. ACETAL (DELRIN, CELCON)
11. ABS (CYCOLAC)
12. BUNA "N" (NITRILE RUBBER)
13. EPR, EPDM (NORDEL)
14. FLOURAZ
15. KALREZ
16. LEATHER-THIOKOL IMPREGNATED
17. NEOPRENE
18. NYLON
19. POLYCARBONATE (LEXAN)
20. POLYETHYLENE LOW DENSITY
21. POLYETHYLENE HIGH DENSITY
22. POLYPROPLENE 120°
23. POLYURETHANE, URATHANE
24. PVC POLYVINYL CHLORIDE
25. PVDF 150° (KYNAR)
26. TEFLON 120°
27. VITON

Material	1	2	3	4	5	6	7	8	9	10	11	12	13	14	15	16	17	18	19	20	21	22	23	24	25	26	27
Chlorbenzol (Conc. Pure)	□	□	☆	☆	□	□	□	□	□	□	□	□	□	□	□	□	□	□	□	□	□	□	□	□	□	□	□
Chlorextol	□	□	□	□	□	□	□	□	□	□	★	■	□	□	□	☆	□	□	□	□	□	■	■	□	☆	☆	
Chloric Acid	■	■	■	■	■	■	■	■	□	□	□	□	□	□	□	□	□	□	□	□	□	□	□	□	□	□	
Chlorinated Glue	■	■	□	☆	□	□	□	□	■	□	★	□	□	□	□	■	□	□	□	□	□	□	□	□	☆	☆	
Chlorinated Lime - 35% (Bleach)	■	■	☆	☆	□	□	□	□	□	□	■	☆	□	□	□	□	□	□	□	□	□	□	□	□	☆	☆	
Chlorinated Salt Brine	□	□	□	□	□	□	□	□	□	□	□	□	□	□	□	□	□	□	□	□	□	□	□	□	☆	☆	
Chlorinated Solvents, Dry	□	□	□	□	□	■	□	□	☆	□	■	■	□	□	□	□	□	□	★	★	■	□	□	☆	☆		
Wet	□	□	□	□	□	■	□	□	☆	□	■	■	□	□	□	□	□	□	★	★	■	□	□	☆	☆		
Chlorinated Water — Saturated	□	□	■	★	□	□	□	□	□	□	■	□	□	□	□	□	□	□	□	□	□	□	□	☆	☆		
Chlorine, Dry	■	■	■	★	■	■	■	☆	□	□	★	□	■	★	□	■	□	★	★	★	★	□	★	☆	☆	☆	
Chlorine, Wet	■	■	■	■	■	■	■	□	■	□	★	■	★	★	□	■	□	★	★	★	★	■	★	☆	☆	☆	
Chlorine Gas — Dry	■	■	■	★	■	■	■	☆	■	□	★	□	□	□	□	□	□	□	□	□	□	□	□	☆	☆	□	
Gas — Moist	■	■	■	■	■	■	■	□	■	□	□	■	□	□	□	□	□	□	□	□	□	□	□	☆	☆	□	
Gas — 100°C.	□	□	■	■	□	□	□	■	□	□	■	□	□	□	□	□	□	□	□	□	□	□	□	☆	☆	□	
Chlorine, Anhydrous Liquid	■	■	★	★	■	■	■	□	□	☆	■	□	□	★	□	■	□	★	★	□	□	□	□	☆	☆	☆	
Chlorine Dioxide	■	■	■	■	■	■	☆	□	□	□	□	□	□	□	□	□	□	□	□	□	□	□	■	□	□	☆	
Chlorine Trifluoride	☆	☆	☆	☆	☆	☆	☆	□	□	■	■	■	□	★	□	■	□	□	■	■	□	□	◦	□	□	■	
Chloroacetic Acid	■	■	■	■	■	■	■	□	□	□	★	□	□	□	□	☆	☆	★	□	☆	☆	☆	■				
Chloracetone	★	★	★	★	★	★	■	★	□	□	■	■	☆	□	■	☆	■	□	□	□	□	□	■	□	□	☆	
Chlorobenzene	★	★	★	★	★	★	★	□	■	■	■	■	□	(☆)	□	■	□	□	■	□	□	■	□	□	☆	☆	
Chlorobromo Methane	★	★	★	★	★	★	■	★	□	□	■	★	□	(☆)	□	■	□	□	☆	☆	□	■	□	□	☆		
Chlorobutadiene	★	★	☆	☆	□	□	■	□	□	■	■	★	□	□	□	■	□	□	□	□	□	□	□	□	☆		
Chlorodane	□	□	□	□	□	□	□	□	□	★	■	■	□	□	□	■	□	□	□	□	□	□	□	□	☆		
Chlorododecane	□	□	□	□	□	□	□	□	□	□	■	■	□	□	□	■	□	□	□	□	□	□	□	□	☆		
Chloroform	★	★	☆	☆	☆	☆	□	☆	☆	☆	■	■	■	(☆)	■	■	☆	■	★	★	★	□	★	☆	☆	☆	
0 — Chloronaphthalene	★	★	★	★	★	★	■	★	□	■	■	■	☆	□	□	■	□	□	□	□	□	□	□	□	□	☆	
1 — Chlorol 1 Nitro Ethane	□	□	□	□	□	□	□	□	□	□	□	□	□	□	□	□	□	□	□	□	□	□	□	□	□	■	
Chloropicrin	□	□	□	□	□	■	□	□	□	□	■	■	□	(☆)	□	□	□	□	□	□	□	□	□	□	☆	□	
Chlorosulfonic Acid (Dry)	★	★	★	★	□	□	■	□	★	■	■	■	★	□	□	□	□	□	□	□	□	□	■	□	☆	★	
Chlorosulfonic Acid (Wet)	■	■	■	★	□	□	■	□	□	□	■	□	□	□	□	□	□	□	□	□	□	□	■	□	☆	■	
Chlorotoluene	★	★	★	★	□	□	■	□	★	□	■	■	□	(☆)	□	■	□	□	□	□	□	□	□	□	□	☆	
Chlorox (Bleach)	■	■	☆	☆	□	□	□	□	□	□	□	★	★	□	□	□	★	□	□	□	□	□	□	□	□	☆	
0 — Chlorphenol	★	★	★	★	★	★	■	★	□	□	■	■	□	(☆)	□	■	□	□	□	□	□	□	□	□	□	☆	
Chocolate Syrup	■	■	☆	☆	□	☆	□	□	☆	☆	□	☆	☆	□	□	☆	□	□	□	□	□	□	□	□	☆	□	
Chrome Alum.	■	■	☆	☆	☆	☆	□	□	□	☆	☆	□	(☆)	□	☆	□	□	□	□	□	□	■	☆	☆			
Chrome Plating Solutions	□	□	☆	☆	□	□	□	□	□	□	■	★	□	□	□	□	■	□	□	□	□	□	□	□	☆	☆	
Chromic Acid — 5%	■	■	★	☆	■	■	■	■	■	■	★	□	☆	□	□	■	■	■	★	☆	☆	☆	■	★	☆	☆	
Chromic Acid — 50%	■	■	★	☆	■	■	■	■	■	■	★	□	☆	□	□	■	■	★	☆	☆	★	■	★	☆	☆		
Chromic Acid — 50% Solution (Hot)	■	■	■	■	■	■	■	■	■	■	□	□	□	□	□	□	■	□	□	□	□	■	□	□	☆	□	
Chromic Oxide .88 Wt. % Aqueous Sol.	□	□	□	□	□	□	□	□	□	□	■	★	□	□	□	□	□	□	□	□	□	□	□	□	□	☆	
Chromium Plating Bath	□	□	☆	☆	□	□	□	□	□	□	□	□	□	□	□	□	□	□	□	□	□	□	□	□	☆	□	
Chromium Sulfate (Basic)	■	■	★	★	□	□	□	□	■	□	□	□	□	(☆)	□	☆	□	□	□	□	□	□	□	□	☆	□	
Cider (Apple Juice)	■	■	☆	☆	□	□	★	□	□	☆	□	☆	□	☆	□	□	☆	☆	□	□	☆	□	□	□	☆	☆	
Cinnamon Oil	■	■	☆	☆	□	□	□	□	□	□	□	□	□	□	□	■	□	■	■	■	■	□	■	□	☆	□	
Circo Light Process Oil	☆	☆	☆	☆	☆	☆	☆	☆	□	☆	□	☆	■	□	□	□	☆	□	□	□	□	□	□	□	☆	☆	

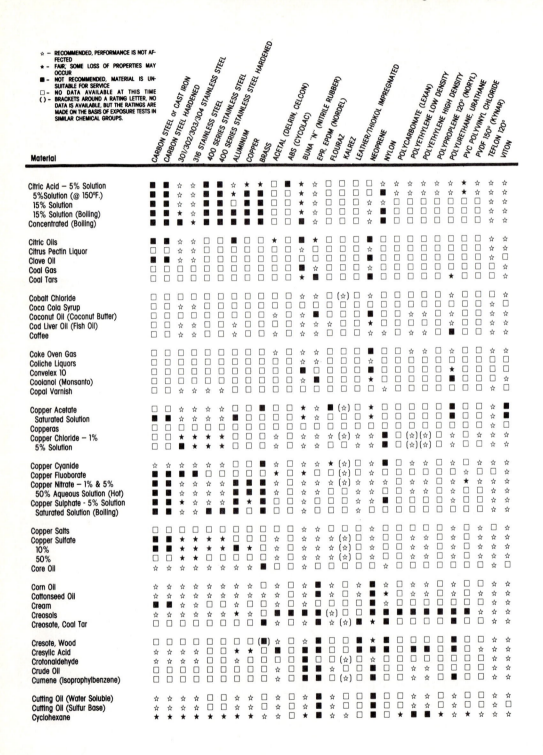

Legend:

☆ – RECOMMENDED, PERFORMANCE IS NOT AFFECTED
★ – FAIR; SOME LOSS OF PROPERTIES MAY OCCUR
■ – NOT RECOMMENDED, MATERIAL IS UNSUITABLE FOR SERVICE
☐ – NO DATA AVAILABLE AT THIS TIME
( ) – BRACKETS AROUND A RATING LETTER, NO DATA IS AVAILABLE, BUT THE RATINGS ARE MADE ON THE BASIS OF EXPOSURE TESTS IN SIMILAR CHEMICAL GROUPS.

Column headers:
CARBON STEEL or CAST IRON · CARBON STEEL HARDENED · 301/302/303/304 STAINLESS STEEL · 316 STAINLESS STEEL · 400 SERIES STAINLESS STEEL · 400 SERIES STAINLESS STEEL HARDENED · ALUMINUM · COPPER · BRASS · ACETAL (DELRIN, CELCON) · ABS (CYCOLAC) · BUNA "N" (NITRILE RUBBER) · EPR, EPDM (NORDEL) · FLOURAZ · KALREZ · LEATHER/THIOKOL IMPREGNATED · NEOPRENE · NYLON · POLYCARBONATE (LEXAN) · POLYETHYLENE LOW DENSITY · POLYETHYLENE HIGH DENSITY · POLYPROPLENE 120° (NORYL) · POLYURETHANE, URATHANE · PVC POLYVINYL CHLORIDE · PVDF 150° (KYNAR) · TEFLON 120° · VITON

Material

- Citric Acid – 5% Solution
- 5% Solution (@ 150°F.)
- 15% Solution
- 15% Solution (Boiling)
- Concentrated (Boiling)

- Citric Oils
- Citrus Pectin Liquor
- Clove Oil
- Coal Gas
- Coal Tars

- Cobalt Chloride
- Coca Cola Syrup
- Coconut Oil (Coconut Butter)
- Cod Liver Oil (Fish Oil)
- Coffee

- Coke Oven Gas
- Coliche Liquors
- Convelex 10
- Coolanol (Monsanto)
- Copal Varnish

- Copper Acetate
- Saturated Solution
- Copperas
- Copper Chloride – 1%
- 5% Solution

- Copper Cyanide
- Copper Fluoborate
- Copper Nitrate – 1% & 5%
- 50% Aqueous Solution (Hot)
- Copper Sulphate - 5% Solution
- Saturated Solution (Boiling)

- Copper Salts
- Copper Sulfate
- 10%
- 50%
- Core Oil

- Corn Oil
- Cottonseed Oil
- Cream
- Creosols
- Creosote, Coal Tar

- Cresote, Wood
- Cresylic Acid
- Crotonaldehyde
- Crude Oil
- Cumene (Isoprophylbenzene)

- Cutting Oil (Water Soluble)
- Cutting Oil (Sulfur Base)
- Cyclohexane

Legend:

- ☆ – RECOMMENDED, PERFORMANCE IS NOT AFFECTED
- ★ – FAIR, SOME LOSS OF PROPERTIES MAY OCCUR
- ■ – NOT RECOMMENDED, MATERIAL IS UNSUITABLE FOR SERVICE
- □ – NO DATA AVAILABLE AT THIS TIME
- ( ) – BRACKETS AROUND A RATING LETTER, NO DATA IS AVAILABLE, BUT THE RATINGS ARE MADE ON THE BASIS OF EXPOSURE TESTS IN SIMILAR CHEMICAL GROUPS.

Material	CARBON STEEL or CAST IRON	CARBON STEEL HARDENED	301/302/303/304 STAINLESS STEEL	316 STAINLESS STEEL	400 SERIES STAINLESS STEEL	400 SERIES STAINLESS STEEL HARDENED	ALUMINUM	COPPER	BRASS	ACETAL (DELRIN CELCON)	ABS (CYCOLAC)	BUNA "N" (NITRILE RUBBER)	EPR, EPDM (NORDEL)	FLOURAZ	KALREZ	LEATHER/THIOKOL IMPREGNATED	NEOPRENE	NYLON	POLYCARBONATE (LEXAN)	POLYETHYLENE LOW DENSITY	POLYETHYLENE HIGH DENSITY	POLYPROPLENE 120°	POLYURETHANE, URATHANE	PVC POLYVINYL CHLORIDE (NORYL)	PVDF 150° (KYNAR)	TEFLON 120°	VITON
Cyclohexanol	★	★	★	★	★	★	■	★	□	□	□	★	■	□	(☆)	□	★	□	□	□	□	□	□	□	□	☆	☆
Cyclohexanone	★	★	★	★	★	★	★	★	□	□	□	■	★	☆	(☆)	□	■	☆	□	□	□	□	□	□	□	☆	■
Cyanic Acid	□	□	□	□	☆	□	□	□	□	□	□	■	★	☆	(☆)	□	■	☆	□	□	□	□	□	□	■	■	☆
P-Cymene	□	□	□	□	□	□	□	□	☆	□	■	☆	□	(☆)	□	■	☆	□	□	□	□	☆	☆	□	■	□	■
D.D.T.	■	■	☆	☆	☆	☆	★	★	□	□	□	☆	☆	■	(☆)	□	■	□	□	□	□	☆	□	□	□	☆	☆
Decalin	□	□	□	□	□	□	□	□	□	■	■	☆	■	□	□	□	■	☆	■	★	★	★	□	★	□	☆	☆
Decanal	□	□	□	□	□	□	□	□	□	■	■	☆	□	□	□	□	■	□	□	□	□	□	□	□	□	☆	■
Decane	□	□	□	□	□	□	□	□	□	□	■	☆	■	□	□	□	■	□	□	□	□	□	★	□	□	☆	☆
Decyl Alcohol (Decanol)	□	□	□	□	□	□	□	□	□	□	□	☆	□	□	□	□	☆	□	□	□	□	□	★	□	□	☆	★
De-Ionized Water	□	□	□	□	□	□	□	□	□	□	□	□	□	□	□	□	□	□	□	□	□	□	□	□	□	□	☆
Degreasing Fluid (Chlorinated)	□	□	□	□	□	□	□	□	□	□	□	□	□	□	□	□	□	□	□	□	□	□	□	□	□	□	☆
Denatured Alcohol	□	□	□	□	□	□	□	□	□	□	□	☆	□	□	□	□	□	□	□	□	□	□	□	□	□	□	☆
Detergent Solutions	□	□	☆	☆	□	□	□	★	□	★	★	☆	□	□	□	□	★	□	□	□	□	□	☆	■	□	□	☆
Developing Fluids (Photo)	■	■	■	★	□	□	☆	☆	□	☆	☆	★	□	□	□	□	★	□	□	□	□	□	☆	■	□	☆	☆
Dextrose	■	■	☆	☆	□	□	□	□	□	★	☆	□	(☆)	□	□	★	□	□	□	□	□	☆	□	□	□	□	☆
Dextron	□	□	□	□	□	□	□	□	□	■	□	☆	□	□	□	★	□	□	□	□	□	★	□	(☆)	□	☆	□
Diacetone	☆	☆	☆	☆	□	□	□	□	□	■	□	■	□	□	□	■	□	□	□	□	□	□	□	□	□	☆	■
Diacetone Alcohol	☆	☆	☆	☆	□	□	☆	☆	☆	■	□	■	(☆)	□	□	■	□	□	□	□	□	★	★	□	□	☆	■
Diamylamine	□	□	□	□	☆	□	□	□	□	■	□	☆	(☆)	□	☆	□	□	□	□	□	□	□	□	□	□	☆	■
Diazinon	□	□	□	□	□	□	□	□	□	■	□	☆	□	□	□	■	□	□	□	□	□	□	□	□	□	□	★
Dibenzyl Ether	☆	☆	☆	☆	☆	☆	☆	☆	☆	■	★	□	(☆)	□	□	■	□	□	□	□	□	★	□	□	□	☆	■
Dibenzyl Sebacate	□	□	□	□	□	□	□	□	□	■	■	□	(☆)	□	□	■	□	□	□	□	□	□	□	□	□	□	★
Dibromoethyl Benzene	□	□	□	□	□	□	□	□	□	■	■	□	□	□	□	■	□	□	□	□	□	□	□	□	□	□	★
Dibutylamine	☆	☆	□	□	□	□	■	□	□	■	■	□	(☆)	□	☆	■	□	□	□	□	□	□	□	□	□	□	☆
Dibutyl Ether	☆	☆	☆	☆	☆	☆	☆	☆	☆	■	□	□	(☆)	□	□	■	□	□	□	□	□	★	□	□	□	☆	☆
Dibutyl Phthalate	☆	☆	☆	☆	☆	☆	□	□	□	■	★	■	(☆)	□	■	□	□	□	□	■	□	□	□	□	□	☆	☆
Dibutyl Sebacate	☆	☆	☆	☆	☆	☆	□	□	□	■	★	□	(☆)	□	■	□	□	□	□	□	□	□	□	□	□	☆	★
Dichloroacetic Acid	□	□	□	□	□	□	□	□	□	□	□	☆	□	□	□	□	☆	□	□	□	□	□	□	□	□	□	★
Dichlorethane	☆	☆	☆	☆	☆	☆	■	☆	□	□	□	■	☆	□	□	□	■	□	□	□	□	□	□	■	★	★	☆
P-Dichlorobenzene	☆	☆	☆	☆	☆	☆	■	☆	□	□	□	■	□	□	□	□	■	□	□	□	□	□	☆	■	□	□	☆
Dichloro-Butane	□	□	□	□	□	□	■	□	□	□	□	★	■	□	□	□	■	□	□	□	□	□	□	■	□	□	☆
Dichloroethyl Ether	□	□	□	□	□	□	■	□	□	□	★	■	□	□	□	□	■	□	□	□	□	□	□	■	□	☆	□
Dichloro-Isopropyl Ether	□	□	□	□	□	□	■	□	□	□	□	■	□	□	□	□	■	□	□	□	□	□	★	□	□	☆	□
Dichloropenthane	□	□	□	□	□	□	■	□	□	□	□	■	□	□	□	□	☆	□	□	□	□	□	□	□	□	□	□
Dichlorophenol	☆	☆	☆	☆	☆	☆	□	□	□	□	□	□	□	(☆)	□	□	□	□	□	□	□	□	□	□	□	□	☆
Dicyclohexylamine	□	□	□	□	□	□	■	□	□	□	□	■	□	□	□	□	□	□	□	□	□	□	□	■	□	□	□
Diesel Oil (Fuel ASTM #2)	☆	☆	☆	☆	☆	☆	☆	☆	☆	□	□	☆	■	□	□	□	■	□	□	□	□	□	□	☆	□	■	☆
Di-Ester Lubricant Mil-L-7808	☆	☆	☆	☆	☆	☆	☆	□	□	□	□	★	■	□	□	□	□	□	□	□	□	□	★	☆	☆	☆	☆
Di-Ester Synthetic Lubricants	☆	☆	☆	☆	☆	☆	☆	□	□	□	□	★	□	□	□	□	□	□	□	□	□	□	□	□	□	☆	□
Diethanolamine	☆	☆	☆	☆	☆	☆	■	□	□	□	■	★	☆	☆	□	□	■	□	□	□	□	□	□	■	□	☆	□
Diethylamine	☆	☆	☆	☆	☆	☆	■	□	□	★	★	■	(☆)	□	★	□	☆	□	□	□	□	□	■	■	□	☆	■
Diethyl Aniline	□	□	□	□	□	□	□	□	☆	□	□	■	□	(☆)	□	□	□	□	□	□	□	□	☆	□	☆	■	★
Diethyl Benzene	□	□	□	□	□	□	□	□	□	■	□	■	□	(☆)	□	■	□	□	■	■	■	□	□	☆	□	☆	☆
Diethyl Carbonate	☆	☆	□	□	□	□	□	□	□	■	□	■	□	(☆)	□	■	□	□	□	□	□	□	□	□	□	☆	□
Diethyl Ether	☆	☆	☆	☆	☆	☆	□	□	□	■	■	■	(☆)	□	■	□	■	□	■	■	■	■	□	★	■	☆	■
Diethyl Phthalate (Dep)	☆	☆	☆	☆	□	☆	☆	□	□	■	□	■	(☆)	□	■	□	■	□	■	□	□	■	□	□	□	☆	■
Diethyl Sebacate	☆	☆	☆	☆	☆	☆	☆	□	□	■	★	□	(☆)	□	■	□	■	□	□	□	□	■	□	□	□	☆	★
Diethyl Sulfate	□	□	□	□	□	□	□	□	□	■	□	□	(☆)	□	■	□	■	□	□	□	□	□	□	□	□	□	□

☆ — RECOMMENDED, PERFORMANCE IS NOT AFFECTED
★ — FAIR, SOME LOSS OF PROPERTIES MAY OCCUR
■ — NOT RECOMMENDED, MATERIAL IS UNSUITABLE FOR SERVICE
□ — NO DATA AVAILABLE AT THIS TIME
( ) — BRACKETS AROUND A RATING LETTER, NO DATA IS AVAILABLE, BUT THE RATINGS ARE MADE ON THE BASIS OF EXPOSURE TESTS IN SIMILAR CHEMICAL GROUPS.

Material	CARBON STEEL or CAST IRON	CARBON STEEL HARDENED	301/302/303/304 STAINLESS STEEL	316 STAINLESS STEEL	400 SERIES STAINLESS STEEL	400 SERIES STAINLESS STEEL HARDENED	ALUMINUM	COPPER	BRASS	ACETAL (DELRIN, CELCON)	ABS (CYCOLAC)	BUNA "N" (NITRILE RUBBER)	EPR, EPDM (NORDEL)	FLOURAZ	KALREZ	LEATHER/THIOKOL IMPREGNATED	NEOPRENE	NYLON	POLYCARBONATE (LEXAN)	POLYETHYLENE LOW DENSITY	POLYETHYLENE HIGH DENSITY	POLYPROPLENE 120° (NORYL)	POLYURETHANE, URATHANE	PVC POLYVINYL CHLORIDE	PVDF 150° (KYNAR)	TEFLON 120°	VITON
Diethylene Ether (Dioxane)	☆	☆	☆	☆	□	□	☆	□	□	□	□	■	☆	□	□	□	■	□	□	□	□	□	□	□	□	☆	■
Diethylene Glycol	□	□	□	☆	□	□	□	□	□	☆	★	☆	★	☆	☆	□	☆	★	☆	☆	☆	☆	□	■	□	☆	☆
Diethylene Triamine	☆	☆	☆	☆	□	□	□	□	□	□	□	★	□	☆	☆	□	☆	□	□	□	□	□	□	□	□	☆	□
Difluorodibromomethane	□	□	□	□	□	■	□	□	□	□	□	■	★	□	□	□	■	□	□	□	□	■	□	□	□	☆	□
Diisobutyl Ketone	☆	☆	☆	☆	□	□	☆	□	□	□	□	■	★	■	□	□	■	□	□	□	□	□	□	□	□	☆	■
Diisobutylene	★	★	★	★	□	□	★	□	□	□	□	★	■	□	□	□	★	□	□	□	□	■	□	□	□	☆	☆
Diisodecyl Adipate (Dida)	□	□	□	□	□	□	□	□	□	□	□	■	□	□	□	□	■	□	□	□	□	□	□	□	□	☆	■
Diisodecyl Phthalate (Diop)	□	□	□	□	□	□	□	□	□	□	□	■	☆	□	□	□	■	□	□	□	□	□	□	□	□	☆	■
Diisooctyl Adipate (Dioa)	□	□	□	□	□	□	□	□	□	□	□	■	□	□	□	□	■	□	□	□	□	□	□	□	□	☆	■
Diisooctyl Phthalate (Diop)	□	□	□	□	□	□	□	□	□	□	□	■	□	□	□	□	■	□	□	□	□	□	□	□	□	☆	■
Diisooctyl Sebacate	□	□	□	□	□	□	□	□	□	□	□	■	■	□	□	□	■	□	□	□	□	□	□	□	□	□	★
Diisopropyl Amine	□	□	□	□	□	□	□	□	□	□	★	□	□	□	□	□	□	□	□	□	□	□	□	□	□	☆	□
Diisopropyl Benzene	□	□	□	□	□	□	□	□	□	□	□	□	□	□	□	□	□	□	□	□	□	□	□	□	□	☆	☆
Diisopropyl Ketone	☆	☆	☆	☆	□	□	☆	□	□	□	□	■	☆	■	□	□	■	□	□	□	□	□	□	□	□	☆	■
Dimethyl Aniline	□	□	□	□	□	□	□	□	□	□	□	■	☆	□	□	□	■	□	□	□	□	□	□	□	□	☆	■
Diemethyl Formamide	☆	☆	☆	☆	□	□	□	□	□	★	■	■	(☆)	□	■	□	■	☆	☆	☆	☆	■	■	■	■	☆	■
Dimethyl Phthalate	□	□	□	★	□	□	□	□	□	★	★	★	(☆)	■	□	□	■	□	□	□	□	□	□	□	□	☆	★
Dimethyl Sulfate	☆	□	□	□	□	□	□	□	□	□	□	■	☆	□	□	□	■	□	□	□	□	□	□	□	□	☆	■
Dimethyl Sulfide	□	□	□	□	□	□	□	□	□	□	□	■	□	□	□	□	■	□	□	□	□	□	□	□	□	☆	■
Dinitro Toluene	□	□	□	□	□	□	□	□	□	□	□	■	□	□	□	□	■	□	□	□	□	■	□	□	□	☆	■
Dioctylphalate	□	□	□	□	□	□	□	□	□	★	□	□	□	□	□	□	□	□	□	□	□	□	□	□	□	□	☆
Dioctyl Phthalate	☆	☆	☆	☆	☆	☆	☆	☆	☆	□	■	★	☆	☆	□	□	■	□	□	□	□	□	□	★	□	□	★
Dioctyl Sebacate	☆	☆	☆	☆	□	□	☆	□	□	□	■	★	□	□	□	□	■	□	□	□	□	□	□	★	□	□	■
Dioxane	☆	☆	☆	☆	☆	☆	★	☆	□	□	□	■	★	□	□	□	■	□	□	□	□	□	□	□	□	☆	■
Dioxolane (Dioxalans)	□	□	□	□	□	□	□	□	□	□	□	■	★	□	□	□	■	□	□	□	□	□	□	□	□	☆	■
Dipentene	☆	☆	☆	□	□	□	☆	□	□	□	☆	□	(☆)	□	□	■	□	□	□	□	□	□	□	□	□	☆	□
Diphenyl	★	★	★	★	□	□	☆	□	□	□	□	■	■	■	(☆)	□	■	□	□	□	□	■	□	□	□	☆	☆
Diphenyl Oxides	☆	☆	☆	☆	☆	☆	☆	☆	□	□	□	■	□	(☆)	□	□	■	□	□	□	□	□	□	□	□	☆	☆
Dipropylamine	□	□	□	□	□	□	□	□	□	□	★	□	□	□	□	□	□	□	□	□	□	□	□	□	□	☆	□
Dipropylene Glycol	□	□	□	☆	□	□	□	□	□	□	☆	★	□	□	□	□	□	□	□	□	□	□	□	□	□	☆	□
Dipropyl Ketone (Butyrone)	□	□	□	□	□	□	□	□	□	□	□	■	☆	□	□	□	■	□	□	□	□	□	□	□	□	☆	■
Dispersing Oil #10	☆	☆	☆	☆	□	□	☆	□	□	□	■	■	□	□	□	□	■	□	□	□	□	□	□	□	□	☆	★
Divinyl Benzene (DVB)	□	□	□	□	□	□	□	□	□	□	□	■	□	□	□	□	□	□	□	□	□	□	□	□	□	☆	☆
Dodecyl Benzene (Alkane)	☆	☆	☆	☆	□	□	☆	□	□	□	□	■	□	□	□	□	■	□	□	□	□	□	□	□	□	☆	☆
Dow (Silicones)	□	□	□	□	□	□	□	□	□	□	□	☆	□	☆	(☆)	□	☆	□	□	□	□	□	□	□	□	☆	☆
Dowtherm A	☆	☆	☆	☆	☆	☆	☆	☆	□	□	□	■	■	★	☆	□	■	□	□	□	□	□	□	□	□	★	☆
Dowtherm E	☆	☆	☆	☆	☆	☆	☆	☆	□	□	□	■	■	☆	☆	□	■	□	□	□	□	□	□	□	□	★	☆
Drinking Water	■	■	☆	☆	☆	☆	☆	☆	☆	□	☆	☆	☆	☆	□	□	★	☆	□	☆	☆	☆	□	☆	□	☆	☆
Drilling Mud (Oil Base)	□	□	□	□	□	□	□	□	□	□	□	☆	□	□	□	□	☆	□	□	□	□	□	□	□	□	☆	□
Drilling Mud (Water Base)	□	□	□	□	□	□	□	□	□	□	□	☆	□	□	□	□	☆	□	□	□	□	□	□	□	□	□	□
Dry Cleaning Fluid	☆	☆	☆	☆	☆	☆	☆	□	□	□	■	■	□	☆	□	□	■	□	□	□	□	■	□	□	□	☆	☆
DTE Light Oil	☆	☆	☆	☆	☆	☆	☆	☆	□	□	☆	☆	■	□	□	□	★	□	□	□	□	☆	□	★	□	☆	☆
Dyes	☆	☆	☆	□	□	☆	☆	☆	☆	□	□	□	□	(☆)	□	□	☆	□	□	□	□	□	□	□	□	☆	☆
Elco 28-EP Lubricant	☆	☆	☆	☆	☆	☆	☆	☆	□	□	☆	☆	■	□	□	□	■	□	□	□	□	■	□	■	□	☆	☆
Epichlorohydrin	★	★	☆	☆	☆	☆	☆	□	□	□	■	★	☆	□	□	□	■	□	□	□	□	■	□	■	□	(☆)	■
Epoxy Resins	☆	☆	☆	☆	☆	☆	☆	□	□	□	☆	☆	□	□	□	□	☆	□	□	□	□	☆	□	□	□	☆	■
Epsom Salts	☆	☆	☆	☆	☆	☆	☆	☆	☆	□	☆	☆	☆	□	□	□	☆	□	□	□	□	☆	□	□	□	☆	☆
Esam-6 Fluid	□	□	□	□	□	□	□	□	□	□	☆	☆	□	□	□	□	★	□	□	□	□	□	□	□	□	☆	■
Esstic 42, 43	□	□	□	□	□	□	□	□	□	□	☆	■	□	□	□	□	★	□	□	□	□	□	□	★	□	☆	☆

☆ – RECOMMENDED, PERFORMANCE IS NOT AFFECTED
★ – FAIR, SOME LOSS OF PROPERTIES MAY OCCUR
■ – NOT RECOMMENDED, MATERIAL IS UNSUITABLE FOR SERVICE
□ – NO DATA AVAILABLE AT THIS TIME
( ) – BRACKETS AROUND A RATING LETTER, NO DATA IS AVAILABLE, BUT THE RATINGS ARE MADE ON THE BASIS OF EXPOSURE TESTS IN SIMILAR CHEMICAL GROUPS.

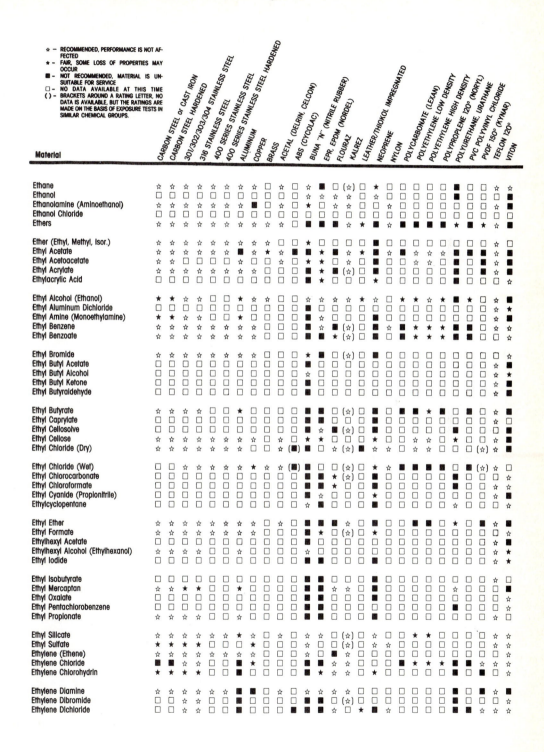

Chemical resistance ratings table. Columns (left to right): Carbon Steel or Cast Iron; Carbon Steel Hardened; 301/302/303/304 Stainless Steel; 316 Stainless Steel; 400 Series Stainless Steel; 400 Series Stainless Steel Hardened; Aluminum; Copper; Brass; Acetal (Delrin, Celcon); ABS (Cycolac); Buna "N" (Nitrile Rubber); EPR, EPDM (Nordel); Flouraz; Kalrez; Leather/Thiokol Impregnated; Neoprene; Nylon; Polycarbonate (Lexan); Polyethylene Low Density; Polyethylene High Density; Polyproplene 120° (Noryl); Polyurethane, Urathane; PVC Polyvinyl Chloride; PVDF 150° (Kynar); Teflon 120°; Viton 120°.

Material

- Ethane
- Ethanol
- Ethanolamine (Aminoethanol)
- Ethanol Chloride
- Ethers

- Ether (Ethyl, Methyl, Isor.)
- Ethyl Acetate
- Ethyl Acetoacetate
- Ethyl Acrylate
- Ethylacrylic Acid

- Ethyl Alcohol (Ethanol)
- Ethyl Aluminum Dichloride
- Ethyl Amine (Monoethylamine)
- Ethyl Benzene
- Ethyl Benzoate

- Ethyl Bromide
- Ethyl Butyl Acetate
- Ethyl Butyl Alcohol
- Ethyl Butyl Ketone
- Ethyl Butyraldehyde

- Ethyl Butyrate
- Ethyl Caprylate
- Ethyl Cellosolve
- Ethyl Cellose
- Ethyl Chloride (Dry)

- Ethyl Chloride (Wet)
- Ethyl Chlorocarbonate
- Ethyl Chloroformate
- Ethyl Cyanide (Propionitrile)
- Ethylcyclopentane

- Ethyl Ether
- Ethyl Formate
- Ethylhexyl Acetate
- Ethylhexyl Alcohol (Ethylhexanol)
- Ethyl Iodide

- Ethyl Isobutyrate
- Ethyl Mercaptan
- Ethyl Oxalate
- Ethyl Pentachlorobenzene
- Ethyl Propionate

- Ethyl Silicate
- Ethyl Sulfate
- Ethylene (Ethene)
- Ethylene Chloride
- Ethylene Chlorohydrin

- Ethylene Diamine
- Ethylene Dibromide
- Ethylene Dichloride

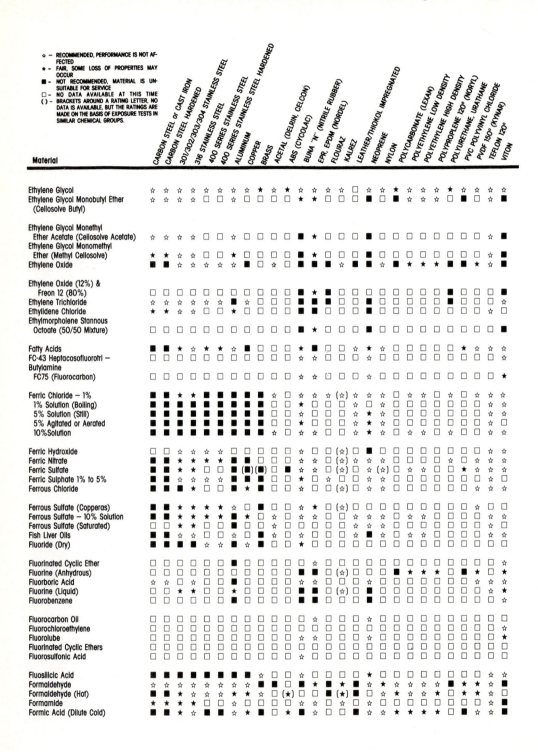

☆ – RECOMMENDED, PERFORMANCE IS NOT AF-FECTED
★ – FAIR, SOME LOSS OF PROPERTIES MAY OCCUR
■ – NOT RECOMMENDED, MATERIAL IS UN-SUITABLE FOR SERVICE
☐ – NO DATA AVAILABLE AT THIS TIME
( ) – BRACKETS AROUND A RATING LETTER, NO DATA IS AVAILABLE, BUT THE RATINGS ARE MADE ON THE BASIS OF EXPOSURE TESTS IN SIMILAR CHEMICAL GROUPS.

Materials listed in this table:

Ethylene Glycol
Ethylene Glycol Monobutyl Ether (Cellosolve Butyl)
Ethylene Glycol Monethyl Ether Acetate (Cellosolve Acetate)
Ethylene Glycol Monomethyl Ether (Methyl Cellosolve)
Ethylene Oxide
Ethylene Oxide (12%) & Freon 12 (80%)
Ethylene Trichloride
Ethylidene Chloride
Ethylmorpholene Stannous Octoate (50/50 Mixture)
Fatty Acids
FC-43 Heptacosofluorotri – Butylamine
FC75 (Fluorocarbon)
Ferric Chloride – 1%
1% Solution (Boiling)
5% Solution (Still)
5% Agitated or Aerated
10% Solution
Ferric Hydroxide
Ferric Nitrate
Ferric Sulfate
Ferric Sulphate 1% to 5%
Ferrous Chloride
Ferrous Sulfate (Copperas)
Ferrous Sulfate – 10% Solution
Ferrous Sulfate (Saturated)
Fish Liver Oils
Fluoride (Dry)
Fluorinated Cyclic Ether
Fluorine (Anhydrous)
Fluorboric Acid
Fluorine (Liquid)
Fluorobenzene
Fluorocarbon Oil
Fluorochloroethylene
Fluorolube
Fluorinated Cyclic Ethers
Fluorosulfonic Acid
Fluosilicic Acid
Formaldehyde
Formaldehyde (Hot)
Formamide
Formic Acid (Dilute Cold)

Legend:

☆ – RECOMMENDED, PERFORMANCE IS NOT AFFECTED
★ – FAIR, SOME LOSS OF PROPERTIES MAY OCCUR
■ – NOT RECOMMENDED, MATERIAL IS UNSUITABLE FOR SERVICE
□ – NO DATA AVAILABLE AT THIS TIME
( ) – BRACKETS AROUND A RATING LETTER, NO DATA IS AVAILABLE, BUT THE RATINGS ARE MADE ON THE BASIS OF EXPOSURE TESTS IN SIMILAR CHEMICAL GROUPS.

Material	CARBON STEEL or CAST IRON	CARBON STEEL HARDENED	301/302/303/304 STAINLESS STEEL	316 STAINLESS STEEL	400 SERIES STAINLESS STEEL	400 SERIES STAINLESS STEEL HARDENED	ALUMINUM	COPPER	BRASS	ACETAL (DELRIN, CELCON)	ABS (CYCOLAC)	BUNA "N" (NITRILE RUBBER)	EPR, EPDM (NORDEL)	FLOURAZ	KALREZ	LEATHER/THIOKOL IMPREGNATED	NEOPRENE	NYLON	POLYCARBONATE (LEXAN)	POLYETHYLENE LOW DENSITY	POLYETHYLENE HIGH DENSITY	POLYPROPLENE 120°	POLYURETHANE, URATHANE	PVC POLYVINYL CHLORIDE	PVDF 150° (KYNAR)	TEFLON 120°	VITON
Formic Acid (Dilute Hot)	■	■	★	☆	■	■	☆	★	★	■	□	(☆)	■	☆	□	□	☆	□	★	★	☆	★	□	★	☆	☆	■
Freon 11	☆	☆	☆	☆	☆	☆	■	☆	□	■	☆	★	■	(★)	□	■	■	□	□	□	□	□	★	□	★	☆	★
Freon 12	☆	☆	☆	☆	☆	☆	■	☆	☆	■	☆	■	★	★	□	☆	■	□	□	□	□	★	☆	■	☆	☆	☆
Freon 13	☆	☆	☆	☆	☆	☆	■	☆	☆	■	☆	□	☆	(★)	□	☆	□	□	□	□	□	□	□	□	☆	☆	☆
Freon 13BI	☆	☆	☆	☆	☆	☆	■	☆	☆	■	□	☆	☆	(★)	□	☆	□	□	□	□	□	□	□	□	☆	☆	☆
Freon 14	☆	☆	☆	☆	☆	☆	■	□	□	■	□	□	□	(★)	□	☆	□	□	□	□	□	□	□	□	☆	☆	☆
Freon 21	☆	☆	☆	☆	☆	☆	■	□	□	■	□	■	☆	★	□	□	■	□	□	□	□	□	□	□	☆	☆	■
Freon 22	☆	☆	☆	☆	☆	☆	■	☆	☆	■	■	■	■	(☆)	□	★	□	★	□	□	□	★	■	■	☆	☆	■
Freon 31	☆	☆	☆	☆	☆	☆	■	□	□	■	□	■	■	☆	□	□	☆	□	□	□	□	□	□	□	☆	☆	■
Freon 32	☆	☆	☆	☆	☆	☆	■	□	□	■	□	☆	☆	☆	□	☆	□	□	□	□	□	□	□	□	☆	☆	■
Freon 112	☆	☆	☆	☆	☆	☆	■	□	□	■	□	★	■	■	□	□	★	□	□	□	★	□	□	□	☆	☆	☆
Freon 113	☆	☆	☆	☆	☆	☆	■	☆	□	■	□	☆	■	★	□	☆	□	□	□	□	☆	□	□	□	☆	☆	★
Freon 114	☆	☆	☆	☆	☆	☆	■	☆	□	■	□	☆	☆	☆	■	(★)	□	□	□	□	□	□	□	□	☆	☆	☆
Freon 114B2	☆	☆	☆	☆	☆	☆	■	☆	□	■	□	□	□	★	■	□	□	□	□	□	□	□	□	□	☆	☆	★
Freon 115	☆	☆	☆	☆	☆	☆	■	□	□	■	□	□	☆	(★)	□	☆	□	□	□	□	□	□	□	□	☆	☆	☆
Freon 142b	□	□	□	□	□	□	■	□	□	■	□	□	□	□	□	☆	□	□	□	□	□	□	□	□	☆	☆	■
Freon 152a	□	□	□	□	□	□	■	□	□	■	□	□	□	□	□	☆	□	□	□	□	□	□	□	□	☆	☆	■
Freon 218	□	□	□	□	□	□	■	□	□	■	□	□	□	□	□	☆	□	□	□	□	□	□	□	□	☆	☆	☆
Freon 502	☆	☆	☆	☆	☆	☆	■	☆	□	□	□	★	☆	(★)	□	☆	□	□	□	□	□	□	□	□	☆	☆	★
Freon, BF	☆	☆	☆	☆	☆	☆	■	☆	□	□	□	★	■	□	□	★	□	□	□	□	□	□	□	□	☆	☆	☆
Freon C316	□	□	□	□	□	□	■	□	□	■	□	□	□	□	□	☆	□	□	□	□	□	□	□	□	☆	☆	☆
Freon, C318	☆	☆	☆	☆	☆	☆	■	□	□	■	□	□	□	□	□	☆	□	□	□	□	□	□	□	□	☆	☆	☆
Freon, K-152A	☆	☆	☆	☆	☆	☆	■	□	□	■	□	□	□	☆	□	☆	□	□	□	□	□	□	□	□	☆	☆	■
Freon, K-142B	☆	☆	☆	☆	☆	☆	■	□	□	■	□	□	□	☆	□	☆	□	□	□	□	□	□	□	□	☆	☆	■
Freon, MF	☆	☆	☆	☆	☆	☆	■	☆	□	□	□	★	■	□	□	★	□	□	□	□	□	□	□	□	☆	☆	★
Freon, PCA	☆	☆	☆	☆	☆	☆	■	□	□	□	■	■	□	□	□	☆	□	□	□	□	□	□	□	☆	☆	☆	★
Freon, TF	☆	☆	☆	☆	☆	☆	■	□	□	□	□	□	□	□	□	☆	■	□	□	□	□	□	□	☆	☆	☆	☆
Freon T-WD602	□	□	□	□	□	□	■	□	□	□	□	★	□	□	□	★	□	□	□	□	□	□	□	☆	☆	☆	☆
Freon TMC	□	□	□	□	□	□	■	□	□	□	□	★	□	□	□	☆	□	□	□	□	□	□	★	□	☆	☆	☆
Freon T-P35	□	□	□	□	□	□	■	□	□	□	□	★	□	□	□	□	□	□	□	□	□	□	□	□	☆	☆	☆
Freon TA	□	□	□	□	□	□	■	□	□	□	□	□	□	□	□	☆	□	□	□	□	□	□	□	□	☆	☆	■
Freon TC	□	□	□	□	□	□	■	□	□	□	□	□	□	□	□	☆	□	□	□	□	□	□	□	□	☆	☆	☆
Fruit Juices	□	□	☆	☆	☆	☆	☆	☆	□	■	★	■	☆	☆	□	☆	□	☆	☆	☆	☆	☆	☆	☆	☆	☆	☆
Fuel Oil	☆	☆	☆	☆	☆	☆	☆	★	☆	■	☆	☆	■	☆	□	☆	★	☆	★	★	★	★	■	☆	★	☆	☆
Fumaric Acid	☆	☆	☆	☆	☆	☆	☆	☆	□	■	□	■	□	■	□	□	★	□	★	★	□	□	□	□	☆	☆	☆
Fuming Sulphuric Acid (20/25% Oleum)	□	□	□	□	□	□	□	□	□	■	□	■	■	□	□	■	□	□	■	□	□	■	□	□	☆	☆	
Furan (Furfuran)	□	□	□	□	□	□	□	□	□	■	□	■	□	□	□	■	□	□	□	□	□	■	□	□	□	□	□
Furan Resin	□	□	□	□	□	□	□	□	□	■	□	□	□	□	□	■	□	□	□	□	□	■	□	□	□	□	□
Furfural (Ant Oil)	★	★	☆	☆	☆	☆	☆	☆	□	☆	■	■	☆	☆	□	(☆)	■	■	□	□	□	■	□	□	□	☆	■
Furfuraldehyde	■	■	☆	☆	☆	☆	☆	☆	□	□	□	■	□	□	□	□	■	□	□	□	□	■	□	□	□	☆	■
Furfuryl Alcohol	☆	☆	☆	☆	☆	☆	☆	☆	□	□	■	★	□	□	□	☆	■	□	□	□	□	■	□	□	□	☆	□
Furyl Carbinol	□	□	□	□	□	□	□	□	□	□	■	★	□	□	□	☆	■	□	□	□	□	■	□	□	□	☆	□
Fusel Oil (Grain Oil)	□	□	□	□	□	□	□	□	□	□	☆	☆	□	□	□	☆	□	□	□	□	□	■	□	□	□	☆	□
Fyrquel A60	□	□	□	□	□	□	□	□	□	□	■	★	□	□	□	☆	□	□	□	□	□	■	□	□	□	□	□
90, 100, 150m 220, 300, 500	□	□	□	□	□	□	□	□	□	□	■	★	□	□	□	☆	□	□	□	□	□	■	□	□	□	□	■
Gallic Acid	■	■	☆	☆	☆	☆	■	☆	□	☆	□	★	★	□	☆	★	☆	□	☆	☆	☆	★	□	■	☆	★	☆
Gasoline (Sour)	★	★	☆	☆	■	■	■	■	☆	□	□	■	■	□	■	(☆)	☆	☆	□	★	☆	☆	★	□	★	☆	□
Gasoline (Meter)	☆	☆	☆	☆	☆	☆	☆	☆	■	☆	■	☆	■	□	■	(☆)	☆	■	■	★	☆	☆	★	☆	★	☆	☆

Legend:

☆ — RECOMMENDED, PERFORMANCE IS NOT AFFECTED
★ — FAIR, SOME LOSS OF PROPERTIES MAY OCCUR
■ — NOT RECOMMENDED, MATERIAL IS UNSUITABLE FOR SERVICE
□ — NO DATA AVAILABLE AT THIS TIME
( ) — BRACKETS AROUND A RATING LETTER, NO DATA IS AVAILABLE, BUT THE RATINGS ARE MADE ON THE BASIS OF EXPOSURE TESTS IN SIMILAR CHEMICAL GROUPS.

Material	CARBON STEEL or CAST IRON	CARBON STEEL HARDENED	301/302/303/304 STAINLESS STEEL	316 STAINLESS STEEL	400 STAINLESS STEEL	400 SERIES STAINLESS STEEL	400 SERIES STAINLESS STEEL HARDENED	ALUMINUM	COPPER	BRASS	ACETAL (DELRIN, CELCON)	ABS (CYCOLAC)	BUNA "N" (NITRILE RUBBER)	EPR, EPDM (NORDEL)	FLOURAZ	KALREZ	LEATHER/THIOKOL IMPREGNATED	NEOPRENE	NYLON	POLYCARBONATE (LEXAN)	POLYETHYLENE LOW DENSITY	POLYETHYLENE HIGH DENSITY	POLYPROPLENE 120°	POLYURETHANE, URATHANE	PVC POLYVINYL CHLORIDE	PVDF 150° (KYNAR)	TEFLON 120°	VITON
Gasoline (Aviation)	☆	☆	☆	☆	☆	☆	☆	☆	□	■	★	□	■	(☆)	☆		☆	☆	□	☆	☆	☆	□	☆	☆	☆	☆	□
Gelatin	☆	☆	☆	☆	■	■	☆	☆	■	☆	□	☆	☆	☆	□	☆	☆	☆	□	☆	☆	☆	□	☆	★	☆	☆	☆
Ginger Oil	■	■	■	■	□	□	□	□	□	□	□	□	□	□	□	☆	□	☆	□	□	□	□	□	□	☆	□	☆	☆
Glacial Acetic Acid	□	□	□	□	□	□	□	□	□	□	★	★	□	□	□	■	□	■	★	☆	★	■	★	□	□	■
Glauber's Salt	□	□	□	□	□	□	□	■	□	■	★	☆	(☆)	□	★	□	☆	☆	□	☆	☆	☆	□	☆	☆	☆
Glucose	☆	☆	☆	☆	☆	☆	☆	☆	☆	★	☆	☆	(☆)	☆	☆	☆	□	☆	☆	■	☆	☆	☆	☆
Glue	□	□	□	☆	□	□	★	□	□	□	☆	☆	(☆)	□	☆	☆	□	☆	☆	□	☆	☆	☆	★
Glutamic Acid	■	■	☆	☆	★	★	■	■	□	□	□	□	□	□	□	☆	□	☆	□	□	□	□	□	☆	□	☆	☆
Glycerine-Glycerol	★	★	☆	☆	☆	☆	☆	★	☆	■	☆	☆	(☆)	☆	☆	☆	☆	☆	■	☆	☆	☆	☆	☆
Glycolic Acid	□	□	☆	☆	☆	☆	□	□	☆	★	☆	□	□	☆	□	☆	☆	□	★	★	☆	☆
Glycol	□	□	☆	☆	☆	☆	□	□	□	☆	□	(☆)	□	☆	☆	□	☆	☆	☆	□	☆	☆
Gold Monocyanide	■	■	☆	☆	☆	☆	□	□	□	☆	□	□	□	□	□	□	□	□	□	□	☆	■	☆
Grape Juice	■	■	☆	☆	□	□	☆	☆	★	□	□	□	□	■	□	□	☆	□	□	☆	□	☆	☆
Grapefruit Oil	■	■	☆	☆	□	□	☆	☆	□	■	□	□	■	□	□	□	□	□	☆	□	☆	☆
Grease (Ester Base)	☆	☆	☆	☆	☆	☆	☆	☆	□	☆	(☆)	(☆)	□	☆	□	□	□	□	□	□	□	☆	☆
Grease (Petroleum Base)	☆	☆	☆	☆	☆	☆	☆	☆	□	■	☆	(☆)	(☆)	■	☆	□	□	□	□	□	☆	☆
Grease (Silicone Base)	☆	☆	☆	☆	☆	☆	☆	☆	□	☆	(☆)	(☆)	□	★	□	□	□	□	□	□	☆	☆
Green Sulfate Liquor	■	■	☆	☆	□	□	□	□	★	☆	☆	□	□	□	□	□	□	□	□	☆
Halothane	□	□	□	□	□	□	□	□	■	■	□	□	□	■	□	□	□	□	■	□	□	☆
Halowax Oil	□	□	□	□	□	□	□	□	■	■	□	□	□	■	□	□	□	□	□	□	□	☆
Hannifin Lube A	□	□	□	□	□	□	□	☆	■	□	□	□	☆	□	□	□	□	☆	□	□	☆
Heavy Water	□	□	□	□	□	□	□	☆	☆	□	□	★	□	□	■	□	□	□	□	□	□
HEF-2 (High Energy Fuel)	□	□	□	□	□	□	□	★	■	□	□	□	□	□	□	□	□	□	□	□	☆
Helium	□	□	□	□	□	□	□	☆	☆	□	□	☆	□	□	□	□	□	□	□	□	☆
Heptanal	☆	☆	☆	☆	□	□	☆	□	☆	☆	□	□	□	□	□	□	□	□	□	□	■
Heptane	☆	☆	☆	☆	☆	☆	☆	☆	□	■	☆	■	■	☆	★	★	☆	★	★	★	☆	★	☆	☆
N-Hexaldehyde	☆	☆	☆	☆	☆	☆	☆	□	☆	■	☆	□	□	□	□	□	□	□	□	□	■
Hexamine	☆	☆	☆	☆	☆	☆	☆	★	□	□	□	(☆)	□	□	□	□	□	□	□	□	□	☆
Hexane	☆	☆	☆	☆	☆	☆	☆	☆	□	☆	■	■	☆	★	★	☆	■	■	★	★	■	☆	☆
Hexanol, Tertiary	☆	☆	☆	☆	□	□	☆	☆	☆	□	☆	☆	□	□	□	□	□	□	□	□	☆	□
N-Hexene-1	□	□	□	□	□	□	□	★	■	□	□	□	□	□	□	□	□	★	□	□	☆
Hexyl Alcohol	☆	☆	☆	☆	□	□	☆	□	☆	■	☆	☆	□	★	□	□	■	□	□	☆
Hexylene Glycol	☆	☆	☆	☆	□	□	☆	□	☆	■	☆	(☆)	□	□	□	□	□	□	□	☆
High Viscosity Lubricant, U4 H2	□	□	□	□	□	□	□	☆	☆	□	□	☆	□	□	□	□	□	■	□	□	☆
Hilo MS #1	□	□	□	□	□	□	□	■	☆	□	□	☆	□	□	■	□	□	□	□	□	☆
Honey	☆	☆	☆	☆	☆	☆	☆	□	□	□	□	□	□	☆	□	□	□	□	□	☆	☆	☆
Houghto-Safe 271 (Water & Glycol Base)																												
620 Water/Glycol	□	□	□	□	□	□	□	☆	☆	□	□	★	□	□	■	□	□	□	□	★
1010, Phosphate Ester	□	□	□	□	□	□	□	■	☆	□	□	★	□	□	■	□	□	□	□	★
1055, Phosphate Ester	□	□	□	□	□	□	□	■	☆	□	□	★	□	□	■	□	□	□	□	☆
1120, Phosphate Ester	□	□	□	□	□	□	□	■	☆	□	□	☆	□	□	■	□	□	□	□	☆
5040 (Water/Oil Emulsion)	□	□	□	□	□	□	□	☆	☆	□	□	☆	□	□	■	□	□	□	□	☆
Hydraulic Oils (Petroleum)	☆	☆	☆	☆	☆	☆	☆	☆	□	☆	■	□	☆	★	☆	□	□	□	□	☆	☆
Hydraulic Oils (Synthetic)	☆	☆	☆	☆	☆	☆	☆	☆	□	☆	☆	□	☆	★	☆	□	□	□	□	☆	☆
Hydrazine (Diamine)	■	■	☆	☆	★	★	☆	☆	□	□	★	☆	☆	☆	□	□	□	□	□	☆	☆	□
Hydrobromic Acid	■	■	■	■	■	■	■	■	(■)	□	■	☆	☆	★	□	□	■	□	★	☆	☆	☆
Hydrobromic Acid 40%	■	■	■	■	■	■	■	■	(■)	□	■	☆	☆	★	□	□	■	□	★	☆	☆	☆

Legend:

☆ – RECOMMENDED, PERFORMANCE IS NOT AFFECTED
★ – FAIR, SOME LOSS OF PROPERTIES MAY OCCUR
■ – NOT RECOMMENDED, MATERIAL IS UNSUITABLE FOR SERVICE
□ – NO DATA AVAILABLE AT THIS TIME
( ) – BRACKETS AROUND A RATING LETTER, NO DATA IS AVAILABLE, BUT THE RATINGS ARE MADE ON THE BASIS OF EXPOSURE TESTS IN SIMILAR CHEMICAL GROUPS.

Material	Carbon Steel or Cast Iron	Carbon Steel Hardened	301/302/303/304 SS	316 SS	400 Series SS	400 Series SS Hardened	Aluminum	Copper	Brass	Acetal (Delrin, Celcon)	ABS (Cycolac)	Buna "N" (Nitrile Rubber)	EPR, EPDM (Nordel)	Flouraz	Kalrez	Leather/Thiokol Impregnated	Neoprene	Nylon	Polycarbonate (Lexan)	Polyethylene Low Density	Polyethylene High Density	Polyproplene 120° (Noryl)	Polyurethane Urathane	PVC Polyvinyl Chloride	PVDF 150° (Kynar)	Teflon 120°	Viton
Hydrocarbons (Saturated)	□	□	□	□	□	□	□	□	□	□	☆	■	■	☆	□	★	□	□	□	□	□	□	★	□	□	□	☆
Hydrocloric Acid	■	■	■	■	■	■	■	■	■	☆	☆	■	☆	☆	□	★	★	■	★	☆	☆	☆	□	★	□	□	☆
3 Molar	■	□	■	■	■	■	■	■	■	☆	☆	■	☆	☆	□	★	☆	■	☆	☆	☆	☆	□	■	□	□	☆
Concentrated	■	□	■	■	■	□	■	■	■	☆	□	■	☆	★	□	■	■	■	☆	☆	☆	☆	□	☆	□	□	☆
Hydrocyanic Acid	■	■	□	□	■	■	☆	■	■	★	★	★	□	□	□	■	★	☆	☆	☆	☆	☆	□	★	☆	☆	☆
Hydrofluoric Acid 65% or less, Cold	■	■	■	■	□	□	□	□	□	■	■	■	☆	□	□	□	★	□	■	□	☆	☆	□	☆	□	☆	☆
65% or More, Cold	■	■	■	■	□	□	□	■	■	■	★	□	□	□	□	□	■	□	□	□	☆	☆	□	☆	□	☆	★
65% or Less, Hot	■	■	■	■	□	□	■	■	■	□	★	□	□	□	□	□	■	□	□	□	☆	☆	□	☆	□	☆	☆
65% or More, Hot	■	■	■	■	□	□	■	■	■	□	□	□	□	□	□	□	☆	□	□	□	☆	☆	□	☆	□	☆	■
Anhydrous	□	□	□	□	□	□	□	□	□	□	■	★	□	□	□	□	□	□	□	□	□	□	□	☆	□	□	■
Hydrofluoric Acid – Anhydrous	□	□	□	□	□	□	□	□	□	□	■	★	□	□	□	□	□	□	□	□	□	□	□	□	□	□	■
Hydrofluosilicic Acid	■	■	■	■	□	□	□	■	□	□	□	□	□	□	□	□	□	□	□	□	□	□	□	□	□	□	□
Hydrogen	□	□	☆	☆	☆	☆	☆	□	□	□	☆	■	□	(☆)	□	□	☆	□	□	☆	☆	☆	□	☆	□	☆	☆
Hydrogen Chloride Gas	□	□	☆	☆	☆	☆	☆	□	□	☆	☆	☆	□	□	□	□	☆	□	□	☆	☆	☆	□	☆	□	☆	☆
Hydrogen Cyanide Gas	★	★	★	☆	☆	★	★	☆	☆	☆	□	★	□	□	□	□	★	□	□	☆	☆	☆	□	☆	□	☆	☆
Hydrogen Fluoride	☆	☆	☆	☆	★	★	□	□	□	□	□	□	□	□	□	□	□	□	□	□	□	□	□	□	□	□	□
Hydrogen Gase (Cold)	★	★	☆	☆	☆	☆	☆	★	□	□	□	☆	□	□	□	□	☆	□	□	☆	☆	☆	□	☆	☆	☆	☆
Hydrogen Gas (Hot)	☆	☆	□	☆	□	■	☆	☆	□	□	□	☆	□	□	□	□	☆	□	□	☆	☆	☆	□	☆	☆	☆	☆
Hydrogen Peroxide	■	■	★	☆	★	★	☆	★	■	■	☆	☆	☆	☆	□	☆	☆	☆	☆	☆	☆	☆	☆	☆	☆	☆	☆
90%	□	□	□	□	□	□	□	□	■	□	★	■	■	□	□	■	□	☆	★	☆	★	□	★	★	□	☆	□
Hydrogen Peroxide (Dilute)	□	□	□	□	□	□	□	□	□	□	☆	□	□	□	□	□	□	□	□	□	□	□	□	□	□	☆	□
Hydrogen Phosphide	□	□	□	□	□	□	□	□	□	□	☆	□	□	□	□	□	□	□	□	□	□	□	□	□	□	☆	□
Hydrogen Sulfide Dry, Cold	☆	☆	■	★	■	■	★	★	■	□	☆	☆	(☆)	★	☆	☆	□	□	□	□	□	□	★	☆	□	■	
Dry, Hot	★	★	★	★	★	★	★	★	□	■	□	☆	☆	(☆)	★	★	☆	□	□	□	□	□	★	□	☆	■	
Wet, Cold	★	★	★	☆	■	★	★	■	□	□	□	☆	☆	(☆)	★	☆	☆	□	□	□	□	□	□	☆	□	■	
Wet, Hot	■	■	□	□	□	□	□	□	□	□	□	☆	☆	(☆)	★	★	☆	□	□	□	□	□	□	☆	□	■	
Hydrogen Sulfide Aqueous Solution	■	■	□	☆	□	□	□	■	□	□	□	□	□	□	□	□	★	□	□	□	□	□	□	□	□	■	■
Hydrolube-Water/Ethylene Glycol	□	□	□	□	□	□	□	□	□	□	□	□	☆	□	□	□	★	□	□	□	□	□	■	□	□	□	☆
Hydroquinone	☆	☆	★	★	□	□	□	★	★	□	□	■	■	(☆)	□	□	■	□	□	□	□	□	□	□	□	□	★
Hydyne	□	□	□	□	□	□	□	□	□	□	□	□	□	★	□	□	★	□	□	□	□	□	□	□	□	□	□
Hydroxyacetic Acid - 10%	★	★	★	★	□	□	□	★	□	□	□	□	□	□	□	□	□	■	□	□	□	□	□	■	□	☆	☆
Hypochlorous Acid	■	■	■	■	■	■	■	☆	☆	□	□	■	★	□	□	□	□	□	□	□	□	□	□	☆	□	☆	☆
Hyporchlorites, Sodium	□	□	□	□	□	□	□	□	□	□	□	☆	□	□	□	□	□	□	□	☆	□	□	□	□	□	☆	☆
Hypoid Grease (Parapoid 10-C)	□	□	□	□	□	□	□	□	□	□	□	☆	□	□	□	□	□	□	□	□	□	□	□	□	□	☆	□
Illuminating Gas	☆	☆	☆	☆	□	□	☆	☆	□	□	□	□	□	□	□	□	□	□	□	□	□	□	□	□	□	□	□
Ink (Printers)	☆	☆	☆	☆	□	□	□	□	□	☆	□	★	☆	☆	■	(☆)	☆	☆	□	☆	☆	☆	□	□	□	☆	☆
Iodine	■	■	■	★	■	■	■	■	■	■	★	★	■	(☆)	■	■	☆	□	□	□	□	□	□	★	☆	■	
Iodine Pentafluoride	□	□	□	□	□	□	□	□	□	□	■	■	□	(☆)	■	■	□	□	□	□	□	□	□	□	□	☆	■
Iodoform	□	□	□	□	□	□	□	□	□	□	☆	□	□	□	★	□	★	□	□	□	□	□	□	□	□	☆	□
Isoamyl Acetate	☆	☆	☆	☆	□	□	□	□	□	□	■	□	□	□	□	■	□	□	□	□	□	□	□	□	□	☆	□
Isoamyl Alcohol	□	□	□	□	□	□	□	□	□	□	☆	□	□	□	□	☆	□	□	□	□	□	□	□	□	□	☆	☆
Isoamyl Butyrate	☆	☆	☆	☆	□	□	□	□	□	□	■	□	□	□	□	■	□	□	□	□	□	□	□	□	□	☆	■
Isoamyl Chloride	□	□	□	□	□	□	□	□	□	□	■	■	□	□	□	■	□	□	□	□	□	□	□	□	□	☆	☆
Iso Butane	□	□	□	☆	□	□	□	□	□	□	☆	■	(☆)	□	■	□	□	□	□	□	□	□	□	□	□	☆	☆
Iso Butyl Acetate	□	□	□	□	□	□	□	□	□	□	☆	■	□	□	■	□	□	□	□	□	□	□	□	□	□	☆	□
Isobutyl Alcohol	□	□	□	□	□	□	□	□	□	★	☆	☆	(☆)	□	☆	□	★	☆	☆	☆	■	★	□	□	□	☆	☆
Isobutyl Amine	□	□	□	□	□	□	□	□	□	☆	☆	☆	□	□	☆	□	□	□	□	□	□	□	□	□	□	☆	☆
Isobutyl Chloride	★	★	★	★	□	□	☆	□	□	□	■	□	□	(☆)	□	□	☆	□	□	□	□	□	□	□	□	☆	★
Isobutyric Acid	□	□	□	□	□	□	☆	□	□	□	■	☆	□	□	□	★	□	□	□	□	□	□	□	□	□	☆	□

☆ – RECOMMENDED, PERFORMANCE IS NOT AFFECTED
★ – FAIR, SOME LOSS OF PROPERTIES MAY OCCUR
■ – NOT RECOMMENDED, MATERIAL IS UNSUITABLE FOR SERVICE
□ – NO DATA AVAILABLE AT THIS TIME
( ) – BRACKETS AROUND A RATING LETTER, NO DATA IS AVAILABLE, BUT THE RATINGS ARE MADE ON THE BASIS OF EXPOSURE TESTS IN SIMILAR CHEMICAL GROUPS.

Material	CARBON STEEL or CAST IRON	CARBON STEEL HARDENED	301/302/303/304 STAINLESS STEEL	316 STAINLESS STEEL	400 SERIES STAINLESS STEEL	400 SERIES STAINLESS STEEL HARDENED	ALUMINUM	COPPER	BRASS	ACETAL (DELRIN CELCON)	ABS (CYCOLAC)	BUNA "N" (NITRILE RUBBER)	EPR, EPDM (NORDEL)	FLOURAZ	KALREZ	LEATHER/THIOKOL IMPREGNATED	NEOPRENE	NYLON	POLYCARBONATE (LEXAN)	POLYETHYLENE LOW DENSITY	POLYETHYLENE HIGH DENSITY	POLYPROPLENE 120°	POLYURETHANE, URATHANE	PVC POLYVINYL CHLORIDE	PVDF 150° (KYNAR)	TEFLON 120°	VITON
Iso-Butyl N-Butyrate	□	□	□	□	□	□	□	□	□	□	□	■	☆	□	□	□	■	□	□	□	□	□	□	□	□	□	☆
Isododecane	★	★	★	★	★	★	★	☆	★	□	□	□	☆	□	(☆)	□	□	★	□	□	□	□	□	□	□	□	☆
Iso Octane	☆	☆	☆	☆	☆	☆	☆	□	□	□	□	☆	□	■	★	(☆)	□	★	□	□	□	□	★	□	□	□	☆
Iso Pentane	□	□	□	□	□	□	□	□	□	□	□	☆	□	■	□	(☆)	□	□	□	□	□	□	□	□	□	☆	☆
Isophorone (Ketone)	★	★	☆	★	★	★	★	☆	★	□	☆	□	■	☆	□	(☆)	□	□	□	□	□	□	□	□	□	□	☆
Isopropanol	☆	☆	★	★	★	★	★	☆	★	□	□	□	☆	☆	☆	(☆)	□	★	□	□	□	□	□	□	■	□	☆
Isopropyl Acetate	★	★	★	☆	□	□	■	□	□	□	□	☆	□	■	☆	□	(☆)	□	■	■	★	★	★	■	■	☆	■
Isopropyl Alcohol	☆	☆	☆	☆	☆	☆	☆	☆	☆	□	□	☆	☆	☆	(☆)	★	★	☆	☆	☆	☆	☆	■	★	□	□	☆
Isopropyl Amine	☆	☆	☆	☆	☆	☆	□	□	□	□	□	□	★	(☆)	□	□	□	□	□	□	□	□	□	□	■	□	☆
Isopropyl Benzene (Cumene)	□	□	□	□	□	□	□	□	□	□	□	■	□	■	□	□	■	□	□	□	□	□	□	□	□	□	☆
Isopropyl Chloride	☆	☆	☆	☆	□	□	☆	□	□	□	■	■	■	□	□	□	■	□	□	□	□	□	■	□	□	□	☆
Isopropyl Ether	☆	☆	☆	☆	□	□	☆	□	□	□	★	■	□	(☆)	□	□	■	□	□	□	□	□	☆	□	★	□	■
JP-1	☆	☆	☆	☆	☆	☆	☆	□	☆	□	□	■	□	□	□	■	□	□	□	□	☆	☆	□	□	□	☆	☆
JP-2	☆	☆	☆	☆	☆	☆	☆	□	☆	□	□	■	□	■	□	□	□	□	□	□	☆	☆	□	□	□	☆	☆
JP-3 (Mil-J-5624)	☆	☆	☆	☆	☆	☆	☆	□	☆	□	□	■	★	(☆)	□	■	□	□	□	□	☆	☆	□	□	□	☆	☆
JP-4 (Mil-J-5624)	☆	☆	☆	☆	☆	☆	☆	□	☆	□	□	■	★	(☆)	□	■	□	□	□	□	☆	☆	□	□	★	□	☆
JP-5 (Mil-J-5624)	☆	☆	☆	☆	☆	☆	☆	□	☆	□	□	■	★	☆	□	■	□	□	□	□	☆	☆	□	□	★	□	☆
JP-6 (Mil-J-25656)	☆	☆	☆	☆	☆	☆	☆	□	☆	□	□	■	★	□	□	★	□	□	□	□	☆	☆	□	□	□	□	☆
JP-X (Mil-F-25604)	☆	☆	☆	☆	☆	☆	☆	□	☆	□	□	■	★	□	□	★	□	□	□	□	☆	☆	□	□	□	□	■
Kel F Liquids	□	□	□	□	□	□	□	□	□	□	□	□	☆	□	☆	□	□	□	□	□	□	□	□	□	□	□	★
Kerosene	☆	☆	☆	☆	☆	☆	☆	☆	■	☆	☆	■	★	☆	☆	■	☆	■	■	★	★	☆	■	☆	☆	☆	☆
Ketchup	□	□	☆	☆	□	□	☆	□	☆	□	□	☆	□	□	□	□	□	□	□	□	□	□	□	□	□	☆	☆
Ketones	☆	☆	☆	☆	☆	★	☆	☆	■	■	☆	■	☆	□	□	★	□	☆	□	□	□	□	□	■	□	☆	■
Keystone #87HX-Grease	☆	☆	☆	☆	☆	☆	☆	☆	☆	□	□	■	□	□	□	■	□	□	□	☆	☆	☆	□	☆	□	☆	■
Lactam-Amino Acids	□	□	□	□	□	□	□	□	□	□	■	★	□	□	□	□	★	□	□	□	□	□	□	□	□	□	□
Lacquers	■	■	☆	☆	☆	☆	☆	□	□	□	■	■	■	□	□	☆	□	■	□	□	□	☆	☆	☆	■	☆	☆
Lacquer Solvents	★	★	☆	☆	☆	☆	☆	□	□	□	■	■	□	□	□	■	□	□	□	☆	☆	☆	□	☆	□	☆	■
Lactic Acid – 5% Solution	■	■	★	☆	■	■	☆	☆	■	■	★	☆	☆	□	★	★	☆	★	☆	★	☆	☆	★	□	★	★	☆
5% Solution	■	■	★	☆	■	■	★	□	■	■	■	☆	☆	□	★	□	(★)	(☆)	(☆)	(★)	■	★	■	☆	★
10% Solution (Boiling)	■	■	★	☆	■	■	■	■	■	■	□	☆	☆	□	★	□	(★)	(☆)	(☆)	(★)	■	★	■	☆	★
Lactol	☆	☆	☆	☆	□	□	☆	□	□	□	□	■	□	□	□	■	□	□	□	□	□	□	□	□	☆	☆
Lard Oil (Cold)	☆	☆	☆	☆	☆	☆	☆	☆	☆	□	■	(☆)	□	☆	★	□	☆	□	□	□	□	□	★	□	(☆)	☆	☆
Lard Oil (Hot)	☆	☆	☆	☆	☆	☆	☆	☆	☆	□	■	(☆)	□	☆	★	□	★	□	□	□	□	□	★	□	(☆)	☆	☆
Latex	□	□	☆	☆	□	□	☆	□	□	★	★	☆	□	□	□	☆	★	□	□	□	□	□	□	□	☆	☆	☆
Lauryl Alcohol (N-Dodecanol)	☆	☆	☆	☆	□	□	□	□	□	□	□	☆	□	□	□	□	☆	□	□	□	□	□	□	□	☆	☆	★
Lavender Oil	□	□	□	□	□	□	□	□	□	□	□	☆	□	★	★	☆	□	□	□	□	□	□	■	□	□	☆	☆
Lead Acetate	☆	☆	★	★	★	★	■	□	□	□	☆	★	★	☆	■	(☆)	★	□	□	□	□	☆	■	★	☆	☆	■
Lead Chloride	□	□	★	★	□	□	□	□	□	□	☆	☆	□	(☆)	□	★	□	□	□	□	□	□	□	☆	☆	☆	□
Lead Nitrate	□	□	★	★	□	□	□	□	□	□	☆	☆	★	(☆)	□	☆	□	□	□	□	□	□	□	☆	□	□	□
Lead Sulfamate	□	□	□	□	□	□	□	□	□	□	★	☆	☆	□	□	□	□	□	□	□	□	□	□	★	☆	★	□
Lead Sulphamate	□	□	□	□	□	□	□	□	□	□	★	☆	□	□	□	☆	□	□	□	□	□	□	□	☆	□	□	□
Lehigh X1169	□	□	□	□	□	□	□	□	□	□	☆	■	□	□	□	☆	□	□	□	□	□	□	□	☆	□	□	□
Lehigh X1170	□	□	□	□	□	□	□	□	□	□	☆	■	□	□	□	★	□	□	□	□	□	□	□	☆	□	□	□
Lemon Oil	□	□	☆	☆	□	□	☆	□	□	□	□	☆	□	□	□	□	□	□	□	□	□	□	■	□	☆	☆	☆
Light Grease	☆	☆	☆	☆	☆	☆	☆	☆	☆	□	□	■	□	□	□	■	□	□	□	□	□	□	★	□	☆	☆	☆
Ligroin (Petroleum Ether or Benzine)	☆	☆	☆	☆	☆	☆	■	☆	☆	□	★	□	□	☆	□	☆	□	□	★	□	□	□	□	□	□	☆	☆
Lignin Liquor	□	□	☆	☆	□	□	□	□	□	□	□	☆	□	□	□	☆	□	□	□	□	□	□	□	□	☆	☆	☆
Lime	■	■	☆	☆	■	■	☆	☆	■	□	□	☆	☆	☆	□	☆	☆	□	□	□	☆	☆	★	★	☆	☆	☆
Lime Bleach	■	■	□	☆	■	■	□	□	□	□	□	☆	☆	☆	□	★	□	□	□	□	□	□	□	☆	☆	☆	☆
Lime Slurries	■	■	☆	☆	■	■	☆	☆	☆	□	□	☆	☆	☆	□	★	□	□	□	□	□	□	★	□	☆	☆	☆

Legend:
- ☆ — RECOMMENDED, PERFORMANCE IS NOT AFFECTED
- ★ — FAIR, SOME LOSS OF PROPERTIES MAY OCCUR
- ■ — NOT RECOMMENDED, MATERIAL IS UNSUITABLE FOR SERVICE
- □ — NO DATA AVAILABLE AT THIS TIME
- ( ) — BRACKETS AROUND A RATING LETTER, NO DATA IS AVAILABLE, BUT THE RATINGS ARE MADE ON THE BASIS OF EXPOSURE TESTS IN SIMILAR CHEMICAL GROUPS.

Material	CARBON STEEL or CAST IRON	CARBON STEEL HARDENED	301/302/303/304 STAINLESS STEEL	316 STAINLESS STEEL	400 SERIES STAINLESS STEEL	400 SERIES STAINLESS STEEL HARDENED	ALUMINUM	COPPER	BRASS	ACETAL (DELRIN, CELCON)	ABS (CYCOLAC)	BUNA "N" (NITRILE RUBBER)	EPR, EPDM (NORDEL)	FLOURAZ	KALREZ	LEATHER/THIOKOL IMPREGNATED	NEOPRENE	NYLON	POLYCARBONATE (LEXAN)	POLYETHYLENE LOW DENSITY	POLYETHYLENE HIGH DENSITY	POLYPROPLENE 120°	POLYURETHANE, URATHANE	PVC POLYVINYL CHLORIDE	PVDF 150° (KYNAR)	TEFLON 120°	VITON
Lime Sulfur	★	★	☆	☆	★	★	■	■	(■)	☆	□	■	☆	□	□	□	☆	☆	□	☆	☆	□	□	□	□	☆	☆
Lime Sulphur	★	★	☆	☆	★	★	■	■	■	☆	□	■	□	□	□	□	☆	☆	□	☆	☆	□	□	□	□	☆	☆
Lindol, Hydraulic Fluid	☆	☆	☆	☆	☆	☆	☆	☆	□	□	□	■	☆	☆	□	□	■	□	□	□	□	□	■	□	□	☆	★
Limonene	□	□	□	□	□	□	□	□	□	□	□	■	■	□	□	□	■	□	□	□	□	□	■	□	□	☆	☆
Linoleic Acid	■	■	★	☆	★	★	☆	■	□	★	□	★	■	□	□	□	★	□	□	☆	☆	☆	□	□	□	☆	★
Linseed Oil	☆	☆	☆	☆	□	□	★	★	★	☆	□	☆	■	□	□	□	☆	☆	■	☆	☆	☆	★	□	☆	☆	☆
Liquid Oxygen	□	□	□	□	□	□	□	□	□	□	□	■	■	□	□	□	■	□	□	□	□	□	□	□	□	☆	□
Liquid Petroleum Gas (LPG)	□	□	□	□	□	□	□	□	□	□	□	■	☆	☆	□	□	★	□	□	□	□	□	★	□	□	☆	☆
Liquimoly	□	□	□	□	□	□	□	□	□	□	□	□	☆	□	□	□	★	□	□	□	□	□	★	□	□	☆	☆
Lithium Bromide	☆	☆	□	□	□	□	□	□	□	□	☆	☆	□	(☆)	■	□	□	□	□	□	□	□	□	□	□	☆	☆
Lithium Chloride	□	□	☆	☆	□	□	□	■	■	□	☆	□	(☆)	□	☆	□	□	□	□	□	□	□	■	□	□	□	■
Lithium Hydroxide	☆	☆	□	□	□	□	■	□	□	□	□	□	(☆)	□	☆	□	□	□	□	□	□	□	□	□	□	□	□
Lubricating Oil Di-Ester	☆	☆	☆	☆	☆	☆	☆	☆	□	□	□	★	■	□	□	(☆)	■	(☆)	□	□	□	□	☆	□	□	☆	☆
Petroleum Base	☆	☆	☆	☆	☆	☆	☆	☆	□	□	□	☆	■	□	□	(☆)	★	(☆)	□	□	□	☆	★	□	□	☆	☆
SAE 10, 20, 30, 40, 50	☆	☆	☆	☆	☆	☆	☆	☆	□	□	□	☆	■	□	□	(☆)	★	(☆)	□	☆	☆	☆	★	□	□	☆	☆
Lye Solutions	□	□	□	□	□	□	□	□	□	□	★	☆	☆	(☆)	□	★	□	□	□	□	□	□	■	□	□	□	★
Lysol	□	□	□	□	□	□	□	□	□	□	□	□	□	□	□	□	□	□	□	□	□	□	□	□	□	☆	☆
Magnesium Bisulfite	□	□	☆	☆	□	□	□	□	□	□	□	□	□	□	□	□	□	□	□	□	□	□	□	□	□	☆	☆
Magnesium Carbonate	★	★	★	★	★	★	☆	★	□	□	★	☆	□	□	□	□	☆	□	□	□	□	□	□	□	★	☆	☆
Magnesium Chloride	★	★	★	☆	□	□	■	★	■	☆	★	☆	☆	□	□	□	☆	☆	□	☆	☆	☆	□	☆	★	☆	☆
Magnesium Hydroxide (Milk of Magnesia)	☆	☆	☆	☆	☆	☆	■	☆	■	☆	★	☆	☆	☆	(☆)	★	☆	☆	□	☆	☆	☆	□	★	☆	☆	☆
Magnesium Hydroxide (Hot)	★	★	☆	☆	☆	☆	■	☆	■	□	□	★	□	□	□	□	☆	☆	□	□	□	□	□	★	☆	☆	☆
Magnesium Nitrate	★	★	★	★	★	★	★	★	☆	□	□	★	☆	★	☆	□	□	☆	□	□	□	□	□	★	☆	☆	☆
Magnesium Oxide	☆	☆	☆	☆	☆	★	□	□	□	□	□	☆	☆	□	□	□	☆	□	□	□	□	□	□	□	☆	★	★
Magnesium Salts	□	□	□	□	□	□	□	□	□	☆	☆	☆	☆	□	□	□	☆	□	□	□	□	□	□	□	□	☆	☆
Magnesium Sulfate	★	★	★	★	□	□	★	★	■	☆	★	☆	☆	(☆)	□	□	☆	☆	□	□	□	□	□	☆	☆	☆	☆
Magnesium Sulfite	□	□	□	□	□	□	□	★	□	☆	□	☆	☆	(☆)	□	□	☆	□	□	□	□	□	□	□	□	□	☆
Malathion	□	□	□	□	□	□	□	□	□	□	□	★	■	□	□	□	□	□	□	□	□	□	□	□	□	☆	☆
Maleic Acid	□	□	★	☆	□	□	□	□	□	□	□	■	■	□	□	□	□	□	□	□	□	□	□	□	☆	☆	☆
Maleic Anhydride	□	□	☆	☆	☆	☆	□	□	□	□	■	■	☆	□	□	□	■	□	□	□	□	□	□	□	□	☆	☆
Malic Acid	□	□	☆	☆	★	★	★	★	★	☆	□	☆	☆	□	□	□	★	★	□	□	□	□	□	■	☆	☆	☆
Manganese Chloride	★	★	★	☆	□	□	■	★	□	☆	□	■	☆	☆	(☆)	★	☆	□	□	□	□	□	□	☆	☆	☆	■
Malt Beverages	■	■	☆	☆	□	□	□	★	□	☆	□	☆	☆	□	□	□	☆	☆	□	□	□	■	□	□	□	☆	☆
Maple Sugar Liquors (Sucrose)	□	□	☆	☆	□	□	□	□	□	☆	□	☆	☆	□	□	□	☆	□	□	□	□	□	□	□	□	☆	☆
Mash	□	□	☆	☆	☆	☆	□	□	□	□	□	□	□	□	□	□	□	□	□	□	□	□	□	□	□	□	☆
Mayonnaise	■	■	★	☆	★	★	□	□	□	□	□	☆	■	□	□	☆	■	☆	□	□	□	□	□	□	□	□	☆
MCS 312	□	□	□	□	□	□	□	□	□	□	□	■	☆	□	□	□	□	□	□	□	□	□	□	□	□	■	□
MCS 352	□	□	□	□	□	□	□	□	□	□	□	■	☆	□	□	□	□	□	□	□	□	□	□	□	□	■	□
MCS 463	□	□	□	□	□	□	□	□	□	□	□	■	☆	□	□	□	□	□	□	□	□	□	□	□	□	■	□
Melamine Resins	□	□	□	■	□	□	□	□	□	☆	□	■	□	(☆)	□	□	□	□	□	□	□	□	□	□	□	☆	□
Mercaptan	☆	☆	☆	☆	☆	☆	□	□	★	□	□	■	□	(☆)	□	□	□	□	□	□	□	□	□	□	□	☆	□
Mercuric Chloride	■	■	■	■	■	■	■	■	■	□	☆	☆	☆	(☆)	□	☆	■	□	☆	☆	☆	☆	□	★	□	☆	☆
Mercuric Cyanide	★	★	★	★	□	□	■	■	□	□	☆	□	(☆)	□	★	□	□	□	□	□	□	□	★	□	☆	☆	☆
Mercurous Nitrate	□	□	□	□	□	□	□	■	□	□	☆	□	(☆)	□	□	□	□	□	□	□	□	□	□	□	□	☆	☆
Mercury	☆	☆	☆	☆	☆	☆	■	■	■	★	★	☆	☆	(☆)	☆	☆	☆	☆	□	☆	☆	☆	★	★	☆	☆	☆
Mercury Salts	□	□	□	□	□	□	□	□	□	□	□	☆	☆	(☆)	□	□	□	□	□	□	□	□	□	□	□	☆	☆
Mercury Vapors	□	□	□	□	□	□	□	□	□	□	□	☆	☆	(☆)	□	★	□	□	□	□	□	□	□	□	□	☆	☆
Mesityl Oxide	☆	☆	☆	☆	☆	☆	☆	☆	□	□	■	★	■	(☆)	□	■	□	□	□	□	□	■	□	□	□	☆	■
Methane	□	□	☆	☆	☆	☆	☆	☆	☆	□	□	■	■	(☆)	□	★	□	□	□	□	□	□	□	□	☆	□	☆

☆ – RECOMMENDED, PERFORMANCE IS NOT AFFECTED
★ – FAIR, SOME LOSS OF PROPERTIES MAY OCCUR
■ – NOT RECOMMENDED, MATERIAL IS UNSUITABLE FOR SERVICE
□ – NO DATA AVAILABLE AT THIS TIME
( ) – BRACKETS AROUND A RATING LETTER, NO DATA IS AVAILABLE, BUT THE RATINGS ARE MADE ON THE BASIS OF EXPOSURE TESTS IN SIMILAR CHEMICAL GROUPS.

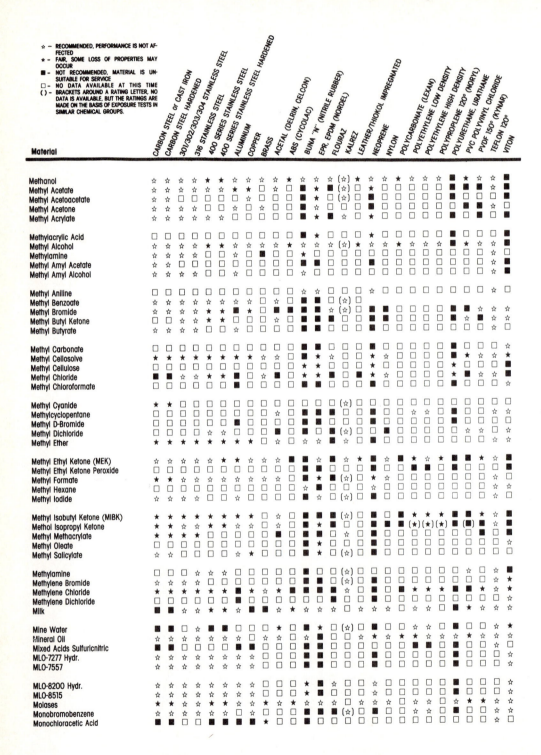

Material	CARBON STEEL or CAST IRON	CARBON STEEL HARDENED	301/302/303/304 STAINLESS STEEL	316 STAINLESS STEEL	400 SERIES STAINLESS STEEL	400 SERIES STAINLESS STEEL HARDENED	ALUMINUM	COPPER	BRASS	ACETAL (DELRIN, CELCON)	ABS (CYCOLAC)	BUNA "N" (NITRILE RUBBER)	EPR, EPDM (NORDEL)	FLOURAZ	KALREZ	LEATHER/THIOKOL IMPREGNATED	NEOPRENE	NYLON	POLYCARBONATE (LEXAN)	POLYETHYLENE LOW DENSITY	POLYETHYLENE HIGH DENSITY	POLYPROPLENE 120° (NORYL)	POLYURETHANE, URATHANE	PVC POLYVINYL CHLORIDE	PVDF 150° (KYNAR)	TEFLON 120°	VITON
Methanol	☆	☆	☆	☆	★	★	☆	☆	☆	☆	★	☆	☆	☆	(☆)	★	☆	☆	★	☆	☆	☆	■	★	☆	☆	■
Methyl Acetate	☆	☆	☆	☆	☆	☆	★	★	□	☆	□	■	★	■	(☆)	□	★	□	□	□	□	□	■	■	■	☆	■
Methyl Acetoacetate	☆	☆	□	□	□	□	□	☆	□	□	□	■	★	□	(☆)	□	■	□	□	□	□	□	■	□	□	□	■
Methyl Acetone	☆	☆	☆	☆	□	□	☆	□	☆	□	□	■	☆	□	□	□	■	□	□	□	□	□	□	■	■	☆	□
Methyl Acrylate	☆	☆	☆	☆	☆	☆	□	□	□	□	□	■	★	■	☆	□	★	□	□	□	□	□	■	□	■	□	■
Methylacrylic Acid	□	□	□	□	□	□	□	□	□	□	□	■	★	□	□	□	★	□	□	□	□	□	■	□	□	□	■
Methyl Alcohol	☆	☆	☆	☆	★	★	☆	☆	☆	☆	★	☆	☆	☆	(☆)	★	☆	☆	★	☆	☆	☆	■	★	☆	☆	■
Methylamine	☆	☆	☆	☆	□	□	☆	□	■	□	□	★	□	□	□	□	□	□	□	□	□	□	□	□	□	☆	□
Methyl Amyl Acetate	☆	☆	□	□	□	□	□	□	□	□	□	■	■	□	□	□	■	□	□	□	□	□	□	□	□	☆	■
Methyl Amyl Alcohol	☆	☆	☆	☆	□	□	☆	□	□	□	□	☆	□	□	□	□	□	□	□	□	□	□	□	□	□	☆	■
Methyl Aniline	□	□	□	□	□	□	□	□	□	□	□	☆	☆	□	□	□	☆	□	□	□	□	□	□	□	□	☆	□
Methyl Benzoate	☆	☆	☆	☆	☆	☆	☆	☆	□	☆	□	■	■	□	(☆)	□											
Methyl Bromide	☆	☆	☆	☆	★	★	■	★	□	■	■	■	■	☆	(☆)	□	■	■	□	□	□	□	■	■	☆	☆	☆
Methyl Butyl Ketone	□	□	☆	☆	★	★	□	□	□	☆	□	■	■	■	□	□	■	■	□	□	□	□	■	☆	■	☆	☆
Methyl Butyrate	☆	☆	☆	☆	□	□	☆	□	□	□	□	■	■	□	□	□	■	□	□	□	□	□	□	□	□	☆	□
Methyl Carbonate	□	□	□	□	□	□	□	□	□	□	□	■	■	□	□	□	■	□	□	□	□	□	■	□	□	□	☆
Methyl Cellosolve	★	★	★	★	★	★	★	★	☆	☆	□	■	★	☆	□	□	★	☆	□	□	□	□	■	★	☆	☆	★
Methyl Cellulose	□	□	□	□	□	□	□	□	□	□	□	★	★	□	□	□	★	□	□	□	□	□	★	□	□	□	■
Methyl Chloride	■	■	☆	☆	★	★	■	□	☆	■	□	★	★	■	□	■	★	☆	□	□	□	□	★	■	☆	☆	■
Methyl Chloroformate	□	□	□	□	□	□	■	□	□	□	□	■	■	□	□	□	■	□	□	□	□	□	■	□	□	□	☆
Methyl Cyanide	★	★	□	□	□	□	□	□	□	□	□	□	□	□	(☆)	□	□	□	□	□	□	□	□	□	□	□	□
Methylcyclopentane	□	□	□	□	□	□	□	□	□	☆	□	■	■	■	□	□	■	□	□	☆	☆	□	■	□	□	☆	☆
Methyl D-Bromide	□	□	□	□	□	□	■	□	□	□	□	■	■	□	□	□	■	□	□	□	□	□	■	□	□	□	□
Methyl Dichloride	□	□	□	□	☆	☆	□	□	□	■	□	■	□	■	(☆)	□	□	■	□	□	□	□	□	☆	☆	□	☆
Methyl Ether	★	★	★	★	★	★	★	★	□	☆	□	☆	☆	■	☆	□	■	□	□	□	□	□	□	□	□	☆	☆
Methyl Ethyl Ketone (MEK)	☆	☆	☆	☆	☆	★	★	☆	☆	☆	■	■	☆	■	☆	★	■	☆	■	★	☆	★	■	■	★	☆	■
Methyl Ethyl Ketone Peroxide	□	□	□	□	□	□	□	□	□	□	□	■	■	□	□	□	■	□	□	■	■	□	■	□	□	□	■
Methyl Formate	★	★	☆	☆	☆	☆	☆	☆	☆	☆	□	■	★	■	(☆)	□	★	☆	□	□	□	□	□	□	□	☆	□
Methyl Hexane	□	□	□	□	□	□	□	□	□	□	□	☆	■	□	□	□	☆	□	□	□	□	□	□	□	□	☆	☆
Methyl Iodide	☆	☆	☆	☆	□	□	☆	□	□	□	□	■	☆	□	(☆)	□	■	□	□	□	□	□	□	□	□	☆	□
Methyl Isobutyl Ketone (MIBK)	★	★	★	★	★	★	★	★	□	☆	□	■	■	■	(☆)	□	■	□	■	★	★	★	■	■	★	☆	■
Methol Isopropyl Ketone	★	★	☆	☆	★	★	☆	☆	□	☆	□	■	★	■	□	□	■	■	■	(★)	(★)	(★)	■	(■)	■	☆	■
Methyl Methacrylate	★	★	★	★	□	□	□	□	□	■	□	■	■	□	☆	□	■	□	□	□	□	□	□	□	■	□	■
Methyl Oleate	□	□	□	□	□	□	□	□	□	□	□	■	★	□	□	□	■	□	□	□	□	□	□	□	□	□	☆
Methyl Salicylate	☆	☆	□	□	□	□	☆	★	□	□	□	■	★	□	(☆)	□	■	□	□	□	□	□	□	□	□	□	□
Methylamine	□	□	□	☆	☆	☆	□	□	□	□	□	■	□	□	(☆)	□	□	□	□	□	□	□	□	☆	□	☆	■
Methylene Bromide	☆	☆	☆	☆	□	□	□	□	□	□	□	■	■	□	(☆)	□	■	□	□	□	□	□	□	□	□	☆	★
Methylene Chloride	★	★	★	★	★	★	■	★	☆	★	■	■	■	■	☆	□	■	□	■	★	★	★	■	■	★	☆	★
Methylene Dichloride	□	□	□	□	□	□	■	□	□	□	□	□	□	■	□	□	■	□	□	□	□	□	□	□	□	□	☆
Milk	■	■	☆	☆	★	★	☆	■	■	☆	★	☆	☆	☆	□	☆	☆	☆	□	☆	☆	□	■	★	☆	☆	☆
Mine Water	■	■	□	☆	■	■	□	□	□	★	□	■	★	□	(☆)	□	■	□	□	☆	☆	□	■	□	□	☆	★
Mineral Oil	☆	☆	☆	☆	☆	☆	☆	□	☆	☆	□	☆	■	□	□	☆	★	☆	★	★	☆	☆	☆	★	☆	☆	☆
Mixed Acids Sulfuricnitric	■	■	□	□	□	□	■	■	□	□	□	■	■	□	□	□	□	□	□	■	■	□	■	□	□	☆	□
MLO-7277 Hydr.	☆	☆	☆	☆	☆	☆	☆	☆	□	□	□	■	■	□	□	□	■	□	□	□	□	□	■	□	□	□	☆
MLO-7557	☆	☆	☆	☆	☆	☆	☆	☆	□	□	□	■	■	□	□	□	■	□	□	□	□	□	■	□	□	□	☆
MLO-8200 Hydr.	☆	☆	☆	☆	☆	☆	☆	☆	□	□	□	★	■	☆	□	□	☆	□	□	□	□	□	■	□	□	□	☆
MLO-8515	☆	☆	☆	☆	☆	☆	☆	☆	□	□	□	★	■	□	□	□	☆	□	□	□	□	□	■	□	□	□	☆
Molases	★	★	☆	☆	★	★	☆	☆	★	☆	★	☆	☆	☆	□	☆	☆	☆	□	☆	☆	□	☆	★	★	☆	☆
Monobromobenzene	☆	☆	☆	☆	☆	☆	□	☆	□	□	□	■	■	■	(☆)	□	■	□	□	☆	☆	□	■	□	□	☆	☆
Monochloracetic Acid	■	■	□	□	■	■	■	■	★	□	□	■	□	□	□	□	□	□	□	□	□	□	□	□	□	☆	□

Legend

- ☆ – RECOMMENDED, PERFORMANCE IS NOT AFFECTED
- ★ – FAIR, SOME LOSS OF PROPERTIES MAY OCCUR
- ■ – NOT RECOMMENDED, MATERIAL IS UNSUITABLE FOR SERVICE
- □ – NO DATA AVAILABLE AT THIS TIME
- ( ) – BRACKETS AROUND A RATING LETTER, NO DATA IS AVAILABLE, BUT THE RATINGS ARE MADE ON THE BASIS OF EXPOSURE TESTS IN SIMILAR CHEMICAL GROUPS.

Material	CARBON STEEL or CAST IRON	CARBON STEEL HARDENED	301/302/303/304 STAINLESS STEEL	316 STAINLESS STEEL	400 SERIES STAINLESS STEEL	400 SERIES STAINLESS STEEL HARDENED	ALUMINUM	COPPER	BRASS	ACETAL (DELRIN CELCON)	ABS (CYCOLAC)	BUNA "N" (NITRILE RUBBER)	EPR, EPDM (NORDEL)	FLOURAZ	KALREZ	LEATHER/THIOKOL IMPREGNATED	NEOPRENE	NYLON	POLYCARBONATE (LEXAN)	POLYETHYLENE LOW DENSITY	POLYETHYLENE HIGH DENSITY	POLYPROLENE 120° HIGH DENSITY	POLYURETHANE, URATHANE	PVC POLYVINYL CHLORIDE (NORYL)	PVDF 150° (KYNAR)	TEFLON 120°	VITON
Monochlorobenzene	□	□	□	□	□	□	■	□	□	■	■	■	■	□	□	■	□	□	□	□	□	■	□	■	□	☆	☆
Molybdic Acid	☆	☆	☆	☆	□	□	□	□	☆	□	☆	☆	□	□	□	□	□	□	☆	☆	☆	☆	□	□	□	☆	☆
Monochlorodifluoro Methane	☆	☆	☆	☆	☆	☆	■	☆	□	□	■	■	□	□	□	■	□	□	□	□	□	□	□	□	□	☆	☆
Monoethanolamine	☆	☆	☆	☆	☆	☆	★	■	■	□	□	★	□	□	■	□	□	□	□	□	□	□	■	□	■	☆	■
Monomethylaniline	★	★	★	★	★	★	★	★	■	□	☆	□	□	□	■	□	□	□	□	□	□	□	☆	☆	□	☆	★
Monomethylether	□	□	□	□	□	□	□	□	□	□	☆	□	□	☆	□	□	□	□	□	□	□	□	□	□	□	□	□
Monomethyl Hydrazine	☆	☆	☆	☆	□	□	□	□	□	★	☆	□	□	□	★	□	□	□	☆	□	□	□	□	□	□	☆	□
Mononitrotoluene & Dinitrotoluene (40/60 Mixture)	□	□	□	□	□	□	□	□	□	◉	■	■	□	□	■	□	■	□	□	□	□	□	□	□	□	□	■
Monosodium Postassium Ammonium Phosphate	□	□	□	□	□	□	□	□	□	□	☆	□	□	□	□	□	□	□	□	□	□	□	□	□	□	□	□
Monovinyl Acetylene	□	□	□	□	□	□	□	☆	☆	□	☆	□	□	□	□	★	□	□	□	□	□	□	□	□	□	□	☆
Muriatic Acid	■	■	■	■	■	■	□	□	□	□	☆	□	□	□	□	□	□	□	□	☆	□	☆	□	□	□	☆	☆
Mustard	■	■	☆	☆	□	□	★	□	□	★	☆	☆	□	□	□	☆	□	□	☆	□	□	☆	□	□	□	☆	☆
Napalm	□	□	□	□	□	□	□	□	□	□	★	□	□	□	□	□	□	□	□	★	□	□	□	□	□	☆	□
Naptha	★	☆	☆	☆	☆	☆	☆	☆	☆	■	■	☆	☆	□	□	■	□	☆	□	■	□	□	□	☆	□	☆	☆
Naptha-Coal Tar (Benzol)	★	★	☆	☆	☆	☆	☆	☆	☆	□	■	■	□	□	□	■	□	□	□	□	□	□	□	☆	□	☆	☆
Napthalene (Tar Camphor)	☆	☆	☆	☆	☆	☆	★	★	☆	■	■	■	□	■	(☆)	■	■	□	☆	□	□	□	□	☆	□	☆	☆
Napthenic Acid	★	★	☆	☆	☆	☆	☆	☆	□	□	□	■	□	□	□	■	□	□	☆	□	□	□	□	☆	□	☆	☆
Natural Gas	☆	☆	☆	☆	☆	☆	★	■	☆	□	☆	■	☆	☆	□	☆	□	□	□	★	□	□	□	☆	□	☆	☆
Neatsfoot Oil	☆	☆	☆	☆	□	□	□	☆	☆	□	☆	★	☆	□	□	☆	□	☆	□	□	□	☆	□	□	□	☆	☆
Neohexane	□	□	□	□	□	□	□	□	□	□	☆	□	□	□	□	□	□	□	□	□	□	□	□	□	□	□	□
Neon	☆	☆	☆	☆	☆	☆	☆	☆	☆	□	☆	□	□	□	□	☆	□	☆	□	☆	□	□	□	☆	□	☆	☆
Neosol	★	★	☆	☆	□	□	★	□	□	□	☆	□	□	□	★	□	□	□	□	□	□	□	□	□	□	☆	□
Neville Acid	□	□	□	□	□	□	□	□	□	■	★	□	□	□	□	■	□	□	□	□	□	□	□	□	□	☆	★
Nickel Acetate	□	★	★	□	□	□	□	★	□	□	★	□	■	(☆)	□	★	□	□	□	□	□	■	□	■	□	□	■
Nickel Ammonium Sulfate	□	□	□	□	□	□	□	□	□	□	☆	□	□	(☆)	□	□	□	□	□	□	□	□	□	□	□	☆	□
Nickel Chloride	■	■	■	★	■	■	■	□	■	□	☆	★	☆	□	(☆)	□	★	☆	□	☆	□	☆	■	★	□	☆	☆
Nickel Nitrate	★	★	★	★	★	★	☆	■	□	□	☆	□	□	(☆)	□	□	□	☆	□	□	□	□	□	□	□	☆	☆
Nickel Salts	□	□	□	□	□	□	□	□	□	☆	☆	□	□	□	□	★	□	□	□	□	□	■	□	□	□	☆	□
Nickel Sulfate	■	■	★	★	(☆)	(☆)	■	□	■	☆	☆	☆	☆	(☆)	□	☆	☆	☆	□	☆	□	☆	☆	□	□	☆	☆
Nicotine	★	★	□	□	□	□	★	□	★	□	□	□	□	(☆)	□	□	□	□	□	□	□	□	□	★	□	☆	□
Nicotinic Acid	□	□	□	□	□	□	★	□	☆	□	☆	□	□	□	□	□	□	□	□	□	□	☆	□	□	□	☆	□
Niter Cake	□	□	□	□	□	□	□	□	□	■	☆	□	□	□	□	□	□	☆	□	□	□	□	□	☆	□	☆	☆
Nitrana (Ammonia Fertilizer)	□	□	☆	☆	□	□	□	□	□	□	☆	□	□	□	□	★	□	□	□	□	□	□	□	□	□	□	■
Nitric Acid																											
3 Molar	■	■	☆	☆	■	■	■	■	■	□	□	■	□	□	□	■	□	■	□	□	□	□	□	□	☆	☆	☆
Concentrated	■	■	☆	☆	■	■	□	■	■	□	□	■	□	□	□	■	□	■	□	□	□	□	□	□	☆	☆	☆
Red Fuming (RFNA)	■	■	☆	☆	☆	☆	☆	■	■	□	□	■	□	□	□	■	□	■	□	□	□	□	□	□	☆	☆	★
Inhibited Red Fuming (IRFNA)	■	■	☆	☆	☆	☆	☆	■	■	□	□	■	□	□	□	■	□	■	□	□	□	□	□	□	☆	☆	★
5% to 10% Solution	■	■	☆	☆	☆	☆	☆	■	■	□	★	■	□	□	□	■	★	■	★	☆	☆	■	□	■	★	☆	☆
20% Solution	■	■	☆	☆	☆	☆	☆	■	■	□	★	■	□	□	□	■	☆	■	☆	☆	☆	☆	□	■	★	☆	☆
50% Solution (Boiling)	■	■	★	★	■	■	☆	■	■	□	□	■	□	□	□	■	★	★	★	★	☆	★	□	★	□	☆	☆
65% Solution (Boiling)	■	■	★	★	■	■	☆	■	■	□	□	■	□	□	□	■	■	☆	■	★	□	■	□	■	□	☆	☆
Concentrated Boiling	■	■	☆	☆	■	■	☆	■	■	□	□	■	□	□	□	■	☆	□	☆	☆	□	■	□	■	□	☆	☆
Nitric Acid & Hydrochloric Acid	■	■	■	■	■	■	■	■	□	■	■	□	■	★	□	□	□	□	□	□	□	□	□	■	□	☆	☆
Nitriding Gases	☆	☆	☆	☆	☆	☆	☆	☆	☆	□	□	■	□	□	□	□	□	□	□	□	□	□	□	☆	□	☆	□
Nitrobenzene	★	★	★	★	★	★	★	★	☆	□	■	■	■	(☆)	□	■	■	■	■	■	■	■	■	■	□	★	★
Nitroethane	☆	☆	☆	☆	☆	□	□	☆	☆	□	■	■	★	(☆)	□	■	★	□	□	☆	□	☆	□	■	□	☆	☆
Nitrogen	□	□	□	□	□	□	□	□	□	□	☆	□	☆	(☆)	□	☆	☆	□	□	☆	□	☆	□	□	□	□	☆

☆ – RECOMMENDED, PERFORMANCE IS NOT AFFECTED
★ – FAIR, SOME LOSS OF PROPERTIES MAY OCCUR
■ – NOT RECOMMENDED, MATERIAL IS UNSUITABLE FOR SERVICE
□ – NO DATA AVAILABLE AT THIS TIME
( ) – BRACKETS AROUND A RATING LETTER, NO DATA IS AVAILABLE, BUT THE RATINGS ARE MADE ON THE BASIS OF EXPOSURE TESTS IN SIMILAR CHEMICAL GROUPS.

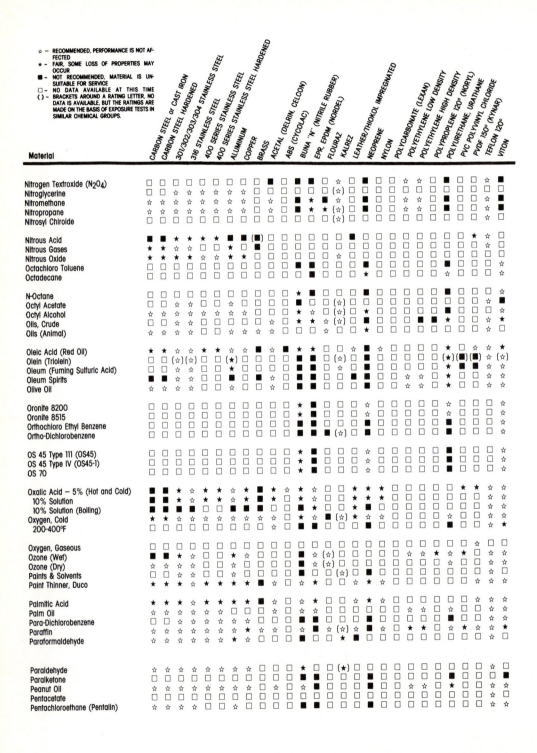

Column headings (left to right):
CARBON STEEL or CAST IRON · CARBON STEEL HARDENED · 301/302/303/304 STAINLESS STEEL · 316 STAINLESS STEEL · 400 SERIES STAINLESS STEEL · 400 SERIES STAINLESS STEEL HARDENED · ALUMINUM · COPPER · BRASS · ACETAL (DELRIN, CELCON) · ABS (CYCOLAC) · BUNA "N" (NITRILE RUBBER) · EPR, EPDM (NORDEL) · FLOURAZ · KALREZ · LEATHER/THIOKOL IMPREGNATED · NEOPRENE · NYLON · POLYCARBONATE (LEXAN) · POLYETHYLENE LOW DENSITY · POLYETHYLENE HIGH DENSITY · POLYPROPLENE 120° (NORYL) · POLYURETHANE, URATHANE · PVC POLYVINYL CHLORIDE · PVDF 150° (KYNAR) · TEFLON 120° · VITON

Material

- Nitrogen Textroxide (N₂O₄)
- Nitroglycerine
- Nitromethane
- Nitropropane
- Nitrosyl Chlroide
- Nitrous Acid
- Nitrous Gases
- Nitrous Oxide
- Octachloro Toluene
- Octadecane
- N-Octane
- Octyl Acetate
- Octyl Alcohol
- Oils, Crude
- Oils (Animal)
- Oleic Acid (Red Oil)
- Olein (Triolein)
- Oleum (Fuming Sulfuric Acid)
- Oleum Spirits
- Olive Oil
- Oronite 8200
- Oronite 8515
- Orthochloro Ethyl Benzene
- Ortho-Dichlorobenzene
- OS 45 Type 111 (OS45)
- OS 45 Type IV (OS45-1)
- OS 70
- Oxalic Acid – 5% (Hot and Cold)
-   10% Solution
-   10% Solution (Boiling)
- Oxygen, Cold
-   200–400°F
- Oxygen, Gaseous
- Ozone (Wet)
- Ozone (Dry)
- Paints & Solvents
- Paint Thinner, Duco
- Palmitic Acid
- Palm Oil
- Para-Dichlorobenzene
- Paraffin
- Paraformaldehyde
- Paraldehyde
- Paralketone
- Peanut Oil
- Pentacetate
- Pentachloroethane (Pentalin)

Legend:

- ☆ – RECOMMENDED, PERFORMANCE IS NOT AFFECTED
- ★ – FAIR, SOME LOSS OF PROPERTIES MAY OCCUR
- ■ – NOT RECOMMENDED, MATERIAL IS UNSUITABLE FOR SERVICE
- □ – NO DATA AVAILABLE AT THIS TIME
- ( ) – BRACKETS AROUND A RATING LETTER, NO DATA IS AVAILABLE, BUT THE RATINGS ARE MADE ON THE BASIS OF EXPOSURE TESTS IN SIMILAR CHEMICAL GROUPS.

Material	CARBON STEEL or CAST IRON	CARBON STEEL HARDENED	301/302/303/304 STAINLESS STEEL	316 STAINLESS STEEL	400 SERIES STAINLESS STEEL	400 SERIES STAINLESS STEEL HARDENED	ALUMINUM	COPPER	BRASS	ACETAL (DELRIN, CELCON)	ABS (CYCOLAC)	BUNA "N" (NITRILE RUBBER)	EPR, EPDM (NORDEL)	FLOURAZ	KALREZ	LEATHER/THIOKOL IMPREGNATED	NEOPRENE	NYLON	POLYCARBONATE (LEXAN)	POLYETHYLENE LOW DENSITY	POLYETHYLENE HIGH DENSITY	POLYPROPLENE 120°	POLYURETHANE 120° (NORYL)	PVC POLYVINYL CHLORIDE	PVDF 150° (KYNAR)	TEFLON 120°	VITON
Pentachlorophenol (PCP)	☆	☆	☆	☆	□	□	☆	□	□	□	□	■	■	□	□	□	■	□	□	□	□	□	□	□	□	☆	☆
Pentane, 2 Methyl	□	□	□	□	□	□	□	□	□	☆	□	☆	□	☆	(☆)	□	★	□	□	☆	☆	□	■	□	□	☆	☆
2-4, Dimethyl	□	□	□	□	□	□	□	□	□	☆	□	☆	□	☆	(☆)	□	★	□	□	☆	☆	□	■	□	□	☆	☆
3-Methyl	□	□	□	□	□	□	□	□	□	☆	□	☆	□	☆	■	□	★	□	□	☆	☆	□	■	□	□	☆	☆
N-Pentane	★	★	★	★	□	□	★	□	□	☆	□	☆	■	☆	(☆)	★	☆	□	□	☆	☆	□	■	□	□	☆	☆
Pentanedione-2,4	□	□	□	□	□	□	□	□	□	□	□	☆	□	☆	□	□	☆	□	□	□	□	□	□	□	□	☆	☆
Peppermint Oil	□	□	☆	☆	□	□	□	□	□	☆	□	■	■	□	□	■	★	□	□	□	□	□	□	□	□	☆	☆
Perchloric Acid	■	■	★	★	□	□	■	■	□	☆	□	■	★	□	□	□	★	□	■	★	★	★	★	★	★	☆	☆
Perchloroethylene (Tetrachloroethylene)	☆	☆	★	☆	☆	☆	■	★	□	☆	□	■	■	★	□	■	■	■	■	■	■	■	■	■	□	☆	☆
Permachlor (Degreasing Fluid)	☆	☆	☆	☆	☆	☆	☆	□	□	☆	□	■	■	□	□	□	□	□	□	□	□	□	□	□	□	☆	☆
Petrolatum	□	□	☆	☆	☆	□	□	□	□	☆	☆	☆	(☆)	□	★	□	■	□	□	□	□	□	□	□	★	☆	★
Petroleum Ether	□	□	☆	☆	☆	□	□	□	□	☆	☆	☆	■	(☆)	□	□	□	■	□	□	■	■	☆	□	□	☆	☆
Petroleum Oils (Refined)	☆	☆	☆	☆	☆	☆	☆	□	□	☆	□	□	□	□	□	□	□	□	□	□	□	□	□	□	□	□	□
Petroleum Oils (Sour)	★	★	☆	☆	■	■	☆	■	■	□	□	■	□	□	□	□	★	□	□	□	□	□	□	□	□	□	□
Petroleum Oil, Crude	□	□	□	□	□	□	□	□	□	☆	■	□	□	□	□	★	□	□	□	□	□	☆	☆	□	□	☆	□
Below 250° F	□	□	□	□	□	□	□	□	□	☆	■	□	□	□	□	★	□	□	□	□	□	★	□	□	□	☆	□
Above 250°F	□	□	□	□	□	□	□	□	□	■	■	□	□	□	□	■	□	□	□	□	□	■	□	□	□	□	★
Phenethyl Alcohol (Benzyl Carbinol)	☆	☆	☆	☆	□	□	☆	□	□	□	□	■	★	□	□	□	■	□	□	□	□	□	□	□	□	☆	■
Phenol	☆	☆	☆	☆	★	★	☆	☆	□	■	■	■	★	☆	☆	■	■	■	■	■	■	■	■	■	■	☆	☆
70%/30% H₂O	□	□	☆	☆	□	□	☆	□	■	■	□	■	□	☆	(☆)	□	□	(☆)	(☆)	(☆)	(☆)	■	(☆)	□	☆	☆	☆
85%/15% H₂O	□	□	☆	☆	□	□	☆	□	■	■	□	■	□	☆	(☆)	□	■	(☆)	(☆)	(☆)	(☆)	■	(☆)	□	☆	☆	☆
Phenolic Sulphonate	□	□	□	□	□	□	□	□	□	□	□	□	(☆)	□	□	□	□	□	□	□	□	□	□	□	□	☆	□
Phenol Sulfonic Acid	★	★	★	★	□	□	□	□	□	□	□	□	□	□	□	□	□	□	□	□	□	□	□	□	□	☆	■
Phenyl Acetate	□	□	□	□	□	□	□	□	□	■	★	□	(☆)	□	■	□	□	□	□	□	□	□	□	■	□	☆	☆
Phenylbenzene	□	□	□	□	□	□	□	□	□	□	□	☆	□	□	□	□	■	□	□	□	□	□	□	■	□	□	☆
Phenyl Hydrazine	□	□	□	□	□	□	□	□	□	□	□	□	□	□	□	□	■	□	□	□	□	□	□	■	□	☆	■
Phorone Diisopropylidene Acetone)	□	□	□	□	□	□	□	□	□	☆	□	■	★	□	□	□	■	□	□	☆	☆	□	■	□	□	☆	■
Phosphate Esters	□	□	□	□	□	□	□	□	□	☆	☆	□	□	□	□	□	□	□	□	□	□	□	□	□	□	□	□
Phosphoric Acid	□	□	□	★	□	□	■	■	□	☆	★	★	(☆)	□	□	★	□	□	★	★	★	★	■	★	☆	☆	☆
3 Molar	□	□	□	□	□	□	■	■	□	☆	☆	★	(☆)	□	■	□	□	■	□	□	□	□	□	□	☆	☆	☆
Concentrated	□	□	□	□	□	□	■	■	□	☆	★	★	(☆)	□	■	□	□	■	□	□	□	□	□	□	☆	☆	☆
1% Solution	□	□	☆	☆	☆	☆	☆	■	□	☆	★	□	☆	(☆)	□	□	★	□	★	★	★	★	★	☆	★	☆	☆
5% Solution	□	□	☆	☆	☆	☆	☆	■	□	☆	★	□	☆	(☆)	□	□	★	□	★	★	★	★	★	☆	★	☆	☆
10% Solution	■	■	☆	☆	■	■	☆	■	□	☆	★	□	☆	(☆)	□	★	★	□	☆	☆	☆	★	★	★	☆	☆	☆
10% Solution (Hot)	■	■	□	☆	☆	■	☆	■	□	☆	★	□	☆	(☆)	□	★	★	□	☆	★	★	★	★	★	☆	☆	☆
50% Solution	■	■	☆	☆	■	■	■	■	□	☆	□	□	☆	(☆)	□	★	□	□	☆	★	★	★	★	★	★	☆	☆
50% Solution (Hot)	■	■	★	☆	■	■	■	■	□	☆	□	□	□	(☆)	□	★	□	□	□	□	□	□	□	□	□	☆	☆
85% Solution	■	■	★	★	■	■	■	■	□	☆	□	□	★	(☆)	□	★	□	★	☆	★	★	★	★	★	★	☆	☆
85% Solution (Hot)	■	■	■	■	■	■	☆	■	□	☆	□	□	□	(☆)	□	★	□	□	□	□	□	□	□	□	□	☆	☆
Phosphoric Acid (Aerated)	■	■	□	★	□	□	□	■	■	□	☆	□	□	□	□	□	□	□	□	□	□	□	□	□	□	☆	□
Phosphoric Acid Air Free	■	■	□	□	□	□	■	■	■	□	☆	□	□	□	□	□	□	□	□	□	□	□	□	□	□	☆	□
Phosphorous	★	★	☆	☆	□	□	★	☆	□	□	□	□	□	□	(☆)	□	□	□	□	□	□	□	□	□	□	☆	□
Phosphorous Oxychloride	■	■	■	■	□	□	★	■	□	□	□	□	□	□	□	□	□	□	□	□	□	□	□	□	□	☆	□
Phosphorous Trichloride	★	★	☆	☆	■	□	□	□	□	□	□	■	☆	☆	(☆)	□	■	□	□	□	□	□	□	□	□	★	☆

☆ – RECOMMENDED, PERFORMANCE IS NOT AFFECTED
★ – FAIR, SOME LOSS OF PROPERTIES MAY OCCUR
■ – NOT RECOMMENDED, MATERIAL IS UNSUITABLE FOR SERVICE
□ – NO DATA AVAILABLE AT THIS TIME
( ) – BRACKETS AROUND A RATING LETTER, NO DATA IS AVAILABLE, BUT THE RATINGS ARE MADE ON THE BASIS OF EXPOSURE TESTS IN SIMILAR CHEMICAL GROUPS.

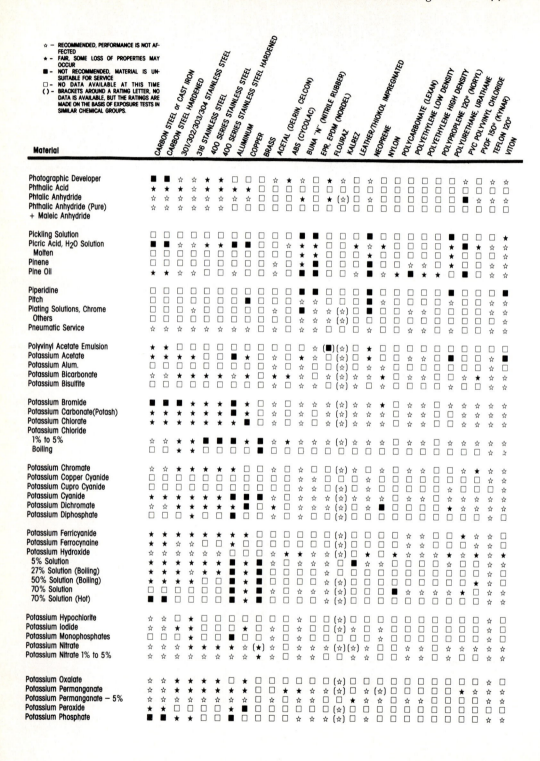

Column headers (left to right):
CARBON STEEL or CAST IRON · CARBON STEEL HARDENED · 301/302/303/304 STAINLESS STEEL · 316 STAINLESS STEEL · 400 SERIES STAINLESS STEEL · 400 SERIES STAINLESS STEEL HARDENED · ALUMINUM · COPPER · BRASS · ACETAL (DELRIN, CELCON) · ABS (CYCOLAC) · BUNA "N" (NITRILE RUBBER) · EPR, EPDM (NORDEL) · FLUORAZ · KALREZ · LEATHER/THIOKOL IMPREGNATED · NEOPRENE · NYLON · POLYCARBONATE (LEXAN) · POLYETHYLENE LOW DENSITY · POLYETHYLENE HIGH DENSITY · POLYPROPLENE 120° · POLYURETHANE, URATHANE · PVC POLYVINYL CHLORIDE · PVDF 150° (KYNAR) · TEFLON 120° · VITON

Material

- Photographic Developer
- Phthalic Acid
- Phtalic Anhydride
- Phthalic Anhydride (Pure) + Maleic Anhydride
- Pickling Solution
- Picric Acid, H₂O Solution
- Molten
- Pinene
- Pine Oil
- Piperidine
- Pitch
- Plating Solutions, Chrome
- Others
- Pneumatic Service
- Polyvinyl Acetate Emulsion
- Potassium Acetate
- Potassium Alum.
- Potassium Bicarbonate
- Potassium Bisulfite
- Potassium Bromide
- Potassium Carbonate(Potash)
- Potassium Chlorate
- Potassium Chloride
  - 1% to 5%
  - Boiling
- Potassium Chromate
- Potassium Copper Cyanide
- Potassium Cupro Cyanide
- Potassium Cyanide
- Potassium Dichromate
- Potassium Diphosphate
- Potassium Ferricyanide
- Potassium Ferrocynaine
- Potassium Hydroxide
  - 5% Solution
  - 27% Solution (Boiling)
  - 50% Solution (Boiling)
  - 70% Solution
  - 70% Solution (Hot)
- Potassium Hypochlorite
- Potassium Iodide
- Potassium Monophosphates
- Potassium Nitrate
- Potassium Nitrate 1% to 5%
- Potassium Oxalate
- Potassium Permanganate
- Potassium Permanganate – 5%
- Potassium Peroxide
- Potassium Phosphate

☆ – RECOMMENDED, PERFORMANCE IS NOT AFFECTED
★ – FAIR, SOME LOSS OF PROPERTIES MAY OCCUR
■ – NOT RECOMMENDED, MATERIAL IS UNSUITABLE FOR SERVICE
□ – NO DATA AVAILABLE AT THIS TIME
( ) – BRACKETS AROUND A RATING LETTER, NO DATA IS AVAILABLE, BUT THE RATINGS ARE MADE ON THE BASIS OF EXPOSURE TESTS IN SIMILAR CHEMICAL GROUPS.

Material	CARBON STEEL or CAST IRON	CARBON STEEL HARDENED	301/302/303/304 STAINLESS STEEL	316 STAINLESS STEEL	400 SERIES STAINLESS STEEL	400 SERIES STAINLESS STEEL HARDENED	ALUMINUM	COPPER	BRASS	ACETAL (DELRIN CELCON)	ABS (CYCOLAC)	BUNA "N" (NITRILE RUBBER)	EPR, EPDM (NORDEL)	FLOURAZ	KALREZ	LEATHER/THIOKOL IMPREGNATED	NEOPRENE	NYLON	POLYCARBONATE (LEXAN)	POLYETHYLENE LOW DENSITY	POLYETHYLENE HIGH DENSITY	POLYPROPLENE 120°	POLYURETHANE, URATHANE	PVC POLYVINYL CHLORIDE	PVDF 150° (KYNAR)	TEFLON 120°	VITON
Potassium Salts	□	□	□	□	□	□	□	□	□	□	□	☆	☆	□	□	□	☆	□	□	□	☆	□	□	□	☆	□	☆
Potassium Sulfate	☆	☆	★	★	★	★	☆	★	■	□	☆	☆	☆	☆	(☆)	□	☆	□	□	☆	☆	□	☆	☆	☆	☆	☆
Potassium Sulfate — 1% and 5%	☆	☆	☆	☆	☆	☆	☆	☆	■	☆	☆	☆	☆	□	☆	□	☆	□	□	☆	☆	□	☆	☆	☆	☆	☆
Potassium Sulfide	★	★	★	★	★	★	☆	□	■	□	☆	☆	□	☆	(☆)	□	☆	□	□	□	□	□	□	□	★	□	☆
Potassium Sulfite	☆	☆	☆	☆	☆	□	□	□	□	□	☆	☆	□	☆	(☆)	□	☆	□	□	□	☆	□	□	□	☆	□	☆
Potassium Sulphate	☆	☆	★	★	★	★	☆	★	□	□	☆	☆	☆	□	□	□	☆	□	□	☆	☆	□	☆	☆	☆	☆	☆
Potassium Sulphite or Nitrate	☆	☆	☆	□	★	★	★	☆	□	□	☆	☆	☆	□	□	□	☆	□	□	☆	☆	□	☆	☆	☆	☆	☆
Potassium Triphosphate	□	□	□	★	☆	□	■	□	□	□	☆	□	□	□	(☆)	□	□	□	□	☆	☆	□	□	□	□	□	□
Prestone	☆	☆	☆	☆	☆	☆	☆	☆	□	☆	☆	☆	□	□	□	□	■	☆	□	☆	☆	□	■	☆	☆	☆	☆
PRL-High Temp. Hydr. Oil	☆	☆	☆	☆	☆	☆	☆	☆	☆	□	☆	■	□	□	□	□	★	☆	□	☆	☆	□	★	□	☆	☆	☆
Producer Gas	☆	☆	☆	☆	☆	☆	☆	☆	☆	□	☆	■	□	□	□	□	★	□	■	□	■	■	□	☆	□	☆	☆
Propane	☆	☆	☆	☆	☆	☆	☆	☆	☆	□	☆	■	☆	(☆)	□	★	□	■	■	■	■	□	□	☆	☆	☆	☆
Propane (Liquified)	☆	☆	☆	☆	☆	☆	☆	☆	☆	□	☆	■	□	□	□	★	□	■	□	★	★	☆	□	☆	☆	☆	☆
Propane Propionitrile	□	□	☆	☆	☆	☆	☆	☆	☆	□	☆	■	□	□	□	★	□	□	□	□	□	■	□	☆	□	☆	☆
Propionaldehyde (Propanal)	☆	☆	☆	☆	☆	☆	☆	☆	☆	□	■	□	☆	□	□	□	□	□	□	□	□	■	□	☆	□	☆	☆
Propionic Acid	☆	☆	□	★	☆	☆	★	★	□	□	■	□	☆	□	□	■	□	□	□	□	□	■	□	☆	☆	☆	□
Propyl Acetate	☆	☆	☆	□	□	□	☆	☆	□	□	■	★	□	(☆)	□	□	□	□	□	□	□	■	□	■	☆	☆	□
N-Propyl Acetone	□	□	□	□	□	□	□	□	□	□	■	☆	□	□	□	□	□	□	□	□	□	□	□	■	□	☆	□
Propyl Alcohol	☆	☆	☆	☆	☆	☆	☆	☆	☆	□	■	☆	☆	(☆)	□	☆	□	☆	□	☆	☆	☆	■	☆	☆	☆	☆
Propylene	☆	☆	☆	☆	☆	☆	☆	☆	☆	□	■	■	☆	(☆)	□	■	□	☆	☆	☆	☆	□	■	☆	☆	☆	☆
Propylene Dichloride	☆	☆	★	☆	☆	☆	■	☆	□	□	□	□	□	(☆)	□	□	□	☆	□	□	□	□	□	□	☆	☆	☆
Propylene Glycol	☆	☆	★	★	☆	☆	☆	☆	□	☆	★	☆	☆	(☆)	□	☆	★	☆	☆	☆	☆	☆	★	☆	☆	☆	☆
Propylene Oxide	★	★	☆	☆	☆	★	★	★	□	□	■	■	☆	(☆)	□	■	□	★	★	☆	★	■	★	■	□	☆	□
Propyl Nitrate	□	□	☆	☆	☆	☆	☆	☆	□	☆	□	■	★	□	□	□	□	□	□	☆	☆	■	□	□	☆	☆	■
Pyranol, Transformer Oil	☆	☆	☆	☆	☆	☆	☆	☆	☆	□	☆	■	□	□	□	□	★	□	□	★	★	☆	□	☆	☆	☆	☆
Pydraul 10E, 29 ELT	□	□	□	□	□	□	□	□	□	□	□	■	☆	□	□	□	■	□	□	□	□	■	□	□	□	□	☆
Pydraul 30E, 50E, 65E, 90E	□	□	□	□	□	□	□	□	□	□	□	■	☆	□	□	□	■	□	□	□	□	■	□	□	□	□	☆
Pydraul 115E	□	□	□	□	□	□	□	□	□	□	□	■	☆	□	□	□	■	□	□	□	□	■	□	□	□	□	☆
Pydraul 230E, 312C, 540C	□	□	□	□	□	□	□	□	□	□	□	■	☆	□	□	□	■	□	□	□	□	■	□	□	□	□	☆
Pyridine	☆	☆	★	★	★	★	★	★	□	☆	■	☆	☆	(☆)	□	□	■	□	■	■	■	□	■	☆	★	☆	☆
Pyridine Oil	□	□	□	□	□	□	□	□	□	□	□	■	☆	□	□	□	■	★	□	□	□	□	□	☆	☆	☆	☆
Pyrogallic Acid	■	■	★	★	★	★	★	★	□	■	□	☆	☆	★	□	□	■	□	□	★	★	□	★	★	★	☆	★
Pyroligneous	□	□	□	□	□	□	□	□	□	☆	■	□	☆	■	□	□	■	□	□	★	★	■	★	□	□	☆	□
Pyroligneous Acid	☆	☆	★	★	☆	☆	☆	★	□	■	□	☆	☆	★	□	□	■	□	□	★	★	■	★	□	□	☆	■
Pyrolube	□	□	□	□	□	□	□	□	□	□	■	★	□	□	□	□	■	□	□	□	□	■	□	□	□	□	☆
Pyrrole	□	□	□	□	□	□	□	□	□	□	■	□	■	□	□	□	■	□	□	□	□	□	□	□	□	□	☆
Quaternary Ammonium Salts	■	■	☆	☆	☆	□	☆	□	□	□	☆	☆	☆	□	□	□	□	□	□	☆	☆	□	□	□	☆	☆	☆
Quench Oil	□	□	☆	☆	□	□	☆	□	□	□	☆	★	□	□	□	□	★	□	□	□	□	□	□	□	☆	☆	☆
Quinine Bisulphate (Dry)	□	□	★	☆	□	□	□	□	□	☆	☆	☆	☆	□	□	□	☆	□	□	□	□	□	□	□	☆	☆	☆
Quinine Sulphate (Dry)	□	□	☆	☆	□	□	□	□	□	☆	☆	☆	☆	□	□	□	☆	□	□	☆	☆	□	☆	□	☆	☆	☆
Radiation	□	□	□	□	□	□	□	□	□	☆	■	■	☆	□	□	□	■	□	□	☆	☆	□	■	□	□	☆	■
Rape Seed Oil	□	□	☆	☆	□	□	□	□	□	☆	■	☆	□	□	□	□	★	□	□	■	■	□	★	□	□	☆	☆
Rayon Spin Bath	□	□	□	■	□	□	□	□	□	□	☆	□	□	□	□	□	☆	□	□	□	□	☆	□	□	□	☆	☆
Red Oil (Mil-H-5606)	□	□	□	□	□	□	□	□	□	□	■	□	□	□	□	□	★	□	□	□	□	□	□	□	□	☆	☆

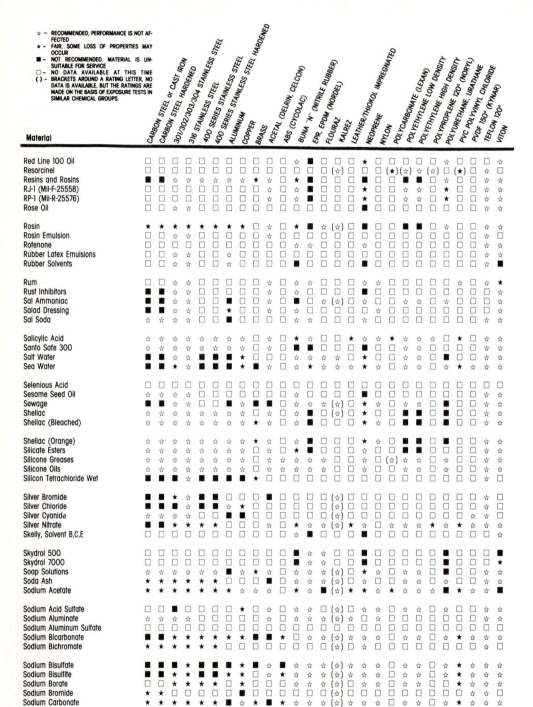

Legend

☆ – RECOMMENDED, PERFORMANCE IS NOT AFFECTED
★ – FAIR, SOME LOSS OF PROPERTIES MAY OCCUR
■ – NOT RECOMMENDED, MATERIAL IS UNSUITABLE FOR SERVICE
☐ – NO DATA AVAILABLE AT THIS TIME
( ) – BRACKETS AROUND A RATING LETTER, NO DATA IS AVAILABLE, BUT THE RATINGS ARE MADE ON THE BASIS OF EXPOSURE TESTS IN SIMILAR CHEMICAL GROUPS.

Column headings (left to right): CARBON STEEL or CAST IRON · CARBON STEEL HARDENED · 301/302/303/304 STAINLESS STEEL · 316 STAINLESS STEEL · 400 SERIES STAINLESS STEEL · 400 SERIES STAINLESS STEEL HARDENED · ALUMINUM · COPPER · BRASS · ACETAL (DELRIN, CELCON) · ABS (CYCOLAC) · BUNA 'N' (NITRILE RUBBER) · EPR, EPDM (NORDEL) · FLOURAZ · KALREZ · LEATHER/THIOKOL IMPREGNATED · NEOPRENE · NYLON · POLYCARBONATE (LEXAN) · POLYETHYLENE LOW DENSITY · POLYETHYLENE HIGH DENSITY · POLYPROPLENE 120° (NORYL) · POLYURETHANE, URATHANE · PVC POLYVINYL CHLORIDE · PVDF 150° (KYNAR) · TEFLON 120° · VITON

Materials (in order):

- Red Line 100 Oil
- Resorcinel
- Resins and Rosins
- RJ-1 (Mil-F-25558)
- RP-1 (Mil-R-25576)
- Rose Oil
- Rosin
- Rosin Emulsion
- Rotenone
- Rubber Latex Emulsions
- Rubber Solvents
- Rum
- Rust Inhibitors
- Sal Ammoniac
- Salad Dressing
- Sal Soda
- Salicylic Acid
- Santo Safe 300
- Salt Water
- Sea Water
- Selenious Acid
- Sesame Seed Oil
- Sewage
- Shellac
- Shellac (Bleached)
- Shellac (Orange)
- Silicate Esters
- Silicone Greases
- Silicone Oils
- Silicon Tetrachloride Wet
- Silver Bromide
- Silver Chloride
- Silver Cyanide
- Silver Nitrate
- Skelly, Solvent B,C,E
- Skydrol 500
- Skydrol 7000
- Soap Solutions
- Soda Ash
- Sodium Acetate
- Sodium Acid Sulfate
- Sodium Aluminate
- Sodium Aluminum Sulfate
- Sodium Bicarbonate
- Sodium Bichromate
- Sodium Bisulfate
- Sodium Bisulfite
- Sodium Borate
- Sodium Bromide
- Sodium Carbonate

Legend:

☆ – RECOMMENDED, PERFORMANCE IS NOT AFFECTED
★ – FAIR, SOME LOSS OF PROPERTIES MAY OCCUR
■ – NOT RECOMMENDED, MATERIAL IS UNSUITABLE FOR SERVICE
□ – NO DATA AVAILABLE AT THIS TIME
( ) – BRACKETS AROUND A RATING LETTER, NO DATA IS AVAILABLE, BUT THE RATINGS ARE MADE ON THE BASIS OF EXPOSURE TESTS IN SIMILAR CHEMICAL GROUPS.

Material	CARBON STEEL or CAST IRON	CARBON STEEL HARDENED	301/302/303/304 STAINLESS STEEL	316 STAINLESS STEEL	400 SERIES STAINLESS STEEL	400 SERIES STAINLESS STEEL HARDENED	ALUMINUM	COPPER	BRASS	ACETAL (DELRIN, CELCON)	ABS (CYCOLAC)	BUNA 'N' (NITRILE RUBBER)	EPR, EPDM (NORDEL)	FLOURAZ	KALREZ	LEATHER/THIOKOL IMPREGNATED	NEOPRENE	NYLON	POLYCARBONATE (LEXAN)	POLYETHYLENE LOW DENSITY	POLYETHYLENE HIGH DENSITY	POLYPROPLENE 120° (NORYL)	POLYURETHANE, URATHANE	PVC POLYVINYL CHLORIDE	PVDF 150° (KYNAR)	TEFLON 120°	VITON
Sodium Chlorate	★	★	★	★	★	★	■	★	□	☆	□	☆	☆	□	☆	☆	☆	□	□	☆	☆	□	★	☆	☆	☆	☆
Sodium Chloride	■	■	★	★	★	★	■	□	□	☆	★	☆	☆	□	(☆)	□	☆	☆	□	☆	☆	□	★	☆	☆	☆	☆
Sodium Chloride – 2% Solution	★	★	★	☆	★	★	■	★	■	☆	□	☆	☆	□	□	☆	☆	☆	□	☆	☆	□	☆	☆	☆	☆	☆
Sodium Chloride – 5% Solution	□	□	★	☆	☆	☆	■	★	■	☆	□	☆	☆	□	□	□	☆	☆	□	☆	☆	□	☆	☆	☆	☆	☆
5% @ 150°F	□	□	★	☆	□	□	■	□	■	☆	□	☆	☆	□	□	□	☆	□	□	☆	☆	□	□	☆	☆	☆	☆
Saturated Solution	□	□	☆	☆	☆	☆	■	□	□	☆	□	☆	☆	□	□	□	☆	□	□	☆	☆	□	□	☆	☆	☆	☆
Saturated Solution (Boiling)	□	□	★	☆	□	□	■	□	□	☆	□	□	□	□	□	□	☆	□	□	□	□	□	□	□	☆	□	□
Sodium Chloride Slurry	□	□	□	□	□	□	□	□	□	□	□	□	□	□	□	□	□	□	□	□	□	□	□	□	□	□	□
Sodium Chromate	☆	☆	★	★	★	★	★	★	□	□	□	☆	☆	□	(☆)	☆	□	□	□	☆	☆	□	□	☆	☆	☆	☆
Sodium Citrate	☆	☆	★	★	□	□	■	■	□	☆	□	☆	☆	□	□	☆	□	□	□	☆	☆	□	□	☆	☆	☆	☆
Sodium Cyanide	☆	☆	☆	☆	☆	☆	■	■	■	☆	□	☆	☆	■	(☆)	☆	□	□	□	☆	☆	□	★	☆	☆	☆	☆
Sodium Dichromate	□	□	□	□	□	□	□	□	□	☆	□	☆	☆	□	□	□	☆	□	□	□	□	□	□	□	□	□	□
Sodium Ferricyanide	★	★	★	★	□	□	□	□	□	☆	□	☆	☆	□	(☆)	□	□	□	□	□	□	□	□	□	☆	□	□
Sodium Fluoride	★	★	★	★	□	□	☆	☆	☆	☆	□	☆	☆	□	(☆)	□	□	□	□	☆	☆	□	□	★	☆	☆	☆
Sodium Fluoride - 5%	★	★	★	★	□	□	☆	☆	□	☆	□	☆	☆	□	□	□	☆	□	□	☆	☆	□	□	☆	☆	☆	☆
Sodium Hydrosulfide	□	□	□	□	□	□	□	□	□	☆	□	■	☆	☆	(☆)	□	☆	□	□	☆	☆	□	□	☆	□	☆	☆
Sodium Hydroxide	★	★	□	□	□	□	□	□	■	■	□	★	☆	☆	□	■	★	□	★	☆	☆	☆	★	☆	★	☆	★
3 Molar	□	□	□	□	□	□	□	□	■	■	□	★	★	☆	□	■	★	□	★	☆	☆	☆	☆	☆	□	□	★
20% (Cold)	☆	☆	☆	☆	☆	☆	■	☆	■	■	□	★	☆	☆	□	■	☆	□	■	☆	☆	★	★	★	★	★	★
20% Solution (Hot)	★	★	☆	☆	★	★	★	★	■	■	□	☆	□	☆	□	■	☆	□	■	☆	☆	□	□	□	□	★	★
50% Solution (Cold)	■	■	☆	☆	★	★	★	★	■	■	□	★	★	☆	□	■	☆	□	■	☆	☆	☆	□	★	★	★	★
80% Solution (Hot)	■	■	■	■	■	■	■	■	■	■	□	☆	□	☆	□	■	□	□	■	□	□	□	□	□	□	□	★
70% Solution (Cold)	□	□	□	★	□	□	■	□	■	■	□	★	★	☆	□	■	☆	□	■	☆	☆	☆	□	□	□	□	★
70% Solution (Hot)	■	■	☆	☆	■	■	■	■	■	□	□	□	□	☆	□	■	□	□	■	☆	☆	☆	□	□	□	□	★
Sodium Hypochlorite	■	■	□	□	■	■	■	■	■	□	□	☆	☆	(☆)	□	□	☆	★	☆	☆	☆	■	★	☆	☆	☆	★
Sodium Hypochlorite 5%	■	■	□	□	■	■	■	■	■	☆	□	☆	☆	(☆)	□	□	☆	★	☆	☆	☆	■	★	☆	☆	☆	★
Sodium Hpochlorite 20%	■	■	■	☆	■	■	■	■	□	□	■	☆	★	☆	(☆)	□	★	☆	★	☆	☆	☆	■	★	☆	☆	☆
Sodium Hyposulfate	■	■	☆	☆	☆	☆	■	□	☆	□	□	☆	□	(☆)	□	■	☆	□	☆	☆	☆	□	□	■	☆	☆	□
Sodium Metaphosphate	★	★	★	★	★	★	★	★	■	□	□	☆	☆	(☆)	□	★	□	□	☆	☆	☆	□	□	★	☆	☆	□
Sodium Metasilicate (Cold)	☆	☆	☆	☆	☆	☆	★	★	☆	□	□	□	□	(☆)	□	□	□	□	☆	☆	☆	□	□	☆	☆	☆	□
Sodium Metasilicate (Hot)	★	★	☆	☆	☆	☆	□	★	□	□	□	□	□	■	(☆)	□	□	□	☆	☆	☆	□	□	☆	☆	☆	□
Sodium Monophosphate	□	□	☆	☆	☆	☆	★	■	□	□	□	□	□	(☆)	□	☆	□	☆	☆	☆	☆	□	□	□	☆	☆	□
Sodium Nitrate	★	★	☆	☆	☆	☆	★	★	■	☆	□	★	☆	■	(☆)	□	★	□	☆	☆	☆	□	☆	☆	☆	☆	☆
Sodium Nitrate – Fused	☆	☆	■	■	■	☆	☆	■	■	□	■	□	□	■	□	☆	■	□	☆	☆	☆	□	□	□	☆	☆	☆
Sodium Orthosilicate	□	□	□	□	□	□	□	□	□	□	□	□	□	(☆)	□	□	□	□	□	□	□	□	□	□	□	□	□
Sodium Perborate	■	■	★	★	★	★	☆	★	■	☆	□	★	☆	☆	(☆)	□	★	□	□	☆	☆	□	□	★	☆	☆	☆
Sodium Peroxide (Sodium Dioxide)	■	■	☆	☆	★	★	☆	★	■	■	□	★	☆	☆	(☆)	■	★	□	☆	☆	☆	□	■	★	☆	☆	☆
Sodium Phosphate	★	★	★	★	★	★	□	★	■	□	□	☆	☆	(☆)	□	★	□	□	☆	☆	☆	□	□	☆	□	☆	☆
Sodium Phosphate (Mono)	□	□	□	□	□	□	□	□	□	□	□	☆	☆	(☆)	□	☆	□	□	☆	☆	☆	□	□	☆	□	☆	☆
Sodium Phosphate (Dibasic)	□	□	□	□	□	□	□	□	□	□	□	☆	☆	(☆)	□	★	□	□	☆	☆	☆	□	□	☆	□	☆	☆
Sodium Phosphate (Tribasic)	★	★	★	★	★	★	□	★	□	□	□	☆	☆	(☆)	□	★	□	□	☆	☆	☆	□	□	☆	□	☆	☆
Sodium Polyphosphate	■	■	★	★	★	★	☆	☆	■	□	□	□	□	(☆)	□	□	□	☆	□	☆	☆	□	□	☆	□	☆	□
Sodium Pyrophosphate	□	□	□	□	□	□	□	□	□	□	□	☆	□	(☆)	□	□	□	☆	□	□	□	□	□	☆	□	☆	□
Sodium Resinate	★	★	★	★	□	□	□	□	□	□	□	☆	☆	(☆)	□	☆	□	☆	□	□	□	□	□	☆	□	☆	□
Sodium Salicylate	★	★	★	★	★	★	□	□	□	□	□	☆	□	(☆)	□	□	□	□	□	□	□	□	□	□	□	☆	☆
Sodium Salts	□	□	□	□	□	□	□	□	□	□	□	☆	☆	□	□	□	★	□	□	□	□	□	□	□	□	☆	□

☆ – RECOMMENDED, PERFORMANCE IS NOT AFFECTED
★ – FAIR, SOME LOSS OF PROPERTIES MAY OCCUR
■ – NOT RECOMMENDED, MATERIAL IS UNSUITABLE FOR SERVICE
□ – NO DATA AVAILABLE AT THIS TIME
( ) – BRACKETS AROUND A RATING LETTER, NO DATA IS AVAILABLE, BUT THE RATINGS ARE MADE ON THE BASIS OF EXPOSURE TESTS IN SIMILAR CHEMICAL GROUPS.

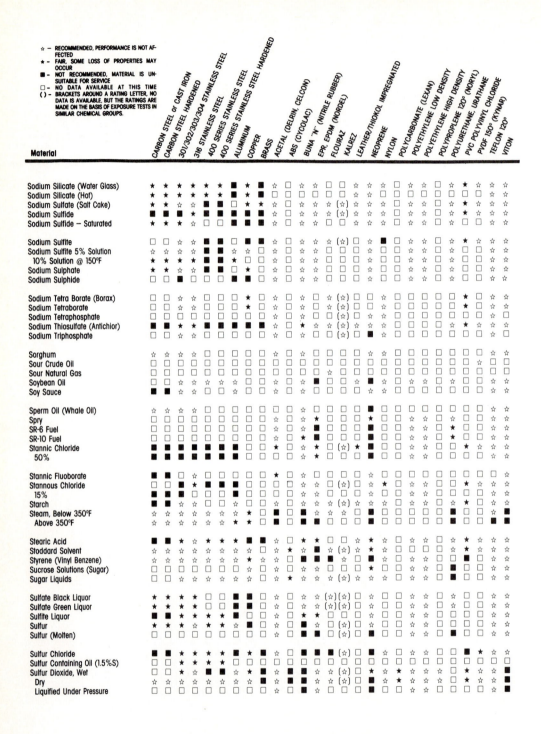

Column headings (left to right): CARBON STEEL or CAST IRON · CARBON STEEL HARDENED · 301/302/303/304 STAINLESS STEEL · 316 STAINLESS STEEL · 400 SERIES STAINLESS STEEL · 400 SERIES STAINLESS STEEL HARDENED · ALUMINUM · COPPER · BRASS · ACETAL (DELRIN, CELCON) · ABS (CYCOLAC) · BUNA "N" (NITRILE RUBBER) · EPR, EPDM (NORDEL) · FLOURAZ · KALREZ · LEATHER/THIOKOL IMPREGNATED · NEOPRENE · NYLON · POLYCARBONATE (LEXAN) · POLYETHYLENE LOW DENSITY · POLYETHYLENE HIGH DENSITY · POLYPROPYLENE 120° (NORYL) · POLYURETHANE, URATHANE · PVC POLYVINYL CHLORIDE · PVDF 150° (KYNAR) · TEFLON 120° · VITON

Material

Sodium Silicate (Water Glass)
Sodium Silicate (Hot)
Sodium Sulfate (Salt Cake)
Sodium Sulfide
Sodium Sulfide – Saturated

Sodium Sulfite
Sodium Sulfite 5% Solution
   10% Solution @ 150°F
Sodium Sulphate
Sodium Sulphide

Sodium Tetra Borate (Borax)
Sodium Tetraborate
Sodium Tetraphosphate
Sodium Thiosulfate (Antichlor)
Sodium Triphosphate

Sorghum
Sour Crude Oil
Sour Natural Gas
Soybean Oil
Soy Sauce

Sperm Oil (Whale Oil)
Spry
SR-6 Fuel
SR-10 Fuel
Stannic Chloride
   50%

Stannic Fluoborate
Stannous Chloride
   15%
Starch
Steam, Below 350°F
   Above 350°F

Stearic Acid
Stoddard Solvent
Styrene (Vinyl Benzene)
Sucrose Solutions (Sugar)
Sugar Liquids

Sulfate Black Liquor
Sulfate Green Liquor
Sulfite Liquor
Sulfur
Sulfur (Molten)

Sulfur Chloride
Sulfur Containing Oil (1.5%S)
Sulfur Dioxide, Wet
   Dry
   Liquified Under Pressure

Legend:

☆ – RECOMMENDED, PERFORMANCE IS NOT AFFECTED
★ – FAIR, SOME LOSS OF PROPERTIES MAY OCCUR
■ – NOT RECOMMENDED, MATERIAL IS UNSUITABLE FOR SERVICE
□ – NO DATA AVAILABLE AT THIS TIME
( ) – BRACKETS AROUND A RATING LETTER, NO DATA IS AVAILABLE, BUT THE RATINGS ARE MADE ON THE BASIS OF EXPOSURE TESTS IN SIMILAR CHEMICAL GROUPS.

Material	Carbon Steel or Cast Iron	Carbon Steel Hardened	301/302/303/304 Stainless Steel	316 Stainless Steel	400 Series Stainless Steel	400 Series Stainless Steel Hardened	Aluminum	Copper	Brass	Acetal (Delrin, Celcon)	ABS (Cycolac)	Buna "N" (Nitrile Rubber)	EPR, EPDM (Nordel)	Flouraz	Kalrez	Leather/Thiokol Impregnated	Neoprene	Nylon	Polycarbonate (Lexan)	Polyethylene Low Density	Polyethylene High Density	Polyproplene 120°	Polyurethane, Urathane	PVC Polyvinyl Chloride (Noryl)	PVDF 150° (Kynar)	Teflon 120°	Viton
Sulfur Hexafluoride	□	□	★	★	□	□	■	□	■	□	☆	□	★	☆	□	★	☆	□	☆	□	□	□	□	□	□	☆	■
Sulfur Trioxide	★	★	★	★	☆	□	★	□	■	□	■	★	☆	☆	□	■	☆	□	★	★	★	□	□	□	□	☆	□
Sulfur Trioxide (Dry)	☆	☆	☆	☆	□	□	☆	★	☆	■	□	□	□	□	□	☆	☆	□	☆	□	□	□	□	□	★	★	☆
Sulfuric Acid – 85%	■	■	★	☆	■	■	■	■	■	□	□	■	□	☆	□	■	☆	☆	★	★	☆	■	★	□	☆		
3 Molar	□	□	□	□	□	□	□	□	□	■	★	□	□	□	□	□	□	□	□	■	□	□	☆	☆	□		
Concentrated	□	□	☆	☆	☆	☆	□	□	□	■	★	□	□	□	□	□	□	□	□	★	□	□	☆	☆	□		
(Aerated) No Velocity	★	★	★	★	■	■	■	■	□	□	□	□	□	□	□	□	□	□	□	□	□	□	☆	☆	□		
(Air Free) No Velocity	■	■	★	★	■	■	■	□	□	□	□	□	□	□	□	□	□	□	□	□	□	□	☆	☆	□		
(Fuming) Oleum	★	★	☆	☆	☆	☆	★	■	□	□	□	□	□	□	□	□	□	□	□	□	□	□	☆	☆	□		
Sulfurous Acid	■	■	■	★	■	■	★	■	□	■	□	★	★	□	□	★	■	□	□	☆	☆	□	■	★	☆	☆	☆
Sulphuric Acid – 5%	★	★	★	☆	■	■	■	★	■	□	□	☆	□	□	□	■	■	□	□	☆	☆	□	□	☆	☆	☆	☆
5% Solution (Boiling)	■	■	■	■	■	■	■	★	■	□	□	■	□	□	□	■	■	□	□	☆	☆	□	□	☆	☆	☆	□
10% Solution	■	■	★	☆	■	■	■	■	■	□	□	★	□	□	□	■	■	□	□	☆	☆	□	□	☆	☆	☆	☆
10% Solution (Boiling)	■	■	■	■	■	■	■	■	■	□	□	■	□	□	□	■	■	□	□	☆	☆	□	□	☆	☆	☆	□
50% Solution (Boiling)	■	■	■	■	■	■	■	■	■	□	□	■	□	□	□	■	□	□	□	☆	☆	□	□	☆	☆	☆	□
Concentrated Solution	□	□	☆	☆	☆	☆	☆	□	□	■	★	□	□	□	□	□	□	□	□	☆	☆	□	■	□	☆	☆	☆
Concentrated Solution (Boiling)	■	■	■	■	■	■	■	■	■	□	□	■	□	□	□	■	□	□	□	☆	☆	□	□	☆	☆	☆	□
Concentrated Solution @ 300%	■	■	■	■	■	■	■	■	■	□	□	■	□	□	□	■	□	□	□	☆	☆	□	□	☆	☆	☆	□
Fuming	★	★	★	★	☆	☆	■	■	□	□	□	□	□	□	□	□	□	□	□	☆	☆	□	□	☆	☆	☆	□
Sulphurous Acid – Saturated	□	□	☆	☆	★	★	■	■	■	□	□	■	□	□	□	■	□	★	□	□	☆	□	□	□	☆	☆	□
Sulphurous Spray	□	■	■	■	■	■	□	□	□	□	□	□	□	□	□	□	□	□	□	□	□	□	□	□	☆	☆	□
Sunsafe (Fire Resist. Hydr. Fluid)	☆	☆	☆	☆	☆	☆	☆	□	□	□	□	□	□	□	□	★	□	□	□	□	□	□	■	□	□	☆	☆
Syrup	□	□	☆	☆	☆	☆	☆	□	□	☆	□	☆	☆	☆	(☆)	☆	☆	□	☆	☆	☆	□	□	☆	☆	☆	☆
Talc Oil	□	□	☆	☆	□	□	★	□	□	□	□	☆	□	☆	□	□	☆	□	□	□	□	□	□	□	□	☆	□
Talc Slurry	□	□	□	□	□	□	□	□	□	☆	□	□	□	□	☆	□	☆	□	□	□	□	□	□	□	☆	☆	□
Tall Oil	★	★	★	★	□	□	■	□	□	□	□	☆	□	☆	□	☆	□	□	□	□	□	□	□	□	☆	☆	☆
Tallow, Molten	□	□	☆	☆	☆	☆	☆	□	□	□	□	☆	□	☆	(☆)	□	☆	□	□	□	□	□	□	□	☆	☆	□
Tannic Acid	★	★	★	★	★	★	■	☆	★	■	□	★	★	□	□	★	★	☆	□	☆	☆	□	■	★	★	☆	☆
Tannin	□	□	□	□	□	□	□	☆	☆	□	□	☆	□	□	(☆)	□	□	□	□	□	□	□	□	□	□	☆	□
Tar and Tar Oil	☆	☆	☆	☆	★	★	☆	☆	★	☆	□	★	☆	(☆)	□	★	☆	□	★	★	□	□	□	□	□	☆	☆
Tar, Bituminous	☆	☆	☆	☆	☆	☆	☆	□	□	□	★	■	☆	□	□	★	■	□	□	□	□	□	□	□	□	☆	☆
Tartaric Acid	■	■	☆	☆	★	★	■	☆	■	☆	□	☆	★	□	□	★	★	☆	□	☆	☆	☆	■	★	★	☆	☆
Terpene Monocylic	□	□	□	□	□	□	□	□	□	☆	□	□	□	□	□	□	□	□	□	□	□	□	□	□	□	☆	□
Terpineol	☆	☆	☆	☆	☆	☆	☆	□	□	★	■	□	(☆)	□	■	□	□	□	□	□	□	□	★	□	□	☆	☆
Teriary Butyl Alcohol	□	□	□	□	□	□	□	□	□	■	★	□	☆	□	★	□	□	□	□	□	□	□	■	□	□	☆	☆
P-Tertiary Butyl Catechol	★	★	★	★	□	□	□	□	□	■	★	□	□	□	□	□	□	□	□	□	□	□	□	□	□	☆	☆
Tertiary Butyl Mercaptan	□	□	□	□	□	□	□	□	□	■	□	□	□	□	□	□	□	□	□	□	□	□	□	□	□	☆	□
Tetrabromoethane	□	□	□	□	□	□	□	□	□	■	□	□	□	□	□	□	□	□	□	□	□	□	□	□	□	☆	□
Tetrabutyl Titanate	□	□	□	□	□	□	□	□	□	★	☆	□	□	□	★	□	□	□	□	□	□	□	□	□	□	☆	☆
Tetrachlorodifluorethane	□	□	□	□	□	□	□	□	□	■	□	□	□	□	□	□	□	□	□	□	□	□	□	□	□	☆	□
Tetrachloroethane	☆	☆	☆	☆	☆	☆	■	☆	□	□	■	★	■	□	★	□	■	□	□	□	□	□	■	■	☆	☆	□
Tetrachloroethylene	☆	☆	☆	☆	☆	☆	■	☆	□	□	■	★	■	□	★	□	■	□	□	□	□	□	□	☆	☆	☆	□
Tetraethyl Lead	□	□	□	□	□	□	□	☆	☆	★	■	■	□	□	★	□	■	□	□	★	★	□	☆	□	☆	☆	□
Tetraethyl Lead "Blend"	□	□	□	□	□	□	□	□	□	★	■	□	□	□	★	□	■	□	□	□	□	□	☆	□	□	☆	☆
Tetraethylene Glycol	□	□	□	□	□	□	□	□	□	☆	□	☆	□	□	☆	□	□	□	□	□	□	□	□	□	□	☆	☆
Tetrahydrofuran	□	□	☆	☆	☆	☆	☆	□	□	■	★	■	☆	□	■	□	■	□	□	□	□	□	□	■	★	☆	★
Tetralin	☆	☆	☆	☆	☆	☆	☆	□	□	■	■	□	(☆)	□	■	□	□	□	□	□	□	□	□	□	□	☆	☆
Tetraphosphoglucosate	□	□	□	□	□	□	□	□	□	☆	□	□	(☆)	□	□	□	□	□	□	□	□	□	□	□	□	☆	□

Legend

☆ – RECOMMENDED, PERFORMANCE IS NOT AFFECTED
★ – FAIR, SOME LOSS OF PROPERTIES MAY OCCUR
■ – NOT RECOMMENDED, MATERIAL IS UNSUITABLE FOR SERVICE
□ – NO DATA AVAILABLE AT THIS TIME
( ) – BRACKETS AROUND A RATING LETTER, NO DATA IS AVAILABLE, BUT THE RATINGS ARE MADE ON THE BASIS OF EXPOSURE TESTS IN SIMILAR CHEMICAL GROUPS.

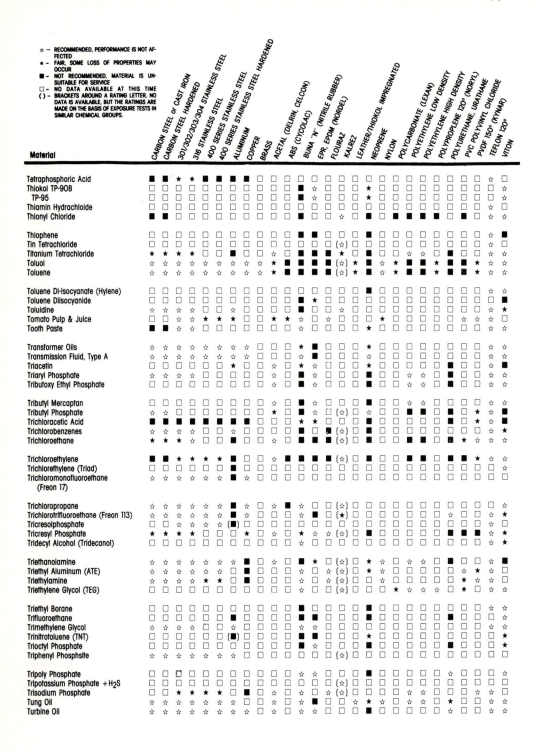

Material	Carbon Steel or Cast Iron	Carbon Steel Hardened	301/302/303/304 Stainless Steel	316 Stainless Steel	400 Series Stainless Steel	400 Series Stainless Steel Hardened	Aluminum	Copper	Brass	Acetal (Delrin, Celcon)	ABS (Cycolac)	Buna "N" (Nitrile Rubber)	EPR, EPDM (Nordel)	Flouraz	Kalrez	Leather/Thiokol Impregnated	Neoprene	Nylon	Polycarbonate (Lexan)	Polyethylene Low Density	Polyethylene High Density	Polyproplene 120° (Noryl)	Polyurethane, Urathane	PVC Polyvinyl Chloride	PVDF 150° (Kynar)	Teflon 120°	Viton
Tetraphosphoric Acid	■	■	★	★	■	■	☆	■	□	□	□	□	□	□	□	□	□	□	□	□	□	□	□	□	□	☆	□
Thiokol TP-90B	□	□	□	□	□	□	□	□	□	□	■	☆	□	□	□	□	★	□	□	□	□	□	□	□	□	☆	☆
TP-95	□	□	□	□	□	□	□	□	□	□	■	☆	□	□	□	□	★	□	□	□	□	□	□	□	□	☆	☆
Thiamin Hydrachloride	□	□	□	□	□	□	□	□	□	□	□	□	□	□	□	□	□	□	□	□	□	□	□	□	□	☆	☆
Thionyl Chloride	■	■	□	□	□	□	□	■	□	□	□	■	□	□	☆	□	■	■	■	■	■	■	■	□	☆	☆
Thiophene	□	□	□	□	□	□	□	□	□	□	■	■	☆	□	□	□	■	□	□	□	□	□	□	□	☆	■
Tin Tetrachloride	□	□	□	□	□	□	□	□	□	□	□	(☆)	□	□	□	□	■	□	□	□	□	□	□	□	☆	☆
Titanium Tetrachloride	★	★	★	★	□	□	■	□	□	☆	■	■	■	★	□	■	■	□	□	□	□	□	☆	□	☆	☆
Toluol	☆	☆	☆	☆	☆	☆	☆	☆	☆	★	■	■	■	■	(☆)	★	■	☆	★	■	■	★	■	★	☆	☆
Toluene	☆	☆	☆	☆	☆	☆	☆	☆	☆	★	■	■	■	■	(☆)	★	■	☆	★	■	■	★	■	★	☆	☆
Toluene Di-Isocyanate (Hylene)	□	□	□	□	□	□	□	□	□	□	■	★	□	□	□	□	■	□	□	□	□	□	□	□	☆	☆
Toluene Diisocyanide	□	□	□	□	□	□	□	□	□	□	■	★	□	□	□	□	■	□	□	□	□	□	□	□	□	■
Toluidine	☆	☆	☆	☆	□	□	□	□	□	□	■	□	□	□	□	□	■	□	□	□	□	□	□	□	☆	★
Tomato Pulp & Juice	□	□	☆	☆	★	★	★	□	□	☆	□	□	□	□	□	□	■	□	□	□	□	□	□	□	☆	☆
Tooth Paste	■	■	☆	☆	□	□	□	□	□	□	□	☆	□	□	□	□	★	□	□	□	□	□	□	□	☆	☆
Transformer Oils	☆	☆	☆	☆	☆	☆	☆	□	□	■	■	☆	□	□	□	□	☆	□	□	□	□	□	□	□	☆	☆
Transmission Fluid, Type A	☆	☆	☆	☆	☆	☆	☆	□	□	☆	■	□	□	□	□	□	☆	□	□	□	□	□	□	□	☆	☆
Triacetin	□	□	□	□	□	□	★	□	□	☆	■	☆	□	□	□	□	★	□	□	□	□	■	□	□	☆	■
Triaryl Phosphate	☆	☆	☆	☆	☆	☆	☆	□	□	■	☆	□	□	□	□	□	■	□	□	□	□	☆	□	☆	☆	☆
Tributoxy Ethyl Phosphate	□	□	□	□	□	□	☆	□	□	■	☆	□	□	□	□	□	■	□	□	□	□	☆	□	☆	☆	☆
Tributyl Mercaptan	□	□	□	□	□	□	□	□	□	□	☆	□	□	□	□	□	■	□	□	□	□	☆	□	□	☆	☆
Tributyl Phosphate	☆	☆	□	□	□	□	☆	□	□	★	■	☆	□	(☆)	☆	□	■	■	□	■	■	☆	□	★	☆	■
Trichloracetic Acid	■	■	■	■	■	■	□	□	□	□	★	□	★	□	□	□	■	□	□	□	□	□	□	★	☆	■
Trichlorobenzenes	☆	☆	☆	☆	□	□	□	■	□	□	■	■	□	■	(☆)	□	■	□	□	□	□	□	□	□	☆	★
Trichloroethane	★	★	★	☆	□	□	■	□	□	☆	■	■	■	■	(☆)	■	■	□	■	■	■	□	■	★	☆	☆
Trichloroethylene	■	■	★	★	★	★	■	□	□	☆	■	■	■	■	(☆)	□	■	□	■	■	■	□	■	★	☆	☆
Trichlorethylene (Triad)	□	□	□	□	□	□	□	□	□	□	□	□	□	□	□	□	□	□	□	□	□	□	□	□	□	☆
Trichloromonofluoroethane (Freon 17)	☆	☆	☆	☆	□	□	□	□	□	□	■	□	□	□	□	□	□	□	□	□	□	□	□	□	☆	☆
Trichloropropane	☆	☆	☆	☆	☆	☆	■	□	□	☆	■	■	□	(☆)	□	□	■	□	□	□	□	□	□	□	☆	☆
Trichlorotrifluoroethane (Freon 113)	☆	☆	☆	☆	☆	☆	■	□	☆	□	□	■	■	□	(★)	□	□	□	□	□	□	□	☆	□	☆	★
Tricresolphosphate	□	□	☆	☆	☆	☆	(■)	□	□	□	■	□	□	□	□	□	■	□	□	□	□	☆	□	☆	☆	☆
Tricresyl Phosphate	★	★	★	★	□	□	□	★	□	☆	■	☆	□	☆	(☆)	□	■	□	□	■	■	☆	■	■	☆	★
Tridecyl Alcohol (Tridecanol)	□	□	□	□	□	□	□	□	□	□	□	☆	□	□	□	□	□	□	□	□	□	□	□	□	☆	★
Triethanolamine	☆	☆	☆	☆	☆	☆	☆	□	□	☆	■	★	□	(☆)	□	★	□	□	☆	□	□	☆	□	☆	☆	■
Triethyl Aluminum (ATE)	☆	☆	☆	☆	☆	☆	□	■	□	☆	■	□	□	(☆)	□	★	□	□	☆	□	□	□	□	★	★	☆
Triethylamine	☆	☆	☆	☆	★	★	☆	■	□	☆	■	□	□	(☆)	□	□	□	□	☆	□	□	□	□	★	☆	■
Triethylene Glycol (TEG)	□	□	□	□	□	□	□	□	□	□	☆	□	□	(☆)	□	□	□	★	☆	☆	☆	□	★	☆	☆	☆
Triethyl Borane	□	□	□	□	□	□	□	□	□	□	■	□	□	□	□	□	■	□	□	□	□	□	□	□	☆	☆
Trifluoroethane	□	□	□	□	□	□	□	□	□	□	■	■	□	□	□	□	■	□	□	□	□	□	□	□	☆	☆
Trimethylene Glycol	□	□	□	☆	□	□	☆	□	□	□	■	☆	□	□	□	□	☆	□	□	□	□	□	□	□	☆	☆
Trinitrotoluene (TNT)	□	□	□	□	□	□	(■)	□	□	□	■	■	□	□	□	□	★	□	□	□	□	□	□	□	☆	★
Trioctyl Phosphate	□	□	□	□	□	□	□	□	□	□	■	☆	□	□	□	□	■	□	□	□	□	■	□	☆	☆	★
Triphenyl Phosphsite	☆	☆	☆	☆	☆	☆	□	□	□	□	□	□	□	(☆)	□	□	□	□	□	□	□	□	□	□	☆	☆
Tripoly Phosphate	□	□	□	□	□	□	□	□	□	☆	☆	□	□	□	□	□	■	□	□	□	□	□	□	☆	□	☆	□
Tripotassium Phosphate +H₂S	□	□	□	□	□	□	□	□	□	□	☆	□	□	(☆)	□	□	□	□	□	□	□	□	□	□	□	☆	□
Trisodium Phosphate	□	□	★	★	★	★	□	□	□	☆	☆	□	□	(☆)	□	□	□	□	□	□	□	□	□	□	□	☆	□
Tung Oil	☆	☆	☆	☆	☆	☆	□	□	□	☆	■	☆	□	□	☆	★	☆	□	□	□	□	□	★	□	☆	☆	☆
Turbine Oil	☆	☆	☆	☆	☆	☆	☆	□	□	☆	■	☆	□	□	□	■	☆	□	□	□	□	□	□	□	☆	☆	☆

Legend:

☆ – RECOMMENDED, PERFORMANCE IS NOT AFFECTED
★ – FAIR, SOME LOSS OF PROPERTIES MAY OCCUR
■ – NOT RECOMMENDED, MATERIAL IS UNSUITABLE FOR SERVICE
□ – NO DATA AVAILABLE AT THIS TIME
( ) – BRACKETS AROUND A RATING LETTER, NO DATA IS AVAILABLE, BUT THE RATINGS ARE MADE ON THE BASIS OF EXPOSURE TESTS IN SIMILAR CHEMICAL GROUPS.

Material	Carbon Steel or Cast Iron	Carbon Steel Hardened	301/302/303/304 Stainless Steel	316 Stainless Steel	400 Series Stainless Steel	400 Series Stainless Steel Hardened	Aluminum	Copper	Brass	Acetal (Delrin, Celcon)	ABS (Cycolac)	Buna "N" (Nitrile Rubber)	EPR, EPDM (Nordel)	Flouraz	Kalrez	Leather/Thiokol Impregnated	Neoprene	Nylon	Polycarbonate (Lexan)	Polyethylene Low Density	Polyethylene High Density	Polyproplene 120° (Noryl)	Polyurethane, Urathane	PVC Polyvinyl Chloride	PVDF 150° (Kynar)	Teflon 120°	Viton
Turbine Oil #15 (Mil-L-7808A)	☆	☆	☆	☆	☆	☆	☆	□	□	□	☆	☆	□	□	□	■	□	□	□	□	☆	□	□	☆	☆		
Turbo Oil #35	☆	☆	☆	☆	☆	☆	☆	□	□	□	★	★	□	□	□	■	□	□	□	□	■	□	□	☆	☆		
Turpentine	■	■	☆	☆	★	★	☆	☆	☆	■	★	□	■	☆	(☆)	★	■	★	■	■	■	■	■	☆	☆		
Type 1 Fuel (Mil-S-3136) (ASTM Ref. Fuel A)	☆	☆	☆	☆	☆	☆	☆	□	□	□	☆	☆	□	□	□	★	□	□	□	□	☆	□	□	☆	☆		
Type 11 Fuel (Mil-S-3136)	☆	☆	☆	☆	☆	☆	☆	□	□	□	★	■	□	□	□	■	□	□	□	□	★	□	□	☆	☆		
Type 111 (Fuel Mil-S-3136) (ASTM Ref. Fuel B)	☆	☆	☆	☆	☆	☆	☆	□	□	□	☆	■	□	□	□	■	□	□	□	□	★	□	□	□	☆		
Univis 40 (Hydr. Fluid)	☆	☆	☆	☆	☆	☆	☆	☆	★	□	☆	■	□	□	□	★	□	□	□	□	☆	□	□	☆	☆		
Univolt #35 (Mineral Oil)	□	□	☆	☆	☆	☆	☆	☆	☆	□	☆	☆	□	□	□	★	□	□	□	□	☆	□	□	☆	☆		
Unsymmetrical Dimethyl Hydrazine (UDMH)	☆	☆	☆	☆	□	□	☆	☆	☆	□	★	☆	□	□	□	★	□	□	□	□	☆	□	□	□	□		■
Urea (Carbamide)	□	□	★	★	□	□	★	□	□	□	★	□	■	□	□	★	□	■	□	☆	☆	☆	■	□	☆	☆	
Uric Acid	■	■	★	★	★	★	■	□	□	□	□	□	□	□	□	□	■	□	□	□	□	□	□	□	□		☆
Urine	☆	☆	☆	☆	☆	☆	☆	□	□	□	☆	☆	□	□	■	□	□	★	□	□	□	□	□	□	□		☆
Valeric Acid	□	□	□	□	□	□	□	□	□	□	■	☆	□	□	□	□	□	□	□	□	□	□	□	☆	□		□
Vanadium Ashes	□	□	☆	☆	☆	☆	□	□	□	□	□	☆	□	□	□	□	□	□	□	□	□	□	□	□	□		□
Vanadium Pentoxide	□	□	☆	☆	☆	☆	□	□	□	□	□	☆	□	□	□	□	□	□	□	□	□	□	□	□	□		□
Vanilla Extract	□	□	☆	☆	□	□	□	□	□	□	□	☆	□	□	□	■	□	□	□	□	□	□	□	□	☆		■
Varnish	★	★	☆	☆	☆	☆	☆	☆	☆	□	★	■	■	(☆)	★	□	☆	□	☆	☆	☆	■	□	☆	☆		
Vegetable Juices	■	■	☆	☆	□	□	■	□	□	□	☆	□	☆	□	□	★	□	★	□	□	□	□	■	□	☆	☆	
Vegetable Oil	★	★	☆	☆	☆	☆	☆	★	★	☆	□	☆	■	□	□	☆	■	☆	□	■	■	□	□	☆	☆		
Vegetable Oil (Hot)	★	★	★	★	☆	☆	☆	★	★	□	□	□	□	□	□	☆	□	☆	□	□	□	□	□	☆	☆		
Versilube	☆	☆	☆	☆	☆	☆	☆	☆	□	□	☆	□	□	□	□	☆	□	☆	□	□	□	□	☆	☆	☆		
Versilube F-50	☆	☆	☆	☆	☆	☆	☆	☆	□	□	☆	□	□	□	□	★	□	☆	□	□	□	☆	□	☆	☆		
Vinegar	■	■	☆	☆	★	★	■	☆	★	■	★	★	★	☆	□	■	★	★	□	☆	☆	☆	■	★	★	☆	★
Vinyl Acetate	★	★	★	★	☆	☆	□	□	□	□	□	■	☆	□	□	□	★	□	□	☆	☆	□	□	☆	□		☆
Vinyl Chloride (Chloroethylene)	★	★	★	★	☆	☆	★	☆	□	□	■	★	☆	☆	□	□	☆	□	□	☆	☆	□	□	■	□		☆
VV-H-910	□	□	□	□	□	□	□	□	□	□	□	□	■	☆	□	□	★	□	□	□	□	□	□	□	□		☆
Walnut Oil	□	□	□	□	□	□	□	□	□	□	□	☆	□	□	□	□	★	□	□	□	□	□	□	□	☆		☆
Water	■	■	☆	☆	☆	☆	☆	☆	☆	□	☆	□	☆	☆	☆	(☆)	□	☆	□	☆	☆	☆	□	☆	★	☆	☆
Water, Acid Mine	■	■	★	★	■	■	■	■	■	☆	★	■	☆	☆	(☆)	□	★	☆	□	☆	☆	☆	■	★	☆	☆	☆
Water, Fresh	■	■	☆	☆	☆	☆	☆	☆	★	□	☆	☆	☆	☆	(☆)	□	☆	☆	□	☆	☆	☆	□	☆	☆	☆	☆
Water, Distilled	■	■	☆	☆	☆	☆	☆	★	☆	☆	☆	☆	☆	☆	(☆)	□	☆	☆	□	☆	☆	☆	□	★	☆	☆	☆
Water, Salt	■	■	★	★	■	■	■	★	☆	☆	☆	☆	☆	☆	☆	★	☆	☆	□	☆	☆	☆	■	★	☆	☆	☆
Water — Brine, Process, Beverage	■	■	★	★	■	■	□	★	□	☆	☆	☆	☆	☆	(☆)	□	☆	□	□	☆	☆	☆	■	□	☆	☆	☆
Waxes	□	□	☆	☆	☆	☆	☆	☆	☆	□	□	☆	□	□	□	□	☆	☆	□	☆	☆	□	□	☆	☆		☆
Weed Killers	□	□	☆	☆	☆	☆	■	□	□	□	□	☆	□	□	□	□	★	□	□	□	□	□	□	☆	□		☆
Wemco C	□	□	□	□	□	□	□	□	□	□	□	☆	■	□	□	□	★	□	□	□	□	□	□	☆	□		☆
Whey	□	□	☆	☆	□	□	★	□	□	□	□	☆	□	□	□	□	□	□	□	□	□	□	□	□	☆		☆
Whiskey and Wines	■	■	☆	☆	■	■	★	★	★	☆	□	☆	☆	☆	□	□	☆	☆	□	☆	☆	☆	■	□	☆	☆	☆
White Pine Oil	□	□	□	□	□	□	□	□	□	□	★	■	□	□	□	□	■	□	□	☆	☆	□	□	☆	□		☆
White Oil	□	□	☆	☆	□	□	□	□	□	□	□	☆	□	□	□	■	□	□	★	□	□	□	□	☆	☆		★
White Sulfate Liquor	★	★	☆	☆	□	□	★	□	□	□	□	★	☆	□	(☆)	□	☆	□	□	☆	☆	□	□	□	□		★

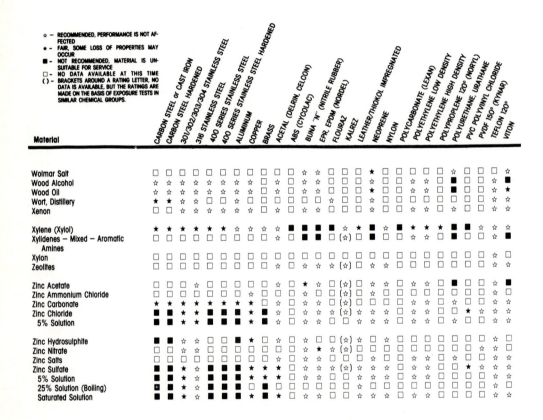

☆ – RECOMMENDED, PERFORMANCE IS NOT AFFECTED
★ – FAIR, SOME LOSS OF PROPERTIES MAY OCCUR
■ – NOT RECOMMENDED, MATERIAL IS UNSUITABLE FOR SERVICE
□ – NO DATA AVAILABLE AT THIS TIME
() – BRACKETS AROUND A RATING LETTER, NO DATA IS AVAILABLE, BUT THE RATINGS ARE MADE ON THE BASIS OF EXPOSURE TESTS IN SIMILAR CHEMICAL GROUPS.

Material	CARBON STEEL or CAST IRON	CARBON STEEL HARDENED	301/302/303/304 STAINLESS STEEL	316 STAINLESS STEEL	400 SERIES STAINLESS STEEL	400 SERIES STAINLESS STEEL HARDENED	ALUMINUM	COPPER	BRASS	ACETAL (DELRIN, CELCON)	ABS (CYCOLAC)	BUNA "N" (NITRILE RUBBER)	EPR, EPDM (NORDEL)	FLOURAZ	KALREZ	LEATHER/THIOKOL IMPREGNATED	NEOPRENE	NYLON	POLYCARBONATE (LEXAN)	POLYETHYLENE LOW DENSITY	POLYETHYLENE HIGH DENSITY	POLYPROPLENE 120°	POLYURETHANE, URATHANE	PVC POLYVINYL CHLORIDE	PVDF 150° (KYNAR)	TEFLON 120°	VITON
Wolmar Salt	□	□	□	□	□	□	□	□	□	□	☆	☆	□	□	□	★	□	□	□	□	□	☆	□	□	□	☆	
Wood Alcohol	☆	☆	☆	☆	☆	☆	☆	☆	☆	☆	☆	☆	□	□	□	☆	□	☆	■	□	□	☆	■	
Wood Oil	☆	☆	☆	☆	□	□	□	☆	□	☆	☆	☆	□	□	□	★	□	□	☆	□	■	□	☆	★	
Wort, Distillery	★	★	☆	☆	□	□	□	☆	□	□	□	□	□	☆	□	□	□	□	□	☆	□	☆	☆	
Xenon	□	□	☆	☆	☆	☆	☆	□	☆	□	☆	□	□	☆	□	□	□	□	☆	□	□	☆	☆	
Xylene (Xylol)	★	★	★	★	★	★	☆	☆	☆	■	■	■	☆	★	■	☆	■	★	★	★	■	■	☆	☆	☆
Xylidenes – Mixed – Aromatic Amines	□	□	□	□	□	□	□	□	□	☆	□	■	■	□	(☆)	■	■	□	☆	☆	■	□	□	☆	■
Xylon	□	□	□	□	□	□	□	□	□	□	□	□	☆	□	□	□	□	□	□	□	□	□	☆	□
Zeolites	□	□	□	□	□	□	□	□	□	☆	□	☆	☆	☆	(☆)	□	☆	☆	□	□	□	□	☆	☆
Zinc Acetate	□	□	☆	☆	□	□	□	□	☆	★	☆	□	(☆)	□	☆	☆	☆	☆	□	■	□	□	☆	■
Zinc Ammonium Chloride	□	□	□	□	□	□	☆	□	□	□	☆	□	(☆)	□	☆	□	☆	☆	□	□	□	□	☆	□
Zinc Carbonate	★	★	★	★	★	★	★	★	□	☆	☆	☆	(☆)	□	☆	☆	☆	☆	□	□	□	□	☆	☆
Zinc Chloride	■	■	★	★	■	■	■	★	■	☆	□	☆	☆	(☆)	☆	☆	☆	☆	□	□	★	☆	☆	☆
5% Solution	■	■	★	★	■	■	■	★	■	☆	□	☆	☆	□	☆	☆	☆	□	□	□	□	☆	☆	☆
Zinc Hydrosulphite	■	■	☆	☆	□	■	■	★	□	□	☆	□	(☆)	☆	□	□	□	□	□	□	□	□	☆	□
Zinc Nitrate	□	□	☆	☆	□	□	□	□	□	☆	★	☆	(☆)	□	□	□	□	□	□	□	□	□	☆	□
Zinc Salts	□	□	□	□	□	□	□	□	□	☆	☆	☆	☆	□	☆	☆	☆	☆	□	□	□	□	☆	☆
Zinc Sulfate	■	■	★	☆	■	■	■	★	★	★	□	☆	☆	(☆)	☆	☆	☆	☆	□	□	★	☆	☆	☆
5% Solution	■	■	★	☆	■	■	■	★	★	☆	□	☆	☆	□	☆	☆	☆	☆	□	□	□	□	☆	☆
25% Solution (Boiling)	c	■	★	☆	■	■	□	■	□	□	□	□	□	☆	□	□	□	□	□	□	□	□	☆	□
Saturated Solution	■	■	★	☆	■	■	■	★	■	★	□	☆	☆	□	☆	☆	☆	☆	□	□	□	□	☆	☆

REFERENCE CHART
(Material Service Temp.)

MATERIAL NAME	MAXIMUM SERVICE TEMPERATURE (1°F) (at zero stress)
ACETAL (DELRIN)	195°
ACETAL (CELCON)	212°
CHLOROPRENE (NEOPRENE)	225°
ETHYLENE-PROPYLENE (EPDM) (EPM)	325°
FLUOROCARBON (TEFLON)	550°
FLUOROELASTOMER (VITON)	450°
KYNAR	302°
NATURAL RUBBER	250°
NITRITE (BUNA N)	200°
NYLON	300°
PHENALIC	550°
POLYCARBONATE (ABS)	220°
POLYETHYLENE	250°
POLYISOBUTYLENE (BUTYL)	200°
POLYPROPYLENE	250°/300°
POLYSILOXANE (SILICONE)	600°
POLYSULFIDE (THIOKOL)	250°
POLYURETHANE	190°/210°
POLYVINYL CHLORIDE (PVC)	175°
SYNTHETIC RUBBER (ISOPRENE)	180°

CHEMICAL	FORMULA	PLASTIC			METAL				ELASTOMER				
		POLYPROPYLENE	RYTON®	TEFLON®	ALUMINUM	CAST IRON	HASTELLOY-C®	STAINLESS STEEL	BUNA-N	E.P.D.M.	NEOPRENE	TEFLON®	VITON®
Acetaldehyde (Ethanal)	CH_3CHO	C	A	A	A	B	A	A	X	A	X	A	X
Acetamide (Acetic Acid Amide)	CH_3CONH_2	A	A	A	A	X	A	X	B	A	B	A	B
Acetate Solvents	CH_3COOR	X	A	A	A	·	·	A	X	·	X	A	X
Acetic Acid — 20%		C	A	A	B	·	A	A	C	A	B	A	C
Acetic Acid — 30%		C		A	B	X	A	A	C	A	B	A	X
Acetic Acid — 50%	CH_3COOH	C		A	B	X	A	A	C	A	C	A	C
Acetic Acid — Glacial	CH_3COOH	B	A	A	B	X	A	A	C	B	X	A	X
Acetic Anhydride (Acetic Oxide)	$(CH_3CO)_2O$	X	A	A	B	B	A	A	C	B	B	A	X
Acetone (Dimethylketone)	CH_3COCH_3	X	A	A	B	A	A	A	X	A	X	A	X
Acetone Cyanohydrin	$(CH_3)_2C(OH)CN$			A	A	B	B	B	X	X	B	A	X
Acetonitrile (Methyl Cyanide)	CH_3CN	B	A	A	A	A	A	A	C	A	A	A	X
Acetophenone (Phenyl Methyl Ketone)	$C_6H_5COCH_3$	A	A	A	B	A	B	A	X	A	X	A	X
Acetyl Acetone (2,4-Pentanedione)	$CH_3COCH_2COCH_3$			A	B	X	B	B	X	A	X	A	X
Acetyl Chloride	CH_3COCl	X	A	A	X	A	A	B	X	C	X	A	B
Acetylene (Ethyne)	$HC{\equiv}CH$	X	A	A	A	A	A	A	A	A	C	A	A
Acetyl Salicylic Acid (Aspirin)	$(CH_3OCO)C_6H_4COOH$			A	A	X	B	B	·	B	X	A	·
Acetylene Tetrabromide (Tetra Bromoethane)	$(CHBr_2)_2$			A	X	X	·	A	X	·	X	A	A
Acrolein (Acrylaldehyde)	$H_2C{=}CHCHO$			A	A	B	B	B	B	·	·	A	A
Acrylonitrile (Vinyl Cyanide)	$CH_2{=}CHCN$	B		A	A	A	A	A	X	X	X	A	X
Adipic Acid (1,4-Butanedicarboxylic Acid)	$HOOC(CH_2)_4COOH$	A		A	B	B	A	B	B	·	X	A	A
Allyl Alcohol (2-Propen-1-ol)	CH_2CHCH_2OH	·		A	B	A	A	A	A	A	A	A	B
Alcohols	R-OH	A	A	A	—	—	A	—	—	—	—	—	—
Amyl (1-Pentanol)	$C_4H_9CH_2OH$	B		A	B	—	A	A	B	—	B	A	B
Benzyl (Phenylcarbinol)	$C_6H_5CH_2OH$	A		A	B	—	A	A	X	—	B	A	A
Butyl (Butanol)	$C_3H_7CH_2OH$	B	A	A	B	—	A	A	A	—	A	A	A
Diacetone (Tyranton)	$(CH_3)_2C(OH)CH_2COCH_3$	X		A	A	A	A	A	X	B	X	A	X
Ethyl (Ethanol)	CH_3CH_2OH	A	A	A	B	A	A	A	A	—	A	A	A
Hexyl (1-Hexanol)	$C_5H_{11}CH_2OH$	A		A	A	—	A	A	A	—	B	A	A
Isobutyl (2-Methyl-1-Propanol)	$C_3H_7CH_2OH$	·		A	B	—	A	A	C	—	A	A	A
Isopropyl (2-Propanol)	$H_3CCH(OH)CH_3$	A		A	B	C	A	A	C	—	B	A	A
Methyl (Methanol)	CH_3OH	A		A	B	A	A	A	A	—	A	A	X
Octyl (Caprylic Alcohol)	$C_7H_{15}{\bullet}CH_2OH$			A	A	—	A	A	B	—	B	A	A
Propyl (Propanol)	$C_2H_6CH_2OH$	A		A	A	—	A	A	A	—	A	A	A
Allyl Bromide (3-Bromopropene)	$H_2C{=}CHCH_2Br$			A	X	A	·	·	X	·	X	A	B
Allyl Chloride (3-Chloropropene)	$CH_2{=}CHCH_2Cl$	A		A	X	C	·	B	X	X	X	A	B

RATING KEY: (A) EXCELLENT   (B) GOOD   (C) FAIR TO POOR   (X) NOT RECOMMENDED
(*) OR (—) NO DATA AVAILABLE

CHEMICAL	FORMULA	POLYPROPYLENE	RYTON®	TEFLON®	ALUMINUM	CAST IRON	HASTELLOY-C®	STAINLESS STEEL	BUNA-N	E.P.D.M.	NEOPRENE	TEFLON®	VITON®
Alkazene® (Chlorethyl or Polyisopropyl benzenes)				A	—	—	—	—	X	—	X	A	A
Almond Oil (Artificial)				A	•	•	•	•	X	B	X	A	X
Alum (Aluminum Potassium Sulfate Dodecahydrate)	KAl(SO₄)₂•12H₂O	A		A	•	•	B	B	A	A	A	A	X
Aluminum Acetate (Burow's Solution)				A	B	C	A	A	C	A	C	A	X
Aluminum Ammonium Sulfate (Alum)	AlNH₄(SO₄)₂			A	•	•	•	•	B	•	B	A	A
Aluminum Bromide	AlBr₃			A	•	•	•	•	A	•	A	A	•
Aluminum Chloride	AlCl₃	A	A	A	X	C	A	B	A	A	A	A	A
Aluminum Fluoride	AlF₃	A	A	A	A	C	A	C	A	B	A	A	A
Aluminum Hydroxide (Alumina Trihydrate)	Al(OH)₃	A		A	B	B	B	B	B	A	A	A	C
Aluminum Nitrate	Al(NO₃)₃•9H₂O	A		A	X	•	B	A	A	A	A	A	A
Aluminum Phosphate	AlPO₄			A	•	•	•	•	A	A	A	A	A
Aluminum Potassium Sulfate (Potash Alum)	KAl(SO₄)₂	A		A	A	X	B	A	A	A	A	A	A
Aluminum Sodium Sulfate (Soda Alum)	NaAl(SO₄)₂			A	•	•	•	•	A	A	A	A	A
Aluminum Sulfate (Cake Alum)	Al₂(SO₄)₃	A	A	A	B	X	A	A	A	A	A	A	A
Amines	R-NH₂	B	A	A	A	—	—	A	X	—	B	—	X
Ammonia Anhydrous, Liquid	NH₃	A	A	A	A	A	A	A	B	A	B	A	X
Ammonia Gas — Cold				A	—	—	—	—	A	—	A	A	A
Ammonia Gas — Hot				A	—	—	—	—	C	—	B	A	X
Ammonia Liquors				A	A	A	•	A	•	•	A	A	X
				A									
Ammonium Cupric Sulfate	(NH₄)₂Cu(SO₄)₂			A	•	•	•	•	A	•	•	A	A
Ammonium Acetate	CH₃CO₂NH₄			A	A	B	•	A	•	•	A	A	A
Ammonium Bicarbonate	NH₄HCO₃			A	B	B	•	B	A	A	A	A	A
Ammonium Bifluoride — 10%	NH₄HF₂	A		A	C	X	B	B	B	•	X	A	•
Ammonium Carbonate	(NH₄)₂CO₃	A	A	A	B	B	B	B	X	A	B	A	A
Ammonium Casenite				A	—	—	—	A	—	—	A	—	—
Ammonium Chloride (Sal Ammoniac)	NH₄Cl	A	A	A	X	X	A	B	A	A	A	A	A
Ammonium Dichromate	(NH₄)₂Cr₂O₇			A	A	A	•	A	A	A	A	A	•
Ammonium Fluoride	NH₄F	B		A	B	B	A	B	B	•	B	A	A
Ammonium Hydroxide (Aqua Ammonia)	NH₄OH	A	A	A	A	B	A	A	B	A	B	A	B
Ammonium Metaphosphate				A	B	B	A	B	A	A	A	A	A
Ammonium Nitrate	NH₄NO₃	A	A	A	B	C	B	A	A	A	A	A	A
Ammonium Nitrite	NH₄NO₂	A		A	—	—	—	—	A	—	A	A	—

RATING KEY: (A) EXCELLENT   (B) GOOD   (C) FAIR TO POOR   (X) NOT RECOMMENDED
(*) OR (—) NO DATA AVAILABLE

CHEMICAL	FORMULA	PLASTIC			METAL				ELASTOMER				
		POLYPROPYLENE	RYTON®	TEFLON®	ALUMINUM	CAST IRON	HASTELLOY-C®	STAINLESS STEEL	BUNA-N	E.P.D.M.	NEOPRENE	TEFLON®	VITON®
Ammonium Oxalate	$(NH_4OOC)_2$			A	—	—	A	A	A	—	A	—	—
Ammonium Persulfate	$(NH_4)_2S_2O_8$	A		A	C	X	*	A	C	B	A	A	A
Ammonium Phosphate, Monobasic	$(NH_4)H_2PO_4$	A		A	X	X	A	B	A	A	A	A	A
Ammonium Phosphate, Di-Basic	$(NH_4)_2HPO_4$	A	A	A	B	*	A	A	A	*	A	A	A
Ammonium Phosphate, Tri-Basic	$(NH_4)_3PO_4 \bullet 3H_2O$	A		A	X	*	B	B	A	*	A	A	A
Ammonium Sulfate	$(NH_4)_2SO_4$	A	A	A	X	B	B	A	A	A	A	A	A
Ammonium Sulfide	$(NH_4)_2S$			A	B	*	A	B	A	*	A	A	A
Ammonium Sulfite	$(NH_4)_2SO_3 \bullet H_2O$	A		A	C	X	A	B	A	*	*	A	A
Ammonium Thiocyanate	NH_4SCN			A	C	C	A	A	A	A	A	A	A
Ammonium Thiosulfate	$(NH_4)_2S_2O_3$			A	A	X	*	A	A	A	A	A	A
Amyl Acetate (Banana Oil)	$CH_3CO_2C_5H_{11}$	X	A	A	A	B	B	A	X	A	X	A	X
Amyl Alcohol (Pentyl Alcohol)	$CH_3(CH_2)_4OH$	A		A	A	A	B	A	B	A	A	A	A
n-Amyl Amine (1-Aminopentane)	$CH_3(CH_2)_4NH_2$			A	*	*	*	*	C	X	X	A	X
Amyl Borate	$C_5H_{11}BO_3$			A	—	—	—	—	A	—	B	A	A
Amyl Chloride (Chloropentane)	$CH_3(CH_2)_4Cl$	X		A	X	A	B	A	X	X	X	A	A
Amyl Chloronaphthalene				A	—	—	—	—	B	—	X	A	A
Amyl Naphthalene	$C_{15}H_{18}$			A	—	—	—	—	X	X	X	A	A
Amyl Phenol	$C_6H_4(OH)C_5H_{11}$			A	A	A	A	A	X	*	*	A	A
Aniline (Aniline Oil) (Amino Benzene)	$C_6H_5NH_2$	A	A	A	B	A	B	A	X	C	X	A	B
Aniline Dyes				A	B	C	*	B	C	C	C	A	B
Aniline Hydrochloride	$C_6H_5NH_2 \bullet HCl$	X		A	X	X	—	X	C	—	X	A	B
Animal Fats & Oils				A	A	X	A	A	A	B	C	A	A
Animal Gelatin				A	*	*	*	A	A	A	A	A	A
Anisole (Methylphenyl Ether)	$C_6H_5OCH_3$			A	B	B	B	B	*	*	X	A	X
Ansul Ether				A	—	—	—	—	C	—	X	A	X
Anthraquinone	$C_{14}H_8O_2$			A	B	B	A	B	*	*	*	A	*
Anti-Freeze (Alcohol Base)				A	A	A	A	A	A	A	A	A	A
Anti-Freeze (Glycol Base) (Prestone® Etc.)				A	A	A	A	A	A	A	B	A	A
Antimony Pentachloride	$SbCl_5$			A	A	A	A	A	X	*	*	A	*
Antimony Trichloride	$SbCl_3$	A		A	B	A	B	A	B	A	*	A	A
Aqua Regia (Nitric & Hydrochloric Acid)		B	X	A	X	X	C	X	X	X	X	A	B
Aroclor®	PCB mixtures	X		A	A	B	A	A	C	X	X	A	A
Aromatic Hydrocarbons	C_6H_5R			A	A	A	—	A	X	—	X	A	A
Aromatic Solvents (Benzene Etc.)				A	A	B	B	A	C	X	X	A	B
Arsenic Acid	AsH_3O_4	A	A	A	A	X	B	B	B	A	A	A	A
Arsenic Trichloride (Arsenic Butter)	$AsCl_3$			A	B	B	B	X	C	X	A	A	X

RATING KEY:　(A) EXCELLENT　(B) GOOD　(C) FAIR TO POOR　(X) NOT RECOMMENDED
(*) OR (—) NO DATA AVAILABLE

CHEMICAL	FORMULA	PLASTIC			METAL				ELASTOMER				
		POLYPROPYLENE	RYTON®	TEFLON®	ALUMINUM	CAST IRON	HASTELLOY-C®	STAINLESS STEEL	BUNA-N	E.P.D.M.	NEOPRENE	TEFLON®	VITON®
Ascorbic Acid	C₆H₈O₆			A	A	X	*	A	*	*	*	A	A
Askarel® (Pyranol®)	PCB mixtures			A	*	*	*	A	B	X	X	A	C
Asphalt	Hydrocarbons	A		A	A	B	*	A	B	X	C	A	A
Asphalt Topping	Hydrocarbons			A	*	A	*	A	C	*	A	A	C
ASTM — Ref Motor Fuel A (Aliphatic)	Hydrocarbons			A	A	A	A	A	A	X	B	A	A
ASTM — Ref Motor Fuel B(30% Aromatic)	Hydrocarbons			A	A	A	A	A	A	X	X	A	A
ASTM — Ref Motor Fuel C (50% Aromatic)	Hydrocarbons			A	A	A	A	A	B	X	X	A	A
ASTM — Ref #1 Oil (High Aniline)	Hydrocarbons			A	A	A	A	A	A	X	B	A	A
ASTM — Ref #2 Oil (Medium Aniline)	Hydrocarbons			A	A	A	A	A	A	X	B	A	A
ASTM — Ref #3 Oil (Low Aniline)	Hydrocarbons			A	A	A	A	A	A	X	C	A	A
ASTM — Ref #4 Oil (High Aniline)	Hydrocarbons			A	A	A	A	A	B	X	X	A	A
Aviation Gasoline	Hydrocarbons			A	A	A	A	A	A	X	C	A	A
Barbeque Sauce	Water, oils, spices			A	*	X	*	A	A	*	A	A	*
Barium Carbonate	BaCO₃	A	A	A	X	B	B	B	A	A	A	A	A
Barium Chloride Dihydrate	BaCl₂ • 2H₂O	A	A	A	B	B	B	B	A	A	A	A	A
Barium Cyanide	Ba(CN)₂	X		A	—	—	—	A	C	—	A	—	A
Barium Hydroxide (Barium Hydrate)	Ba(OH)₂	A	A	A	X	B	B	A	A	A	A	A	A
Barium Nitrate	Ba(NO₃)₂	A		A	B	A	A	A	A	*	A	A	*
Barium Sulfate (Blanc Fixe)	BaSO₄	A	A	A	B	B	*	B	A	A	A	A	A
Barium Sulfide	BaS	A	A	A	X	*	A	B	A	A	A	A	A
Beef Extract				A	*	X	*	A	A	*	A	A	A
Beer	Water, carbonate	A	A	A	A	X	A	A	C	A	A	A	A
Beet Sugar Liquors (Sucrose)		A		A	A	B	*	A	A	A	A	A	A
Benzaldehyde	C₆H₅CHO	X	A	A	A	A	A	A	X	B	X	A	X
Benzene (Benzol)	C₆H₆	X	A	A	B	B	B	A	X	X	X	A	B
Benzene Sulfonic Acid	C₆H₅SO₃H	B	A	A	C	A	A	A	C	C	A	A	A
Benzoic Acid (Benzene Carboxylic Acid)	C₆H₅COOH	X	A	A	B	X	A	B	X	B	B	A	A
Benzoyl Chloride	C₆H₅COCl			A	X	A	B	B	X	X	X	A	B
Benzyl Acetate	CH₃CO₂CH₂C₆H₅			A	A	A	B	A	X	*	*	A	X
Benzyl Alcohol	C₆H₅CH₂OH	A	A	A	A	A	B	A	X	C	C	A	A
Benzyl Benzoate	C₆H₅CO₂CH₂C₆H₅			A	A	B	B	B	X	B	X	A	A
Benzyl Chloride (Chlorotoluene)	C₆H₅CH₂Cl	X	A	A	X	A	A	B	X	X	X	A	A
Benzyl Dichloride (Benzal Chloride)	C₆H₅CHCl₂			A	X	B	B	A	X	*	*	A	*

RATING KEY: (A) EXCELLENT (B) GOOD (C) FAIR TO POOR (X) NOT RECOMMENDED
(*) OR (—) NO DATA AVAILABLE

| | PLASTIC | | | METAL | | | | ELASTOMER | | | | |
CHEMICAL	POLYPROPYLENE	RYTON®	TEFLON®	ALUMINUM	CAST IRON	HASTELLOY-C®	STAINLESS STEEL	BUNA-N	E.P.D.M.	NEOPRENE	TEFLON®	VITON®
Benzol (Benzene) C_6H_6	X	A	A	B	B	B	A	X	X	X	A	B
Biphenyl (Diphenyl) $C_6H_5C_6H_5$			A	A	A	*	*	X	X	X	A	A
Bismuth Subcarbonate (Bismuth Carbonate) $(BiO)_2CO_3$			A	*	*	*	B	A	A	A	A	A
Black Sulfate Liquor			A	C	B	B	A	B	A	A	A	A
Blast Furnace Gas CO,H_2,CH_4,CO_2,N_2			A	—	—	—	—	C	—	A	A	A
Bleach Solutions Water, chlorine, oxygen	B		A	X	*	A	B	X	A	X	A	B
Borax (Sodium Borate) $B_4Na_2O_7$	A	A	A	B	B	A	A	B	A	A	A	A
Bordeaux Mixture Copper sulfate salts			A	*	*	A	A	A	A	A	A	B
Boric Acid (Boracic Acid) H_3BO_3	A	A	A	A	X	A	A	A	A	A	A	A
Brake Fluid (Non-Petroleum Base) Silicones or glycols	X		A	A	A	A	A	X	A	A	A	*
Brewery Slop			A	*	A	*	A	A	*	A	A	A
Brine (Sodium Chloride) Salt water	A		A	*	X	A	A	A	A	B	A	A
Bromine — Anhydrous Br_2	X		A	B	C	A	X	X	C	X	A	A
Bromine Trifluoride BrF_3	X		A	A	*	*	B	X	X	X	A	X
Bromine Water	X		A	X	X	A	X	X	X	B	A	B
Bromobenzene C_6H_5Br	X		A	X	B	B	A	X	X	X	A	B
Bromochloromethane $BrCH_2Cl$			A	X	B	B	B	X	B	X	A	C
Bromotoluene $C_6H_4BrCH_3$			A	X	A	A	A	X	*	*	A	B
Bronzing Liquid			A	*	*	A	A	X	B	X	A	X
Bunker Oil (Fuel) #5, #6 & C Hydrocarbons			A	A	A	A	A	A	X	B	A	A
Butadiene C_4H_6	X	A	A	A	A	*	A	X	C	C	A	C
Butane (LPG) (Butyl Hydride) C_4H_{10}	X	A	A	A	A	A	A	A	X	B	A	A
Butter Fats			A	A	X	*	A	A	A	C	A	A
Buttermilk Fats, water	A		A	A	—	—	A	A	—	A	—	A
Butyl Acetate $CH_3CO_2(CH_2)_3CH_3$	X	A	A	A	A	A	A	X	B	X	A	X
n-Butyl Acetate $CH_3CO_2(CH_2)_3CH_3$			A	A	A	A	A	X	X	X	A	X
Butyl Acetyl Ricinoleate $C_{24}H_{44}O_5$			A	*	*	A	*	C	C	X	A	B
Butyl Acrylate $CH_2CHCO_2C_4H_9$			A	*	*	*	*	X	X	X	A	X
Butyl Alcohol (Butanol) $CH_3(CH_2)_3OH$			A	A	B	A	A	A	B	A	A	A
Butyl Amine (Aminobutane) $CH_3(CH_2)_2CH_2NH_2$	B	A	A	A	A	*	A	B	X	X	A	X
Butyl Benzoate C_6H_5COO $(CH_2)_3CH_3$			A	B	B	B	B	*	B	X	A	A
Butyl Bromide $CH_3(CH_2)_2CH_2Br$			A	*	*	*	*	X	*	*	A	B
Butyl Butyrate $CH_3(CH_2)_2$ $CH_2CO_2C_4H_9$			A	A	A	A	A	X	*	*	A	X
Butyl Carbitol® $CH_3(CH_2)_3OCH_2$ $CH_2OCH_2CH_2OH$			A	*	*	*	*	A	A	B	A	A
Butyl Cellosolve® $HOCH_2CH_2OC_4H_9$			A	—	—	—	—	B	—	C	A	C
Butyl Chloride (Chlorobutane) $CH_3(CH_2)_3Cl$	X		A	X	B	B	B	X	*	*	A	A

RATING KEY: (A) EXCELLENT   (B) GOOD   (C) FAIR TO POOR   (X) NOT RECOMMENDED
(*) OR (—) NO DATA AVAILABLE

CHEMICAL	FORMULA	PLASTIC			METAL				ELASTOMER				
		POLYPROPYLENE	RYTON®	TEFLON®	ALUMINUM	CAST IRON	HASTELLOY-C®	STAINLESS STEEL	BUNA-N	E.P.D.M.	NEOPRENE	TEFLON®	VITON®
Butyl Ether (Dibutyl Ether)	(CH₃(CH₂)₃)₂O	X	A	A	A	B	A	A	A	·	B	A	C
Butyl Oleate	C₂₂H₄₂O₂			A	·	·	·	·	·	C	X	A	A
Butyl Stearate	CH₃(CH₂)₁₆ CO₂(CH₂)₃CH₃			A	B	B	B	B	A	C	X	A	B
Butylene (Butene)	C₄H₈	X	A	A	A	·	·	A	B	X	X	A	B
Butyraldehyde	CH₃(CH₂)₂CHO			A	A	A	A	A	X	C	X	A	X
Butyric Acid	CH₃(CH₂)₂CO₂H	A	A	A	A	X	A	B	C	C	X	A	C
Butyric Anhydride	(CH₃CH₂CH₂CO)₂O			A	A	A	A	A	C	·	·	A	A
Butyronitrile	CH₃CH₂CH₂CN			A	·	·	·	·	X	A	X	A	A
Calcium Acetate Hydrate	Ca(CH₃COO)₂ • H₂O			A	C	C	B	B	B	A	C	A	X
Calcium Bisulfite	Ca(HSO₃)₂	A	A	A	X	X	A	A	A	X	A	A	A
Calcium Carbonate (Chalk)	CaCO₃	A		A	C	B	B	B	A	A	A	A	A
Calcium Chlorate	Ca(ClO₃)₂			A	B	B	B	B	A	A	A	A	A
Calcium Chloride (Brine)	CaCl₂ • 6H₂O	A	A	A	A	A	A	A	A	A	A	A	A
Calcium Hydrosulfide (Calcium Sulfhydrate)	Ca(HS)₂ • 6H₂O			A	·	·	·	·	A	·	A	A	A
Calcium Hydroxide (Slaked Lime)	Ca(OH)₂	A		A	X	B	A	B	A	A	A	A	A
Calcium Hypochlorite 20% (Calcium Oxichloride)	Ca(ClO)₂	A	A	A	X	X	B	B	C	B	X	A	B
Calcium Nitrate	Ca(NO₃)₂	A	A	A	B	B	B	B	A	A	A	A	A
Calcium Oxide (Unslaked Lime)	CaO			A	A	A	A	A	A	A	A	A	·
Calcium Silicate	Ca₂SiO₄			A	A	B	A	A	A	·	·	A	A
Calcium Sulfate (Gypsum)	CaSO₄	A	A	A	C	B	A	A	A	A	A	A	A
Calcium Sulfide	CaS	A		A	A	B	A	B	A	A	B	A	A
Calcium Sulfite	CaSO₃ • 2H₂O			A	B	B	·	A	A	·	·	A	A
Calgon®	(NaPO₃)₆	A		A	—	X	—	A	A	—	A	—	A
Cane Juice	Sucrose, water	X		A	B	A	—	A	A	—	A	—	—
Cane Sugar Liquors	Sucrose, water	A		A	A	A	·	A	A	A	A	A	A
Capryl Alcohol (Octanol)	CH₃(CH₂)₆CH₂OH			A	A	A	A	A	A	C	B	A	B
Caprylic Acid (Octanoic Acid)	CH₃(CH₂)₆ COOH			A	A	·	A	A	C	·	·	A	·
Carbamate	H₂NCO₂R			A	·	·	·	·	C	C	C	A	A
Carbitol®	CH₃CH₂OCH₂CH₂ OCH₂CH₂OH			A	A	A	A	A	B	C	C	A	C
Carbolic Acid (see Phenol)	C₆H₅OH	C	A	A	B	A	A	B	X	C	C	A	A
Carbon Dioxide (Carbonic Acid Gas)	CO₂	A	A	A	A	A	A	A	A	B	A	A	A
Carbon Disulfide (Carbon Bisulfide)	CS₂	X	A	A	A	B	·	A	X	X	X	A	A
Carbon Monoxide	CO	A		A	A	A	A	A	C	C	A	A	C
Carbon Tetrachloride (Tetrachloromethane)	CCl₄	X	A	A	X	C	A	B	C	X	X	A	A
Carbonated Beverages	CO₂/H₂O	A		A	C	·	A	A	A	·	A	A	·

RATING KEY: (A) EXCELLENT (B) GOOD (C) FAIR TO POOR (X) NOT RECOMMENDED (*) OR (—) NO DATA AVAILABLE

CHEMICAL	FORMULA	PLASTIC			METAL				ELASTOMER				
		POLYPROPYLENE	RYTON®	TEFLON®	ALUMINUM	CAST IRON	HASTELLOY-C®	STAINLESS STEEL	BUNA-N	E.P.D.M.	NEOPRENE	TEFLON®	VITON®
Carbonic Acid (liquid)	H_2CO_3	A	A	A	A	X	A	B	B	—	A	A	A
Casein	a phosphoprotein			A	B	*	B	B	A	A	A	A	A
Castor Oil	a mixture of fatty acids			A	A	B	A	A	A	B	A	A	A
Catsup (Ketchup)		A		A	B	X	A	A	A	*	C	A	A
Cellosolve® (Glycol Ethers)	$HOCH_2CH_2OR$	A	A	A	A	*	A	A	C	C	C	A	B
Cellulose Acetate	$C_8H_{12}O_5$			A	B	B	A	A	B	*	B	A	C
Cellulube® Hydraulic Fluids (Phosphate Esters)				A	A	A	A	A	X	A	X	A	B
Chlorinated Lime — 35% Bleach	$Ca(ClO)_2$			A	*	X	*	A	C	A	X	A	A
Chlorinated Water		B	X	A	C	*	A	B	C	*	C	A	A
Chlorine, Dry	Cl_2	X	X	A	X	X	—	—	C	—	C	A	A
Chlorine, Wet	Cl_2/H_2O	X	X	A	B	C	A	A	C	X	X	A	A
Chlorine, Anhydrous Liquid	Cl_2	X		A	X	X	A	X	X	—	X	A	A
Chlorine Dioxide	ClO_2			A	B	*	B	X	X	C	X	A	B
Chlorine Trifluoride	ClF_3	X		A	A	*	*	A	X	X	X	A	B
Chloroacetic Acid (Monochloroacetic Acid)	$ClCH_2COOH$	A	A	A	X	X	A	X	X	B	C	A	C
Chloroacetone (Monochloroacetone)	$ClCH_2COCH_3$	X		A	X	B	B	B	X	A	C	A	C
Chlorobenzene (Monochlorobenzene)	C_6H_5Cl	X	A	A	X	B	B	B	X	X	X	A	A
Chlorobutadiene (Chloroprene)	C_4H_5Cl	X		A	X	B	B	B	X	X	X	A	A
Chlorobromomethane	$ClCH_2Br$	X		A	X	B	—	B	X	—	X	A	A
Chloroform	$CHCl_3$	X	A	A	X	A	A	A	X	X	X	A	A
1-Chloronaphthalene	$C_{10}H_7Cl$	X		A	X	B	A	B	X	X	X	A	C
Chlorosulfonic Acid	HSO_3Cl	X	X	A	B	B	A	B	X	X	X	A	X
o-Chlorophenol	C_6H_5ClO		A	A	B	B	B	B	X	X	X	A	B
Chlorothene® (Chlorinated Solvents)	CH_3CCl_3			A	X	X	A	A	X	*	X	A	C
Chlorotrifluoroethylene	$C_2H_2ClF_3$			A	B	B	B	B	X	*	*	A	*
Chlorox®		B		A	*	X	B	A	C	*	B	A	A
Chocolate Syrup	Corn syrup, water, sugar	A		A	*	X	*	A	A	*	A	A	*
Chromic Acid — To 25%	H_2CrO_4	A	A	A	B	B	B	X	X	A	X	A	A
Chromic Acid — Over 25%	H_2CrO_4	A	A	A	X	B	B	X	X	C	X	A	A
Cider (Apple Juice)	Sucrose, water			A	B	X	A	A	A	*	A	A	A
Cinnamon Oil	Cinnamic acid esters			A	*	X	*	A	*	*	C	A	*
Citric Acid	$C_6H_8O_7 \bullet H_2O$	B	A	A	B	X	A	A	B	A	A	A	A
Citric Oils	Citric acid esters	A		A	*	X	*	A	C	B	X	A	A
Citrus Pectin Liquor				A	*	*	*	A	A	*	A	A	A
Clove Oil (Eugenol)	$C_{10}H_{12}O_2$			A	*	X	*	A	*	*	C	A	*
Cobalt Chloride	$CoCl_2 \bullet 6H_2O$	A		A	X	*	*	*	A	C	A	A	A

RATING KEY: (A) EXCELLENT (B) GOOD (C) FAIR TO POOR (X) NOT RECOMMENDED (*) OR (—) NO DATA AVAILABLE

CHEMICAL	FORMULA	POLYPROPYLENE	RYTON®	TEFLON®	ALUMINUM	CAST IRON	HASTELLOY-C®	STAINLESS STEEL	BUNA-N	E.P.D.M.	NEOPRENE	TEFLON®	VITON®
Coconut Oil (Coconut Butter)	Fatty acid mixture			A	B	A	•	A	B	A	B	A	A
Cod Liver Oil (Fish Oil)	Glycerides, acids, esters			A	A	X	•	A	B	A	B	A	A
Coffee	Fatty oils, acids, cellulose, water	A		A	A	•	A	A	A	•	A	A	•
Coke Oven Gas	$H_2(53\%),CH_4(26\%),$ $N_2(11\%),CO(7\%)\&$ hydrocarbons(3%)			A	—	—	—	—	C	—	C	A	A
Copper Acetate	$Cu(C_2H_3O_2)_2 \bullet$ $CuO \bullet 6H_2O$			A	X	A	B	B	B	A	C	A	•
Copper Chloride	$CuCl_2 \bullet 2H_2O$	A		A	X	X	B	X	A	A	A	A	A
Copper Cyanide	CuCN	A	A	A	X	A	A	A	A	A	A	A	A
Copper Fluoroborate				A	X	X	B	X	B	—	A	—	A
Copper Nitrate Hexahydrate	$Cu(NO_3)_2 \bullet 6H_2O$	A	A	A	X	X	B	A	A	A	A	A	A
Copper Sulfate (Blue Copperas)	$CuSO_4 \bullet 5H_2O$	A	A	A	X	X	A	A	A	A	A	A	A
Copper Sulfide	CuS			A	•	•	•	•	A	•	•	A	A
Corn Oil (Maize oil)	Glycerides of fatty acids	A		A	B	C	•	B	A	C	C	A	A
Cotton Seed Oil		A		A	A	C	•	A	A	A	C	A	A
Cream		A		A	•	X	•	A	A	•	C	A	A
Creosote, Coal-Tar (Tar Oil)	Hydrocarbon mixture	X		A	B	B	B	B	A	X	C	A	A
Creosote, Wood-Tar	Mixture of phenols	X		A	•	•	•	B	A	X	B	A	A
Cresylic Acid (Cresol)	$C_8H_{10}O_2$	X		A	B	C	B	A	C	X	X	A	A
Crotonaldehyde	$CH_3CHCHCHO$			A	A	A	A	A	X	•	A	A	A
Cumeme (Isopropylbenzene)	$C_6H_5CH(CH_3)_2$			A	B	B	B	B	X	X	X	A	A
Cutting Oil (Water Soluble)				A	A	A	A	A	C	•	X	A	A
Cutting Oil (Sulfur Base)				A	A	A	A	A	A	•	C	A	•
Cyclohexane	C_6H_{12}	X	A	A	B	B	B	B	B	X	X	A	A
Cyclohexanol	$C_6H_{11}OH$	B	A	A	C	B	A	A	B	X	A	A	A
Cyclohexanone	$C_6H_{10}O$	X	A	A	B	B	B	B	X	C	X	A	X
Cyclopentane	C_5H_{10}			A	B	B	B	B	B	X	A	A	A
Cymene (Isopropyltoluene)	$C_{10}H_{14}$			A	•	•	•	•	C	X	X	A	A
Decahydronaphthalene (Decalin®)	$C_{10}H_{18}$			A	•	•	•	•	X	X	X	A	A
Decanal	$CH_3(CH_2)_8CHO$			A	•	•	•	•	X	X	•	A	X
Decane	$CH_3(CH_2)_8CH_3$	A		A	•	•	•	•	B	C	X	A	A
Decyl Alcohol (Decanol)	$C_{10}H_{21}OH$			A	•	•	•	•	A	•	X	A	B
Denatured Alcohol	Ethanol and denaturant	A		A	B	B	A	A	A	A	B	A	B
Detergent Solutions		A	A	A	B	•	•	A	A	A	A	A	A
Developing Fluids & Solutions				A	•	X	A	A	A	C	A	A	A
Dextrose	$C_6H_{12}O_6$			A	A	X	A	A	B	A	B	A	A
Diacetone Alcohol (Diacetone)	$(CH_3)_2COHCH_2$ $COCH_3$	X		A	A	A	A	A	X	B	X	A	X

RATING KEY: (A) EXCELLENT (B) GOOD (C) FAIR TO POOR (X) NOT RECOMMENDED (•) OR (—) NO DATA AVAILABLE

CHEMICAL	FORMULA	PLASTIC			METAL				ELASTOMER				
		POLYPROPYLENE	RYTON	TEFLON	ALUMINUM	CAST IRON	HASTELLOY-C	STAINLESS STEEL	BUNA-N	E.P.D.M.	NEOPRENE	TEFLON	VITON
Dibenzyl Ether	(C₆H₅CH₂)₂O			A	B	B	B	B	X	C	X	A	C
Dibenzyl Sebecate	C₂₄H₃₀O₄			A	•	•	•	•	X	C	X	A	B
Dibutyl Amine	(C₄H₉)₂NH	X		A	•	A	A	A	C	X	X	A	X
Dibutyl Phthalate (DBP)	C₆H₄(CO₂C₄H₉)₂	X	A	A	A	A	A	A	X	A	X	A	B
Dibutyl Sebecate (DBS)	C₁₈H₃₄O₄	C		A	•	A	•	A	X	C	X	A	C
Dichloroacetic Acid	Cl₂CHCOOH			A	•	•	•	•	X	•	X	A	X
o-Dichlorobenzene	C₆H₄Cl₂	B		A	X	B	A	B	X	X	X	A	A
Dichlorobutane	C₄H₆Cl₂			A	X	B	•	B	X	•	•	A	A
Dichloroethyl Ether	[ClCH₂CH₂]₂O			A	B	•	•	•	X	•	•	A	•
Dichloro Isopropyl Ether	C₆H₁₂OCl₂	X		A	•	•	•	•	X	X	X	A	X
Dicyclohexylamine	(C₆H₁₁)₂NH			A	•	•	•	•	X	X	X	A	B
Diesel Oil (Fuel ASTM #2)	Hydrocarbons	B		A	A	A	A	A	A	X	C	A	A
Diester Synthetic Oils				A	A	A	A	A	B	X	X	A	A
Diethanol Amine	(HOCH₂CH₂)₂NH	A		A	•	A	A	A	B	•	A	A	•
Diethyl Amine	(CH₃CH₂)₂NH	A		A	B	B	A	A	C	C	C	A	X
Diethyl Benzene	C₆H₄(C₂H₅)₂			A	•	•	•	•	X	X	X	A	A
Diethyl Carbonate	(C₂H₅O)₂CO			A	•	A	•	•	X	•	X	A	•
Diethyl Ether (Ether)	(CH₃CH₂)₂O	X	A	A	B	A	A	A	B	X	C	A	X
Diethyl Phthalate (DEP)	C₆H₄(CO₂C₂H₅)₂			A	A	A	A	A	X	•	•	A	C
Diethyl Sebecate	C₁₄H₂₆O₄	A		A	A	A	A	A	X	C	X	A	B
Diethylene Ether (Dioxane)	C₄H₈O₂			A	A	A	•	A	X	A	X	A	X
Diethylene Glycol (DEG)	HOCH₂CH₂OCH₂CH₂OH	A		A	A	A	A	A	A	A	A	A	A
Diethylene Triamine	(NH₂C₂H₄)₂NH			A	A	A	A	A	B	•	•	A	•
Diisobutyl Ketone	C₄H₉COC₄H₉			A	A	A	A	A	X	B	X	A	X
Diisobutylene	[HC=C(CH₃)₂]₂	A	A	A	•	•	•	•	B	•	C	A	C
Diisodecyl Adipate (DIDA)	C₂₆H₅₀O₄			A	•	•	•	•	X	•	•	A	C
Diisodecyl Phthalate (DIDP)	C₂₈H₄₇O₄			A	•	•	•	•	X	A	X	A	C
Diisooctyl Adipate (DIOA)	C₂₂H₄₂O₄			A	A	A	A	A	X	•	•	A	C
Diisooctyl Phthalate (DIOP)	C₂₄H₃₉O₄			A	•	•	•	•	X	•	•	A	C
Diisooctyl Sebecate (DIOS)	C₂₆H₄₆O₄			A	•	•	•	•	•	B	•	A	A
Diisopropyl Amine	[(CH₃)₂CH]₂NH			A	•	•	•	•	B	•	•	A	•
Diisopropyl Benzene	C₆H₄ • [CH(CH₃)₂]₂			A	•	•	•	•	X	X	X	A	A
Diisopropyl Ketone	[(CH₃)₂CH]₂CO			A	•	•	•	A	X	A	X	A	X
N,N-Dimethylaniline	C₆H₅N(CH₃)₂	X	A	A	B	B	•	•	X	C	X	A	X
Dimethyl Ether	CH₃OCH₃			A	B	B	B	B	A	•	B	A	A
N,N-Dimethyl Formamide (DMF)	HCON(CH₃)₂	A	A	A	A	•	A	A	C	•	X	A	X
Dimethyl Phthalate	C₆H₄(CO₂CH₃)₂		A	A	•	•	•	•	X	C	X	A	C

RATING KEY: (A) EXCELLENT (B) GOOD (C) FAIR TO POOR (X) NOT RECOMMENDED
(*) OR (—) NO DATA AVAILABLE

	PLASTIC			METAL				ELASTOMER				
CHEMICAL / FORMULA	POLYPROPYLENE	RYTON®	TEFLON®	ALUMINUM	CAST IRON	HASTELLOY-C®	STAINLESS STEEL	BUNA-N	E.P.D.M.	NEOPRENE	TEFLON®	VITON®
Dimethyl Sulfate — $(CH_3)_2SO_4$			A	•	A	•	•	X	•	•	A	X
Dimethyl Sulfide — $(CH_3)_2S$			A	A	A	A	A	X	•	•	A	•
Dinitrotoluene (DNT) — $CH_3C_6H_3(NO_2)_2$			A	•	•	•	A	X	X	X	A	C
Dioctyl Phthalate (DOP) — $C_{24}H_{38}O_4$			A	A	A	A	A	X	B	X	A	B
Dioctyl Sebecate — $C_{26}H_{50}O_4$			A	A	A	A	A	X	C	X	A	C
Dioxolanes (Dioxolans) — Glycol ethers			A	•	•	•	•	X	B	X	A	C
Dipentene (Limonene) — $C_{10}H_{16}$			A	A	A	A	A	C	X	X	A	A
Diphenyl Oxides (Phenyl Ether) — $C_6H_5OC_6H_5$			A	B	A	A	A	X	C	X	A	A
Dipropylamine — $(CH_3CH_2CH_2)_2NH$			A	•	•	•	•	B	•	•	A	•
Dipropylene Glycol — $(C_3H_6OH)_2O$	A		A	•	•	•	•	A	•	•	A	A
Dipropyl Ketone (Butyrone) — $(C_3H_7)_2CO$			A	•	•	•	•	X	•	•	A	•
Dispersing Oil #10			A	A	A	A	A	X	X	X	A	C
Divinyl Benzene (DVB) — $C_6H_4(CH=CH_2)_2$			A	•	•	•	•	X	•	•	A	A
Dodecyl Benzene (Alkane) — $C_6H_5(CH_2)_{11}CH_3$			A	A	A	•	A	X	•	•	A	A
Dow Corning® (Silicones) — $[(CH_3)_2SiO]_2$			A	A	•	•	•	A	•	A	A	A
Dowtherm® (Biphenyl & Phenyl Ether) — $(C_6H_5)_2$ and $(C_6H_5)_2O$			A	A	B	A	A	X	X	X	A	A
Drycleaning Fluids — Chlorinated hydrocarbons	X		A	A	A	—	A	C	—	X	A	A
Dyes			A	B	—	—	A	—	—	C	—	A
Epichlorohydrin — C_3H_5ClO	A	A	A	A	A	A	A	X	B	X	A	X
Epsom Salts (Magnesium Sulfate) — $MgSO_4 \bullet 7H_2O$	A		A	A	—	B	A	A	—	A	A	A
Ethane — C_2H_6	C		A	A	A	A	A	A	X	C	A	A
Ethanolamine (Aminoethanol) — $H_2NCH_2CH_2OH$	X	A	A	B	A	•	A	B	B	C	A	X
Ethyl Acetate — $CH_3COOCH_2CH_3$	C	A	A	A	A	A	A	X	B	X	A	X
Ethyl Acetoacetate (Acetoacetic Ester) — $CH_3COCH_2COOCH_2CH_3$			A	A	A	A	A	X	C	X	A	X
Ethyl Acrylate — $CH_2CHCO_2CH_2CH_3$	B		A	A	A	A	A	X	C	X	A	X
Ethyl Alcohol (Ethanol) — CH_3CH_2OH	A	A	A	B	B	A	A	A	A	A	A	B
Ethyl Aluminum Dichloride — $CH_3CH_2AlCl_2$			A	•	•	•	•	X	•	•	A	B
Ethyl Amine (Monoethylamine) — $CH_3CH_2NH_2$			A	B	B	•	A	X	A	C	A	X
Ethyl Benzene — $CH_3CH_2C_6H_5$	X		A	B	B	A	B	X	X	X	A	A
Ethyl Benzoate — $C_6H_5CO_2CH_2CH_3$	B		A	A	A	A	A	X	C	X	A	A
Ethyl Bromide (Bromoethane) — CH_3CH_2Br			A	X	A	A	A	X	B	B	A	•
Ethyl Butyl Acetate — $CH_3CO_2CH_2CH(C_2H_5)_2$			A	•	•	•	•	X	•	•	A	X
Ethyl Butyl Alcohol — $CH_3CH(C_2H_5)(CH_2)_2OH$			A	•	•	•	•	A	•	•	A	B

RATING KEY: (A) EXCELLENT (B) GOOD (C) FAIR TO POOR (X) NOT RECOMMENDED
(•) OR (—) NO DATA AVAILABLE

		PLASTIC			METAL				ELASTOMER				
CHEMICAL	FORMULA	POLYPROPYLENE	RYTON®	TEFLON®	ALUMINUM	CAST IRON	HASTELLOY-C®	STAINLESS STEEL	BUNA-N	E.P.D.M.	NEOPRENE	TEFLON®	VITON®
Ethyl Butyl Ketone	$CH_3CH_2COC_4H_9$			A	•	•	•	•	X	•	•	A	X
Ethyl Butyraldehyde	$C_6H_{12}O$			A	•	•	•	•	X	•	•	A	X
Ethyl Butyrate	$CH_3CH_2CH_2CO_2C_2H_5$	B		A	B	A	A	A	X	X	X	A	C
Ethyl Caprylate	$CH_3(CH_2)_6CO_2C_2H_5$			A	•	•	•	•	X	X	X	A	•
Ethyl Cellosolve®	$C_2H_5O(CH_2)_2OH$			A	•	•	•	•	C	B	C	A	X
Ethyl Cellulose (Ethocel®)		C		A	B	A	B	B	B	B	B	A	C
Ethyl Chloride (Chloroethane)	C_2H_5Cl	X	A	A	X	B	B	A	A	A	C	A	A
Ethyl Chlorocarbonate (Ethyl Chloroformate)	$ClCO_2C_2H_5$			A	•	•	•	•	•	•	C	A	A
Ethyl Cyanide (Propionitrile)	C_2H_5CN			A	•	•	•	•	X	A	B	A	X
Ethyl Formate	$HCOOCH_2CH_3$			A	B	A	B	B	X	C	B	A	A
Ethylhexyl Acetate	$CH_3CO_2CH_2CH(C_2H_5)C_4H_9$			A	•	•	•	•	X	•	•	A	X
Ethylhexyl Alcohol (Ethylhexanol)	$C_8H_{17}OH$			A	A	A	A	A	A	•	•	A	B
Ethyl Iodide	CH_3CH_2I			A									
Ethyl Isobutyrate	$(CH_3)_2CHCOOCH_2CH_3$			A	•	•	•	•	X	X	X	A	•
Ethyl Mercaptan (Ethanethiol)	CH_3CH_2SH			A	B	A	B	B	X	X	C	A	B
Ethyl Oxalate	$C_2H_5O_2CCO_2C_2H_5$			A	•	•	•	•	X	A	X	A	B
Ethyl Pentachlorobenzene	$C_2H_5C_6Cl_5$	X		A	X	—	—	—	X	—	X	A	A
Ethyl Propionate	$CH_3CH_2COOCH_2CH_3$			A	A	A	A	A	X	X	X	A	•
Ethyl Silicate	$Si(OCH_2CH_3)_4$			A	B	A	A	A	A	A	A	A	A
Ethyl Sulfate	$C_2H_5OSO_2OH$			A	—	—	—	X	A	—	—	A	A
Ethylene (Ethene)	C_2H_4			A	A	A	•	A	B	C	A	A	A
Ethylene Chlorohydrin	$ClCH_2CH_2OH$	X		A	•	B	A	A	X	A	B	A	B
Ethylene Diamine	$(CH_2)_2(NH_2)_2$	A	A	A	C	A	A	A	B	A	A	A	X
Ethylene Dibromide (Ethylene Bromide)	$Br(CH_2)_2Br$	X		A	X	X	B	B	X	C	X	A	B
Ethylene Dichloride (Dutch Oil)	$Cl(CH_2)_2Cl$	X	A	A	X	B	B	B	X	X	X	A	B
Ethylene Glycol (Ethylene Alcohol) (Glycol)	$(CH_2OH)_2$	A	A	A	A	A	A	A	A	A	A	A	A
Ethylene Glycol Monobutyl Ether (Butyl Cellosolve®)	$C_4H_9OCH_2CH_2OH$			A	A	A	A	A	B	B	X	A	C
Ethylene Glycol Monoethyl Ether Acetate (Cellosolve Acetate®)	$C_2H_5O(CH_2)_2O_2CCH_3$			A	A	A	A	A	C	B	X	A	C
Ethylene Glycol Monomethyl Ether (Methyl Cellosolve®)	$CH_3O(CH_2)_2OH$			A	B	B	A	A	C	B	C	A	X
Ethylene Oxide	$(CH_2)_2O$	C	X	A	A	B	A	A	X	X	X	A	C

RATING KEY: (A) EXCELLENT   (B) GOOD   (C) FAIR TO POOR   (X) NOT RECOMMENDED (*) OR (—) NO DATA AVAILABLE

CHEMICAL	FORMULA	PLASTIC			METAL				ELASTOMER				
		POLYPROPYLENE	RYTON®	TEFLON®	ALUMINUM	CAST IRON	HASTELLOY-C®	STAINLESS STEEL	BUNA-N	E.P.D.M.	NEOPRENE	TEFLON®	VITON®
Ethylene Trichloride (Trichloroethene)	ClCHCCl₂	X		A	X	A	•	A	X	X	X	A	A
Ethylidene Chloride	CH₃CHCl₂			A	X	B	B	A	X	X	X	A	•
Fatty Acids	CₙH₂ₙ₊₁COOH	B		A	A	X	A	A	B	X	C	A	A
Ferric Chloride	FeCl₃	A	A	A	X	X	A	X	A	A	A	A	A
Ferric Hydroxide	FeHO₂			A	•	•	B	A	B	•	•	A	C
Ferric Nitrate	Fe(NO₃)₃	A	A	A	X	X	A	B	A	A	A	A	A
Ferric Sulfate	Fe₂(SO₄)₃	A	A	A	C	X	A	B	A	A	A	A	A
Ferrous Chloride	FeCl₂	A	A	A	X	X	B	B	A	A	A	A	A
Ferrous Sulfate	FeSO₄	A	A	A	A	C	A	B	A	A	A	A	A
Fish Oil				A	—	—	—	—	A	—	—	A	A
Fluoboric Acid (Fluoroboric Acid)	HBF₄	A	A	A	X	X	•	A	A	A	B	A	C
Fluorine (Liquid)	F₂	X		A	A	•	•	A	X	C	C	A	B
Fluorobenzene	FC₆H₅	X		A	•	•	•	•	X	X	X	A	A
Fluorolube (Fluorocarbon Oils)	FₓCᵧHᵤ	X		A	A	A	A	A	C	A	A	A	B
Fluosilicic Acid (Sand Acid)	H₂SiF₆	A	A	A	X	X	B	A	B	B	A	A	A
Formaldehyde (Formalin)	HCHO	A	A	A	A	C	A	A	B	A	C	A	A
Formamide	HCONH₂			A	A	B	B	B	A	A	A	A	X
Formic Acid	HCOOH	A	A	A	X	X	A	C	C	B	B	A	C
Freon 11 (Trichlorofluoromethane)	CCl₃F	B	A	A	B	A	•	A	C	X	B	A	B
Freon 12 (Dichlorodifluoromethane)	Cl₂CF₂			A	A	A	•	A	B	B	B	A	B
Freon 13 (Chlorotrifluoromethane)	ClCF₃			A	A	A	A	A	A	A	A	A	A
Freon 13B1 (Bromotrifluoromethane)	BrCF₃			A	•	•	•	•	A	A	A	A	A
Freon 14 (Tetrafluoromethane)	CF₄			A	•	•	•	•	X	B	X	A	•
Freon 21 (Dichlorofluoromethane)	FCHCl₂			A	A	•	•	•	X	X	B	A	X
Freon 22 (Chlorodifluoromethane)	HCClF₂			A	A	A	A	A	X	C	B	A	X
Freon 113 (Trichlorotrifluoroethane)	Cl₃CCF₃			A	B	•	•	A	B	X	A	A	B
Freon 114 (Dichlorotetrafluoroethane)	C₂Cl₂F₄			A	B	•	•	A	A	C	A	A	A
Freon 114B2 (Dibromotetrafluoroethane)	C₂Br₂F₄			A	•	•	•	•	B	X	A	A	B
Freon 115 (Chloropentafluoroethane)	C₂ClF₅			A	A	•	•	•	A	A	A	A	B
Fruit Juices	Water, sucrose	A	A	A	A	X	A	A	A	•	A	A	A
Fuel Oils (ASTM #1 thru #9)	Hydrocarbons	C	A	A	A	A	A	A	A	X	C	A	A
Fumaric Acid (Boletic Acid)	HOOCCH = CHCOOH			A	•	•	•	•	C	•	B	A	A

RATING KEY: (A) EXCELLENT   (B) GOOD   (C) FAIR TO POOR   (X) NOT RECOMMENDED
(*) OR (—) NO DATA AVAILABLE

		PLASTIC			METAL				ELASTOMER				
CHEMICAL	FORMULA	POLYPROPYLENE	RYTON®	TEFLON®	ALUMINUM	CAST IRON	HASTELLOY-C®	STAINLESS STEEL	BUNA-N	E.P.D.M.	NEOPRENE	TEFLON®	VITON®
Furan (Furfuran)	C_4H_4O	C	A	A	•	•	•	•	X	X	X	A	C
Furfural (Ant Oil)	$C_5H_4O_2$	X	A	A	A	B	B	A	X	B	B	A	C
Furfuryl Alcohol	$C_5H_6O_2$			A	A	A	A	A	•	•	•	A	X
Fusel Oil (Grain Oil)	$(CH_3)_2$ $CHCH_2CH_2OH$			A	•	•	•	•	A	A	A	A	A
Gallic Acid	$C_6H_2(OH)_3$ $COOH$	A	A	A	A	X	B	B	B	B	C	A	A
Gasoline (Unleaded)	C_4 to C_{12} hydrocarbons	C	A	A	A	A	A	A	X	X	X	A	A
Gasoline (Petrol)	Hydrocarbons	C	A	A	A	A	A	A	A	X	C	A	A
Gelatin	Water soluble proteins	A		A	A	A	•	A	A	A	A	A	B
Ginger Oil	$C_{17}H_{26}O_4$			A	•	X	•	A	•	•	A	A	A
Glauber's Salt (Sodium Sulfate Decahydrate)	Na_2SO_4•$10H_2O$			A	•	•	•	•	A	B	A	A	A
Gluconic Acid	$C_6H_{12}O_7$			A	B	C	A	A	C	•	•	A	A
Glucose (Corn Syrup)	$C_6H_{12}O_6$	A		A	A	A	•	A	A	A	A	A	A
Glue		A		A	A	A	A	B	A	B	A	A	A
Glycerol (Glycerine)	$C_3H_8O_3$	A	A	A	A	B	A	A	A	A	A	A	A
Glycolic Acid	$HOCH_2COOH$	A	A	A	—	—	A	—	A	—	A	—	A
Glycols		A	A	A	B	B	—	B	A	—	A	A	A
Gold Monocyanide	AuCN	—		A	—	—	A	X	A	—	A	—	A
Grape Juice	Water, sucrose	A		A	•	X	•	A	C	•	X	A	A
Grapefruit Oil				A	•	X	•	A	X	•	X	A	•
Grease	Hydrocarbons			A	A	—	—	A	A	—	X	A	A
Green Sulfate Liquor		A		A	B	C	B	A	B	A	B	A	A
Halowax Oil	Chlorinated naphthalenes			A	X	•	•	•	X	X	X	A	A
Heptanal	$CH_3(CH_2)_5CHO$	A		A	A	A	A	A	A	•	•	•	A
Heptane	C_7H_{16}	C	A	A	A	A	A	A	A	X	C	A	A
Hexanal	$CH_3(CH_2)_4CHO$			A	A	B	B	A	X	B	A	A	C
Hexalin (Cyclohexanol)	$C_6H_{11}OH$			A	•	•	•	•	B	C	A	A	A
n-Hexane	C_6H_{14}	C	A	A	A	A	A	A	A	X	B	A	A
n-Hexene 1 (Hexylene)	$H_2CCH(CH_2)_3CH_3$			A	•	•	•	•	A	X	B	A	A
Hexyl Alcohol (1-Hexanol)	$C_6H_{13}OH$			A	A	A	•	A	A	C	B	A	A
Hexylene Glycol (Brake Fluid)	$C_6H_{12}(OH)_2$			A	A	A	A	A	A	C	A	A	A
Honey		A		A	A	A	•	A	•	•	A	A	•
Hydraulic Oil (Petroleum Base)	Hydrocarbons	X		A	A	A	A	A	A	X	B	A	A
Hydrazine (Diamine)	H_2NNH_2	A		A	A	X	A	A	C	A	C	A	X
Hydrobromic Acid	HBr	B	A	A	A	A	•	A	X	A	C	A	A
Hydrochloric Acid 10%	HCl	A	A	A	X	C	B	X	B	A	B	A	A

RATING KEY: (A) EXCELLENT   (B) GOOD   (C) FAIR TO POOR   (X) NOT RECOMMENDED
(•) OR (—) NO DATA AVAILABLE

CHEMICAL	FORMULA	PLASTIC			METAL				ELASTOMER				
		POLYPROPYLENE	RYTON®	TEFLON®	ALUMINUM	CAST IRON	HASTELLOY-C®	STAINLESS STEEL	BUNA-N	E.P.D.M.	NEOPRENE	TEFLON®	VITON®
Hydrochloric Acid 20%	HCl	A	A	A	X	C	A	X	B	A	B	A	A
Hydrochloric Acid 30% (Conc.)	HCl	B	A	A	X	X	A	X	C	A	C	A	B
Hydrocyanic Acid (Formonitrile)	HCN	A		A	A	X	B	A	B	A	C	A	A
Hydrogen Fluoride — Anhydrous	HF	A		A	X	*	A	X	X	C	C	A	A
Hydrofluoric Acid (Conc.) Cold	HF	X	A	A	C	X	B	A	*	C	C	A	B
Hydrogen Peroxide — 3%	H_2O_2	A	X	A	A	*	*	*	B	B	B	A	A
Hydrogen Peroxide — 10%	H_2O_2	A	X	A	A	B	A	A	C	B	C	A	A
Hydrogen Peroxide — 30%	H_2O_2	A	X	A	A	X	A	B	C	B	X	A	A
Hydrogen Peroxide — 90%	H_2O_2		X	A	A	X	*	A	X	C	B	A	A
Hydrogen Sulfide (Wet)	H_2S	A	A	A	A	X	A	A	X	A	C	A	X
Hydroquinone	$C_6H_4(OH)_2$			A	A	B	B	A	C	*	X	A	C
Hydroxyacetic Acid — 10%	$HOCH_2COOH$			A	B	*	*	B	X	*	X	A	*
Hypochlorous Acid	HClO	A		A	X	X	A	X	X	B	X	A	A
Ink				A	C	X	A	A	A	*	A	A	A
Iodine	I_2	A	X	A	A	X	A	X	B	B	B	A	A
Iodoform	CHI_3			A	A	A	A	A	*	A	*	A	*
Isoamyl Acetate	$CH_3CO_2CH_2CH_2CH(CH_3)_2$			A	A	A	A	A	X	B	X	A	X
Isoamyl Alcohol	$(CH_3)_2CHCH_2CH_2OH$			A	*	*	*	*	A	A	A	A	A
Isoamyl Butyrate	$C_9H_{18}O_2$			A	A	A	A	A	X	*	*	A	X
Isoamyl Chloride	$(CH_3)_2CHCH_2CH_2Cl$			A	X	*	*	*	X	X	X	A	A
Isobutyl Acetate				A	A	A	A	A	X	C	X	A	X
Isobutyl Alcohol (Isobutanol)	$(CH_3)_2CHCH_2OH$	A	A	A	A	*	*	*	B	A	B	A	A
Isobutyl Amine	$(CH_3)_2CHCH_2NH_2$			A	*	*	*	*	X	*	*	A	X
Isobutyl Chloride	$(CH_3)_2CHCH_2Cl$			A	X	B	A	B	X	*	*	A	B
Isobutyric Acid	$(CH_3)_2CHCOOH$			A	A	*	*	*	X	A	B	A	*
Isododecane	$(CH_3)_2CH(CH_2)_8CH_3$			A	B	B	B	B	B	X	A	A	A
Isooctane (Trimethylpentane)	C_8H_{18}	A	A	A	A	A	A	A	A	X	B	A	A
Isopentane	$(CH_3)_2CHCH_2CH_3$			A	*	*	*	*	A	*	*	A	A
Isophorone	$C_9H_{14}O$			A	A	A	A	A	X	C	X	A	X
Isopropyl Acetate	$CH_3COOCH(CH_3)_2$	B		A	A	A	A	A	X	B	X	A	X
Isopropyl Alcohol (Isopropanol)	$CH_3CH(OH)CH_3$	A	A	A	A	A	A	A	B	B	A	A	A
Isopropyl Amine	$C_3H_7NH_2$			A	*	A	*	A	X	*	*	A	X

RATING KEY: (A) EXCELLENT   (B) GOOD   (C) FAIR TO POOR   (X) NOT RECOMMENDED
(*) OR (—) NO DATA AVAILABLE

CHEMICAL	FORMULA	PLASTIC			METAL				ELASTOMER				
		POLYPROPYLENE	RYTON®	TEFLON®	ALUMINUM	CAST IRON	HASTELLOY-C®	STAINLESS STEEL	BUNA-N	E.P.D.M.	NEOPRENE	TEFLON®	VITON®
Isopropyl Chloride	(CH₃)₂CHCl	X		A	X	A	A	A	X	X	X	A	B
Isopropyl Ether	(CH₃)₂CHOCH(CH₃)₂	X		A	B	*	*	A	C	X	C	A	C
Jet Fuels (JP1 to JP6) (ASTM-A, A1 & B)		X	A	A	A	A	A	A	A	X	C	A	A
Kerosine (Kerosene)	Hydrocarbons	X	A	A	A	A	A	A	A	X	C	A	A
Lacquers				A	X	B	A	A	X	X	X	A	X
Lacquer Solvents		C		A	X	B	A	A	X	X	X	A	X
Lactic Acid	CH₃CHOH COOH	A	A	A	A	X	A	A	B	A	B	A	A
Lactol (Aliphatic Naptha Solvent)	CH₃CHOH CO₂C₁₀H₇			A	A	A	A	A	C	*	X	A	A
Lard (Lard Oil)	Olein, stearin	A		A	A	A	A	B	A	X	C	A	A
Latex	Rubber emulsion	A		A	A	*	*	A	A	*	A	A	*
Lauryl Alcohol (n-Dodecanol)	CH₃(CH₂)₁₀CH₂OH			A	A	A	A	A	A	*	*	A	B
Lavender Oil	Ester mixture			A	*	*	*	*	B	X^	X	A	B
Lead Acetate (Sugar of Lead)	Pb(CH₃CO₂)₂	A	A	A	X	*	B	B	B	A	A	A	X
Lead Chloride	PbCl₂			A	X	*	B	B	*	*	B	A	*
Lead Nitrate	Pb(NO₃)₂	A		A	X	B	B	B	B	A	A	A	A
Lead Sulfamate		A		A	—	—	—	—	B	—	A	A	A
Lemon Oil (Cedro Oil)	Hydrocarbons			A	A	*	*	A	*	*	C	A	A
Ligroin (Ligroine) (Benzine)	Petroleum fraction	X		A	*	A	*	A	A	X	B	A	A
Lignin Liquor	Blend of natural aromatic oils			A	*	*	*	A	A	*	A	A	A
Lime, Soda (Slaked Lime & Soda Ash)	CaO			A	*	*	*	*	B	A	B	A	B
Lime Bleach		B		A	X	*	*	*	A	A	C	A	A
Lime Slurries				A	B	*	*	B	B	*	A	A	B
Lime Sulfur	CaS + CaSO₄	A		A	X	*	*	A	A	A	A	A	A
Limonene	C₁₀H₁₆			A	*	*	*	*	C	X	X	A	A
Linoleic Acid	C₁₈H₃₂O₂	A		A	A	*	A	A	B	X	X	A	B
Linseed Oil (Flaxseed Oil)	Glycerides	A	A	A	A	A	A	A	A	C	A	A	A
Lindol (Tritolyl Phosphate)	C₂₁H₂₁O₄P			A	—	—	—	—	X	—	C	A	B
Lithium Bromide	LiBrH₂O			A	*	A	*	*	A	*	X	A	A
Lubricating Oils (Petroleum)	Hydrocarbons	C	A	A	A	A	A	A	A	X	B	A	A
Lye (Potassium Hydroxide)	KOH	A	A	A	—	—	—	A	C	—	B	A	B
Magnesium Carbonate	MgCO₃	A		A	A	B	B	B	A	C	A	A	A
Magnesium Chloride	MgCl₂O	A	A	A	A	B	A	B	A	A	A	A	A
Magnesium Hydroxide (Milk of Magnesia)	Mg(OH)₂	A	A	A	A	A	A	A	B	A	B	A	A

RATING KEY: (A) EXCELLENT   (B) GOOD   (C) FAIR TO POOR   (X) NOT RECOMMENDED   (*) OR (—) NO DATA AVAILABLE

	PLASTIC			METAL				ELASTOMER				
	POLYPROPYLENE	RYTON	TEFLON	ALUMINUM	CAST IRON	HASTELLOY-C	STAINLESS STEEL	BUNA-N	E.P.D.M.	NEOPRENE	TEFLON	VITON
CHEMICAL / **FORMULA**												
Magnesium Nitrate — $Mg(NO_3)_2 \cdot 6H_2O$	A	A	A	B	B	B	A	A	A	A	A	A
Magnesium Oxide — MgO			A	A	A	A	A	A	·	A	A	B
Magnesium Sulfate (Epsom Salts) — $MgSO_4 \cdot 7H_2O$	A	A	A	A	A	A	A	A	A	A	A	A
Maleic Acid — $(CHCOOH)_2$	A			A	B	A	B	X	X	A	A	A
Maleic Anhydride — $C_4H_2O_3$			A	A	B	A	A	·	X	·	A	A
Malic Acid (Apple Acid) — $C_4H_6O_5$			A	B	·	B	A	B	X	C	A	A
Maple Sugar Liquors (Sucrose) — Water, sucrose			A	·	·	·	A	A	A	A	A	A
Mayonnaise — Water, fats, oils	A		A	X	X	A	A	A	·	A	A	·
Mercuric Chloride — $HgCl_2$	A		A	X	X	B	X	A	A	B	A	A
Mercuric Cyanide — $Hg(CN)_2$	A		A	X	B	B	B	B	A	B	A	A
Mercurous Nitrate — $Hg_2(NO_3)_2 \cdot 2H_2O$	A		A	X	B	B	B	B	A	B	A	A
Mercury — Hg	A		A	X	A	A	A	A	A	A	A	A
Mesityl Oxide — $(CH_3)_2C = CHCOCH_3$			A	A	A	A	A	X	B	X	A	X
Methane — CH_4	A		A	A	A	A	A	A	X	B	A	A
Methyl Acetate — $CH_3CO_2CH_3$	C		A	A	A	A	A	X	C	C	A	X
Methyl Acetoacetate — $CH_3COCH_2COOCH_3$			A	·	A	A	A	X	·	·	A	X
Methyl Acrylate — $CH_2CHCO_2CH_3$			A	·	A	·	A	·	C	C	A	X
Methyl Acrylic Acid (Crotonic Acid) — $CH_3(CH)_2COOH$			A	·	·	·	·	·	C	C	A	X
Methyl Alcohol (Methanol) — CH_3OH	A	A	A	B	A	A	A	A	A	A	A	B
Methyl Amine (Monomethylamine) — CH_3NH_2	A		A	B	B	B	A	B	A	A	A	A
Methyl Amyl Acetate — $C_8H_{16}O_2$			A	A	A	A	A	A	·	·	A	X
Methyl Amyl Alcohol — $C_6H_{13}OH$			A	A	A	A	A	A	·	·	A	X
Methyl Aniline — $C_6H_5NH(CH_3)$			A	·	·	·	·	A	A	A	A	·
Methyl Bromide (Bromo Methane) — CH_3Br	X		A	X	A	B	A	C	A	X	A	A
Methyl Butyl Ketone (2-hexanone) — $CH_3COC_4H_9$	X		A	·	·	·	A	X	B	X	A	X
Methyl Butyrate — $CH_3(CH_2)_2CO_2CH_3$			A	A	A	A	A	X	X	X	A	·
Methyl Cellosolve® — $CH_3OCH_2CH_2OH$	A		A	A	—	—	—	X	—	X	A	X
Methyl Chloride — CH_3Cl	X	A	A	X	A	A	A	X	C	X	A	B
Methyl Cyclopentane — C_6H_{12}			A	·	·	·	A	B	X	X	A	A
Methyl Dichloride — CH_2Cl_2	X		A	X	—	—	—	X	—	X	—	A
Methyl Ethyl Ketone (Butanone) — $CH_3CO \cdot CH_2CH_3$	X	A	A	A	A	A	A	X	A	X	A	X
Methyl Formate — $HCOOCH_3$			A	A	A	·	A	X	C	B	A	X
Methyl Hexane — C_7H_{16}			A	·	·	·	·	A	X	A	A	A

RATING KEY: (A) EXCELLENT   (B) GOOD   (C) FAIR TO POOR   (X) NOT RECOMMENDED
(*) OR (—) NO DATA AVAILABLE

CHEMICAL	FORMULA	POLYPROPYLENE	RYTON®	TEFLON®	ALUMINUM	CAST IRON	HASTELLOY-C®	STAINLESS STEEL	BUNA-N	E.P.D.M.	NEOPRENE	TEFLON®	VITON®
Methyl Iodide	CH_3I			A	X	A	A	A	X	A	X	A	*
Methyl Isobutyl Ketone (Hexone)	$CH_3COCH_2CH(CH_3)_2$	C	A	A	A	B	A	B	X	C	X	A	X
Methyl Isopropyl Ketone	$CH_3COCH(CH_3)_2$	C		A	*	*	*	A	X	C	X	A	X
Methyl Methacrylate	$CH_2C(CH_3)CO_2CH_3$			A	B	*	*	A	X	X	X	A	C
Methyl Oleate	$C_{19}H_{36}O_2$			A	*	*	*	*	X	C	X	A	B
Methyl Propyl Ketone	$CH_3CH_2CH_2COCH_3$			A	*	*	*	*	X	B	X	A	X
Methyl Salicylate (Betula Oil)	$HOC_6H_4COOCH_3$			A	A	A	*	*	X	C	X	A	B
Methylacrylic Acid	$CH_3CHCHCO_2H$			A	—	—	—	—	—	—	B	A	B
Methylamine	CH_3NH_2	A		A	B	B	B	A	B	A	A	A	A
Methylene Bromide	CH_2Br_2			A	X	A	A	A	X	*	X	A	B
Methylene Chloride	CH_2Cl_2	X	A	A	X	B	A	A	X	X	X	A	B
Milk		A		A	A	X	A	A	B	A	A	A	A
Mine Water				A	B	*	A	B	A	*	*	A	*
Mineral Oil (Petroleum)	Hydrocarbons	B	A	A	A	A	A	A	A	X	B	A	A
Mixed Acids (Sulfuric & Nitric)	H_2SO_4, HNO_3	X		A	X	X	B	B	X	B	X	A	B
Molasses		A		A	A	A	A	A	A	A	A	A	A
Monochlorobenzene	C_6H_5Cl	X	A	A	X	A	—	A	X	—	X	A	A
N-Methyl Aniline	$C_6H_5NHCH_3$	C		A	—	—	—	—	X	—	X	A	C
Monoethanolamine	$NH_2C_2H_4OH$	X	A	A	B	A	—	A	B	—	C	A	C
				A									
				A									
Mustard		A		A	B	X	A	A	C	*	A	A	X
Naphtha (Petroleum Spirits) (Thinner)	Petroleum fractions	X	A	A	A	B	A	A	A	X	X	A	A
Naphtha Coal Tar (Benzol)	Hydrocarbons			A	A	B	A	A	X	X	X	A	A
Naphthalene (Tar Camphor)	$C_{10}H_8$	A	A	A	B	A	A	A	X	X	X	A	A
Naphthoic Acid	$C_{11}H_8O_2$			A	B	B	B	A	B	X	*	A	A
Neatsfoot Oil				A	*	*	*	*	A	C	*	A	A
Neohexane (2,2-dimethylbutane)	C_6H_{14}			A	*	*	*	*	A	*	*	A	A
Neosol				A	B	B	A	A	A	B	A	A	C
Neville Acid				A	*	*	*	*	C	C	C	A	B
Nickel Acetate	$Ni(CH_3CO_2)_2$			A	B	*	*	A	B	A	B	A	X
Nickel Chloride	$NiCl_2$	A	A	A	X	X	A	B	A	A	A	A	A
Nickel Nitrate	$Ni(NO_3)_2 \bullet 6H_2O$	A	A	A	X	*	B	A	A	A	A	A	A
Nickel Sulfate	$NiSO_4$	A	A	A	X	X	B	A	A	A	A	A	A
Nitrana (Ammonia Fertilizer)				A	*	*	*	A	B	*	B	A	C

RATING KEY: (A) EXCELLENT   (B) GOOD   (C) FAIR TO POOR   (X) NOT RECOMMENDED
(*) OR (—) NO DATA AVAILABLE

CHEMICAL	FORMULA	PLASTIC			METAL				ELASTOMER				
		POLYPROPYLENE	RYTON®	TEFLON®	ALUMINUM	CAST IRON	HASTELLOY-C®	STAINLESS STEEL	BUNA-N	E.P.D.M.	NEOPRENE	TEFLON®	VITON®
Nitric Acid — 10%	HNO_3	A	X	A	A	X	A	A	X	B	B	A	A
Nitric Acid — 25%	HNO_3	A	X	A	X	X	A	A	X	B	C	A	A
Nitric Acid — 35%	HNO_3		X	A	X	X	A	A	X	C	X	A	A
Nitric Acid — 50%	HNO_3	C	X	A	X	X	X	A	X	X	X	A	A
Nitric Acid — 70%	HNO_3		X	A	*	X	X	A	X	X	X	A	A
Nitric Acid (Conc.)	HNO_3	X	X	A	A	X	A	A	X	X	X	A	B
Nitric Acid (Red Fuming)		X		A	A	X	B	A	X	X	X	A	B
Nitrobenzene	$C_6H_5NO_2$	A	A	A	A	A	B	A	X	X	X	A	B
Nitroethane	$C_2H_5NO_2$	C		A	A	A	A	A	X	C	C	A	X
Nitrogen Tetroxide	N_2O_4	X		A	A	B	A	A	X	X	X	A	C
Nitromethane	CH_3NO_2	C	A	A	A	A	A	A	X	C	C	A	X
1-Nitropropane	$CH_3(CH_2)_2NO_2$			A	A	A	A	A	X	A	C	A	X
Octadecane	$CH_3(CH_2)_{16}CH_3$			A	*	*	*	*	A	X	B	A	A
n-Octane	C_8H_{18}	X		A	*	*	*	*	A	X	*	A	A
Octyl Acetate	CH_3COO $(CH_2)_7CH_3$			A	A	*	*	A	X	*	*	A	X
Oleic Acid (Red Oil)	$C_{18}H_{34}O_2$	B	A	A	A	C	A	B	C	C	X	A	B
Octachlorotoluene	C_7Cl_8	X		A	X	—	—	—	X	—	X	A	A
Oleum (Fuming Sulfuric Acid)	H_2SO_4/SO_3	X		A	X	X	—	A	C	—	X	A	A
Olein (Triolene)	$C_{57}H_{104}O_6$			A	*	*	*	*	B	*	C	A	A
o-Dichlorobenzene	$C_6H_4Cl_2$	X		A	X	A	—	A	X	—	X	A	A
Olive Oil	mixed glycerides of acids	A		A	A	A	A	A	A	C	C	A	A
Oxalic Acid	$(COOH)_2$	A	A	A	B	X	B	B	C	A	B	A	C
Ozone	O_3	X		A	A	A	A	A	X	A	B	A	A
Paints & Solvents				A	X	*	A	A	X	*	X	A	*
Paint Thinner, DUCO	Hydrocarbons	X		A	X	*	A	A	A	X	C	A	B
Palm Oil	Mixture of terpenes			A	*	A	A	A	A	*	C	A	A
Palmitic Acid	$CH_3(CH_2)_{14}$ $COOH$	A		A	B	B	*	A	B	B	C	A	B
Paraffins (Paraffin Oil)	Hydrocarbons	A		A	A	*	A	A	A	*	*	A	*
Paraformaldehyde	$(CH_2O)_n$			A	A	A	A	A	B	*	B	A	C
Paraldehyde	$C_6H_{12}O_3$			A	A	A	A	A	C	A	B	A	X
Peanut Oil	Glycerides of fatty acids			A	*	A	A	A	A	X	B	A	A
Pentachloroethane (Pentalin)	Cl_2 $CHCCl_3$			A	X	A	A	A	X	*	X	A	A
Pentachlorophenol (PCP)	C_6Cl_5OH			A	A	A	A	A	X	X	X	A	A
Pentane (Amyl Hydride)	C_5H_{12}			A	A	B	*	B	A	X	B	A	A
Peppermint Oil				A	*	*	*	*	X	*	X	A	A
Perchloric Acid	$HClO_4$		A	A	X	X	*	B	X	B	B	A	A

RATING KEY: (A) EXCELLENT (B) GOOD (C) FAIR TO POOR (X) NOT RECOMMENDED (*) OR (—) NO DATA AVAILABLE

CHEMICAL	FORMULA	PLASTIC			METAL				ELASTOMER				
		POLYPROPYLENE	RYTON®	TEFLON®	ALUMINUM	CAST IRON	HASTELLOY-C®	STAINLESS STEEL	BUNA-N	E.P.D.M.	NEOPRENE	TEFLON®	VITON®
Perchloroethylene (Tetrachloroethylene)	C_2Cl_4	X	A	A	X	B	B	A	X	X	X	A	A
Petroleum (Crude Oil) (Sour)	Hydrocarbons	B		A	B	B	A	A	B	X	C	A	A
Phenethyl Alcohol (Benzyl Carbinol)	$C_6H_5(CH_2)_2OH$			A	A	A	A	A	X	B	X	A	X
Phenol (Carbolic Acid)	C_6H_5OH	C	A	A	B	A	A	B	X	C	C	A	A
Phenol Sulfonic Acid	$C_6H_4(OH)SO_3H$			A	B	B	*	B	X	*	*	A	X
Phenyl Acetate	$CH_3COOC_6H_5$			A	*	*	*	*	X	B	X	A	X
Phenylbenzene	C_6H_5			A	—	—	—	—	X	—	X	A	A
Phenyl Ethyl Ether (Phenetole)	$C_6H_5OC_2H_5$			A	*	*	*	*	X	X	X	A	C
Phenyl Hydrazine	$C_6H_5NHNH_2$	X		A	A	X	*	*	X	X	X	A	A
Phorone (Diisopropylidene Acetone)	$C_9H_{14}O$			A	*	*	*	*	X	C	X	A	A
Phosphoric Acid — 10%	H_3PO_4	A	A	A	X	X	*	A	A	A	B	A	A
Phosphoric Acid — 20%	H_3PO_4	A	A	A	X	X	A	A	C	A	B	A	A
Phosphoric Acid — 50%	H_3PO_4	A	A	A	X	X	C	A	X	B	B	A	A
Phosphoric Acid (Conc.)	H_3PO_4	A	A	A	X	X	*	A	X	B	B	A	A
Phosphorus Oxychloride	$POCl_3$			A	B	B	B	B	*	*	X	A	*
Phosphorus Trichloride	PCl_3	A	A	A	C	B	A	A	X	A	X	A	A
Photographic Developer		A	A	A	C	X	A	A	A	—	A	—	A
Pickling Solution				A	*	*	A	*	*	X	X	A	B
Picric Acid (Carbazotic Acid)	$(NO_2)_3$ C_6H_2OH	B		A	A	C	B	A	B	B	B	A	A
Pine Oil (Yarmor)	Cyclic terpene alcohols			A	A	B	*	*	B	X	X	A	A
Pinene	$C_{10}H_{16}$			A	*	*	*	*	B	X	X	A	A
Piperidine	$C_5H_{11}N$			A	*	*	*	*	X	X	X	A	X
Plating Solution — Cadmium		A		A	*	*	*	A	B	*	B	A	*
Plating Solution — Chrome		A		A	*	*	A	*	X	C	X	A	A
Plating Solution — Lead		A		A	*	*	*	*	B	*	B	A	*
Plating Solution — Others				A	*	*	*	*	A	A	C	A	B
Polyvinyl Acetate Emulsion	$PVac + H_2O$			A	*	B	*	*	*	A	C	A	*
Potassium Acetate	CH_3CO_2K	A		A	B	A	B	B	B	A	B	A	X
Potassium Bicarbonate	$KHCO_3$	A	A	A	B	B	B	A	A	*	A	A	A
Potassium Bisulfate	$KHSO_4$			A	A	X	*	A	A	*	A	A	A
Potassium Bisulfite	$KHSO_3$			A	B	*	B	B	A	*	A	A	A
Potassium Bromide	KBr	A	A	A	A	B	A	B	A	A	A	A	A
Potassium Carbonate (Potash)	K_2CO_3	A	A	A	X	B	A	B	A	A	A	A	A
Potassium Chlorate	$KClO_3$	A	A	A	X	B	A	B	A	A	A	A	A
Potassium Chloride	KCl	A	A	A	X	B	A	A	A	A	A	A	A

RATING KEY: (A) EXCELLENT   (B) GOOD   (C) FAIR TO POOR   (X) NOT RECOMMENDED (*) OR (—) NO DATA AVAILABLE

CHEMICAL	FORMULA	PLASTIC			METAL				ELASTOMER				
		POLYPROPYLENE	RYTON®	TEFLON®	ALUMINUM	CAST IRON	HASTELLOY-C®	STAINLESS STEEL	BUNA-N	E.P.D.M.	NEOPRENE	TEFLON®	VITON®
Potassium Chromate	K_2CrO_4	A		A	A	A	*	A	A	*	A	A	A
Potassium Copper Cyanide	$K_3[Cu(CN)_4]$			A	*	*	*	*	A	A	A	A	A
Potassium Cyanide	KCN	A	A	A	C	B	B	B	A	A	A	A	A
Potassium Dichromate	$K_2Cr_2O_7$	A	A	A	A	A	B	A	A	A	A	A	A
Potassium Hydroxide (Caustic Potash) (Lye)	KOH	A	A	A	X	B	B	A	B	A	B	A	B
Potassium Iodide	KI	A		A	B	*	B	B	A	A	A	A	A
Potassium Nitrate (Saltpeter)	KNO_3	A	A	A	A	B	B	B	A	A	A	A	A
Potassium Nitrite	KNO_2			A	B	B	B	B	A	A	A	A	A
Potassium Permanganate (Purple Salt)	$KMnO_4$	B	A	A	A	B	A	B	C	A	C	A	B
Potassium Phosphate	KH_2PO_4			A	X	X	B	B	A	A	A	A	A
Potassium Silicate	$K_2Si_2O_5$			A	B	B	B	B	A	A	A	A	A
Potassium Sulfate	K_2SO_4	A	A	A	B	B	A	A	A	A	A	A	A
Potassium Sulfide	K_2S	A		A	X	B	B	B	A	A	A	A	A
Potassium Sulfite	$K_2SO_3 2H_2O$	A		A	A	X	*	B	A	A	A	A	A
Propane (LPG)	C_3H_8	X		A	A	A	A	A	A	X	B	A	A
Propionaldehyde (Propanal)	C_2H_5CHO			A	A	A	A	A	X	*	*	A	X
Propionic Acid (Methylacetic Acid)	$CH_3CH_2CO_2H$			A	A	X	A	B	X	A	X	A	A
n-Propyl Acetate	$CH_3COO(CH_2)_2CH_3$	C		A	A	*	A	A	X	A	X	A	X
Propyl Alcohol (1-Propanol)	$CH_3CH_2CH_2OH$	A	A	A	A	A	A	A	B	A	B	A	A
n-Propyl Nitrate (NPN)	$CH_3(CH_2)_2NO_3$			A	A	X	*	*	A	B	*	A	C
Propylene	C_3H_6			A	A	A	A	A	X	X	X	A	A
Propylene Dichloride	$CH_3CH(Cl)CH_2Cl$			A	X	A	B	A	X	X	X	A	B
Propylene Glycol (Methyl Glycol)	$C_3H_6(OH)_2$	A	A	A	A	A	A	A	A	A	C	A	A
Propylene Oxide	C_3H_6O	C		A	B	B	*	A	*	C	X	A	X
Pydraul (Phosphate Ester Base Fluid)				A	*	A	A	A	X	B	X	A	A
Pyranol				A	—	—	—	—	A	—	X	A	A
Pyridine	$N(CH)_4CH$	C	A	A	A	B	A	A	X	C	X	A	X
Pyroligneous Acid (Wood Vinegar)		A	A	A	B	X	*	A	C	C	C	A	A
Pyrrole (Azole)	C_4H_5N			A	*	*	*	*	X	X	X	A	C
Quaternary Ammonium Salts	$NH_4(X)$			A	*	X	*	A	A	*	A	A	A
Quench Oil				A	A	*	A	A	B	*	B	A	A
Rape-Seed Oil (Colza Oil)				A	*	A	A	A	B	A	C	A	A
Rose Oil	Geraniol, citronellol			A	*	*	*	A	*	*	C	A	A
Rosin	$C_{20}H_{30}O_2$	A		A	A	*	A	A	A	*	C	A	*
Rosin Oil (Rosinol)				A	*	*	*	*	A	*	A	A	A

RATING KEY: (A) EXCELLENT (B) GOOD (C) FAIR TO POOR (X) NOT RECOMMENDED (*) OR (—) NO DATA AVAILABLE

CHEMICAL	FORMULA	PLASTIC			METAL				ELASTOMER				
		POLYPROPYLENE	RYTON	TEFLON	ALUMINUM	CAST IRON	HASTELLOY-C	STAINLESS STEEL	BUNA-N	E.P.D.M.	NEOPRENE	TEFLON	VITON
Rotenone	$C_{23}H_{22}O_6$			A	·	·	·	·	A	A	A	A	A
Rubber Latex Emulsions	$(C_5H_8)_n/H_2O$			A	A	·	A	A	·	·	·	A	A
Rubber Solvents (Petroleum Distillate)	Hydrocarbons			A	A	·	A	A	X	·	C	A	X
Rum	Alcoholic liquor from molasses			A	·	·	A	A	A	A	A	A	B
Rust Inhibitors		A		A	—	—	—	A	A	—	C	—	A
Salad Dressing	Fats, oils, water	A		A	B	X	—	A	A	—	—	—	A
Sal Ammoniac (Ammonium Chloride)	NH_4Cl			A	X	X	—	A	A	—	A	A	A
Sal Soda (Sodium Carbonate)	$NaCO_3$			A	X	A	A	A	A	A	A	A	A
Salicylic Acid	HOC_6H_4COOH	A		A	A	X	A	B	B	A	B	A	B
Salt Water (Brine)	$NaCl/H_2O$	A		A	B	X	A	A	A	A	B	A	A
Sea Water	(Brine)	A	A	A	A	C	A	A	A	A	B	A	A
Sesame Seed Oil	Olein, stearin, palmitin			A	·	A	·	A	A	·	C	A	A
Sewage		A		A	B	B	A	A	A	C	B	A	A
Silicate Esters	$Si(OR)_4$			A	·	·	·	·	B	X	A	A	A
Silicone Oils (Versilube Etc.)	$[(CH_3)_2SiO_2]_n$	A	A	A	B	B	A	A	A	C	C	A	A
Silver Cyanide	AgCN	A		A	X	A	A	A	·	·	A	A	·
Silver Nitrate	$AgNO_3$	A	A	A	X	X	A	A	B	A	A	A	A
Skydrol Hydraulic Fluid® (Phosphate Ester Base)				A	·	·	A	A	X	A	X	A	C
Soap Solutions	Salt of fatty acid in H_2O	A	A	A	C	X	A	A	A	A	B	A	A
Soda Ash (Sodium Carbonate)	Na_2CO_3			A	X	A	A	A	A	A	A	A	A
Sodium Acetate	CH_3COONa	A	A	A	A	A	A	A	C	A	C	A	X
Sodium Aluminate	$Na_2Al_2O_4$	A		A	·	A	B	A	A	·	A	A	A
Sodium Bicarbonate (Baking Soda)	$NaHCO_3$	A	A	A	B	C	A	A	A	A	A	A	A
Sodium Bisulfite (Niter Cake)	$NaHSO_4$	A	A	A	B	C	B	B	A	A	A	A	A
Sodium Bisulfite	$NaHSO_3$	A		A	B	B	B	A	C	A	A	A	A
Sodium Borate	$Na_2B_4O_7$	A	A	A	B	·	A	A	A	A	A	A	A
Sodium Bromide	NaBr	A		A	C	C	B	B	·	·	·	A	·
Sodium Chlorate	$NaClO_3$	A	A	A	B	B	B	B	A	A	B	A	A
Sodium Chloride (Table Salt)	NaCl	A	A	A	B	B	A	A	A	A	A	A	A
Sodium Chromate	Na_2CrO_4	A		A	A	A	A	A	A	·	A	A	A
Sodium Cyanide	NaCN	A	A	A	X	A	·	A	A	A	A	A	A
Sodium Dichromate (Sodium Bichromate)	$Na_2Cr_2O_7 \cdot 2H_2O$	A	A	A	·	·	·	·	·	A	B	A	A
Sodium Fluoride	NaF	A		A	B	·	B	B	A	A	A	A	A

RATING KEY: (A) EXCELLENT (B) GOOD (C) FAIR TO POOR (X) NOT RECOMMENDED (*) OR (—) NO DATA AVAILABLE

CHEMICAL	FORMULA	PLASTIC			METAL				ELASTOMER				
		POLYPROPYLENE	RYTON®	TEFLON®	ALUMINUM	CAST IRON	HASTELLOY-C®	STAINLESS STEEL	BUNA-N	E.P.D.M.	NEOPRENE	TEFLON®	VITON®
Sodium Hexametaphosphate (Calgon)	(NaPO$_3$)$_6$			A	C	B	A	B	B	B	B	A	A
Sodium Hydroxide (Caustic Soda) (Lye)	NaOH	A	X	A	X	B	B	A	B	A	B	A	X
Sodium Hypochlorite	NaClO	B	X	A	X	X	B	X	X	C	B	A	B
Sodium Metaphosphate (Kurrol's Salt)	Na(PO$_3$)H	X		A	X	*	A	B	B	A	C	A	A
Sodium Metasilicate	Na$_2$SiO$_3$			A	B	—	A	A	A	—	A	—	A
Sodium Nitrate (Chile Saltpeter)	NaNO$_3$	A	A	A	A	A	A	A	C	A	B	A	A
Sodium Nitrite	NaNO$_2$			A	A	A	A	A	A	*	X	A	A
Sodium Perborate	NaBO$_3$	A		A	X	B	B	A	C	A	B	A	A
Sodium Peroxide (Sodium Dioxide)	Na$_2$O$_2$	B		A	B	A	B	B	B	B	B	A	A
Sodium Phosphate (Tribasic) (TSP)	Na$_3$PO$_4$	A		A	X	B	B	B	B	A	B	A	A
Sodium Silicates (Water Glass)	Na$_2$O • SiO$_2$	A	A	A	A	A	B	A	A	A	A	A	A
Sodium Sulfate (Salt Cake) (Thenardite)	Na$_2$SO$_4$	A	A	A	B	B	A	A	A	A	B	A	A
Sodium Sulfide (Pentahydrate)	Na$_2$S • 5H$_2$O	A	A	A	A	B	B	A	A	A	A	A	A
Sodium Sulfite	Na$_2$SO$_3$	A	A	A	A	X	B	A	A	A	A	A	A
Sodium Tetraborate	Na$_2$B$_4$O$_7$ 10H$_2$O	C	A	A	—	—	—	A	A	—	—	A	A
Sodium Thiosulfate (Antichlor)	Na$_2$S$_2$O$_3$	A	A	A	A	C	B	A	A	A	A	A	A
Sorgum				A	*	A	A	A	A	*	A	A	*
Soybean Oil	Triglycerides of acids	B	A	A	A	A	A	A	A	C	A	A	A
Soy Sauce	Fermented soya bean/wheat			A	*	X	*	A	A	*	A	A	*
Sperm Oil (Whale Oil)	Fatty acid esters			A	*	A	A	A	A	*	X	A	A
Stannic Chloride (Tin Chloride)	SnCl$_4$	A	A	A	X	C	B	A	A	B	B	A	A
Stannous Chloride (Tin Salt)	SnCl$_2$	A	A	A	X	B	A	A	A	B	A	A	A
Starch	C$_6$H$_{10}$O$_5$	A	A	A	A	C	A	A	A	B	A	A	C
Stearic Acid	CH$_3$(CH$_2$)$_{16}$ CO$_2$H	A		A	C	C	B	A	B	B	B	A	A
Stoddard Solvent	Petroleum distillate	A		A	A	A	X	A	A	XA	C	A	*
Styrene (Vinylbenzene)	C$_6$H$_5$CHCH$_2$			A	A	A	A	A	X	X	X	A	A
Sucrose Solution (Sugar)	C$_{12}$H$_{22}$O$_{11}$/H$_2$O			A	A	A	A	A	A	A	A	A	A
Sulfamic Acid	H$_2$NSO$_3$H			A	A	X	*	X	B	*	A	A	*
Sulfite Liquors				A	*	*	A	*	A	C	B	A	A
Sulfur	S	A	A	A	A	A	B	A	X	A	B	A	A

RATING KEY: (A) EXCELLENT (B) GOOD (C) FAIR TO POOR (X) NOT RECOMMENDED (*) OR (—) NO DATA AVAILABLE

		PLASTIC			METAL				ELASTOMER				
CHEMICAL	FORMULA	POLYPROPYLENE	RYTON®	TEFLON®	ALUMINUM	CAST IRON	HASTELLOY-C®	STAINLESS STEEL	BUNA-N	E.P.D.M.	NEOPRENE	TEFLON®	VITON®
Sulfur Chloride	S_2Cl_2	X		A	B	X	A	B	C	X	X	A	A
Sulfur Dioxide	SO_2	A	A	A	A	B	A	A	X	B	A	A	A
Sulfur Hexafluoride	SF_6			A	*	*	*	*	B	A	A	A	A
Sulfur Trioxide	SO_3	X		A	B	B	B	B	C	C	C	A	A
Sulfuric Acid 10%	H_2SO_4	A	X	A	X	X	A	A	B	A	A	A	A
Sulfuric Acid 25%	H_2SO_4	A	X	A	X	X	A	B	C	B	B	A	A
Sulfuric Acid 50%	H_2SO_4	A	X	A	X	X	A	X	C	B	B	A	A
Sulfuric Acid 60%	H_2SO_4	A	X	A	X	X	A	X	X	B	C	A	A
Sulfuric Acid 75%	H_2SO_4	A	X	A	X	C	A	C	X	C	X	A	A
Sulfuric Acid 95%	H_2SO_4	X	X	A	X	B	A	A	X	C	X	A	A
Sulfuric Acid (Conc.)	H_2SO_4	X		A	X	B	A	B	X	C	X	A	A
Sulfuric Acid (Fuming)	H_2SO_4			A	C	X	B	B	X	X	X	A	B
Sulfurous Acid	H_2SO_3	A	A	A	B	X	B	B	B	C	X	A	A
Tall Oil (Liquid Rosin)	Rosin acids	A		A	X	B	A	A	A	B	B .	A	A
Tallow	Fat from cattle, sheep	B		A	A	—	—	A	A	—	—	A	A
Tannic Acid	$C_{76}H_{52}O_{46}$	A	A	A	A	A	B	A	C	C	B	A	A
Tanning Liquors	Tannic acid	A		A	A	*	A	A	A	*	B	A	*
Tar, Bituminous (Coal Tar) (Pitch)	Mixture of aromatic and phenolic hydrocarbons	A		A	A	*	A	A	B	X	C	A	A
Tartaric Acid	$C_4H_6O_6$	A	A	A	A	X	A	A	B	B	A	A	A
Terpenes	C_{10} hydrocarbons			A	A	X	*	*	C	X	X	A	A
Terpineol (Terpilenol)	$C_{10}H_{18}O$	X		A	A	A	A	A	C	C	X	A	A
Tertiary Butyl Alcohol	$(CH_3)_3COH$	B		A	—	—	—	—	A		A	A	B
Tertiary Butyl Catechol	$C_9H_{14}O_2$			A	C	B	—	B	X		B	A	A
Tertiary Butyl Mercaptan	$C_4H_{10}S$			A	—	—	—	—	X		X	A	A
Tetra Bromomethane	CBr_4	X		A	X	—	—	—	X	—	X	A	A
Tetrabutyl Titanate	$Ti(C_4H_9)$			A	*	*	*	*	B	B	A	A	A
Tetrachloroethylene	$Cl_2C = CCl_2$			A									
Tetrachlorodifluoroethane	$(Cl_2FC)_2$			A	*	*	*	*	X	*	X	A	*
Tetrachloroethane (Acetylene Tetrachloride)	$(Cl_2HC)_2$	X		A	X	A	A	C	X	X	X	A	A
Tetraethyl Lead	$Pb(C_2H_5)_4$	A		A	B	A	*	A	B	X	X	A	B
Tetraethylene Glycol (TEG)	$HOCH_2(CH_2OCH_2)_3CH_2OH$			A	*	*	*	*	A	*	*	A	A
Tetrahydrofuran (THF)	C_4H_8O	C	A	A	*	*	*	*	X	C	X	A	X
Tetrahydronaphthalene (Tetralin)	$C_{10}H_{12}$	C	A	A	A	A	A	A	X	X	X	A	A
Thionyl Chloride	$SOCl_2$	B		A	C	A	A	A	X	X	X	A	B
Thiophene	C_4H_4S			A	*	*	*	*	X	X	X	A	C
Titanium Tetrachloride	$TiCl_4$	B		A	X	A	B	B	C	X	X	A	A

RATING KEY: (A) EXCELLENT (B) GOOD (C) FAIR TO POOR (X) NOT RECOMMENDED
(*) OR (—) NO DATA AVAILABLE

CHEMICAL	FORMULA	PLASTIC			METAL				ELASTOMER				
		POLYPROPYLENE	RYTON®	TEFLON®	ALUMINUM	CAST IRON	HASTELLOY-C®	STAINLESS STEEL	BUNA-N	E.P.D.M.	NEOPRENE	TEFLON®	VITON®
Toluene (Toluol)	C_7H_8	X	A	A	A	A	A	A	C	X	X	A	X
Toluene Diisocyanate	$CH_3C_6H_3(NCO)_2$			A	•	•	•	•	•	A	X	A	•
Toluidine	$CH_3C_6H_4NH_2$			A	A	A	A	A	X	•	•	A	B
Tomato Pulp & Juice		A	A	A	B	•	A	A	A	•	•	A	•
Toothpaste				A	•	X	A	A	A	•	C	A	A
Transformer Oil (Petroleum)	Hydrocarbons	B		A	A	A	A	A	B	X	C	A	A
Transmission Fluid (Type A)				A	A	A	A	A	A	X	C	A	A
Triacetin	$C_3H_5(OCOCH_3)_3$			A	B	•	•	•	A	A	B	A	X
Triallyl Phosphate	$P(OC_3H_5)_3$	B		A	•	•	•	•	X	A	C	A	A
Triaryl Phosphate	$(C_6H_5O)_3PO$			A	—	—	—	—	X	—	C	A	A
Tributoxyl Ethyl Phosphate	$(C_4H_9O)_3P(C_2H_5)$			A	•	•	•	•	X	•	X	A	B
Tributyl Phosphate (TBP)	$(C_4H_9)_3PO_4$	A		A	A	A	•	A	X	C	X	A	X
Dibutyl Mercaptan	$(C_4H_9)_2S$			A	—	—	—	—	X	•	X	A	A
Trichloroacetic Acid (TCA)	CCl_3COOH	B	A	A	X	X	B	X	C	C	B	A	B
Trichlorobenzenes	$C_6H_5Cl_3$			A	X	A	B	A	X	•	X	A	B
Trichloroethane	$C_2H_3Cl_3$	X	A	A	X	A	A	A	X	X	X	A	B
Trichloroethylene (Ex-Tri) (Hi-Tri)®	C_2HCl_3	X	A	A	X	B	A	A	X	X	X	A	C
Trichloropropane	$CH_2ClCHClCH_2Cl$	X		A	X	A	A	A	X	•	A	A	B
Tricresyl Phosphate (Lindol) (TCP)	$(CH_3C_6H_4O)_3PO$	B		A	•	A	A	B	X	A	C	A	C
Tridecyl Alcohol (Tridecanol)	$C_{12}H_{25}CH_2OH$			A	•	•	•	•	A	•	•	A	B
Triethanol Amine (TEA)	$N(C_2H_4OH)_3$	A	A	A	A	A	A	A	X	B	A	A	C
Triethyl Aluminum (ATE)	$Al(C_2H_5)_3$			A	•	•	•	•	X	•	X	A	B
Triethyl Amine	$(CH_3CH_2)_3N$	C		A	•	A	A	A	A	•	B	A	•
Triethyl Borane	$(C_2H_5)_3B$			A	•	•	•	•	X	•	X	A	A
Triethylene Glycol (TEG)	$(CH_2OCH_2CH_2OH)_2$	A		A	•	•	•	•	A	•	•	A	A
Trimethylene Glycol	$HO(CH_2)_3OH$			A	A	A	A	A	A	•	•	A	A
Trinitrotoluene (TNT)	$CH_3C_6H_2(NO_2)_3$			A	•	•	•	•	X	X	B	A	C
Trioctyl Phosphate	$(C_8H_{17}O)_3PO$			A	•	•	•	•	X	A	X	A	B
Tung Oil (Wood Oil)	Fatty acids	A		A	A	•	A	A	A	X	C	A	A
Turpentine	$C_{10}H_{16}$	X	A	A	A	A	A	A	A	X	X	A	A
Unsymmetrical Dimethyl Hydrazine (UDMN)	$H_2NN(CH_3)_2$			A	•	•	•	•	C	A	C	A	X
Urea (Carbamide)	$CO(NH_2)_2$	A	A	A	B	•	•	B	B	•	B	A	A
Urine		A		A	A	A	A	A	A	•	X	A	A
Valeric Acid	$CH_3(CH_2)_3COOH$			A	A	•	A	•	X	A	X	A	•

RATING KEY: (A) EXCELLENT (B) GOOD (C) FAIR TO POOR (X) NOT RECOMMENDED
(*) OR (—) NO DATA AVAILABLE

CHEMICAL	FORMULA	PLASTIC			METAL				ELASTOMER				
		POLYPROPYLENE	RYTON®	TEFLON®	ALUMINUM	CAST IRON	HASTELLOY-C®	STAINLESS STEEL	BUNA-N	E.P.D.M.	NEOPRENE	TEFLON®	VITON®
Vanilla Extract (Vanillin)	C$_6$H$_5$(CHO)(OCH$_3$)(OH)			A	*	*	*	A	A	*	X	A	X
Varnish	Oil, gum resins, oil of turpentine	A		A	A	*	A	A	B	X	C	A	A
Vegetable Juices				A	C	*	*	A	A	*	C	A	*
Vegetable Oils		X	A	A	A	B	A	A	B	A	C	A	A
Vinegar	Dilute acetic acid	A	A	A	C	X	A	A	C	A	B	A	A
Vinyl Acetate	CH$_3$COOCHCH$_2$	B		A	B	A	A	A	X	*	B	A	X
Vinyl Chloride (Chloroethylene)	CH$_2$CHCl	X		A	X	A	A	A	X	C	X	A	A
Walnut Oil				A	*	*	*	*	A	*	B	A	A
Water — Distilled	H$_2$O	A	A	A	A	C	A	A	A	A	C	A	A
Water — Fresh	H$_2$O	A	A	A	A	A	A	A	A	A	B	A	A
Waxes	Hydrocarbons			A	A	*	A	A	A	X	A	A	*
Weed Killers				A	X	—	—	A	B	—	C	—	A
Whiskey	Ethanol, esters, acids	A		A	A	X	A	A	B	A	A	A	A
White Oil (Mineral) (Petroleum)	Mixture of liquid hydrocarbons			A	*	*	A	A	A	X	C	A	A
White Sulfate Liquor		A		A	B	C	B	A	B	A	A	A	B
Wines		A		A	C	X	A	A	A	A	A	A	B
Wort, Distillery	Sugar solution from malt			A	A	B	A	A	*	*	A	A	A
Xylene (Xylol)	C$_6$H$_4$(CH$_3$)$_2$	X	A	A	A	B	A	B	X	X	X	A	A
Xylidines (Xylidin)	(CH$_3$)$_2$C$_6$H$_3$NH$_2$			A	B	B	*	*	*	X	X	A	X
Zeolite	Hydrated alkali aluminum silicates			A	*	*	A	A	C	A	C	A	A
Zinc Acetate	Zn(C$_2$H$_3$O$_2$)$_2$			A	C	*	*	*	C	A	B	A	X
Zinc Carbonate	ZnCO$_3$			A	B	B	B	B	A	*	*	A	A
Zinc Chloride	ZnCl$_2$	A	A	A	A	B	A	A	B	A	B	A	A
Zinc Hydrosulfite	ZnHSO$_3$			A	X	*	*	A	A	*	A	A	A
Zinc Sulfate	ZnSO$_4$	A	A	A	B	X	B	B	A	A	A	A	B

RATING KEY: (A) EXCELLENT   (B) GOOD   (C) FAIR TO POOR   (X) NOT RECOMMENDED
(*) OR (—) NO DATA AVAILABLE

PIPE CAPACITIES FLOW DATA (gpm)

Pipe size (in.)	Water		70 SSU		100 SSU		150 SSU		200 SSU		300 SSU		500 SSU	
	Gravity	Pressure	Gravity	Pressure	Gravity	Pressure	Gravity	Pressure	Gravity	Pressure	Gravity	Pressure	Gravity	Pressure
$\frac{3}{4}$	1.42	4.70	0.58	4.21	0.39	4.01	0.25	3.79	0.18	3.64	0.12	3.45	0.068	3.21
1	2.65	9.20	1.04	8.27	0.692	7.85	0.432	7.43	0.324	7.14	0.216	6.75	0.130	5.66
$1\frac{1}{4}$	5.30	18.9	3.24	16.9	2.02	16.1	1.30	15.3	0.943	14.7	0.634	13.9	0.389	12.9
$1\frac{1}{2}$	8.10	28.4	5.90	25.5	3.74	24.2	2.38	22.9	1.73	20.9	1.17	19.8	0.706	19.4
2	15.60	54.7	11.95	49.0	10.22	46.7	6.55	44.2	4.86	42.4	3.24	40.1	1.94	37.4
$2\frac{1}{2}$	25.10	87.4	23.04	78.3	19.50	77.0	13.32	17.5	9.65	67.8	6.55	64.2	3.92	59.8
3	44.50	154	33.2	138	29.5	131	27.1	124	23.2	119	15.7	113	9.22	105
4	91.00	317	80.5	284	77.0	270	72	256	69.8	246	54.0	233	32.3	217
5	164	573	139	514	131	489	123	463	116	445	99.10	421	71.0	392
6	267	930	212	834	202	794	187	751	176	722	159	683	143	570
8	550	1910	469	1710	436	1630	401	1540	379	1480	350	1400	312	1310
10	1010	3480	940	3120	885	2970	825	2810	786	2700	735	2550	670	2380
12	1610	5590	1398	5010	1305	4770	1220	4519	1106	4340	1085	4100	995	3820
14	2160	7250	1880	6500	1780	6190	1650	5850	1580	5620	1470	5320	1350	4960
16	3020	10490	2610	9410	2470	8760	2300	8480	2180	8150	2020	7700	1840	7180
18	4100	14500	3580	13200	3350	12400	3100	13200	2920	12600	2720	11000	2500	10100
20	5500	19180	4850	17200	4600	16400	4300	15500	4100	14900	3860	14100	3580	13100

PRESSURE DROP OF WATER THROUGH SCHEDULE 40 STEEL PIPE

Pressure Drop of Water per 100 ft of Schedule 40 Steel Pipe (psi)

Flow		1/8 in.		1/4 in.		3/8 in.		1/2 in.		3/4 in.		1 in.		1¼ in.		1½ in.		2 in.		2½ in.		3 in.	
gpm	ft³/s	u fps	p psi	u fps	p psi	u fps	p psi	u fps	p psi	u fps	p psi	u fps	p psi	u fps	p psi	u fps	p psi	u fps	p psi	u fps	p psi	u fps	p psi
0.1	0.00022	0.56	0.677																				
0.2	0.00045	1.14	2.48	0.62	0.548																		
0.3	0.00067	1.70	5.26	0.93	1.16	0.50	0.255																
0.5	0.00111	2.82	13.58	1.55	3.00	0.84	0.656	0.53	0.205	0.30	0.050												
0.6	0.00134	3.38	19.12	1.85	4.22	1.01	0.925	0.63	0.290	0.36	0.071												
0.8	0.00178	4.52	32.62	2.47	7.17	1.34	1.58	0.84	0.494	0.48	0.121	0.30	0.036										
1	0.00223			3.09	10.91	1.68	2.39	1.06	0.749	0.60	0.183	0.37	0.055	0.21	0.014								
2	0.00446			6.18	39.60	3.36	8.68	2.11	2.72	1.20	0.665	0.74	0.199	0.43	0.051								
3	0.00668					5.04	18.46	3.17	5.77	1.80	1.41	1.11	0.424	0.64	0.107								
4	0.00891					6.72	31.55	4.22	9.86	2.40	2.42	1.49	0.724	0.86	0.183								
5	0.01114							5.28	14.92	3.01	3.64	1.86	1.09	1.07	0.276								
6	0.01337							6.33	20.95	3.61	5.13	2.23	1.54	1.29	0.390								
8	0.01782									4.81	8.76	2.97	2.62	1.71	0.667	1.26	0.308						
10	0.02228									6.01	13.28	3.713	3.97	2.142	1.01	1.58	0.466						
15	0.03342											5.57	8.46	3.21	2.14	2.36	0.992	1.43	0.285				
20	0.04456											7.43	14.42	4.28	3.66	3.15	1.69	1.91	0.486				
25	0.05570													5.36	5.54	3.94	2.54	2.39	0.736				
30	0.06684													6.43	7.79	4.73	3.60	2.87	1.03	2.01	0.424		
35	0.07798													7.50	10.38	5.51	4.79	3.35	1.37	2.35	0.566		
40	0.08912													8.57	13.28	6.30	6.14	3.82	1.76	2.68	0.724		
50	0.1114															7.88	9.31	4.78	2.67	3.35	1.10	2.17	0.371

										3½ in.		4 in.		5 in.		6 in.		8 in.		10 in.		12 in.	
60	0.1337	9.45	13.08	5.74	3.75	4.02	1.54	2.61	0.520														
70	0.1560			6.70	4.99	4.70	2.05	3.04	0.693	2.27	0.335												
80	0.1782			7.65	6.40	5.37	2.63	3.47	0.890	2.59	0.430												
90	0.2005			8.60	7.96	6.04	3.28	3.91	1.10	2.92	0.535												
100	0.2228			9.56	9.69	6.71	3.98	4.34	1.34	3.24	0.650	2.52	0.346										
125	0.2785					8.38	6.03	5.43	2.01	4.05	0.984	3.15	0.523										
150	0.3342					10.1	8.46	6.52	2.86	4.87	1.38	3.78	0.734										
175	0.3899					11.7	11.3	7.60	3.81	5.68	1.84	4.41	0.978	2.81	0.316								
200	0.4456					13.4	14.4	8.69	4.89	6.49	2.36	5.04	1.25	3.21	0.405								
225	0.5013							9.77	6.09	7.30	2.94	5.67	1.56	3.61	0.505	2.78	0.245						
250	0.5570							10.9	7.41	8.11	3.58	6.30	1.90	4.01	0.616	3.06	0.292						
275	0.6127							11.9	8.84	8.92	4.27	6.93	2.27	4.41	0.734	3.33	0.344						
300	0.6684							13.0	10.4	9.73	5.02	7.56	2.67	4.81	0.863	3.89	0.457						
350	0.7798							15.2	13.8	11.4	6.87	8.82	3.55	5.62	1.15								
400	0.8912									13.0	8.58	10.1	4.56	6.41	1.47	4.44	0.587	2.57	0.149				
450	1.003									14.6	10.7	11.3	5.66	7.22	1.83	5.00	0.731	2.89	0.185				
500	1.114									16.2	13.0	12.6	6.89	8.02	2.23	5.55	0.887	3.21	0.225				
550	1.225									17.8	15.5	13.9	8.25	8.82	2.67	6.11	1.07	3.53	0.270				
600	1.337									19.5	18.2	15.1	9.68	9.62	3.13	6.66	1.25	3.85	0.316				
650	1.449											16.4	11.2	10.4	3.62	7.22	1.45	4.17	0.367	2.65	0.118		
700	1.560											17.6	12.9	11.2	4.16	7.78	1.66	4.49	0.420	2.85	0.135		
750	1.671											18.9	14.7	12.0	4.75	8.33	1.89	4.81	0.480	3.05	0.154		
800	1.782											20.2	16.5	12.8	5.35	8.89	2.13	5.13	0.540	3.26	0.173		
850	1.894											21.4	18.5	13.6	5.98	9.44	2.38	5.45	0.605	3.46	0.194		

(continued)

PRESSURE DROP OF WATER THROUGH SCHEDULE 40 STEEL PIPE Continued

Pressure Drop of Water per 100 ft of Schedule 40 Steel Pipe (psi)

| Flow | | | | | | | | | | | | 14 in. | | 16 in. | | 18 in. | |
gpm	ft³/s	u fps	p psi	u fps	p psi	u fps	p psi	u fps	p psi	u fps	p psi	u fps	p psi	u fps	p psi
900	2.005	10.0	2.66	5.77	0.627	3.66	0.216	2.58	0.090			22.7	20.6	14.4	6.65
950	2.117	10.6	2.93	6.09	0.744	3.87	0.238	2.72	0.099			23.9	22.8	15.2	7.36
1000	2.228	11.1	3.23	6.41	0.817	4.07	0.262	2.87	0.109					16.0	8.10
1100	2.451	12.2	3.85	7.06	0.975	4.48	0.313	3.15	0.130					17.6	9.66
1200	2.674	13.3	4.53	7.70	1.15	4.88	0.368	3.44	0.153	2.85	0.096			19.2	11.4
1300	2.896	14.4	5.26	8.34	1.33	5.29	0.427	3.73	0.178	3.08	0.111			20.8	13.2
1400	3.119	15.6	6.01	8.98	1.53	5.70	0.490	4.01	0.204	3.32	0.127			22.4	15.1
1500	3.342	16.7	6.84	9.62	1.74	6.10	0.556	4.30	0.232	3.56	0.145			24.1	17.2
1600	3.565	17.8	7.73	10.3	1.96	6.51	0.628	4.59	0.262	3.79	0.163	2.91	0.084		
1800	4.010	20.0	9.64	11.5	2.46	7.32	0.782	5.16	0.329	4.27	0.203	3.27	0.104		
2000	4.456	22.2	11.6	12.8	2.97	8.14	0.953	5.73	0.396	4.74	0.247	3.63	0.127		
2500	5.570	27.8	17.6	16.0	4.49	10.2	1.44	7.17	0.601	5.93	0.374	4.54	0.192		
3000	6.684			19.2	6.30	12.2	2.02	8.60	0.842	7.11	0.525	5.45	0.270	4.30	0.149
3500	7.798			22.4	8.41	14.2	2.70	10.0	1.12	8.30	0.700	6.36	0.358	5.02	0.199
4000	8.912			25.7	10.8	16.3	3.46	11.5	1.44	9.48	0.896	7.26	0.459	5.74	0.255
4500	10.03			28.9	13.4	18.3	4.31	12.9	1.76	10.7	1.12	8.17	0.671	6.45	0.317
5000	11.14					20.4	5.20	14.3	2.18	11.9	1.36	9.08	0.695	7.17	0.386
6000	13.37					24.4	7.35	17.2	3.06	14.2	1.91	10.9	0.977	8.60	0.542
7000	15.60					28.5	9.80	20.1	4.08	16.6	2.54	12.7	1.30	10.0	0.723
8000	17.82							22.9	5.22	19.0	3.25	14.5	1.67	11.5	0.926
9000	20.05							25.8	6.51	21.3	4.06	16.3	2.08	12.9	1.15
10000	22.28							28.7	7.91	23.7	4.92	18.2	2.53	14.3	1.40
12000	26.74									28.5	6.92	21.8	3.55	17.2	1.97
14000	31.19											25.4	4.72	20.1	2.62
16000	35.65											29.1	6.06	22.9	3.36
18000	40.10											32.7	7.55	25.8	4.18
20000	44.56													28.7	5.08

REPRESENTATIVE EQUIVALENT LENGTHS OF DIFFERENT VALVES IN PIPE DIAMETERS (L/D)

Type valve	Description	Valve-stem position	Equivalent length in pipe diameters (L/D)
Globe Valves			
Stem Perpendicular to Run	With no obstruction in flat, bevel or plug-type seat	Fully open	340
	With wing or pin guided disc	Fully open	450
Y-Pattern	(No obstruction in flat, bevel or plug-type seat)		
	With stem 60 degrees from run of pipe line	Fully open	175
	With stem 45 degrees from run of pipe line	Fully open	145
Angle Valves	With no obstruction in flat, bevel or plug-type seat	Fully open	145
	With wing or pin guided disc	Fully open	200
Gate Valves			
Wedge, Disc, Double Disc or Plug Disc		Fully open	13
		Three-quarters open	35
		One-half open	160
		One-quarter open	900
Pulp Stock		Fully open	17
		Three-quarters open	50
		One-half open	260
		One-quarter open	1200
Check Valves			
Conventional Swing		0.5[a] Fully open	135
Clearway Swing		0.5[a] Fully open	50
Globe Lift or Stop; Stem Perpendicular to Run or Y-Pattern		2.0[a] Fully open	Same as globe
Angle Lift or Stop		2.0[a] Fully open	Same as angle
In-Line Ball		2.5 vertical and 0.25 horizontal[a]	
		Fully open	150
Butterfly Valves (8-inch and larger)		Fully open	
Cocks			
Straight-Through	Rectangular plug port area equal to 100% of pipe area	Fully open	18
Three-Way	Rectangular plug port area equal to 80% of pipe area (fully open)	Flow straight through	44
		Flow through branch	140

[a] Minimum calculated pressure drop in psi across valve to provide sufficient flow to lift disc fully. Note L/D valves are based on turbulent flow.

B

Tables of
Engineering Data

CONTENTS

Conversion: Kinematic Viscosity to Saybolt Universal Viscosity

Kin Vis, cSt	Equivalent Saybolt Universal Viscosity, SSU		Kin Vis, cSt	Equivalent Saybolt Universal Viscosity, SSU		Kin Vis, cSt	Equivalent Saybolt Universal Viscosity, SSU		Kin Vis, cSt	Equivalent Saybolt Universal Viscosity, SSU	
	At 100°F	At 210°F		At 100°F	At 210°F		At 100°F	At 210°F		At 100°F	At 210°F
			2.45	34.1	34.4	3.15	36.4	36.7	3.85	38.7	39.0
			2.46	34.2	34.4	3.16	36.5	36.7	3.86	38.7	39.0
1.77	...	32.0	2.47	34.2	34.4	3.17	36.5	36.8	3.87	38.8	39.0
1.78	...	32.1	2.48	34.2	34.5	3.18	36.5	36.8	3.88	38.8	39.1
1.79	...	32.1	2.49	34.3	34.5	3.19	36.6	36.8	3.89	38.8	39.1
1.80	...	32.1	2.50	34.3	34.5	3.20	36.6	36.9	3.90	38.9	39.1
1.81	32.0	32.2	2.51	34.3	34.6	3.21	36.6	36.9	3.91	38.9	39.2
1.82	32.0	32.2	2.52	34.4	34.6	3.22	36.7	36.9	3.92	38.9	39.2
1.83	32.0	32.2	2.53	34.4	34.6	3.23	36.7	37.0	3.93	39.0	39.2
1.84	32.1	32.3	2.54	34.4	34.7	3.24	36.7	37.0	3.94	39.0	39.3
1.85	32.1	32.3	2.55	34.5	34.7	3.25	36.8	37.0	3.95	39.0	39.3
1.86	32.1	32.3	2.56	34.5	34.7	3.26	36.8	37.0	3.96	39.1	39.3
1.87	32.2	32.4	2.57	34.5	34.8	3.27	36.8	37.1	3.97	39.1	39.4
1.88	32.2	32.4	2.58	34.6	34.8	3.28	36.9	37.1	3.98	39.1	39.4
1.89	32.2	32.4	2.59	34.6	34.8	3.29	36.9	37.1	3.99	39.2	39.4
1.90	32.3	32.5	2.60	34.6	34.9	3.30	36.9	37.2	4.00	39.2	39.5
1.91	32.3	32.5	2.61	34.7	34.9	3.31	37.0	37.2	4.01	39.2	39.5
1.92	32.3	32.5	2.62	34.7	34.9	3.32	37.0	37.2	4.02	39.3	39.5
1.93	32.4	32.6	2.63	34.7	35.0	3.33	37.0	37.3	4.03	39.3	39.6
1.94	32.4	32.6	2.64	34.8	35.0	3.34	37.1	37.3	4.04	39.3	39.6
1.95	32.4	32.6	2.65	34.8	35.0	3.35	37.1	37.3	4.05	39.4	39.6
1.96	32.5	32.7	2.66	34.8	35.1	3.36	37.1	37.4	4.06	39.4	39.7
1.97	32.5	32.7	2.67	34.9	35.1	3.37	37.2	37.4	4.07	39.4	39.7
1.98	32.5	32.8	2.68	34.9	35.1	3.38	37.2	37.4	4.08	39.5	39.7
1.99	32.6	32.8	2.69	34.9	35.2	3.39	37.2	37.5	4.09	39.5	39.8
2.00	32.6	32.8	2.70	35.0	35.2	3.40	37.3	37.5	4.10	39.5	39.8
2.01	32.6	32.9	2.71	35.0	35.2	3.41	37.3	37.5	4.11	39.6	39.8
2.02	32.7	32.9	2.72	35.0	35.3	3.42	37.3	37.6	4.12	39.6	39.8
2.03	32.7	32.9	2.73	35.1	35.3	3.43	37.4	37.6	4.13	39.6	39.9
2.04	32.7	33.0	2.74	35.1	35.3	3.44	37.4	37.6	4.14	39.6	39.9
2.05	32.8	33.0	2.75	35.1	35.4	3.45	37.4	37.7	4.15	39.7	39.9
2.06	32.8	33.0	2.76	35.2	35.4	3.46	37.5	37.7	4.16	39.7	40.0
2.07	32.8	33.1	2.77	35.2	35.4	3.47	37.5	37.7	4.17	39.7	40.0
2.08	32.9	33.1	2.78	35.2	35.5	3.48	37.5	37.8	4.18	39.8	40.0
2.09	32.9	33.1	2.79	35.3	35.5	3.49	37.6	37.8	4.19	39.8	40.1
2.10	32.9	33.2	2.80	35.3	35.5	3.50	37.6	37.8	4.20	39.8	40.1
2.11	33.0	33.2	2.81	35.3	35.6	3.51	37.6	37.9	4.21	39.9	40.1
2.12	33.0	33.2	2.82	35.4	35.6	3.52	37.6	37.9	4.22	39.9	40.2
2.13	33.0	33.3	2.83	35.4	35.6	3.53	37.7	37.9	4.23	39.9	40.2
2.14	33.1	33.3	2.84	35.4	35.7	3.54	37.7	38.0	4.24	40.0	40.2
2.15	33.1	33.3	2.85	35.5	35.7	3.55	37.7	38.0	4.25	40.0	40.3
2.16	33.1	33.4	2.86	35.5	35.7	3.56	37.8	38.0	4.26	40.0	40.3
2.17	33.2	33.4	2.87	35.5	35.8	3.57	37.8	38.1	4.27	40.1	40.3
2.18	33.2	33.4	2.88	35.6	35.8	3.58	37.8	38.1	4.28	40.1	40.4
2.19	33.2	33.5	2.89	35.6	35.8	3.59	37.9	38.1	4.29	40.1	40.4
2.20	33.3	33.5	2.90	35.6	35.9	3.60	37.9	38.2	4.30	40.2	40.4
2.21	33.3	33.5	2.91	35.7	35.9	3.61	37.9	38.2	4.31	40.2	40.5
2.22	33.3	33.6	2.92	35.7	35.9	3.62	38.0	38.2	4.32	40.2	40.5
2.23	33.4	33.6	2.93	35.7	36.0	3.63	38.0	38.3	4.33	40.3	40.5
2.24	33.4	33.6	2.94	35.8	36.0	3.64	38.0	38.3	4.34	40.3	40.6
2.25	33.5	33.7	2.95	35.8	36.0	3.65	38.1	38.3	4.35	40.3	40.6
2.26	33.5	33.7	2.96	35.8	36.1	3.66	38.1	38.4	4.36	40.4	40.6
2.27	33.5	33.7	2.97	35.9	36.1	3.67	38.1	38.4	4.37	40.4	40.7
2.28	33.6	33.8	2.98	35.9	36.1	3.68	38.2	38.4	4.38	40.4	40.7
2.29	33.6	33.8	2.99	35.9	36.2	3.69	38.2	38.5	4.39	40.4	40.7
2.30	33.6	33.8	3.00	36.0	36.2	3.70	38.2	38.5	4.40	40.5	40.8
2.31	33.7	33.9	3.01	36.0	36.2	3.71	38.3	38.5	4.41	40.5	40.8
2.32	33.7	33.9	3.02	36.0	36.3	3.72	38.3	38.6	4.42	40.5	40.8
2.33	33.7	33.9	3.03	36.0	36.3	3.73	38.3	38.6	4.43	40.6	40.8
2.34	33.8	34.0	3.04	36.1	36.3	3.74	38.4	38.6	4.44	40.6	40.9
2.35	33.8	34.0	3.05	36.1	36.4	3.75	38.4	38.7	4.45	40.6	40.9
2.36	33.8	34.0	3.06	36.1	36.4	3.76	38.4	38.7	4.46	40.7	40.9
2.37	33.9	34.1	3.07	36.2	36.4	3.77	38.5	38.7	4.47	40.7	41.0
2.38	33.9	34.1	3.08	36.2	36.5	3.78	38.5	38.7	4.48	40.7	41.0
2.39	33.9	34.2	3.09	36.2	36.5	3.79	38.5	38.8	4.49	40.8	41.0
2.40	34.0	34.2	3.10	36.3	36.5	3.80	38.6	38.8	4.50	40.8	41.1
2.41	34.0	34.2	3.11	36.3	36.6	3.81	38.6	38.8	4.51	40.8	41.1
2.42	34.0	34.3	3.12	36.3	36.6	3.82	38.6	38.9	4.52	40.9	41.1
2.43	34.1	34.3	3.13	36.4	36.6	3.83	38.7	38.9	4.53	40.9	41.2
2.44	34.1	34.3	3.14	36.4	36.7	3.84	38.7	38.9	4.54	40.9	41.2

Conversion: Kinematic Viscosity to Saybolt Universal Viscosity . . . continued

Kin Vis, cSt	Equivalent Saybolt Universal Viscosity, SSU		Kin Vis, cSt	Equivalent Saybolt Universal Viscosity, SSU		Kin Vis, cSt	Equivalent Saybolt Universal Viscosity, SSU		Kin Vis, cSt	Equivalent Saybolt Universal Viscosity, SSU	
	At 100°F	At 210°F		At 100°F	At 210°F		At 100°F	At 210°F		At 100°F	At 210°F
4.55	41.0	41.2	5.25	43.2	43.5	5.95	45.4	45.7	6.65	47.7	48.0
4.56	41.0	41.3	5.26	43.2	43.5	5.96	45.4	45.8	6.66	47.7	48.0
4.57	41.0	41.3	5.27	43.3	43.5	5.97	45.5	45.8	6.67	47.7	48.0
4.58	41.1	41.3	5.28	43.3	43.6	5.98	45.5	45.8	6.68	47.8	48.1
4.59	41.1	41.4	5.29	43.3	43.6	5.99	45.5	45.9	6.69	47.8	48.1
4.60	41.1	41.4	5.30	43.3	43.6	6.00	45.6	45.9	6.70	47.8	48.1
4.61	41.2	41.4	5.31	43.4	43.7	6.01	45.6	45.9	6.71	47.9	48.2
4.62	41.2	41.5	5.32	43.4	43.7	6.02	45.6	45.9	6.72	47.9	48.2
4.63	41.2	41.5	5.33	43.4	43.7	6.03	45.7	46.0	6.73	47.9	48.2
4.64	41.2	41.5	5.34	43.5	43.8	6.04	45.7	46.0	6.74	47.9	48.3
4.65	41.3	41.6	5.35	43.5	43.8	6.05	45.7	46.0	6.75	48.0	48.3
4.66	41.3	41.6	5.36	43.5	43.8	6.06	45.8	46.1	6.76	48.0	48.3
4.67	41.3	41.6	5.37	43.6	43.9	6.07	45.8	46.1	6.77	48.0	48.4
4.68	41.4	41.7	5.38	43.6	43.9	6.08	45.8	46.1	6.78	48.1	48.4
4.69	41.4	41.7	5.39	43.6	43.9	6.09	45.9	46.2	6.79	48.1	48.4
4.70	41.4	41.7	5.40	43.7	44.0	6.10	45.9	46.2	6.80	48.1	48.5
4.71	41.5	41.7	5.41	43.7	44.0	6.11	45.9	46.2	6.81	48.2	48.5
4.72	41.5	41.8	5.42	43.7	44.0	6.12	46.0	46.3	6.82	48.2	48.5
4.73	41.5	41.8	5.43	43.8	44.1	6.13	46.0	46.3	6.83	48.2	48.6
4.74	41.6	41.8	5.44	43.8	44.1	6.14	46.0	46.3	6.84	48.3	48.6
4.75	41.6	41.9	5.45	43.8	44.1	6.15	46.1	46.4	6.85	48.3	48.6
4.76	41.6	41.9	5.46	43.9	44.2	6.16	46.1	46.4	6.86	48.3	48.7
4.77	41.7	41.9	5.47	43.9	44.2	6.17	46.1	46.4	6.87	48.4	48.7
4.78	41.7	42.0	5.48	43.9	44.2	6.18	46.2	46.5	6.88	48.4	48.7
4.79	41.7	42.0	5.49	44.0	44.2	6.19	46.2	46.5	6.89	48.4	48.8
4.80	41.8	42.0	5.50	44.0	44.3	6.20	46.2	46.5	6.90	48.5	48.8
4.81	41.8	42.1	5.51	44.0	44.3	6.21	46.2	46.6	6.91	48.5	48.8
4.82	41.8	42.1	5.52	44.0	44.3	6.22	46.3	46.6	6.92	48.5	48.9
4.83	41.9	42.1	5.53	44.1	44.4	6.23	46.3	46.6	6.93	48.6	48.9
4.84	41.9	42.2	5.54	44.1	44.4	6.24	46.3	46.7	6.94	48.6	48.9
4.85	41.9	42.2	5.55	44.1	44.4	6.25	46.4	46.7	6.95	48.6	49.0
4.86	41.9	42.2	5.56	44.2	44.5	6.26	46.4	46.7	6.96	48.7	49.0
4.87	42.0	42.3	5.57	44.2	44.5	6.27	46.4	46.8	6.97	48.7	49.0
4.88	42.0	42.3	5.58	44.2	44.5	6.28	46.5	46.8	6.98	48.7	49.1
4.89	42.0	42.3	5.59	44.3	44.6	6.29	46.5	46.8	6.99	48.8	49.1
4.90	42.1	42.4	5.60	44.3	44.6	6.30	46.5	46.8	7.00	48.8	49.1
4.91	42.1	42.4	5.61	44.3	44.6	6.31	46.6	46.9	7.01	48.8	49.1
4.92	42.1	42.4	5.62	44.4	44.7	6.32	46.6	46.9	7.02	48.9	49.2
4.93	42.2	42.5	5.63	44.4	44.7	6.33	46.6	46.9	7.03	48.9	49.2
4.94	42.2	42.5	5.64	44.4	44.7	6.34	46.7	47.0	7.04	48.9	49.2
4.95	42.2	42.5	5.65	44.5	44.8	6.35	46.7	47.0	7.05	49.0	49.3
4.96	42.3	42.5	5.66	44.5	44.8	6.36	46.7	47.0	7.06	49.0	49.3
4.97	42.3	42.6	5.67	44.5	44.8	6.37	46.8	47.1	7.07	49.0	49.3
4.98	42.3	42.6	5.68	44.6	44.9	6.38	46.8	47.1	7.08	49.0	49.4
4.99	42.4	42.6	5.69	44.6	44.9	6.39	46.8	47.1	7.09	49.1	49.4
5.00	42.4	42.7	5.70	44.6	44.9	6.40	46.9	47.2	7.10	49.1	49.4
5.01	42.4	42.7	5.71	44.7	45.0	6.41	46.9	47.2	7.11	49.1	49.5
5.02	42.5	42.7	5.72	44.7	45.0	6.42	46.9	47.2	7.12	49.2	49.5
5.03	42.5	42.8	5.73	44.7	45.0	6.43	47.0	47.3	7.13	49.2	49.5
5.04	42.5	42.8	5.74	44.7	45.0	6.44	47.0	47.3	7.14	49.2	49.6
5.05	42.6	42.8	5.75	44.8	45.1	6.45	47.0	47.3	7.15	49.3	49.6
5.06	42.6	42.9	5.76	44.8	45.1	6.46	47.0	47.4	7.16	49.3	49.6
5.07	42.6	42.9	5.77	44.8	45.1	6.47	47.1	47.4	7.17	49.3	49.7
5.08	42.6	42.9	5.78	44.9	45.2	6.48	47.1	47.4	7.18	49.4	49.7
5.09	42.7	43.0	5.79	44.9	45.2	6.49	47.1	47.5	7.19	49.4	49.7
5.10	42.7	43.0	5.80	44.9	45.2	6.50	47.2	47.5	7.20	49.4	49.8
5.11	42.7	43.0	5.81	45.0	45.3	6.51	47.2	47.5	7.21	49.5	49.8
5.12	42.8	43.1	5.82	45.0	45.3	6.52	47.2	47.6	7.22	49.5	49.8
5.13	42.8	43.1	5.83	45.0	45.3	6.53	47.3	47.6	7.23	49.5	49.9
5.14	42.8	43.1	5.84	45.1	45.4	6.54	47.3	47.6	7.24	49.6	49.9
5.15	42.9	43.2	5.85	45.1	45.4	6.55	47.3	47.7	7.25	49.6	49.9
5.16	42.9	43.2	5.86	45.1	45.4	6.56	47.4	47.7	7.26	49.6	50.0
5.17	42.9	43.2	5.87	45.2	45.5	6.57	47.4	47.7	7.27	49.7	50.0
5.18	43.0	43.3	5.88	45.2	45.5	6.58	47.4	47.8	7.28	49.7	50.0
5.19	43.0	43.3	5.89	45.2	45.5	6.59	47.5	47.8	7.29	49.7	50.1
5.20	43.0	43.3	5.90	45.3	45.6	6.60	47.5	47.8	7.30	49.8	50.1
5.21	43.1	43.3	5.91	45.3	45.6	6.61	47.5	47.8	7.31	49.8	50.1
5.22	43.1	43.4	5.92	45.3	45.6	6.62	47.6	47.9	7.32	49.8	50.2
5.23	43.1	43.4	5.93	45.4	45.7	6.63	47.6	47.9	7.33	49.9	50.2
5.24	43.2	43.4	5.94	45.4	45.7	6.64	47.6	47.9	7.34	49.9	50.2

Conversion: Kinematic Viscosity to Saybolt Universal Viscosity ... continued

Kin Vis, cSt	Equivalent Saybolt Universal Viscosity, SSU		Kin Vis, cSt	Equivalent Saybolt Universal Viscosity, SSU		Kin Vis, cSt	Equivalent Saybolt Universal Viscosity, SSU		Kin Vis, cSt	Equivalent Saybolt Universal Viscosity, SSU	
	At 100°F	At 210°F		At 100°F	At 210°F		At 100°F	At 210°F		At 100°F	At 210°F
7.35	49.9	50.3	8.05	52.2	52.6	8.75	54.6	54.9	9.45	56.9	57.3
7.36	50.0	50.3	8.06	52.3	52.6	8.76	54.6	55.0	9.46	57.0	57.4
7.37	50.0	50.3	8.07	52.3	52.6	8.77	54.6	55.0	9.47	57.0	57.4
7.38	50.0	50.4	8.08	52.3	52.7	8.78	54.7	55.0	9.48	57.0	57.4
7.39	50.1	50.4	8.09	52.4	52.7	8.79	54.7	55.1	9.49	57.1	57.5
7.40	50.1	50.4	8.10	52.4	52.7	8.80	54.7	55.1	9.50	57.1	57.5
7.41	50.1	50.5	8.11	52.4	52.8	8.81	54.8	55.1	9.52	57.2	57.6
7.42	50.2	50.5	8.12	52.5	52.8	8.82	54.8	55.2	9.54	57.2	57.6
7.43	50.2	50.5	8.13	52.5	52.8	8.83	54.8	55.2	9.56	57.3	57.7
7.44	50.2	50.6	8.14	52.5	52.9	8.84	54.9	55.2	9.58	57.4	57.8
7.45	50.3	50.6	8.15	52.6	52.9	8.85	54.9	55.3	9.60	57.5	57.8
7.46	50.3	50.6	8.16	52.6	52.9	8.86	54.9	55.3	9.62	57.5	57.9
7.47	50.3	50.7	8.17	52.6	53.0	8.87	55.0	55.3	9.64	57.6	58.0
7.48	50.3	50.7	8.18	52.7	53.0	8.88	55.0	55.4	9.66	57.7	58.0
7.49	50.4	50.7	8.19	52.7	53.0	8.89	55.0	55.4	9.68	57.7	58.1
7.50	50.4	50.8	8.20	52.7	53.1	8.90	55.1	55.4	9.70	57.8	58.2
7.51	50.4	50.8	8.21	52.8	53.1	8.91	55.1	55.5	9.72	57.9	58.3
7.52	50.5	50.8	8.22	52.8	53.1	8.92	55.1	55.5	9.74	57.9	58.3
7.53	50.5	50.9	8.23	52.8	53.2	8.93	55.2	55.5	9.76	58.0	58.4
7.54	50.5	50.9	8.24	52.9	53.2	8.94	55.2	55.6	9.78	58.1	58.5
7.55	50.6	50.9	8.25	52.9	53.2	8.95	55.2	55.6	9.80	58.1	58.5
7.56	50.6	51.0	8.26	52.9	53.3	8.96	55.3	55.6	9.82	58.2	58.6
7.57	50.6	51.0	8.27	53.0	53.3	8.97	55.3	55.7	9.84	58.3	58.7
7.58	50.7	51.0	8.28	53.0	53.3	8.98	55.3	55.7	9.86	58.4	58.7
7.59	50.7	51.0	8.29	53.0	53.4	8.99	55.4	55.7	9.88	58.4	58.8
7.60	50.7	51.1	8.30	53.1	53.4	9.00	55.4	55.8	9.90	58.5	58.9
7.61	50.8	51.1	8.31	53.1	53.4	9.01	55.4	55.8	9.92	58.6	59.0
7.62	50.8	51.1	8.32	53.1	53.5	9.02	55.5	55.8	9.94	58.6	59.0
7.63	50.8	51.2	8.33	53.2	53.5	9.03	55.5	55.9	9.96	58.7	59.1
7.64	50.9	51.2	8.34	53.2	53.5	9.04	55.5	55.9	9.98	58.8	59.2
7.65	50.9	51.2	8.35	53.2	53.6	9.05	55.6	55.9	10.00	58.8	59.2
7.66	50.9	51.3	8.36	53.3	53.6	9.06	55.6	56.0	10.02	58.9	59.3
7.67	51.0	51.3	8.37	53.3	53.6	9.07	55.6	56.0	10.04	59.0	59.4
7.68	51.0	51.3	8.38	53.3	53.7	9.08	55.7	56.0	10.06	59.0	59.4
7.69	51.0	51.4	8.39	53.4	53.7	9.09	55.7	56.1	10.08	59.1	59.5
7.70	51.1	51.4	8.40	53.4	53.7	9.10	55.7	56.1	10.10	59.2	59.6
7.71	51.1	51.4	8.41	53.4	53.8	9.11	55.8	56.1	10.12	59.3	59.7
7.72	51.1	51.5	8.42	53.5	53.8	9.12	55.8	56.2	10.14	59.3	59.7
7.73	51.2	51.5	8.43	53.5	53.8	9.13	55.8	56.2	10.16	59.4	59.8
7.74	51.2	51.5	8.44	53.5	53.9	9.14	55.9	56.3	10.18	59.5	59.9
7.75	51.2	51.6	8.45	53.6	53.9	9.15	55.9	56.3	10.20	59.5	59.9
7.76	51.3	51.6	8.46	53.6	53.9	9.16	55.9	56.3	10.22	59.6	60.0
7.77	51.3	51.6	8.47	53.6	54.0	9.17	56.0	56.4	10.24	59.7	60.1
7.78	51.3	51.7	8.48	53.7	54.0	9.18	56.0	56.4	10.26	59.7	60.1
7.79	51.4	51.7	8.49	53.7	54.0	9.19	56.0	56.4	10.28	59.8	60.2
7.80	51.4	51.7	8.50	53.7	54.1	9.20	56.1	56.5	10.30	59.9	60.3
7.81	51.4	51.8	8.51	53.8	54.1	9.21	56.1	56.5	10.32	60.0	60.4
7.82	51.5	51.8	8.52	53.8	54.1	9.22	56.2	56.5	10.34	60.0	60.4
7.83	51.5	51.8	8.53	53.8	54.2	9.23	56.2	56.6	10.36	60.1	60.5
7.84	51.5	51.9	8.54	53.9	54.2	9.24	56.2	56.6	10.38	60.2	60.6
7.85	51.6	51.9	8.55	53.9	54.2	9.25	56.3	56.6	10.40	60.2	60.6
7.86	51.6	51.9	8.56	53.9	54.3	9.26	56.3	56.7	10.42	60.3	60.7
7.87	51.6	52.0	8.57	54.0	54.3	9.27	56.3	56.7	10.44	60.4	60.8
7.88	51.7	52.0	8.58	54.0	54.3	9.28	56.4	56.7	10.46	60.4	60.9
7.89	51.7	52.0	8.59	54.0	54.4	9.29	56.4	56.8	10.48	60.5	60.9
7.90	51.7	52.1	8.60	54.1	54.4	9.30	56.4	56.8	10.50	60.6	61.0
7.91	51.8	52.1	8.61	54.1	54.5	9.31	56.5	56.8	10.52	60.7	61.1
7.92	51.8	52.1	8.62	54.1	54.5	9.32	56.5	56.9	10.54	60.7	61.1
7.93	51.8	52.2	8.63	54.2	54.5	9.33	56.5	56.9	10.56	60.8	61.2
7.94	51.9	52.2	8.64	54.2	54.6	9.34	56.6	56.9	10.58	60.9	61.3
7.95	51.9	52.2	8.65	54.2	54.6	9.35	56.6	57.0	10.60	60.9	61.4
7.96	51.9	52.3	8.66	54.3	54.6	9.36	56.6	57.0	10.62	61.0	61.4
7.97	52.0	52.3	8.67	54.3	54.7	9.37	56.7	57.0	10.64	61.1	61.5
7.98	52.0	52.3	8.68	54.3	54.7	9.38	56.7	57.1	10.66	61.2	61.6
7.99	52.0	52.4	8.69	54.4	54.7	9.39	56.7	57.1	10.68	61.2	61.6
8.00	52.1	52.4	8.70	54.4	54.8	9.40	56.8	57.1	10.70	61.3	61.7
8.01	52.1	52.4	8.71	54.4	54.8	9.41	56.8	57.2	10.72	61.4	61.8
8.02	52.1	52.5	8.72	54.5	54.8	9.42	56.8	57.2	10.74	61.4	61.9
8.03	52.2	52.5	8.73	54.5	54.9	9.43	56.9	57.2	10.76	61.5	61.9
8.04	52.2	52.5	8.74	54.5	54.9	9.44	56.9	57.3	10.78	61.6	62.0

Conversion: Kinematic Viscosity to Saybolt Universal Viscosity . . . continued

Kin Vis, cSt	At 100°F	At 210°F	Kin Vis, cSt	At 100°F	At 210°F	Kin Vis, cSt	At 100°F	At 210°F	Kin Vis, cSt	At 100°F	At 210°F
10.80	61.7	62.1	12.20	66.7	67.2	13.60	72.0	72.5	15.00	77.4	77.9
10.82	61.7	62.1	12.22	66.8	67.2	13.62	72.1	72.5	15.02	77.5	78.0
10.84	61.8	62.6	12.24	66.9	67.3	13.64	72.1	72.6	15.04	77.6	78.1
10.86	61.9	62.3	12.26	66.9	67.4	13.66	72.2	72.7	15.06	77.6	78.2
10.88	61.9	62.4	12.28	67.0	67.5	13.68	72.3	72.8	15.08	77.7	78.2
10.90	62.0	62.4	12.30	67.1	67.5	13.70	72.4	72.8	15.10	77.8	78.3
10.92	62.1	62.5	12.32	67.2	67.6	13.72	72.4	72.9	15.12	77.9	78.4
10.94	62.2	62.6	12.34	67.2	67.7	13.74	72.5	73.0	15.14	78.0	78.5
10.96	62.2	62.6	12.36	67.3	67.8	13.76	72.6	73.1	15.16	78.0	78.6
10.98	62.3	62.7	12.38	67.4	67.8	13.78	72.7	73.2	15.18	78.1	78.6
11.00	62.4	62.8	12.40	67.5	67.9	13.80	72.7	73.2	15.20	78.2	78.7
11.02	62.4	62.9	12.42	67.5	68.0	13.82	72.8	73.3	15.22	78.3	78.8
11.04	62.5	62.9	12.44	67.6	68.1	13.84	72.9	73.4	15.24	78.3	78.9
11.06	62.6	63.0	12.46	67.7	68.1	13.86	73.0	73.5	15.26	78.4	79.0
11.08	62.7	63.1	12.48	67.8	68.2	13.88	73.1	73.5	15.28	78.5	79.0
11.10	62.7	63.1	12.50	67.8	68.3	13.90	73.1	73.6	15.30	78.6	79.1
11.12	62.8	63.2	12.52	67.9	68.4	13.92	73.2	73.7	15.32	78.7	79.2
11.14	62.9	63.3	12.54	68.0	68.4	13.94	73.3	73.8	15.34	78.7	79.3
11.16	62.9	63.4	12.56	68.1	68.5	13.96	73.4	73.9	15.36	78.8	79.3
11.18	63.0	63.4	12.58	68.1	68.6	13.98	73.4	73.9	15.38	78.9	79.4
11.20	63.1	63.5	12.60	68.2	68.7	14.00	73.5	74.0	15.40	79.0	79.5
11.22	63.2	63.6	12.62	68.3	68.7	14.02	73.6	74.1	15.42	79.1	79.6
11.24	63.2	63.7	12.64	68.4	68.8	14.04	73.7	74.2	15.44	79.1	79.7
11.26	63.3	63.7	12.66	68.4	68.9	14.06	73.7	74.2	15.46	79.2	79.7
11.28	63.4	63.8	12.68	68.5	69.0	14.08	73.8	74.3	15.48	79.3	79.8
11.30	63.4	63.9	12.70	68.6	69.0	14.10	73.9	74.4	15.50	79.4	79.9
11.32	63.5	63.9	12.72	68.7	69.1	14.12	74.0	74.5	15.52	79.5	80.0
11.34	63.6	64.0	12.74	68.7	69.2	14.14	74.1	74.6	15.54	79.5	80.1
11.36	63.7	64.1	12.76	68.8	69.3	14.16	74.1	74.6	15.56	79.6	80.1
11.38	63.7	64.2	12.78	68.9	69.3	14.18	74.2	74.7	15.58	79.7	80.2
11.40	63.8	64.2	12.80	69.0	69.4	14.20	74.3	74.8	15.60	79.8	80.3
11.42	63.9	64.3	12.82	69.0	69.5	14.22	74.4	74.9	15.62	79.8	80.4
11.44	63.9	64.4	12.84	69.1	69.6	14.24	74.4	74.9	15.64	79.9	80.5
11.46	64.0	64.5	12.86	69.2	69.6	14.26	74.5	75.0	15.66	80.0	80.5
11.48	64.1	64.5	12.88	69.3	69.7	14.28	74.6	75.1	15.68	80.1	80.6
11.50	64.2	64.6	12.90	69.3	69.8	14.30	74.7	75.2	15.70	80.2	80.7
11.52	64.2	64.7	12.92	69.4	69.9	14.32	74.7	75.3	15.72	80.2	80.8
11.54	64.3	64.7	12.94	69.5	69.9	14.34	74.8	75.3	15.74	80.3	80.9
11.56	64.4	64.8	12.96	69.6	70.0	14.36	74.9	75.4	15.76	80.4	80.9
11.58	64.5	64.9	12.98	69.6	70.1	14.38	75.0	75.5	15.78	80.5	81.0
11.60	64.5	65.0	13.00	69.7	70.2	14.40	75.1	75.6	15.80	80.6	81.1
11.62	64.6	65.0	13.02	69.8	70.3	14.42	75.1	75.6	15.82	80.6	81.2
11.64	64.7	65.1	13.04	69.9	70.3	14.44	75.2	75.7	15.84	80.7	81.3
11.66	64.7	65.2	13.06	69.9	70.4	14.46	75.3	75.8	15.86	80.8	81.3
11.68	64.8	65.3	13.08	70.0	70.5	14.48	75.4	75.9	15.88	80.9	81.4
11.70	64.9	65.3	13.10	70.1	70.6	14.50	75.4	76.0	15.90	81.0	81.5
11.72	65.0	65.4	13.12	70.2	70.6	14.52	75.5	76.0	15.92	81.0	81.6
11.74	65.0	65.5	13.14	70.2	70.7	14.54	75.6	76.1	15.94	81.1	81.7
11.76	65.1	65.5	13.16	70.3	70.8	14.56	75.7	76.2	15.96	81.2	81.7
11.78	65.2	65.6	13.18	70.4	70.9	14.58	75.8	76.3	15.98	81.3	81.8
11.80	65.3	65.7	13.20	70.5	70.9	14.60	75.8	76.3	16.00	81.4	81.9
11.82	65.3	65.8	13.22	70.5	71.0	14.62	75.9	76.4	16.02	81.4	82.0
11.84	65.4	65.8	13.24	70.6	71.1	14.64	76.0	76.5	16.04	81.5	82.1
11.86	65.5	65.9	13.26	70.7	71.2	14.66	76.1	76.6	16.06	81.6	82.2
11.88	65.6	66.0	13.28	70.8	71.2	14.68	76.1	76.7	16.08	81.7	82.2
11.90	65.6	66.1	13.30	70.8	71.3	14.70	76.2	76.7	16.10	81.8	82.3
11.92	65.7	66.1	13.32	70.9	71.4	14.72	76.3	76.8	16.12	81.8	82.4
11.94	65.8	66.2	13.34	71.0	71.5	14.74	76.4	76.9	16.14	81.9	82.5
11.96	65.8	66.3	13.36	71.1	71.5	14.76	76.5	77.0	16.16	82.0	82.6
11.98	65.9	66.4	13.38	71.1	71.6	14.78	76.5	77.1	16.18	82.1	82.6
12.00	66.0	66.4	13.40	71.2	71.7	14.80	76.6	77.1	16.20	82.2	82.7
12.02	66.1	66.5	13.42	71.3	71.8	14.82	76.7	77.2	16.22	82.2	82.8
12.04	66.1	66.6	13.44	71.4	71.9	14.84	76.8	77.3	16.24	82.3	82.9
12.06	66.2	66.7	13.46	71.4	71.9	14.86	76.9	77.4	16.26	82.4	83.0
12.08	66.3	66.7	13.48	71.5	72.0	14.88	76.9	77.4	16.28	82.5	83.0
12.10	66.4	66.8	13.50	71.6	72.1	14.90	77.0	77.5	16.30	82.6	83.1
12.12	66.4	66.9	13.52	71.7	72.2	14.92	77.1	77.6	16.32	82.6	83.2
12.14	66.5	67.0	13.54	71.8	72.2	14.94	77.2	77.7	16.34	82.7	83.3
12.16	66.6	67.0	13.56	71.8	72.3	14.96	77.2	77.8	16.36	82.8	83.4
12.18	66.7	67.1	13.58	71.9	72.4	14.98	77.3	77.8	16.38	82.9	83.4

Conversion: Kinematic Viscosity to Saybolt Universal Viscosity . . . continued

Kin Vis, cSt	Equivalent Saybolt Universal Viscosity, SSU At 100°F	At 210°F	Kin Vis, cSt	Equivalent Saybolt Universal Viscosity, SSU At 100°F	At 210°F	Kin Vis, cSt	Equivalent Saybolt Universal Viscosity, SSU At 100°F	At 210°F	Kin Vis, cSt	Equivalent Saybolt Universal Viscosity, SSU At 100°F	At 210°F
16.40	83.0	83.5	17.80	88.7	89.3	19.20	94.5	95.1	21.50	104.2	104.9
16.42	83.0	83.6	17.82	88.7	89.3	19.22	94.5	95.2	21.55	134.4	105 1
16.44	83.1	83.7	17.84	88.8	89.4	19.24	94.6	95.3	21.60	104.6	105.3
16.46	83.2	83.8	17.86	88.9	89.5	19.26	94.7	95.4	21.65	104.8	105.5
16.48	83.3	83.8	17.88	89.0	89.6	19.28	94.8	95.4	21.70	105.0	105.8
16.50	83.4	83.9	17.90	89.1	89.7	19.30	94.9	95.5	21.75	105.3	106.0
16.52	83.5	84.0	17.92	89.2	89.8	19.32	95.0	95.6	21.80	105.5	106.2
16.54	83.5	84.1	17.94	89.2	89.8	19.34	95.0	95.7	21.85	105.7	106.4
16.56	83.6	84.2	17.96	89.3	89.9	19.36	95.1	95.8	21.90	105.9	106.6
16.58	83.7	84.3	17.98	89.4	90.0	19.38	95.2	95.9	21.95	106.1	106.8
16.60	83.8	84.3	18.00	89.5	90.1	19.40	95.3	95.9	22.00	106.3	107.0
16.62	83.9	84.4	18.02	89.6	90.2	19.42	95.4	96.0	22.05	106.6	107.3
16.64	83.9	84.5	18.04	89.6	90.2	19.44	95.5	96.1	22.10	106.8	107.5
16.66	84.0	84.6	18.06	89.7	90.3	19.46	95.6	96.2	22.15	107.0	107.7
16.68	84.1	84.7	18.08	89.8	90.4	19.48	95.6	96.3	22.20	107.2	107.9
16.70	84.2	84.7	18.10	89.9	90.5	19.50	95.7	96.4	22.25	107.4	108.1
16.72	84.3	84.8	18.12	90.0	90.6	19.52	95.8	96.4	22.30	107.6	108.3
16.74	84.3	84.9	18.14	90.1	90.7	19.54	95.9	96.5	22.35	107.8	108.6
16.76	84.4	85.0	18.16	90.1	90.7	19.56	96.0	96.6	22.40	108.1	108.8
16.78	84.5	85.1	18.18	90.2	90.8	19.58	96.1	96.7	22.45	108.3	109.0
16.80	84.6	85.1	18.20	90.3	90.9	19.60	96.1	96.8	22.50	108.5	109.2
16.82	84.7	85.2	18.22	90.4	91.0	19.62	96.2	96.9	22.55	108.7	109.4
16.84	84.7	85.3	18.24	90.5	91.1	19.64	96.3	97.0	22.60	108.9	109.6
16.86	84.8	85.4	18.26	90.6	91.2	19.66	96.4	97.0	22.65	109.1	109.9
16.88	84.9	85.5	18.28	90.6	91.2	19.68	96.5	97.1	22.70	109.4	110.1
16.90	85.0	85.6	18.30	90.7	91.3	19.70	96.6	97.2	22.75	109.6	110.3
16.92	85.1	85.6	18.32	90.8	91.4	19.72	96.6	97.3	22.80	109.8	110.5
16.94	85.1	85.7	18.34	90.9	91.5	19.74	96.7	97.4	22.85	110.0	110.7
16.96	85.2	85.8	18.36	91.0	91.6	19.76	96.8	97.5	22.90	110.2	111.0
16.98	85.3	85.9	18.38	91.1	91.7	19.78	96.9	97.5	22.95	110.4	111.2
17.00	85.4	86.0	18.40	91.1	91.7	19.80	97.0	97.6	23.00	110.6	111.4
17.02	85.5	86.0	18.42	91.2	91.8	19.82	97.1	97.7	23.05	110.9	111.6
17.04	85.6	86.1	18.44	91.3	91.9	19.84	97.1	97.8	23.10	111.1	111.8
17.06	85.6	86.2	18.46	91.4	92.0	19.86	97.2	97.9	23.15	111.3	112.0
17.08	85.7	86.3	18.48	91.5	92.1	19.88	97.3	98.0	23.20	111.5	112.3
17.10	85.8	86.4	18.50	91.5	92.2	19.90	97.4	98.1	23.25	111.7	112.5
17.12	85.9	86.5	18.52	91.6	92.2	19.92	97.5	98.1	23.30	111.9	112.7
17.14	86.0	86.5	18.54	91.7	92.3	19.94	97.6	98.2	23.35	112.2	112.9
17.16	86.0	86.6	18.56	91.8	92.4	19.96	97.7	98.3	23.40	112.4	113.1
17.18	86.1	86.7	18.58	91.9	92.5	19.98	97.7	98.4	23.45	112.6	113.4
17.20	86.2	86.8	18.60	92.0	92.6	20.00	97.8	98.5	23.50	112.8	113.6
17.22	86.3	86.9	18.62	92.0	92.7	20.05	98.0	98.7	23.55	113.0	113.8
17.24	86.4	86.9	18.64	92.1	92.7	20.10	98.2	98.9	23.60	113.2	114.0
17.26	86.5	87.0	18.66	92.2	92.8	20.15	98.5	99.1	23.65	113.5	114.2
17.28	86.5	87.1	18.68	92.3	92.9	20.20	98.7	99.3	23.70	113.7	114.4
17.30	86.6	87.2	18.70	92.4	93.0	20.25	98.9	99.5	23.75	113.9	114.7
17.32	86.7	87.3	18.72	92.5	93.1	20.30	99.1	99.8	23.80	114.1	114.9
17.34	86.8	87.4	18.74	92.5	93.2	20.35	99.3	100.0	23.85	114.3	115.1
17.36	86.9	87.4	18.76	92.6	93.3	20.40	99.5	100.2	23.90	114.6	115.3
17.38	86.9	87.5	18.78	92.7	93.3	20.45	99.7	100.4	23.95	114.8	115.5
17.40	87.0	87.6	18.80	92.8	93.4	20.50	99.9	100.6	24.00	115.0	115.8
17.42	87.1	87.7	18.82	92.9	93.5	20.55	100.1	100.8	24.05	115.2	116.0
17.44	87.2	87.8	18.84	93.0	93.6	20.60	100.4	101.0	24.10	115.4	116.2
17.46	87.3	87.9	18.86	93.0	93.7	20.65	100.6	101.2	24.15	115.6	116.4
17.48	87.3	87.9	18.88	93.1	93.8	20.70	100.8	101.5	24.20	115.9	116.6
17.50	87.4	88.0	18.90	93.2	93.8	20.75	101.0	101.7	24.25	116.1	116.9
17.52	87.5	88.1	18.92	93.3	93.9	20.80	101.2	101.9	24.30	116.3	117.1
17.54	87.6	88.2	18.94	93.4	94.0	20.85	101.4	102.1	24.35	116.5	117.3
17.56	87.7	88.3	18.96	93.5	94.1	20.90	101.6	102.3	24.40	116.7	117.5
17.58	87.8	88.3	18.98	93.5	94.2	20.95	101.8	102.5	24.45	117.0	117.7
17.60	87.8	88.4	19.00	93.6	94.3	21.00	102.1	102.7	24.50	117.2	118.0
17.62	87.9	88.5	19.02	93.7	94.3	21.05	102.3	103.0	24.55	117.4	118.2
17.64	88.0	88.6	19.04	93.8	94.4	21.10	102.5	103.2	24.60	117.6	118.4
17.66	88.1	88.7	19.06	93.9	94.5	21.15	102.7	103.4	24.65	117.8	118.6
17.68	88.2	88.8	19.08	94.0	94.6	21.20	102.9	103.6	24.70	118.0	118.8
17.70	88.3	88.8	19.10	94.0	94.7	21.25	103.1	103.8	24.75	118.3	119.1
17.72	88.3	88.9	19.12	94.1	94.8	21.30	103.3	104.0	24.80	118.5	119.3
17.74	88.4	89.0	19.14	94.2	94.8	21.35	103.6	104.2	24.85	118.7	119.5
17.76	88.5	89.1	19.16	94.3	94.9	21.40	103.8	104.5	24.90	118.9	119.7
17.78	88.6	89.2	19.18	94.4	95.0	21.45	104.0	104.7	24.95	119.1	119.9

Conversion: Kinematic Viscosity to Saybolt Universal Viscosity ... continued

Kin Vis, cSt	Equivalent Saybolt Universal Viscosity, SSU At 100°F	At 210°F	Kin Vis, cSt	Equivalent Saybolt Universal Viscosity, SSU At 100°F	At 210°F	Kin Vis, cSt	Equivalent Saybolt Universal Viscosity, SSU At 100°F	At 210°F	Kin Vis, cSt	Equivalent Saybolt Universal Viscosity, SSU At 100°F	At 210°F
25.00	119.4	120.2	28.50	134.8	135.7	32.00	150.5	151.5	35.50	166.3	167.4
25.05	119.6	120.4	28.55	135.0	135.9	32.05	150.7	151.7	35.55	166.5	167.6
25.10	119.8	120.6	28.60	135.3	136.2	32.10	150.9	152.0	35.60	166.7	167.9
25.15	120.0	120.8	28.65	135.5	136.4	32.15	151.2	152.2	35.65	167.0	168.1
25.20	120.2	121.0	28.70	135.7	136.6	32.20	151.4	152.4	35.70	167.2	168.3
25.25	120.5	121.3	28.75	135.9	136.8	32.25	151.6	152.6	35.75	167.4	168.6
25.30	120.7	121.5	28.80	136.2	137.1	32.30	151.8	152.9	35.80	167.7	168.8
25.35	120.9	121.7	28.85	136.4	137.3	32.35	152.1	153.1	35.85	167.9	169.0
25.40	121.1	121.9	28.90	136.6	137.5	32.40	152.3	153.3	35.90	168.1	169.2
25.45	121.3	122.1	28.95	136.8	137.7	32.45	152.5	153.5	35.95	168.3	169.5
25.50	121.6	122.4	29.00	137.0	138.0	32.50	152.7	153.8	36.00	168.6	169.7
25.55	121.8	122.6	29.05	137.3	138.2	32.55	153.0	154.0	36.05	168.8	169.9
25.60	122.0	122.8	29.10	137.5	138.4	32.60	153.2	154.2	36.10	169.0	170.1
25.65	122.2	123.0	29.15	137.7	138.6	32.65	153.4	154.4	36.15	169.2	170.4
25.70	122.4	123.3	29.20	137.9	138.9	32.70	153.6	154.7	36.20	169.5	170.6
25.75	122.6	123.5	29.25	138.2	139.1	32.75	153.9	154.9	36.25	169.7	170.8
25.80	122.9	123.7	29.30	138.4	139.3	32.80	154.1	155.1	36.30	169.9	171.1
25.85	123.1	123.9	29.35	138.6	139.5	32.85	154.3	155.4	36.35	170.1	171.3
25.90	123.3	124.1	29.40	138.8	139.8	32.90	154.5	155.6	36.40	170.4	171.5
25.95	123.5	124.4	29.45	139.1	140.0	32.95	154.8	155.8	36.45	170.6	171.7
26.00	123.7	124.6	29.50	139.3	140.2	33.00	155.0	156.0	36.50	170.8	172.0
26.05	124.0	124.8	29.55	139.5	140.4	33.05	155.2	156.3	36.55	171.1	172.2
26.10	124.2	125.0	29.60	139.7	140.7	33.10	155.4	156.5	36.60	171.3	172.4
26.15	124.4	125.2	29.65	140.0	140.9	33.15	155.7	156.7	36.65	171.5	172.7
26.20	124.6	125.5	29.70	140.2	141.1	33.20	155.9	156.9	36.70	171.7	172.9
26.25	124.9	125.7	29.75	140.4	141.3	33.25	156.1	157.2	36.75	172.0	173.1
26.30	125.1	125.9	29.80	140.6	141.6	33.30	156.3	157.4	36.80	172.2	173.3
26.35	125.3	126.1	29.85	140.8	141.8	33.35	156.6	157.6	36.85	172.4	173.6
26.40	125.5	126.4	29.90	141.1	142.0	33.40	156.8	157.8	36.90	172.6	173.8
26.45	125.7	126.6	29.95	141.3	142.2	33.45	157.0	158.1	36.95	172.9	174.0
26.50	126.0	126.8	30.00	141.5	142.5	33.50	157.2	158.3	37.00	173.1	174.3
26.55	126.2	127.0	30.05	141.7	142.7	33.55	157.5	158.5	37.05	173.3	174.5
26.60	126.4	127.2	30.10	142.0	142.9	33.60	157.7	158.8	37.10	173.6	174.7
26.65	126.6	127.5	30.15	142.2	143.1	33.65	157.9	159.0	37.15	173.8	174.9
26.70	126.8	127.7	30.20	142.4	143.4	33.70	158.2	159.2	37.20	174.0	175.2
26.75	127.1	127.9	30.25	142.6	143.6	33.75	158.4	159.4	37.25	174.2	175.4
26.80	127.3	128.1	30.30	142.9	144.0	33.80	158.6	159.7	37.30	174.5	175.6
26.85	127.5	128.4	30.35	143.1	144.0	33.85	158.8	159.9	37.35	174.7	175.9
26.90	127.7	128.6	30.40	143.3	144.3	33.90	159.1	160.1	37.40	174.9	176.1
26.95	127.9	128.8	30.45	143.5	144.5	33.95	159.3	160.3	37.45	175.1	176.3
27.00	128.2	129.0	30.50	143.8	144.7	34.00	159.5	160.6	37.50	175.4	176.5
27.05	128.4	129.2	30.55	144.0	144.9	34.05	159.7	160.8	37.55	175.6	176.8
27.10	128.6	129.5	30.60	144.2	145.2	34.10	160.0	161.0	37.60	175.8	177.0
27.15	128.8	129.7	30.65	144.4	145.4	34.15	160.2	161.3	37.65	176.1	177.2
27.20	129.0	129.9	30.70	144.6	145.6	34.20	160.4	161.5	37.70	176.3	177.5
27.25	129.3	130.1	30.75	144.9	145.8	34.25	160.6	161.7	37.75	176.5	177.7
27.30	129.5	130.4	30.80	145.1	146.1	34.30	160.9	161.9	37.80	176.7	177.9
27.35	129.7	130.6	30.85	145.3	146.3	34.35	161.1	162.2	37.85	177.0	178.1
27.40	129.9	130.8	30.90	145.5	146.5	34.40	161.3	162.4	37.90	177.2	178.4
27.45	130.2	131.0	30.95	145.8	146.7	34.45	161.5	162.6	37.95	177.4	178.6
27.50	130.4	131.3	31.00	146.0	147.0	34.50	161.8	162.9	38.00	177.6	178.8
27.55	130.6	131.5	31.05	146.2	147.2	34.55	162.0	163.1	38.05	177.9	179.1
27.60	130.8	131.7	31.10	146.4	147.4	34.60	162.2	163.3	38.10	178.1	179.3
27.65	131.0	131.9	31.15	146.7	147.7	34.65	162.4	163.5	38.15	178.3	179.5
27.70	131.3	132.1	31.20	146.9	147.9	34.70	162.7	163.8	38.20	178.6	179.8
27.75	131.5	132.4	31.25	147.1	148.1	34.75	162.9	164.0	38.25	178.8	180.0
27.80	131.7	132.6	31.30	147.3	148.3	34.80	163.1	164.2	38.30	179.0	180.2
27.85	131.9	132.8	31.35	147.6	148.6	34.85	163.3	164.4	38.35	179.2	180.4
27.90	132.2	133.0	31.40	147.8	148.8	34.90	163.6	164.7	38.40	179.5	180.7
27.95	132.4	133.3	31.45	148.0	149.0	34.95	163.8	164.9	38.45	179.7	180.9
28.00	132.6	133.5	31.50	148.2	149.2	35.00	164.0	165.1	38.50	179.9	181.1
28.05	132.8	133.7	31.55	148.5	149.5	35.05	164.3	165.4	38.55	180.1	181.4
28.10	133.0	133.9	31.60	148.7	149.7	35.10	164.5	165.6	38.60	180.4	181.6
28.15	133.3	134.2	31.65	148.9	149.9	35.15	164.7	165.8	38.65	180.6	181.8
28.20	133.5	134.4	31.70	149.1	150.1	35.20	164.9	166.0	38.70	180.8	182.0
28.25	133.7	134.6	31.75	149.4	150.4	35.25	165.2	166.3	38.75	181.1	182.3
28.30	133.9	134.8	31.80	149.6	150.6	35.30	165.4	166.5	38.80	181.3	182.5
28.35	134.2	135.1	31.85	149.8	150.8	35.35	165.6	166.7	38.85	181.5	182.7
28.40	134.4	135.3	31.90	150.0	151.0	35.40	165.8	167.0	38.90	181.7	183.0
28.45	134.6	135.5	31.95	150.3	151.3	35.45	166.1	167.2	38.95	182.0	183.2

Conversion: Kinematic Viscosity to Saybolt Universal Viscosity ... continued

Kin Vis, cSt	Equivalent Saybolt Universal Viscosity, SSU	
	At 100°F	At 210°F
39.00	182.2	183.4
39.05	182.4	183.6
39.10	182.7	183.9
39.15	182.9	184.1
39.20	183.1	184.3
39.25	183.3	184.6
39.30	183.6	184.8
39.35	183.8	185.0
39.40	184.0	185.3
39.45	184.2	185.5
39.50	184.5	185.7
39.55	184.7	185.9
39.60	184.9	186.2
39.65	185.2	186.4
39.70	185.4	186.6
39.75	185.6	186.9
39.80	185.8	187.1
39.85	186.1	187.3
39.90	186.3	187.5
39.95	186.5	187.8
40.00	186.8	188.0
40.05	187.0	188.2
40.10	187.2	188.5
40.15	187.4	188.7
40.20	187.7	188.9
40.25	187.9	189.2
40.30	188.1	189.4
40.35	188.4	189.6
40.40	188.6	189.8
40.45	188.8	190.1
40.50	189.0	190.3
40.55	189.3	190.5
40.60	189.5	190.8
40.65	189.7	191.0
40.70	189.9	191.2
40.75	190.2	191.5
40.80	190.4	191.7
40.85	190.6	191.9
40.90	190.9	192.1
40.95	191.1	192.4
41.00	191.3	192.6
41.05	191.5	192.8
41.10	191.8	193.1
41.15	192.0	193.3
41.20	192.2	193.5
41.25	192.5	193.7
41.30	192.7	194.0
41.35	192.9	194.2
41.40	193.1	194.4
41.45	193.4	194.7
41.50	193.6	194.9
41.55	193.8	195.1
41.60	194.1	195.4
41.65	194.3	195.6
41.70	194.5	195.8
41.75	194.7	196.0
41.80	195.0	196.3
41.85	195.2	196.5
41.90	195.4	196.7
41.95	195.7	197.0
42.00	195.9	197.2
42.05	196.1	197.4
42.10	196.3	197.7
42.15	196.6	197.9
42.20	196.8	198.1
42.25	197.0	198.3
42.30	197.3	198.6
42.35	197.5	198.8
42.40	197.7	199.0
42.45	197.9	199.3

Kin Vis, cSt	Equivalent Saybolt Universal Viscosity, SSU	
	At 100°F	At 210°F
42.50	198.2	199.5
42.55	198.4	199.7
42.60	198.6	200
42.65	198.9	200
42.70	199.1	200
42.75	199.3	200
42.80	199.5	200
42.85	199.8	201
42.90	200	201
42.95	200	201
43.00	200	201
43.05	201	202
43.10	201	202
43.15	201	202
43.20	201	202
43.25	202	202
43.30	202	203
43.35	202	203
43.40	202	203
43.45	203	203
43.50	203	204
43.55	203	204
43.60	203	204
43.65	203	204
43.70	204	205
43.75	204	205
43.80	204	205
43.85	204	205
43.90	205	205
43.95	205	206
44.00	205	206
44.05	205	206
44.10	205	206
44.15	206	207
44.20	206	207
44.25	206	207
44.30	206	207
44.35	207	208
44.40	207	208
44.45	207	208
44.50	207	208
44.55	208	208
44.60	208	209
44.65	208	209
44.70	208	209
44.75	208	209
44.80	209	210
44.85	209	210
44.90	209	210
44.95	209	210
45.00	210	211
45.05	210	211
45.10	210	211
45.15	210	211
45.20	211	211
45.25	211	212
45.30	211	212
45.35	211	212
45.40	211	212
45.45	212	213
45.50	212	213
45.55	212	213
45.60	212	213
45.65	213	214
45.70	213	214
45.75	213	214
45.80	213	214
45.85	214	214
45.90	214	215
45.95	214	215

Kin Vis, cSt	Equivalent Saybolt Universal Viscosity, SSU	
	At 100°F	At 210°F
46.00	214	215
46.05	214	215
46.10	215	216
46.15	215	216
46.20	215	216
46.25	215	216
46.30	216	217
46.35	216	217
46.40	216	217
46.45	216	217
46.50	216	217
46.55	217	218
46.60	217	218
46.65	217	218
46.70	217	218
46.75	218	219
46.80	218	219
46.85	218	219
46.90	218	219
46.95	219	220
47.00	219	220
47.05	219	220
47.10	219	220
47.15	219	220
47.20	220	221
47.25	220	221
47.30	220	221
47.35	220	221
47.40	221	222
47.45	221	222
47.50	221	222
47.55	221	222
47.60	222	223
47.65	222	223
47.70	222	223
47.75	222	223
47.80	222	223
47.85	223	224
47.90	223	224
47.95	223	224
48.00	223	224
48.05	224	225
48.10	224	225
48.15	224	225
48.20	224	225
48.25	225	226
48.30	225	226
48.35	225	226
48.40	225	226
48.45	225	226
48.50	226	227
48.55	226	227
48.60	226	227
48.65	226	227
48.70	227	228
48.75	227	228
48.80	227	228
48.85	227	228
48.90	227	229
48.95	228	229
49.00	228	229
49.05	228	229
49.10	228	229
49.15	229	230
49.20	229	230
49.25	229	230
49.30	229	230
49.35	230	231
49.40	230	231
49.45	230	231

Kin Vis, cSt	Equivalent Saybolt Universal Viscosity, SSU	
	At 100°F	At 210°F
49.50	230	231
49.55	230	232
49.60	231	232
49.65	231	232
49.70	231	232
49.75	231	232
49.80	232	233
49.85	232	233
49.90	232	233
49.95	232	233
50.0	233	234
50.1	233	234
50.2	233	235
50.3	234	235
50.4	234	235
50.5	235	236
50.6	235	236
50.7	236	237
50.8	236	237
50.9	237	238
51.0	237	238
51.1	238	239
51.2	238	239
51.3	239	240
51.4	239	240
51.5	239	241
51.6	240	241
51.7	240	241
51.8	241	242
51.9	241	242
52.0	242	243
52.1	242	243
52.2	243	244
52.3	243	244
52.4	244	245
52.5	244	245
52.6	245	246
52.7	245	246
52.8	245	247
52.9	246	247
53.0	246	247
53.1	247	248
53.2	247	248
53.3	248	249
53.4	248	249
53.5	249	250
53.6	249	250
53.7	250	251
53.8	250	251
53.9	250	252
54.0	251	252
54.1	251	253
54.2	252	253
54.3	252	254
54.4	253	254
54.5	253	254
54.6	254	255
54.7	254	255
54.8	255	256
54.9	255	256
55.0	256	257
55.1	256	257
55.2	256	258
55.3	257	258
55.4	257	259
55.5	258	259
55.6	258	260
55.7	259	260
55.8	259	260
55.9	260	261

Kin Vis, cSt	Equivalent Saybolt Universal Viscosity, SSU	
	At 100°F	At 210°F
56.0	260	261
56.1	261	262
56.2	261	262
56.3	262	263
56.4	262	263
56.5	262	264
56.6	263	264
56.7	263	265
56.8	264	265
56.9	264	266
57.0	265	266
57.1	265	266
57.2	266	267
57.3	266	267
57.4	267	268
57.5	267	268
57.6	268	269
57.7	268	269
57.8	268	270
57.9	269	270
58.0	269	271
58.1	270	271
58.2	270	272
58.3	271	272
58.4	271	273
58.5	272	273
58.6	272	273
58.7	273	274
58.8	273	274
58.9	274	275
59.0	274	275
59.1	274	276
59.2	275	276
59.3	275	277
59.4	276	277
59.5	276	278
59.6	277	278
59.7	277	279
59.8	278	279
59.9	278	279
60.0	279	280
60.1	279	280
60.2	280	281
60.3	280	281
60.4	280	282
60.5	281	282
60.6	281	283
60.7	282	283
60.8	282	284
60.9	283	284
61.0	283	285
61.1	284	285
61.2	284	286
61.3	285	286
61.4	285	286
61.5	286	287
61.6	286	287
61.7	286	288
61.8	287	288
61.9	287	289
62.0	288	289
62.1	288	290
62.2	289	290
62.3	289	291
62.4	290	291
62.5	290	292
62.6	291	292
62.7	291	292
62.8	291	293
62.9	292	293

Kin Vis, cSt	Equivalent Saybolt Universal Viscosity, SSU	
	At 100°F	At 210°F
63.0	292	294
63.1	293	294
63.2	293	295
63.3	294	295
63.4	294	296
63.5	295	296
63.6	295	297
63.7	296	297
63.8	296	298
63.9	297	298
64.0	297	299
64.1	297	299
64.2	298	299
64.3	298	300
64.4	299	300
64.5	299	301
64.6	300	301
64.7	300	302
64.8	301	302
64.9	301	303
65.0	302	303
65.1	302	304
65.2	303	304
65.3	303	305
65.4	303	305
65.5	304	305
65.6	304	306
65.7	305	306
65.8	305	307
65.9	306	307
66.0	306	308
66.1	307	308
66.2	307	309
66.3	308	309
66.4	308	310
66.5	309	310
66.6	309	311
66.7	309	311
66.8	310	312
66.9	310	312
67.0	311	312
67.1	311	313
67.2	312	313
67.3	312	314
67.4	313	314
67.5	313	315
67.6	314	315
67.7	314	316
67.8	315	316
67.9	315	317
68.0	315	317
68.1	316	318
68.2	316	318
68.3	317	319
68.4	317	319
68.5	318	319
68.6	318	320
68.7	319	320
68.8	319	321
68.9	320	321
69.0	320	322
69.1	321	322
69.2	321	323
69.3	322	323
69.4	322	324
69.5	322	324
69.6	323	325
69.7	323	325
69.8	324	325
69.9	324	326

Kin Vis, cSt	Equivalent Saybolt Universal Viscosity, SSU	
	At 100°F	At 210°F
70.0	325	326
70.1	325	327
70.2	326	327
70.3	326	328
70.4	327	328
70.5	327	329
70.6	328	329
70.7	328	330
70.8	328	330
70.9	329	331
71.0	329	331
71.1	330	332
71.2	330	332
71.3	331	332
71.4	331	333
71.5	332	333
71.6	332	334
71.7	333	334
71.8	333	335
71.9	334	335
72.0	334	336
72.1	334	336
72.2	335	337
72.3	335	337
72.4	336	338
72.5	336	338
72.6	337	339
72.7	337	339
72.8	338	339
72.9	338	340
73.0	339	340
73.1	339	341
73.2	340	341
73.3	340	342
73.4	340	342
73.5	341	343
73.6	341	343
73.7	342	344
73.8	342	344
73.9	343	345
74.0	343	345
74.1	344	345
74.2	344	346
74.3	345	346
74.4	345	347
74.5	346	347
74.6	346	348
74.7	346	348
74.8	347	349
74.9	347	349
75.0	348	350
75.1	348	350
75.2	349	351
75.3	349	351
75.4	350	352
75.5	350	352
75.6	351	352
75.7	351	353
75.8	352	353
75.9	352	354
76.0	352	354
76.1	353	355
76.2	353	355
76.3	354	356
76.4	354	356
76.5	355	357
76.6	355	357
76.7	356	358
76.8	356	358
76.9	357	359

Kin Vis, cSt	Equivalent Saybolt Universal Viscosity, SSU	
	At 100°F	At 210°F
77.0	357	359
77.1	358	359
77.2	358	360
77.3	358	360
77.4	359	361
77.5	359	361
77.6	360	362
77.7	360	362
77.8	361	363
77.9	361	363
78.0	362	364
78.1	362	364
78.2	363	365
78.3	363	365
78.4	364	365
78.5	364	366
78.6	364	366
78.7	365	367
78.8	365	367
78.9	366	368
79.0	366	368
79.1	367	369
79.2	367	369
79.3	368	370
79.4	368	370
79.5	369	371
79.6	369	371
79.7	370	372
79.8	370	372
79.9	370	372
80.0	371	373
80.1	371	373
80.2	372	374
80.3	372	374
80.4	373	375
80.5	373	375
80.6	374	376
80.7	374	376
80.8	375	377
80.9	375	377
81.0	376	378
81.1	376	378
81.2	376	379
81.3	377	379
81.4	377	379
81.5	378	380
81.6	378	380
81.7	379	381
81.8	379	381
81.9	380	382
82.0	380	382
82.1	381	383
82.2	381	383
82.3	382	384
82.4	382	384
82.5	383	385
82.6	383	385
82.7	383	386
82.8	384	386
82.9	384	386
83.0	385	387
83.1	385	387
83.2	386	388
83.3	386	388
83.4	387	389
83.5	387	389
83.6	388	390
83.7	388	390
83.8	389	391
83.9	389	391

Conversion: Kinematic Viscosity to Saybolt Universal Viscosity ... continued

Kin Vis, cSt	Equivalent Saybolt Universal Viscosity, SSU		Kin Vis, cSt	Equivalent Saybolt Universal Viscosity, SSU		Kin Vis, cSt	Equivalent Saybolt Universal Viscosity, SSU		Kin Vis, cSt	Equivalent Saybolt Universal Viscosity, SSU	
	At 100°F	At 210°F		At 100°F	At 210°F		At 100°F	At 210°F		At 100°F	At 210°F
84.0	389	392	91.0	422	424	98.0	454	457	110.0	510	513
84.1	390	392	91.1	422	425	98.1	455	457	110.2	511	514
84.2	390	392	91.2	423	425	98.2	455	458	110.4	512	515
84.3	391	393	91.3	423	426	98.3	456	458	110.6	513	515
84.4	391	393	91.4	424	426	98.4	456	459	110.8	513	516
84.5	392	394	91.5	424	426	98.5	457	459	111.0	514	517
84.6	392	394	91.6	425	427	98.6	457	460	111.2	515	518
84.7	393	395	91.7	425	427	98.7	457	460	111.4	516	519
84.8	393	395	91.8	426	428	98.8	458	460	111.6	517	520
84.9	394	396	91.9	426	428	98.9	458	461	111.8	518	521
85.0	394	396	92.0	426	429	99.0	459	461	112.0	519	522
85.1	395	397	92.1	427	429	99.1	459	462	112.2	520	523
85.2	395	397	92.2	427	430	99.2	460	462	112.4	521	524
85.3	395	398	92.3	428	430	99.3	460	463	112.6	522	525
85.4	396	398	92.4	428	431	99.4	461	463	112.8	523	526
85.5	396	399	92.5	429	431	99.5	461	464	113.0	524	527
85.6	397	399	92.6	429	432	99.6	462	464	113.2	525	528
85.7	397	399	92.7	430	432	99.7	462	465	113.4	525	529
85.8	398	400	92.8	430	433	99.8	463	465	113.6	526	529
85.9	398	400	92.9	431	433	99.9	463	466	113.8	527	530
86.0	399	401	93.0	431	433	100.0	463	466	114.0	528	531
86.1	399	401	93.1	432	434	100.2	464	467	114.2	529	532
86.2	400	402	93.2	432	434	100.4	465	468	114.4	530	533
86.3	400	402	93.3	432	435	100.6	466	469	114.6	531	534
86.4	401	403	93.4	433	435	100.8	467	470	114.8	532	535
86.5	401	403	93.5	433	436	101.0	468	471	115.0	533	536
86.6	401	404	93.6	434	436	101.2	469	472	115.2	534	537
86.7	402	404	93.7	434	437	101.4	470	473	115.4	535	538
86.8	402	405	93.8	435	437	101.6	471	474	115.6	536	539
86.9	403	405	93.9	435	438	101.8	472	474	115.8	537	540
87.0	403	406	94.0	436	438	102.0	473	475	116.0	538	541
87.1	404	406	94.1	436	439	102.2	474	476	116.2	538	542
87.2	404	406	94.2	437	439	102.4	475	477	116.4	539	542
87.3	405	407	94.3	437	440	102.6	475	478	116.6	540	543
87.4	405	407	94.4	438	440	102.8	476	479	116.8	541	544
87.5	406	408	94.5	438	440	103.0	477	480	117.0	542	545
87.6	406	408	94.6	438	441	103.2	478	481	117.2	543	546
87.7	407	409	94.7	439	441	103.4	479	482	117.4	544	547
87.8	407	409	94.8	439	442	103.6	480	483	117.6	545	548
87.9	407	410	94.9	440	442	103.8	481	484	117.8	546	549
88.0	408	410	95.0	440	443	104.0	482	485	118.0	547	550
88.1	408	411	95.1	441	443	104.2	483	486	118.2	548	551
88.2	409	411	95.2	441	444	104.4	484	487	118.4	549	552
88.3	409	412	95.3	442	444	104.6	485	488	118.6	550	553
88.4	410	412	95.4	442	445	104.8	486	488	118.8	550	554
88.5	410	413	95.5	443	445	105.0	487	489	119.0	551	555
88.6	411	413	95.6	443	446	105.2	488	490	119.2	552	556
88.7	411	413	95.7	444	446	105.4	488	491	119.4	553	556
88.8	412	414	95.8	444	447	105.6	489	492	119.6	554	557
88.9	412	414	95.9	444	447	105.8	490	493	119.8	555	558
89.0	413	415	96.0	445	447	106.0	491	494	120.0	556	559
89.1	413	415	96.1	445	448	106.2	492	495	120.2	557	560
89.2	413	416	96.2	446	448	106.4	493	496	120.4	558	561
89.3	414	416	96.3	446	449	106.6	494	497	120.6	559	562
89.4	414	417	96.4	447	449	106.8	495	498	120.8	560	563
89.5	415	417	96.5	447	450	107.0	496	499	121.0	561	564
89.6	415	418	96.6	448	450	107.2	497	500	121.2	562	565
89.7	416	418	96.7	448	451	107.4	498	501	121.4	563	566
89.8	416	419	96.8	449	451	107.6	499	501	121.6	563	567
89.8	417	419	96.9	449	452	107.8	500	502	121.8	564	568
90.0	417	419	97.0	450	452	108.0	500	503	122.0	565	569
90.1	418	420	97.1	450	453	108.2	501	504	122.2	566	570
90.2	418	420	97.2	451	453	108.4	502	505	122.4	567	570
90.3	419	421	97.3	451	453	108.6	503	506	122.6	568	571
90.4	419	421	97.4	451	454	108.8	504	507	122.8	569	572
90.5	420	422	97.5	452	454	109.0	505	508	123.0	570	573
90.6	420	422	97.6	452	455	109.2	506	509	123.2	571	574
90.7	420	423	97.7	453	455	109.4	507	510	123.4	572	575
90.8	421	423	97.8	453	456	109.6	508	511	123.6	573	576
90.9	421	424	97.9	454	456	109.8	509	512	123.8	574	577

Conversion: Kinematic Viscosity to Saybolt Universal Viscosity ... continued

Kin Vis, cSt	Equivalent Saybolt Universal Viscosity, SSU		Kin Vis, cSt	Equivalent Saybolt Universal Viscosity, SSU		Kin Vis, cSt	Equivalent Saybolt Universal Viscosity, SSU		Kin Vis, cSt	Equivalent Saybolt Universal Viscosity, SSU	
	At 100°F	At 210°F		At 100°F	At 210°F		At 100°F	At 210°F		At 100°F	At 210°F
124.0	575	578	128.0	593	597	132.0	612	615	136.0	630	634
124.2	575	579	128.2	594	597	132.2	612	616	136.2	631	635
124.4	576	580	128.4	595	598	132.4	613	617	136.4	632	636
124.6	577	581	128.6	596	599	132.6	614	618	136.6	633	637
124.8	578	582	128.8	597	600	132.8	615	619	136.8	634	638
125.0	579	583	129.0	598	601	133.0	616	620	137.0	635	639
125.2	580	584	129.2	599	602	133.2	617	621	137.2	636	639
125.4	581	584	129.4	600	603	133.4	618	622	137.4	637	640
125.6	582	585	129.6	600	604	133.6	619	623	137.6	638	641
125.8	583	586	129.8	601	605	133.8	620	624	137.8	638	642
126.0	584	587	130.0	602	606	134.0	621	625	138.0	639	643
126.2	585	588	130.2	603	607	134.2	622	625	138.2	640	644
126.4	586	589	130.4	604	608	134.4	623	626	138.4	641	645
126.6	587	590	130.6	605	609	134.6	624	627	138.6	642	646
126.8	588	591	130.8	606	610	134.8	625	628	138.8	643	647
127.0	588	592	131.0	607	611	135.0	625	629	139.0	644	648
127.2	589	593	131.2	608	611	135.2	626	630	139.2	645	649
127.4	590	594	131.4	609	612	135.4	627	631	139.4	646	650
127.6	591	595	131.6	610	613	135.6	628	632	139.6	647	651
127.8	592	596	131.8	611	614	135.8	629	633	139.8	648	652

Note: The relationship between Saybolt seconds and kinematic viscosities is linear above 75 cSt. To convert kinematic viscosities above 75 cSt to SSU at 210°F, multiply by 4.664 — to convert centistokes to SSU at 100°F, multiply by 4.632

Kin Vis, cSt	SSU At 100°F	Kin Vis, cSt	SSU At 100°F	Kin Vis, cSt	SSU At 100°F	Kin Vis, cSt	SSU At 100°F	Kin Vis, cSt	SSU At 100°F	Kin Vis, cSt	SSU At 100°F
140.0	649	148.0	686	156.0	723	164.0	760	172.0	797	180.0	834
140.2	650	148.2	687	156.2	724	164.2	761	172.2	798	180.2	835
140.4	650	148.4	688	156.4	725	164.4	762	172.4	799	180.4	836
140.6	651	148.6	688	156.6	726	164.6	763	172.6	800	180.6	837
140.8	652	148.8	689	156.8	726	164.8	763	172.8	801	180.8	838
141.0	653	149.0	690	157.0	727	165.0	764	173.0	801	181.0	839
141.2	654	149.2	691	157.2	728	165.2	765	173.2	802	181.2	839
141.4	655	149.4	692	157.4	729	165.4	766	173.4	803	181.4	840
141.6	656	149.6	693	157.6	730	165.6	767	173.6	804	181.6	841
141.8	657	149.8	694	157.8	731	165.8	768	173.8	805	181.8	842
142.0	658	150.0	695	158.0	732	166.0	769	174.0	806	182.0	843
142.2	659	150.2	696	158.2	733	166.2	770	174.2	807	182.2	844
142.4	660	150.4	697	158.4	734	166.4	771	174.4	808	182.4	845
142.6	661	150.6	698	158.6	735	166.6	772	174.6	809	182.6	846
142.8	662	150.8	699	158.8	736	166.8	773	174.8	810	182.8	847
143.0	663	151.0	700	159.0	737	167.0	774	175.0	811	183.0	848
143.2	663	151.2	701	159.2	738	167.2	775	175.2	812	183.2	849
143.4	664	151.4	701	159.4	738	167.4	776	175.4	813	183.4	850
143.6	665	151.6	702	159.6	739	167.6	776	175.6	814	183.6	851
143.8	666	151.8	703	159.8	740	167.8	777	175.8	814	183.8	851
144.0	667	152.0	704	160.0	741	168.0	778	176.0	815	184.0	852
144.2	668	152.2	705	160.2	742	168.2	779	176.2	816	184.2	853
144.4	669	152.4	706	160.4	743	168.4	780	176.4	817	184.4	854
144.6	670	152.6	707	160.6	744	168.6	781	176.6	818	184.6	855
144.8	671	152.8	708	160.8	745	168.8	782	176.8	819	184.8	856
145.0	672	153.0	709	161.0	746	169.0	783	177.0	820	185.0	857
145.2	673	153.2	710	161.2	747	169.2	784	177.2	821	185.2	858
145.4	674	153.4	711	161.4	748	169.4	785	177.4	822	185.4	859
145.6	675	153.6	712	161.6	749	169.6	786	177.6	823	185.6	860
145.8	676	153.8	713	161.8	750	169.8	787	177.8	824	185.8	861
146.0	676	154.0	713	162.0	751	170.0	788	178.0	825	186.0	862
146.2	677	154.2	714	162.2	751	170.2	789	178.2	826	186.2	863
146.4	678	154.4	715	162.4	752	170.4	789	178.4	826	186.4	864
146.6	679	154.6	716	162.6	753	170.6	790	178.6	827	186.6	864
146.8	680	154.8	717	162.8	754	170.8	791	178.8	828	186.8	865
147.0	681	155.0	718	163.0	755	171.0	792	179.0	829	187.0	866
147.2	682	155.2	719	163.2	756	171.2	793	179.2	830	187.2	867
147.4	683	155.4	720	163.4	757	171.4	794	179.4	831	187.4	868
147.6	684	155.6	721	163.6	758	171.6	795	179.6	832	187.6	869
147.8	685	155.8	722	163.8	759	171.8	796	179.8	833	187.8	870

Kin Vis, cSt	Equivalent Saybolt Universal Viscosity, SSU At 100°F	Kin Vis, cSt	Equivalent Saybolt Universal Viscosity, SSU At 100°F	Kin Vis, cSt	Equivalent Saybolt Universal Viscosity, SSU At 100°F	Kin Vis, cSt	Equivalent Saybolt Universal Viscosity, SSU At 100°F	Kin Vis, cSt	Equivalent Saybolt Universal Viscosity, SSU At 100°F	Kin Vis, cSt	Equivalent Saybolt Universal Viscosity, SSU At 100°F
188.0	871	205.0	950	240.0	1112	275.0	1274	310.0	1436	345.0	1598
188.2	872	205.5	952	240.5	1114	275.5	1276	310.5	1438	345.5	1601
188.4	873	206.0	954	241.0	1116	276.0	1279	311.0	1441	346.0	1603
188.6	874	206.5	957	241.5	1119	276.5	1281	311.5	1443	346.5	1605
188.8	875	207.0	959	242.0	1121	277.0	1283	312.0	1445	347.0	1607
189.0	876	207.5	961	242.5	1123	277.5	1286	312.5	1448	347.5	1610
189.2	877	208.0	964	243.0	1126	278.0	1288	313.0	1450	348.0	1612
189.4	877	208.5	966	243.5	1128	278.5	1290	313.5	1452	348.5	1614
189.6	878	209.0	968	244.0	1130	279.0	1292	314.0	1455	349.0	1617
189.8	879	209.5	971	244.5	1133	279.5	1295	314.5	1457	349.5	1619
190.0	880	210.0	973	245.0	1135	280.0	1297	315.0	1459	350.0	1621
190.2	881	210.5	975	245.5	1137	280.5	1299	315.5	1462	350.5	1624
190.4	882	211.0	977	246.0	1140	281.0	1302	316.0	1464	351.0	1626
190.6	883	211.5	980	246.5	1142	281.5	1304	316.5	1466	351.5	1628
190.8	884	212.0	982	247.0	1144	282.0	1306	317.0	1468	352.0	1631
191.0	885	212.5	984	247.5	1147	282.5	1309	317.5	1471	352.5	1633
191.2	886	213.0	987	248.0	1149	283.0	1311	318.0	1473	353.0	1635
191.4	887	213.5	989	248.5	1151	283.5	1313	318.5	1475	353.5	1638
191.6	888	214.0	991	249.0	1154	284.0	1316	319.0	1478	354.0	1640
191.8	889	214.5	994	249.5	1156	284.5	1318	319.5	1480	354.5	1642
192.0	889	215.0	996	250.0	1158	285.0	1320	320.0	1482	355.0	1645
192.2	890	215.5	998	250.5	1160	285.5	1323	320.5	1485	355.5	1647
192.4	891	216.0	1001	251.0	1163	286.0	1325	321.0	1487	356.0	1649
192.6	892	216.5	1003	251.5	1165	286.5	1327	321.5	1489	356.5	1651
192.8	893	217.0	1005	252.0	1167	287.0	1330	322.0	1492	357.0	1654
193.0	894	217.5	1008	252.5	1170	287.5	1332	322.5	1494	357.5	1656
193.2	895	218.0	1010	253.0	1172	288.0	1334	323.0	1496	358.0	1658
193.4	896	218.5	1012	253.5	1174	288.5	1336	323.5	1499	358.5	1661
193.6	897	219.0	1015	254.0	1177	289.0	1339	324.0	1501	359.0	1663
193.8	898	219.5	1017	254.5	1179	289.5	1341	324.5	1503	359.5	1665
194.0	899	220.0	1019	255.0	1181	290.0	1343	325.0	1506	360.0	1668
194.2	900	220.5	1021	255.5	1184	290.5	1346	325.5	1508	360.5	1670
194.4	901	221.0	1024	256.0	1186	291.0	1348	326.0	1510	361.0	1672
194.6	902	221.5	1026	256.5	1188	291.5	1350	326.5	1512	361.5	1675
194.8	902	222.0	1028	257.0	1191	292.0	1353	327.0	1515	362.0	1677
195.0	903	222.5	1031	257.5	1193	292.5	1355	327.5	1517	362.5	1679
195.2	904	223.0	1033	258.0	1195	293.0	1357	328.0	1519	363.0	1682
195.4	905	223.5	1035	258.5	1198	293.5	1360	328.5	1522	363.5	1684
195.6	906	224.0	1038	259.0	1200	294.0	1362	329.0	1524	364.0	1686
195.8	907	224.5	1040	259.5	1202	294.5	1364	329.5	1526	364.5	1689
196.0	908	225.0	1042	260.0	1204	295.0	1367	330.0	1529	365.0	1691
196.2	909	225.5	1045	260.5	1207	295.5	1369	330.5	1531	365.5	1693
196.4	910	226.0	1047	261.0	1209	296.0	1371	331.0	1533	366.0	1695
196.6	911	226.5	1049	261.5	1211	296.5	1374	331.5	1536	366.5	1698
196.8	912	227.0	1052	262.0	1214	297.0	1376	332.0	1538	367.0	1700
197.0	913	227.5	1054	262.5	1216	297.5	1378	332.5	1540	367.5	1702
197.2	914	228.0	1056	263.0	1218	298.0	1380	333.0	1543	368.0	1705
197.4	914	228.5	1059	263.5	1221	298.5	1383	333.5	1545	368.5	1707
197.6	915	229.0	1061	264.0	1223	299.0	1385	334.0	1547	369.0	1709
197.8	916	229.5	1063	264.5	1225	299.5	1387	334.5	1550	369.5	1712
198.0	917	230.0	1065	265.0	1228	300.0	1390	335.0	1552	370.0	1714
198.2	918	230.5	1068	265.5	1230	300.5	1392	335.5	1554	370.5	1716
198.4	919	231.0	1070	266.0	1232	301.0	1394	336.0	1557	371.0	1719
198.6	920	231.5	1072	266.5	1235	301.5	1397	336.5	1559	371.5	1721
198.8	921	232.0	1075	267.0	1237	302.0	1399	337.0	1561	372.0	1723
199.0	922	232.5	1077	267.5	1239	302.5	1401	337.5	1563	372.5	1726
199.2	923	233.0	1079	268.0	1242	303.0	1404	338.0	1566	373.0	1728
199.4	924	233.5	1082	268.5	1244	303.5	1406	338.5	1568	373.5	1730
199.6	925	234.0	1084	269.0	1246	304.0	1408	339.0	1570	374.0	1733
199.8	926	234.5	1086	269.5	1248	304.5	1411	339.5	1573	374.5	1735
200.0	927	235.0	1089	270.0	1251	305.0	1413	340.0	1575	375.0	1737
200.5	929	235.5	1091	270.5	1253	305.5	1415	340.5	1577	375.5	1739
201.0	931	236.0	1093	271.0	1255	306.0	1418	341.0	1580	376.0	1742
201.5	933	236.5	1096	271.5	1258	306.5	1420	341.5	1582	376.5	1744
202.0	936	237.0	1098	272.0	1260	307.0	1422	342.0	1584	377.0	1746
202.5	938	237.5	1100	272.5	1262	307.5	1424	342.5	1587	377.5	1749
203.0	940	238.0	1103	273.0	1265	308.0	1427	343.0	1589	378.0	1751
203.5	943	238.5	1105	273.5	1267	308.5	1429	343.5	1591	378.5	1753
204.0	945	239.0	1107	274.0	1269	309.0	1431	344.0	1594	379.0	1756
204.5	947	239.5	1109	274.5	1272	309.5	1434	344.5	1596	379.5	1758

Conversion: Kinematic Viscosity to Saybolt Universal Viscosity ... continued

Kin Vis, cSt	Equivalent Saybolt Universal Viscosity, SSU At 100 F	Kin Vis, cSt	Equivalent Saybolt Universal Viscosity, SSU At 100 F	Kin Vis, cSt	Equivalent Saybolt Universal Viscosity, SSU At 100 F	Kin Vis, cSt	Equivalent Saybolt Universal Viscosity, SSU At 100 F	Kin Vis, cSt	Equivalent Saybolt Universal Viscosity, SSU At 100 F	Kin Vis, cSt	Equivalent Saybolt Universal Viscosity, SSU At 100 F
380.0	1760	400.0	1853	420.0	1946	440.0	2038	460.0	2131	480.0	2224
380.5	1763	400.5	1855	420.5	1948	440.5	2041	460.5	2133	480.5	2226
381.0	1765	401.0	1858	421.0	1950	441.0	2043	461.0	2136	481.0	2228
381.5	1767	401.5	1860	421.5	1953	441.5	2045	461.5	2138	481.5	2231
382.0	1770	402.0	1862	422.0	1955	442.0	2048	462.0	2140	482.0	2233
382.5	1772	402.5	1865	422.5	1957	442.5	2050	462.5	2142	482.5	2235
383.0	1774	403.0	1867	423.0	1960	443.0	2052	463.0	2145	483.0	2237
383.5	1777	403.5	1869	423.5	1962	443.5	2054	463.5	2147	483.5	2240
384.0	1779	404.0	1872	424.0	1964	444.0	2057	464.0	2149	484.0	2242
384.5	1781	404.5	1874	424.5	1966	444.5	2059	464.5	2152	484.5	2244
385.0	1783	405.0	1876	425.0	1969	445.0	2061	465.0	2154	485.0	2247
385.5	1786	405.5	1878	425.5	1971	445.5	2064	465.5	2156	485.5	2249
386.0	1788	406.0	1881	426.0	1973	446.0	2066	466.0	2159	486.0	2251
386.5	1790	406.5	1883	426.5	1976	446.5	2068	466.5	2161	486.5	2254
387.0	1793	407.0	1885	427.0	1978	447.0	2071	467.0	2163	487.0	2256
387.5	1795	407.5	1888	427.5	1980	447.5	2073	467.5	2166	487.5	2258
388.0	1797	408.0	1890	428.0	1983	448.0	2075	468.0	2168	488.0	2261
388.5	1800	408.5	1892	428.5	1985	448.5	2078	468.5	2170	488.5	2263
389.0	1802	409.0	1895	429.0	1987	449.0	2080	469.0	2173	489.0	2265
389.5	1804	409.5	1897	429.5	1990	449.5	2082	469.5	2175	489.5	2268
390.0	1807	410.0	1899	430.0	1992	450.0	2085	470.0	2177	490.0	2270
390.5	1809	410.5	1902	430.5	1994	450.5	2087	470.5	2180	490.5	2272
391.0	1811	411.0	1904	431.0	1997	451.0	2089	471.0	2182	491.0	2275
391.5	1814	411.5	1906	431.5	1999	451.5	2092	471.5	2184	491.5	2277
392.0	1816	412.0	1909	432.0	2001	452.0	2094	472.0	2187	492.0	2279
392.5	1818	412.5	1911	432.5	2004	452.5	2096	472.5	2189	492.5	2281
393.0	1821	413.0	1913	433.0	2006	453.0	2098	473.0	2191	493.0	2284
393.5	1823	413.5	1916	433.5	2008	453.5	2101	473.5	2193	493.5	2286
394.0	1825	414.0	1918	434.0	2010	454.0	2103	474.0	2196	494.0	2288
394.5	1827	414.5	1920	434.5	2013	454.5	2105	474.5	2198	494.5	2291
395.0	1830	415.0	1922	435.0	2015	455.0	2108	475.0	2200	495.0	2293
395.5	1832	415.5	1925	435.5	2017	455.5	2110	475.5	2203	495.5	2295
396.0	1834	416.0	1927	436.0	2020	456.0	2112	476.0	2205	496.0	2298
396.5	1837	416.5	1929	436.5	2022	456.5	2115	476.5	2207	496.5	2300
397.0	1839	417.0	1932	437.0	2024	457.0	2117	477.0	2210	497.0	2302
397.5	1841	417.5	1934	437.5	2027	457.5	2119	477.5	2212	497.5	2305
398.0	1844	418.0	1936	438.0	2029	458.0	2122	478.0	2214	498.0	2307
398.5	1846	418.5	1939	438.5	2031	458.5	2124	478.5	2217	498.5	2312
399.0	1848	419.0	1941	439.0	2034	459.0	2126	479.0	2219	499.0	2312
399.5	1851	419.5	1943	439.5	2036	459.5	2129	479.5	2221	499.5	2314
										500.0	2316

Note: To convert kinematic viscosities above 500 cSt to SSU at 100°F, refer to the note on page 49.

Viscosity: Approximate Equivalents at Same Temperature

Kinematic (Centistokes)	Saybolt Universal (Seconds)	Redwood No. 1 (Seconds)	Engler (Degrees)	Saybolt Furol (Seconds)	Redwood No. 2 (Seconds)	Kinematic (Centistokes)	Saybolt Universal (Seconds)	Redwood No. 1 (Seconds)	Engler (Degrees)	Saybolt Furol (Seconds)	Redwood No. 2 (Seconds)
1.8	32	30.8	1.14	–	–						
2.7	35	32.2	1.18	–	–	102.2	475	419	13.5	49	–
4.2	40	36.2	1.32	–	–	107.6	500	441	14.2	51	–
5.8	45	40.6	1.46	–	–	118.4	550	485	15.6	56	–
7.4	50	44.9	1.60	–	–	129.2	600	529	17.0	61	–
8.9	55	49.1	1.75	–	–	140.3	650	573	18.5	66	–
10.3	60	53.5	1.88	–	–	151	700	617	19.9	71	–
11.7	65	57.9	2.02	–	–	162	750	661	21.3	76	–
13.0	70	62.3	2.15	–	–	173	800	705	22.7	81	–
14.3	75	67.6	2.31	–	–	183	850	749	24.2	86	–
15.6	80	71.0	2.42	–	–	194	900	793	25.6	91	–
16.8	85	75.1	2.55	–	–	205	950	837	27.0	96	–
18.1	90	79.6	2.68	–	–	215	1,000	882	28.4	100	–
19.2	95	84.2	2.81	–	–	259	1,200	1,058	34.1	121	104
20.4	100	88.4	2.95	–	–	302	1,400	1,234	39.8	141	122
22.8	110	97.1	3.21	–	–	345	1,600	1,411	45.5	160	138
25.0	120	105.9	3.49	–	–	388	1,800	1,587	51	180	153
27.4	130	114.8	3.77	–	–	432	2,000	1,763	57	200	170
29.6	140	123.6	4.04	–	–	541	2,500	2,204	71	250	215
31.8	150	132.4	4.32	–	–	650	3,000	2,646	85	300	255
34.0	160	141.1	4.59	–	–	758	3,500	3,087	99	350	300
36.0	170	150.0	4.88	–	–	866	4,000	3,526	114	400	345
38.4	180	158.8	5.15	–	–	974	4,500	3,967	128	450	390
40.6	190	167.5	5.44	–	–	1,082	5,000	4,408	142	500	435
42.8	200	176.4	5.72	23.0	–	1,190	5,500	4,849	156	550	475
47.2	220	194.0	6.28	25.3	–	1,300	6,000	5,290	170	600	515
51.6	240	212	6.85	27.0	–	1,405	6,500	5,730	185	650	560
55.9	260	229	7.38	28.7	–	1,515	7,000	6,171	199	700	600
60.2	280	247	7.95	30.5	–	1,625	7,500	6,612	213	750	645
64.5	300	265	8.51	32.5	–	1,730	8,000	7,053	227	800	690
69.9	325	287	9.24	35.0	–	1,840	8,500	7,494	242	850	730
75.3	350	309	9.95	37.2	–	1,950	9,000	7,934	256	900	770
80.7	375	331	10.7	39.5	–	2,055	9,500	8,375	270	950	815
86.1	400	353	11.4	42.0	–	2,165	10,000	8,816	284	1,000	855
91.5	425	375	12.1	44.2	–						
96.8	450	397	12.8	47.0	–						

Viscosity Conversion Table (122°F/50°C)

Saybolt Furol, SSF	Kinematic, cSt	Saybolt, Universal, SSU	Saybolt Furol, SSF	Kinematic, cSt	Saybolt, Universal, SSU	Saybolt Furol, SSF	Kinematic, cSt	Saybolt, Universal, SSU	Saybolt Furol, SSF	Kinematic, cSt	Saybolt, Universal, SSU
25	47.6	222	60	124.8	579	95	200.1	928	160	338.8	1572
26	50.0	232	61	127.0	589	96	202.2	938	162	343.0	1592
27	52.1	242	62	129.2	599	97	204.4	948	164	347.2	1611
28	54.2	252	63	131.3	609	98	206.6	958	166	351.5	1631
29	56.5	262	64	133.5	619	99	208.7	968	168	355.8	1651
30	58.7	273	65	135.6	629	100	210.9	978	170	360.0	1671
31	60.9	283	66	137.7	639	102	215.1	998	172	364.2	1691
32	63.1	293	67	139.8	649	104	219.3	1018	174	368.5	1711
33	65.4	303	68	140.0	659	106	223.6	1038	176	372.8	1730
34	67.6	314	69	144.1	669	108	227.8	1057	178	377.0	1750
35	69.8	324	70	146.3	679	110	232.1	1077	180	381.2	1770
36	72.0	334	71	148.5	689	112	236.5	1097	182	385.6	1790
37	74.2	344	72	150.6	699	114	240.8	1117	184	389.8	1809
38	76.4	355	73	152.8	709	116	245.0	1137	186	394.0	1829
39	78.7	365	74	155.0	719	118	249.3	1157	188	398.4	1849
40	80.9	375	75	157.1	729	120	253.5	1176	190	402.6	1870
41	83.1	386	76	159.3	739	122	257.7	1196	192	406.8	1890
42	85.3	396	77	161.5	749	124	262.0	1216	194	411.0	1910
43	87.5	406	78	163.6	759	126	266.2	1235	196	415.2	1929
44	89.7	416	79	165.8	769	128	270.5	1255	198	419.6	1949
45	91.9	427	80	168.0	779	130	274.8	1275	200	423.8	1969
46	94.2	437	81	170.1	789	132	279.2	1295	210	444	2071
47	96.4	447	82	172.2	799	134	283.4	1315	220	466	2174
48	98.6	458	83	174.3	809	136	285.6	1335	230	487	2274
49	100.8	468	84	176.4	819	138	292.0	1355	240	508	2373
50	103.0	478	85	178.6	829	140	296.3	1375	250	529	2473
51	105.1	488	86	180.8	839	142	300.6	1395	260	551	2578
52	107.3	498	87	182.9	849	144	304.8	1414	270	592	2677
53	109.5	508	88	185.1	859	146	309.1	1434	280	593	2777
54	111.7	518	89	187.3	869	148	313.3	1454			
55	114.0	529	90	189.4	879	150	317.6	1474			
56	116.2	539	91	191.5	889	152	321.8	1493			
57	118.3	549	92	193.6	898	154	326.0	1513			
58	120.5	559	93	195.7	908	156	330.3	1533			
59	122.7	569	94	197.9	918	158	334.5	1552			

Equations for Converting Kinematic Viscosity, Centistokes, to Flow Times

$$T=DV+\frac{1+EV}{F+GV+HV^2+IV^3}$$

Where T=Flow time in seconds (or Engler degrees)
V=Kinematic viscosity in centistokes
D, E, F, G, H, I=Constants given below

Equation No.	For	D	E	F	G	H	I	Range, cs
1	Saybolt seconds universal at 100°F	4.6324	0.03264	0.039302	0.002627	0.0002397	0.00001646	> 1.8
1a	Saybolt seconds universal at 100°F	4.6324	0.0	0.039911	0.000938	0.000280	0.00000274	> 1.8
1b	Saybolt seconds universal at 210°F	4.6635	0.00677	0.039911	0.000938	0.000280	0.00000274	> 1.8
2[a]	Saybolt seconds furol at 122°F	0.47170	0.0	0.4895	−0.005213	0.0000718	0.0	> 48
3[b]	Saybolt seconds furol at 210°F	0.47916	0.0	0.3797	0.0	0.0001783	0.0	> 48
4	Redwood No. 1 seconds at 140°F	4.0984	0.0	0.038014	0.001919	0.0000278	0.00000521	> 4.0
5	Redwood No. 2 seconds	0.40984	0.0	0.38014	0.01919	0.000278	0.0000521	> 73
6	Engler degrees	0.13158	0.0	1.1326	0.01040	0.00656	0.0	> 1.0

Note: For Saybolt Seconds Universal at other temperatures.

$$SSU_t=cs_t\,[1+0.000061\,(t-100°F)]\left[\frac{SSU_{100°F}}{cs_{100°F}}\right]$$

[a] Equivalent to Eq 7 in ASTM D 2161.
[b] Equivalent to Eq 8 in ASTM D 2161.

No. 1 and No. 1a are alternative equations.

Equations for Converting Flow Times to Kinematic Viscosity, Centistokes

$$V=AT-\frac{BT}{T^3+C}$$

Where V =Kinematic viscosity in centistokes
T =Flow time in seconds (or Engler degrees)
A, B, C =Constants given below

Equation No.	For	A	B	C	Range	
11	Saybolt seconds universal at 100°F	0.21587	11,069	37,003	SSU	> 32
11b	Saybolt seconds universal at 210°F	0.21443	11,219	37,755	SSU	> 32
12	Saybolt seconds furol at 122°F	2.120	8,920	27,100	SSF	> 25
13	Saybolt seconds furol at 210°F	2.087	2,460	8,670	SSF	> 25
14	Redwood No. 1 seconds at 140°F	0.244	8,000	12,500	R₁	> 35
15	Redwood No. 2 seconds	2.44	3,410	9,550	R₂	> 31
16	Engler degrees	7.60	18.0	1.7273	E	> 1,000

Equation for Converting Kinematic Viscosity to Absolute Viscosity

Absolute viscosity in centipoises = (Kinematic viscosity in centistokes) (Density)

Note: For most purposes, density may be expressed as specific gravity in this equation without significant loss of accuracy.
The specific gravity must be measured at the same temperature as the kinematic viscosity.

TABLES FOR ESTIMATING VISCOSITIES IN KINEMATIC UNITS @ 50°C FROM GIVEN VISCOSITIES @ 50°C, 40°C & 100°F

Based on University of Athens Tables
(Viscosity Index About 70)

| 50°C = 122°F | | | 100°F = 37.78°C | | | 40°C = 104°F |
Centistokes (KvcSt)	Saybolt Furol (SSF)	Engler (Degrees)	Centistokes (KvcSt)	Redwood No. 1 (RW1)	Saybolt Univ. (SSU)	Centistokes (KvcSt)
			1.7	29.9	31.5	1.6
			1.8	30.1	31.9	1.7
			1.9	30.4	32.3	1.8
			2.0	30.7	32.6	1.9
1.7	—	—	2.1	31.0	33.1	2.0
1.8	—	—	2.2	31.2	33.3	2.1
1.9	—	—	2.3	31.5	33.6	2.2
2.0	—	1.141	2.4	31.8	34.0	2.3
2.1	—	1.149	2.6	32.1	34.6	2.5
2.2	—	1.158	2.7	32.4	35.0	2.6
2.3	—	1.166	2.9	32.7	35.6	2.7
2.4	—	1.175	3.0	33.0	36.0	2.8
2.5	—	1.183	3.2	33.4	36.6	3.0
2.6	—	1.191	3.3	33.7	36.9	3.1
2.7	—	1.200	3.4	34.0	37.3	3.2
2.8	—	1.208	3.5	34.3	37.6	3.3
2.9	—	1.217	3.7	34.6	38.2	3.5
3.0	—	1.225	3.8	34.9	38.6	3.6
3.2	—	1.242	3.9	35.4	38.9	3.7
3.4	—	1.259	4.1	36.0	39.5	4.0
3.6	—	1.275	4.4	36.6	40.5	4.2
3.8	—	1.292	4.6	37.3	41.1	4.4
4.0	—	1.309	4.9	38.0	42.1	4.7
4.2	—	1.326	5.3	38.7	43.3	4.9
4.4	—	1.343	5.6	39.4	44.3	5.3
4.6	—	1.361	5.8	40.1	44.9	5.5
4.8	—	1.381	6.0	40.9	46.2	5.7
5.0	—	1.401	6.3	41.8	46.5	6.0
5.2	—	1.418	6.6	42.6	47.5	6.2
5.4	—	1.435	6.8	43.2	48.1	6.3
5.6	—	1.451	7.0	43.9	48.8	6.5
5.8	—	1.467	7.4	44.6	50.1	6.9
6.0	—	1.483	7.8	45.4	51.4	7.2
6.5	—	1.523	8.0	46.1	52.1	7.4
7.0	—	1.573	8.8	47.3	54.7	8.2
7.5	—	1.608	9.5	50.5	57.1	8.8
8.0	—	1.656	10.3	53.0	59.9	9.6
8.5	—	1.703	11.1	55.0	62.7	10.2
9.0	—	1.749	11.9	57.3	65.6	10.9
9.5	—	1.794	12.7	59.5	68.6	11.7
			13.4	62	71	12.3

TABLES FOR ESTIMATING VISCOSITIES IN KINEMATIC UNITS @ 50ºC . . . Continued

50ºC = 122ºF			100ºF = 37.78ºC			40ºC = 104ºF
Centistokes (KvcSt)	Saybolt Furol (SSF)	Engler (Degrees)	Centistokes (KvcSt)	Redwood No. 1 (RW1)	Saybolt Univ. (SSU)	Centistokes (KvcSt)
10.0	—	1.840	14.2	65	74	13.1
10.5	—	1.885	15.1	67	78	13.9
11.0	—	1.932	15.9	70	81	14.6
11.5	—	1.977	16.7	73	84	15.4
12.0	—	2.024	17.6	76	88	16.2
12.5	—	2.074	18.5	79	92	17.0
13.0	—	2.125	19.3	82	95	17.8
13.5	—	2.175	20.1	85	98	18.3
14.0	—	2.224	20.9	88	102	19.0
14.5	—	2.275	21.8	91	106	19.8
15.0	—	2.329	22.7	94	109	20.7
15.5	—	2.384	23.5	97	113	21.4
16.0	—	2.440	24.3	100	116	22.1
16.5	—	2.496	25.1	103	120	22.9
17.0	—	2.547	26.0	106	124	23.7
17.5	—	2.597	26.9	109	128	24.5
18.0	—	2.651	27.8	112	132	25.3
18.5	—	2.707	28.7	116	136	26.1
19.0	—	2.762	29.6	120	140	26.7
19.5	—	2.820	30.2	124	142	26.9
20.0	—	2.878	31.2	128	147	27.2
21	—	2.993	33.0	135	155	30
22	—	3.109	34.5	143	162	31.5
23	—	3.225	36.0	150	169	33
24	—	3.345	38.5	157	180	34.5
25	—	3.465	40.5	164	189	36.5
26	—	3.586	42	171	196	38
27	—	3.706	44	178	205	39
28	—	3.831	45.5	185	212	40
29	—	3.957	47	192	219	42
30	—	4.082	49	200	228	43
31	—	4.208	51	207	237	45
32	18	4.333	53	215	246	47
33	18	4.458	55	223	256	49
34	18	4.584	57	231	265	51
35	19	4.709	59	239	274	52
36	19	4.839	61	247	283	54
37	20	4.970	63	255	292	56
38	20	5.095	65	263	302	58
39	21	5.221	66	270	306	59
40	21	5.351	68	278	315	61

TABLES FOR ESTIMATING VISCOSITIES IN KINEMATIC UNITS @ 50°C . . . Continued

50°C = 122°F			100°F = 37.78°C			40°C = 104°F
Centistokes (KvcSt)	Saybolt Furol (SSF)	Engler (Degrees)	Centistokes (KvcSt)	Redwood No. 1 (RW1)	Saybolt Univ. (SSU)	Centistokes (KvcSt)
42	22	5.607	72	292	334	63
44	23	5.863	76	308	352	66
46	24	6.123	80	323	371	70
48	25.1	6.384	84	338	389	74
50	26.0	6.650	88	354	408	78
52	27.0	6.911	92	370	426	81
54	27.9	7.127	96	391	445	85
56	28.8	7.392	100	407	463	88
58	29.7	7.656	104	423	482	91
60	30.6	7.920	108	439	500	94
62	31.5	8.183	112	455	519	98
64	32.4	8.447	116	471	538	101
66	33.3	8.712	120	487	556	104
68	34.2	8.976	124	503	575	108
70	35.1	9.240	128	520	593	112
72	36.0	9.503	132	537	612	116
74	36.9	9.767	136	555	630	119
76	37.8	10.165	141	573	653	123
78	38.7	10.432	147	592	681	128
80	39.6	10.700	152	612	704	132
90	44.1	12.037	172	700	797	150
100	48.6	13.374	192	780	889	167
110	53.2	14.71	212	870	982	182
120	57.7	16.05	232	965	1075	197
130	62.3	17.39	252	1055	1167	214
140	66.9	18.72	275	1150	1274	236
150	71.5	19.98	298	1240	1380	253
160	76.1	21.32	321	1330	1487	272
170	80.8	22.65	345	1425	1598	293
180	85.4	23.98	369	1520	1709	313
190	90.1	25.31	393	1620	1821	326
200	94.8	26.64	417	1720	1932	346
210	99.4	27.98	441	1820	2043	361
220	104.1	29.31	465	1920	2154	386
230	108.8	30.64	489	2020	2265	406
240	113.5	31.97	513	2120	2376	426
250	118.2	33.31	537	2220	2487	446
260	122.9	34.64	561	2320	2598	465
270	127.6	35.97	585	2420	2709	485
280	132.3	37.30	610	2525	2825	506
290	137.0	38.64	635	2630	2941	527

TABLES FOR ESTIMATING VISCOSITIES IN KINEMATIC UNITS @ 50°C . . . Continued

50°C = 122°F			100°F = 37.78°C			40°C = 104°F
Centistokes (KvcSt)	Saybolt Furol (SSF)	Engler (Degrees)	Centistokes (KvcSt)	Redwood No. 1 (RW1)	Saybolt Univ. (SSU)	Centistokes (KvcSt)
300	141.7	39.97	660	2735	3057	548
310	146.4	41.30	685	2840	3173	568
320	151.1	42.63	710	2945	3288	589
330	155.8	43.96	735	3050	3404	610
340	160.5	45.30	760	3160	3520	631
350	165.2	46.63	785	3270	3636	651
360	169.9	47.96	810	3380	3752	672
370	174.6	49.29	836	3490	3872	694
380	179.4	50.63	862	3600	3993	715
390	184.1	51.96	888	3710	4113	737
400	188.8	53.29	914	3820	4234	758
410	193.5	54.62	940	3930	4354	780
420	198.2	55.96	966	4040	4474	802
430	203	57.29	992	4150	4595	823
440	208	58.62	1018	4260	4715	845
450	212	59.95	1044	4370	4836	866
460	217	61.28	1070	4408	4956	888
470	222	62.62	1096	4590	5077	909
480	226	63.94	1122	4700	5197	931
490	231	65.28	1148	4810	5317	953
500	236	66.61	1174	4920	5438	974
510	241	67.95	1200	5030	5558	996
520	245	69.28	1226	5140	5679	1017
530	250	70.61	1252	5250	5799	1039
540	255	71.92	1278	5360	5920	1061
550	259	73.27	1304	5470	6040	1082
560	264	74.61	1330	5580	6161	1104
570	269	75.94	1357	5690	6286	1126
580	274	77.27	1384	5800	6411	1149
590	278	78.60	1412	5910	6540	1172
600	283	79.94	1440	6020	6670	1195
610	288	81.27	1468	6130	6800	1218
620	292	82.60	1496	6240	6929	1241
630	297	83.93	1524	6350	7059	1265
640	302	85.26	1552	6460	7189	1288
650	307	86.60	1580	6570	7318	1311
700	330	93.30	1715	7150	7944	1423
750	354	99.90	1850	7800	8569	1535
800	377	106.60	1985	8400	9194	1647
850	401	113.20	2120	9000	9820	1759
900	425	119.90	2265	9600	10491	1880
700	330	93.30	1715	7150	7944	1423
750	354	99.90	1850	7800	8569	1535
800	377	106.60	1985	8400	9194	1647
850	401	113.20	2120	9000	9820	1759
900	425	119.90	2265	9600	10491	1880

TABLES FOR ESTIMATING VISCOSITIES IN SAYBOLT UNIVERSAL @ 100°F (Tables IA & IB)
KINEMATIC UNITS @ 40°C (Tables IIA & IIB) FROM GIVEN VISCOSITIES IN ENGLER DEGREES @ 20°C & 50°C

TABLE IA

Engler Viscosity °E @ 20°C (68°F)	Saybolt Universal Viscosity SSU @ 100°F for:		
	0 V.I. Oil	30 V.I. Oil	100 V.I. Oil
1.5	38.6	38.9	39.5
2.0	47	48	49
2.5	54	55	57
3.0	60	61	64
4.0	71	73	77
5.0	84	86	90
6.0	93	96	100
7.0	106	109	113
8.0	114	117	125
9.0	127	131	140
10.	137	141	150
15.	188	193	205
20.	230	237	255
30.	300	315	350
40.	380	398	440
50.	440	467	530
75.	630	675	775
100.	775	842	1,000
150.	1,050	1,155	1,400
200.	1,350	1,485	1,800
300.	2,000	2,180	2,600
400.	2,480	2,710	3,250
500.	3,000	3,300	4,000

TABLE IB

Engler Viscosity E° @ 50°C (122°F)	Saybolt Universal Viscosity SSU @ 100°F for:		
	0 V.I. Oil	30 V.I. Oil	100 V.I. Oil
1.5	56	55	54
2.0	93	92	89
2.5	132	130	125
3.0	175	171	162
4.0	257	251	238
5.0	340	332	312
6.0	420	410	385
7.0	515	497	455
8.0	612	584	550
9.0	700	680	635
10.	798	772	710
15.	1,200	1,155	1,050
20.	1,680	1,625	1,500
30.	2,600	2,495	2,250
40.	3,950	3,725	3,200
50.	5,000	4,700	4,000
75.	8,000	7,580	6,600
100.	11,500	10,750	9,000

TABLE IIA

Engler Viscosity °E @ 20°C (68°F)	Kinematic Viscosity cSt @ 40°C for:		
	0 V.I. Oil	30 V.I. Oil	100 V.I. Oil
1.5	3.59	3.72	3.90
2.0	6.02	6.31	6.63
2.5	7.95	8.26	8.85
3.0	9.54	9.83	10.67
4.0	12.25	12.77	13.8
5.0	15.3	15.7	16.7
6.0	17.2	17.9	18.9
7.0	19.9	20.6	21.5
8.0	21.6	22.2	24.1
9.0	24.2	25.1	27.1
10.	26.1	27.0	29.0
15.	35.9	37.0	39.8
20.	43.9	45.4	49.4
30.	56.9	60.0	67.6
40.	71.6	75.3	84.7
50.	82.7	88.1	101.7
75.	117.1	126.1	147.8
100.	143.3	156.7	189.9
150.	192.9	213.3	264.4
200.	246.6	272.7	338.8
300.	361.8	397.1	500.0
400.	460.0	500.0	600.0
500.	540.0	600.0	750.0

TABLE IIB

Engler Viscosity °E @ 50°C (122°F)	Kinematic Viscosity cSt @ 40°C for:		
	0 V.I. Oil	30 V.I. Oil	100 V.I. Oil
1.5	8.48	8.26	8.03
2.0	17.2	16.9	16.5
2.5	25.1	24.7	24.1
3.0	33.4	32.6	31.3
4.0	48.9	48.0	46.1
5.0	64.2	63.1	60.4
6.0	78.8	77.6	74.3
7.0	96.3	93.6	87.5
8.0	113.8	109.6	105.4
9.0	129.7	127.1	121.5
10.	147.4	143.9	135.6
15.	219.6	213.4	199.5
20.	305.4	297.7	282.9
30.	483.0	464.0	422.0
40.	710.0	680.0	600.0
50.	900.0	850.0	750.0
75.	1,460.0	1,400.0	1,280.0
100.	1,930.0	1,830.0	1,610.0

Volume Correction Data
For Toluene and Mixed Xylene

°F	Mixed Xylene	Toluene
−5	1.0383	
0	1.0353	
5	1.0324	1.0293
10	1.0294	1.0266
15	1.0265	1.0240
20	1.0235	1.0214
25	1.0206	1.0187
30	1.0177	1.0161
35	1.0147	1.0134
40	1.0118	1.0107
45	1.0088	1.0081
50	1.0059	1.0054
55	1.0029	1.0027
60	1.0000	1.0000
65	0.9971	0.9973
70	0.9941	0.9946
75	0.9912	0.9919
80	0.9882	0.9891
85	0.9853	0.9864
90	0.9823	0.9837
95	0.9794	0.9809
100	0.9765	0.9782
105	0.9735	0.9754
110	0.9706	0.9726
115	0.9676	0.9698
120	0.9647	0.9671

Comparison of Barometric Pressure
to Boiling Point of Water

Barometric Pressure		Boiling Point	
mm	in	°C	°F
736	28.976	99.105	210.387
737	29.016	99.142	210.456
738	29.055	99.180	210.524
739	29.094	99.218	210.592
740	29.134	99.255	210.659
741	29.173	99.293	210.727
742	29.213	99.331	210.795
743	29.252	99.368	210.862
744	29.291	99.406	210.931
745	29.331	99.443	210.997
746	29.370	99.481	211.066
747	29.409	99.518	211.132
748	29.449	99.555	211.199
749	29.488	99.593	211.266
750	29.528	99.630	211.334
751	29.567	99.667	211.401
752	29.606	99.704	211.467
753	29.646	99.741	211.534
754	29.685	99.778	211.600
755	29.724	99.815	211.667
756	29.764	99.852	211.734
757	29.803	99.889	211.800
758	29.842	99.926	211.867
759	29.882	99.963	211.930
760	29.921	100.000	212.000
761	29.961	100.037	212.067
762	30.000	100.074	212.133
763	30.039	100.110	212.198
764	30.079	100.147	212.264
765	30.118	100.184	212.331
766	30.157	100.220	212.396
767	30.196	100.257	212.462
768	30.236	100.293	212.527
769	30.276	100.330	212.594
770	30.315	100.366	212.658
771	30.354	100.403	212.725
772	30.394	100.439	212.790
773	30.433	100.475	212.855
774	30.472	100.511	212.919
775	30.512	100.548	212.986
776	30.551	100.584	213.051
777	30.590	100.620	213.116
778	30.630	100.656	213.181
779	30.669	100.692	213.245
780	30.709	100.728	213.310
781	30.748	100.764	213.375
782	30.787	100.800	213.440
783	30.827	100.836	213.504
784	30.866	100.872	213.569
785	30.905	100.908	213.634

Relative Density of Water

Temp. °C	Rel. Density g/ml
0	.99987
3	.99999
4	1.0
5	.99999
10	.99973
15	.99913
15.56	.99904
16	.99897
17	.99880
18	.99862
19	.99843
20	.99823
21	.99802
22	.99780
23	.99756
24	.99732
25	.99707
26	.99681
27	.99654
28	.99626
29	.99597
30	.99567
35	.99406
37.78	.99307
40	.99224
45	.99025
50	.98807
55	.98573
60	.98324
65	.98059
70	.97781
75	.97489
80	.97183
85	.96865
90	.96534
100	.95838

Typical Compositions of Petroleum Products

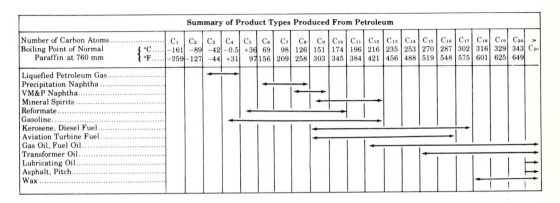

Summary of Product Types Produced From Petroleum																					
Number of Carbon Atoms		C_1	C_2	C_3	C_4	C_5	C_6	C_7	C_8	C_9	C_{10}	C_{11}	C_{12}	C_{13}	C_{14}	C_{15}	C_{16}	C_{17}	C_{18}	C_{19}	C_{20} >
Boiling Point of Normal	°C	−161	−89	−42	−0.5	+36	69	98	126	151	174	196	216	235	253	270	287	302	316	329	343 C_{20}
Paraffin at 760 mm	°F	−259	−127	−44	+31	97	156	209	258	303	345	384	421	456	488	519	548	575	601	625	649

Liquefied Petroleum Gas
Precipitation Naphtha
VM&P Naphtha
Mineral Spirits
Reformate
Gasoline
Kerosene, Diesel Fuel
Aviation Turbine Fuel
Gas Oil, Fuel Oil
Transformer Oil
Lubricating Oil
Asphalt, Pitch
Wax

Quantities for Various Depths of Cylindrical Tanks in Horizontal Position

% Depth Filled	% of Capacity	% Depth Filled	% of Capacity	% Depth Filled	% of Capacity	% Depth Filled	% of Capacity
1	0.20	26	20.73	51	51.27	76	81.50
2	0.50	27	21.86	52	52.55	77	82.60
3	0.90	28	23.00	53	53.81	78	83.68
4	1.34	29	24.07	54	55.08	79	84.74
5	1.87	30	25.31	55	56.34	80	85.77
6	2.45	31	26.48	56	57.60	81	86.77
7	3.07	32	27.66	57	58.86	82	87.76
8	3.74	33	28.84	58	60.11	83	88.73
9	4.45	34	30.03	59	61.36	84	89.68
10	5.20	35	31.19	60	62.61	85	90.60
11	5.98	36	32.44	61	63.86	86	91.50
12	6.80	37	33.66	62	65.10	87	92.36
13	7.64	38	34.90	63	66.34	88	93.20
14	8.50	39	36.14	64	67.56	89	94.02
15	9.40	40	37.39	65	68.81	90	94.80
16	10.32	41	38.64	66	69.97	91	95.55
17	11.27	42	39.89	67	71.16	92	96.26
18	12.24	43	41.14	68	72.34	93	96.93
19	13.23	44	42.40	69	73.52	94	97.55
20	14.23	45	43.66	70	74.69	95	98.13
21	15.26	46	44.92	71	75.93	96	98.66
22	16.32	47	46.19	72	77.00	97	99.10
23	17.40	48	47.45	73	78.14	98	99.50
24	18.50	49	48.73	74	79.27	99	99.80
25	19.61	50	50.00	75	80.39	100	100.00

Volume Calibration of a Vertical 55-gallon Drum

The inside diameter of a standard 55-gallon drum is 22.5 inches.

Therefore, each vertical inch of an upright drum contains 1.72 gallons.

Physical Constants of Hydrocarbons

	Formula	Molec. Wt	Boiling Point °F	Melting Point °F	Density °API	Density Sp Gr 60°/60°	Density lb/gal	Heat of Combustion at 60°F – Btu/lb Gross	Heat of Combustion at 60°F – Btu/lb Net
NORMAL PARAFFINS C_nH_{2n+2}									
Methane	CH_4	16.0	−258.9	−296.5	340	0.30	2.50	23,860[1]	21,500[1]
Ethane	C_2H_6	30.1	−128.0	−297.8	247	.374	3.11	22,300[1]	20,420[1]
Propane	C_3H_8	44.1	− 43.8	−305.7	147	.508	4.23	21,650[1]	19,930[1]
Butane	C_4H_{10}	58.1	+ 31.1	−216.9	111	.584	4.86	21,290[1]	19,670[1]
Pentane	C_5H_{12}	72.1	96.9	−201.5	92.7	.631	5.25	21,070[1]	19,500[1]
Hexane	C_6H_{14}	86.2	155.7	−139.5	81.6	.664	5.53	20,780	19,240
Heptane	C_7H_{16}	100.2	209.2	−131.1	74.2	.688	5.73	20.670	19,160
Octane	C_8H_{18}	114.2	258.2	− 70.3	68.6	.707	5.89	20,590	19,100
Nonane	C_9H_{20}	128.2	303.4	− 64.5	64.5	.722	6.01	20,530	19,050
Decane	$C_{10}H_{22}$	142.3	345.2	− 21.5	61.3	.734	6.11	20,480	19,020
Undecane	$C_{11}H_{24}$	156.3	384.4	− 14.1	58.7	.744	6.19	20,450	19,000
Dodecane	$C_{12}H_{26}$	170.3	421.3	+ 14.7	56.4	.753	6.27	20,420	18,980
ISO-PARAFFINS C_nH_{2n+2}									
Isobutane	C_4H_{10}	58.1	10.9	−255.0	120	.563	4.69	21,240[1]	19,610[1]
2-Methylbutane (Isopentane)	C_5H_{12}	72.1	82.2	−255.5	94.9	.625	5.20	21,030[1]	19,450[1]
2,2-Dimethylpropane (Neopentane)	C_5H_{12}	72.1	49.0	+ 2.1	105	.597	4.97	20,960[1]	19,330[1]
2-Methylpentane (Isohexane)	C_6H_{14}	86.2	140.5	−245	83.5	.658	5.48	20,750	19,210
3-Methylpentane	C_6H_{14}	86.2	145.9	−180	80.0	.669	5.57	20,760	19,220
2,2-Dimethylbutane (Neohexane)	C_6H_{14}	86.2	121.5	−147.6	84.9	.654	5.44	20,700	19,160
2,3-Dimethylbutane (Diisopropyl)	C_6H_{14}	86.2	136.4	−198.8	81.0	.666	5.54	20,740	19,200
2-Methylhexane (Isoheptane)	C_7H_{16}	100.2	194.1	−180.8	75.7	.683	5.68	20,650	19,140
3-Methylhexane	C_7H_{16}	100.2	197.5	−182.9	73.0	.692	5.76	20,660	19,150
3-Ethylpentane	C_7H_{16}	100.2	200.2	−181.5	69.8	.703	5.85	20,670	19,160
2,2-Dimethylpentane	C_7H_{16}	100.2	174.6	−190.8	77.2	.678	5.64	20,600	19,090
2,3-Dimethylpentane	C_7H_{16}	100.2	193.6	−	70.6	.700	5.83	20,640	19,130
2,4-Dimethylpentane	C_7H_{16}	100.2	176.9	−183.1	77.2	.678	5.64	20,620	19,110
3,3-Dimethylpentane	C_7H_{16}	100.2	186.9	−211.0	71.2	.698	5.81	20,620	19,110
2,2,3-Trimethylbutane (Triptane)	C_7H_{16}	100.2	177.6	− 13.0	72.1	0.695	5.78	20,620	19,110
2-Methylheptane (Isooctane)	C_8H_{18}	114.2	243.8	−165.1	70.1	.702	5.84	20,570	19,080
3-Ethylhexane	C_8H_{18}	114.2	245.4	−	65.6	.718	5.98	20,570	19,080
2,5-Dimethylhexane (Diisobutyl)	C_8H_{18}	114.2	228.4	−130	71.2	.698	5.81	20,550	19,060
2,2,4-Trimethylpentane ("Isooctane")	C_8H_{18}	114.2	210.6	−161.2	71.8	.696	5.79	20,540	19,050
OLEFINS C_nH_{2n}									
Ethylene	C_2H_4	28.0	−154.7	−272.5	273	.35	2.91	21,640[1]	20,290[1]
Propylene	C_3H_6	42.1	−53.9	−301.4	140	.522	4.35	21,040[1]	19,690[1]
Butene-1	C_4H_8	56.1	20.7	−	104	.601	5.00	20,840[1]	19,490[1]
Cis-Butene-2	C_4H_8	56.1	38.6	−218.0	94.2	.627	5.22	20,780[1]	19,430[1]
Trans-Butene-2	C_4H_8	56.1	33.6	−157.7	100	.610	5.08	20,750[1]	19,400[1]
Isobutene (Isobutylene)	C_4H_8	56.1	19.6	−220.5	104	.600	4.99	20,720[1]	19,370[1]
Pentene-1 (Amylene)	C_5H_{10}	70.1	86.2	−216.4	87.2	.647	5.38	20,710[1]	19,360[1]
Cis-Pentene-2	C_5H_{10}	70.1	86.2	−290.2	87.2	.661	5.50	20,660[1]	19,310[1]
Trans-Pentene-2	C_5H_{10}	70.1	98.6	−211.0	84.9	.654	5.44	20,640[1]	19,290[1]
2-Methylbutene-1	C_5H_{10}	70.1	88.0	−	84.5	.655	5.45	20,610[1]	19,260[1]
3-Methylbutene-1 (Isoamylene)	C_5H_{10}	70.1	68.4	−292.0	92.0	.633	5.27	20,660[1]	19,310[1]
2-Methylbutene-2	C_5H_{10}	70.1	101.2	−207.0	80.6	.667	5.55	20,570[1]	19,220[1]
Hexene-1	C_6H_{12}	84.2	146.4	−218.0	77.2	.678	5.64	20,450	19,100
Cis-Hexene-2	C_6H_{12}	84.2	155.4	−231.0	73.9	.689	5.73	20,420	19,070
Trans-Hexene-2	C_6H_{12}	84.2	154.2	−207.0	75.7	.683	5.68	20,400	19,050
Cis-Hexene-3	C_6H_{12}	84.2	153.7	−211.0	75.4	.684	5.69	20,420	19,070
Trans-Hexene-3	C_5H_{12}	84.2	154.6	−171	76.0	.682	5.68	20,400	19,050
DIOLEFINS C_nH_{2n-2}									
Propadiene	C_3H_4	40.1	− 30.1	−213.0	106	.595	4.95	20,880[1]	19,930
Butadiene-1,2	C_4H_6	54.1	+ 50.5	−	83.5	.658	5.48	−	−
Butadiene-1,3	C_4H_6	54.1	24.1	−164.0	94.2	.627	5.22	20,230[1]	19,180[1]
Pentadiene-1,2	C_5H_8	68.1	112.8	− 85.0	71.5	.697	5.80	−	−
Cis-Pentadiene-1,3	C_5H_8	68.1	111.6	−	71.8	.696	5.79	20,150[1]	19,040[1]
Trans-Pentadiene-1,3	C_5H_8	68.1	108.1	−	76.0	.682	5.68	20,150[1]	19,040[1]
Pentadiene-1,4	C_5H_8	68.1	78.9	−234.0	81.3	.665	5.53	20,320[1]	19,210[1]
3-Methylbutadiene-1,2	C_5H_8	68.1	104	−184.0	82.9	.685	5.70	−	−
2-Methylbutadiene-1,3 (Isoprene)	C_5H_8	68.1	93.3	−231.0	74.8	.686	5.71	20,060[1]	18,950[1]

[1] Heat of combustion as a gas — otherwise as a liquid.   *Mixture of cis- and trans-isomers.   **Sublimes.

Physical Constants of Hydrocarbons ... continued

	Formula	Molec. wt	Boiling Point °F	Melting Point °F	Density °API	Density Sp Gr 60°/60°	Density lb/gal	Heat of Combustion at 60°F – Btu/lb Gross	Heat of Combustion at 60°F – Btu/lb Net
DIOLEFINS (*Cont.*)									
Hexadiene-1,2	C6H10	82.1	172	—	64.5	0.722	6.01	—	—
Hexadiene-1,3*	C6H10	82.1	163	—	67.8	.710	5.91	—	—
Hexadiene-1,4*	C6H10	82.1	149	—	70.6	.700	5.83	—	—
Hexadiene-1,5	C6H10	82.1	139.3	−221.4	71.8	.696	5.79	20,130	18,980
Hexadiene-2,3	C6H10	82.1	154.4	—	75.1	.685	5.70	—	—
Hexadiene-2,4*	C6H10	82.1	176	—	63.7	.725	6.03	—	—
3-Methylpentadiene-1,2	C6H10	82.1	158	—	65.0	.720	5.99	—	—
4-Methylpentadiene-1,2	C6H10	82.1	158.0	—	67.0	.713	5.93	—	—
2-Methylpentadiene-1,3*	C6H10	82.1	169	—	63.9	.724	6.03	—	—
3-Methylpentadiene-1,3*	C6H10	82.1	171	—	59.7	.740	6.16	—	—
4-Methylpentadiene-1,3	C6H10	82.1	169.3	− 94.0	63.9	.724	6.03	—	—
2-Methylpentadiene-1,4	C6H10	82.1	133	—	70.9	.699	5.82	—	—
2-Methylpentadiene-2,3	C6H10	82.1	162.0	—	66.1	.716	5.96	—	—
2,3-Dimethylbutadiene-1,3	C6H10	82.1	155.7	−105	62.1	.731	6.08	19,880	18,730
2-Ethylbutadiene-1,3	C6H10	82.1	167	—	61.0	.735	6.12	—	—
ACETYLENES CnH2n-2									
Acetylene	C2H2	26.0	−119**	−114	209	.416	3.46	21,470[1]	20,740[1]
Methylacetylene	C3H4	40.1	− 9.8	−153	94.9	.625	5.20	20,810[1]	19,860[1]
Butyne-1 (Ethylacetylene)	C4H6	54.1	+ 47.7	−188.5	86.2	.650	5.41	20,650[1]	19,600[1]
Butyne-2 (Dimethylacetylene)	C4H6	54.1	80.4	− 26.0	71.2	.698	5.81	20,510[1]	19,460[1]
Pentyne-1 (Propylacetylene)	C5H8	68.1	104.4	−159	71.8	.696	5.79	20,550[1]	19,440[1]
Pentyne-2	C5H8	68.1	132.8	−148	66.1	.716	5.96	20,450[1]	19,340[1]
3-Methylbutyne-1 (Isopropylacetylene)	C5H8	68.1	82	—	79.7	.670	5.58	20,500[1]	19,390[1]
Hexyne-1 (Butylacetylene)	C6H10	82.1	160.9	−205.6	65.0	.720	5.99	—	—
Hexyne-2	C6H10	82.1	184.1	−126.4	60.8	.736	6.13	—	—
Hexyne-3	C6H10	82.1	179.2	−149.8	63.1	.727	6.05	—	—
4-Methylpentyne-1	C6H10	82.1	142.1	−157.1	67.5	.711	5.92	—	—
4-Methylpentyne-2	C6H10	82.1	162	—	65.3	.719	5.98	—	—
3,3-Dimethylbutyne-1	C6H10	82.1	100.0	−114.2	78.7	.673	5.60	—	—
OLEFINS-ACETYLENES CnH2n-4									
Buten-3-yne-1 (Vinylacetylene)	C4H4	52.1	42	—	73.9	.689	5.73	—	—
Penten-1-yne-3	C5H6	66.1	138.6	—	58.7	0.744	6.19	—	—
Penten-1-yne-4 (Allylacetylene)	C5H6	66.1	107	—	49.4	.782	6.51	—	—
2-Methylbuten-1-yne-3	C5H6	66.1	90	—	—	—	—	—	—
Hexen-1-yne-3	C6H8	80.1	185	—	56.4	.753	6.27	—	—
Hexen-1-yne-5	C6H8	80.1	158	—	32.8	.861	7.17	—	—
2-Methylpenten-1-yne-3	C6H8	80.1	169	—	—	-—	—	—	—
3-Methylpenten-3-yne-1*	C6H8	80.1	156	—	—	—	—	—	—
AROMATICS CnH2n-6									
Benzene	C6H6	78.1	176.2	41.9	28.6	.884	7.36	17,990	17,270
Toluene	C7H8	92.1	231.1	−139.0	30.8	.872	7.26	18,270	17,450
o-Xylene	C8H10	106.2	292.0	− 13.3	28.4	.885	7.37	18,500	17,610
m-Xylene	C8H10	106.2	282.4	− 54.2	31.3	.869	7.24	18,500	17,610
p-Xylene	C8H10	106.2	281.0	+ 55.9	31.9	.866	7.21	18,430	17,540
Ethylbenzene	C8H10	106.2	277.1	−138.9	30.8	.872	7.26	18,490	17,600
1,2,3-Trimethylbenzene	C9H12	120.2	349.0	− 13.8	25.7	.900	7.49	—	—
1,2,4-Trimethylbenzene (Pseudocumene) . . .	C9H12	120.2	336.5	− 47.3	29.1	.881	7.34	18,570	17,620
1,3,5-Trimethylbenzene (Mesitylene)	C9H12	120.2	328.3	− 48.6	31.1	.870	7.24	18,620	17,670
Propylbenzene	C9H12	120.2	318.6	−147.1	31.9	.866	7.21	18,660	17,710
Isopropylbenzene (Cumene)	C9H12	120.2	306.3	−140.8	31.9	.866	7.21	18,670	17,720
1-Methyl-2-Ethylbenzene	C9H12	120.2	329.2	−126.6	28.7	.883	7.35	—	—
1-Methyl-3-Ethylbenzene	C9H12	120.2	322.7	—	31.1	.870	7.24	—	—
1-Methyl-4-Ethylbenzene	C9H12	120.2	324.5	− 82.7	31.5	.868	7.23	—	—
CYCLOPARAFFINS CnH2n									
Cyclopropane	C3H6	42.1	− 27.0	−196.6	98.6	.615	5.12		
Cyclobutane	C4H8	56.1	+ 54.7	− 58.0	74.8	.686	5.71	—	—
Cyclopentane	C5H10	70.1	120.7	−136.7	56.9	.751	6.25	20,350[1]	19,000
Methylcyclopentane	C6H12	84.2	161.3	−224.4	56.2	.754	6.28	20,110	18,760
1,1-Dimethylcyclopentane	C7H14	98.2	189.5	−105	54.7	.760	6.33	—	—
1,2-Dimethylcyclopentane-cis.	C7H14	98.2	210.7	− 62	50.4	.778	6.48	20,020	18,670
1,2-Dimethylcyclopentane-trans. . . .	C7H14	98.2	197.4	−182	55.4	.757	6.30	20,020	18,670
1,3-Dimethylcyclopentane-trans. . . .	C7H14	98.2	195.4	−213	57.2	.750	6.24	—	—
Ethylcyclopentane	C7H14	98.2	218.2	−217	52.0	.771	6.42	20,110	18,760
Cyclohexane	C6H12	84.2	177.3	+ 44	49.0	.784	6.53	20,030	18,680
Methylcyclohexane	C7H14	98.2	213.6	−195.6	51.3	.774	6.44	20,000	18,650

[1]Heat of combustion as a gas — otherwise as a liquid.   *Mixture of cis- and trans- isomers.   **Sublimes.

421

Temperature Conversions

		Degrees			
	Centi-grade	Fahren-heit	Abso-lute (Kelvin)	Rankine	
Absolute zero	−273.16	−459.58	0	0	
Boiling point of water	100	212	373.2	671.6	
Freezing point of water	0	32	273.2	491.6	
Room temperature	25	77	298.2	536.6	

$°F = \frac{9}{5}°C + 32$

$°F = °R - 459.58$

$°K = °C + 273.16$

$°R = °F + 459.58$

$°C = \frac{5}{9}(°F - 32)$

$°K = \frac{5}{9}°R$

Temperature Conversion Table

To °F	From	To °C
−148.0	−100	−73.33
−146.2	−99	−72.78
−144.4	−98	−72.22
−142.6	−97	−71.67
−140.8	−96	−71.11
−139.0	−95	−70.56
−137.2	−94	−70.00
−135.4	−93	−69.44
−133.6	−92	−68.89
−131.8	−91	−68.33
−130.0	−90	−67.78
−128.2	−89	−67.22
−126.4	−88	−66.67
−124.6	−87	−66.11
−122.8	−86	−65.56
−121.0	−85	−65.00
−119.2	−84	−64.44
−117.4	−83	−63.89
−115.6	−82	−63.33
−113.8	−81	−62.78
−112.0	−80	−62.22
−110.2	−79	−61.67
−108.4	−78	−61.11
−106.6	−77	−60.56
−104.8	−76	−60.00
−103.0	−75	−59.44
−101.2	−74	−58.89
− 99.4	−73	−58.33
− 97.6	−72	−57.78
− 95.8	−71	−57.22
− 94.0	−70	−56.67
− 92.2	−69	−56.11
− 90.4	−68	−55.56
− 88.6	−67	−55.00
− 86.8	−66	−54.44
− 85.0	−65	−53.89
− 83.2	−64	−53.33
− 81.4	−63	−52.78
− 79.6	−62	−52.22
− 77.8	−61	−51.67
− 76.0	−60	−51.11
− 74.2	−59	−50.56
− 72.4	−58	−50.00
− 70.6	−57	−49.44
− 68.8	−56	−48.89
− 67.0	−55	−48.33
− 65.2	−54	−47.78
− 63.4	−53	−47.22
− 61.6	−52	−46.67
− 59.8	−51	−46.11

To °F	From	To °C
−58.0	−50	−45.56
−56.2	−49	−45.00
−54.4	−48	−44.44
−52.6	−47	−43.89
−50.8	−46	−43.33
−49.0	−45	−42.78
−47.2	−44	−42.22
−45.4	−43	−41.67
−43.6	−42	−41.11
−41.8	−41	−40.56
−40.0	−40	−40.00
−38.2	−39	−39.44
−36.4	−38	−38.89
−34.6	−37	−38.33
−32.8	−36	−37.78
−31.0	−35	−37.22
−29.2	−34	−36.67
−27.4	−33	−36.11
−25.6	−32	−35.56
−23.8	−31	−35.00
−22.0	−30	−34.44
−20.2	−29	−33.89
−18.4	−28	−33.33
−16.6	−27	−32.78
−14.8	−26	−32.22
−13.0	−25	−31.67
−11.2	−24	−31.11
− 9.4	−23	−30.56
− 7.6	−22	−30.00
− 5.8	−21	−29.44
− 4.0	−20	−28.89
− 2.2	−19	−28.33
− 0.4	−18	−27.78
+ 1.4	−17	−27.22
+ 3.2	−16	−26.67
+ 5.0	−15	−26.11
+ 6.8	−14	−25.56
+ 8.6	−13	−25.00
+10.4	−12	−24.44
+12.2	−11	−23.89
+14.0	−10	−23.33
+15.8	− 9	−22.78
+17.6	− 8	−22.22
+19.4	− 7	−21.67
+21.2	− 6	−21.11
+23.0	− 5	−20.56
+24.8	− 4	−20.00
+26.6	− 3	−19.44
+28.4	− 2	−18.89
+30.2	− 1	−18.33

To °F	From	To °C
+ 32.0	0	−17.78
+ 33.8	+1	−17.22
+ 35.6	+2	−16.67
+ 37.4	+3	−16.11
+ 39.2	+4	−15.56
+ 41.0	+5	−15.00
+ 42.8	+6	−14.44
+ 44.6	+7	−13.89
+ 46.4	+8	−13.33
+ 48.2	+9	−12.78
+ 50.0	+10	−12.22
+ 51.8	+11	−11.67
+ 53.6	+12	−11.11
+ 55.4	+13	−10.56
+ 57.2	+14	−10.00
+ 59.0	+15	− 9.44
+ 60.8	+16	− 8.89
+ 62.6	+17	− 8.33
+ 64.4	+18	− 7.78
+ 66.2	+19	− 7.22
+ 68.0	+20	− 6.67
+ 69.8	+21	− 6.11
+ 71.6	+22	− 5.56
+ 73.4	+23	− 5.00
+ 75.2	+24	− 4.44
+ 77.0	+25	− 3.89
+ 78.8	+26	− 3.33
+ 80.6	+27	− 2.78
+ 82.4	+28	− 2.22
+ 84.2	+29	− 1.67
+ 86.0	+30	− 1.11
+ 87.8	+31	− 0.56
+ 89.6	+32	0.00
+ 91.4	+33	+ 0.56
+ 93.2	+34	+ 1.11
+ 95.0	+35	+ 1.67
+ 96.8	+36	+ 2.22
+ 98.6	+37	+ 2.78
+100.4	+38	+ 3.33
+102.2	+39	+ 3.89
+104.0	+40	+ 4.44
+105.8	+41	+ 5.00
+107.6	+42	+ 5.56
+109.4	+43	+ 6.11
+111.2	+44	+ 6.67
+113.0	+45	+ 7.22
+114.8	+46	+ 7.78
+116.6	+47	+ 8.33
+118.4	+48	+ 8.89
+120.2	+49	+ 9.44

To °F	From	To °C
+122.0	+50	+10.00
+123.8	+51	+10.56
+125.6	+52	+11.11
+127.4	+53	+11.67
+129.2	+54	+12.22
+131.0	+55	+12.78
+132.8	+56	+13.33
+134.6	+57	+13.89
+136.4	+58	+14.44
+138.2	+59	+15.00
+140.0	+60	+15.56
+141.8	+61	+16.11
+143.6	+62	+16.67
+145.4	+63	+17.22
+147.2	+64	+17.78
+149.0	+65	+18.33
+150.8	+66	+18.89
+152.6	+67	+19.44
+154.4	+68	+20.00
+156.2	+69	+20.56
+158.0	+70	+21.11
+159.8	+71	+21.67
+161.6	+72	+22.22
+163.4	+73	+22.78
+165.2	+74	+23.33
+167.0	+75	+23.89
+168.8	+76	+24.44
+170.6	+77	+25.00
+172.4	+78	+25.56
+174.2	+79	+26.11
+176.0	+80	+26.67
+177.8	+81	+27.22
+179.6	+82	+27.78
+181.4	+83	+28.33
+183.2	+84	+28.89
+185.0	+85	+29.44
+186.8	+86	+30.00
+188.6	+87	+30.56
+190.4	+88	+31.11
+192.2	+89	+31.67
+194.0	+90	+32.22
+195.8	+91	+32.78
+197.6	+92	+33.33
+199.4	+93	+33.89
+201.2	+94	+34.44
+203.0	+95	+35.00
+204.8	+96	+35.56
+206.6	+97	+36.11
+208.4	+98	+36.67
+210.2	+99	+37.22

Temperature Conversion Table . . . continued

To °F	From	To °C	To °F	From	To °C	To °F	From	To °C	To °F	From	To °C
212.0	100	37.78	347.0	175	79.44	482.0	250	121.11	617.0	325	162.78
213.8	101	38.33	348.8	176	80.00	483.8	251	121.67	618.8	326	163.33
215.6	102	38.89	350.6	177	80.56	485.6	252	122.22	620.6	327	163.89
217.4	103	39.44	352.4	178	81.11	487.4	253	122.78	622.4	328	164.44
219.2	104	40.00	354.2	179	81.67	489.2	254	123.33	624.2	329	165.00
221.0	105	40.56	356.0	180	82.22	491.0	255	123.89	626.0	330	165.56
222.8	106	41.11	357.8	181	82.78	492.8	256	124.44	627.8	331	166.11
224.6	107	41.67	359.6	182	83.33	494.6	257	125.00	629.6	332	166.67
226.4	108	42.22	361.4	183	83.89	496.4	258	125.56	631.4	333	167.22
228.2	109	42.78	363.2	184	84.44	498.2	259	126.11	633.2	334	167.78
230.0	110	43.33	365.0	185	85.00	500.0	260	126.67	635.0	335	168.33
231.8	111	43.89	366.8	186	85.56	501.8	261	127.22	636.8	336	168.89
233.6	112	44.44	368.6	187	86.11	503.6	262	127.78	638.6	337	169.44
235.4	113	45.00	370.4	188	86.67	505.4	263	128.33	640.4	338	170.00
237.2	114	45.56	372.2	189	87.22	507.2	264	128.89	642.2	339	170.56
239.0	115	46.11	374.0	190	87.78	509.0	265	129.44	644.0	340	171.11
240.8	116	46.67	375.8	191	88.33	510.8	266	130.00	645.8	341	171.67
242.6	117	47.22	377.6	192	88.89	512.6	267	130.56	647.6	342	172.22
244.4	118	47.78	379.4	193	89.44	514.4	268	131.11	649.4	343	172.78
246.2	119	48.33	381.2	194	90.00	516.2	269	131.67	651.2	344	173.33
248.0	120	48.89	383.0	195	90.56	518.0	270	132.22	653.0	345	173.89
249.8	121	49.44	384.8	196	91.11	519.8	271	132.78	654.8	346	174.44
251.6	122	50.00	386.6	197	91.67	521.6	272	133.33	656.6	347	175.00
253.4	123	50.56	388.4	198	92.22	523.4	273	133.89	658.4	348	175.56
255.2	124	51.11	390.2	199	92.78	525.2	274	134.44	660.2	349	176.11
257.0	125	51.67	392.0	200	93.33	527.0	275	135.00	662.0	350	176.67
258.8	126	52.22	393.8	201	93.89	528.8	276	135.56	663.8	351	177.22
260.6	127	52.78	395.6	202	94.44	530.6	277	136.11	665.6	352	177.78
262.4	128	53.33	397.4	203	95.00	532.4	278	136.67	667.4	353	178.33
264.2	129	53.89	399.2	204	95.56	534.2	279	137.22	669.2	354	178.89
266.0	130	54.44	401.0	205	96.11	536.0	280	137.78	671.0	355	179.44
267.8	131	55.00	402.8	206	96.67	537.8	281	138.33	672.8	356	180.00
269.6	132	55.56	404.6	207	97.22	539.6	282	138.89	674.6	357	180.56
271.4	133	56.11	406.4	208	97.78	541.4	283	139.44	676.4	358	181.11
273.2	134	56.67	408.2	209	98.33	543.2	284	140.00	678.2	359	181.67
275.0	135	57.22	410.0	210	98.89	545.0	285	140.56	680.0	360	182.22
276.8	136	57.78	411.8	211	99.44	546.8	286	141.11	681.8	361	182.78
278.6	137	58.33	413.6	212	100.00	548.6	287	141.67	683.6	362	183.33
280.4	138	58.89	415.4	213	100.56	550.4	288	142.22	685.4	363	183.89
282.2	139	59.44	417.2	214	101.11	552.2	289	142.78	687.2	364	184.44
284.0	140	60.00	419.0	215	101.67	554.0	290	143.33	689.0	365	185.00
285.8	141	60.56	420.8	216	102.22	555.8	291	143.89	690.8	366	185.56
287.6	142	61.11	422.6	217	102.78	557.6	292	144.44	692.6	367	186.11
289.4	143	61.67	424.4	218	103.33	559.4	293	145.00	694.4	368	186.67
291.2	144	62.22	426.2	219	103.89	561.2	294	145.56	696.2	369	187.22
293.0	145	62.78	428.0	220	104.44	563.0	295	146.11	698.0	370	187.78
294.8	146	63.33	429.8	221	105.00	564.8	296	146.67	699.8	371	188.33
296.6	147	63.89	431.6	222	105.56	566.6	297	147.22	701.6	372	188.89
298.4	148	64.44	433.4	223	106.11	568.4	298	147.78	703.4	373	189.44
300.2	149	65.00	435.2	224	106.67	570.2	299	148.33	705.2	374	190.00
302.0	150	65.56	437.0	225	107.22	572.0	300	148.89	707.0	375	190.56
303.8	151	66.11	438.8	226	107.78	573.8	301	149.44	708.8	376	191.11
305.6	152	66.67	440.6	227	108.33	575.6	302	150.00	710.6	377	191.67
307.4	153	67.22	442.4	228	108.89	577.4	303	150.58	712.4	378	192.22
309.2	154	67.78	444.2	229	109.44	579.2	304	151.11	714.2	379	192.78
311.0	155	68.33	446.0	230	110.00	581.0	305	151.67	716.0	380	193.33
312.8	156	68.89	447.8	231	110.56	582.8	306	152.22	717.8	381	193.89
314.6	157	69.44	449.6	232	111.11	584.6	307	152.78	719.6	382	194.44
316.4	158	70.00	451.4	233	111.67	586.4	308	153.33	721.4	383	195.00
318.2	159	70.56	453.2	234	112.22	588.2	309	153.89	723.2	384	195.56
320.0	160	71.11	455.0	235	112.78	590.0	310	154.44	725.0	385	196.11
321.8	161	71.67	456.8	236	113.33	591.8	311	155.00	726.8	386	196.67
323.6	162	72.22	458.6	237	113.89	593.6	312	155.56	728.6	387	197.22
325.4	163	72.78	460.4	238	114.44	595.4	313	156.11	730.4	388	197.78
327.2	164	73.33	462.2	239	115.00	597.2	314	156.67	732.2	389	198.33
329.0	165	73.89	464.0	240	115.56	599.0	315	157.22	734.0	390	198.89
330.8	166	74.44	465.8	241	116.11	600.8	316	157.78	735.8	391	199.44
332.6	167	75.00	467.6	242	116.67	602.6	317	158.33	737.6	392	200.00
334.4	168	75.56	469.4	243	117.22	604.4	318	158.89	739.4	393	200.56
336.2	169	76.11	471.2	244	117.78	606.2	319	159.44	741.2	394	201.11
338.0	170	76.67	473.0	245	118.33	608.0	320	160.00	743.0	395	201.67
339.8	171	77.22	474.8	246	118.89	609.8	321	160.56	744.8	396	202.22
341.6	172	77.78	476.6	247	119.44	611.6	322	161.11	746.6	397	202.78
343.4	173	78.33	478.4	248	120.00	613.4	323	161.67	748.4	398	203.33
345.2	174	78.89	480.2	249	120.56	615.2	324	162.22	750.2	399	203.89

Temperature Conversion Table ... continued

To °F	From	To °C	To °F	From	To °C	To °F	From	To °C	To °F	From	To °C
752.0	400	204.44	887.0	475	246.11	1022.0	550	287.78	1337.0	725	385.00
753.8	401	205.00	888.8	476	246.67	1023.8	551	288.33	1346.0	730	387.78
755.6	402	205.56	890.6	477	247.22	1025.6	552	288.89	1355.0	735	390.56
757.4	403	206.11	892.4	478	247.78	1027.4	553	289.44	1364.0	740	393.33
759.2	404	206.67	894.2	479	248.33	1029.2	554	290.00	1373.0	745	396.11
761.0	405	207.22	896.0	480	248.89	1031.0	555	290.56	1382.0	750	398.89
762.8	406	207.78	897.8	481	249.44	1032.8	556	291.11	1391.0	755	401.67
764.6	407	208.33	899.6	482	250.00	1034.6	557	291.67	1400.0	760	404.44
766.4	408	208.89	901.4	483	250.56	1036.4	558	292.22	1409.0	765	407.22
768.2	409	209.44	903.2	484	251.11	1038.2	559	292.78	1418.0	770	410.00
770.0	410	210.00	905.0	485	251.67	1040.0	560	293.33	1427.0	775	412.78
771.8	411	210.56	906.8	486	252.22	1041.8	561	293.89	1436.0	780	415.56
773.6	412	211.11	908.6	487	252.78	1043.6	562	294.44	1445.0	785	418.33
775.4	413	211.67	910.4	488	253.33	1045.4	563	295.00	1454.0	790	421.11
777.2	414	212.22	912.2	489	253.89	1047.2	564	295.56	1463.0	795	423.89
779.0	415	212.78	914.0	490	254.44	1049.0	565	296.11	1472.0	800	426.67
780.8	416	213.33	915.8	491	255.00	1050.8	566	296.67	1481.0	805	429.44
782.6	417	213.89	917.6	492	255.56	1052.6	567	297.22	1490.0	810	432.22
784.4	418	214.44	919.4	493	256.11	1054.4	568	297.78	1499.0	815	435.00
786.2	419	215.00	921.2	494	256.67	1056.2	569	298.33	1508.0	820	437.78
788.0	420	215.56	923.0	495	257.22	1058.0	570	298.89	1517.0	825	440.56
789.8	421	216.11	924.8	496	257.78	1059.8	571	299.44	1526.0	830	443.33
791.6	422	216.67	926.6	497	258.33	1061.6	572	300.00	1535.0	835	446.11
793.4	423	217.22	928.4	498	258.89	1063.4	573	300.56	1544.0	840	448.89
795.2	424	217.78	930.2	499	259.44	1065.2	574	301.11	1553.0	845	451.67
797.0	425	218.33	932.0	500	260.00	1067.0	575	301.67	1562.0	850	454.44
798.8	426	218.89	933.8	501	260.56	1068.8	576	302.22	1571.0	855	457.22
800.6	427	219.44	935.6	502	261.11	1070.6	577	302.78	1580.0	860	460.00
802.4	428	220.00	937.4	503	261.67	1072.4	578	303.33	1589.0	865	462.78
804.2	429	220.56	939.2	504	262.22	1074.2	579	303.89	1598.0	870	465.56
806.0	430	221.11	941.0	505	262.78	1076.0	580	304.44	1607.0	875	468.33
807.8	431	221.67	942.8	506	263.33	1077.8	581	305.00	1616.0	880	471.11
809.6	432	222.22	944.6	507	263.89	1079.6	582	305.56	1625.0	885	473.89
811.4	433	222.78	946.4	508	264.44	1081.4	583	306.11	1634.0	890	476.67
813.2	434	223.33	948.2	509	265.00	1083.2	584	306.67	1643.0	895	479.44
815.0	435	223.89	950.0	510	265.56	1085.0	585	307.22	1652.0	900	482.22
816.8	436	224.44	951.8	511	266.11	1086.8	586	307.78	1661.0	905	485.00
818.6	437	225.00	953.6	512	266.67	1088.6	587	308.33	1670.0	910	487.78
820.4	438	225.56	955.4	513	267.22	1090.4	588	308.89	1679.0	915	490.56
822.2	439	226.11	957.2	514	267.78	1092.2	589	309.44	1688.0	920	493.33
824.0	440	226.67	959.0	515	268.33	1094.0	590	310.00	1697.0	925	496.11
825.8	441	227.22	960.8	516	268.89	1095.8	591	310.56	1706.0	930	498.89
827.6	442	227.78	962.6	517	269.44	1097.6	592	311.11	1715.0	935	501.67
829.4	443	228.33	964.4	518	270.00	1099.4	593	311.67	1724.0	940	504.44
831.2	444	228.89	966.2	519	270.56	1101.2	594	312.22	1733.0	945	507.22
833.0	445	229.44	968.0	520	271.11	1103.0	595	312.78	1742.0	950	510.00
834.8	446	230.00	969.8	521	271.67	1104.8	596	313.33	1751.0	955	512.78
836.6	447	230.56	971.6	522	272.22	1106.8	597	313.89	1760.0	960	515.56
838.4	448	231.11	973.4	523	272.78	1108.4	598	314.44	1769.0	965	518.33
840.2	449	231.67	975.2	524	273.33	1110.2	599	315.00	1778.0	970	521.11
842.0	450	232.22	977.0	525	273.89	1112.0	600	315.56	1787.0	975	523.89
843.8	451	232.78	978.8	526	274.44	1121.0	605	318.33	1796.0	980	526.67
845.6	452	233.33	980.6	527	275.00	1130.0	610	321.11	1805.0	985	529.44
847.4	453	233.89	982.4	528	275.56	1139.0	615	323.89	1814.0	990	532.22
849.2	454	234.44	984.2	529	276.11	1148.0	620	326.67	1823.0	995	535.00
851.0	455	235.00	986.0	530	276.67	1157.0	625	329.44	1832.0	1000	537.78
852.8	456	235.56	987.8	531	277.22	1166.0	630	332.22	1841.0	1005	540.56
854.6	457	236.11	989.6	532	277.78	1175.0	635	335.00	1850.0	1010	543.33
856.4	458	236.67	991.4	533	278.33	1184.0	640	337.78	1859.0	1015	546.11
858.2	459	237.22	993.2	534	278.89	1193.0	645	340.56	1868.0	1020	548.89
860.0	460	237.78	995.0	535	279.44	1202.0	650	343.33	1877.0	1025	551.67
861.8	461	238.33	996.8	536	280.00	1211.0	655	346.11	1886.0	1030	554.44
863.6	462	238.89	998.6	537	280.56	1220.0	660	348.89	1895.0	1035	557.22
865.4	463	239.44	1000.4	538	281.11	1229.0	665	351.67	1904.0	1040	560.00
867.2	464	240.00	1002.2	539	281.67	1238.0	670	354.44	1913.0	1045	562.78
869.0	465	240.56	1004.0	540	282.22	1247.0	675	357.22	1922.0	1050	565.56
870.8	466	241.11	1005.8	541	282.78	1256.0	680	360.00	1931.0	1055	568.33
872.6	467	241.67	1007.6	542	283.33	1265.0	685	362.78	1940.0	1060	571.11
874.4	468	242.22	1009.4	543	283.89	1274.0	690	365.56	1949.0	1065	573.89
876.2	469	242.78	1011.2	544	284.44	1283.0	695	368.33	1958.0	1070	576.67
878.0	470	243.33	1013.0	545	285.00	1292.0	700	371.11	1967.0	1075	579.44
879.8	471	243.89	1014.8	546	285.56	1301.0	705	373.89	1976.0	1080	582.22
881.6	472	244.44	1016.6	547	286.11	1310.0	710	376.67	1985.0	1085	585.00
883.4	473	245.00	1018.4	548	286.67	1319.0	715	379.44	1994.0	1090	587.78
885.2	474	245.56	1020.2	549	287.22	1328.0	720	382.22	2003.0	1095	590.56

Temperature Conversion Table . . . continued

To °F	From	To °C	To °F	From	To °C	To °F	From	To °C	To °F	From	To °C
2012.0	1100	593.33	2237.0	1225	662.78	2462.0	1350	732.22	2687.0	1475	801.67
2021.0	1105	596.11	2246.0	1230	665.56	2471.0	1355	735.00	2696.0	1480	804.44
2030.0	1110	598.89	2255.0	1235	668.33	2480.0	1360	737.78	2705.0	1485	807.22
2039.0	1115	601.67	2264.0	1240	671.11	2489.0	1365	740.56	2714.0	1490	810.00
2048.0	1120	604.44	2273.0	1245	673.89	2498.0	1370	743.33	2723.0	1495	812.78
2057.0	1125	607.22	2282.0	1250	676.67	2507.0	1375	746.11	2732.0	1500	815.56
2066.0	1130	610.00	2291.0	1255	679.44	2516.0	1380	748.89	2741.0	1505	818.33
2075.0	1135	612.78	2300.0	1260	682.22	2525.0	1385	751.67	2750.0	1510	821.11
2084.0	1140	615.56	2309.0	1265	685.00	2534.0	1390	754.44	2759.0	1515	823.89
2093.0	1145	618.33	2318.0	1270	687.78	2543.0	1395	757.22	2768.0	1520	826.67
2102.0	1150	621.11	2327.0	1275	690.56	2552.0	1400	760.00	2777.0	1525	829.44
2111.0	1155	623.89	2336.0	1280	693.33	2561.0	1405	762.78	2786.0	1530	832.22
2120.0	1160	626.67	2345.0	1285	696.11	2570.0	1410	765.56	2795.0	1535	835.00
2129.0	1165	629.44	2354.0	1290	698.89	2579.0	1415	768.33	2804.0	1540	837.78
2138.0	1170	632.22	2363.0	1295	701.67	2588.0	1420	771.11	2813.0	1545	840.56
2147.0	1175	635.00	2372.0	1300	704.44	2597.0	1425	773.89	2822.0	1550	843.33
2156.0	1180	637.78	2381.0	1305	707.22	2606.0	1430	776.67	2831.0	1555	846.11
2165.0	1185	640.56	2390.0	1310	710.00	2615.0	1435	779.44	2840.0	1560	848.89
2174.0	1190	643.33	2399.0	1315	712.78	2624.0	1440	782.22	2849.0	1565	851.67
2183.0	1195	646.11	2408.0	1320	715.56	2633.0	1445	785.00	2858.0	1570	854.44
2192.0	1200	648.89	2417.0	1325	718.33	2642.0	1450	787.78	2867.0	1575	857.22
2201.0	1205	651.67	2426.0	1330	721.11	2651.0	1455	790.56	2876.0	1580	860.00
2210.0	1210	654.44	2435.0	1335	723.89	2660.0	1460	793.33	2885.0	1585	862.78
2219.0	1215	657.22	2444.0	1340	726.67	2669.0	1465	796.11	2894.0	1590	865.56
2228.0	1220	660.00	2453.0	1345	729.44	2678.0	1470	798.89	2903.0	1595	868.33
									2912.0	1600	871.11

Weight-Volume Relationships

Specific Gravities

Specific Gravity at 60°/60°F	Degrees API	Pounds Per Gallon	Gallons Per Pound	Kilograms per Cubic Meter
0.650	86.19	5.410	.1848	******
0.651	85.86	5.418	.1846	******
0.652	85.52	5.426	.1843	******
0.653	85.19	5.435	.1840	******
0.654	84.86	5.443	.1837	653.9
0.655	84.53	5.452	.1834	654.9
0.656	84.20	5.460	.1832	655.9
0.657	83.87	5.468	.1829	656.9
0.658	83.55	5.476	.1826	657.9
0.659	83.22	5.485	.1823	658.9
0.660	82.89	5.493	.1820	659.9
0.661	82.57	5.502	.1818	660.9
0.662	82.25	5.510	.1815	661.9
0.663	81.92	5.518	.1812	662.9
0.664	81.60	5.526	.1810	663.9
0.665	81.28	5.535	.1807	664.9
0.666	80.96	5.543	.1804	665.9
0.667	80.64	5.552	.1801	666.9
0.668	80.33	5.560	.1799	667.9
0.669	80.01	5.568	.1796	668.9
0.670	79.69	5.577	.1793	669.9
0.671	79.38	5.585	.1790	670.9
0.672	79.07	5.593	.1788	671.9
0.673	78.75	5.602	.1785	672.9
0.674	78.44	5.610	.1782	673.9
0.675	78.13	5.618	.1780	674.9
0.676	77.82	5.627	.1777	675.9
0.677	77.51	5.635	.1775	676.9
0.678	77.20	5.643	.1772	677.9
0.679	76.89	5.652	.1769	678.9
0.680	76.59	5.660	.1767	679.9
0.681	76.28	5.668	.1764	680.9
0.682	75.98	5.677	.1762	681.9
0.683	75.67	5.685	.1759	682.9
0.684	75.37	5.693	.1756	683.8
0.685	75.07	5.702	.1754	684.8
0.686	74.77	5.710	.1751	685.8
0.687	74.47	5.718	.1749	686.8
0.688	74.17	5.727	.1746	687.8
0.689	73.87	5.735	.1744	688.8
0.690	73.57	5.743	.1741	689.8
0.691	73.28	5.752	.1739	690.8
0.692	72.98	5.760	.1736	691.8
0.693	72.68	5.768	.1734	692.8
0.694	72.39	5.777	.1731	693.8
0.695	72.10	5.785	.1729	694.8
0.696	71.80	5.793	.1726	695.8
0.697	71.51	5.802	.1724	696.8
0.698	71.22	5.810	.1721	697.8
0.699	70.93	5.818	.1719	698.8
0.700	70.64	5.827	.1716	699.8
0.701	70.35	5.835	.1714	700.8
0.702	70.07	5.843	.1711	701.8
0.703	69.78	5.852	.1709	702.8
0.704	69.49	5.860	.1706	703.8
0.705	69.21	5.868	.1704	704.8
0.706	68.92	5.877	.1702	705.8
0.707	68.64	5.885	.1699	706.8
0.708	68.36	5.894	.1697	707.8
0.709	68.08	5.902	.1694	708.8
0.710	67.80	5.910	.1692	709.8
0.711	67.52	5.918	.1690	710.8
0.712	67.24	5.927	.1687	711.8
0.713	66.96	5.935	.1685	712.8
0.714	66.68	5.944	.1682	713.8
0.715	66.40	5.952	.1680	714.8
0.716	66.13	5.960	.1678	715.8
0.717	65.85	5.968	.1676	716.8
0.718	65.58	5.977	.1673	717.8
0.719	65.30	5.985	.1671	718.8
0.720	65.03	5.994	.1668	719.8
0.721	64.76	6.002	.1666	720.8
0.722	64.48	6.010	.1664	721.8
0.723	64.21	6.018	.1662	722.8
0.724	63.94	6.027	.1659	723.8
0.725	63.67	6.035	.1657	724.8
0.726	63.40	6.044	.1655	725.8
0.727	63.14	6.052	.1652	726.8
0.728	62.87	6.060	.1650	727.8
0.729	62.60	6.068	.1645	728.8
0.730	62.34	6.077	.1646	729.8
0.731	62.07	6.085	.1643	730.8
0.732	61.81	6.094	.1641	731.8
0.733	61.54	6.102	.1639	732.8
0.734	61.28	6.110	.1637	733.8
0.735	61.02	6.119	.1634	734.8
0.736	60.76	6.127	.1632	735.8
0.737	60.49	6.135	.1630	736.8
0.738	60.23	6.144	.1628	737.8
0.739	59.97	6.152	.1626	738.8
0.740	59.72	6.160	.1623	739.8
0.741	59.46	6.169	.1621	740.8
0.742	59.20	6.177	.1619	741.8
0.743	58.94	6.185	.1617	742.8
0.744	58.69	6.194	.1615	743.8
0.745	58.43	6.202	.1612	744.8
0.746	58.18	6.210	.1610	745.8
0.747	57.92	6.219	.1608	746.8
0.748	57.67	6.227	.1606	747.8
0.749	57.42	6.235	.1604	748.8
0.750	57.17	6.244	.1602	749.8
0.751	56.92	6.252	.1600	750.8
0.752	56.66	6.260	.1597	751.8
0.753	56.41	6.269	.1595	752.8
0.754	56.17	6.277	.1593	753.8
0.755	55.92	6.285	.1591	754.8
0.756	55.67	6.294	.1589	755.8
0.757	55.42	6.302	.1587	756.8
0.758	55.18	6.310	.1585	757.7
0.759	54.93	6.319	.1583	758.7
0.760	54.68	6.327	.1580	759.7
0.761	54.44	6.335	.1578	760.7
0.762	54.20	6.344	.1576	761.7
0.763	53.95	6.352	.1574	762.7
0.764	53.71	6.360	.1572	763.7
0.765	53.47	6.369	.1570	764.7
0.766	53.23	6.377	.1568	765.7
0.767	52.98	6.386	.1566	766.7
0.768	52.74	6.394	.1564	767.7
0.769	52.51	6.402	.1562	768.7
0.770	52.27	6.410	.1560	769.7
0.771	52.03	6.419	.1558	770.7
0.772	51.79	6.427	.1556	771.7
0.773	51.55	6.436	.1554	772.7
0.774	51.32	6.444	.1552	773.7
0.775	51.08	6.452	.1550	774.7
0.776	50.85	6.460	.1548	775.7
0.777	50.61	6.469	.1546	776.7
0.778	50.38	6.477	.1544	777.7
0.779	50.14	6.486	.1542	778.7
0.780	49.91	6.494	.1540	779.7
0.781	49.68	6.502	.1538	780.7
0.782	49.45	6.510	.1536	781.7
0.783	49.22	6.519	.1534	782.7
0.784	48.98	6.527	.1532	783.7
0.785	48.75	6.536	.1530	784.7
0.786	48.53	6.544	.1528	785.7
0.787	48.30	6.552	.1526	786.6
0.788	48.07	6.560	.1524	787.6
0.789	47.84	6.569	.1522	788.6
0.790	47.61	6.577	.1520	789.6
0.791	47.39	6.586	.1518	790.6
0.792	47.16	6.594	.1517	790.6
0.793	46.94	6.602	.1515	792.6
0.794	46.71	6.611	.1513	793.6
0.795	46.49	6.619	.1511	794.6
0.796	46.26	6.627	.1509	795.6
0.797	46.04	6.636	.1507	796.6
0.798	45.82	6.644	.1505	797.6
0.799	45.60	6.652	.1503	798.6
0.800	45.38	6.661	.1501	799.6
0.801	45.15	6.669	.1500	800.6
0.802	44.93	6.677	.1498	801.6
0.803	44.71	6.686	.1496	802.6
0.804	44.49	6.694	.1494	803.6
0.805	44.28	6.702	.1492	804.6
0.806	44.06	6.711	.1490	805.6
0.807	43.84	6.719	.1488	806.6
0.808	43.62	6.727	.1486	807.6
0.809	43.41	6.736	.1485	808.6
0.810	43.19	6.744	.1483	809.6
0.811	42.98	6.752	.1481	810.6
0.812	42.76	6.761	.1479	811.6
0.813	42.55	6.769	.1477	812.6
0.814	42.33	6.777	.1476	813.6
0.815	42.12	6.786	.1474	814.6
0.816	41.91	6.794	.1472	815.6
0.817	41.69	6.802	.1470	816.6
0.818	41.48	6.811	.1468	817.6
0.819	41.27	6.819	.1466	818.6
0.820	41.06	6.827	.1465	819.6
0.821	40.85	6.836	.1463	820.6
0.822	40.64	6.844	.1461	821.6
0.823	40.43	6.852	.1459	822.6
0.824	40.22	6.861	.1458	823.6
0.825	40.02	6.869	.1456	824.6
0.826	39.81	6.877	.1454	825.6
0.827	39.60	6.886	.1452	826.6
0.828	39.39	6.894	.1450	827.6
0.829	39.19	6.902	.1449	828.6
0.830	38.98	6.911	.1447	829.6
0.831	38.78	6.919	.1445	830.6
0.832	38.57	6.927	.1444	831.6
0.833	38.37	6.936	.1442	832.6
0.834	38.16	6.944	.1440	833.6
0.835	37.96	6.952	.1438	834.6
0.836	37.76	6.961	.1437	835.6
0.837	37.56	6.969	.1435	836.6
0.838	37.35	6.978	.1433	837.6
0.839	37.15	6.986	.1432	838.6
0.840	36.95	6.994	.1430	839.6
0.841	36.75	7.002	.1428	840.6
0.842	36.55	7.011	.1426	841.6
0.843	36.35	7.019	.1425	842.6
0.844	36.15	7.028	.1423	843.6
0.845	35.96	7.036	.1421	844.6
0.846	35.76	7.044	.1420	845.6
0.847	35.56	7.052	.1418	846.6
0.848	35.36	7.061	.1416	847.6
0.849	35.17	7.069	.1415	848.6
0.850	34.97	7.078	.1413	849.6
0.851	34.77	7.086	.1411	850.6
0.852	34.58	7.094	.1410	851.6
0.853	34.39	7.103	.1408	852.5
0.854	34.19	7.111	.1406	853.5
0.855	34.00	7.119	.1405	854.5
0.856	33.80	7.128	.1403	855.5
0.857	33.61	7.136	.1401	856.5
0.858	33.42	7.144	.1400	857.5
0.859	33.23	7.153	.1398	858.5
0.860	33.03	7.161	.1396	859.5
0.861	32.84	7.169	.1395	860.5
0.862	32.65	7.178	.1393	861.5
0.863	32.46	7.186	.1392	862.5
0.864	32.27	7.194	.1390	863.5
0.865	32.08	7.203	.1388	864.5
0.866	31.89	7.211	.1387	865.5
0.867	31.71	7.219	.1385	866.5
0.868	31.52	7.228	.1384	867.5
0.869	31.38	7.236	.1382	868.5
0.870	31.14	7.244	.1380	869.5
0.871	30.96	7.253	.1379	870.5
0.872	30.77	7.261	.1377	871.5
0.873	30.58	7.269	.1376	872.5
0.874	30.40	7.278	.1374	873.5

Weight-Volume Relationships

Specific Gravities

Specific Gravity at 60°/60°F	Degrees API	Pounds Per Gallon	Gallons Per Pound	Kilograms per Cubic Meter
0.875	30.21	7.286	.1372	874.5
0.876	30.03	7.284	.1371	875.5
0.877	29.85	7.303	.1369	876.5
0.878	29.66	7.311	.1368	877.5
0.879	29.48	7.319	.1366	878.5
0.880	29.30	7.328	.1365	879.5
0.881	29.11	7.336	.1363	880.5
0.882	28.93	7.344	.1362	881.5
0.883	28.75	7.353	.1360	882.5
0.884	28.57	7.361	.1358	883.5
0.885	28.39	7.369	.1357	884.5
0.886	28.21	7.378	.1355	885.5
0.887	28.03	7.386	.1354	886.5
0.888	27.85	7.394	.1352	887.5
0.889	27.67	7.403	.1351	888.5
0.890	27.49	7.411	.1349	889.5
0.891	27.31	7.419	.1348	890.5
0.892	27.13	7.428	.1346	891.5
0.893	26.95	7.436	.1345	892.5
0.894	26.78	7.444	.1343	893.5
0.895	26.60	7.453	.1342	894.5
0.896	26.42	7.461	.1340	895.5
0.897	26.25	7.469	.1339	896.5
0.898	26.07	7.478	.1337	897.5
0.899	25.90	7.486	.1336	898.5
0.900	25.72	7.494	.1334	899.5
0.901	25.55	7.503	.1333	900.5
0.902	25.37	7.511	.1331	901.5
0.903	25.20	7.519	.1330	902.5
0.904	25.03	7.528	.1328	903.5
0.905	24.85	7.536	.1327	904.5
0.906	24.68	7.544	.1326	905.5
0.907	24.51	7.553	.1324	906.5
0.908	24.34	7.561	.1323	907.5
0.909	24.17	7.569	.1321	908.5
0.910	23.99	7.578	.1320	909.5
0.911	23.82	7.586	.1318	910.5
0.912	23.65	7.594	.1317	911.5
0.913	23.48	7.603	.1315	912.5
0.914	23.31	7.611	.1314	913.5
0.915	23.14	7.620	.1312	914.5
0.916	22.98	7.628	.1311	915.5
0.917	22.81	7.636	.1310	916.5
0.918	22.64	7.645	.1308	917.5
0.919	22.47	7.653	.1307	918.5
0.920	22.30	7.661	.1305	919.5
0.921	22.14	7.670	.1304	920.5
0.922	21.97	7.678	.1302	921.5
0.923	21.80	7.686	.1301	922.5
0.924	21.64	7.695	.1300	923.5
0.925	21.47	7.703	.1298	924.5
0.926	21.31	7.711	.1297	925.5
0.927	21.14	7.720	.1295	926.5
0.928	20.98	7.728	.1294	927.5
0.929	20.81	7.736	.1293	928.5
0.930	20.65	7.745	.1291	929.5
0.931	20.49	7.753	.1290	930.5
0.932	20.32	7.761	.1288	931.5
0.933	20.16	7.770	.1287	932.5
0.934	20.00	7.778	.1286	933.5
0.935	19.84	7.786	.1284	934.5
0.936	19.68	7.795	.1283	935.5
0.937	19.51	7.803	.1282	936.5
0.938	19.35	7.811	.1280	937.5
0.939	19.19	7.820	.1279	938.5
0.940	19.03	7.828	.1278	939.5
0.941	18.87	7.836	.1276	940.5
0.942	18.71	7.845	.1275	941.4
0.943	18.55	7.853	.1273	942.4
0.944	18.39	7.861	.1272	943.4
0.945	18.24	7.870	.1271	944.4
0.946	18.08	7.878	.1269	945.4
0.947	17.92	7.886	.1268	946.4
0.948	17.76	7.895	.1267	947.4
0.949	17.60	7.903	.1265	948.4
0.950	17.45	7.911	.1264	949.4
0.951	17.29	7.920	.1263	950.4
0.952	17.13	7.928	.1261	951.4
0.953	16.98	7.937	.1260	952.4
0.954	16.82	7.945	.1259	953.4
0.955	16.67	7.953	.1257	954.4
0.956	16.51	7.962	.1256	955.4
0.957	16.36	7.970	.1255	956.4
0.958	16.20	7.978	.1253	957.4
0.959	16.05	7.987	.1252	958.4
0.960	15.90	7.995	.1251	959.4
0.961	15.74	8.003	.1250	960.4
0.962	15.59	8.012	.1248	961.4
0.963	15.44	8.020	.1247	962.4
0.964	15.28	8.028	.1246	963.4
0.965	15.13	8.036	.1244	964.4
0.966	14.98	8.045	.1243	965.4
0.967	14.83	8.053	.1242	966.4
0.968	14.68	8.062	.1240	967.4
0.969	14.53	8.070	.1239	968.4
0.970	14.38	8.078	.1238	969.4
0.971	14.23	8.087	.1237	970.4
0.972	14.08	8.095	.1235	971.4
0.973	13.93	8.103	.1234	972.4
0.974	13.78	8.112	.1233	973.4
0.975	13.63	8.120	.1232	974.4
0.976	13.48	8.128	.1230	975.4
0.977	13.33	8.137	.1229	976.4
0.978	13.18	8.145	.1228	977.4
0.979	13.04	8.153	.1227	978.4
0.980	12.89	8.162	.1225	979.4
0.981	12.74	8.170	.1224	980.4
0.982	12.59	8.178	.1223	981.4
0.983	12.45	8.187	.1221	982.4
0.984	12.30	8.195	.1220	983.4
0.985	12.15	8.203	.1219	984.4
0.986	12.01	8.212	.1218	985.4
0.987	11.86	8.220	.1217	986.4
0.988	11.72	8.228	.1215	987.4
0.989	11.57	8.237	.1214	988.4
0.990	11.43	8.245	.1213	989.4
0.991	11.29	8.253	.1212	990.4
0.992	11.14	8.262	.1210	991.4
0.993	11.00	8.270	.1209	992.4
0.994	10.85	8.278	.1208	993.4
0.995	10.71	8.287	.1207	994.4
0.996	10.57	8.295	.1206	995.4
0.997	10.43	8.303	.1204	996.4
0.998	10.28	8.312	.1203	997.4
0.999	10.14	8.320	.1202	998.4
1.000	10.00	8.328	.1201	999.4
1.001	9.86	8.337	.1199	1000.4
1.002	9.72	8.345	.1198	1001.4
1.003	9.58	8.353	.1197	1002.4
1.004	9.44	8.362	.1196	1003.4
1.005	9.30	8.370	.1195	1004.4
1.006	9.16	8.378	.1194	1005.4
1.007	9.02	8.387	.1192	1006.4
1.008	8.88	8.395	.1191	1007.4
1.009	8.74	8.403	.1190	1008.4
1.010	8.60	8.412	.1189	1009.4
1.011	8.46	8.420	.1188	1010.4
1.012	8.32	8.428	.1186	1011.4
1.013	8.18	8.437	.1185	1012.4
1.014	8.05	8.445	.1184	1013.4
1.015	7.91	8.453	.1183	1014.4
1.016	7.77	8.462	.1182	1015.4
1.017	7.63	8.470	.1181	1016.4
1.018	7.50	8.478	.1180	1017.4
1.019	7.36	8.487	.1178	1018.4
1.020	7.33	8.495	.1177	1019.4
1.021	7.09	8.503	.1176	1020.4
1.022	6.95	8.512	.1175	1021.4
1.023	6.82	8.520	.1174	1022.4
1.024	6.68	8.528	.1173	1023.4
1.025	6.55	8.537	.1171	1024.4
1.026	6.41	8.545	.1170	1025.4
1.027	6.28	8.553	.1169	1026.4
1.028	6.15	8.562	.1168	1027.4
1.029	6.01	8.570	.1167	1028.4
1.030	5.88	8.578	.1166	1029.4
1.031	5.75	8.587	.1165	1030.4
1.032	5.61	8.595	.1163	1031.4
1.033	5.48	8.603	.1162	1032.4
1.034	5.35	8.612	.1161	1033.4
1.035	5.21	8.620	.1160	1034.3
1.036	5.08	8.628	.1159	1035.3
1.037	4.95	8.637	.1158	1036.3
1.038	4.82	8.645	.1157	1037.3
1.039	4.69	8.653	.1156	1038.3
1.040	4.56	8.662	.1154	1039.3
1.041	4.43	8.670	.1153	1040.3
1.042	4.30	8.678	.1152	1041.3
1.043	4.17	8.687	.1151	1042.3
1.044	4.04	8.695	.1150	1043.3
1.045	3.91	8.704	.1149	1044.3
1.046	3.78	8.712	.1148	1045.3
1.047	3.65	8.720	.1147	1046.3
1.048	3.52	8.729	.1146	1047.3
1.049	3.39	8.737	.1145	1048.3
1.050	3.26	8.745	.1143	1049.3
1.051	3.13	8.754	.1142	1050.3
1.052	3.01	8.762	.1141	1051.3
1.053	2.88	8.770	.1140	1052.3
1.054	2.75	8.779	.1139	1053.3
1.055	2.62	8.787	.1138	1054.3
1.056	2.50	8.795	.1137	1055.3
1.057	2.37	8.804	.1136	1056.3
1.058	2.24	8.812	.1135	1057.3
1.059	2.12	8.820	.1134	1058.3
1.060	1.99	8.829	.1133	1059.3
1.061	1.86	8.837	.1132	1060.3
1.062	1.74	8.845	.1131	1061.3
1.063	1.61	8.854	.1129	1062.3
1.064	1.49	8.862	.1128	1063.3
1.065	1.36	8.870	.1127	1064.3
1.066	1.24	8.879	.1126	1065.3
1.067	1.11	8.887	.1125	1066.3
1.068	.99	8.895	.1124	1067.3
1.069	.87	8.904	.1123	1068.3
1.070	.74	8.912	.1122	1069.3
1.071	.62	8.920	.1121	1070.3
1.072	.50	8.929	.1120	1071.3
1.073	.37	8.937	.1119	1072.3
1.074	.25	8.945	.1118	1073.3

Weight-Volume Relationships

Degrees API

Degrees API	Specific gravity at 60°/60°F	Pounds Per Gallon	Gallons Per Pound	Kilograms Per Cubic Meter
0.0	1.0760	8.962	0.1116	1075.3
0.1	1.0752	8.956	0.1117	1074.5
0.2	1.0744	8.949	0.1117	1073.7
0.3	1.0736	8.942	0.1118	1072.9
0.4	1.0728	8.935	0.1119	1072.1
0.5	1.0720	8.928	0.1120	1071.1
0.6	1.0712	8.922	0.1121	1070.5
0.7	1.0703	8.915	0.1122	1069.5
0.8	1.0695	8.908	0.1123	1069.8
0.9	1.0687	8.901	0.1123	1068.0
1.0	1.0679	8.895	0.1124	1067.2
1.1	1.0671	8.888	0.1125	1066.4
1.2	1.0663	8.881	0.1126	1065.6
1.3	1.0655	8.875	0.1127	1064.8
1.4	1.0647	8.868	0.1128	1064.0
1.5	1.0639	8.861	0.1129	1063.2
1.6	1.0631	8.855	0.1129	1062.4
1.7	1.0623	8.848	0.1130	1061.6
1.8	1.0615	8.841	0.1131	1060.8
1.9	1.0607	8.835	0.1132	1060.0
2.0	1.0599	8.828	0.1133	1059.2
2.1	1.0591	8.821	0.1134	1058.5
2.2	1.0583	8.815	0.1134	1057.7
2.3	1.0575	8.808	0.1135	1056.9
2.4	1.0568	8.802	0.1136	1056.1
2.5	1.0560	8.795	0.1137	1055.3
2.6	1.0552	8.788	0.1138	1054.5
2.7	1.0544	8.782	0.1139	1053.7
2.8	1.0536	8.775	0.1140	1052.9
2.9	1.0528	8.769	0.1140	1052.2
3.0	1.0520	8.762	0.1141	1051.4
3.1	1.0513	8.756	0.1142	1050.6
3.2	1.0505	8.749	0.1143	1049.8
3.3	1.0497	8.743	0.1144	1049.0
3.4	1.0489	8.736	0.1145	1048.3
3.5	1.0481	8.730	0.1145	1047.5
3.6	1.0474	8.723	0.1146	1046.7
3.7	1.0466	8.717	0.1147	1045.9
3.8	1.0458	8.710	0.1148	1045.2
3.9	1.0451	8.704	0.1149	1044.4
4.0	1.0443	8.698	0.1150	1043.6
4.1	1.0435	8.691	0.1151	1042.8
4.2	1.0427	8.685	0.1151	1042.1
4.3	1.0420	8.678	0.1152	1041.3
4.4	1.0412	8.672	0.1153	1040.5
4.5	1.0404	8.666	0.1154	1039.8
4.6	1.0397	8.659	0.1155	1039.0
4.7	1.0389	8.653	0.1156	1038.3
4.8	1.0382	8.646	0.1157	1037.5
4.9	1.0374	8.640	0.1157	1036.7
5.0	1.0366	8.634	0.1158	1036.0
5.1	1.0359	8.627	0.1159	1035.2
5.2	1.0351	8.621	0.1160	1034.5
5.3	1.0344	8.615	0.1161	1033.7
5.4	1.0336	8.608	0.1162	1033.0
5.5	1.0328	8.602	0.1163	1032.2
5.6	1.0321	8.596	0.1163	1031.4
5.7	1.0313	8.590	0.1164	1030.7
5.8	1.0306	8.583	0.1165	1029.9
5.9	1.0298	8.577	0.1166	1029.2
6.0	1.0291	8.571	0.1167	1028.4
6.1	1.0283	8.565	0.1168	1027.7
6.2	1.0276	8.558	0.1168	1027.0
6.3	1.0269	8.552	0.1169	1026.2
6.4	1.0261	8.546	0.1170	1025.5
6.5	1.0254	8.540	0.1171	1024.7
6.6	1.0246	8.534	0.1172	1024.0
6.7	1.0239	8.527	0.1173	1023.2
6.8	1.0231	8.521	0.1174	1022.5
6.9	1.0224	8.515	0.1174	1021.8
7.0	1.0217	8.509	0.1175	1021.0
7.1	1.0209	8.503	0.1176	1020.3
7.2	1.0202	8.497	0.1177	1019.6
7.3	1.0195	8.490	0.1178	1018.8
7.4	1.0187	8.484	0.1179	1018.1

Degrees API	Specific gravity at 60°/60°F	Pounds Per Gallon	Gallons Per Pound	Kilograms Per Cubic Meter
7.5	1.0180	8.478	0.1180	1017.4
7.6	1.0173	8.472	0.1180	1016.6
7.7	1.0165	8.466	0.1181	1015.9
7.8	1.0158	8.460	0.1182	1015.2
7.9	1.0151	8.454	0.1183	1014.2
8.0	1.0143	8.448	0.1184	1013.7
8.1	1.0136	8.442	0.1185	1013.0
8.2	1.0129	8.436	0.1185	1012.3
8.3	1.0122	8.430	0.1186	1011.5
8.4	1.0114	8.424	0.1187	1010.8
8.5	1.0107	8.418	0.1188	1010.1
8.6	1.0100	8.412	0.1189	1009.4
8.7	1.0093	8.406	0.1190	1008.6
8.8	1.0086	8.400	0.1190	1007.9
8.9	1.0078	8.394	0.1191	1007.2
9.0	1.0071	8.388	0.1192	1006.5
9.1	1.0064	8.382	0.1193	1005.8
9.2	1.0057	8.376	0.1194	1005.1
9.3	1.0050	8.370	0.1195	1004.4
9.4	1.0043	8.364	0.1196	1003.6
9.5	1.0035	8.358	0.1196	1002.9
9.6	1.0028	8.352	0.1197	1002.2
9.7	1.0021	8.346	0.1198	1001.5
9.8	1.0014	8.340	0.1199	1000.8
9.9	1.0007	8.334	0.1200	1000.1
10.0	1.0000	8.328	0.1201	999.4
10.1	.9993	8.322	0.1202	998.7
10.2	.9986	8.317	0.1202	998.0
10.3	.9979	8.311	0.1203	997.3
10.4	.9972	8.305	0.1204	996.6
10.5	.9965	8.299	0.1205	995.9
10.6	.9958	8.293	0.1206	995.2
10.7	.9951	8.287	0.1207	994.5
10.8	.9944	8.282	0.1207	993.8
10.9	.9937	8.276	0.1208	993.1
11.0	.9930	8.270	0.1209	992.4
11.1	.9923	8.264	0.1210	991.7
11.2	.9916	8.258	0.1211	991.0
11.3	.9909	8.252	0.1212	990.3
11.4	.9902	8.246	0.1213	989.6
11.5	.9895	8.241	0.1213	988.9
11.6	.9888	8.235	0.1214	988.2
11.7	.9881	8.229	0.1215	987.5
11.8	.9874	8.223	0.1216	986.8
11.9	.9868	8.218	0.1217	986,2
12.0	.9861	8.212	0.1218	985.5
12.1	.9854	8.206	0.1219	984.8
12.2	.9847	8.201	0.1219	984.1
12.3	.9840	8.195	0.1220	983.4
12.4	.9833	8.189	0.1221	982.7
12.5	.9826	8.183	0.1222	982.0
12.6	.9820	8.178	0.1223	981.4
12.7	.9813	8.172	0.1224	980.7
12.8	.9806	8.166	0.1225	980.0
12.9	.9799	8.161	0.1225	979.3
13.0	.9792	8.155	0.1226	978.6
13.1	.9786	8.150	0.1227	978.0
13.2	.9779	8.144	0.1228	977.3
13.3	.9772	8.138	0.1229	976.6
13.4	.9765	8.132	0.1230	975.9
13.5	.9759	8.127	0.1230	975.3
13.6	.9752	8.122	0.1231	974.6
13.7	.9745	8.116	0.1232	973.9
13.8	.9738	8.110	0.1233	973.3
13.9	.9732	8.105	0.1234	972.6
14.0	.9725	8.099	0.1235	971.9
14.1	.9718	8.093	0.1236	971.3
14.2	.9712	8.088	0.1236	970.6
14.3	.9705	8.082	0.1237	969.9
14.4	.9698	8.076	0.1238	969.3
14.5	.9692	8.071	0.1239	968.6
14.6	.9685	8.066	0.1240	967.9
14.7	.9679	8.061	0.1241	967.3
14.8	.9672	8.055	0.1241	966.6
14.9	.9665	8.049	0.1242	966.0

Degrees API	Specific gravity at 60°/60°F	Pounds Per Gallon	Gallons Per Pound	Kilograms Per Cubic Meter
15.0	.9659	8.044	0.1243	965.3
15.1	.9652	8.038	0.1244	964.6
15.2	.9646	8.033	0.1245	964.0
15.3	.9639	8.027	0.1246	963.3
15.4	.9632	8.021	0.1247	962.7
15.5	.9626	8.016	0.1248	962.0
15.6	.9619	8.011	0.1248	961.4
15.7	.9613	8.006	0.1249	960.7
15.8	.9606	8.000	0.1250	960.1
15.9	.9600	7.995	0.1251	959.4
16.0	.9593	7.989	0.1252	958.8
16.1	.9587	7.984	0.1253	958.1
16.2	.9580	7.978	0.1253	957.5
16.3	.9574	7.973	0.1254	956.8
16.4	.9567	7.967	0.1255	956.2
16.5	.9561	7.962	0.1256	955.5
16.6	.9554	7.956	0.1257	954.9
16.7	.9548	7.951	0.1258	954.2
16.8	.9541	7.946	0.1258	953.6
16.9	.9535	7.940	0.1259	952.9
17.0	.9529	7.935	0.1260	952.3
17.1	.9522	7.930	0.1261	951.7
17.2	.9516	7.925	0.1262	951.0
17.3	.9509	7.919	0.1263	950.4
17.4	.9503	7.914	0.1264	949.7
17.5	.9497	7.909	0.1264	949.1
17.6	.9490	7.903	0.1265	948.5
17.7	.9484	7.898	0.1266	947.8
17.8	.9478	7.893	0.1267	947.2
17.9	.9471	7.887	0.1268	946.6
18.0	.9465	7.882	0.1269	945.9
18.1	.9459	7.877	0.1270	945.3
18.2	.9452	7.871	0.1270	944.7
18.3	.9446	7.866	0.1271	944.0
18.4	.9440	7.861	0.1272	943.4
18.5	.9433	7.856	0.1273	942.8
18.6	.9427	7.851	0.1274	942.2
18.7	.9421	7.846	0.1275	941.5
18.8	.9415	7.841	0.1275	940.9
18.9	.9408	7.835	0.1276	940.3
19.0	.9402	7.830	0.1277	939.7
19.1	.9396	7.825	0.1278	939.0
19.2	.9390	7.820	0.1279	938.4
19.3	.9383	7.814	0.1280	937.8
19.4	.9377	7.809	0.1281	937.2
19.5	.9371	7.804	0.1281	936.5
19.6	.9365	7.799	0.1282	935.9
19.7	.9358	7.793	0.1283	935.3
19.8	.9352	7.788	0.1284	934.7
19.9	.9346	7.783	0.1285	934.1
20.0	.9340	7.778	0.1286	933.5
20.1	.9334	7.773	0.1287	932.8
20.2	.9328	7.768	0.1287	932.2
20.3	.9321	7.762	0.1288	931.6
20.4	.9315	7.757	0.1289	931.0
20.5	.9309	7.752	0.1290	930.4
20.6	.9303	7.747	0.1291	929.8
20.7	.9297	7.742	0.1292	929.2
20.8	.9291	7.737	0.1292	928.6
20.9	.9285	7.732	0.1293	927.9
21.0	.9279	7.727	0.1294	927.3
21.1	.9273	7.722	0.1295	926.7
21.2	.9267	7.717	0.1296	926.1
21.3	.9260	7.711	0.1297	925.5
21.4	.9254	7.706	0.1298	924.9
21.5	.9248	7.701	0.1299	924.3
21.6	.9242	7.696	0.1299	923.7
21.7	.9236	7.691	0.1300	923.1
21.8	.9230	7.686	0.1301	922.5
21.9	.9224	7.681	0.1302	921.9
22.0	.9218	7.676	0.1303	921.3
22.1	.9212	7.671	0.1304	920.7
22.2	.9206	7.666	0.1304	920.1
22.3	.9200	7.661	0.1305	919.5
22.4	.9194	7.656	0.1306	918.9

Weight-Volume Relationships
Degrees API

Degrees API	Specific gravity at 60°/60°F	Pounds Per Gallon	Gallons Per Pound	Kilograms Per Cubic Meter
22.5	.9188	7.651	0.1307	918.3
22.6	.9182	7.646	0.1308	917.7
22.7	.9176	7.641	0.1309	917.1
22.8	.9170	7.636	0.1310	916.5
22.9	.9165	7.632	0.1310	915.9
23.0	.9159	7.627	0.1311	915.3
23.1	.9153	7.622	0.1312	914.7
23.2	.9147	7.617	0.1312	914.2
23.3	.9141	7.612	0.1314	913.6
23.4	.9135	7.607	0.1315	913.0
23.5	.9129	7.602	0.1315	912.4
23.6	.9123	7.597	0.1316	911.8
23.7	.9117	7.592	0.1317	911.2
23.8	.9111	7.587	0.1318	910.6
23.9	.9106	7.583	0.1319	910.0
24.0	.9100	7.578	0.1320	909.5
24.1	.9094	7.573	0.1320	908.9
24.2	.9088	7.568	0.1321	908.3
24.3	.9082	7.563	0.1322	907.7
24.4	.9076	7.558	0.1323	907.1
24.5	.9071	7.554	0.1324	906.5
24.6	.9065	7.549	0.1325	906.0
24.7	.9059	7.544	0.1326	905.4
24.8	.9053	7.539	0.1326	904.8
24.9	.9047	7.534	0.1327	904.2
25.0	.9042	7.529	0.1328	903.6
25.1	.9036	7.524	0.1329	903.1
25.2	.9030	7.519	0.1330	902.5
25.3	.9024	7.514	0.1331	901.9
25.4	.9018	7.509	0.1332	901.3
25.5	.9013	7.505	0.1332	900.8
25.6	.9007	7.500	0.1333	900.2
25.7	.9001	7.495	0.1334	899.6
25.8	.8996	7.491	0.1335	899.1
25.9	.8990	7.486	0.1336	898.5
26.0	.8984	7.481	0.1337	897.9
26.1	.8978	7.476	0.1338	897.3
26.2	.8973	7.472	0.1338	896.8
26.3	.8967	7.467	0.1339	896.2
26.4	.8961	7.462	0.1340	895.6
26.5	.8956	7.458	0.1341	895.1
26.6	.8950	7.453	0.1342	894.5
26.7	.8944	7.448	0.1343	893.9
26.8	.8939	7.443	0.1344	893.4
26.9	.8933	7.438	0.1344	892.8
27.0	.8927	7.434	0.1345	892.2
27.1	.8922	7.429	0.1346	891.7
27.2	.8916	7.424	0.1347	891.1
27.3	.8911	7.420	0.1348	890.6
27.4	.8905	7.415	0.1349	890.0
27.5	.8899	7.410	0.1350	889.4
27.6	.8894	7.406	0.1350	888.9
27.7	.8888	7.401	0.1351	888.3
27.8	.8883	7.397	0.1352	887.8
27.9	.8877	7.392	0.1353	887.2
28.0	.8871	7.387	0.1354	886.7
28.1	.8866	7.383	0.1354	886.1
28.2	.8860	7.378	0.1355	885.5
28.3	.8855	7.373	0.1356	885.0
28.4	.8849	7.368	0.1357	884.4
28.5	.8844	7.364	0.1358	883.9
28.6	.8838	7.360	0.1359	883.3
28.7	.8833	7.355	0.1360	882.8
28.8	.8827	7.351	0.1360	882.2
28.9	.8822	7.346	0.1361	881.7
29.0	.8816	7.341	0.1362	881.1
29.1	.8811	7.337	0.1363	880.6
29.2	.8805	7.332	0.1364	880.0
29.3	.8800	7.328	0.1365	879.5
29.4	.8794	7.323	0.1366	878.9
29.5	.8789	7.318	0.1366	878.4
29.6	.8783	7.314	0.1367	877.9
29.7	.8778	7.309	0.1368	877.3
29.8	.8772	7.305	0.1369	876.8
29.9	.8767	7.300	0.1370	876.2
30.0	.8762	7.296	0.1371	875.7
30.1	.8756	7.291	0.1372	875.1
30.2	.8751	7.287	0.1372	874.6
30.3	.8745	7.282	0.1373	874.1
30.4	.8740	7.278	0.1374	873.5
30.5	.8735	7.273	0.1375	873.0
30.6	.8729	7.269	0.1376	872.4
30.7	.8724	7.264	0.1377	871.9
30.8	.8718	7.259	0.1378	871.4
30.9	.8713	7.255	0.1378	870.8
31.0	.8708	7.251	0.1379	870.3
31.1	.8702	7.246	0.1380	869.8
31.2	.8697	7.242	0.1381	869.2
31.3	.8692	7.238	0.1382	868.7
31.4	.8686	7.233	0.1383	868.2
31.5	.8681	7.228	0.1384	867.6
31.6	.8676	7.224	0.1384	867.1
31.7	.8670	7.219	0.1385	866.6
31.8	.8665	7.215	0.1386	866.0
31.9	.8660	7.211	0.1387	865.5
32.0	.8654	7.206	0.1388	865.0
32.1	.8649	7.202	0.1389	864.5
32.2	.8644	7.198	0.1389	863.9
32.3	.8639	7.193	0.1390	863.4
32.4	.8633	7.188	0.1391	862.9
32.5	.8628	7.184	0.1392	862.3
32.6	.8623	7.180	0.1393	861.8
32.7	.8618	7.176	0.1394	861.3
32.8	.8612	7.171	0.1395	860.8
32.9	.8607	7.167	0.1395	860.2
33.0	.8602	7.163	0.1396	859.7
33.1	.8597	7.158	0.1397	859.2
33.2	.8591	7.153	0.1398	858.7
33.3	.8586	7.149	0.1399	858.2
33.4	.8581	7.145	0.1400	857.6
33.5	.8576	7.141	0.1400	857.1
33.6	.8571	7.137	0.1401	856.6
33.7	.8565	7.132	0.1402	856.1
33.8	.8560	7.128	0.1403	855.6
33.9	.8555	7.123	0.1404	855.0
34.0	.8550	7.119	0.1405	854.5
34.1	.88545	7.115	0.1405	854.0
34.2	.8540	7.111	0.1406	853.5
34.3	.8534	7.106	0.1407	853.0
34.4	.8529	7.102	0.1408	852.5
34.5	.8524	7.098	0.1409	852.0
34.6	.8519	7.093	0.1410	851.4
34.7	.8514	7.089	0.1411	850.9
34.8	.8509	7.085	0.1411	850.4
34.9	.8504	7.081	0.1412	849.9
35.0	.8498	7.076	0.1413	849.4
35.1	.8493	7.072	0.1414	848.9
35.2	.8488	7.068	0.1415	848.4
35.3	.8483	7.063	0.1416	847.9
35.4	.8478	7.059	0.1417	847.4
35.5	.8473	7.055	0.1417	846.9
35.6	.8468	7.051	0.1418	846.4
35.7	.8463	7.047	0.1419	845.8
35.8	.8458	7.042	0.1420	845.3
35.9	.8453	7.038	0.1421	844.8
36.0	.8448	7.034	0.1422	844.3
36.1	.8443	7.030	0.1422	843.8
36.2	.8438	7.026	0.1423	843.3
36.3	.8433	7.022	0.1424	842.8
36.4	.8428	7.018	0.1425	842.3
36.5	.8423	7.013	0.1426	841.8
36.6	.8418	7.009	0.1427	841.3
36.7	.8413	7.005	0.1428	840.8
36.8	.8408	7.001	0.1428	840.3
36.9	.8403	6.997	0.1429	839.8
37.0	.8398	6.993	0.1430	839.3
37.1	.8393	6.989	0.1431	838.8
37.2	.8388	6.985	0.1432	838.3
37.3	.8383	6.980	0.1433	837.8
37.4	.8378	6.976	0.1433	837.3
37.5	.8373	6.972	0.1434	836.8
37.6	.8368	6.968	0.1435	836.4
37.7	.8363	6.964	0.1436	835.9
37.8	.8358	6.960	0.1437	835.4
37.9	.8353	6.955	0.1438	834.9
38.0	.8348	6.951	0.1439	834.4
38.1	.8343	6.947	0.1439	833.9
38.2	.8338	6.943	0.1440	833.4
38.3	.8333	6.938	0.1441	832.9
38.4	.8328	6.935	0.1442	832.4
38.5	.8324	6.930	0.1443	831.9
38.6	.8319	6.926	0.1444	831.4
38.7	.8314	6.922	0.1445	831.0
38.8	.8309	6.918	0.1446	830.5
38.9	.8304	6.914	0.1446	830.0
39.0	.8299	6.910	0.1447	829.5
39.1	.8294	6.906	0.1448	829.0
39.2	.8289	6.902	0.1449	828.5
39.3	.8285	6.898	0.1450	828.0
39.4	.8280	6.894	0.1451	827.6
39.5	.8275	6.890	0.1451	827.1
39.6	.8270	6.886	0.1452	826.6
39.7	.8265	6.882	0.1453	826.1
39.8	.8260	6.878	0.1454	825.6
39.9	.8256	6.874	0.1455	825.1
40.0	.8251	6.870	0.1456	824.7
40.1	.8246	6.866	0.1456	824.2
40.2	.8241	6.862	0.1457	823.7
40.3	.8236	6.858	0.1458	823.2
40.4	.8232	6.854	0.1459	822.7
40.5	.8227	6.850	0.1460	822.3
40.6	.8222	6.846	0.1461	821.8
40.7	.8217	6.842	0.1462	821.3
40.8	.8212	6.838	0.1462	820.8
40.9	.8208	6.834	0.1463	820.4
41.0	.8203	6.830	0.1464	819.9
41.1	.8198	6.826	0.1465	819.4
41.2	.8193	6.822	0.1466	818.9
41.3	.8189	6.818	0.1467	818.5
41.4	.8184	6.814	0.1468	818.0
41.5	.8179	6.810	0.1468	817.5
41.6	.8174	6.806	0.1469	817.0
41.7	.8170	6.802	0.1470	816.6
41.8	.8165	6.798	0.1471	816.1
41.9	.8160	6.794	0.1472	815.6
42.0	.8155	6.790	0.1473	815.2
42.1	.8151	6.786	0.1474	814.7
42.2	.8146	6.782	0.1474	814.2
42.3	.8142	6.779	0.1475	813.8
42.4	.8137	6.775	0.1476	813.3
42.5	.8132	6.771	0.1477	812.8
42.6	.8128	6.767	0.1478	812.4
42.7	.8123	6.763	0.1479	811.9
42.8	.8118	6.759	0.1480	811.4
42.9	.8114	6.756	0.1480	811.0
43.0	.8109	6.752	0.1481	810.5
43.1	.8104	6.748	0.1482	810.0
43.2	.8100	6.744	0.1483	809.6
43.3	.8095	6.740	0.1484	809.1
43.4	.8090	6.736	0.1485	808.6
43.5	.8086	6.732	0.1485	808.2
43.6	.8081	6.728	0.1486	807.7
43.7	.8076	6.724	0.1487	807.3
43.8	.8072	6.720	0.1488	806.8
43.9	.8067	6.716	0.1489	806.3
44.0	.8063	6.713	0.1490	805.9
44.1	.8058	6.709	0.1491	805.4
44.2	.8054	6.705	0.1491	805.0
44.3	.8049	6.701	0.1492	804.5
44.4	.8044	6.697	0.1493	804.1
44.5	.8040	6.694	0.1494	803.6
44.6	.8035	6.690	0.1495	803.1
44.7	.8031	6.686	0.1496	802.7
44.8	.8026	6.682	0.1497	802.2
44.9	.8022	6.679	0.1497	801.8

Weight-Volume Relationships
Degrees API

Degrees API	Specific gravity at 60°/60°F	Pounds Per Gallon	Gallons Per Pound	Kilograms Per Cubic Meter	Degrees API	Specific gravity at 60°/60°F	Pounds Per Gallon	Gallons Per Pound	Kilograms Per Cubic Meter	Degrees API	Specific gravity at 60°/60°F	Pounds Per Gallon	Gallons Per Pound	Kilograms Per Cubic Meter
45.0	.8017	6.675	0.1498	801.3	52.5	.7690	6.402	0.1562	768.8	60.0	.7389	6.151	0.1626	738.7
45.1	.8012	6.671	0.1499	800.9	52.6	.7686	6.399	0.1563	768.3	60.1	.7385	6.148	0.1627	738.3
45.2	.8008	6.667	0.1500	800.4	52.7	.7682	6.396	0.1563	767.9	60.2	.7381	6.144	0.1628	737.9
45.3	.8003	6.663	0.1501	800.0	52.8	.7678	6.392	0.1564	767.5	60.3	.7377	6.141	0.1628	737.5
45.4	.7999	6.660	0.1502	799.5	52.9	.7674	6.389	0.1565	767.1	60.4	.7374	6.138	0.1629	737.1
45.5	.7994	6.656	0.1502	799.1	53.0	.7669	6.385	0.1566	766.7	60.5	.7370	6.135	0.1630	736.8
45.6	.7990	6.652	0.1503	798.6	53.1	.7665	6.381	0.1567	766.3	60.6	.7366	6.132	0.1631	736.4
45.7	.7995	6.648	0.1504	798.2	53.2	.7661	6.378	0.1568	765.8	60.7	.7362	6.129	0.1632	736.0
45.8	.7981	6.645	0.1505	797.5	53.3	.7657	6.375	0.1569	765.4	60.8	.7358	6.125	0.1633	735.6
45.9	.7976	6.641	0.1506	797.3	53.4	.7653	6.371	0.1570	765.0	60.9	.7354	6.122	0.1633	735.2
46.0	.7972	6.637	0.1507	796.8	53.5	.7649	6.368	0.1570	764.6	61.0	.7351	6.119	0.1634	734.8
46.1	.7967	6.633	0.1508	796.4	53.6	.7645	6.365	0.1571	76.2	61.1	.7347	6.116	0.1635	734.5
46.2	.7963	6.630	0.1508	795.9	53.7	.7640	6.361	0.1572	763.8	61.2	.7343	6.113	0.1636	734.1
46.3	.7958	6.626	0.1509	795.5	53.8	.7636	6.357	0.1573	763.4	61.3	.7339	6.109	0.1637	733.7
46.4	.7954	6.622	0.1510	795.0	53.9	.7632	6.354	0.1574	763.0	61.4	.7335	6.106	0.1638	733.3
46.5	.7949	6.618	0.1511	794.6	54.0	.7628	6.350	0.1575	762.5	61.5	.7332	6.103	0.1639	732.9
46.6	.7945	6.615	0.1512	794.1	54.1	.7624	6.347	0.1576	762.1	61.6	.7328	6.100	0.1639	732.6
46.7	.7941	6.611	0.1513	793.7	54.2	.7620	6.344	0.1576	761.7	61.7	.7324	6.097	0.1640	732.2
46.8	.7936	6.607	0.1514	793.2	54.3	.7616	6.340	0.1577	761.3	61.8	.7320	6.094	0.1641	731.8
46.9	.7932	6.604	0.1514	792.8	54.4	.7612	6.337	0.1578	760.9	61.9	.7316	6.090	0.1642	731.4
47.0	.7927	6.600	0.1515	792.4	54.5	.7608	6.334	0.1579	760.5	62.0	.7313	6.087	0.1643	731.1
47.1	.7923	6.596	0.1516	791.9	54.6	.7603	6.330	0.1580	760.1	62.1	.7309	6.084	0.1644	730.7
47.2	.7918	6.592	0.1517	791.5	54.7	.7599	6.326	0.1581	759.7	62.2	.7305	6.081	0.1644	730.3
47.3	.7914	6.589	0.1518	791.0	54.8	.7595	6.323	0.1582	759.3	62.3	.7301	6.078	0.1645	729.9
47.4	.7909	6.585	0.1519	790.6	54.9	.7591	6.320	0.1582	758.9	62.4	.7298	6.075	0.1646	729.5
47.5	.7905	6.582	0.1519	790.1	55.0	.7587	6.316	0.1583	758.5	62.5	.7294	6.072	0.1647	729.2
47.6	.7901	6.578	0.1520	789.7	55.1	.7583	6.313	0.1584	758.1	62.6	.7290	6.068	0.1648	728.8
47.7	.7896	6.574	0.1521	789.3	55.2	.7579	6.310	0.1585	757.6	62.7	.7286	6.065	0.1649	728.4
47.8	.7892	6.571	0.1522	788.8	55.3	.7575	6.306	0.1586	757.2	62.8	.7283	6.062	0.1650	728.0
47.9	.7887	6.567	0.1523	788.4	55.4	.7571	6.303	0.1587	756.8	62.9	.7279	6.059	0.1650	727.7
48.0	.7883	6.563	0.1524	787.9	55.5	.7567	6.300	0.1587	756.4	63.0	.7275	6.056	0.1651	727.3
48.1	.7879	6.560	0.1524	787.5	55.6	.7563	6.296	0.1588	756.0	63.1	.7271	6.053	0.1652	726.9
48.2	.7874	6.556	0.1525	787.1	55.7	.7559	6.293	0.1589	755.6	63.2	.7268	6.050	0.1653	726.5
48.3	.7870	6.552	0.1526	786.6	55.8	.7555	6.290	0.1590	755.2	63.3	.7264	6.047	0.1654	726.2
48.4	.7865	6.548	0.1527	786.2	55.9	.7551	6.287	0.1591	754.8	63.4	.7260	6.044	0.1655	725.8
48.5	.7861	6.545	0.1528	785.8	56.0	.7547	6.283	0.1592	754.4	63.5	.7256	6.040	0.1656	725.4
48.6	.7857	6.541	0.1529	785.3	56.1	.7543	6.280	0.1592	754.0	63.6	.7253	6.037	0.1656	725.1
48.7	.7852	6.537	0.1530	784.9	56.2	.7539	6.276	0.1593	753.6	63.7	.7249	6.034	0.1657	724.7
48.8	.7848	6.534	0.1530	784.5	56.3	.7535	6.273	0.1594	753.2	63.8	.7245	6.031	0.1658	724.3
48.9	.7844	6.530	0.1531	784.0	56.4	.7531	6.270	0.1595	752.8	63.9	.7242	6.028	0.1659	723.9
49.0	.7839	6.526	0.1532	783.6	56.5	.7527	6.266	0.1596	752.4	64.0	.7238	6.025	0.1660	723.6
49.1	.7835	6.523	0.1533	783.2	56.6	.7523	6.263	0.1597	752.0	64.1	.7234	6.022	0.1661	723.2
49.2	.7831	6.520	0.1534	782.7	56.7	.7519	6.259	0.1598	751.6	64.2	.7230	6.019	0.1661	722.8
49.3	.7826	6.516	0.1535	782.3	56.8	.7515	6.256	0.1598	751.2	64.3	.7227	6.016	0.1662	722.5
49.4	.7822	6.512	0.1536	781.9	56.9	.7511	6.253	0.1599	750.8	64.4	.7223	6.013	0.1663	722.1
49.5	.7818	6.509	0.1536	781.4	57.0	.7507	6.249	0.1600	750.4	64.5	.7219	6.010	0.1664	721.7
49.6	.7813	6.505	0.1537	781.0	57.1	.7503	6.246	0.1601	750.0	64.6	.7216	6.007	0.1665	721.4
49.7	.7809	6.501	0.1538	780.6	57.2	.7499	6.243	0.1602	749.6	64.7	.7212	6.004	0.1666	721.0
49.8	.7805	6.498	0.1539	780.2	57.3	.7495	6.240	0.1603	749.2	64.8	.7208	6.000	0.1667	720.6
49.9	.7800	6.494	0.1540	779.7	57.4	.7491	6.236	0.1604	748.8	64.9	.7205	5.997	0.1668	720.3
50.0	.7796	6.490	0.1541	779.3	57.5	.7487	6.233	0.1604	748.4	65.0	.7201	5.994	0.1668	719.9
50.1	.7792	6.487	0.1542	778.9	57.6	.7483	6.229	0.1605	748.0	65.1	.7197	5.991	0.1669	719.5
50.2	.7788	6.484	0.1542	778.4	57.7	.7479	6.226	0.1606	747.6	65.2	.7194	5.988	0.1670	719.2
50.3	.7783	6.480	0.1543	778.0	57.8	.7475	6.223	0.1607	747.3	65.3	.7190	5.985	0.1671	718.8
50.4	.7779	6.476	0.1544	777.6	57.9	.7471	6.219	0.1608	746.9	65.4	.7186	5.982	0.1672	718.4
50.5	.7775	6.473	0.1545	777.2	58.0	.7467	6.216	0.1609	746.5	65.5	.7183	5.979	0.1673	718.1
50.6	.7770	6.469	0.1546	776.7	58.1	.7463	6.213	0.1610	746.1	65.6	.7179	5.976	0.1673	717.7
50.7	.7766	6.466	0.1547	776.3	58.2	.7459	6.209	0.1611	745.7	65.7	.7175	5.973	0.1674	717.3
50.8	.7762	6.462	0.1548	775.9	58.3	.7455	6.206	0.1611	745.3	65.8	.7172	5.970	0.1675	717.0
50.9	.7758	6.459	0.1548	775.5	58.4	.7451	6.203	0.1612	744.9	65.9	.7168	5.967	0.1676	716.6
51.0	.7753	6.455	0.1549	775.1	58.5	.7447	6.199	0.1613	744.5	66.0	.7165	5.964	0.1677	716.3
51.1	.7749	6.451	0.1550	774.6	58.6	.7443	6.196	0.1614	744.1	66.1	.7161	5.961	0.1678	715.9
51.2	.7745	6.448	0.1551	774.2	58.7	.7440	6.193	0.1615	743.7	66.2	.7157	5.958	0.1678	715.5
51.3	.7741	6.445	0.1552	773.8	58.8	.7436	6.190	0.1616	743.3	66.3	.7154	5.955	0.1679	715.2
51.4	.7736	6.441	0.1553	773.4	58.9	.7432	6.187	0.1616	742.9	66.4	.7150	5.952	0.1680	714.8
51.5	.7732	6.437	0.1554	772.9	59.0	.7428	6.184	0.1617	742.6	66.5	.7146	5.949	0.1681	714.5
51.6	.7728	6.434	0.1554	772.5	59.1	.7424	6.180	0.1618	742.2	66.6	.7143	5.946	0.1682	714.1
51.7	.7724	6.430	0.1555	772.1	59.2	.7420	6.177	0.1619	741.8	66.7	.7139	5.943	0.1683	713.7
51.8	.7720	6.427	0.1556	771.7	59.3	.7416	6.174	0.1620	741.4	66.8	.7136	5.940	0.1684	713.4
51.9	.7715	6.423	0.1557	771.3	59.4	.7412	6.170	0.1621	741.0	66.9	.7132	5.937	0.1684	713.0
52.0	.7711	6.420	0.1558	770.8	59.5	.7408	6.167	0.1622	740.6	67.0	.7128	5.934	0.1685	712.7
52.1	.7707	6.416	0.1559	770.4	59.6	.7405	6.164	0.1622	740.2	67.1	.7125	5.931	0.1686	712.3
52.2	.7703	6.413	0.1559	770.0	59.7	.7401	6.161	0.1623	739.8	67.2	.7121	5.928	0.1687	711.9
52.3	.7699	6.410	0.1560	769.6	59.8	.7397	6.158	0.1624	739.4	67.3	.7118	5.925	0.1688	711.6
52.4	.7694	6.406	0.1561	769.2	59.9	.7393	6.154	0.1625	739.1	67.4	.7114	5.922	0.1689	711.2

Weight-Volume Relationships

Degrees API

Degrees API	Specific gravity at 60°/60°F	Pounds Per Gallon	Gallons Per Pound	Kilograms Per Cubic Meter
67.5	.7111	5.919	0.1689	710.9
67.6	.7107	5.916	0.1690	710.5
67.7	.7103	5.913	0.1691	710.2
67.8	.7100	5.910	0.1692	709.8
67.9	.7096	5.907	0.1693	709.4
68.0	.7093	5.904	0.1694	709.1
68.1	.7089	5.901	0.1695	708.7
68.2	.7086	5.898	0.1695	708.4
68.3	.7082	5.895	0.1696	708.0
68.4	.7079	5.892	0.1697	707.7
68.5	.7075	5.889	0.1698	707.3
68.6	.7071	5.886	0.1699	707.0
68.7	.7068	5.883	0.1700	706.6
68.8	.7064	5.880	0.1701	706.3
68.9	.7061	5.877	0.1702	705.9
69.0	.7057	5.874	0.1702	705.6
69.1	.7054	5.871	0.1703	705.2
69.2	.7050	5.868	0.1704	704.9
69.3	.7047	5.866	0.1705	704.5
69.4	.7043	5.863	0.1706	704.2
69.5	.7040	5.860	0.1706	703.8
69.6	.7036	5.857	0.1707	703.5
69.7	.7033	5.854	0.1708	703.1
69.8	.7029	5.851	0.1709	702.8
69.9	.7026	5.848	0.1710	702.4
70.0	.7022	5.845	0.1711	702.1
70.1	.7019	5.842	0.1712	701.7
70.2	.7015	5.839	0.1713	701.4
70.3	.7012	5.836	0.1714	701.0
70.4	.7008	5.833	0.1714	700.7
70.5	.7005	5.831	0.1715	700.3
70.6	.7001	5.828	0.1716	700.0
70.7	.6998	5.825	0.1717	699.6
70.8	.6995	5.823	0.1717	699.3
70.9	.6991	5.820	0.1718	698.9
71.0	.6988	5.817	0.1719	698.6
71.1	.6984	5.814	0.1720	698.2
71.2	.6981	5.811	0.1721	697.9
71.3	.6977	5.808	0.1722	697.6
71.4	.6974	5.805	0.1723	697.2
71.5	.6970	5.802	0.1724	696.9
71.6	.6967	5.799	0.1724	696.5
71.7	.6964	5.796	0.1725	696.2
71.8	.6960	5.793	0.1726	695.8
71.9	.6957	5.791	0.1727	695.5
72.0	.6953	5.788	0.1728	695.2
72.1	.6950	5.785	0.1729	694.8
72.2	.6946	5.782	0.1730	694.5
72.3	.6943	5.779	0.1730	694.1
72.4	.6940	5.776	0.1731	693.8
72.5	.6936	5.773	0.1732	693.5
72.6	.6933	5.771	0.1733	693.1
72.7	.6929	5.768	0.1734	692.8
72.8	.6926	5.765	0.1735	692.4
72.9	.6923	5.762	0.1736	692.1
73.0	.6919	5.759	0.1736	691.8
73.1	.6916	5.757	0.1737	691.4
73.2	.6913	5.754	0.1738	691.1
73.3	.6909	5.751	0.1739	690.8
73.4	.6906	5.748	0.1740	690.4
73.5	.6902	5.745	0.1741	690.1
73.6	.6899	5.743	0.1741	689.7
73.7	.6896	5.740	0.1742	689.4
73.8	.6892	5.737	0.1743	689.1
73.9	.6889	5.734	0.1744	688.7
74.0	.6886	5.731	0.1745	688.4
74.1	.6882	5.728	0.1746	688.1
74.2	.6879	5.726	0.1746	687.7
74.3	.6876	5.723	0.1747	687.4
74.4	.6872	5.720	0.1748	687.1
74.5	.6869	5.718	0.1749	686.7
74.6	.6866	5.715	0.1750	686.4
74.7	.6862	5.712	0.1751	686.1
74.8	.6859	5.709	0.1752	685.7
74.9	.6856	5.706	0.1753	685.4
75.0	.6852	5.703	0.1753	685.1
75.1	.6849	5.701	0.1754	684.7
75.2	.6846	5.689	0.1755	684.4
75.3	.6842	5.695	0.1756	684.1
75.4	.6839	5.693	0.1757	683.8
75.5	.6836	5.690	0.1757	683.4
75.6	.6832	5.687	0.1758	683.1
75.7	.6829	5.685	0.1759	682.8
75.8	.6826	5.682	0.1760	682.4
75.9	.6823	5.679	0.1761	682.1
76.0	.6819	5.676	0.1762	681.8
76.1	.6816	5.673	0.1763	681.5
76.2	.6813	5.671	0.1763	681.1
76.3	.6809	5.668	0.1764	680.8
76.4	.6806	5.665	0.1765	680.5
76.5	.6803	5.662	0.1766	680.1
76.6	.6800	5.660	0.1767	679.8
76.7	.6796	5.657	0.1768	679.5
76.8	.6793	5.654	0.1769	679.2
76.9	.6790	5.652	0.1769	678.8
77.0	.6787	5.649	0.1770	678.5
77.1	.6783	5.646	0.1771	678.2
77.2	.6780	5.643	0.1772	677.9
77.3	.6777	5.641	0.1773	677.5
77.4	.6774	5.638	0.1774	677.2
77.5	.6770	5.635	0.1775	676.9
77.6	.6767	5.632	0.1776	676.6
77.7	.6764	5.630	0.1776	676.2
77.8	.6761	5.627	0.1777	675.9
77.9	.6757	5.624	0.1778	675.6
78.0	.6754	5.622	0.1779	675.3
78.1	.6751	5.619	0.1780	675.0
78.2	.6748	5.617	0.1780	674.6
78.3	.6745	5.614	0.1781	674.3
78.4	.6741	5.611	0.1782	674.0
78.5	.6738	5.608	0.1783	673.7
78.6	.6735	5.606	0.1784	673.4
78.7	.6732	5.603	0.1785	673.0
78.8	.6728	5.600	0.1786	672.7
78.9	.6725	5.598	0.1786	672.4
79.0	.6722	5.595	0.1787	672.1
79.1	.6719	5.592	0.1788	671.8
79.2	.6716	5.590	0.1789	671.4
79.3	.6713	5.587	0.1790	671.1
79.4	.6709	5.584	0.1791	670.8
79.5	.6706	5.582	0.1791	670.5
79.6	.6703	5.579	0.1792	670.2
79.7	.6700	5.577	0.1793	669.9
79.8	.6697	5.574	0.1794	669.5
79.9	.6693	5.571	0.1795	669.2
80.0	.6690	5.568	0.1796	668.9
80.1	.6687	5.566	0.1797	668.6
80.2	.6684	5.563	0.1798	668.3
80.3	.6681	5.561	0.1798	668.0
80.4	.6678	5.558	0.1799	667.6
80.5	.6675	5.556	0.1800	667.3
80.6	.6671	5.553	0.1801	667.0
80.7	.6668	5.550	0.1802	666.7
80.8	.6665	5.548	0.1802	666.4
80.9	.6662	5.545	0.1803	666.1
81.0	.6659	5.542	0.1804	665.8
81.1	.6656	5.540	0.1805	665.4
81.2	.6653	5.537	0.1806	665.1
81.3	.6649	5.534	0.1807	664.8
81.4	.6646	5.532	0.1808	664.5
81.5	.6643	5.529	0.1809	664.2
81.6	.6640	5.526	0.1810	663.9
81.7	.6637	5.524	0.1810	663.6
81.8	.6634	5.522	0.1811	663.3
81.9	.6631	5.519	0.1812	663.0
82.0	.6628	5.516	0.1813	662.6
82.1	.6625	5.514	0.1814	662.3
82.2	.6621	5.511	0.1815	662.0
82.3	.6618	5.508	0.1816	661.7
82.4	.6615	5.506	0.1816	661.4
82.5	.6612	5.503	0.1817	661.1
82.6	.6609	5.501	0.1818	660.8
82.7	.6606	5.498	0.1819	660.5
82.8	.6603	5.496	0.1820	660.2
82.9	.6600	5.493	0.1820	659.9
83.0	.6597	5.491	0.1821	659.6
83.1	.6594	5.489	0.1822	659.3
83.2	.6591	5.486	0.1823	658.9
83.3	.6588	5.483	0.1824	658.6
83.4	.6584	5.480	0.1825	658.3
83.5	.6581	5.477	0.1826	658.0
83.6	.6578	5.475	0.1826	657.7
83.7	.6575	5.472	0.1827	657.4
83.8	.6572	5.470	0.1828	657.1
83.9	.6569	5.467	0.1829	656.8
84.0	.6566	5.465	0.1830	656.6
84.1	.6563	5.462	0.1831	656.2
84.2	.6560	5.460	0.1832	655.9
84.3	.6557	5.458	0.1832	655.6
84.4	.6554	5.455	0.1833	655.3
84.5	.6551	5.453	0.1834	655.0
84.6	.6548	5.450	0.1835	654.7
84.7	.6545	5.448	0.1836	654.4
84.8	.6542	5.445	0.1837	654.1
84.9	.6539	5.443	0.1837	653.8
85.0	.6536	5.440	0.1838	653.5

Miscellaneous Measurement Conversions

Multiply This	By This	To Obtain This
acre	43.560	ft²
	1.562×10^{-3}	sq mi
	160	sq rod
	0.4047	hectare
Angstrom unit (Å)	3.937×10^{-9}	in
	1×10^{-4}	μ
	0.1	mμ
	1×10^{-8}	cm
	1×10^{-10}	m
are	3.954	sq rod
	100	m²
	119.6	yd²
	0.02471	acre
atmosphere (atm)	760	torr
	760	mm Hg at 0°C
	29.92	in Hg at 0°C
	406.79	in H₂O at 4°C
	33.899	ft H₂O at 4°C
	407.16	in H₂O at 15°C
	33.93	ft H₂O at 15°C
	1.0333	kg/cm²
	10.333	kg/m²
	1.01325×10^{6}	dyne/cm²
	14.696	lb/in²
	2116.32	lb/ft²
	1.0133	bar
	101.3	k/Pa
bar	0.9869	atm
	750	mm Hg (0°C)
	10 197	kg/m²
	1×10^{6}	dyne/cm²
	14.50	lb/in²
	100	k/Pa
barrel, petroleum (bbl)	42	gal
	0.15918	m³
British Thermal Units (Btu)	777.98	ft-lb
	3.930×10^{-4}	hp-hr
	2.931×10^{-4}	kwh
	2.520×10^{-1}	cal (kg)
	107.6	kg-m
	1055	joule
Btu/ft³	37.26	kJ/m³
Btu/sec	1055	watt
Btu/gal	278.7	kJ/m³
Btu/min	0.0236	hp
	17.58	watt
Btu/lb	0.55556	cal/gm
Btu/lb °F	2.326	kJ/kg
bushel	35.238	l
	0.3524	hectoliter
	4	peck
	1.2444	cu ft
calorie gram (cal)	3.968×10^{-3}	Btu
calorie/gm	1.8	Btu/lb
	9.184	kJ/kg
carat	0.2	g
cental	100	lb
centare	1	m²
centimeter (cm)	0.032808	ft
	0.393700	in
	10,000	μ
	1×10^{8}	Å
	0.01	m
	10	mm
centimeter of Hg (0°C)	0.01316	atm
	0.1934	lb/in²
	27.845	lb/ft²
	0.44604	ft H₂O at 4°C
	135.95	kg/m²
	13332	dyne/cm²
	1.332	k/Pa
centimeter²	100	mm²
	1×10^{-4}	m²
	0.1550	in²
	0.00108	ft²
	1.196×10^{-4}	yd²

Multiply This	By This	To Obtain This
centimeter³	0.061	in³
	3.531×10^{-5}	ft³
	1.3079×10^{-6}	yd³
	1×10^{-6}	m³
	0.99997	ml
	1×10^{-3}	l
	0.0338	oz (US fl)
	0.0351	oz (Brit fl)
	10.567×10^{-4}	qt (US fl)
	8.7988×10^{-4}	qt (Brit fl)
	2.6417×10^{-4}	gal (US)
	2.1997×10^{-4}	gal (Brit)
centimeter/second	0.0328	ft/sec
	1.9685	ft/min
	0.60	m/min
	3.728×10^{-4}	mi/min
	0.02237	mi/hr
	0.03600	km/hr
centipoise (cP)	6.72×10^{-4}	lb/sec-ft
	3.60	kg/hr-m
	0.001	Pa/sec
centistokes (cSt)	1×10^{-6}	m²/sec
chain (surveyors)	100	link
	66	ft
	20.117	m
cheval-vapeur	735.499	watt
	0.9863	hp
cord	128	ft³
	3.625	stere
day	86400	sec
	1440	min
	0.143	wk
	0.0028	yr
decimeter (dm)	3.937	in
	0.328	ft
degrees, angular	0.0175	radians (rad)
	60	minutes
dram (dry)	0.0625	oz (mass)
	1.7718	g
dram (fluid)	3.6967	cc
	0.125	oz (fl)
dyne	1.020×10^{-3}	g
	2.248×10^{-4}	lb
dyne/centimeter²	1×10^{-6}	bar
	10.197×10^{-4}	g/cm²
	1.4504×10^{-5}	lb/in²
ell	45	in
erg	9.4805×10^{10}	Btu
	1	dyne-cm
	7.376×10^{-8}	ft-lb
	2.38810×10^{-11}	cal (kg)
fathom	6	ft
	1.826	m
firkin	9	gal
foot (ft)	30.48	cm
	0.3048	m
foot H₂O (4°C)	0.0295	atm
	0.883	in Hg (0°C)
	2.419	cm Hg (0°C)
	0.4335	lb/in²
	62.427	lb/ft²
	304.79	kg/m²
foot²	0.111	yd²
	3.587×10	mi²
	2.296×10^{-5}	acre
	0.0929	m²
	929	cm²
foot³	1728	in³
	0.037	yd³
	28.316	cc
	0.02832	m³
	28.316	l
	7.4805	gas (US)
	6.2288	gal (Brit)

Miscellaneous Measurement Conversions. . . continued

Multiply This	By This	To Obtain This
foot³ H₂O	62.42	lb (4°C)
	62.36	lb (15°C)
foot³/min (cfm)	0.4720	l/sec
	472	cc/sec
	448.831	gal/min
foot-pound	0.1383	kg-m
	1.356	J
foot-pound/min	3.030x10⁻⁵	hp
foot-pound/sec	1.356	W
foot/min	0.01667	ft/sec
	0.0114	mph
	0.508	cm/sec
	0.005	m/sec
	0.3048	m/min
	0.0183	km/hr
foot/second	0.01136	mi/min
	0.6818	mph
	30.48	cm/sec
	18.288	m/min
	1.097	km/hr
foot/second²	0.3048	m/second²
furlong	660	ft
	220	yd
	40	rod
gallon, U.S. (gal)	0.8327	gal (Brit.)
	128	oz
	8	pt
	3785.4	cc
	3.785	l
	0.00379	m³
	231	in³
	0.1337	ft³
	0.00495	yd³
	8.3378	lb H₂O (60°F)
	0.0238	bbl
gallon, British	1.2009	gal (U.S.)
	4546	cc
	4.546	l
	277.419	in³
	0.16054	ft³
	10	lb H₂O (60°F)
	160	oz (Brit fl)
	0.00455	m³
gallon/minute (U.S.)	8.0208	ft³/hr
	0.06309	l/sec
gill (U.S.)	4	oz
grain	0.00229	oz (avoir.)
	0.0648	g
gram (g)	0.0353	oz (avoir.)
	0.0022	lb
	15.432	grain
gram/centimeter²	9.6784x10⁻⁴	atm
	0.7356	mm Hg (0°C)
	0.0289	in Hg (0°C)
	0.3284	ft H₂O (60°F)
	0.0142	lb/in²
	2.0482	lb/ft²
	980.665	dyne/cm²
	10	kg/m²
gram/centimeter³	62.428	lb/ft³
	8.345	lb/gal (U.S.)
	0.0361	lb/in³
	1	g/ml
gram/liter	1000	ppm
	0.0624	lb/ft³
gravity constant	32.174	ft/sec²
	980.665	cm/sec²
hand	4	in
hectare (ha)	2.471	acre
	100	are
	107 640	ft²
	10 000	m²
hogshead	63	gal

Multiply This	By This	To Obtain This
horsepower (hp)	42.418	Btu/min
	33 000	ft-lb/min
	550	ft-lb/sec
	0.7457	kW
	10.688	cal (kg)/min
	1.014	cheval-vapeur
hour (h)	3600	sec
	0.04167	day
	0.0059	wk
horsepower hour (hp-h)	0.7457	mJ
hundredweight (cwt)	100	lb (short)
	112	lb (long)
inch (in)	1000	mil
	0.083	ft
	0.02778	yd
	2.54	cm
	0.0254	m
	2.54x10⁸	Å
inch Hg (0°C)	13.595	in H₂O
	1.133	ft H₂O
	0.0334	atm
	25.4	mm Hg
	34.5	g/cm²
	345.3	kg/m²
	33 864	dyne/cm²
	0.4912	lb/in²
	70.73	lb/ft²
inch H₂O (4°C)	0.00245	atm
	0.07355	in Hg
	25.399	kg/m²
	5.2022	lb/ft²
	0.0361	lb/in²
inch²	6.4516	cm²
	6.4516x10⁻⁴	m²
	0.0069	ft²
	0.00077	yd²
inch³	16.387	cc
	5.7870x10⁻⁴	ft³
	1.6387x10⁻⁵	m³
	2.14335x10⁵	yd³
	0.003606	gal (Brit)
	0.004329	gal (U.S.)
	0.01639	l
	0.5541	oz (U.S. fl)
	0.01488	qt (dry)
	0.01732	qt (U.S. fl)
joule (J) (abs)	0.737562	ft-lb
	0.1019716	kg-m
kilogram (kg)	35.274	oz (avoir.)
	2.2046	lb
	9.842x10⁻⁴	ton (long)
	11.02x10⁻⁴	ton (short)
kilogram-meter	7.2330	ft-lb
kilogram/meter	0.67197	lb/ft
kilogram/meter²	0.07356	mm Hg (0°C)
	0.00142	lb/in²
	9.6777x10⁻⁵	atm
	0.20482	lb/ft²
kilogram/meter³	0.06243	lb/ft³
	0.001	g/cc
kiloliter (kl)	35.317	ft³
kilometer (km)	3280.8	ft
	0.53959	mi (naut)
	0.62137	mi
	1093.6	yd
kilometer/hour	27.7778	cm/sec
	54.68	ft/min
	0.9113	ft/sec
	0.5396	knot
	16 667	m/min
	0.27778	m/sec
kilometer²	1.076x10⁷	ft²
	1.196x10⁶	yd²
	0.386	mi²
	1x10⁶	m²
	247.1	acre

Multiply This	By This	To Obtain This
kilopascal (kPa)	0.14503774	lb/in^2
kilowatt (kw)	56 884	Btu/min
	1.3410	hp
	3.600	mJ
kilowatt-hour	3413	Btu
	1.3410	hp-hr
knot	51.479	cm/sec
	6080.2	ft/hr
	1.15155	mph
league	3	mi (naut.)
light-year	9.4637x10^{12}	km
	5.88x10^{12}	mi
link	0.66	ft
liter	61.025	in^3
	0.035	ft^3
	0.2642	gal (U.S.)
	33.814	oz (U.S. fl)
	1.05668	qt (U.S. fl)
	0.8799	qt (Brit. fl)
liter/second	15.8507	gal/min
meter	1x10^{10}	Å
	3.2808	ft
	39.370	in
	5.359x10^{-4}	mi (naut.)
	6.2137x10^{-4}	mi
	1.0936	yd
	1x10^9	mμ
meter/minute	0.05468	ft/sec
	0.06	km/hr
	0.03728	mph
meter^2	0.01	are
	2.471x10^{-4}	acre
	10.7639	ft^2
	1550	in^2
	3.8610x10^{-7}	mi^2
	1.19598	yd^2
meter^3	35.314	ft^3
	61023	in^3
	1.3079	yd^3
	264.173	gal
	999.973	l
	1056.7	qt (U.S. fl)
micron (μm)	1x10^4	Å
	1x10^{-4}	cm
	3.937x10^{-5}	in
mil	25.4	μm
mile (statute)	8	furlong
	63360	in
	1.60935	km
	1609.35	m
	0.8684	mi (naut)
	320	rod
	1760	yd
mile^2	2.59	km^2
mile (nautical)	6080.2	ft
	1.8520	km
	1.1516	mi (statute)
miles/hour	44.704	cm/sec
	88	ft/min
	0.8684	knot
	26.82	m/min
	1.609	km/hr
milligram (mg)	3.5274x10^{-5}	oz (avoir.)
	5.6438x10^{-4}	dram
	2.2046x10^{-6}	lb
millimeter (mm)	0.03937	in
	1000	μ
millimeter Hg (0°C)	1.316x10^{-3}	atm
	1333.22	dyne/cm^2
	1.3595	g/cm^2
	13.595	kg/m^2
	2.7845	lb/ft^2
	0.1333	kPa

Multiply This	By This	To Obtain This
millimicron (mμ)	10	Å
	1x10^{-7}	cm
minute (min)	6.94444x10^{-4}	day
	9.9206x10^{-5}	wk
mole (lb-mol)	0.4536	kmol
month (mean calendar)	30.4202	day
	730.085	hr
	43.085	min
	2.6283x10^6	sec
ounce (avoirdupois) (oz)	16	dram
	437.5	grain
	0.91146	oz (troy)
	3.125x10^{-5}	ton (short)
	28.35	g
ounce (fluid)	29.5735	cc
	1.8047	in^3
	8	dram
	0.25	gill
	0.029573	l
	0.03125	qt
ounce (British fl)	28.413	cc
pace	2.5	ft
parts per million (ppm)	0.0584	grain/gal
peck	537.6	in^3
	8.8096	l
	16	pint
pint (dry)	550.61	cc
	33.6003	in^3
	0.5506	l
pint (fluid)	473.179	cc
	28.875	in^3
	128	dram
	4	gill
	0.473168	l
pound (lb)	256	dram
	7000	grain
	453.5924	g
	0.45359	kg
	1.2153	lb (troy)
pound/barrel	2.853	kg/m^3
pound/foot	1.48816	kg/m
pounds/foot^2	4.7252x10^{-4}	atm
	4.7880x10^{-4}	bar
	478.78	dyne/cm^2
	0.48824	g/cm^2
	4.8824	kg/m^2
	0.35913	mm Hg (0°C)
	6.9445x10^{-3}	lb/in^2
pound-force	4.448	N
pound-inch^2	0.068046	atm
	70.307	g/cm^2
	703.07	kg/m^2
	51.715	mm Hg (0°C)
	6.894757	kPa
pound/inch^3	27.68	g/cc
pound/foot^3	0.016018	g/cc
	16.018	kg/m^3
	5.787x10^{-4}	lb/in^3
quad	1015	Btu
quart (dry) (qt)	0.03125	bushel
	1101.23	cc
	0.03889	ft^3
	1.1012	l
quart (liquid)	946.358	cc
	57.749	in^3
	0.03342	ft^3
	256	dram
	8	gill
	0.946333	l
quire	24	sheet

Miscellaneous Measurement Conversions. . . continued

Multiply This	By This	To Obtain This
rod	0.25	chain
	16.5	feet
	0.025	furlong
	198	in
	25	link
	5.029216	m
	3.125×10^{-3}	mi
rood	0.25	acre
second (sec or S)	0.01667	min
	2.7778×10^{-4}	hr
	1.1574×10^{-5}	day
slug	32.174	lb
	14.594	kg
span	9	in
stere	1	m^3
	999.973	l
stone	14	lb
	6.3503	kg
ton	20	cwt
	907.1846	kg
	0.89286	ton (long)
	0.9072	tonne (metric)

Multiply This	By This	To Obtain This
ton (long)	22.4	cwt
	1016.047	kg
	1.12	ton (short)
tonne (metric)	1000	kg
	2204.62	lb
torr	1.316×10^{-3}	atm
	1.0	mm Hg at 0°C
watt (W)	44.254	ft-lb/min
	1.340×10^{-3}	hp
	3.41304	Btu/hr
week	168	hr
	10 080	min
	604 800	sec
yard (yd)	91.4402	cm
	5.68182×10^{-4}	mi
yard²	0.881	m^2
yard³	0.7646	m^3
year	365.256	day
	8766.144	hr

The "Système International (SI)" Of Metric Units

Conversion Tables

Note: "E" (exponent) implies 10 raised to a power:
$2.0 E+03 = 2.0 \times 10^3 = 2000$

Customary Unit × Conversion Factor = Preferred Metric Unit

SI Unit	Customary Unit	Preferred Metric Unit	Conversion Factor
Length m			
	mi	km	1.609E+00
	yd	m	9.144E−01
	ft	m	3.048E−01
	in	cm	2.540E+00
	mil	μm	2.540E+01
	micron	μm	1.000E+00
Area m^2			
	mi²	km²	2.590E+00
	acre	ha (hectare)	4.047E−01
	yd²	m²	8.361E−01
	ft²	m²	9.290E−02
	in²	cm²	6.452E+00
Volume, Capacity m^3			
	yd³	m³	7.646E−01
	bbl (42 gal)	m³	1.590E−01
	ft³	m³	2.832E−02
	gal	m³	3.785E−03
	qt	dm³	9.464E−01
	fl oz	cm³	2.957E+01
	in³	cm³	1.639E+01
	ml	cm³	1.000E+00
Mass kg			
	ton	t(tonne)	9.072E−01
	lb	kg	4.536E−01
	oz	g	2.835E+01
	grain	mg	6.480E+01
Amount of substance mol			
	lb-mol	kmol	4.536E−01
	std m³ (0°C, 1 atm)	kmol	4.462E−02
	std ft³ (60°F, 1 atm)	kmol	1.195E−03

The "Système International (SI)" Of Metric Units ... continued

SI Unit	Customary Unit	Preferred Metric Unit	Conversion Factor
Calorific value (mass basis)			
J/kg	Btu/lb	kJ/kg	2.326E+00
	cal/g	kJ/kg	4.184E+00
Calorific value (mole basis)			
J/mol	Btu/lb-mol	kJ/kmol	2.326E+00
	kcal/g-mol	kJ/kmol	4.184E+03
Calorific value (volume basis)			
J/m³	Btu/gal	kJ/m³	2.787E+02
	Btu/ft³	kJ/m³	3.726E+01
	Cal/ml	J/cm³	4.184E+00
Specific Entropy			
J/kg·K	Btu/lb·°R	kJ/kg·K	4.187E+00
Specific Heat Capacity (Mass Basis)			
J/kg·K	Btu/lb·°F	kJ/kg·°C	4.187E+00
	kcal/kg·°C	kJ/kg·°C	4.184E+00
Specific Heat Capacity (Mole Basis)			
J/mol·K	Btu/lb-mol·°F	kJ/kmol·°C	4.187E+00
	cal/g-mol·°C	kJ/kmol·°C	4.184E+00
Temperature (Absolute)			
K	°R	K	5/9°R
Temperature (Customary)			
K	°F	°C	5/9 (°F−32)
	°F	K	5/9 (°F+459.6)
Pressure			
Pa	atm	kPa	1.013E+02
	bar	kPa	1.000E+02
	lb/in²	kPa	6.895E+00
	inHg (60°F)	kPa	3.377E+00
	inH₂O (60°F)	kPa	2.488E−01
	mmHg (0°C)	kPa	1.333E−01
Density (Gases)			
kg/m³	lb/ft³	kg/m³	1.602E+01
Density (Liquids)			
kg/m³	lb/gal	kg/m³	1.198E+02
Density (Solids)			
kg/m³	lb/ft³	kq/m³	1.602E+01
Concentration (Mass/Volume)			
kg/m³	lb/bbl	kg/m³	2.853E+00
	g/gal	kg/m³	2.642E−01
	grains/gal	g/m³	1.712E+01
Concentration (Volume/Volume)			
m³/m³	ml/gal	dm³/m³	2.642E−01
Concentration (Mole/Volume)			
mol/m³	lb-mol/gal	kmol/m³	1.198E+02
	lb-mol/ft³	kmol/m³	1.602E+01
Energy, Work			
J	hp-h	MJ	2.685E+00
	kwh	MJ	3.600E+00
	Btu	kJ	1.055E+00
	kcal	kJ	4.184E+00
	ft-lb	J	1.356E+00
	erg	J	1.000E−07
Power			
W	hp	kW	7.457E−01
	ft–lb/s	W	1.359E+00
	kcal/h	W	1.162E+00
	Btu/h	W	2.931E−01
Fuel Consumption (Automotive)			
m³/m	gal/mi	dm³/100km	2.352E+02
	mi/gal	km/dm³	4.251E−01
Acceleration			
m/s²	ft/s²	m/s²	3.048E−01
Force			
N	lb	N	4.448E+00
	dyne	μN	1.000E+01

The "Système International (SI)" Of Metric Units ... continued

SI Unit	Customary Unit	Preferred Metric Unit	Conversion Factor
Velocity			
m/s	ft/s	m/s	3.048E−01
	mi/h	km/h	1.609E+00
Thermal Resistance			
K·m²/W	°F·ft²·h/Btu	°C·m²/W	1.761E−01
Heat Transfer Coefficient			
W/m²·K	Btu/h·ft²·°F	W/m²·°C	5.678E+00
Thermal Conductivity			
W/m·K	Btu/h·ft²·°F/ft	W/m·°C	1.162E+00
Viscosity (Dynamic)			
Pa·s	cP	Pa·s	1.000E−03
Viscosity (Kinematic)			
m²/s	cSt	m²/s	1.000E−06
Surface Tension			
N/m	dyne/cm	mN/m	1.000E+00

SI Nomenclature

Symbol	Name	Quantity	Remarks
A	ampere	electric current	base unit
a	annum (year)	time	equals 365 d
bar	bar	pressure	equals 10^5 Pa
°C	degree Celsius	temperature	equals K−273.15
d	day	time	equals 24 h
g	gram	mass	equals 10^{-3} kg
h	hour	time	equals 3600 s
ha	hectare	area	equals $10^4 m^2$
J	joule	work, energy	equals 1.0 N·m
K	kelvin	temperature (absolute)	base unit
kg	kilogram	mass	base unit
m	meter	length	base unit
mol	mole	amount of substance	base unit
N	newton	force	equals 1.0 kg·m/s²
P	poise	dynamic viscosity	equals 10^{-1} Pa·s
Pa	pascal	pressure	equals 1.0 N/m²
s	second	time	base unit
St	stoke	kinematic viscosity	equals 10^{-4} m²/s
t	tonne	mass	equals 10^3 kg
W	watt	power	equals 1.0 J/s

Decimal Multiples and Submultiples of SI Units

Multiplication Factors	Prefix	SI Symbol	Multiplication Factors	Prefix	SI Symbol
10^{18}	exa	E	10^{-1}	deci	d
10^{15}	pcta	P	10^{-2}	centi	c
10^{12}	tera	T	10^{-3}	milli	m
10^{9}	giga	G	10^{-6}	micro	u
10^{6}	mega	M	10^{-9}	nano	n
10^{3}	kilo	k	10^{-12}	pico	p
10^{2}	hecto	h	10^{-15}	femto	f
10^{1}	deka	da	10^{-16}	atto	a

C

Glossary of Engineering Terms

absolute humidity — see *humidity.*

absolute pressure — see *pressure.*

Absolute scale — see *temperature scales.*

absolute viscosity — the ratio of *shear stress* to *shear rate.* It is a fluid's internal resistance to flow. The common unit of absolute viscosity is the poise (see *viscosity).* Absolute viscosity divided by the fluid's density equals *kinematic viscosity.*

absorber oil — oil used to selectively absorb heavier hydrocarbon components from a gas mixture. Also called **wash oil** or **scrubber oil.**

absorption — the assimilation of one material into another; in petroleum refining, the use of an absorptive liquid to selectively remove components from a *process stream.*

AC — see *asphalt cement.*

acetylene — highly flammable hydrocarbon gas (C_2H_2), used in welding and cutting, and in plastics manufacture. Also a term for a series of unsaturated *aliphatic hydrocarbons,* each containing at least one triple carbon bond, the simplest member of the series being acetylene. The triple carbon bond makes acetylenes highly reactive. See *hydrocarbon, unsaturated hydrocarbon.*

$$H - C \equiv C - H$$
acetylene

acid — hydrogen-containing compound that reacts with metals to form salts, and with metallic oxides and bases to form a salt and water. The strength of an acid depends on the extent to which its molecules ionize, or dissociate, in water, and on the resulting concentration of hydrogen ions (H^+) in solution. Petroleum hydrocarbons, in the presence of oxygen and heat, can oxidize to form weak acids, which attack metals. See *corrosion.*

acidizing — treatment of underground oil-bearing formations with acid in order to increase production. Hydrochloric or other acid is injected into the formation and held there under pressure until it etches the rock, thereby enlarging the pore spaces and passages through which the oil flows. The acid is then pumped out and the well is swabbed and put back into production.

acid number — see *neutralization number.*

acid treating — refining process for improving the color, odor, and other properties of white oils or lube stocks, whereby the unfinished product is contacted with sulfuric acid to remove the less stable hydrocarbon molecules.

acid wash color — an indication of the presence of *olefins* and *polar compounds* in petroleum *solvents.* A sample of solvent is mixed with sulfuric acid and let stand until formation of an acid layer, the color of which is compared against color standards.

acrylic resin — any of a group of thermoplastic *resins* formed from the polymerization (see *polymer)* of acrylic acid, methac-

438

rylic acid, *esters* of these acids, or acrylonitrile. It is used in the manufacture of lightweight, weather-resistant, exceptionally clear plastics.

acute effect — toxic effect in mammals and aquatic life that rapidly follows exposure to a toxic substance. An acute effect is usually evident after a single oral intake, a single contact with the skin or eyes, or a single exposure to contaminated air lasting any period up to eight hours. Also known as **acute toxicity.**

acute toxicity — see *acute effect.*

additive — chemical substance added to a petroleum product to impart or improve certain properties. Common petroleum product additives are: *anti-foam agent, anti-icing additive, anti-wear additive, corrosion inhibitor, demulsifier, detergent, dispersant, emulsifier, EP additive, oiliness agent, oxidation inhibitor, pour point depressant, rust inhibitor, tackiness agent, viscosity index (V.I.) improver.*

adiabatic compression — compression of a gas without extraction of heat, resulting in increased temperature. The temperature developed in compression of a gas is an important factor in lubrication since oil deteriorates more rapidly at elevated temperatures. *Oxidation inhibitors* help prevent rapid lubricant breakdown under these conditions.

adjuvant — a part of a pesticide formulation that helps or adds to the action of the active ingredient. Petroleum products are sometimes used as adjuvants.

adsorption — adhesion of the molecules of gases, liquids, or dissolved substances to a solid surface, resulting in relatively high concentration of the molecules at the place of contact; e.g. the plating out of an *anti-wear additive* on metal surfaces. Also, any refining process in which a gas or a liquid is contacted with a solid, causing some compounds of the gas or liquid to adhere to the solid; e.g. contacting of lube oils with activated clay to improve color. See *clay filtration.*

aerosol — a highly dispersed suspension of fine solid or liquid particles in a gas. Petroleum solvents are commonly used either as carriers or as vapor pressure depressants in packaged aerosol specialty products.

aftercooling — the process of cooling compressed gases under constant pressure after the final stage of compression. See *intercooling.*

afterrunning — the continued running of a spark-ignited engine after the ignition is turned off; also known as **dieseling.** There are two basic causes of afterrunning: **surface ignition** and **compression ignition.** In surface ignition, the surfaces of the combustion chamber remain hot enough to provide a source of ignition after the spark ignition is terminated. In compression ignition, the conditions of temperature, pressure, fuel composition, and engine idle speed allow ignition to continue.

age hardening — increase in the consistency of a lubricating grease with storage time.

AGMA — American Gear Manufacturers Association, which as one of its activities establishes and promotes standards for gears and lubricants.

AGMA lubricant numbers — AGMA specifications covering gear lubricants. The *viscosity* ranges of the AGMA numbers conform to the International Standards Organization (ISO) viscosity classification system (see *ISO viscosity classification system*). AGMA numbers and their viscosity ranges are as follows:

AGMA Lubricant Number	Corresponding ISO Grade	Viscosity Range cSt @ 40°C
1	46	41.4 - 50.6
2	68	61.2 - 74.8
3	100	90 - 110
4	150	135 - 165
5	220	198 - 242
6	320	288 - 352
6 Compounded	460	414 - 506
8 Compounded	680	612 - 748
8A Compounded	1000	900 - 1000
2 EP	68	61.2 - 74.8
4 EP	150	135 - 165
5 EP	220	198 - 242
6 EP	320	288 - 352
6 EP	460	414 - 506
8 EP	680	612 - 748

alcohol — any of a class of chemical compounds containing an hydroxyl (OH) group and having the general formula $C_n H_{2n+1}$ OH; e.g., *methanol*, CH_3 OH; *ethanol*, $C_2 H_5$ OH.

aliphatic hydrocarbon — hydrocarbon in which the carbon atoms are joined in open chains, rather than rings. See *hydrocarbon, normal paraffin.*

alkali — a hydroxide or carbonate of an alkali metal (e.g. lithium, sodium, potassium, etc.), the aqueous solution of which is characteristically basic in chemical reactions. The term may be extended to apply to hydroxides and carbonates of barium, calcium, magnesium, and the ammonium ion. See *base.*

alkyl — any of a series of monovalent *radicals* having the general formula C_nH_{2n+1}, derived from *aliphatic hydrocarbons* by the removal of a hydrogen atom; for example, CH_3- (methyl radical, from methane).

alkylate — product of an *alkylation* process.

alkylated aromatic — benzene-derived *synthetic lubricant* base with good hydrolytic stability (resistance to chemical reaction with water) and good compatibility with *mineral oils.* Used in turbines, compressors, jet engines, and hydraulic power steering.

alkylation — in refining, the chemical reaction of a low-molecular-weight *olefin* with an *isoparaffin* to form a liquid product, **alkylate,** that has a high *octane number* and is used to improve the *antiknock* properties of gasoline. The reaction takes place in the presence of a strong acid *catalyst,* and at controlled temperature and pressure. Alkylation less commonly describes certain other reactions, such as that of an olefin with an *aromatic* hydrocarbon.

ambient — pertaining to any localized conditions, such as temperature, humidity, or atmospheric pressure, that may affect the operating characteristics of equipment or the performance of a petroleum product; e.g. a high ambient temperature may cause gasoline vapor lock in an automobile engine.

American Gear Manufacturers Association — *see AGMA.*

American National Standards Institute — see *ANSI.*

American Petroleum Institute — see *API*.

American Society for Testing and Materials — see *ASTM*.

American Society of Lubrication Engineers — see *ASLE*.

amphoteric — having the capacity to behave as either an *acid or base*; e.g., aluminum hydroxide, Al(OH)₃, which neutralizes acids to form aluminum salts and reacts with strong bases to form aluminates.

anesthetic effect — the loss of sensation with or without the loss of consciousness. It can be caused by the inhalation of volatile hydrocarbons.

anhydrous — devoid of water.

aniline point — lowest temperature at which a specified quantity of aniline (a benzene derivative) is soluble in a specified quantity of a petroleum product, as determined by test method ASTM D 611; hence, an empirical measure of the solvent power of a hydrocarbon — the lower the aniline point, the greater the solvency. Paraffinic hydrocarbons have higher aniline points than *aromatic* types. See *paraffin*.

anionic emulsified asphalt — see *emulsified anionic asphalt*.

ANSI (American National Standards Institute) — organization of industrial firms, trade associations, technical societies, consumer organizations, and government agencies, intended to establish definitions, terminologies, and symbols; improve methods of rating, testing, and analysis; coordinate national safety, engineering and industrial standards; and represent U.S. interests in international standards work.

anti-foam agent — one of two types of *additives* used to reduce foaming in petroleum products: silicone oil to break up large surface bubbles, and various kinds of *polymers* that decrease the amount of small bubbles entrained in the oils. See *foaming, entrainment*.

anti-icing additive — substance added to gasoline to prevent ice formation on the throttle plate of a *carburetor*. Anti-icing *additives* are of two types: those that lower the freezing point of water, and those that alter the growth of ice crystals so that they remain small enough to be carried away in the air stream. See *carburetor icing*.

antiknock — resistance of a gasoline to *detonation* in a *combustion chamber*. See *knock, octane number*.

antiknock compounds — substances which raise the antiknock quality of a gasoline, as expressed by *octane number*. Historically, tetraethyl lead (see *lead alkyl*) has been the most common antiknock compound, but its use is being phased out under Environmental Protection Agency *(EPA)* regulations. Other additives of the oxygenated organic type — e.g., tertiary butyl alcohol (TBA) and methyl tertiary-butyl ether (MTBE) — are coming into increasing use as octane boosters in gasoline.

anti-oxidant — see *oxidation inhibitor*.

anti-seize compound — grease-like substance containing graphite or metallic solids, which is applied to threaded joints, particularly those subjected to high temperatures, to facilitate separation when required.

anti-wear additive — *additive* in a lubricant that reduces friction and excessive wear. See *boundary lubrication*.

API (American Petroleum Institute) — trade association of petroleum producers, refiners, marketers, and transporters, organized for the advancement of the petroleum industry by conducting research, gathering and disseminating information, and maintaining cooperation between government and the industry on all matters of mutual interest. One API technical activity has been the establishment of *API Engine Service Categories* for lubricating oils.

API Engine Service Categories — gasoline and diesel engine oil quality levels established jointly by *API, SAE,* and *ASTM,* and sometimes called SAE or API/SAE categories; formerly called **API Engine Service Classifications.** API Service Categories are as follows:

Service Station Oils

SA	straight mineral oil (no additives)
SB	anti-oxidant, anti-scuff, but non-detergent
SC	protection against high- and low-temperature deposits, wear, rust, and corrosion; meets car makers' warranty requirements for 1964-1967 models
SD	improved protection over SC oils; meets warranty requirements for 1968-1971 models
SE	improved protection over SD oils; meets warranty requirements for 1972-1980 models
SF	improved anti-wear and anti-oxidation; meets warranty requirements for 1980 and later models

Commercial Oils (Diesel Engine)

CA	light-duty service; meets obsolete Military Specification MIL-L-2104A
CB	moderate-duty; meets MIL-L-2104A, Supplement 1
CC	moderate-to-severe duty; meets obsolete Military Specification MIL-L-2104B
CD	severe duty; highest protection against high- and low-temperature deposits, wear, rust, corrosion; meets Military Specification MIL-L-2104C

API gravity — see *specific gravity*.

apparent viscosity — viscosity of a fluid that holds only for the *shear rate* (and temperature) at which the viscosity is determined. See *shear stress, Brookfield viscosity*.

Aromatic Structures

Basic Aromatic Ring

Two aromatic rings joined together with a paraffinic side chain. Larger molecules are formed by adding more rings and side chains.

aromatic — *unsaturated hydrocarbon* identified by one or more *benzene* rings or by chemical behavior similar to benzene. The benzene ring is characterized by three double bonds alternating with single bonds between carbon atoms (compare with *olefins*). Because of these multiple bonds, aromatics are usually more reactive and have higher solvency than *paraffins* and *naphthenes*. Aromatics readily undergo electrophylic substitution; that is, they react to add other active molecular groups, such as nitrates, *sulfonates*, etc. Aromatics are used extensively as *petrochemical* building blocks in the manufacture of pharmaceuticals, dyes, plastics, and many other chemicals.

aryl — any organic group derived from an *aromatic* hydrocarbon by the removal of a hydrogen atom, for example C_6H_5- (phenyl *radical*, from *benzene*).

ash content — noncombustible residue of a lubricating oil or fuel, determined in accordance with test methods ASTM D 482 and D 874 (sulfated ash). Lubricating oil detergent additives contain metallic derivatives, such as barium, calcium, and magnesium sulfonates, that are common sources of ash. Ash deposits can impair engine efficiency and power. See *detergent*.

ashless disperant — see *dispersant*.

askarel — generic term for a group of synthetic, fire-resistant, chlorinated *aromatic* hydrocarbons used as electrical insulating liquids. Gases produced in an askarel by arcing conditions consist predominantly of noncombustible hydrogen chloride, with lesser amounts of combustible gases. Manufacture of askarels has been discontinued in the U.S. because of their toxicity. See *PCB*.

ASLE (American Society of Lubrication Engineers) — organization intended to advance the knowledge and application of lubrication and related sciences.

asperities — microscopic projections on metal surfaces resulting from normal surface-finishing processes. Interference between opposing asperities in sliding or rolling applications is a source of friction, and can lead to metal welding and *scoring*. Ideally, the lubricating film between two moving surfaces should be thicker than the combined height of the opposing asperities. See *boundary lubrication, EP additive*.

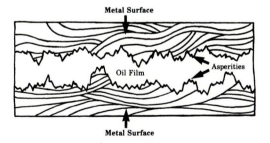

Metal Surface

Oil Film — Asperities

Metal Surface

asphalt — brown-to-black, bituminous material (see *bitumen)* of high molecular weight, occurring naturally or as a residue from the distillation of crude oil; used as a bonding agent in road building, and in numerous industrial applications, including the manufacture of roofing. **Blown,** or **oxidized, asphalt** is produced by blowing air through asphalt at high temperatures, producing a tougher, more durable asphalt. See also *asphalt cement, asphaltenes, emulsified anionic asphalt, emulsified cationic asphalt, penetration grading (asphalt), reclaimed asphalt pavement, recycling of asphalt paving, viscosity (asphalt), viscosity grading (asphalt)*.

asphalt cement (AC) — asphalt refined to meet specifications for paving and special purposes. Specifications are established by *ASTM* and the American Association of State Highway Transportation Officials (AASHTO).

asphaltenes — high-molecular-weight hydrocarbon components of *asphalt* and heavy residual stocks (see *bottoms*) that are soluble in carbon disulfide but not in paraffinic naphtha.

asphalt grading — see *penetration grading (asphalt), viscosity grading (asphalt)*.

asphaltic — containing significant amounts of *asphaltenes*.

Asphalt Institute — an international, non-profit association sponsored by members of the petroleum asphalt industry that serves both users and producers of asphaltic materials through programs of engineering service, research, and education.

asphalt paving — see *asphalt cement*.

asphalt recycling — see *recycling of asphalt paving*.

aspiration — drawing of air at atmospheric pressure into a *combustion chamber;* as opposed to supercharging or turbocharging. See *supercharger*.

ASTM (American Society for Testing and Materials) — organization devoted to "the promotion of knowledge of the materials of engineering, and the standardization of specifications and methods of testing". A preponderance of the data used to describe, identify, or specify petroleum products is determined in accordance with ASTM test methods.

ASTM scale (D 1500) — see *color scale*.

ATF — see *automatic transmission fluid*.

atmosphere — unit of *pressure* equal to 101.3 kilopascals (kPa), or 14.7 pounds per square inch (psi), or 760 mm (29.9 in) of mercury; standard atmospheric pressure at sea level.

atmospheric pollutants — see *pollutants*.

atmospheric pressure — see *pressure*.

atomization — the reduction of a liquid into fine particles or spray. Atomization of a fuel is necessary for efficient combustion. Atomization is accomplished by the *carburetor* or the *fuel injection* system in *internal combustion engines,* and by special steam or air atomizers in furnaces and boilers.

austempering — see *quenching*.

auto-ignition temperature — lowest temperature at which a flammable gas or vaporized liquid will ignite in the absence of a spark or flame, as determined by test method ASTM D 2155; not to be confused with *flash point* or *fire point*, which is typically lower. Auto-ignition temperature is a critical factor in heat transfer oils and transformer oils, and in solvents used in high-temperature applications.

automatic transmission fluid (ATF) — fluid for automatic hydraulic transmissions in motor vehicles; also called **hydraulic transmission fluid.** Automatic transmission fluids must have a suitable *coefficient of friction,* good low-temperature viscosity, and anti-wear properties. Other necessary properties are: high *oxida-*

tion stability, anti-corrosion, anti-foaming, and compatibility with *synthetic rubber* seals. See *corrosion, foaming.*

automotive emissions — see *emissions (automotive).*

aviation gasoline (avgas) — high-quality gasoline manufactured under stringent controls to meet the rigorous performance and safety requirements of piston-type aircraft engines. Volatility of aviation gasoline is closely controlled since, in most aircraft engines, excessive volatility can lead to *vapor lock.* Aviation gasolines are formulated to resist chemical degradation and to prevent fuel system corrosion. There are two basic grades of aviation gaso-

lines (based on their *antiknock* value): 80 (80 lean/87 rich) and 100 (100 lean/130 rich). Aviation gasoline has different properties than *turbo fuel,* which fuels gas-turbine-powered aircraft. See *lean and rich octane numbers.*

azeotrope — liquid mixture of two or more components that boils at a temperature either higher or lower than the boiling point of any of the individual components. In refining, if the components of a solution are very close in boiling point and cannot be separated by conventional *distillation,* a substance can be added that forms an azeotrope with one component, modifying its boiling point and making it separable by distillation.

bactericide — additive included in the formulations of water-mixed *cutting fluids* to inhibit the growth of bacteria promoted by the presence of water, thus preventing the unpleasant odors that can result from bacterial action.

barrel — standard unit of measurement in the petroleum industry, equivalent to 42 standard U.S. gallons.

base — any of a broad class of compounds, including *alkalis,* that react with *acids* to form salts, plus water. Also known as hydroxides. Hydroxides ionize in solution to form hydroxyl ions (OH-); the higher the concentration of these ions, the stronger the base. Bases are used extensively in petroleum refining in *caustic washing* of *process streams* to remove acidic impurities, and are components in certain *additives* that neutralize weak acids formed during *oxidation.*

base coat — see *launching lubricant.*

base number — see *neutralization number.*

base stock — a primary refined petroleum fraction, usually a lube oil, into which *additives* and other oils are blended to produce finished products. See *distillation.*

basin — trough-like geological area, the former bed of an ancient sea. Because basins consist of sedimentary rock and have contours that provide traps for petroleum, they are considered good prospects for exploration.

bearing — basic machine component designed to reduce friction between moving parts and to support moving loads. There are two main types of bearings: (1) **rolling contact bearings** (commonly ball or roller), and (2) **sliding (plain) bearings,** either plain journal (a metal jacket fully or partially enclosing a rotating inner shaft) or pad-type bearings, for linear motion. Rolling-contact bearings are more effective in reducing friction. With few exceptions, bearings require lubrication to reduce wear and extend bearing life.

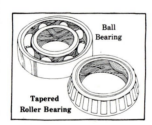

Ball Bearing

Tapered Roller Bearing

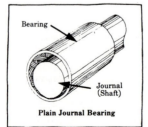

Bearing

Journal (Shaft)

Plain Journal Bearing

benzene — aromatic hydrocarbon consisting of six carbon atoms and six hydrogen atoms arranged in a hexagonal ring structure. See *aromatic, hydrocarbon.* It is used extensively in the *petrochemical* industry as a chemical intermediate and reaction *diluent* and in some applications as a *solvent.* Benzene is a toxic substance, and proper safety precautions should be observed in handling it.

benzene

bevel gear — see *gears.*

bhp — brake horsepower, the effective or available power of an engine or turbine, measured at the output shaft. It is equivalent to the calculated horsepower, less the power lost in friction.

bitumen — any of various mixtures of viscous, brown-to-black hydrocarbons, such as *asphalt* and tar, together with any accompanying non-metallic derivatives, such as sulfur or nitrogen compounds; may occur naturally or may be obtained as residues from refining processes.

black oil — lubricant containing *asphaltic* materials, used in heavy-duty equipment applications, such as mining and quarrying, where extra adhesiveness is desired.

bleeding — the separation of liquid lubricant from a lubricating grease. See *syneresis.*

block grease — very firm grease manufactured in block form to be applied to certain large open plain bearings, generally operating at slow speeds and moderate temperatures. See *bearing.*

blocking point — lowest temperature at which waxed papers stick together, or block, sufficiently to injure the surface films and performance properties, as determined by test method ASTM D 1465.

block penetration — see *penetration (grease).*

blow-by — in an *internal combustion engine,* seepage of fuel and gases past the piston rings and cylinder wall into the crankcase, resulting in crankcase oil dilution and *sludge* formation. See *positive crankcase ventilation, dilution of engine oil.*

blown asphalt — see *asphalt.*

blown rapeseed oil — see *rapeseed oil.*

blowout — uncontrolled eruption of gas, oil, or other fluids from a well to the atmosphere.

blowout preventer — equipment installed at the *wellhead* to prevent the escape of pressure, or pressurized material, from the drill hole.

BMEP — brake mean effective pressure, the theoretical average pressure that would have to be imposed on the pistons of a frictionless engine (of the same dimensions and speed) to produce the same power output as the engine under consideration; a measure of how effectively an engine utilizes its piston displacement to do work.

boiling range — temperature spread between the initial boiling point and final boiling point. See *distillation test*.

bomb oxidation stability — measure of the *oxidation stability* of greases and lubricating oils in separate tests: ASTM D 942 (grease), ASTM D 2272 (oil) and ASTM D 2112 (insulating oils). In all tests, the sample is placed in a container, or bomb, which is then charged with oxygen and pressurized; a constant elevated temperature is maintained. ASTM D 2272 utilizes a rotating bomb, which is placed in a heated bath; the test therefore is commonly called the **rotary bomb oxidation test.** Oxidation stability is expressed in terms of pressure drop in a given time period (D 942) or in terms of the time required to achieve a specified pressure drop (D 2272, D 2112).

borehole — the hole made by drilling, or boring, a well; also called **wellbore.** See *rotary drilling*.

bottled gas — a gas pressurized and stored in a transportable metal container. See *LPG*.

bottoms — in refining, the high-boiling residual liquid (also called **residuum**) — including such components as heavy fuels and asphaltic substances — that collects at the bottom of a distillation column, such as a pipe still. See *distillation, asphalt, fuel oil*.

boundary lubrication — form of lubrication between two rubbing surfaces without development of a full-fluid lubricating film. See *full-fluid-film lubrication, ZN/P curve*. Boundary lubrication can be made more effective by including additives in the lubricating oil that provide a stronger oil film, thus preventing excessive friction and possible *scoring*. There are varying degrees of boundary lubrication, depending on the severity of service. For mild conditions, *oiliness agents* may be used; these are *polar compounds* that have an exceptionally high affinity for metal surfaces. By plating out on these surfaces in a thin but durable film, oiliness agents prevent scoring under some conditions that are too severe for a *straight mineral oil. Compounded oils,* which are formulated with polar *fatty oils,* are sometimes used for this purpose. *Anti-wear additives* are commonly used in more severe boundary lubrication applications. High quality motor oils contain anti-wear additives to protect heavily loaded engine components, such as the valve train. The more severe cases of boundary lubrication are defined as extreme pressure conditions; they are met with lubricants containing *EP additives* that prevent sliding surfaces from fusing together at high local temperatures and pressures.

BR — see *polybutadiene rubber*.

brake horsepower — see *bhp*.

breakdown voltage — see *dielectric strength*.

bright stock — high viscosity oil, highly refined and dewaxed, produced from residual stocks, or *bottoms;* used for blending with lower viscosity oils.

British thermal unit — see *Btu*.

bromine index — number of milligrams of bromine that will react with 100 grams of a petroleum product (test method ASTM D 2710). Bromine index is essentially equivalent to *bromine number* × 1000.

bromine number — number of grams of bromine that react with 100 grams of a sample of a petroleum *distillate,* giving an indication of its relative degree of reactivity, as determined by test method ASTM D 1159; it can be used as an indicator of the relative amount of *olefins* and *diolefins,* which are double-bonded straight-chain or cyclic hydrocarbons. See *hydrocarbon*.

Brookfield viscosity — *apparent viscosity* of an oil, as determined under test method ASTM D 2983. Since the apparent viscosity of a *non-Newtonian fluid* holds only for the *shear rate* (as well as temperature) at which it is determined, the Brookfield viscometer provides a known rate of shear by means of a spindle of specified configuration that rotates at a known constant speed in the fluid. The torque imposed by fluid friction can be converted to absolute viscosity units (centipoises) by a multiplication factor. See *viscosity, shear stress*. The viscosities of certain petroleum waxes and wax-*polymer* blends in the molten state can also be determined by the Brookfield test method ASTM D 2669.

BS&W — abbreviation of "bottoms sediment and water", the water and other extraneous material present in crude oil. Normally, the BS&W content must be quite low before the oil is accepted for pipeline delivery to a refinery.

Btu (British thermal unit) — quantity of heat required to raise the temperature of one pound of water one degree Fahrenheit, at 60°F and at a pressure of one *atmosphere*. See *energy*.

bulk appearance — visual appearance of grease when the undisturbed surface is viewed in an opaque container. Bulk appearance should be described in the following terms: **smooth** — a surface relatively free of irregularities; **rough** — a surface composed of many small irregularities; **grainy** — a surface composed of small granules or lumps of soap particles; **cracked** — showing surface cracks of appreciable number and magnitude; **bleeding** — showing free oil on the surface of the grease (or in the cracks of a cracked grease). See *texture*.

bulk delivery — large quantity of unpackaged petroleum product delivered directly from a tank truck, tank car, or barge into a consumer's storage tank.

bulk modulus — measure of a fluid's resistance to compressibility; the reciprocal of compressibility.

bulk odor — odor of vapor emanating from bulk liquid quantities of a petroleum product; also referred to as **impact odor.** The odor remaining after the product has evaporated is called **residual odor.**

Buna-N — See *nitrile rubber*.

Bunker C fuel oil — see *fuel oil*.

butadiene rubber — see *polybutadiene rubber*.

butane — gaseous paraffinic hydrocarbon ($C_4 H_{10}$), usually a mixture of iso- and normal butane (see *isomer, normal paraffin*); also called, along with *propane,* liquefied petroleum gas *(LPG).*

$$H-\overset{\displaystyle H}{\underset{\displaystyle H}{C}}-\overset{\displaystyle H}{\underset{\displaystyle H}{C}}-\overset{\displaystyle H}{\underset{\displaystyle H}{C}}-\overset{\displaystyle H}{\underset{\displaystyle H}{C}}-H$$

normal butane

butylene — any of three isomeric (see *isomer*) flammable, gaseous hydrocarbons of the molecular structure $C_4 H_8$; commonly derived from hydrocarbon *cracking.*

butyl rubber (IIR) — *synthetic rubber,* produced by copolymerization of isobutylene with isoprene or butadiene (see *polymer*). It is resistant to weather and heat, has low air-permeability and low resiliency; used in the manufacture of cable insulation, tubeless tire innerliners and other applications requiring good weather resistance and air retention.

°C (Celsius) — see *temperature scales.*

calcium soap grease — see *grease.*

calorie — term applicable either to the **gram calorie** or the **kilocalorie.** The gram calorie is defined as the amount of heat required at a pressure of one *atmosphere* to raise the temperature of one gram of water one degree Celsius at 15°C. The kilocalorie is the unit used to express the energy value of food; it is defined as the amount of heat required at a pressure of one atmosphere to raise the temperature of one kilogram of water one degree Celsius; it is equal to 1000 gram calories. See *energy.*

calorific value — see *heat of combustion.*

carbonizable substances — petroleum components which can be detected in *white oil, petrolatum,* and *paraffin wax* when any of these products is mixed with concentrated sulfuric acid, causing the acid to discolor, as outlined in test method ASTM D 565 (white oil and petrolatum) or ASTM D 612 (paraffin wax).

carbon monoxide (CO) — colorless, odorless, poisonous gas, formed by the incomplete combustion of any carbonaceous material (e.g. gasoline, wood, coal). CO is the most widely distributed and most commonly occurring air pollutant, with motor vehicles being the primary source of man-made emissions, although emission controls are reducing the automobile's contribution. It is estimated that more than 90% of atmospheric CO comes from natural sources, such as decaying organic matter. See *catalytic converter, emissions (automotive), pollutants.*

carbon residue — percent of coked material remaining after a sample of oil has been exposed to high temperatures under test method ASTM D 189 (**Conradson**) or D 524 (**Ramsbottom**); hence, a measure of coke-forming tendencies. Results should be interpreted cautiously, as there may be little similarity between test conditions and actual service conditions.

carbon type analysis — empirical analysis of rubber process oil composition that expresses the percentages of carbon atoms in aromatic, naphthenic and paraffinic components, respectively. See *rubber oil, aromatic , naphthene, paraffin.*

carbonyl — the divalent *radical* CO, which occurs in various organic substances, such as organic acids; also, any metal compound containing this radical. Carbonyls are highly reactive and considered to be catalyst poisons when present in solvents used as reaction *diluents* in *polyolefin* plastics manufacture.

carburetor — device in an internal combustion engine that atomizes and mixes fuel with air in the proper proportion for efficient combustion at all engine speeds, and controls the engine's power output by throttling, or metering, the air-fuel mixture admitted to the cylinders. The automobile carburetor is a complex mechanism designed to compensate for many variables over a wide range of speeds and loads. Intake air is drawn through the **venturi,** a

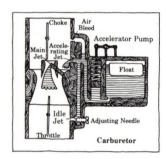

Carburetor

constricted throat in the air passage that causes a pressure reduction in the air stream, which draws fuel from the carburetor bowl through either the **main jet** or the **idle jet.** The fuel is atomized by the high-velocity air, and the resulting air-fuel mixture is piped through the intake manifold to the individual cylinders, where it is burned. A **throttle plate** between the venturi and the cylinders controls power and speed by controlling the volume of air-fuel mixture reaching the cylinders. In most carburetors, closing of this (venturi) throttle valve shuts down the main jet and activates the idle jet, which provides the fuel-rich mixture that idling requires. An **accelerator pump** in the carburetor provides momentary fuel enrichment when the accelerator pedal is depressed rapidly, to compensate for the sudden influx of air. During cold starting, a **choke** (or butterfly valve) restricts airflow to the carburetor, thus enriching the mixture for faster starting. The choke on most automotive engine carburetors is operated automatically by a thermostatic spring, which opens the choke as the engine warms up. See *fuel injection, supercharger.*

carburetor icing — freezing of the moisture in humid air inside the *carburetor,* restricting air supply to the engine and causing it to stall. The air is brought to freezing by the chilling effect of vaporizing fuel. Carburetor icing is most likely to occur when the air temperature is between 3° and 13°C (38° to 55°F); if the *ambient* temperature were higher, the moisture would not freeze, and if it were lower, the absolute humidity would not provide sufficient moisture. Carburetor icing can be prevented by using a gasoline with an *anti-icing additive.*

carcinogen — cancer-causing substance.

carrier — a liquid, such as water, solvent, or oil, in which an active ingredient is dissolved or dispersed.

CAS (Chemical Abstract Service) Registry Numbers — identifying numbers assigned to chemical substances by the Chemical Abstract Service of the American Chemical Society and used by the Environmental Protection Agency *(EPA)* to aid in registering chemicals under the federal Toxic Substances Control Act (TSCA) of 1976. CAS numbers are assigned to generic refinery *process streams,* such as *kerosene* and lube *base stocks,* that contain no *additives.* Petroleum products containing additives are termed "mixtures" by the TSCA and, as such, do not have CAS numbers. All chemical substances used in such mixtures are assigned CAS numbers and must be listed with the EPA by the refiner or the additive supplier.

casing — steel pipe placed in a *borehole* as drilling progresses, to prevent the wall of the hole from caving in. See *rotary drilling.*

casinghead gasoline — see *natural gasoline.*

catalyst — substance that causes or speeds up a chemical reaction without itself undergoing an associated change; catalysts are important in a number of refining processes.

catalytic converter — an emissions control device, incorporated into an automobile's exhaust system, containing catalysts — such as platinum, palladium, or rhodium — that reduce the levels of *hydrocarbons* (HC), *carbon monoxide* (CO), and — in more recent designs — *nitrogen oxides* (NOx) emitted to the air. In the catalytic converter, HC and CO are oxidized to form carbon dioxide (CO_2), and NOx are reduced to nitrogen and oxygen. Three-way catalytic converters that control all three substances require associated electronic controls for precise regulation of oxygen levels in the exhaust gas. Catalytic converters are also effective in removing *PNA* (polynuclear aromatic) hydrocarbons. Cars equipped with catalytic converters require unleaded gasoline, since the lead in tetraethyl lead, an *antiknock compound,* is a catalyst "poison". See *emissions (automotive), hydrocarbon emissions, pollutants, lead alkyl.*

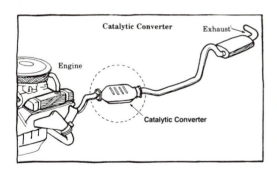

catalytic cracking — in refining, the breaking down at elevated temperatures of large, high-boiling hydrocarbon molecules into smaller molecules in the presence of a *catalyst.* The principal application of catalytic cracking is the production of high-octane gasoline, to supplement the gasoline produced by *distillation* and other processes. Catalytic cracking also produces heating oil components and hydrocarbon *feedstocks,* such as *propylene* and *butylene,* for *polymerization, alkylation,* and *petrochemical* operations.

cationic emulsified asphalt — see *emulsified cationic asphalt.*

caustic washing (scrubbing) — treatment of a petroleum liquid or gas with a caustic alkaline material (e.g., sodium hydroxide) to remove *hydrogen sulfide,* low-weight *mercaptans,* and other acidic impurities.

cavitation — formation of an air or vapor pocket (or bubble) due to lowering of pressure in a liquid, often as a result of a solid body, such as a propeller or piston, moving through the liquid; also, the pitting or wearing away of a solid surface as a result of the collapse of a vapor bubble. Cavitation can occur in a *hydraulic system* as a result of low fluid levels that draw air into the system, producing tiny bubbles that expand explosively at the pump outlet, causing metal erosion and eventual pump destruction. Cavitation can also result when reduced pressure in lubricating grease dispensing systems forms a void, or cavity, which impedes suction and prevents the flow of greases.

Celsius (°C) — see *temperature scales.*

centigrade — see *temperature scales.*

centipoise — see *viscosity.*

centistoke — see *viscosity.*

centralized lubrication — automatic dispensing of grease or oil from a reservoir to the lubricated parts of one or more machines. Flow is maintained by one or more pumps, and the amount of lubricant supplied to each point can be regulated by individual metering devices. Such a system provides *once-through lubrication.*

cetane — colorless liquid hydrocarbon, $C_{15}H_{34}$, used as a standard in determining *diesel fuel* ignition performance. See *cetane number.*

cetane improver — additive for raising the *cetane number* of a *diesel fuel.*

cetane index — an approximation of *cetane number* based on API gravity (see *specific gravity*) and mid-boiling point (see *distillation test*) of a *diesel fuel.* See *diesel index.*

cetane number — measure of the ignition quality of a *diesel fuel,* expressed as the percentage of cetane that must be mixed with liquid methylnaphthalene to produce the same ignition performance as the diesel fuel being rated, as determined by test method ASTM D 613. A high cetane number indicates shorter ignition lag and a cleaner burning fuel. See *cetane, cetane index, diesel index.*

channeling — formation of a channel in lubricating grease by a lubricated element, such as a gear or rolling contact bearing, leaving shoulders of grease that serve as a seal and reservoir. This phenomenon is usually desirable, although a channel that is too deep or permanent could cause lubrication failure.

Chemical Abstract Service Registry Numbers — see *CAS Registry Numbers*

chlorine — see *halogen.*

Christmas tree — structure of valves, fittings, and pressure gauges at the top of a well to control the flow of oil and gas.

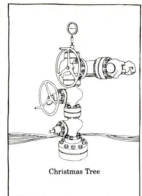

Christmas Tree

chronic effect — cumulative physiological damage resulting from prolonged exposure or series of exposures to a toxic substance. Also known as **chronic toxicity.**

chronic toxicity — see *chronic effect.*

circulating lubrication system — system in which oil is recirculated from a sump or tank to the lubricated parts, in most cases requiring a pump to maintain circulation. Circulating lubrication makes possible extended lubricant use, and usually requires a high-quality rust-and-oxidation-inhibited *(R&O)* oil.

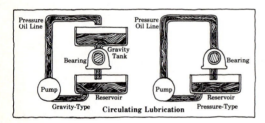

Gravity-Type   **Circulating Lubrication**   Pressure-Type

clay filtration — refining process using fuller's earth (activated clay) or bauxite to adsorb minute solids from lubricating oil, as well as remove traces of water, acids, and *polar compounds*. See *adsorption.*

clay/silica gel analysis — composition analysis test (ASTM D 2007), for determining weight percent of *asphaltenes, saturated hydrocarbons, aromatics,* and *polar compounds* in a petroleum product. The test material is mixed with pentane, asphaltenes are extracted as pentane insolubles, and the remainder of the sample is washed through a column. Active clay at the top of the column separates polar compounds, silica gel at the bottom separates aromatics, and saturates *(naphthenes* and *paraffins)* pass through with the pentane.

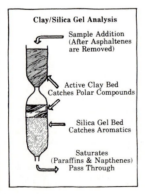

Clay/Silica Gel Analysis

Sample Addition (After Asphaltenes are Removed)

Active Clay Bed Catches Polar Compounds

Silica Gel Bed Catches Aromatics

Saturates (Paraffins & Napthenes) Pass Through

Cleveland Open Cup (COC) — test (ASTM D 92) for determining the *flash point* and *fire point* of all petroleum products except fuel oil and products with flash points below 79°C (175°F). The oil sample is heated in a precisely specified brass cup containing a thermometer. At specified intervals a small flame is passed across the cup. The lowest temperature at which the vapors above the cup briefly ignite is the flash point; the temperature at which the vapors sustain combustion for at least five seconds is the fire point. See *Tag open cup.*

closed cup — method for determining the *flash point* of fuels, *solvents,* and *cutback asphalts,* utilizing a covered container in which the test sample is heated and periodically exposed to a small flame introduced through a shuttered opening. The lowest temperature at which the vapors above the sample briefly ignite is the flash point. See *Pensky-Martens closed tester, Tag closed tester.*

cloud point — temperature at which a cloud or haze of wax crystals appears at the bottom of a sample of lubricating oil in a test jar, when cooled under conditions prescribed by test method ASTM D 2500. Cloud point is an indicator of the tendency of the oil to plug filters or small orifices at cold operating temperatures. It is very similar to *wax appearance point.*

CMS asphalt — see *emulsified cationic asphalt.*

coastal oil — common term for any predominately *naphthenic* crude derived from fields in the Texas Gulf Coast area.

coefficient of friction — see *friction.*

cohesion — molecular attraction causing substances to stick together; a factor in the resistance of a lubricant, especially a grease, to flow.

cold-end corrosion — corrosion due to the acid-forming condensation of sulfur trioxide (SO_3) on cool surfaces of a boiler, especially the cooler parts of the chimney and the air heater; also called **low-temperature corrosion.** It can be prevented or minimized by using resistant alloys, by operating at low excess air levels (which reduces SO_3 production), or by operating at higher stack temperatures.

cold-flow improver — additive to improve flow of *diesel fuel* in cold weather. In some instances, a cold-flow improver may improve operability by modifying the size and structure of the wax crystals that precipitate out of the fuel at low temperatures, permitting their passage through the fuel filter. In most cases, the additive depresses the *pour point*, which delays agglomeration of the wax crystals, but usually has no significant effect on diesel engine performance. A preferred means of improving cold flow is to blend kerosene with the diesel fuel, which lowers the *wax appearance point* by about 1°C (2°F) for each 10% increment of kerosene added.

cold sett grease — see *sett grease.*

colloid — suspension of finely divided particles, 5 to 5000 angstroms in size, in a gas or liquid, that do not settle and are not easily filtered. Colloids are usually ionically stabilized by some form of surface charge on the particles to reduce the tendency to agglomerate. A lubricating grease is a colloidal system, in which metallic soaps or other thickening agents are dispersed in, and give structure to, the liquid lubricant.

color scale — standardized range of colors against which the colors of petroleum products may be compared. There are a number of widely used systems of color scales, including: **ASTM scale** (test method ASTM D 1500), the most common scale, used extensively for industrial and process oils; **Tag-Robinson colorimeter,** used with solvents, waxes, industrial and process oils; **Saybolt chromometer** (ASTM D 156), used with white oils, naphthas, waxes, fuels, kerosene, solvents; **Lovibond tintometer,** for USP petrolatums, sulfonates, chemicals; and the **Platinum-Cobalt (APHA) system** (ASTM D 1209), for lacquer solvents, diluents, petrochemicals. These scales serve primarily as indicators of product uniformity and freedom from contamination.

combustion — rapid oxidation of a fuel (burning). The products of an ideal combustion process are water (H_2O) and carbon dioxide (CO_2); if combustion is incomplete, some carbon is not fully oxidized, yielding *carbon monoxide* (CO). A *stoichiometric* combustible mixture contains the exact quantities of air (oxygen) and fuel required for complete combustion. For gasoline, this air-fuel ratio is about 15:1 by weight. If the fuel concentration is too rich or too lean relative to the oxygen in the mixture, combustion cannot take place. See *explosive limits.*

combustion chamber — in an *internal combustion engine,* the volume, bounded by the top of the piston and the inner surface of the cylinder head, in which the air-fuel charge ignites and burns. Valves and spark plugs are fitted into the combustion chamber.

commercial oils — see *API Engine Service Categories.*

complex soap — see *grease.*

compounded oil — mixture of a petroleum oil with animal or vegetable fat or oil. Compounded oils have a strong affinity for metal surfaces; they are particularly suitable for wet-steam conditions and for applications where *lubricity* and extra load-carrying ability are needed. They are not generally recommended where long-term *oxidation stability* is required.

compression ignition — see *afterrunning.*

compression-ignition engine — diesel engine. See *internal combustion engine.*

compression ratio — in an *internal combustion engine,* the ratio of the volume of the combustion space in the cylinder at the bottom of the piston stroke to the volume at the top of the stroke. High-compression-ratio gasoline engines require high octane fuels. Not to be confused with the *pressure ratio* of a compressor.

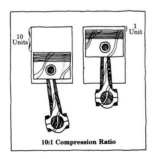

10:1 Compression Ratio

compressor — any of a wide variety of mechanisms designed to compress air or other gas to produce useful work. There are two basic types of compressors: **positive displacement** and **kinetic** (or dynamic). Positive displacement compressors increase pressure directly by reducing the volume of the chamber in which the gas is confined. Kinetic compressors are designed to impart velocity to the gas by means of a bladed rotor, then convert velocity energy to pressure by diverting the gas stream through stationary blades that change its direction of flow.

concrete form coating — an oil, wax, or grease applied to wooden or metal concrete forms to keep the hardened concrete from adhering to the forms. The liquid materials are also called **form oil.**

condensate — in refining, the liquid produced when hydrocarbon vapors are cooled. In oil and gas production, the term applies to hydrocarbons that exist in gaseous form under *reservoir* conditions, but condense to a liquid when brought to the surface.

congealing point — the temperature at which molten wax ceases to flow, as measured by test method ASTM D 938; of importance where storage or application temperature is a critical factor.

coning oil — lubricant, containing *emulsifiers* and anti-static agents, applied to synthetic-fiber yarn to reduce snagging and pulling as the yarn is run off a cone, and to facilitate further processing. See *fiber lubricant.*

Conradson carbon residue — see *carbon residue.*

conservation — see *energy conservation.*

consistency (grease) — a basic property describing the softness or hardness of a *grease,* i.e., the degree to which a grease resists deformation under the application of force. Consistency is measured by means of a cone penetration test. See *penetration (grease).* The consistency of a grease depends on the viscosity of the base oil and the type and proportion of the thickener. It may also be affected by recent agitation; to take this phenomenon into consideration, a grease may be subjected to working (a standard churning process) prior to measuring its penetration value. See *NLGI consistency grades.*

copolymer — see *polymer.*

copper strip corrosion — the tendency of a petroleum product to corrode cuprous metals, as determined by test method ASTM D 130; the corrosion stains on a test copper strip are matched against standardized corroded strips.

corrosion — chemical attack on a metal or other solid by contaminants in a lubricant. Common corrosive contaminants are: (1) water, which causes rust, and (2) acids, which may form as oxidation products in a deteriorating oil, or may be introduced into the oil as combustion by-products in piston engines.

corrosion inhibitor — *additive* for protecting lubricated metal surfaces against chemical attack by water or other contaminants. There are several types of corrosion inhibitors. *Polar compounds* wet the metal surface preferentially, protecting it with a film of oil. Other compounds may absorb water by incorporating it in a water-in-oil emulsion so that only the oil touches the metal surface. Another type of corrosion inhibitor combines chemically with the metal to present a non-reactive surface. See *rust inhibitor.*

cp (centipoise) — see *viscosity.*

CPSC (Consumer Products Safety Commission) — federal government commission that administers legislation on consumer product safety, such as the Consumer Product Safety Act, the Federal Hazardous Substances Act, the Flammable Fabrics Act, the Poison Prevention Packaging Act, and the Refrigeration Safety Act.

CR — see *neoprene rubber.*

cracking — petroleum refining process in which large-molecule liquid hydrocarbons are converted to small-molecule, lower-boiling liquids or gases; the liquids leave the reaction vessel as unfinished *gasoline, kerosene,* and *gas oils.* At the same time, certain unstable, more reactive molecules combine into larger molecules to form tar or coke *bottoms*. The cracking reaction may be carried out under heat and pressure alone *(thermal cracking)*, or in the presence of a catalyst *(catalytic cracking)*.

crankcase oil — see *engine oil*.

CRS asphalt — see *emulsified cationic asphalt*.

crude oil — complex, naturally occurring fluid mixture of petroleum hydrocarbons, yellow to black in color, and also containing small amounts of oxygen, nitrogen, and sulfur derivatives and other impurities. Crude oil was formed by the action of bacteria, heat, and pressure on ancient plant and animal remains, and is usually found in layers of porous rock such as *limestone* or *sandstone*, capped by an impervious layer of shale or clay that traps the oil (see *reservoir*). Crude oil varies in appearance and hydrocarbon composition depending on the locality where it occurs, some crudes being predominately naphthenic, some paraffinic, and others asphaltic. Crude is refined to yield petroleum products. See *distillation, hydrocarbon, sour crude, sweet crude, asphalt, naphthene, paraffin*.

CSS asphalt — see *emulsified cationic asphalt*.

cSt (centistoke) — see *viscosity*.

cut — segregated part, or fraction, separated from crude in the distillation process. See *distillation*.

cutback asphalt — a solution of *asphalt cement* and a petroleum *diluent*. Upon exposure to the atmosphere, the diluent evaporates, leaving the asphalt cement to perform its function. There are three grades: **RC asphalt** — rapid curing cutback asphalt, composed of asphalt cement and a *naphtha*-type diluent of high volatility; **MC**

asphalt — medium-curing cutback asphalt, composed of asphalt cement and a *kerosene*-type diluent of medium volatility; **SC asphalt** — slow-curing cutback asphalt, composed of asphalt cement and a low-volatility oil. For industrial uses, oxidized asphalts can be blended with a petroleum diluent to meet the specific requirements of coatings, mastics, etc. (see *asphalt*).

cutting fluid — fluid, usually of petroleum origin, for cooling and lubricating the tool and work in machining and grinding. Some fluids are fortified with *EP additives* to facilitate cutting of hard metals, to improve finishes, and to lengthen tool life. Some cutting fluids are transparent to provide a better view of the work. Soluble cutting oils are emulsifiable with water to improve cooling. Since the resulting *emulsions* are subject to bacterial action and the development of odors, soluble cutting fluids may contain *bactericides*.

cyclic hydrocarbon — *hydrocarbon* in which the carbon atoms are joined in rings.

cycloalkane — see *naphthene*.

cycloparaffin — see *naphthene*.

cylinder oil — lubricant for independently lubricated cylinders, such as those of steam engines and air compressors; also for lubrication of valves and other elements in the cylinder area. Steam cylinder oils are available in a range of grades with high viscosities to compensate for the thinning effect of high temperatures; of these, the heavier grades are formulated for super-heated and high-pressure steam, and the less heavy grades for wet, saturated, or low-pressure steam. Some grades are compounded for service in excessive moisture; see *compounded oil*. Cylinder oils lubricate on a once-through basis (see *once-through lubrication*).

deasphalting — refining step for removal of *asphaltic* compounds from heavy lubricating oils. Liquid propane, liquid butane, or a mixture of the two is used to dilute the oil and precipitate the asphalt.

demerit rating — arbitrary graduated numerical rating sometimes used in evaluating engine deposit levels following testing of an engine oil's *detergent-dispersant* characteristics. On a scale of 0-10, the higher the number, the heavier the deposits. A more commonly used method of evaluating engine cleanliness is *merit rating*. See *engine deposits*.

demulsibility — ability of an oil to separate from water, as determined by test method ASTM D 1401 or D 2711. Demulsibility is an important consideration in lubricant maintenance in many *circulating lubrication systems*.

demulsifier — *additive* that promotes oil-water separation in lubricants that are exposed to water or steam. See *demulsibility*.

denaturing oil — unpalatable oil, commonly *kerosene* or No. 2 *heating oil*, required to be added to food substances condemned by the U.S. Department of Agriculture, to ensure that these substances will not be sold as food or consumed as such.

density — see *specific gravity*.

de-oiling — removal of oil from petroleum wax; a refinery process usually involving filtering or pressing a chilled mixture of *slack wax* and a solvent that is miscible in the oil, to lower the oil content of the wax.

depletion allowance — a reduction in U.S. taxes on producers of minerals, including petroleum, to compensate for the exhaustion of an irreplaceable capital asset.

deposits — see *engine deposits*.

dermatitis — inflammation of the skin; can be caused by contact with many commercial substances, including petroleum products. Oil and grease in contact with the skin can result in plugging of sweat glands and hair follicles or defatting of the skin, which can lead to dermatitis. Dermatitis can be prevented in such cases by avoiding contact with the causative substances, or, if contact occurs, by promptly washing the skin with soap, water, and a soft skin brush. Clothes soaked with the substance should be removed.

detergent — important component of engine oils that helps control *varnish*, ring zone deposits, and rust by keeping insoluble particles in colloidal suspension (see *colloid*) and in some cases, by neutralizing acids. A detergent is usually a metallic (commonly barium, calcium, or magnesium) compound, such as a sulfonate, phosphonate, thiophosphonate, phenate, or salicylate.

Because of its metallic composition, a detergent leaves a slight ash when the oil is burned. A detergent is normally used in conjunction with a *dispersant.* See *ash content.*

detergent-dispersant — engine oil *additive* that is a combination of a *detergent* and a *dispersant;* important in preventing the formation of *sludge* and other engine deposits.

detonation — see *knock.*

dewaxing — removal of *paraffin wax* from lubricating oils to improve low temperature properties, especially to lower the *cloud point* and *pour point.*

dibasic acid ester (diester) — *synthetic lubricant* base; an organic ester, formed by reacting a dicarboxylic acid and an *alcohol;* properties include a high *viscosity index* (V.I.) and low *volatility.* With the addition of specific additives, it may be used as a lubricant in compressors, hydraulic systems, and internal combustion engines.

dielectric loss — see *power factor.*

dielectric — non-conductor of electricity, such as *insulating oil* for transformers. See *power factor.*

dielectric strength (breakdown voltage) — minimum voltage required to produce an electric arc through an oil sample, as measured by test method ASTM D 877; hence, an indication of the insulating (arc preventive) properties of a transformer oil. A low dielectric strength may indicate contamination, especially by water. See *insulating oil, power factor.*

diesel engine — see *internal combustion engine.*

diesel fuel — that portion of crude oil that distills out within the temperature range of approximately 200°C (392°F) to 370°C (698°F), which is higher than the boiling range of gasoline. See *distillation.* Diesel fuel is ignited in an *internal combustion engine* cylinder by the heat of air under high compression — in contrast to motor gasoline, which is ignited by an electrical spark. Because of the mode of ignition, a high *cetane number* is required in a good diesel fuel. Diesel fuel is close in boiling range and composition to the lighter *heating oils.* There are two grades of diesel fuel, established by the American Society for Testing and Materials *(ASTM):* **Diesel 1** and **Diesel 2.** Diesel 1 is a *kerosene*-type fuel, lighter, more volatile, and cleaner burning than Diesel 2, and is used in engine applications where there are frequent changes in speed and load. Diesel 2 is used in industrial and heavy mobile service.

diesel index — an approximation of the *cetane number* of a *distillate* fuel: the product of the API gravity (see *specific gravity)* and the *aniline point* (°F) divided by 100. Fuels of unusual composition may show erroneous cetane numbers by this method. See *cetane index.*

dieseling — see *afterrunning.*

diester — see *dibasic acid ester.*

diluent — a usually inert (unreactive) liquid or *solvent,* used to dilute, carry, or increase the bulk of some other substance. Petroleum oils and solvents are commonly used as diluents in such products as paints, pesticides, and additives. See *reaction diluent.*

dilution of engine oil — thinning of *engine oil* by seepage of fuel into the crankcase, as measured by test method ASTM D 322, which indicates the volume percentage of fuel in the sample. Dilution is detrimental to lubrication, and may indicate defective en-

gine components — such as worn piston rings — or improper fuel system adjustment.

dimer — see *polymerization.*

diolefin — highly reactive straight-chain hydrocarbon with two double bonds between adjacent carbon atoms. See *olefin.*

dispersant — engine oil *additive* that helps prevent *sludge, varnish,* and other engine deposits by keeping particles suspended in a colloidal state (see *colloid).* Dispersants are normally used in conjunction with *detergents.* A dispersant is commonly distinguished from a detergent in that the former is nonmetallic and, thus, does not leave an ash when the oil is burned; hence, the term **ashless dispersant.** Also, a dispersant can keep appreciably larger quantities of contaminants in suspension than a detergent.

dispersion — minute discrete particles suspended in a liquid, a gas, or a solid. Though it may have the general characteristics of a *colloid,* a dispersion is not necessarily a truly homogeneous mixture.

dissipation factor — in an electrical system with an applied alternating voltage, it is expressed as the tangent of the loss angle (δ) or the cotangent of the phase angle (θ); a measure of the degree of electrical loss due to the imperfect nature of an insulating liquid surrounding the electrical system, as determined by test method ASTM D 924. Dissipation factor is related to *power factor.*

distillate — any of a wide range of petroleum products produced by distillation, as distinct from *bottoms,* cracked stock (see *cracking),* and *natural gas liquids.* In fuels, a term referring specifically to those products in the mid-boiling range, which include *kerosene, turbo fuel,* and *heating oil* — also called **middle distillates** and **distillate fuels.** In lubricating oils, a term applied to the various fractions separated under vacuum in a distillation tower for further processing (lube distillate). See *distillation.*

distillate fuels — see *distillate.*

distillation (fractionation) — the primary refining step, in which crude is separated into **fractions,** or components, in a **distillation tower,** or **pipe still.** Heat, usually applied at the bottom of the tower, causes the oil vapors to rise through progressively cooler levels of the tower, where they condense onto plates and are drawn off in order of their respective condensation temperatures, or boiling points — the lighter-weight, lower-boiling-point fractions, exiting higher in the tower. The primary fractions, from low to high boiling point, are: hydrocarbon gases (e.g., *ethane, propane);* naphtha (e.g. *gasoline);* kerosene, diesel fuel *(heating oil);* and heavy *gas oil* for *cracking.* Heavy materials remaining at the bottom are called the *bottoms,* or residuum, and include such components as heavy fuel oil (see *fuel oil)* and asphaltic substances (see *asphalt).* Those fractions taken in liquid form from any level other than the very top or bottom are called *sidestream* products; a product, such as propane, removed in vapor form from the top of the distillation tower is called *overhead* product. Distillation may take place in two stages: first, the lighter fractions — gases, naphtha, and kerosene — are recovered at essentially atmospheric pressure; next, the remaining crude is distilled at reduced pressure in a **vacuum tower,** causing the heavy lube fractions to distill at much lower temperatures than possible at atmospheric pressure, thus permitting more lube oil to be distilled without the molecular cracking that can occur at excessively high temperatures.

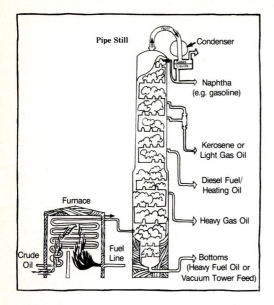

distillation test — method for determining the full range of *volatility* characteristics of a hydrocarbon liquid by progressively boiling off (evaporating) a sample under controlled heating. **Initial boiling point** (IBP) is the fluid temperature at which the first drop falls into a graduated cylinder after being condensed in a condenser connected to a distillation flask. **Mid-boiling point** (MBP) is the temperature at which 50% of the fluid has collected in the cylinder. **Dry point** is the temperature at which the last drop of fluid disappears from the bottom of the distillation flask. **Final boiling point** (FBP) is the highest temperature observed. **Front-end volatility** and **tail-end volatility** are the amounts of test sample that evaporate, respectively, at the low and high temperature ranges. If the boiling range is small, the fluid is said to be **narrow cut,** that is, having components with similar volatilities; if the boiling range is wide, the fluid is termed **wide cut.** Distillation may be carried out by several ASTM test methods, including ASTM D 86, D 850, D 1078, and D 1160.

distillation tower — see *distillation.*

dN factor — also called **speed factor,** determined by multiplying the bore of a rolling-contact bearing (d), in millimeters, by the speed (N), in rpm, of the journal (the shaft or axle supported by the bearings); used in conjunction with operating temperature to help determine the appropriate *viscosity* of the bearing lubricating oil. See *bearing.*

dolomite — sedimentary rock similar to *limestone,* but rich in magnesium carbonate. Dolomite is sometimes a *reservoir* rock for petroleum.

drawing compound — in metal forming, a lubricant for the die or blank used to shape the metal; often contains *EP additives* to increase die life and to improve the surface finish of the metal being drawn.

drilling — see *rotary drilling.*

drilling fluid — also called **drilling mud.** See *mud.*

drilling mud — see *mud.*

dropping point — lowest temperature at which a grease is sufficiently fluid to drip, as determined by test method ASTM D 566 or D 2265; hence, an indication of whether a grease will flow from a bearing at operating temperatures. The test is of limited significance in predicting overall service performance.

dry point — see *distillation test.*

dumbbell blend — mixture of hydrocarbons, usually two components, that have markedly different volatilities, viscosities, or other properties. See *volatility, viscosity.*

dynamic demulsibility — test of water separation properties of an oil, involving continuous mixture of oil and water at elevated temperatures in an apparatus that simulates a lubricating oil circulating system. Samples are then drawn off both the top and bottom of the test apparatus. Ideally, the top sample should be 100% oil, and the bottom 100% water. Because of the severity of the test conditions, separation is virtually never complete.

elasto-hydrodynamic (EHD) lubrication — lubrication phenomenon occurring during elastic deformation of two non-conforming surfaces under high load. A high load carried by a small area (as between the ball and race of a rolling contact bearing) causes a temporary increase in lubricant viscosity as the lubricant is momentarily trapped between slightly deformed opposing surfaces.

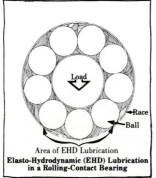

Area of EHD Lubrication

Elasto-Hydrodynamic (EHD) Lubrication in a Rolling-Contact Bearing

elastomer — rubber or rubber-like material, both natural and synthetic, used in making a wide variety of products, such as tires, seals, hose, belting, and footwear. In oil seals, an elastomer's chemical composition is a factor in determining its compatibility with a lubricant, particularly a *synthetic lubricant.* See *natural rubber, synthetic rubber.*

electronic emission controls (EEC) — in automobiles, computerized engine operating controls that reduce automotive exhaust emissions of *carbon monoxide* (CO), *nitrogen oxides* (NOx), and *hydrocarbons* (HC), primarily by optimizing combustion efficiency. This is accomplished by automatic monitoring and control of key engine functions and operating parameters, such as air-fuel ratio (see *combustion*), spark timing, and *exhaust gas recirculation.* See *emissions (automotive), hydrocarbon emissions.*

electrostatic precipitation — removal of particles suspended in a gas — as in a furnace flue — by electrostatic charging of the particles, and subsequent precipitation onto a collector in a strong electrical field. See *emissions (stationary source), particulates, pollutants.*

emission controls — see *catalytic converter, emissions (automotive), emissions (stationary source), electronic emission controls, exhaust gas recirculation, positive crankcase ventilation.*

emissions (automotive) — the three major pollutant emissions for which gasoline-powered vehicles are controlled are: unburned *hydrocarbons* (HC), *carbon monoxide* (CO), and *nitrogen oxides* (NOx). Diesel-powered vehicles primarily emit NOx and *particulates.* Motor vehicles contribute only a small percentage of total man-made emissions of other atmospheric pollutants, such as *sulfur oxides.* Evaporative HC emissions from the fuel tank and carburetor are adsorbed by activated carbon contained in a canister installed on the vehicle. *Blow-by* HC emissions from the crankcase are controlled by *positive crankcase ventilation* (PCV). Exhaust emissions of HC, CO, and NOx — the products of incomplete combustion — are controlled primarily by a *catalytic converter,* in conjunction with *exhaust gas recirculation* and increasingly sophisticated technology for improving combustion efficiency, including *electronic emission controls.* See *emissions (stationary source), hydrocarbon emissions, pollutants.*

emissions (stationary source) — atmospheric pollutants from fossil fuel combustion in furnaces and boilers. Stationary combustion sources contribute significantly to total man-made emissions of *sulfur oxides* — predominantly sulfur dioxide (SO_2), with some sulfur trioxide (SO_3) — *nitrogen oxides* (NOx), and *particulates,* but emit comparatively minor amounts of *carbon monoxide (CO)* and *hydrocarbons* (HC). Means of controlling stationary source emissions include: flue gas scrubbing with a chemical substance such as sodium hydroxide to remove sulfur, burning naturally low-sulfur fuels, lowering combustion temperatures to reduce NOx formation, and using *electrostatic precipitation* to reduce particulate emissions. See *emissions (automotive), pollutants.*

emulsified anionic asphalt (anionic emulsified asphalt) — emulsified *asphalt* in which the asphalt globules are electronegatively charged. Common grades of anionic asphalt are: **RS asphalt** — anionic rapid-setting emulsified asphalt; **MS asphalt** — anionic medium-setting emulsified asphalt; **SS asphalt** — anionic slow-setting emulsified asphalt. See *emulsified cationic asphalt.*

emulsified cationic asphalt (cationic emulsified asphalt) — emulsified *asphalt* in which the asphalt globules are electropositively charged. Common grades of cationic emulsified asphalt are: **CRS asphalt** — cationic rapid-setting emulsified asphalt; **CMS asphalt** — cationic medium-setting emulsified asphalt; **CSS asphalt** — cationic slow-setting emulsified asphalt. Emulsified cationic asphalt is superior to *emulsified anionic asphalt* in its ability to mix with wet stones, or aggregate, because its electropositive charge aids in rapidly replacing the water adhering to the stones.

emulsifier — *additive* that promotes the formation of a stable mixture, or *emulsion,* of oil and water. Common emulsifiers are: metallic soaps, certain animal and vegetable oils, and various *polar compounds* (having molecules that are water-soluble at one extremity of their structures and oil-soluble at the other).

emulsion — intimate mixture of oil and water, generally of a milky or cloudy appearance. Emulsions may be of two types: oil-in-water (where water is the continuous phase) and water-in-oil (where water is the discontinuous phase). Oil-in-water emulsions are used as *cutting fluids* because of the need for the cool-

ing effect of the water. Water-in-oil emulsions are used where the oil, not the water, must contact a surface — as in *rust preventives,* non-flammable *hydraulic fluids,* and compounded steam *cylinder oils* (see *compounded oil);* such emulsions are sometimes referred to as **inverse emulsions.** Emulsions are produced by adding an *emulsifier.* Emulsibility is not a desirable characteristic in certain lubricating oils, such as crankcase or turbine oils, that must separate from water readily. Unwanted emulsification can occur as a result of oxidation products — which are usually *polar compounds* — or other contaminants in the oil. See illustration of an oil-in-water emulsion at *polar compound.*

energy — capacity to do work. There are many forms of energy, any of which can be converted into any other form of energy. To produce electrical power in a steam turbine-generator system, the chemical energy in coal is converted into heat energy, which (through steam) is converted to the mechanical energy of the turbine, and in turn, converted into electrical energy. Electrical energy may then be converted into the mechanical energy of a vacuum cleaner, the radiant and heat energy of a light bulb, the chemical energy of a charged battery, etc. Conversion from one form of energy to another results in some energy being lost in the process (usually as heat). There are two kinds of mechanical energy: **kinetic energy,** imparted by virtue of a body's motion, and **potential energy,** imparted by virtue of a body's position (e.g. a coiled spring, or a stone on the edge of a cliff). Solar (radiant) energy is the basis of all life through the process of photosynthesis, by which green plants convert solar energy into chemical energy. Nuclear energy is the result of the conversion of a small amount of the mass of an unstable (radioactive) atom into energy. The fundamental unit of energy in the *Système International* is the *joule.* It can be expressed in other energy units, such as the *calorie, British thermal unit (Btu), kilowatt-hour,* etc. by use of appropriate conversion factors.

energy conservation — employment of less energy to accomplish the same amount of useful work; also, the reduction or elimination of any energy-consuming activity. Energy conservation is a vital goal in U.S. and world efforts to adjust to the declining availability of conventional oil and gas resources. Conservation will extend the time available to make the transition to alternative energy sources, such as solar energy and *synthetic oil and gas,* and will also reduce the economic hardship imposed by rising energy prices.

engine deposits — hard or persistent accumulations of *sludge, varnish,* and carbonaceous residues due to *blow-by* of unburned and partially burned (partially oxidized) fuel, or from partial breakdown of the crankcase lubricant. Water from condensation of combustion products, carbon, residues from fuel or lubricating oil additives, dust, and metal particles also contribute. Engine deposits can impair engine performance and damage engine components by causing valve and ring sticking, clogging of the oil screen and oil passages, and excessive wear of pistons and cylinders. Hot, glowing deposits in the *combustion chamber* can also cause *pre-ignition* of the air-fuel mix. Engine deposits are increased by short trips in cold weather, high-temperature operation, heavy loads (such as pulling a trailer), and over-extended oil drain intervals.

engine oil (crankcase oil, motor oil) — oil carried in the crankcase, sump, or oil pan of a reciprocating *internal combustion engine* to lubricate all major engine parts; also used in reciprocating compressors and in steam engines of crankcase design. In automotive applications, it is the function of the engine oil not only to lubricate, but to cool hot engine parts, keep the engine free of rust and deposits (see *engine deposits),* and seal the rings and valves against leakage of combustion gases. Oil-feed to the engine parts is generally under pressure developed by a gear pump (forced feed). The oil circulates through passages formed by tubing and drilling (rifling) through the engine parts, and through an oil filter to remove metallic contaminants and other foreign particles. In some engines, lubrication may also be accomplished in part by splashing resulting from the rotation of the crankshaft in the oil in the sump. Modern engine oils are formulated with *additives* to improve performance. Additive content in a single-viscosity-grade oil is typically around 15 mass percent, and in a multi-grade oil, 20 percent or more. See *API Engine Service Categories, military specifications for engine oils, SAE viscosity grades.*

Engler viscosity — method for determining the *viscosity* of petroleum products; it is widely used in Europe, but has limited use in the U.S. The test method is similar to *Saybolt Universal viscosity;* viscosity values are reported as "Engler degrees."

enhanced recovery — in crude oil production, any method used to produce the oil remaining in a reservoir that has largely been depleted. See *secondary recovery, tertiary recovery, reservoir.*

entrainment — the state of a liquid or gas that is dispersed, but undissolved, in a liquid or gaseous medium.

Environmental Protection Agency — see *EPA.*

EPA (Environmental Protection Agency) — agency of the federal executive branch, established in 1970 to abate and control pollution through monitoring, regulation, and enforcement, and to coordinate and support environmental research.

EP additive — lubricant *additive* that prevents sliding metal surfaces from seizing under conditions of extreme pressure (EP). At the high local temperatures associated with metal-to-metal contact, an EP additive combines chemically with the metal to form a surface film that prevents the welding of opposing *asperities,* and the consequent *scoring* that is destructive to sliding surfaces under high loads. Reactive compounds of sulfur, chlorine, or phosphorus are used to form these inorganic films.

EPDM rubber — see *ethylene-propylene rubber.*

EP oil — lubricating oil formulated to withstand extreme pressure (EP) operating conditions. See *EP additive.*

ester — chemical compound formed by the reaction of an organic or inorganic *acid* with an *alcohol* or with another organic compound containing the hydroxyl (-OH) *radical.* The reaction involves replacement of the hydrogen of the acid with a hydrocarbon group. The name of an ester indicates its derivation; e.g., the ester resulting from the reaction of ethyl alcohol and acetic acid is called ethyl acetate. Esters have important uses in the formulation of some petroleum additives and *synthetic lubricants.* See *dibasic acid ester, phosphate ester.*

ethane — gaseous paraffinic hydrocarbon (C_2H_6) present in natural gas and petroleum; used as a fuel, and as a *feedstock* in *petrochemical* manufacture. See *natural gas, distillation.*

$$
\begin{array}{ccc}
 & H & H \\
 & | & | \\
H - & C - & C - H \\
 & | & | \\
 & H & H
\end{array}
$$
ethane

ethanol — also known as ethyl alcohol (C_2H_5OH). Obtained principally from the fermentation of grains or blackstrap molasses; also obtained from *ethylene,* by absorption in sulfuric acid and hydrolyzing with water. Widely used as an industrial solvent, *extraction* medium, chemical intermediate, and in many proprietary products; a component of *gasohol.*

ethylene — flammable gas (C_2H_4) derived from natural gas and petroleum; the lowest molecular weight member of the generic family of *olefins.* Ethylene is widely used as a *feedstock* in the manufacture of *petrochemicals,* including *polyethylene* and other plastics.

$$
\begin{array}{ccc}
H & & H \\
| & & | \\
C & = & C \\
| & & | \\
H & & H
\end{array}
$$
ethylene

ethylene-propylene rubber (EPM and EPDM) — *synthetic rubber:* EPM is a *polymer* of *ethylene* and *propylene;* EPDM is a polymer of ethylene and propylene with a small amount of a third *monomer* (usually a *diolefin*) to permit vulcanization with sulfur. EPM and EPDM possess excellent resistance to ozone, sunlight, and weathering, have good flexibility at low temperatures, and good electrical insulation properties. Used in the manufacture of tires, hoses, auto parts, coated fabrics, and electrical insulation.

evaporation — conversion of a liquid into a vapor; also, a test procedure that yields data on the *volatility* of a petroleum product. Evaporation testing of solvents may be performed in accordance with the Federation of Societies for Paint Technology Method II. In this method, a small sample of product is applied by hypodermic syringe to a filter paper on a sensitive spring balance, then allowed to evaporate under controlled conditions of temperature, relative humidity, air movement, etc. The loss of sample weight is plotted with respect to time. See *distillation test.*

exhaust gas recirculation (EGR) — system designed to reduce automotive exhaust emissions of *nitrogen oxides (NOx).* The system routes exhaust gases into the *carburetor* or intake manifold; the gases dilute the air-fuel mixture (see *combustion*) which lowers peak combustion temperatures, thus reducing the tendency for NOx to form.

explosive limits (flammability limits) — upper and lower limits of petroleum vapor concentration in air outside of which combustion will not occur. As a general rule, below one volume percent concentration in air (lower explosive limit) the mixture is too lean to support combustion; above six volume percent (upper explosive limit), the mixture is too rich to burn. See *combustion.*

extender — material added to a formulation to improve quality and processability, or to reduce costs by substituting for a more expensive material (e.g., a low-cost solvent partially replacing a higher-cost solvent; a petroleum oil added to a rubber formulation).

extraction — use of a solvent to remove edible and commercial oils from seeds (e.g. soybeans), or oils and fats from meat scraps; also, the removal of reactive components from lube distillates (see *solvent extraction)* or other refinery *process streams.*

extreme pressure (EP) additive — see *EP additive.*

°F (Fahrenheit) — see *temperature scales.*

Falex test — a method for determining the extreme-pressure (EP) or anti-wear properties of oils and greases. Vee blocks (with a large "V"-shaped notch) are placed on opposite sides of a rotating steel shaft, and the apparatus is immersed in a bath of the test lubricant. Load is automatically increased until seizure occurs. Measurable wear scars are formed on the blocks.

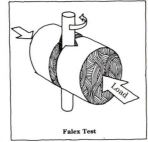

Falex Test

false brinelling — see *fretting.*

fatty acid — any monobasic (one displaceable hydrogen atom per molecule) organic acid having the general formula $CnH_{2n+1}COOH$. Fatty acids derived from natural fats and oils are used to make soaps used in the manufacture of greases and other lubricants. See *grease.*

fatty oil — organic oil of animal or vegetable origin; can be added to petroleum oils to increase load-carrying ability, or oiliness. See *compounded oil, saponification number.*

FDA (Food and Drug Administration) — agency administered under the U.S. Department of Health and Human Services (formerly Health, Education and Welfare) "to enforce the Federal Food, Drug, and Cosmetic Act and thereby carry out the purpose of Congress to insure that foods are safe, pure, and wholesome, and made under sanitary conditions; drugs and therapeutic devices are safe and effective for their intended uses; cosmetics are safe and prepared from appropriate ingredients; and that all of these products are honestly and informatively labeled and packaged."

feedstock — any material to be processed; e.g., *gas oil* for *cracking, ethylene* for *petrochemical* manufacture.

FIA analysis — see *fluorescent indicator adsorption.*

fiber — in *grease,* form in which soap thickeners occur. On the average, soap fibers are about 20 times as long as they are thick; most are microscopic, so that the grease appears smooth.

fiber lubricant — an oil containing *emulsifiers* and anti-static agents, applied to synthetic fibers to lubricate them during processing into yarn; also called **spin finish.** See *coning oil.*

film strength — See *lubricity.*

final boiling point — see *distillation test.*

fingerprint neutralizer — *polar compound* in some *rust preventives* that places a barrier between the metal surface and perspiration deposited during handling of metal parts. In this way, corrosive activity of the salts and acids in perspiration is suppressed.

fire point — temperature at which the vapor concentration of a combustible liquid is sufficient to sustain combustion, as determined by test method ASTM D 92, *Cleveland Open Cup.* See *flash point.*

fire-resistant fluid — lubricant used especially in high-temperature or hazardous hydraulic applications. Three common types of fire-resistant fluids are: (1) water-petroleum oil *emulsions,* in which the water prevents burning of the petroleum constituent; (2) water-glycol fluids; and (3) non-aqueous fluids of low volatility, such as *phosphate esters, silicones,* and halogenated (see *halogen*) hydrocarbon-type fluids. See *flame propagation, synthetic lubricant.*

flame propagation — self-sustaining burning of fuel after heat of combustion has been reached. Many fire-resistant hydraulic fluids — though they can be made to burn if subjected to sufficiently intense heat — do not generate sufficient heat of combustion of themselves to continue burning once the external source of heat is removed. See *fire-resistant fluids.*

flammability limits — see *explosive limits.*

flash point — lowest temperature at which the vapor of a combustible liquid can be made to ignite momentarily in air, as distinct from *fire point.* Flash point is an important indicator of the fire and explosion hazards associated with a petroleum product. There are a number of ASTM tests for flash point, e.g., *Cleveland open cup, Pensky-Martens closed tester, Tag closed tester, Tag open cup.* See illustration on facing page.

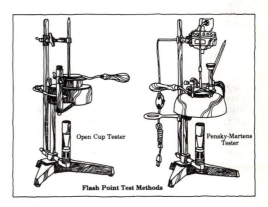

Open Cup Tester   Pensky-Martens Tester

Flash Point Test Methods

floc point — temperature at which waxy materials in a lubricating oil separate from a mixture of oil and Freon* R-12 refrigerant, giving a cloudy appearance to the mixture; also called **Freon floc point.** Generally used to evaluate the tendency of *refrigeration oils* to plug expansion valves or capillaries in refrigerant systems. Not to be confused with *cloud point,* the temperature at which wax precipitates from an undiluted oil.
*Registered trademark of E. I. DuPont de Nemours, Inc.

fluid friction — see *friction.*

fluorescent indicator adsorption (FIA) — method of measuring the relative concentration of *saturated hydrocarbons, aromatics,* and *olefins* in a petroleum product (usually a *solvent* or light *distillate),* as determined by test method ASTM D 1319. The sample is passed through a column where it reacts with three dyes, each sensitive to one of the components. The relative concentration of the component is indicated by the length of the respective dyed zones, viewed under ultraviolet light, which brings out the coloration of the dyes.

flux (flux oil) — a relatively non-volatile fraction of petroleum used as a *diluent* to soften asphalt to a desired consistency; also, a *base stock* for the manufacture of roofing asphalts.

foaming — occurrence of frothy mixture of air and a petroleum product (e.g. lubricant, fuel oil) that can reduce the effectiveness of the product, and cause sluggish hydraulic operation, air binding of oil pumps, and overflow of tanks or sumps. Foaming can result from excessive agitation, improper fluid levels, air inclusion, or contamination with water or other foreign materials. Foaming can be inhibited with an *anti-foam agent*. The foaming characteristics of a lubricating oil can be determined by blowing air through a sample at a specified temperature and measuring the volume of foam, as described in test method ASTM D 892.

fogging oil — light *mineral seal oil* injected in small quantities into a gas transmission line to settle dust or to seal joints by soaking the fiber or jute gasket materials. Also, oil used to generate smoke or fog to obscure the movement of troops and naval vessels, or as a carrier for insecticides applied to large outdoor areas.

follower plate — heavy disc in a grease container which rests on the surface of the grease, and assists its downward movement toward a dispensing pump located at the bottom of the container.

food additive — non-nutritional substance added directly or indirectly to food during processing or packaging. Petroleum food additives are usually *refined waxes* or *white oils* that meet applicable *FDA* standards. Applications include direct additives, such as coatings for fresh fruits and vegetables, and indirect additives, such as impregnating oils for fruit and vegetable wrappers, dough divider oils, defoamers for yeast and beet sugar manufacture, release and polishing agents in confectionery manufacture, and rust preventives for meat processing equipment.

Food and Drug Administration — see *FDA*

form oil — see *concrete form coating.*

fossil fuel — any fuel, such as crude oil and coal, derived from remains of ancient organisms that have been transformed over the ages by heat, pressure, and chemical action.

four-ball method — either of two lubricant test procedures, the Four-Ball Wear Method (ASTM D 2266) and Four-Ball EP (extreme pressure) Method (ASTM D 2596), based on the same principle. Three steel balls are clamped together to form a cradle upon which a fourth ball rotates on a vertical axis. The balls are immersed in the lubricant under investigation. The **Four Ball Wear Method** is used to determine the anti-wear properties of lubricants operating under *boundary lubrication* conditions. The test is carried out at a specified speed, temperature and load. At the end of a specified test time, the average diameter of the wear scars on the three lower balls is reported. The **Four-Ball EP Method** is designed to evaluate performance under much higher unit loads. The loading is increased at specified intervals until the rotating ball seizes and welds to the other balls. At the end of each interval the average scar diameter is recorded. Two values are generally re-

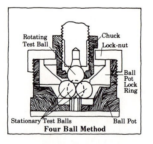

Rotating Test Ball — Chuck — Lock-nut

Ball Pot Lock Ring

Stationary Test Balls — Ball Pot
Four Ball Method

ported — *load wear index* (formerly **mean Hertz load**) and *weld point.*

four-square gear oil tester — device consisting of two automotive drive-axle systems to test the load-carrying capacity of hypoid gear (see *gear*) lubricants.

four-stroke-cycle — see *internal combustion engine.*

fraction, fractionation — see *cut, distillation.*

freezing point — a specific temperature that can be defined in two ways, depending on the ASTM test used. In ASTM D 1015, which measures the freezing point of high-purity petroleum products (such as *nitration-grade* toluene), freezing point is the temperature at which a liquid solidifies. In ASTM D 2386, which measures the freezing point of aviation fuel, freezing point is that temperature at which hydrocarbon crystals formed on cooling disappear when the temperature of the fuel is allowed to rise.

Freon floc point — see *floc point.*

fretting — form of wear resulting from small-amplitude oscillations or vibrations that cause the removal of very finely divided particles from rubbing surfaces (e.g., the vibrations imposed on the wheel bearings of an automobile when transported by rail car). With ferrous metals the wear particles oxidize to a reddish, abrasive iron oxide, which has the appearance of rust or corrosion, and is therefore sometimes called **fretting corrosion;** other terms applied to this phenomenon are **false Brinelling** (localized fretting involving the rolling elements of a bearing) and **friction oxidation.**

fretting corrosion — see *fretting.*

friction — resistance to the motion of one surface over another. The amount of friction is dependent on the smoothness of the contacting surfaces, as well as the force with which they are pressed together. Friction between unlubricated solid bodies is independent of speed and area. The **coefficient of friction** is obtained by dividing the force required to move one body over a horizontal surface at constant speed by the weight of the body; e.g., if a force of 4 kilograms is required to move a body weighing 10 kilograms, the coefficient of friction is 4/10, or 0.4. Coefficients of rolling friction (e.g., the motion of a tire or ball bearing) are much less than coefficients of sliding friction (back and forth motion over two flat surfaces). Sliding friction is thus more wasteful of energy and can cause more wear. **Fluid friction** occurs between the molecules of a gas or liquid in motion, and is expressed as *shear stress.* Unlike solid friction, fluid friction varies with speed and area. In general, lubrication is the substitution of low fluid friction in place of high solid-to-solid friction. See *tribology.*

friction oxidation — see *fretting.*

front-end volatility — see *distillation test.*

fuel-economy oil — engine oil specially formulated to increase fuel efficiency. A fuel-economy oil works by reducing the *friction* between moving engine parts that wastefully consumes fuel energy. There are two known means of accomplishing this goal: 1) by reducing the *viscosity* of the oil to decrease fluid friction and 2) by using friction-reducing additives in the oil to prevent metal-to-metal contact, or rubbing friction, between surface *asperities.*

fuel injection — method of introducing fuel under pressure through a small nozzle into the intake system of cylinders of an engine. Fuel injection is essential to the diesel cycle, and an alternative to conventional carburetion in the gasoline engine. In some designs, each cylinder has a cam-operated injector, which is a plunger pump that delivers precisely metered quantities of fuel at precise intervals. The fuel is injected in a minutely divided spray at high discharge pressures. The amount of the charge is controlled by the throttle pedal. A combination of fuel injection and carburetion is used in advanced emission-control systems, involving fuel injection into the throttle body of the carburetor. Fuel injection offers certain advantages over carburetion, including: more balanced fuel distribution in the cylinders for improved combustion, more positive delivery of fuel to the cylinder (hence, easier starting and faster acceleration), and higher power output because of improved *volumetric efficiency*. See *carburetor*.

fuel oil — term encompassing a broad range of *distillate* and residual fuels identified by ASTM grades 1 through 6. Grade No. 1, a *kerosene*-type fuel, is a light distillate fuel that has the lowest boiling range. No. 2 fuel oil, popularly called *heating oil*, has a higher boiling range and is commonly used in home heating. It is comparable in boiling range to *diesel fuel*. Grades 4, 5, and 6 are called **heavy fuel oils** (HFO), or **residual fuel oils**; they are composed largely of heavy pipe still *bottoms*. Because of their high viscosity, No. 5 and No. 6 fuel oils require preheating to facilitate pumping and burning. No. 6 fuel oil is also called **Bunker C fuel oil.**

fuel pump — mechanism for delivering fuel from the tank to the *carburetor* of a gasoline engine, to the fuel injectors of a diesel engine, or to the fuel atomizers of an oil-fired boiler.

gas blanket — atmosphere of inert gas (usually nitrogen or carbon dioxide) lying above a lubricant in a tank and preventing contact with air. In the absence of such a covering, the lubricant would be subject to oxidation. Gas blankets are commonly used with *heat transfer fluids* and electrical *insulating oils*.

gas engine — *internal combustion engine*, either two- or four-stroke cycle, powered by natural gas or *LPG*. Commonly used to drive compressors on gas pipelines, utilizing as fuel a portion of the gas being compressed.

gasohol — blend of 10 volume percent anhydrous *ethanol* (ethyl alcohol) and 90 volume percent unleaded gasoline.

gas oil — liquid petroleum *distillate,* higher boiling than *naphtha;* initial boiling point may be as low as 204°C (400°F). Gas oil is called light or heavy, depending on its final boiling point. It is used in blending *fuel oil* and as refinery *feedstock* in *cracking* operations.

gasoline — blend of light hydrocarbon fractions of relatively high *antiknock* value. Finished motor and *aviation gasolines* may consist of the following components: straight-run *naphthas,* obtained by the primary *distillation* of crude oil; *natural gaso-*

full-fluid-film lubrication —presence of a continuous lubricating film sufficient to completely separate two surfaces, as distinct from *boundary lubrication.* Full-fluid-film lubrication is normally **hydrodynamic lubrication,** whereby the oil adheres to the moving part and is drawn into the area between the sliding surfaces, where it forms a pressure, or hydrodynamic, wedge. See *ZN/P curve.* A less

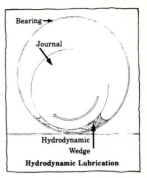

Hydrodynamic Lubrication

common form of full-fluid-lubrication is **hydrostatic lubrication,** wherein the oil is supplied to the bearing area under sufficient external pressure to separate the sliding surfaces.

fully refined wax — see *refined wax.*

furfural — colorless liquid, C_4H_3OCHO, employed in petroleum refining as a solvent to extract *mercaptans, polar compounds, aromatics,* and other impurities from oils and waxes. Also used in the manufacture of dyes and plastics.

Furol viscosity — viscosity of a petroleum oil measured with a Saybolt Furol viscometer; see *viscosity.*

FZG four-square gear oil test — test used in developing industrial gear lubricants to meet equipment manufacturers' specifications. The FZG test equipment consists of two gear sets, arranged in a four-square configuration, driven by an electric motor. The test gear set is run in the lubricant at gradually increased load stages until failure, which is the point at which a 10 milligram weight loss by the gear set is recorded. Also called **Niemann four-square gear oil test.**

line, which is "stripped", or condensed, out of natural gas; cracked naphthas; reformed naphthas; and alkylate. (See *alkylation, catalytic cracking, reforming.*). A high-quality gasoline has the following properties: (1) proper *volatility* to ensure easy starting and rapid warm-up; (2) clean-burning characteristics to prevent harmful *engine deposits;* (3) *additives* to prevent rust, *oxidation,* and carburetor icing; (4) sufficiently high *octane number* to prevent engine knock.

gas turbine — see *internal combustion engine, turbine.*

gauge pressure — see *pressure.*

gear — machine part which transmits motion and force from one rotary shaft to another by means of successively engaging projections, called teeth. The smaller gear of a pair is called the **pinion;** the larger, the gear. When the pinion is on the driving shaft, the gear set acts as a

Helical Gear

speed reducer; when the gear drives, the set acts as a speed multiplier. The basic gear type is the **spur gear,** or **straight-tooth gear,** with teeth cut parallel to the gear axis. Spur gears transmit power in applications utilizing parallel shafts. In this type of gear, the teeth mesh along their full length, creating a sudden shift in load from one tooth to the next, with consequent noise and vibration. This problem is overcome by the **helical gear,** which has teeth cut at an angle to the center of rotation, so that the load is transferred progressively along the length of the tooth from one edge of the gear to the other. When the shafts are not parallel, the most common gear type used is the **bevel gear,** with teeth cut on a sloping gear face, rather than parallel to the

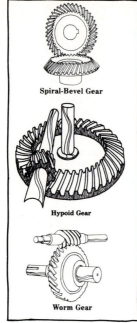

Spiral-Bevel Gear

Hypoid Gear

Worm Gear

shaft. The **spiral bevel gear** has teeth cut at an angle to the plane of rotation, which, like the helical gear, reduces vibration and noise. A **hypoid gear** resembles a spiral bevel gear, except that the pinion is offset so that its axis does not intersect the gear axis; it is widely used in automobiles between the engine driveshaft and the rear axle. Offset of the axes of hypoid gears introduces additional sliding between the teeth, which, when combined with high loads, requires a high-quality EP oil. A **worm gear** consists of a spirally grooved screw moving against a toothed wheel; in this type of gear, where the load is transmitted across sliding, rather than rolling, surfaces, *compounded oils* or *EP oils* are usually necessary to maintain effective lubrication.

gear oil (automotive) — long-life oil of relatively high viscosity for the lubrication of rear axles and some manual transmissions. Most final drives and many accessories in agricultural and construction equipment also require gear oils. Straight (non-additive) mineral gear oils are suitable for most spiral-bevel rear axles (see *gear)* and for some manual transmissions. Use of such oils is declining, however, in favor of EP (extreme pressure) gear oils (see *EP oil)* suitable both for hypoid gears (see *gear)* and for all straight mineral oil applications. An EP gear oil is also appropriate for off-highway and other automotive applications for which the lubricant must meet the requirements of Military Specification MIL-L-2105C.

gear oil (industrial) — high-quality oil with good *oxidation stability,* rust protection, and resistance to *foaming,* for service in gear housings and enclosed chain drives. A *turbine oil* or *R&O* oil is the usual gear oil recommendation. Specially formulated industrial EP gear oils (see *EP oil)* are used where highly loaded gear sets or excessive sliding action (as in worm gears) is encountered. See *gear.*

gear shield — highly adhesive lubricant of heavy consistency, formulated with *asphaltic* compounds or *polymers* for protection of exposed gears and wire rope in circumstances where the lubricant cannot readily be replenished. Many gear shield lubricants must be softened with heat or cut back with solvents before they can be applied.

general purpose oils — see *once-through lubrication.*

gilsonite — a naturally occurring *asphalt* mined from rock fissures. It is hard and brittle, and has a high melting point. It is used in the manufacture of *rust preventives,* paints, sealants, and lacquers.

gloss — property of wax determinable by measuring light reflected from a wax-treated paper surface. Gloss stability is evaluated after a sample of treated paper has been held at an elevated temperature for a specified period.

gram calorie — see *calorie.*

graphite — a soft form of elemental carbon, gray to black, in color. It occurs naturally or is synthesized from coal or other carbon sources. It is used in the manufacture of paints, lead pencils, crucibles, and electrodes, and is also widely used as a lubricant, either alone or added to conventional lubricants.

grease — mixture of a fluid lubricant (usually a petroleum oil) and a thickener (usually a soap) dispersed in the oil. Because greases do not flow readily, they are used where extended lubrication is required and where oil would not be retained. The thickener may play as important a role as the oil in lubrication. **Soap thickeners** are formed by reacting (saponifying) a metallic hydroxide, or *alkali,* with a fat, *fatty acid,* or *ester.* The type of soap used depends on the grease properties desired. **Calcium** (lime) **soap greases** are highly resistant to water, but unstable at high temperatures. **Sodium soap greases** are stable at high temperatures, but wash out in moist conditions. **Lithium soap greases** resist both heat and moisture. A mixed-base soap is a combination of soaps, offering some of the advantages of each type. A **complex soap** is formed by the reaction of an alkali with a high-molecular-weight fat or fatty acid to form a soap, and the simultaneous reaction of the alkali with a short-chain organic or inorganic acid to form a metallic salt (the complexing agent). Complexing agents usually increase the *dropping point* of grease. Lithium, calcium, and aluminum greases are common alkalis in complex-soap greases. **Non-soap thickeners,** such as clays, silica gels, carbon black, and various synthetic organic materials are also used in grease manufacture. A *multipurpose grease* is designed to provide resistance to heat, as well as to water, and may contain additives to increase load-carrying ability and inhibit rust. See *consistency (grease), penetration (grease).*

grease gun injury — see *high-pressure-injection injury.*

gum in gasoline — oily, viscous contaminant that may form due to oxidation during storage. Gum formation in gasoline can cause serious fuel system problems, such as carburetor malfunctioning and intake valve sticking. See *oxidation.* The amount of gum in motor gasoline, aviation gasoline, and aircraft turbine fuel can be determined by evaporating a measured sample by means of air or steam flow at controlled temperature, and weighing the residue, as described in test method ASTM D 381.

halogen — any of a group of five chemically related nonmetallic elements: chlorine, bromine, fluorine, iodine, and astatine. Chlorine compounds are used as *EP additives* in certain lubricating oils, and as constituents of certain *petrochemicals* (e.g. vinyl chloride, chlorinated waxes). Chlorine and fluorine compounds are also used in some *synthetic lubricants*.

heating oil — see *fuel oil.*

heating value — see *heat of combustion.*

heat of combustion — measure of the available energy content of a fuel, under controlled conditions specified by test method ASTM D 240 or D 2382. Heat of combustion is determined by burning a small quantity of a fuel in an oxygen bomb calorimeter and measuring the heat absorbed by a specified quantity of water within the calorimeter. Heat of combustion is expressed either as *calories* per gram or *British thermal units* per pound. Also called **thermal value, heating value, calorific value.**

heat transfer fluid — circulating medium (often a petroleum oil) that absorbs heat in one part of a system (e.g. a solar heating system or a remote oil-fired system) and releases it to another part of the system. Heat transfer fluids require high resistance to cracking (molecular breakdown) when used in systems with fluid temperatures above 260°C (500°F). Systems can be either closed or open to the atmosphere. To prevent oxidation in a closed system inert gas is sometimes used in the expansion tank (or reservoir) to exclude air (oxygen). See *gas blanket.* If the system is open and the fluid is exposed simultaneously to air and to temperatures above 66°C (150°F), the fluid must also have good *oxidation stability,* since a protective gas blanket cannot be contained.

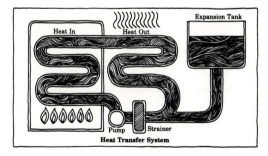

Heat Transfer System

heat treating oil — see *quenching oil.*

heavy crude naphtha — see *naphtha.*

heavy ends — highest boiling portion in a distilled petroleum fraction or finished product. In motor gasoline, the heavy ends do not fully volatilize until the engine has warmed. See *light ends.*

heavy fuel oil — see *fuel oil.*

helical gear — see *gear.*

heptane — liquid paraffinic (see *paraffin)* hydrocarbon containing seven carbon atoms in the molecule, which may be straight-chain (normal) or branched-chain (iso). Heptane can be used in place of *hexane* where a less volatile solvent is desired, as in the manufac-

ture of certain adhesives and lacquers, and in *extraction* of edible and commercial oils. Heptane is blended with *isooctane* to create a standard reference fuel in laboratory determinations of *octane number.*

hexane — highly volatile paraffinic (see *paraffin)* hydrocarbon containing six carbon atoms in the molecule; it may also contain six-carbon *isoparaffins.* Widely used as a solvent in adhesive and rubber solvent formulations, and in the *extraction* of a variety of edible and commercial oils. Hexane is a neurotoxin and must be handled with adequate precautions.

high-pressure-injection injury — injury caused by the accidental injection of grease or oil under pressure through the skin and into the underlying tissue; also called a **grease gun injury.** Such an injury requires immediate medical attention.

homogenization — intimate mixing of a lubricating grease or an *emulsion* by intensive shearing action to obtain more uniform dispersion of the components.

horsepower — unit of power equal to 33 000 foot-pounds per minute, equivalent to 745.7 watts.

hot-dip rust preventive — petroleum-base *rust preventive,* consisting of a blend of oil, wax or asphalt, and rust-inhibiting additives, that must be melted before application.

humidity — water vapor in the atmosphere. **Absolute humidity** is the amount of water vapor in a given quantity of air; it is not a function of temperature. **Relative humidity** is a ratio of actual atmospheric moisture to the maximum amount of moisture that could be carried at a given temperature, assuming constant atmospheric pressure. The higher the temperature — other factors remaining constant — the lower the relative humidity (i.e., the drier the air).

hydrated soap — grease thickener that has water incorporated into its structure to improve structural stability of the grease. See *grease.*

hydraulic fluid — fluid serving as the power transmission medium in a *hydraulic system.* The most commonly used fluids are petroleum oils, *synthetic lubricants,* oil-water emulsions, and water-glycol mixtures. The principal requirements of a premium hydraulic fluid are proper *viscosity,* high *viscosity index,* anti-wear protection (if needed), good *oxidation stability,* adequate *pour point,* good *demulsibility,* rust inhibition (see *rust inhibitor),* resistance to *foaming,* and compatibility with seal materials. Anti-wear oils are frequently used in compact, high-pressure, and high-capacity pumps that require extra lubrication protection. Certain *synthetic lubricants* and water-containing fluids are used where fire resistance is needed. See *fire-resistant fluids.*

hydraulic system — system designed to transmit power through a liquid medium, permitting multiplication of force in accordance with Pascal's law, which states that "a pressure exerted on a confined liquid is transmitted undiminished in all directions and acts with equal force on all equal areas." Hydraulic systems have six basic components: (1) a reservoir to hold the fluid supply; (2) a fluid to transmit the power; (3) a pump to move the fluid; (4) a valve to regulate pressure; (5) a directional valve to control the flow, and (6) a working component — such as a cylinder and piston or a shaft rotated by pressurized fluid — to turn hydraulic power into mechanical motion. Hydraulic systems offer several advantages over mechanical systems: they eliminate complicated mechanisms such as cams, gears, and levers; are less subject to wear; are usually more easily adjusted for control of speed and force; are easily adaptable to both rotary and linear transmission of power; and can transmit power over long distances and in any direction with small losses.

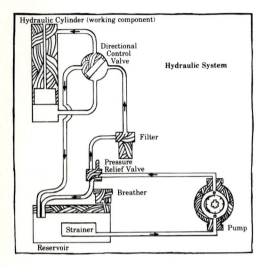

hydraulic transmission fluid — see *automatic transmission fluid.*

hydraulic turbine — see *turbine.*

hydrocarbon — chemical compound of hydrogen and carbon; also called an *organic compound.* Hydrogen and carbon atoms can be combined in virtually countless ways to make a diversity of products. Carbon atoms form the skeleton of the hydrocarbon molecule, and may be arranged in chains (aliphatic) or rings (cyclic). There are three principal types of hydrocarbons that occur naturally in petroleum: *paraffins, naphthenes,* and *aromatics,* each with distinctive properties. Paraffins are aliphatic, the others cyclic. Paraffins and naphthenes are saturated; that is, they have a full complement of hydrogen atoms and, thus, only single bonds between carbon atoms. Aromatics are unsaturated, and have as part of their molecular structure at least one *benzene* ring, i.e., six carbon atoms in a ring configuration with alternating single and double bonds. Because of these double bonds, aromatics are usually more reactive than paraffins and naphthenes, and are thus prime starting materials for chemical synthesis. Other types of hydrocarbons are formed during the petroleum refining process. Important among these are *olefins* and *acetylenes.* Olefins are unsaturated hydrocarbons with at least one double bond in the molecular structure, which may be in either an open chain or ring configuration; olefins are highly reactive. Acetylenes are also unsaturated and contain at least one triple bond in the molecule. See *saturated hydrocarbons, unsaturated hydrocarbons.*

hydrocarbon (HC) emissions — substances considered to be atmospheric pollutants because the more reactive *hydrocarbons* (e.g., *aromatics)* undergo a photochemical reaction with *nitrogen oxides (NOx)* to form oxidants, components of smog that can cause eye irritation and respiratory problems. Motor vehicles account for about one-third of man-made hydrocarbon emissions, although automotive emission controls are reducing this amount. The greatest portion of total atmospheric hydrocarbons is from natural sources, such as pine trees. See *catalytic converter, emissions (automotive), pollutants.*

hydrocracking — refining process in which middle and heavy *distillate* fractions are cracked (broken into smaller molecules) in the presence of hydrogen at high pressure and moderate temperature to produce high-octane gasoline, turbine fuel components, and middle distillates with good flow characteristics and *cetane* ratings. The process is a combination of *hydrogenation* and *cracking.*

hydrodynamic lubrication — see *full-fluid-film lubrication.*

Hydrofining® — form of *hydrogen treating* in which refinery distillate, lube, and wax streams are treated with hydrogen at elevated temperatures and moderate pressures in the presence of a *catalyst,* to improve color and stability, and reduce sulfur content. The patented process was developed by Exxon in 1951.

hydroforming — a dehydrogenation process in which *naphthas* are passed over a solid *catalyst* at elevated temperatures and moderate pressures in the presence of hydrogen to form high-octane motor gasoline, high-grade aviation gasoline, or aromatic solvents. The process is a net producer of hydrogen.

hydrogenation — in refining, the chemical addition of hydrogen to a *hydrocarbon* in the presence of a *catalyst;* a severe form of *hydrogen treating.* Hydrogenation may be either destructive or non-destructive. In the former case, hydrocarbon chains are ruptured (cracked) and hydrogen is added where the breaks have occurred. In the latter, hydrogen is added to a molecule that is unsaturated (see *unsaturated hydrocarbon)* with respect to hydrogen. In either case, the resulting molecules are highly stable. Temperature and pressures in the hydrogenation process are usually greater than in *Hydrofining®.*

hydrogen sulfide (H₂S) — gaseous compound of sulfur and hydrogen commonly found in crude oil; it is extremely poisonous, corrosive, and foul-smelling.

hydrogen treating — refining process in which *hydrocarbons* are treated with hydrogen in the presence of a *catalyst* at relatively low temperatures to remove *mercaptans* and other sulfur compounds, and improve color and stability. See *Hydrofining®.*

hydrolytic stability — ability of additives and certain *synthetic lubricants* to resist chemical decomposition (hydrolysis) in the presence of water.

hydrometer — see *specific gravity.*

hydrophilic — also **hygroscopic,** having an affinity for water. Some *polar compounds* are simultaneously hydrophilic and oil soluble.

hydrophobic — the opposite of *hydrophilic.*

hydrostatic lubrication — see *full-fluid-film lubrication.*

hygroscopic — see *hydrophilic.*

hypoid gear — see *gear.*

IBP — initial boiling point; see *distillation test.*

IFT — see *interfacial tension.*

immiscible — incapable of being mixed without separation of phases. Water and petroleum oil are immiscible under most conditions, although they can be made miscible with the addition of an *emulsifier.* See *miscible.*

impact odor — see *bulk odor.*

induction period — the time period in an *oxidation* test during which oxidation proceeds at a constant and relatively low rate. It ends at a point at which the rate of oxidation increases sharply.

industrial asphalt — oxidized asphalt (see *asphalt*) used in the manufacture of roofing, asphaltic paints, *mastics,* and adhesives for laminating paper and foil. Industrial asphalt is generally harder than *asphalt cement,* which is used for paving.

industrial lubricant — any petroleum or synthetic-base fluid (see *synthetic lubricant)* or grease commonly used in lubricating industrial equipment, such as gears, turbines, compressors.

inhibitor — additive that improves the performance of a petroleum product through the control of undesirable chemical reactions. See *corrosion inhibitor, oxidation inhibitor, rust inhibitor.*

initial boiling point (IBP) — see *distillation test.*

inorganic compound — chemical compound, usually mineral, that does not include *hydrocarbons* and their derivatives. However, some relatively simple carbon compounds, such as carbon dioxide, metallic carbonates, and carbon disulfide are regarded as inorganic compounds.

insoluble resins — see *insolubles.*

insolubles — test for contaminants in used lubricating oils, under conditions prescribed by test method ASTM D 893. The oil is first diluted with *pentane,* causing the oil to lose its solvency for certain oxidation *resins,* and also causing the precipitation of such extraneous materials as dirt, soot, and wear metals. These contaminants are called **pentane insolubles.** The pentane insolubles may then be treated with *toluene,* which dissolves the oxidation resins (benzene was formerly used). The remaining solids are called **toluene insolubles.** The difference in weight between the pentane insolubles and the toluene insolubles is called **insoluble resins.**

insulating oil — high-quality oxidation-resistant oil refined to give long service as a *dielectric* and coolant for transformers and other electrical equipment. Its most common application is as a **transformer oil.** An insulating oil must resist the effects of elevated temperatures, electrical stress, and contact with air, which can lead to sludge formation and loss of insulation properties. It must be kept dry, as water is detrimental to *dielectric strength.*

intercooling — cooling of a gas at constant pressure between stages in a compressor. It permits reduced work in the compression phase because cooler gas is more easily compressed. *Aftercooling* is the final cooling following the last compression stage.

interfacial tension (IFT) — the force required to rupture the interface between two liquid phases. The interfacial tension between water and a petroleum oil can be determined by measuring the force required to move a platinum ring upward through the interface, under conditions specified by test method ASTM D 971. Since the interface can be weakened by oxidation products in the oil, this measurement may be evidence of oil deterioration. The lower the surface tension below the original value, the greater the extent of oxidation. ASTM D 971 is not widely used with additive-containing oils, since additives may affect surface tension, thus reducing the reliability of the test as an indicator of oxidation.

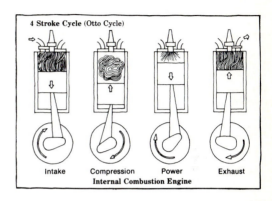

4 Stroke Cycle (Otto Cycle)

Intake — Compression — Power — Exhaust
Internal Combustion Engine

internal combustion engine — heat engine driven directly by the expansion of combustion gases, rather than by an externally produced medium, such as steam. Basic versions of the internal combustion engine are: **gasoline engine** and **gas engine** (spark ignition), **diesel engine** (compression ignition), and **gas turbine** (continuous combustion). Diesel compression-ignition engines are more fuel-efficient than gasoline engines because *compression ratios* are higher, and because the absence of air throttling improves *volumetric efficiency.* Gasoline, gas (natural gas, propane), and diesel engines operate either on a **four-stroke cycle (Otto cycle)** or a **two-stroke cycle.** Most gasoline engines are of the four-stroke type, with operation as follows: (1) *intake* – piston moves down the cylinder, drawing in a fuel-air mixture through the intake valve; (2) *compression* – all valves closed, piston moves up, compressing the fuel-air mixture, and spark ignites mixture near top of stroke; (3) *power* – rapid expansion of hot combustion gases drives piston down, all valves remain closed; (4) *exhaust* – exhaust valve opens and piston returns, forcing out spent gases. The diesel four-stroke cycle differs in that only air is admitted on the intake stroke, fuel is injected at the top of the compression stroke, and the fuel-air mixture is ignited by the heat of compression rather than by an electric spark. The four-stroke-cycle engine has certain advantages over a two-stroke, which include: higher piston speeds, wider variation in speed and load, cooler pistons, no fuel lost through the exhaust, and lower fuel consumption. ● The two-stroke cycle eliminates the intake and exhaust strokes of the four-stroke cycle. As the piston ascends, it compresses the charge in the cylinder, while simultaneously drawing a new fuel-air charge into the crankcase, which is air-tight. (In the diesel two-stroke cycle, only air is drawn in; the fuel is injected at the top of the compression stroke.) After ignition, the piston descends on the power stroke, simultaneously compressing the fresh charge in the crankcase. Toward the end of the power stroke, intake ports in the piston skirt admit a new fuel-air charge that sweeps exhaust products from the cylinder through exhaust ports; this means of flushing out exhaust gases is called "scavenging". Because the crankcase is needed to contain the intake charge, it cannot double as an oil reservoir. Therefore, lubrication is generally supplied by oil that is pre-mixed with the fuel.   An important advantage of the two-stroke-cycle engine is that it offers twice as many power strokes per cycle and, thus, greater output for the same displacement and speed. Because two-stroke engines are light in relation to their output, they are frequently used where small engines are desirable, as in chain saws, outboard motors, and lawn mowers. Many commercial, industrial, and railroad diesel engines are also of the two-stroke type. ● Gas turbines differ from conventional internal combustion engines in that a continuous stream of hot gases is directed at the blades of a rotor. A compressor section supplies air to a combustion chamber into which fuel is sprayed, maintaining continuous combustion. The resulting hot gases expand through the turbine unit, turning the rotor and driveshaft. See *turbine*.

International System of Units — see *SI*.

inverse emulsion — see *emulsion*.

ion — electrically charged atom, or group of atoms, that has lost or gained electrons. Electron loss makes the resulting particle positive, while electron gain makes the particle negative.

isomer — molecule having the same molecular formula as another molecule, but having a different structure and, therefore, different properties. As the carbon atoms in a molecule increase, the number of possible combinations, or isomers, increases sharply. For example, octane, an 8-carbon-atom molecule, has 18 isomers; decane, a 10-carbon-atom molecule, has 75 isomers.

jet fuel — see *turbo fuel*.

joule — unit of energy in the *Système International*, equal to the work done when the point of application of a force of one *newton* is

°K (Kelvin) — see *temperature scales*.

kauri-butanol (KB) value — a measure of the solvency of a *hydrocarbon;* the higher the kauri-butanol value, the greater the general solvent power of the hydrocarbon. Under test conditions prescribed in test method ASTM D 1133, a hydrocarbon sample is added to a standard solution of kauri gum in butyl alcohol (butanol) until sufficient kauri gum precipitates to blur vision of 10-point type viewed through the flask. When used in varnish, lacquer, and enamel formulations, a hydrocarbon *diluent* with a high kauri-butanol value dissolves relatively large quantities of solids.

Kelvin (°K) — see *temperature scales*.

kerosene — relatively colorless light *distillate*, heavier than gasoline (see *distillation*). It is used for lighting and heating, and as a fuel for some *internal combustion engines*.

kilocalorie — see *calorie*.

kilowatt-hour — unit of work or energy, equivalent to the energy expended in one hour at a steady rate of one kilowatt. Total

laminating strength — bonding strength of a petroleum wax used as an adhesive between layers of paper or foil, measured in terms of

normal butane

isobutane

isooctane — an *isomer* of octane (C_8H_{18}) having very good *anti-knock* properties. With a designated octane number of 100, isooctane is used as a standard for determining the *octane number* of gasolines.

isoparaffin — branched *isomer* of a straight-chain *paraffin* molecule.

isoprene rubber — see *polyisoprene rubber*.

isothermal — pertaining to the conduct of a process or operation of equipment under conditions of constant temperature.

ISO viscosity classification system — international system, approved by the International Standards Organization (ISO), for classifying industrial lubricants according to *viscosity*. Each ISO viscosity grade number designation corresponds to the mid-point of a viscosity range expressed in centistokes (cSt) at 40°C. For example, a lubricant with an ISO grade of 32 has a viscosity within the range of 28.8 — 35.2 cSt, the mid-point of which is 32.

displaced a distance of one meter in the direction of the force.

journal — that part of a shaft or axle which rotates in or against a *bearing*.

kilowatt-hours (kwh) consumed by one 100-watt light bulb burning for 150 hours can be calculated as follows: 100 watts × 150 hours = 15,000 watt-hours = 15 kwh.

kinematic viscosity — *absolute viscosity* of a fluid divided by its density at the same temperature of measurement. It is the measure of a fluid's resistance to flow under gravity, as determined by test method ASTM D 445. To determine kinematic viscosity, a fixed volume of the test fluid is allowed to flow through a calibrated capillary tube *(viscometer)* that is held at a closely controlled temperature. The kinematic viscosity, in centistokes (cSt), is the product of the measured flow time in seconds and the calibration constant of the viscometer. See *viscosity*.

knock — in the cylinder of a spark-ignited *internal combustion engine*, premature explosion of a portion of the air-fuel mixture, independent of spark plug ignition, as a result of excessive heat buildup during compression. The high local pressures resulting from the explosion are the source of the objectionable clatter or ping associated with knock. Knock reduces efficiency and can be destructive to engine parts. High-octane gasolines resist knocking. Also called **detonation.** See *octane number, pre-ignition*.

the specific number of grams of force required to peel the layers apart. It is expressed in grams per inch of width of the layers.

Sealing strength, similar in meaning to laminating strength, may also be measured in this manner.

lard oil — *fatty oil* used for compounding. See *compounded oil.*

latent heat — quantity of heat absorbed or released by a substance undergoing a change of state (e.g., ice changing to liquid water, or water to steam) without change of temperature. At standard atmospheric pressure, the latent heat of vaporization of water is 2256 kJ/kg (970 Btu/lb); this is the amount of energy required to convert water at 100°C (212°F) to steam at the same temperature. Conversely, when steam condenses at 100°C, the same amount of heat is released.

launching lubricant — lubricant applied to inclined launching guides, or ways, to facilitate launching of a ship. Two separate lubricants are usually used: a firm, abrasion-resistant base coat, plus a softer, low-friction slip coat.

LC50 — lethal concentration, 50% mortality; a measure of inhalation toxicity. It is the concentration in air of a volatile chemical compound at which half the test population of an animal species dies when exposed to the compound. It is expressed as parts per million by volume of the toxicant per million parts of air for a given exposure period.

LCN — light crude naphtha. See *naphtha.*

LD50 — lethal dose, 50% mortality, a general measurement of toxicity. It is the dose of a chemical compound that, when administered to laboratory animals, causes death in one-half the test population. It is expressed in milligrams of toxicant per kilogram of animal weight. The route of administration, may be oral, epidermal, or intraperitoneal.

lead alkyl — any of several lead compounds used to improve *octane number* in a gasoline. The best known is **tetraethyl lead** (TEL), Pb (C$_2$H$_5$)$_4$. Another is **tetramethyl lead** (TML), Pb (CH$_3$)$_4$. Other compounds have varying proportions of methyl *radicals* (CH$_3$) and ethyl radicals (C$_2$H$_5$). Use of lead compounds in motor gasoline is being phased out for environmental reasons. Beginning with the 1980-model year, all new U.S. and foreign-made cars sold in the U.S. required *unleaded gasoline.*

lead naphthenate, lead oleate — lead soaps that serve as mild *EP additives.* These additives are seldom used in modern lubricants because of environmental considerations.

lead scavenger — see *scavenger.*

lean and rich octane number — expression of the *antiknock* value of an *aviation gasoline* at lean air-fuel mixtures (relatively low concentration of fuel) and rich air-fuel mixtures, respectively. A grade designation of 80/87 means that at lean mixtures the fuel performs like an 80-octane gasoline and at rich mixtures, like an 87-octane gasoline. See *performance number.*

light crude naphtha — see *naphtha.*

light ends — low-boiling-point *hydrocarbons* in gasoline having up to five carbon atoms, e.g., *butanes,* butenes, *pentanes,* pentenes, etc. Also, any extraneous low-boiling fraction in a refinery process stream.

limestone — porous, sedimentary rock composed chiefly of calcium carbonate; sometimes serves as a *reservoir* rock for petroleum.

linear paraffin — see *normal paraffin.*

liquefied natural gas — see *LNG.*

liquefied petroleum gas — see *LPG.*

lithium soap grease — see *grease.*

LNG (liquefied natural gas) — natural gas that has been liquefied at extremely low temperature. It is stored or transported in insulated tanks capable of sustaining the high pressure developed by the product at normal ambient temperatures.

load wear index (LWI) — measure of the relative ability of a lubricant to prevent wear under applied loads; it is calculated from data obtained from the Four Ball EP Method. Formerly called **mean Hertz load.** See *four-ball method.*

local effect — toxic effect that is limited to the area of the body (commonly the skin and eyes) that has come into contact with a toxicant.

Lovibond tintometer — device for measuring the color of a petroleum product, particularly *petrolatums.* The melted petrolatum is contained in a cell and the color is compared with a series of yellow and red Lovibond glasses. The length of the cell and the color standards that give the best match are reported. See *color scale.*

lower flammable limit — the concentration of a flammable vapor mixed with air that will just propagate flame, that is, continue to burn. See *explosive limits.*

low-temperature corrosion — see *cold-end corrosion.*

LP gas — see *LPG.*

LPG (liquefied petroleum gas) — *propane* or (less commonly) *butane,* obtained by extraction from *natural gas* or from refinery processes. LPG has a *vapor pressure* sufficiently low to permit compression and storage in a liquid state at moderate pressures and normal ambient temperatures. Pressurized in metal bottles or tanks, LPG is easily handled and readily lends itself to a variety of applications as a fuel, refrigerant, and propellant in packaged aerosols. LPG is also called **LP gas** and *bottled gas.* See *natural gas liquids.*

lubrication — control of friction and wear by the introduction of a friction-reducing film between moving surfaces in contact. The lubricant used may be a fluid, solid, or plastic substance. For principles of lubrication, see *boundary lubrication, full-fluid-film lubrication, ZN/P curve.* For methods of lubrication, see *centralized lubrication, circulating lubrication, oil mist lubrication, once-through lubrication.*

lubricity — ability of an oil or grease to lubricate; also, called **film strength.** Lubricity can be enhanced by *additive* treatment. See *compounded oil.*

luminometer number — measure of the flame radiation characteristics of a turbine fuel, as determined by test method ASTM D 1740. A sample of fuel is burned in a luminometer lamp, and the temperature rise, at a specified flame radiation value, is compared with the corresponding temperature rise of reference fuels. The higher the luminometer number, the lower the flame radiation and the better the combustion characteristics.

LWI — see *load wear index.*

machine oil — see *once-through lubrication.*

marquenching — see *quenching.*

martempering — see *quenching.*

mass spectrometer — apparatus for rapid quantitative and qualitative analysis of hydrocarbon compounds in a petroleum sample. It utilizes the principle of accelerating molecules in a circular path in an electrical field. The compounds are separated by centrifugal force, with the molecules having a greater mass (weight) being thrown to the outer periphery of the path. Quantitative measurements are accomplished by use of either a photographic plate or electronic determination of the relative proportions of each type of particle of a given mass.

mastic — any of various semi-solid substances, usually formulated with rubber, other *polymers,* or oxidized *asphalt;* commonly used as a tile adhesive caulking, and a sound-reducing treatment on various surfaces.

MC asphalt — see *cutback asphalt.*

mean Hertz load — see *load wear index.*

mechanical stability — see *structural stability.*

melting point of wax — temperature at which a sample of wax either melts or solidifies from the solid or liquid state, respectively, depending on the *ASTM* test used. Low melting point generally indicates low *viscosity,* low *blocking point,* and relative softness.

mercaptan — any of a generic series of malodorous, toxic sulfur compounds occurring in crude oil. Mercaptans are removed from most petroleum products by refining, but may be added to natural gas and *LPG* in very low concentrations to give a distinctive warning odor.

merit rating — arbitrary graduated numerical rating commonly used in evaluating engine deposit levels when testing the *detergent-dispersant* characteristics of an engine oil. On a scale of 10-0, the lower the number, the heavier the deposits. A less common method of evaluating engine cleanliness is *demerit rating.* See *engine deposits.*

metal wetting — see *polar compound.*

methane — a light, odorless, flammable gas (CH_4); the chief constituent of natural gas.

methanol — the lowest molecular weight *alcohol* (CH_3OH). Also called **methyl alcohol** and **wood alcohol.**

methyl alcohol — see *methanol.*

metric system — international decimal system of weights and measures based on the meter and kilogram. The following table presents common metric units and their U.S. equivalents:

Metric Unit		Equivalent in U.S. Customary Unit
1 meter (m)	=	3.28 feet
1 centimeter (cm)	=	0.394 inches
1 kilometer (km)	=	0.621 miles
1 kilogram (kg)	=	2.205 pounds
1 gram (g)	=	0.353 ounces
1 liter (L)	=	1.056 quarts (liquid)

microcrystalline wax, microwax — see *wax (petroleum).*

mid-boiling point — see *distillation test.*

middle distillate — see *distillate.*

military specifications for engine oils — There are six military specifications for engine oils. Some of these specifications are obsolete, but are still commonly used to designate required engine oil performance levels. To qualify under a military specification, an oil must meet minimum requirements in laboratory engine tests. In most cases these are the same *ASTM* tests used to define *API Engine Service Categories.* The following is a listing of the military specifications and API Service equivalents for engine oils in order of ascending quality (the API Service equivalents of obsolete military specifications remain valid):

Mil. Spec.	Description
MIL-L-2104A (API CA)	obsolete, superseded by MIL-L-2104B; describes an oil type still used by some fleet owners for older, less critical equipment
Supplement I (API CB)	obsolete; describes oil type similar to MIL-L-2104A, but tested under more severe conditions
MIL-L-2104B (API CC)	obsolete, superseded by MIL-L-46152, but still widely used as a standard
MIL-L-46152 (API SE-CC, SF-CC)	supersedes MIL-L-2104B
MIL-L-45199B (API CD)	obsolete, superseded by MIL-L-2104C; the specification is essentially the same as the *Series 3* specification established by Caterpillar Tractor Company
MIL-L-2104C (API CD)	supersedes MIL-L-45199B; covers oil requirements for heavy trucks, tanks, and other tactical military vehicles, and includes gasoline engine performance level between API SC and SD

mineral oil — any petroleum oil, as contrasted to animal or vegetable oils. Also, a highly refined petroleum *distillate,* or *white oil,* used medicinally as a laxative.

mineral seal oil — *distillation* fraction between *kerosene* and *gas oil,* widely used as a solvent oil in gas absorption processes (see *absorber oil),* as a lubricant for the rolling of metal foil, and as a base oil in many specialty formulations. Mineral seal oil takes its name — not from any sealing function — but from the fact that it originally replaced oil derived from seal blubber for use as an illuminant for signal lamps and lighthouses.

mineral spirits — *naphthas* of mixed hydrocarbon composition and intermediate volatility, within the boiling range of 149°C (300°F) to 204°C (400°F) and with a *flash point* greater than 38°C (100°F); widely used as solvents or thinners in the manufacture of cleaning products, paints, lacquers, inks, and rubber. Also used uncompounded for cleaning metal and fabrics.

miscible — capable of being mixed in any concentration without separation of phases; e.g., water and ethyl alcohol are miscible. See *immiscible.*

mist lubrication — see *oil mist lubrication.*

mobilometer — device for measuring the relative consistency or resistance to flow of fluid grades of grease too soft to be tested in the *penetrometer.* See *consistency (grease).*

mold lubricant — a compound, often of petroleum origin, for coating the interiors of molds for glass and ceramic products. The mold lubricant facilitates removal of the molded object from the mold, protects the surface of the mold, and reduces or eliminates the need for cleaning it. Also called **release agent.**

moly, molysulfide — see *molybdenum disulfide.*

molybdenum disulfide — a black, lustrous powder (MoS_2) that serves as a dry-film lubricant in certain high-temperature and high-vacuum applications. It is also used in the form of pastes to prevent *scoring* when assembling press-fit parts, and as an additive to impart residual lubrication properties to oils and greases. Molybdenum disulfide is often called **moly** or **molysulfide.**

monomer — see *polymer.*

Mooney scorch value — see *Mooney viscosity.*

Mooney viscosity — measure of the resistance of raw or unvulcanized rubber to deformation, as measured in a Mooney viscometer. A steel disc is embedded in a heated rubber specimen and slowly rotated. The resistance to the shearing action of the disc is measured and expressed as a Mooney viscosity value. Viscosity increases with continued rotation, and the time required to produce a specified rise in Mooney viscosity is known as the **Mooney scorch value,** which is an indication of the tendency of a rubber mixture to cure, or vulcanize, prematurely during processing.

Motor Octane Number — see *octane number.*

naphtha — generic, loosely defined term covering a range of light petroleum *distillates* (see *distillation*). Included in the naphtha classification are: *gasoline blending stocks, mineral spirits,* and a broad selection of petroleum *solvents.* In refining, the term **light crude naphtha** (LCN) usually refers to the first liquid distillation fraction, boiling range 32° to 100°C (90° to 175°F), while **heavy crude naphtha** is usually the second distillation fraction, boiling range 163° to 218°C (325° to 425°F).

naphthene — hydrocarbon characterized by saturated carbon atoms in a ring structure, and having the general formula $CnH2n$; also called **cycloparaffin** or **cycloalkane.** Naphthenic lubricating oils have low *pour points,* owing to their very low wax content, and good solvency properties. See *hydrocarbon, saturated hydrocarbon.*

naphthenic — see *naphthene.*

narrow cut — see *distillation test.*

National Formulary — see *NF.*

natural gas — naturally occurring mixture of gaseous *saturated hydrocarbons,* consisting of 80-95% *methane* (CH_4), lesser amounts of *propane, ethane* and *butane,* and small quantities of nonhydrocarbon gases (e.g., nitrogen, helium). Natural gas is found in sandstone, limestone, and other porous rocks beneath the earth's surface, often in association with crude oil. Because of its high heating value and clean-burning characteristics, natural gas is widely used as a fuel. The heavier hydrocarbons in natural gas can be extracted, through compression or absorption processes, to yield *LPG* (propane or butane), *natural gasoline,* and raw materials for *petrochemical* manufacture. See *natural gas liquids.*

natural gas liquids — hydrocarbons extracted from *natural gas:* primarily *LPG* (propane or butane) and *natural gasoline,* the latter

motor oil — see *engine oil.*

MS asphalt — see *emulsified anionic asphalt.*

mud — liquid circulated through the *borehole* during *rotary drilling;* it is used to bring cuttings to the surface, cool and lubricate the drill stem, protect against blowouts by holding back subsurface pressures, and prevent fluid loss by plastering the borehole wall. Mud formulations originally were suspensions of clay or other earth solid in water, but today are more complex mixtures of liquids (not necessarily water), reactive solids, and inert solids.

multi-grade oil — engine oil that meets the requirements of more than one *SAE* (Society of Automotive Engineers) viscosity grade classification (see *SAE viscosity grades),* and may therefore be suitable for use over a wider temperature range than a single-grade oil. Multi-grade oils have two viscosity grade numbers indicating their lowest and highest classification, e.g., SAE 10W-40. The lower grade number indicates the relative fluidity of the oil in cold weather for easy starting and immediate oil flow. The higher grade number indicates the relative viscosity of the oil at high operating temperatures for adequate wear protection. The "W" means "winter" grade. Multi-grade oils generally contain *viscosity index (V.I.) improvers* that reduce the tendency of an oil to lose viscosity, or thin out, at high temperatures.

multi-purpose grease — high-quality grease that can be used in a variety of applications. See *grease.*

Naphthenic Structures

Basic naphthenic structure for a six carbon ring.
Single bonds exist between carbon atoms.

Six and five carbon rings joined together.

being commonly blended with crude-derived gasoline to improve volatility. Natural gas liquids can be separated from the lighter hydrocarbons of natural gas by compression (the gas is compressed and cooled until the heavier hydrocarbons liquefy) or by absorption (the gas is mixed with a petroleum distillate, such as kerosene, which absorbs, or dissolves, the heavier hydrocarbons).

natural gasoline — liquid hydrocarbons recovered from wet natural gas; also called **casinghead gasoline.** See *natural gas liquids.*

natural rubber — resilient *elastomer* generally prepared from the milky sap, or latex, of the rubber tree *(hevea brasilensis)*. Natural rubber possesses a degree of tack (adhesive properties) not inherent in most *synthetic rubbers*. It may be used unblended in large tires for construction and agricultural equipment and airplanes, where low rolling resistance and low heat buildup are of greater importance than wear resistance. In the manufacture of tires for highway vehicles, natural rubber may be added to synthetic rubber to provide the necessary tack.

NBR — see *nitrile rubber.*

neoprene rubber (CR) — *synthetic rubber,* a chloroprene *polymer,* with excellent resistance to weather, oil, chemicals, and flame. Widely used for electrical cable insulation, industrial hose, adhesives, shoe soles, and paints.

neutralization number — also called **neut number,** an indication of the acidity or alkalinity of an oil; the number is the weight in milligrams of the amount of acid (hydrochloric acid [HCl]) or base (potassium hydroxide [KOH]) required to neutralize one gram of the oil, in accordance with test method ASTM D 664 (potentiometric method) or ASTM D 974 (colorimetric method). **Strong acid number** is the weight in milligrams of base required to titrate a one-gram sample up to a *pH* of 4; **total acid number** is the weight in milligrams of base required to neutralize all acidic constituents. **Strong base number** is the quantity of acid, expressed in terms of the equivalent number of milligrams of KOH, required to titrate a one-gram sample to a pH of 11; **total base number** is the milligrams of acid, expressed in equivalent milligrams of KOH, to neutralize all basic constituents. If the neutralization number indicates increased acidity (i.e., high **acid number**) of a used oil, this may indicate that oil oxidation, additive depletion, or a change in the oil's operating environment has occurred.

newton — in the *Système International,* the unit of force required to accelerate a mass of one kilogram one meter per second.

Newtonian fluid — fluid, such as a *straight mineral oil,* whose *viscosity* does not change with rate of flow. See *shear stress.*

NF (National Formulary) — listing of drugs, drug formulas, quality standards, and tests published by the United States Pharmacopeial Convention, Inc., which also publishes the *USP* (United States Pharmacopeia). The purpose of the NF is to ensure the uniformity of drug products and to maintain and upgrade standards of drug quality, packaging, labeling, and storage. In 1980, all NF responsibility for *white oil* classification was transferred to the USP.

Niemann four-square gear oil test — see *FZG four-square gear oil test.*

nitration grade — term for *toluene, xylene,* or *benzene* refined under close controls for very narrow boiling range and high purity.

occupational exposure limit (OEL) — the *time-weighted average* concentration of a material in air for an eight-hour workday, 40-hour workweek to which nearly all workers may be exposed repeatedly without adverse effect. Also called **threshhold limit value (TLV).**

Occupational Safety and Health Act of 1970 — the main legisla-

Nitration-grade specifications are given in test methods ASTM D 841, D 843, and D 835, respectively.

nitrile rubber (NBR) — *synthetic rubber* made by the copolymerization of butadiene and acrylonitrile. It resists heat, oil, and fuels; hence, is used in gasoline and oil hose, and in tank linings. Originally called **Buna-N.** See *polymer.*

nitrogen blanket — see *gas blanket.*

nitrogen oxides (NOx) — emissions, from man-made and natural sources, of nitric oxide (NO), with minor amounts of nitrogen dioxide (NO_2). NOx are formed whenever fuel is burned at high temperatures in air, from nitrogen in the air as well as in the fuel. Motor vehicles and stationary combustion sources (furnaces and boilers) are the primary man-made sources, although automotive emission controls are reducing the automobile's contribution. Natural emissions of NOx arise from bacterial action in the soil. NOx can react with hydrocarbons to produce smog. See *catalytic converter, emissions (automotive), emissions (stationary source), pollutants, hydrocarbon emissions.*

NLGI (National Lubricating Grease Institute) — trade association whose main interest is grease and grease technology. NLGI is best known for its system of rating greases by penetration. See *NLGI consistency grades, penetration (grease).*

NLGI consistency grades — simplified system established by the National Lubricating Grease Institute (NLGI) for rating the consistency of grease. See *penetration (grease).* The following are the NLGI consistency grades:

NLGI Consistency Grades	
NLGI Grade	**ASTM Worked Penetration, mm/10**
000	445-475
00	400-430
0	355-385
1	310-340
2	265-295
3	220-250
4	175-205
5	130-160
6	85-115

non-Newtonian fluid — fluid, such as a grease or a *polymer-*containing oil (e.g. multi-grade oil), in which *shear stress* is not proportional to *shear rate.* See *Brookfield viscosity.*

non-soap thickener — see *grease.*

non-volatiles — see *solids content.*

normal paraffin — hydrocarbon consisting of unbranched molecules in which any carbon atom is attached to no more than two other carbon atoms; also called **straight chain paraffin** and **linear paraffin.** See *isoparaffin, paraffin.*

tion affecting health and safety in the workplace. It created the Occupational Safety and Health Administration (OSHA) in the Department of Labor, and the National Institute for Occupational Safety and Health in the Department of Health and Human Services (formerly Department of Health, Education, and Welfare).

OCS — see *outer continental shelf.*

octane number — expression of the *antiknock* properties of a gasoline, relative to that of a standard reference fuel. There are two distinct types of octane number measured in the laboratory: **Research Octane Number (RON)** and **Motor Octane Number (MON)**, determined in accordance with ASTM D 2699 and D 2700, respectively. Both the RON and MON tests are conducted in the same laboratory engine, but RON is determined under less severe conditions, and is therefore numerically greater than MON for the same fuel. The average of the two numbers — (RON + MON)/2 — is commonly used as the indicator of a gasoline's road antiknock performance. The gasoline being tested is run in a special single-cylinder engine, whose compression ratio can be varied (the higher the compression ratio, the higher the octane requirement). The knock intensity of the test fuel, as measured by a knockmeter, is compared with the knock intensities of blends of *isooctane* (assigned a knock rating of 100) and *heptane* (with a knock rating of zero), measured under the same conditions as the test fuel. The percentage, by volume, of the isooctane in the blend that matches the characteristics of the test fuel is designated as the octane number of the fuel. For example, if the matching blend contained 90% isooctane, the octane number of the test fuel would be 90. In addition to the laboratory tests for RON and MON, there is a third method, **Road Octane Number**, which is conducted in a specially equipped test car by individuals trained to hear trace levels of engine knock. See *antiknock compounds, knock.*

odorless solvents — *solvents,* generally *mineral spirits,* that are synthesized by *alkylation* and refined to remove odorous *aromatics* and sulfur compounds; there remains, however, a relatively low level of odor inherent in the hydrocarbons. Odorless solvent applications include dry cleaning and odorless paint manufacture.

odor panel — a group of individuals trained to identify and rate odors, in order to check the odor quality of solvents, waxes, etc.

OEL — see *occupational exposure limit.*

oil content of petroleum wax — a measure of wax refinement, under conditions prescribed by test method ASTM D 721. The sample is dissolved in methyl ethyl ketone, and cooled to −32°C (−26°F) to precipitate the wax, which is then filtered out. The oil content of the remaining filtrate is determined by evaporating the solvent and weighing the residue. Waxes with an oil content generally of 1.0 mass percent or less are known as *refined waxes.* Refined waxes are harder and have greater resistance to blocking (see *blocking point*) and staining than waxes with higher oil content. Waxes with an oil content up to 3.0 mass percent are generally referred to as *scale waxes,* and are used in applications where the slight color, odor, and taste imparted by the higher oil content can be tolerated. Semi-refined *slack waxes* may have oil contents up to 30 mass percent, and are used in non-critical applications. The distinction between scale and slack waxes at intermediate oil content levels (2-4 mass percent) is not clearly defined, and their suitability for particular applications depends upon properties other than oil content alone. See *wax.*

oiliness agent — *polar compound* used to increase the *lubricity* of a lubricating oil and aid in preventing wear and *scoring* under conditions of *boundary lubrication.*

oil mist lubrication — type of *centralized lubrication* that employs compressed air to transform liquid oil into a mist that is then distributed at low pressure to multiple points of application. The oil mist is formed in a "generator", where compressed air is passed across an orifice, creating a pressure reduction that causes oil to be drawn from a reservoir into the airstream. The resulting mist (composed of fine droplets on the average of 1.5 microns) is distrib-

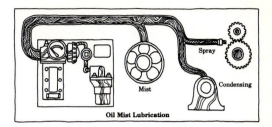

Oil Mist Lubrication

uted through feed lines to various application points. Here, it is reclassified, or condensed, to a liquid, spray, or coarser mist by specialized fittings, depending on the lubrication requirements. Oils for use in a mist lubrication system are formulated with carefully selected *base stocks* and *additives* for maximum delivery of oil to the lubrication points and minimal coalescence of oil in the feed lines.

oil shale — shale containing a rubbery hydrocarbon known as kerogen. When shale is heated, the kerogen vaporizes and condenses as a tar-like oil called **shale oil**, which can be upgraded and refined into products in much the same way as liquid petroleum. There are large oil shale deposits in the U.S., the richest being in Colorado, Utah, and Wyoming.

olefin — any of a series of unsaturated, relatively unstable hydrocarbons characterized by the presence of a double bond between two carbon atoms in its structure, which is commonly straight-chain or branched. The double bond is chemically active and provides a focal point for the addition of other reactive elements, such as oxygen. Due to their ease of oxidation, olefins are undesirable in petroleum solvents and lube oils. Examples of olefins are: *ethylene* and *propylene.* See *hydrocarbon, unsaturated hydrocarbon.*

olefin oligomer — *synthetic lubricant* base, formed by the polymerization of *olefin* monomers (see *polymer*); properties include good *oxidation stability* at high temperatures, good *hydrolytic stability,* good compatibility with *mineral oils,* and low *volatility.* Used in turbines, compressors, gears, automotive engines, and electrical applications.

once-through lubrication — system of lubrication in which the lubricant is supplied to the lubricated part at a minimal rate and is not returned or recirculated. Lubrication by oil can, mechanical lubricator, centralized grease system, lubricating device, oil mist, etc., is done on a once-through basis. Since the lubricant is not recovered, high *oxidation stability* and long service life are usually not necessary, but *viscosity* and other properties may be very important. Oils that meet the moderate requirements of once-through lubrication are known variously as **machine oils** and **general-purpose oils**. See *centralized lubrication, oil mist lubrication.*

OPEC (Organization of Petroleum Exporting Countries) — group of oil-producing nations founded in 1960 to advance member interests in dealings with industrialized oil-consuming nations. The 13 OPEC members are: Algeria, Ecuador, Gabon, Indonesia, Iran, Iraq, Kuwait, Libya, Nigeria, Qatar, Saudi Arabia, United Arab Emirates, and Venezuela. Rising world oil demand, tight world oil supplies, and declining U.S. oil and gas production have enabled OPEC to dramatically increase the price of its oil exports since 1973.

open cup — see *Cleveland open cup, Tag open cup.*

orchard spray oil — petroleum oil suitable for emulsifying with water to form an insecticide spray that kills orchard pests by suffocation. When applied to fruit trees as directed, it has proved highly effective in the control of certain insects that attack citrus, apples, pears, peaches, nuts, and other orchard crops. The phytotoxicity (harmfulness to plants) depends on the boiling range and purity of the oil. Purity is broadly defined by the *unsulfonated residue* of the oil. Oils with an unsulfonated residue of 92% or higher can be used in sensitive applications, such as verdant, or summer sprays when trees are in leaf. These are known as "superior" spray oils. Oils with lower unsulfonated residues — at least 80% — are called "regular" spray oils, and are limited to application only in the dormant phase of plant growth.

organic compound — chemical substance containing carbon and hydrogen; other elements, such as nitrogen or oxygen, may also be present. See *hydrocarbon, inorganic compound.*

Organization of Petroleum Exporting Countries — see *OPEC.*

organosol — see *plastisols and organosols.*

OSHA (Occupational Safety and Health Administration) — See *Occupational Safety and Health Act of 1970.*

Otto cycle — four-stroke engine cycle. See *internal combustion engine.*

outer continental shelf — the part of the continental margin that slopes gradually away from the shore to a point where a much steeper drop begins. The outer continental shelf may extend from a few miles to several hundred miles from the shore. It is much wider off the U.S. East and Gulf coasts, for example, than off the West Coast. Much of the remaining U.S. oil and gas resources are believed to lie beneath the outer continental shelf.

overhead — the distillation fraction removed as vapor or liquid from the top of a distillation column, e.g., a pipe still. See *distillation.*

pale oil — straight naphthenic mineral oil, straw or pale yellow in color, used as a once-through lubricant and in the formulation of *process oils.* See *naphthene, straight mineral oil, once-through lubrication.*

paraffin — hydrocarbon identified by saturated straight (normal) or branched (iso) carbon chains. The generalized paraffinic molecule can be symbolized by the formula CnH_{2n+2}. Paraffins are relatively non-reactive and have excellent *oxidation stability.* In contrast to naphthenic (see *naphthene)* oils, paraffinic lube oils have relatively high wax content and *pour point,* and generally have a high *viscosity index (V.I.).* Paraffinic solvents are generally lower in solvency than naphthenic or *aromatic* solvents. See *hydrocarbon, normal paraffin, isoparaffin, saturated hydrocarbon.* (See illustration on facing page.)

paraffinic — see *paraffin.*

paraffin wax — petroleum-derived wax usually consisting of high-molecular-weight *normal paraffins;* distinct from other natural waxes, such as beeswax and carnauba wax (palm tree), which are composed of high-molecular-weight *esters,* in combination with

oxidation — the chemical combination of a substance with oxygen. All petroleum products are subject to oxidation, with resultant degradation of their composition and performance. The process is accelerated by heat, light, metal catalysts (e.g., copper), and the presence of water, acids, or solid contaminants. The first reaction products of oxidation are organic peroxides. Continued oxidation, catalyzed by peroxides, forms alcohols, aldehydes, ketones, and organic acids, which can be further oxidized to form high-molecular-weight, oil-insoluble *polymers;* these settle out as sludges, varnishes, and gums that can impair equipment operation. Also, the organic acids formed from oxidation are corrosive to metals. Oxidation resistance of a product can be improved by careful selection of base stocks (*paraffins* have greater oxidation resistance than *naphthenes*), special refining methods, and addition of *oxidation inhibitors.* Also, oxidation can be minimized by good maintenance of oil and equipment to prevent contamination and excessive heat. See *oxidation stability.*

oxidation inhibitor — substance added in small quantities to a petroleum product to increase its oxidation resistance, thereby lengthening its service or storage life; also called **anti-oxidant.** An oxidation inhibitor may work in one of three ways: (1) by combining with and modifying peroxides (initial oxidation products) to render them harmless, (2) by decomposing the peroxides, or (3) by rendering an oxidation catalyst (metal or metal ions) inert. See *oxidation.*

oxidation stability — resistance of a petroleum product to oxidation; hence, a measure of its potential service or storage life. There are a number of ASTM tests to determine the oxidation stability of a lubricant or fuel, all of which are intended to simulate service conditions on an accelerated basis. In general, the test sample is exposed to oxygen or air at an elevated temperature, and sometimes to water or catalysts (usually iron or copper). Depending on the test, results are expressed in terms of the time required to produce a specified effect (such as a pressure drop), the amount of sludge or gum produced, or the amount of oxygen consumed during a specified period.

oxidized asphalt — also called **blown asphalt.** See *asphalt.*

high-molecular-weight acids, alcohols, and hydrocarbons. See *wax (petroleum).*

partial pressure — pressure exerted by a single component of a gaseous mixture. The sum of the partial pressures in a gaseous mixture equals the total pressure. The partial pressure of a substance is a function both of its *volatility,* or *vapor pressure,* and its concentration.

particulates — atmospheric particles made up of a wide range of natural materials (e.g., pollen, dust, resins), combined with man-made pollutants (e.g., smoke particles, metallic ash); in sufficient concentrations, particulates can be a respiratory irritant. Primary sources of man-made particulate emissions are industrial process losses (e.g. from cement plants) and stationary combustion sources. Motor vehicles contribute a relatively minor amount of particulates. See *emissions (stationary source), pollutants.*

pascal (Pa) — in the *Système International,* a unit of pressure equivalent to a force of one *newton (n)* applied to an area of one square meter.

paving asphalt — see *asphalt cement.*

Paraffinic Chain Structures

Basic straight paraffinic chain showing all bonds.
Single bonds exist between carbon atoms.

Branched paraffinic chain with single carbon bonds

PCB — polychlorinated biphenyl, a class of synthetic chemicals consisting of an homologous series of compounds beginning with monochlorobiphenyl and ending with decachlorobiphenyl. PCB's do not occur naturally in petroleum, but have been found as contaminants in used oil. PCB's have been legally designated as a health hazard, and any oil so contaminated must be handled in strict accordance with state and federal regulations.

PCV — see *positive crankcase ventilation.*

PE — see *polyethylene.*

penetration (asphalt) — method for determining the penetration, or consistency, of semi-solid and solid bituminous (see *bitumen)* materials, in a *penetrometer* under conditions prescribed by test method ASTM D 5. The test sample is melted and cooled under controlled conditions, then a weighted standard needle is positioned at the surface of the sample and allowed to penetrate it, by means of gravity, for a specified time. The penetration is measured in tenths of a millimeter. See *viscosity (asphalt).*

penetration (grease) — measure of the consistency of a grease, utilizing a *penetrometer.* Penetration is reported as the tenths of a millimeter (penetration number) that a standard cone, acting under the influence of gravity, will penetrate the grease sample under test conditions prescribed by test method ASTM D 217. Standard test temperature is 25°C (77°F). The higher the penetration number, the softer the grease. **Undisturbed penetration** is the penetration of a grease sample as originally received in its container. **Unworked penetration** is the penetration of a grease sample that has received only minimal handling in transfer from its original container to the test apparatus. **Worked penetration** is the penetration of a sample immediately after it has been subjected to 60 double strokes in a standard grease worker; other penetration measurements may utilize more than 60 strokes. **Block penetration** is the penetration of *block grease* (grease sufficiently hard to hold its shape without a container).

penetration (wax) — measure of the consistency of a petroleum wax, utilizing a *penetrometer.* Penetration is reported as the depth, in tenths of a millimeter, to which a standard needle penetrates the wax under conditions described in test method ASTM D 1321. Prior to penetration, the wax sample is heated to 17°C (30°F) above its *congealing point,* air cooled, then conditioned at test temperature in a water bath, where the sample remains during the penetration test. The test temperature may range from 25° to 75°C (77° to 130°F).

penetration grading (asphalt) — classification system for *asphalt cement,* defined in AASHTO (American Association of State Highway Transportation Officials) Specification M20, and based on tests for *penetration (asphalt), flash point,* ductility, purity, etc., as specified in test method ASTM D 946. There are five standard grades, ranging from hard to soft: 40-50, 60-70, 85-100, 120-150, and 200-300. Asphalt cement is also classified by *viscosity grading.*

penetrometer — apparatus for measuring the consistency of lubricating grease or asphalt. A standard cone (for grease) or needle (for wax or asphalt) is lowered onto a test sample, under prescribed conditions, and the depth of penetration is measured. See *mobilometer, penetration (asphalt), penetration (grease), penetration (wax).*

Penetrometer

Pensky-Martens closed tester — apparatus used in determining the *flash point* of *fuel oils* and *cutback asphalt,* under conditions prescribed by test method ASTM D 93. The test sample is slowly heated in a closed cup, at a specified constant rate, with continual stirring. A small flame is introduced into the cup at specified intervals through shuttered openings. The lowest temperature at which the vapors above the sample briefly ignite is the flash point. See *Tag closed tester.*

pentane — saturated paraffinic hydrocarbon (C_5H_{12}); it is a colorless, volatile liquid, normally blended into gasoline. See *paraffin, saturated hydrocarbon.*

pentane insolubles — see *insolubles.*

performance number — expression of the *antiknock* properties of an *aviation gasoline* with a Motor Octane Number higher than 100. The laboratory procedure for determining performance number is the same as that for Motor Octane Number (see *octane number),* except that performance number is based on the number of milliliters of tetraethyl lead (see *lead alkyl)* in the reference *isooctane* blend that matches the antiknock characteristics of the test fuel. A 100/130 grade aviation gasoline has an octane number of 100 at lean fuel mix and a performance number of 130 at rich mix. See *lean and rich octane numbers.*

peroxide — any compound containing two linked oxygen atoms (e.g. Na_2O_2) that yields hydrogen peroxide (H_2O_2) when reacted with acid; also, H_2O_2 itself. Relatively unstable, peroxides are strong oxidizing agents and, when present in lubricating oils, can accelerate oil oxidation and promote bearing corrosion. See *oxidation.*

pesticide — any chemical substance intended to kill or control pests. Common pesticides are: insecticides, rodenticides, herbicides, fungicides, and bactericides. Petroleum products and their petrochemical derivatives are important in the formulation of many types of pesticides. Specialized petroleum oils are used to kill insects by suffocation (see *orchard spray oil*); other petroleum products serve as solvents or *diluents* for the active component.

petrochemical — any chemical derived from crude oil, crude products, or natural gas. A petrochemical is basically a compound of carbon and hydrogen, but may incorporate many other elements. Petrochemicals are used in the manufacture of numerous products such as *synthetic rubber,* synthetic fibers (such as nylon and polyester), plastics, fertilizers, paints, detergents, and pesticides.

petrolatum — semi-solid, noncrystalline hydrocarbon, pale to yellow in color, composed primarily of high-molecular-weight waxes; used in lubricants, rust preventives, and medicinal ointments. See *wax (petroleum).*

petroleum — term applied to crude oil and its products; also called **rock oil.** See *crude oil.*

pH — measure of the acidity or alkalinity of an aqueous solution. The pH scale ranges from 0 (very acidic) to 14 (very alkaline), with a pH of 7 indicating a neutral solution equivalent to the pH of distilled water. See *neutralization number.*

phenol — white, crystalline compound (C_6H_5OH) derived from *benzene;* used in the manufacture of phenolic *resins,* weed killers, plastics, disinfectants; also used in *solvent extraction,* a petroleum refining process. Phenol is a toxic material; skin contact must be avoided.

phosphate ester — any of a group of *synthetic lubricants* having superior fire resistance. A phosphate ester generally has poor *hydrolytic stability,* poor compatibility with *mineral oil,* and a relatively low *viscosity index (V.I.).* It is used as a fire-resistant *hydraulic fluid* in high-temperature applications.

phytotoxic — injurious to plants.

pig — solid plug inserted into pipelines and pushed through by fluid pressure. It may be used for separating two fluids being pumped through the line, or for cleaning foreign materials from a line.

pinion — see *gear.*

pipe still — see *distillation.*

piston sweep — see *sweep (of a piston).*

plain bearing — see *bearing.*

plasticity — the property of an apparently solid material that enables it to be permanently deformed under the application of force, without rupture. (Plastic flow differs from fluid flow in that the *shear stress* must exceed a *yield point* before any flow occurs.)

plasticizer — any *organic compound* used in modifying plastics, *synthetic rubber,* and similar materials to incorporate flexibility and toughness.

plastisols and organosols — coating materials composed of *resins* suspended in a hydrocarbon liquid. An organosol is a plastisol with an added solvent, which swells the resin particles, thereby increasing *viscosity.* Applications include spray coating, dipping, and coatings for aluminum, fabrics, and paper.

Platinum-Cobalt system — see *color scale.*

PNA (polynuclear aromatic) — any of numerous complex hydrocarbon compounds consisting of three or more *benzene* rings in a compact molecular arrangement. Some types of PNA's are known to be carcinogenic (cancer causing). PNA's are formed in *fossil fuel* combustion and other heat processes, such as *catalytic cracking.* They can also form when foods or other organic substances are charred. PNA's occur naturally in many foods, including leafy vegetables, grain cereals, fruits, and meats.

poise — see *viscosity.*

polar compound — a chemical compound whose molecules exhibit electrically positive characteristics at one extremity and negative characteristics at the other. Polar compounds are used as additives in many petroleum products. Polarity gives certain molecules a strong affinity for solid surfaces; as lubricant additives (*oiliness agents*), such molecules plate out to form a tenacious, friction-reducing film. Some polar molecules are oil-soluble at one end and water-soluble at the other end; in lubricants, they act as *emulsifiers,* helping to form stable oil-water emulsions. Such lubricants are said to have good **metal-wetting** properties. Polar compounds with a strong attraction for solid contaminants act as *detergents* in engine oils by keeping contaminants finely dispersed.

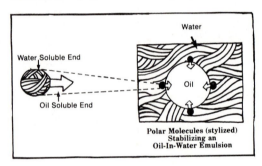

**Polar Molecules (stylized)
Stabilizing an
Oil-In-Water Emulsion**

pollutants (atmospheric) — any substances released to the environment that threaten health or damage vegetation if present in sufficient concentration. The major pollutants emitted as a result of man's industrial activity (largely through the combustion of *fossil fuels*) are: *sulfur oxides* – predominantly sulfur dioxide (SO_2) — *nitrogen oxides (NOx), carbon monoxide (CO), hydrocarbons (HC),* and *particulates.* Such pollutants have a relatively short residence in the atmosphere before being removed by natural scavenging processes. SO_2, for example, has an atmospheric residence time of about four days. There has thus been no evidence of a global buildup of these pollutants. In a given locality, however, pollutants can reach high concentrations in the atmosphere, causing respiratory ailments, as well as inhibiting growth of vegetation, turning soil acid, eroding masonry in buildings, and corroding metals. See *emissions (automotive), emissions (stationary source).*

polybutadiene rubber (BR) — one of the *stereo rubbers,* a term designating high uniformity of composition. High in abrasion resistance, BR is blended with *styrene-butadiene rubber (SBR)* for tire tread manufacture. See *synthetic rubber.*

polyester — any of a number of synthetic *resins* usually produced by the polymerization of dibasic acid with a dihydric alcohol (see *polymer*). Polyester resins have high sealing strength, and are weather resistant. They are used in the manufacture of boat hulls, waterproof fibers, and adhesives.

polyethylene (PE) — polymerized (see *polymer)* ethylene, ranging from a colorless liquid to a white solid; used in the manufacture of plastic films and sheets, and a wide variety of containers, kitchenware, tubing, etc.

polyglycols — *polymers* of ethylene or propylene oxides used as a *synthetic lubricant* base. Properties include very good *hydrolytic stability, high viscosity index (VI)*, and low *volatility*. Used particularly in water *emulsion* fluids.

polyisoprene rubber (IR) — one of the *stereo rubbers*, a term designating a high uniformity of composition. Sometimes called "synthetic natural rubber" because of its similar chemical composition, high tack, resiliency, and heat resistance. It can replace *natural rubber* in many applications. See *synthetic rubber*.

polymer — substance formed by the linkage (polymerization) of two or more simple, unsaturated molecules (see *unsaturated hydrocarbon*), called **monomers**, to form a single heavier molecule having the same elements in the same proportions as the original monomers; i.e. each monomer retains its structural identity. A polymer may be liquid or solid; solid polymers may consist of millions of repeated linked units. A polymer made from two or more dissimilar monomers is called a **copolymer**; a copolymer composed of three different types of monomers is a **terpolymer**. *Natural rubber* and *synthetic rubbers* are polymers.

Major Monomers Used in Synthetic Rubbers

Monomer	Physical State	Chemical Structure
Butadiene	Gas	
Styrene	Liquid	
Isoprene	Liquid	
Isobutylene	Gas	
Ethylene	Gas	
Propylene	Gas	
Chloroprene	Liquid	
Acrylonitrile	Liquid	

(Cl) Chlorine Atom
(N) Nitrogen Atom

---

polymerization — in petroleum refining, polymerization refers to the combination of light, gaseous hydrocarbons, usually *olefins*, into high-molecular-weight hydrocarbons that are used in manufacturing motor gasoline and aviation fuel. The product formed by combining two identical olefin molecules is called a **dimer**, and by three such molecules, a **trimer**. See *polymer*.

polynuclear aromatic — see *PNA*.

polyolefin — *polymer* derived by polymerization of relatively simple *olefins*. *Polyethylene* and *polyisoprene* are important polyolefins.

polyol ester — *synthetic lubricant* base, formed by reacting *fatty acids* with a polyol (such as a glycol) derived from petroleum. Properties include good *oxidation stability* at high temperatures and low *volatility*. Used in formulating lubricants for turbines, compressors, jet engines, and automotive engines.

polystyrene — hard, clear thermoplastic *polymer* of *styrene*, easily colored and molded for a variety of applications, including structural materials. It is a good thermal and electrical insulator and, in the form of expanded foam, extremely buoyant.

positive crankcase ventilation (PCV) — system for removing *blow-by* gases from the crankcase and returning them, through the *carburetor* intake manifold, to the combustion chamber, where the recirculated hydrocarbons are burned, thus reducing hydrocarbon emissions to the atmosphere. A PCV valve, operated by engine vacuum, controls the flow of gases from the crankcase. PCV systems have been standard equipment in all U.S. cars since 1963, replacing the simpler vent, or breather, that allowed crankcase vapors to be emitted to the atmosphere.

pour point — lowest temperature at which an oil or *distillate* fuel is observed to flow, when cooled under conditions prescribed by test method ASTM D 97. The pour point is 3°C (5°F) above the temperature at which the oil in a test vessel shows no movement when the container is held horizontally for five seconds. Pour point is lower than *wax appearance point* or *cloud point*. It is an indicator of the ability of an oil or distillate fuel to flow at cold operating temperatures.

pour point depressant — *additive* used to lower the *pour point* of a petroleum product.

power — rate at which *energy* is used, or at which work is done. Power is commonly measured in terms of the watt (one joule per second) or horsepower (33 000 foot-pounds per minute, or 745.7 watts).

power factor — ratio of the power in watts (W) dissipated in an insulating medium to the product of the effective values of voltage (V) and current (I) in volt-amperes; a measure of the tendency of an insulating oil, which is a *dielectric* (nonconductor of electricity), to permit leakage of current through the oil, as determined by test method ASTM D 924. Such current leakage is called **dielectric loss**. The lower the power factor, the lower the dielectric loss. Determination of power factor can be used to indicate not only the inherent dielectric properties of an oil, but the extent of deterioration of a used oil, since oxidation products and other polar contaminants reduce dielectric strength, causing the power factor to rise. Power factor is related to *dissipation factor*.

Powerforming® — catalytic *reforming* process patented by Exxon.

ppb — parts per billion.

ppm — parts per million.

pre-ignition — ignition of a fuel-air mixture in an *internal combustion engine* (gasoline) before the spark plug fires. It can be caused by a hot spot in the combustion chamber or a very high *compression ratio*. Pre-ignition reduces power and can damage the engine.

pressure — force per unit area, measured in kilopascals (kPa) or pounds per square inch (psi). Standard **atmospheric pressure** at sea level is 101.3 kPa (14.7 psi), or one *atmosphere*. **Gauge pressure,** as indicated by a conventional pressure gauge, is the pressure in excess of atmospheric pressure. **Absolute pressure** is the sum of atmospheric and gauge pressures. Pressure is also expressed in terms of the height of a column of mercury that would exert the same pressure. One atmosphere is equal to 760 mm (29.9 in) of mercury.

pressure maintenance — method for increasing ultimate oil recovery by injecting gas, water, or other fluids into an oil *reservoir*, usually early in the life of the field in order to maintain or slow the decline of the reservoir pressures that force the oil to the surface.

pressure ratio (of a compressor) — the ratio (r) of the absolute discharge pressure to the absolute pressure at the inlet. This is mathematically expressed as:

$$r = P_2/P_1$$

where P_2 is the discharge pressure and P_1 is the inlet pressure.

process oil — oil that serves as a temporary or permanent component of a manufactured product. *Aromatic* process oils have good solvency characteristics; their applications include proprietary chemical formulations, ink oils, and *extenders* in *synthetic rubbers*. Naphthenic (see *naphthene)* process oils are characterized by low *pour points* and good solvency properties; their applications include rubber compounding, printing inks, textile conditioning, leather tanning, shoe polish, rustproofing compounds, and dust suppressants. Paraffinic (see *paraffin)* process oils are characterized by low aromatic content and light color; their applications include furniture polishes, ink oils, and proprietary chemical formulations.

process stream — general term applied to a partially finished petroleum product moving from one refining stage to another; less commonly applied to a finished petroleum product. See *CAS Registry Numbers*.

propane — gaseous paraffinic hydrocarbon (C_3H_8) present in natural gas and crude oil; also termed, along with *butane*, liquefied petroleum gas *(LPG)*. See *paraffin.*

propellant — volatile gas or liquid which, when permitted to escape from a pressurized container, carries with it particles or droplets of another material mixed or suspended in it. *Propane* and *butane* are common petroleum-derived propellants.

propylene — flammable gas (CH_3CHCH_2), derived from hydrocarbon *cracking*; used in the manufacture of polypropylene plastics.

psi — pounds per square inch.

psia — pounds per square inch absolute, equivalent to the gauge pressure plus atmospheric pressure. See *pressure*.

quality of steam — see *saturated steam.*

quenching — immersion of a heated manufactured steel part, such as a gear or axle, in a fluid to achieve rapid and uniform cooling. Petroleum oils are often used for this purpose. Quenching provides hardness superior to that possible if the heat-treated part were allowed to cool slowly in air. **Marquenching** is a slower cooling process that minimizes distortion and cracking. There are two types of marquenching: **martempering** and **austempering;** the latter is the slower process and helps improve ductility. See

quenching oil.

quenching oil — also called **heat treating oil;** it is used to cool metal parts during their manufacture, and is often preferred to water because the oil's slower heat transfer lessens the possibility of cracking or warping of the metal. A quenching oil must have excellent *oxidation stability* and *thermal stability*, and should yield clean parts, essentially free of residue. See *quenching*. In refining terms, a quenching oil is an oil introduced into high temperature vapors of cracked (see *cracking*) petroleum fractions to cool them.

°R (Rankine) — see *temperature scales.*

°R (Reaumur) — see *temperature scales.*

radical — atom or group of atoms with one or more unpaired electrons. A group of atoms functioning as a radical acts as a single atom, remaining intact during a chemical reaction.

raffinate — in *solvent extraction,* that portion of the oil which remains undissolved and is not removed by the selective solvent.

Ramsbottom carbon residue — see *carbon residue.*

R&O — rust- and-oxidation inhibited. A term applied to highly refined industrial lubricating oils formulated for long service in circulating systems, compressors, hydraulic systems, bearing housing, gear cases, etc. The finest R&O oils are often referred to as *turbine oils.*

Rankine (°R) — see *temperature scales.*

rapeseed oil (blown rapeseed oil) — *fatty oil* used for compounding petroleum oil. See *compounded oil.*

rate of shear — see *shear rate.*

RC asphalt — see *cutback asphalt.*

reaction diluent — a material (usually a light *saturated hydrocarbon*, e.g., *pentane, hexane*) that is used as a *carrier* for the polymerization catalyst in the manufacture of *polyolefins* (see *polymer)*. The material must be very pure, since impurities "poison" the catalyst or hinder the polymerization by reacting with the *olefins*.

Reaumur (°R) — see *temperature scales.*

reclaimed aggregate material (RAM) — reprocessed pavement materials containing no reusable binding agent such as *asphalt*.

reclaimed asphalt pavement (RAP) — reprocessed pavement materials containing *asphalt* and aggregate (e.g. pebbles, crushed stone, shells).

recycling of asphalt paving — the reprocessing of old *asphalt* paving and associated materials for reuse as paving. There are three basic methods of recycling: (1) hot mix recycling, a process in which reclaimed asphalt and aggregate materials are combined with new asphalt, recycling agents (petroleum oils), and new aggregate (e.g. crushed stone) in a central plant to produce hot-mix paving mixtures; (2) cold mix recycling, the recombination of reclaimed asphalt and aggregate materials either in place, or at a central plant to produce a cold mix; (3) surface recycling, a process in which the old asphalt pavement surface is heated in place, scarified, remixed with new asphalt or recycling agents as necessary, relaid, and rolled.

Redwood viscosity — method for determining the *viscosity* of petroleum products; it is widely used in Europe, but has limited use in the U.S. The method is similar to *Saybolt Universal viscosity*; viscosity values are reported as "Redwood seconds."

refined wax — low-oil-content wax, generally with an oil content of 1.0 mass percent or less, white in color, and meeting Food and Drug Administration standards for purity and safety. Refined waxes are suitable for the manufacture of drugs and cosmetics, for coating paper used in food packaging, and for other critical applications. Also called **fully refined wax**. See *wax (petroleum)*.

refining — series of processes for converting crude oil and its fractions to finished petroleum products. Following *distillation*, a petroleum fraction may undergo one or more additional steps to purify or modify it. These refining steps include: *thermal cracking, catalytic cracking, polymerization, alkyation, reforming, hydrocracking, hydroforming, hydrogenation, hydrogen treating, Hydrofining®, solvent extraction, dewaxing, de-oiling, acid treating, clay filtration,* and *deasphalting*. Refined lube oils may be blended with other lube stocks, and *additives* may be incorporated, to impart special properties; refined *naphthas* may be blended with *alkylates*, cracked stock or *reformates* to improve *octane number* and other properties of gasolines.

reformate — product of the *reforming* process.

reforming — thermal or catalytic refining process in which the hydrocarbon molecules of a *naphtha* are rearranged to improve its *octane number;* the resulting product is used in blending high-octane gasoline.

refractive index — ratio of the velocity of light at a specified wave length in air to its velocity in a substance under examination. The refractive index of light-colored petroleum liquids can be determined by test method ASTM D 1218, using a refractometer and a monochromatic light source. Refractive index is an excellent test for uniform composition of *solvents*, rubber *process oils,* and other petroleum products. It may also be used in combination with other simple tests to estimate the distribution of naphthenic, paraffinic, (see *naphthene, paraffin*), and *aromatic* carbon atoms in a process oil.

refrigeration oil — lubricant for refrigeration compressors. It should be free of moisture to avoid reaction (hydrolysis) with halogenated refrigerants (see *halogen*) and prevent freezing of water particles that could impede refrigerant flow. It should have a low wax content to minimize wax precipitation on dilution with the refrigerant, which could block capillary-size passages in the circulating system.

Reid vapor pressure — see *vapor pressure*.

relative humidity — see *humidity*.

release agent — see *mold lubricant*.

Research Octane Number (RON) — see *octane number*.

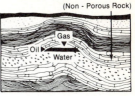

(Non - Porous Rock)

Oil and Gas Reservoir

reservoir — subsurface formation of porous, permeable rock (usually *sandstone, limestone,* or *dolomite*) containing oil or gas within the rock pores. A typical oil reservoir contains gas, oil, and water, which occupy the upper, middle, and lower regions of the reservoir, respectively. The flow of reservoir fluids from the rock to the *borehole* is driven by gas pressure or water pressure.

residual fuel oil — see *fuel oil*.

residual odor — see *bulk odor*.

residuum — see *bottoms*.

resins — solid or semi-solid materials, light yellow to dark brown, composed of carbon, hydrogen, and oxygen. Resins occur naturally in plants, and are common in pines and firs, often appearing as globules on the bark. Synthetic resins, such as *polystyrene, polyesters,* and acrylics (see *acrylic resin*), are derived primarily from petroleum. Resins are widely used in the manufacture of lacquers, varnishes, plastics, adhesives, and rubber. See *plastisols and organosols*.

rheology — study of the deformation and flow of matter in terms of stress, strain, temperature, and time. The rheological properties of a grease are commonly measured by *penetration* and *apparent viscosity*.

rheopectic grease — grease that thickens, or hardens, upon being subjected to shear. The phenomenon is the opposite of *thixotropy*.

rich octane number — see *lean and rich octane numbers*.

ring oil — low-viscosity *R&O* oil for lubricating high-speed textile twister rings. It is normally light in color to prevent staining, and many are compounded with *fatty oils* to prevent wear under conditions of high-speed start-up. See *compounded oil*.

ring oiler — simple device for lubricating a *journal* bearing. A metal ring rides loosely on the journal shaft and the lower part of the ring dips into a small oil reservoir. The rotation of the shaft turns the ring, which carries oil up to the point of contact with the shaft and into the bearing. Though not ordinarily considered a *circulating lubrication system*, the ring oiler is similar in principle and generally requires a long-life oil of the *R&O* type.

ring-sticking — freezing of a piston ring in its groove, in a piston engine or reciprocating compressor, due to heavy deposits in the piston ring zone. This prevents proper action of the ring and tends to increase *blow-by* into the crankcase and to increase oil consumption by permitting oil to flow past the ring zone into the combustion chamber. See *engine deposits*.

Road Octane Number — see *octane number*.

road oil — a heavy petroleum oil, usually one of the slow-curing (SC) grades of liquid asphalt. See *cutback asphalt*.

rock oil — see *petroleum*.

rolling contact bearing — see *bearing*.

roll oil — oil used in hot- and cold-rolling of ferrous and non-ferrous metals to facilitate feed of the metal between the work rolls, improve the plastic deformation of the metal, conduct heat from the metal, and extend the life of the work rolls. Because of the pressures involved, a roll oil may be compounded (see *compounded oil*) or contain *EP additives*. In hot rolling, the oil may also be emulsifiable (see *emulsion*).

RON (Research Octane Number) — see *octane number*.

rotary bomb oxidation test — see *bomb oxidation stability*.

rotary drilling — drilling method utilizing a rotating bit, or cutting element, fastened to and rotated by a drill pipe, which also provides a passageway through which the drilling fluid or *mud* is circulated. Additional lengths of drill stem are added as drilling progresses.

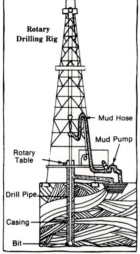

Rotary Drilling Rig

Mud Hose

Mud Pump

Rotary Table

Drill Pipe

Casing

Bit

RS asphalt — see *emulsified anionic asphalt*.

rubber — see *natural rubber, synthetic rubber*.

rubber oil — any petroleum *process oil* used in the manufacture of rubber and rubber products. Rubber oils may be used either as rubber extender oils or as rubber process oils. **Rubber extender oils** are used by the synthetic rubber manufacturer to soften stiff elastomers and reduce their unit volume cost while improving performance characteristics of the rubber. **Rubber process oils** are used by the manufacturer of finished rubber products (tires, footwear, tubing, etc.) to speed mixing and compounding, modify the physical properties of the elastomer, and facilitate processing of the final product.

rubber oil classification — system of four standard classifications for *rubber oils*, based on content of *saturated hydrocarbons, polar compounds*, and *asphaltenes*, as described by ASTM D 2226. The classifications are as follows:

ASTM Type Oil	Hydrocarbon Type, mass %		
	Saturates	Polar Compounds	Asphaltenes
101 (highly aromatic)	20.0 max.	25 max.	0.75 max.
102 (aromatic)	20.1-35.0	12 max.	0.50 max.
103 (naphthenic)	35.1-65.0	6 max.	0.30 max.
104 naphthenic or paraffinic	65.1 min.	1 max.	0.10 max.

Type 104 oils are subclassified into types 104A and 104B for *styrene-butadiene rubber* (SBR) only. Type 104A oils have a *viscosity-gravity constant* (VGC) greater than 0.820 (ASTM D 2501), and are naphthenic; Type 104B oils have a VGC of 0.820 max., and are paraffinic. See *aromatic, naphthene, paraffin*.

rubber swell — see *seal swell*.

rust inhibitor — type of *corrosion inhibitor* used in lubricants to protect the lubricated surfaces against rusting. See *R&O*.

rust preventive — compound for coating metal surfaces with a film that protects against rust; commonly used for the preservation of equipment in storage. The base material of a rust preventive may be a petroleum oil, solvent, wax, or asphalt, to which a *rust inhibitor* is added. A formulation consisting largely of a solvent and additives is commonly called a **thin-film rust preventive** because of the thin coating that remains after evaporation of the solvent. Rust preventives are formulated for a variety of conditions of exposure, e.g., short-time "in-process" protection, indoor storage, exposed outdoor storage, etc.

SAE (Society of Automotive Engineers) — organization responsible for the establishment of many U.S. automotive and aviation standards, including the viscosity classifications of engine oils and gear oils. See *SAE viscosity grades*.

SAE service classification — see *API Engine Service Categories*.

SAE viscosity grades — engine oil classification system developed by the Society of Automotive Engineers (SAE), based on the measured *viscosity* of the oil at either −18°C (−0.4°F), using test method ASTM D 2602, or at 100°C (212°F), using ASTM D 445. If the viscosity is measured at −18°C, the grade number of the oil includes the suffix "W" (e.g. SAE 20W), denoting suitability for winter use. The grade number of an oil tested at 100°C is written without a suffix. It is common for an SAE 20 oil to meet the viscosity requirements at both −18°C and 100°C; such oils are designated 20W-20. Many oils have a high *viscosity index (V.I.)* and therefore may fall into more than one SAE grade classification; these are called *multi-grade oils*, and they are designated by two grade numbers indicating their lowest and highest classification (e.g. SAE 10W-40). The following table shows the viscosity range represented by each SAE grade:

SAE Viscosity Grade	centipoises (cp) at −18°C (ASTM D 2602)	centistokes (cSt) at 100°C (ASTM D 445)	
	Max.	Min.	Max.
5W	1,250	3.8	—
10W	2,500	4.1	—
20W[a]	10,000	5.6	—
20	—	5.6	Less than 9.3
30	—	9.3	Less than 12.5
40	—	12.5	Less than 16.3
50	—	16.3	Less than 21.9

[a]The designation SAE 15W may be used to identify SAE 20W oils that have a maximum viscosity of 5000 cp at −18°C.

After 1981, the low-temperature portion of the SAE viscosity classification system is to be superseded by a revised system, following an 18-month (1980-81) phase-in. The new system introduces the following key changes: 1) two new grades are added — a 0W grade for very low temperature (e.g., arctic) operation, and a 25W grade for temperate climates; 2) low-temperature viscosity is measured at a different temperature for each grade, rather than measuring all winter grades at −18°C; 3) a "borderline pumping temperature" is added to define the ability of a cold oil to flow to the oil pump inlet and through the pump discharge under cold-starting conditions. The revised SAE definitions for "W" grades are as follows:

Revised Definitions for "W" Grades

SAE Viscosity Grade	Viscosity (cp)* at Temperature (°C) Max	Borderline Pumping Temperature (°C)** Max	Viscosity (cSt)*** at 100°C	
			Min	Max
0W	3250 at −30°	−35°	3.8	—
5W	3500 at −25°	−30°	3.8	—
10W	3500 at −20°	−25°	4.1	—
15W	3500 at −15°	−20°	5.6	—
20W	4500 at −10°	−15°	5.6	—
25W	6000 at − 5°	−10°	9.3	—

*By proposed modification of ASTM D 2602, **ASTM D 3829, ***ASTM D 445

sandstone — sedimentary rock usually consisting of grains of quartz cemented by lime, silica, iron oxide, or other materials. Petroleum deposits are commonly found in sandstone formations. See *reservoir.*

saponification — process of converting certain chemicals into soaps, which are the metallic salts of organic acids. It is usually accomplished through reaction of a fat, *fatty acid,* or *ester* with an *alkali* — an important process in *grease* manufacture.

saponification number — number of milligrams of potassium hydroxide (KOH) that combines with 1 gram of oil under conditions specified by test method ASTM D 94. Saponification number is an indication of the amount of fatty saponifiable material in a *compounded oil.* Caution must be used in interpreting test results if certain substances — such as sulfur compounds or *halogens* —are present in the oil, since these also react with KOH, thereby increasing the apparent saponification number.

saturated hydrocarbon — *hydrocarbon* with the basic formula $C_n H_{2n+2}$; it is saturated with respect to hydrogen and cannot combine with the atoms of other elements without giving up hydrogen. Saturates are more chemically stable than *unsaturated hydrocarbons.*

saturated steam — The equilibrium condition at which the temperature of the steam is the same as that of the liquid water from which it is formed. Under this condition, steam containing no unvaporized water particles is called "100% quality." If the temperature of the steam is higher (at the same pressure), the steam is said to be "superheated." If, on the other hand, heat is removed from saturated steam (at the same pressure), as in a steam radiator, the quality drops, and the steam becomes wet, or condenses. At 50% quality, half of the weight represents water, the other half vapor. Saturated steam at atmospheric pressure has a temperature of 100°C (212°F).

Saybolt chromometer — see *color scale.*

Saybolt Furol viscosity — the efflux time in seconds required for 60 milliliters of a petroleum product to flow through the calibrated orifice of a Saybolt Furol *viscometer,* under carefully controlled temperature, as prescribed by test method ASTM D 88. The method differs from *Saybolt Universal viscosity* only in that the viscometer has a larger orifice to facilitate testing of very viscous oils, such as fuel oil (the word "Furol" is a contraction of "fuel and road oils"). The Saybolt Furol method has largely been supplanted by the *kinematic viscosity* method. See *viscosity.*

Saybolt Universal viscosity — the efflux time in seconds required for 60 milliliters of a petroleum product to flow through the calibrated orifice of a Saybolt Universal *viscometer,* under carefully controlled temperature, as prescribed by test method ASTM D 88. This method has largely been supplanted by the *kinematic viscosity* method. See *Saybolt Furol viscosity, viscosity.*

SBR — see *styrene-butadiene rubber.*

SC asphalt — see *cutback asphalt.*

scale wax — soft, semi-refined wax, distinguished from *slack wax* by having a generally lower oil content; usually derived from slack wax by extracting most of the oil from the wax. Used in candle manufacture, coating of carbon paper, and in rubber compounds to prevent surface cracking from sunlight exposure. See *oil content of petroleum wax, wax (petroleum).*

scavenger — a component of lead *antiknock compounds* that reacts with the lead *radical* to form volatile lead compounds that can be easily scavenged from the engine through the exhaust system. Also, an individual who collects used lubricating oils for some secondary use.

scoring — distress marks on sliding metallic surfaces in the form of long, distinct scratches in the direction of motion. Scoring is an advanced stage of *scuffing.*

scrubber oil — see *absorber oil.*

scuffing — localized distress marks on sliding metallic surfaces, appearing as a matte-finished area rather than as individual score marks. See *scoring.*

scuff resistance — property of a wax coating that enables it to withstand abrasion. Scuff resistance is an indication of the extent to which paper-carton-coating machine operation affects the appearance of the coated carton. Poor scuff resistance of a wax can also cause excessive wax deposition on machine parts and adversely affect machine operations.

sealing strength — effectiveness of a coating wax in forming a tight, strong, heat-sealed package closure. See *laminating strength.*

seal oil — see *mineral seal oil.*

seal swell (rubber swell) — swelling of rubber (or other *elastomer*) *gaskets,* or seals, when exposed to petroleum, *synthetic lubricants,* or *hydraulic fluids.* Seal materials vary widely in their resistance to the effect of such fluids. Some seals are designed so that a moderate amount of swelling improves sealing action.

secondary production — see *secondary recovery.*

secondary recovery — restoration of an essentially depleted oil *reservoir* to production by injecting liquids or gases into the reservoir to flush out oil or to increase reservoir pressure. Also called **secondary production.** See *enhanced recovery, tertiary recovery.*

Series 3 — obsolete specification for heavy-duty engine oils used in Caterpillar Tractor Company diesel engines. Caterpillar now specifies that the oil for its engines comply with Military Specification MIL-L-2104C or *API Engine Service Category* CD.

sett grease — any grease that changes from a fluid to a semifluid or plastic state after combination of the components, and often after packaging.

shale oil — see *oil shale*.

shear rate — rate at which adjacent layers of a fluid move with respect to each other, usually expressed as reciprocal seconds (also see *shear stress*). When the fluid is placed between two parallel surfaces moving relative to each other:

$$\text{shear rate} = \frac{\text{relative velocity of surface (meters/ second)}}{\text{distance between surfaces (meters)}} = (\text{seconds})^{-1}$$

shear stress — frictional force overcome in sliding one "layer" of fluid along another, as in any fluid flow. The shear stress of a petroleum oil or other *Newtonian fluid* at a given temperature varies directly with *shear rate* (velocity). The ratio between shear stress and shear rate is constant; this ratio is termed *viscosity*. The higher the viscosity of a Newtonian fluid, the greater the shear stress as a function of rate of shear. In a *non-Newtonian fluid* — such as a *grease* or a *polymer*-containing oil (e.g. *multi-grade oil*) — shear stress is not proportional to the rate of shear. A non-Newtonian fluid may be said to have an *apparent viscosity*, a viscosity that holds only for the shear rate (and temperature) at which the viscosity is determined. See *Brookfield viscosity*.

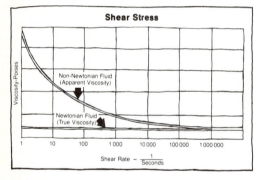

SI (Système International, International System of Units) — metric-based system of weights and measures adopted in 1960 by the 11th General Conference on Weights and Measures, in which 36 countries, including the U.S., participated. SI consists of seven base units:

Unit	Quantity
meter (m)	length
kilogram (kg)	mass
second (s)	time
ampere (A)	electric current
Kelvin (K)	thermodynamic temperature
mole (mol)	amount of substance
candela (cd)	luminous intensity

There are two supplemental units:

Unit	Quantity
radian (rad)	plane angle
steradian (sr)	solid angle

There are many derived units, each defined in terms of the base units: for example, the *newton (N)* — a unit of force — is defined by the formula kg × m/s², and the *joule (J)*, by the relationship N×m. See *metric system*.

sidestream — distillation fraction taken from any level of a distillation tower other than as overhead or *bottoms*. See *distillation*.

sight fluid — transparent liquid in a sight-feed oiler through which the passage of the oil drops can be observed. The sight fluid must be immiscible with the oil. Water and glycerin are often used for this purpose.

silicate esters — class of *synthetic lubricants*, possessing good *thermal stability* and low *volatility*. Commonly used in military applications as high-temperature hydraulic fluids, weapons lubricants, and low-volatility greases.

silicone — generic term for a family of relatively inert liquid organosiloxane *polymers* used as *synthetic lubricants*; properties include high *viscosity index (V.I.)*, good high temperature *oxidation stability*, good *hydrolytic stability* and low *volatility*. Silicones generally have poor *lubricity*, however. Applications include brake fluids, electric motors, oven and kiln preheater fans, automotive fans, plastic bearings, and electrical insulating fluids.

single-grade oil — engine oil that meets the requirements of a single SAE viscosity grade classification. See *SAE viscosity grades*.

slack wax — a semi-refined wax, distinguished from *scale wax* by having a generally higher oil content. Slack waxes with oil content below 10 mass percent are used for manufacture of religious candles, as a feedstock for chlorination processes, and in non-critical paper-making applications. Slack waxes with higher oil content are used in the manufacture of building materials, such as particle board. See *oil content of petroleum wax, wax (petroleum)*.

sleeve bearing — journal bearing. See *bearing*.

sliding bearing — see *bearing*.

sludge — in gasoline engines, a soft, black, mayonnaise-like emulsion of water, other combustion by-products, and oil formed during low-temperature engine operation. Sludge plugs oil lines and screens, and accelerates wear of engine parts. Sludge deposits can be controlled with a *dispersant* additive that keeps the sludge constituents finely suspended in the oil. See *engine deposits*.

slumping — ability of grease to settle to the bottom of a container and form a level surface. When being pumped from the bottom of a container, grease must slump rapidly enough toward the pump suction to maintain flow.

slushing oil — non-drying petroleum-base *rust preventive* used in steel mills to protect the surfaces of steel sheets and strips after rolling.

smoke point (of aviation turbo fuel) — maximum flame height obtainable in a test lamp without causing smoking, as determined by test method ASTM D 1322. It is a measure of burning characteristics; the higher the numerical rating, the cleaner burning the fuel.

SO₂, SO₃ — see *sulfur oxide*.

soap thickener — see *grease*.

Society of Automotive Engineers — see *SAE*.

sodium soap grease — see *grease*.

SOD lead corrosion — test developed by Exxon to measure the corrosiveness of lubricating oils. A small lead panel of known weight and a copper panel (as catalyst) are attached to a spindle, which is immersed in a tube of the lubricant and rotated. Air is introduced at the bottom of the tube and allowed to bubble up through the sample. The weight loss by the lead panel after a specified period of time is a measure of the corrosivity of the oil.

softening point (asphalt) — temperature at which the harder asphalts used in applications other than paving reach an arbitrary degree of softening; usually determined by the ring and ball test method, ASTM D 36.

solids content — that portion of a protective coating material that remains on the surface after drying, often identified as **non-volatiles.**

soluble oil — emulsifiable *cutting fluid*.

solvent — compound with a strong capability to dissolve a given substance. The most common petroleum solvents are *mineral spirits, xylene, toluene, hexane, heptane,* and *naphthas.* Aromatic-type solvents have the highest solvency for organic chemical materials, followed by *naphthenes* and *paraffins.* In most applications the solvent disappears, usually by evaporation, after it has served its purpose. The evaporation rate of a solvent is very important in manufacture: rubber cements often require a fast-drying solvent, whereas rubber goods that must remain tacky during processing require a slower-drying solvent. Solvents have a wide variety of industrial applications, including the manufacture of paints, inks, cleaning products, adhesives, and *petrochemicals.* Other types of solvents have important applications in refining. See *solvent extraction.*

solvent extraction — refining process used to separate reactive components (*unsaturated hydrocarbons*) from lube *distillates* in order to improve the oil's *oxidation stability, viscosity index (V.I.),* and response to *additives.* Commonly used extraction media (solvents) are: *phenol,* N-methylpyrrolidone (NMP), *furfural,* liquid sulfur dioxide, and nitrobenzene. The oil and solvent are mixed in an extraction tower, resulting in the formation of two liquid phases: a heavy phase consisting of the undesirable unsaturates dissolved in the solvent, and a light phase consisting of high quality oil with some solvent dissolved in it. The phases are separated and the solvent recovered from each by distillation. The unsaturates portion, or extract, while undesirable in lubricating oils, is useful in other applications, such as rubber extender oils (see *rubber oil*) and *plasticizer* oils.

solvent neutral — high-quality *paraffin*-base oil refined by *solvent extraction*.

sour crude — crude oil containing appreciable quantities of *hydrogen sulfide* or other *sulfur* compounds, as contrasted to *sweet crude*.

spark-ignition engine — see *internal combustion engine.*

specific gravity — for petroleum products, the ratio of the mass of a given volume of product and the mass of an equal volume of water, at the same temperature. The standard reference temperature is 15.6°C(60°F). Specific gravity is determined by test method ASTM D 1298: the higher the specific gravity, the heavier the product. Specific gravity of a liquid can be determined by means of a **hydrometer,** a graduated float weighted at one end, which provides a direct reading of specific gravity, depending on the depth to which it sinks in the liquid. A related measurement is **density,** an absolute unit defined as mass per unit volume — usually expressed as kilograms per cubic meter (kg/m³). Petroleum products may also be defined in terms of **API gravity** (also determinable by ASTM D 1298), in accordance with the formula:

Hydrometer

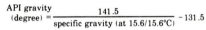

$$\text{API gravity (degree)} = \frac{141.5}{\text{specific gravity (at } 15.6/15.6°C)} - 131.5$$

Hence, the higher the API gravity value, the lighter the material, or the lower its specific gravity.

specific heat — ratio of the quantity of heat required to raise the temperature of a substance one degree Celsius (or Fahrenheit) and the heat required to raise an equal mass of water one degree.

speed factor — see *dN factor.*

spin finish — see *fiber lubricant.*

spindle oil — low-*viscosity* oil of high quality for the lubrication of high-speed textile and metal-working (grinding) machine spindles. In addition to the rust and oxidation (*R&O*) inhibitors needed for prolonged service in humid environments, spindle oils are often fortified with anti-wear agents to reduce torque load and wear, especially at start-up.

spindle test — test to determine the performance life of a grease. Under test method ASTM D 3336, a grease-lubricated SAE No. 204-size ball bearing on a spindle, or shaft, is rotated at 10 000 rpm under light loads and elevated temperatures. The test continues until bearing failure or until completion of a specified number of hours of running time.

spiral bevel gear — see *gear.*

spray oil — see *orchard spray oil.*

spur gear — see *gear.*

squeeze lubrication — phenomenon that occurs when surfaces, such as two gear teeth, move toward each other rapidly enough to develop fluid pressure within the lubricant that will support a load of short duration. The lubricant's viscosity prevents it from immediately flowing away from the area of contact.

SS asphalt — see *emulsified anionic asphalt*.

SSF (Saybolt Furol seconds) — see *Saybolt Furol viscosity*.

SSU, SUS (Saybolt Universal seconds) — see *Saybolt Universal viscosity*.

stationary source emissions — see *emissions (stationary source)*.

steam cylinder oil — see *cylinder oil*.

steam turbine — see *turbine*.

stereo rubber — *elastomer* with a highly uniform arrangement of repeating molecular units (stereoisomers) in its structure. *Polybutadiene rubber* and *polyisoprene rubber* are stereo rubbers.

stick-slip motion — erratic, noisy motion characteristic of some machine *ways*, due to the starting friction encountered by a machine part at each end of its back-and-forth (reciprocating) movement. This undesirable effect can be overcome with a way lubricant, which reduces starting friction.

Stoddard solvent — *mineral spirits* with a minimum *flash point* of 37.8°C (100°F), relatively low odor level, and other properties conforming to Stoddard solvent specifications, as described in test method ASTM D 484. Though formulated to meet dry cleaning requirements, Stoddard solvents are widely used wherever this type of mineral spirits is suitable.

stoichiometric — the exact proportion of two or more substances that will permit a chemical reaction with none of the individual reactants left over. See *combustion*.

stoke — see *viscosity*.

straight-chain paraffin — see *normal paraffin*.

straight mineral oil — petroleum oil containing no *additives*. Straight mineral oils include such diverse products as low-cost once-through lubricants (see *once-through lubrication*) and thoroughly refined *white oils*. Most high-quality lubricants, however, do contain additives. See *mineral oil*.

straight-tooth gear — see *gear*.

strike through — undesirable migration of wax or oil through a paper substrate. A commonly used term in paper laminating operations, but also encountered in partial impregnation of corrugated board with wax. In the printing industry, the term refers to migration of printing ink, formulated with oil or solvent, to the reverse side of the web before setting.

strong acid number — see *neutralization number*.

strong base number — see *neutralization number*.

structural stability — resistance of a grease to change in *consistency* when severely worked in service. Also called **shear stability** and **mechanical stability**.

styrene — colorless liquid (C_8H_8) used as the monomer (see *polymer*) for *polystyrene* and *styrene-butadiene rubber*.

styrene-butadiene rubber (SBR) — general-purpose *synthetic rubber* with good abrasion resistance and tensile properties. SBR can be greatly extended with oil without degrading quality. Applications include automobile tires and wire insulation.

sulfonate — hydrocarbon in which a hydrogen atom has been replaced with the highly polar (SO_2OX) group, where X is a metallic ion or *alkyl* radical. Petroleum sulfonates are refinery by-products of the sulfuric acid treatment of *white oils*. Sulfonates have important applications as *emulsifiers* and chemical intermediates in *petrochemical* manufacture. Synthetic sulfonates can be manufactured from special *feedstocks* rather than from white oil base stocks. See *polar compound*.

sulfur — common natural constituent of petroleum and petroleum products. While certain sulfur compounds are commonly used to improve the EP, or load-carrying, properties of an oil (see *EP oil*), high sulfur content in a petroleum product may be undesirable as it can be corrosive and create an environmental hazard when burned (see *sulfur oxide*). For these reasons, sulfur limitations are specified in the quality control of fuels, solvents, etc. Sulfur content can be determined by *ASTM* tests.

sulfur oxide — major atmospheric pollutant, predominantly sulfur dioxide (SO_2) with some sulfur trioxide (SO_3), primarily emitted from stationary combustion sources (furnaces and boilers). Sulfur oxides are formed whenever fuels containing sulfur are burned. SO_2 is also present in the air from natural land and marine fermentation processes. See *emissions (stationary source)*, *pollutants*.

supercharger — device utilizing a blower or pump to provide intake air to the *carburetor* of an *internal combustion engine* at pressures above atmospheric. Supercharging provides a greater air charge to the cylinders at high crankshaft speeds and at high altitudes, thereby boosting engine power without increasing engine size. Because supercharging maintains maximum intake charge, it offers particular advantages at high altitudes, where the atmosphere contains less oxygen. Some supercharger systems utilize aftercooling to further increase the density of the charge. The blower may be geared to the crankshaft or, in the case of the **turbocharger**, it may consist of a turbine driven by the exhaust gases to operate the centrifugal blower. See *volumetric efficiency*.

supertanker — oil tanker with capacity over 100 000 deadweight tons (dwt); also called **Very Large Crude Carrier** (VLCC). A supertanker with a capacity over 500 000 dwt is called an **Ultra Large Crude Carrier** (ULCC).

surface-active agent — chemical compound that reduces interfacial tension between oil and water and is thus useful as an *emulsifier* in cutting oils (see *cutting fluid*). Sodium *sulfonates* or soaps of *fatty acids* are commonly used for this purpose.

surface ignition — see *afterrunning*.

surfactant — surface-active agent that reduces interfacial tension of a liquid. A surfactant used in a petroleum oil may increase the oil's affinity for metals and other materials.

susceptibility — the tendency of a gasoline toward an increase in *octane number* by addition of a specific amount of a particular *lead alkyl* antiknock compound. See *antiknock compound*.

sweep (of a piston) — internal cylinder surface area over which a piston of a reciprocating compressor moves during its stroke. Total piston sweep is a consideration in the determination of oil-feed rates for some reciprocating compressor cylinders, and may be determined as:

length of stroke x cylinder circumference diameter x 2 x no. of cylinders x rpm x minutes of operation

sweet crude — crude oil containing little or no *sulfur*. See *sour crude*.

syneresis — loss of liquid component from a lubricating grease caused by shrinkage or rearrangement of the structure due to either physical or chemical changes in the thickener; a form of *bleeding*.

synlube — see *synthetic lubricant*.

synthetic gas — see *synthetic oil and gas*.

synthetic lubricant — lubricating fluid made by chemically reacting materials of a specific chemical composition to produce a compound with planned and predictable properties; the resulting *base stock* may be supplemented with *additives* to improve specific properties. Many synthetic lubricants — also called **synlubes** — are derived wholly or primarily from *petrochemicals*; other synlube raw materials are derived from coal and oil shale, or are lipochemicals (from animal and vegetable oils). Synthetic lubricants may be superior to petroleum oils in specific performance areas. Many exhibit higher *viscosity index* (V.I.), better *thermal stability* and *oxidation stability*, and low *volatility* (which reduces oil consumption). Individual synthetic lubricants offer specific outstanding properties: *phosphate esters*, for example, are fire resistant, *diesters* have good oxidation stability and *lubricity*, and *silicones* offer exceptionally high V.I. Most synthetic lubricants can be converted to grease by adding thickeners. Because synthetic lubricants are higher in cost than petroleum oils, they are used selectively where performance or safety requirements may exceed the capabilities of a conventional oil. The following table lists the principal classes of synthetic lubricants:

alkylated aromatics[1]	polyol esters[2]	silicones
olefin oligomers[1]	polyglycols	silicate esters
dibasic acid esters[2]	phosphate esters	halogenated hydrocarbons
[1]organic hydrocarbon	[2]organic ester	

synthetic oil and gas — any oil or gas suitable as fuel, but not produced by the conventional means of pumping from underground reserves. Synthetic gas may be derived from coal, *naphtha*, or liquid petroleum gas (*LPG*); and synthetic oil may be derived from coal, *oil shale*, and tar sands.

synthetic rubber — any *petrochemical*-based *elastomer*. Like *natural rubber*, synthetic rubbers are *polymers*, consisting of a series of simple molecules, called monomers, linked together to form large chain-like molecules. The chain forms a loose coil that returns to its coiled form after it is extended. See under individual listings: *butyl rubber, ethylene-propylene rubber, natural rubber, neoprene rubber, nitrile rubber, polybutadiene rubber, polyisoprene rubber, styrene-butadiene rubber.*

synthetic turbo oil — non-petroleum lubricant for aircraft gas *turbines* generally made from an *ester* base. It is characterized by high *oxidation stability* and *thermal stability*, good load-carrying capacity, and the extreme low *volatility* required to prevent excessive evaporation under wide operating-temperature conditions.

Système International — see *SI*.

systemic effect — toxic effect that is produced in any of the organs of the body after a toxicant has been absorbed into the bloodstream.

tackiness agent — *additive* used to increase the adhesive properties of a lubricant, improve retention, and prevent dripping and splattering.

Tag closed tester — apparatus for determining the *flash point* of petroleum liquids having a *viscosity* below 5.8 centistokes (cSt) at 37.8°C (100°F) and a flash point below 93°C (200°F), under test methods prescribed in ASTM D 56. The test sample is heated in a closed cup at a specified constant rate. A small flame of specified size is introduced into the cup through a shuttered opening at specified intervals. The lowest temperature at which the vapors above the sample briefly ignite is the flash point. See *Pensky-Martens closed tester.*

Tag open cup — apparatus for determining the *flash point* of hydrocarbon liquids, usually *solvents*, having flash points between −17.8° and 168°C (0° to 325°F), under test methods prescribed in ASTM D 1310. The test sample is heated in an open cup at a slow, constant rate. A small flame is passed over the cup at specified intervals. The lowest temperature at which the vapors above the sample briefly ignite is the flash point. See *Cleveland open cup.*

Tag-Robinson colorimeter — see *color scale*.

tail-end volatility — see *distillation test*.

technical white oil — see *white oil*.

TEL (tetraethyl lead) — see *lead alkyl*.

tempering — hardening or strengthening of metal by application of heat or by alternate heating and cooling.

tempering oil — oil used as a medium for heating metals to their *tempering* temperature to relieve stress and improve toughness and ductility.

temperature scales — arbitrary thermometric calibrations that serve as convenient references for temperature determination. There are two thermometric scales based on the freezing and boiling point of water at a pressure of one **atmosphere:** the **Fahrenheit** (F) scale (32° = freezing, 212° = boiling) and the **Celsius** (C), or **Centigrade,** scale (0° = freezing, 100° = boiling).

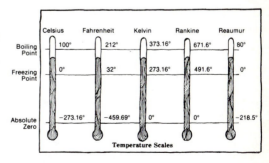

	Celsius	Fahrenheit	Kelvin	Rankine	Reaumur
Boiling Point	100°	212°	373.16°	671.6°	80°
Freezing Point	0°	32°	273.16°	491.6°	0°
Absolute Zero	−273.16°	−459.69°	0°	0°	−218.5°

Temperature Scales

Additionally, there are two scales in which 0° = absolute zero, the temperature at which all molecular movement theoretically ceases: the **Kelvin** (K), or **Absolute** (°A), scale and the **Rankine** (°R) scale, which are related to the Celsius and the Fahrenheit scales, respectively (0°K = −273.16°C; 0°R = −459.69°F). The four scales can be related to each other by the following formulas:

$$°C = 5/9 \,(°F - 32) \qquad °F = 9/5 \,°C + 32$$
$$°K = °C + 273.16 \qquad °R = °F + 459.69$$

terpolymer — *copolymer* formed by the polymerization of three different *monomers*. An example of a terpolymer is *EPDM rubber*, made from *ethylene, propylene,* and a third monomer (usually a *diolefin*). See *polymer*.

tertiary recovery — any method employed to increase removal of hydrocarbons from a *reservoir* after *secondary recovery* methods have been applied.

tetraethyl lead (TEL) — see *lead alkyl*.

tetramethyl lead (TML) — see *lead alkyl*.

texture — that property of a lubricating grease which is observed when a small portion of it is compressed and the pressure slowly released. Texture should be described in the following terms: **brittle** — has a tendency to rupture or crumble when separated; **buttery** — separates in short peaks with no visible fibers; **long fiber** — shows tendency to stretch or string out into bundles of fibers; **short fiber** — shows short break-off with evidence of fibers; **resilient** — capable of withstanding moderate compression without permanent deformation or rupture; **stringy** — shows tendency to stretch or string out into long fine threads, but with no visible evidence of fiber structure.

thermal cracking — in refining, the breaking down of large, high-boiling hydrocarbon molecules into smaller molecules in the presence of heat and pressure. See *cracking*.

thermal stability — ability to resist chemical degradation at high temperatures.

thermal value — see *heat of combustion*.

thin-film rust preventive — see *rust preventive*.

thixotropy — tendency of grease or other material to soften or flow when subjected to shearing action. Grease will usually return to its normal consistency when the action stops. The phenomenon is the opposite of that which occurs with *rheopectic grease*. Thixotropy is also an important characteristic of *drilling fluids,* which must thicken when not in motion so that the cuttings in the fluid will remain in suspension.

threshold limit value (TLV) — see *occupational exposure limit*.

throttle plate — see *carburetor*.

time-weighted average (TWA) — atmospheric concentration of a substance, in parts per million by volume, measured over a seven or eight-hour workday and 40-hour week.

Timken EP test — measure of the extreme-pressure properties of a lubricating oil (see *EP oil*). The test utilizes a Timken machine, which consists of a stationary block pushed upward, by means of a lever arm system, against the rotating outer race of a roller bearing, which is lubricated by the product under test. The test continues under increasing load (pressure) until a measurable wear scar is formed on the block. Timken OK load is the heaviest load that a lubricant can withstand before the block is scored (see *scoring*).

Another scale based on the thermometric properties of water is the **Reaumur** scale, in which the freezing point is set at zero degrees and the boiling point at 80 degrees. This scale has only limited application.

TLV (threshold limit value) — see *occupational exposure limit*.

TML (tetramethyl lead) — see *lead alkyl*.

toluene — aromatic hydrocarbon ($C_6H_5CH_3$) with good solvent properties; used in the manufacture of lacquers and other industrial coatings, adhesives, printing ink, insecticides, and chemical raw materials. Also called **toluol**.

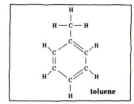

toluene insolubles — see *insolubles*.

toluol — see *toluene*.

torque fluid — lubricating and power-transfer medium for commercial automotive torque converters and transmissions. It possesses the low viscosity necessary for torque transmission, the lubricating properties required for associated gear assemblies, and compatibility with seal materials.

total acid number — see *neutralization number*.

total base number — see *neutralization number*.

transformer oil — see *insulating oil*.

tribology — science of the interactions between surfaces moving relative to each other. Such interactions usually involve the interplay of two primary factors: the load, or force, perpendicular to the surfaces, and the frictional force that impedes movement. Tribological research on friction reduction has important energy conservation applications, since friction increases energy consumption. See *friction*.

trimer — see *polymerization*.

turbine — device that converts the force of a gas or liquid moving across a set of rotor and fixed blades into rotary motion. There are three basic types of turbines: gas, steam, and hydraulic. **Gas turbines** are powered by the expansion of compressed gases generated by the combustion of a fuel. (See *internal combustion engine*.) Some of the power thus produced is used to drive an air compressor, which provides the air necessary for combustion of the fuel. In a turbo-jet aircraft engine, the turbine's only function is to drive the compressor: the plane is propelled by the force of the expanding gases escaping from the rear of the engine. In other applications, however, the rotor shaft provides the driving thrust to some other mechanism, such as a propeller or generator. Thus, gas turbines power not only turbo-jet aircraft, but also turbo-prop aircraft, locomotives, ships, compressors, and small-to-medium-size electric utility generators. Gas turbine-powered aircraft present severe lubrication demands that are best met with a *synthetic turbo oil*. **Steam turbines** employ steam that enters the turbine at high temperature and pressure and expands across both rotating and fixed blades (the latter serving to direct the steam). Steam turbines, which power large electric generators, produce most of the world's electricity. Only the highest-quality lubricants are able to withstand the wet conditions and high temperatures associated with steam turbine operation. The term *turbine oil* has thus become synonymous with quality. **Hydraulic turbines** (water turbines) are either impulse type, in which falling water hits blades or buckets on the periphery of a wheel that turns a shaft, or reaction type, where water under pressure emerges from nozzles on the wheel, causing it to turn. Hydraulic turbines can be used to produce electric power near reservoir or river dams.

turbine oil — top-quality rust- and oxidation-inhibited *(R&O)* oil that meets the rigid requirements traditionally imposed on steam-turbine lubrication. (See *turbine.*) Quality turbine oils are also distinguished by good *demulsibility,* a requisite of effective oil-water separation. Turbine oils are widely used in other exacting applications for which long service life and dependable lubrication are mandatory. Such applications include circulating systems, compressors, hydraulic systems, gear drives, and other equipment. Turbine oils can also be used as *heat transfer fluids* in open systems, where *oxidation stability* is of primary importance.

Ultra Large Crude Carrier (ULCC) — see *supertanker.*

ultraviolet absorbance — measurement of the ultraviolet absorption of petroleum products, determined by standardized tests, such as ASTM D 2008. *Aromatics* absorb more ultraviolet light than do *naphthenes* and *paraffins,* and the amount of absorbance can be used as an indication of the amount of aromatics in a product. Certain polynuclear aromatics *(PNA's)* are known carcinogens (cancer-causing substances), with peaks of absorbance generally between 280 and 400 millimicrons. The Food and Drug Administration (FDA) has therefore imposed limits on the amount of ultraviolet absorbance at these wavelengths for materials classified as food additives. However, not all materials with ultraviolet absorbance at these wavelengths are carcinogenic.

undisturbed penetration — see *penetration (grease).*

United States Geological Survey — see *USGS.*

United States Pharmacopeia — see *USP.*

unleaded gasoline — gasoline that derives its *antiknock* properties from high-octane hydrocarbons or from non-lead *antiknock compounds,* rather than from a lead additive. See *lead alkyl.*

unsaturated hydrocarbon — hydrocarbon lacking a full complement of hydrogen atoms, and thus characterized by one or more double or triple bonds between carbon atoms. Hydrocarbons having only one double bond between adjacent carbon atoms in the molecule are called *olefins*; those having two double bonds in the molecule are *diolefins*. Hydrocarbons having alternating single and double bonds between adjacent carbon atoms in a benzene-ring configuration are called *aromatics*. Hydrocarbons with a triple bond between carbon atoms are called *acetylenes*. Unsaturated

vacuum tower — see *distillation.*

valve beat-in — wear on the valve face or valve seat in *internal combustion engines* resulting from the pounding of the valve on the seat. Also called **valve sink** or **valve recession.**

vapor lock — disruption of fuel movement to a *gasoline* engine *carburetor* caused by excessive vaporization of gasoline. Vapor lock occurs when the fuel pump, which is designed to pump liquid, loses

turbocharger — see *supercharger.*

turbo fuel (jet fuel) — *kerosene*-type fuel used in gas-turbine-powered aircraft. Important properties of a turbo fuel include: clean and efficient burning, ability to provide adequate energy for thrust, resistance to chemical degradation in storage or when used as a heat transfer medium on the aircraft, non-corrosiveness, ability to be transferred and metered under all conditions, volatility high enough for burning but low enough to prevent excessive losses from tank vents, and freedom from dirt, rust, water, and other contaminants. Turbo fuel has different properties than *aviation gasoline,* which fuels piston-engine aircraft.

turbo oil — see *synthetic turbo oil.*

TWA — see *time-weighted average.*

two-stroke cycle — see *internal combustion engine.*

hydrocarbons readily attract additional hydrogen, oxygen, or other atoms, and are therefore highly reactive. See *hydrocarbon, saturated hydrocarbon, hydrogenation.*

unsulfonated residue — measure of the volume of unsulfonated residue in plant spray oils of petroleum origin, as determined by test method ASTM D 483. Unsulfonated residue consists of those components of the oil that will not react with concentrated sulfuric acid. This determination is useful for distinguishing between oils suitable for various types of spraying applications. An oil with a high unsulfonated residue (92 volume percent minimum) is required for spraying orchard crops in the leaf or bud stage. See *orchard spray oil.*

unworked penetration — see *penetration (grease).*

USGS (United States Geological Survey) — bureau of the Department of the Interior, responsible for performing surveys and investigations covering U.S. topography, geology, and mineral and water resources and for enforcing departmental regulations applicable to oil, gas, and other mining activities.

USP (United States Pharmacopeia) — compendium of drugs, drug formulas, quality standards and tests published by the United States Pharmacopeial Convention, Inc., which also publishes the *NF* (National Formulary). The purpose of the USP is to ensure drug uniformity and to maintain and upgrade standards of drug quality and purity, as well as establish packaging, labeling, and storage requirements. The USP includes standards for *white oils* under two classifications: "Mineral Oil" for heavy grades, and "Mineral Oil Light" for lighter grades.

suction as it tries to pump fuel vapor. The engine will usually stall, but in less severe cases may accelerate sluggishly or *knock* due to an excessively lean fuel mixture. Automotive engines are more likely to experience vapor lock during an acceleration that follows a short shutdown period. Vapor lock problems are most likely to occur in the late spring on unseasonably warm days, before the more volatile winter grades of gasoline have been replaced by the less volatile spring and summer grades (see *volatility*). Vapor lock can also occur in other types of pumping systems where volatile liquids are being handled.

vapor pressure — pressure of a confined vapor in equilibrium with its liquid at a specified temperature; thus, a measure of a liquid's *volatility*. Vapor pressure of gasoline and other volatile petroleum products is commonly measured in accordance with test method ASTM D 323 (**Reid vapor pressure**). The apparatus is essentially a double-chambered bomb. One chamber, fitted with a pressure gauge, contains air at atmospheric pressure; the other chamber is filled with the liquid sample. The bomb is immersed in a 37.8°C (100°F) bath, and the resulting vapor pressure of the sample is recorded in pounds per square inch (psi). Reid vapor pressure is useful in predicting seasonal gasoline performance (e.g., higher volatility is needed in cold weather, and lower volatility in hot weather), as well as the tendencies of gasolines, solvents, and other volatile petroleum products toward evaporative loss and fire hazard.

varnish — hard, varnish-like coating formed from oil oxidation products, that bakes on to pistons during high-temperature automotive engine operation. Varnish can accelerate cylinder wear. Varnish formation can be reduced with the use of a *detergent-dispersant* and an *oxidation inhibitor* in the oil. See *engine deposits*.

Very Large Crude Carrier (VLCC) — see *supertanker*.

V.I. — see *viscosity index (V.I.)*.

V.I. (viscosity index) improver — see *viscosity index (V.I.) improver*.

viscometer — device for measuring *viscosity*; commonly in the form of a calibrated capillary tube through which a liquid is allowed to pass at a controlled temperature in a specified time period. See *kinematic viscosity*.

viscosimeter — see *viscometer*.

viscosity — measurement of a fluid's resistance to flow. The common metric unit of *absolute viscosity* is the **poise**, which is defined as the force in dynes required to move a surface one square centimeter in area past a parallel surface at a speed of one centimeter per second, with the surfaces separated by a fluid film one centimeter thick. For convenience, the **centipoise** (cp) — one one-hundredth of a poise — is the unit customarily used. Laboratory measurements of viscosity normally use the force of gravity to produce flow through a capillary tube (*viscometer*) at a controlled temperature. This measurement is called *kinematic viscosity*. The unit of kinematic viscosity is the **stoke**, expressed in square centimeters per second. The more customary unit is the **centistoke** (cSt) — one one-hundredth of a stoke. Kinematic viscosity can be related to absolute viscosity by the equation:

$$cSt = cp \div \text{fluid density}.$$

In addition to kinematic viscosity, there are other methods for determining viscosity, including *Saybolt Universal viscosity*, *Saybolt Furol viscosity*, *Engler viscosity*, and *Redwood viscosity*. Since viscosity varies inversely with temperature, its value is meaningless unless the temperature at which it is determined is reported.

viscosity (asphalt) — determined by any of several *ASTM* test methods. Two common methods are ASTM D 2170 and D 2171. The former method measures *kinematic viscosity*, that is, viscosity under the force of gravity, by allowing a test sample to flow down a capillary tube (*viscometer*) at a temperature of 135°C (275°F); the viscosity is expressed in centistokes. The latter method measures

absolute viscosity. The liquid, at a temperature of 60°C (140°F), is drawn up a tube by applying a vacuum; the viscosity is expressed in poises. In both tests the viscosity is obtained by multiplying the flow time in seconds by the viscometer calibration factor. See *penetration (asphalt)*, *viscosity*.

viscosity grading (asphalt) — classification system for *asphalt cement*, defined in American Association of State Highway Transportation Officials (AASHTO) Specification M 226, and based on tests for *viscosity*, *flash point*, ductility, purity, etc., as outlined in test method ASTM D 3381. There are five standard grades, from softest to hardest: AC-2.5, AC-5, AC-10, AC-20, AC-40. Asphalt cement is also classified by *penetration grading*. See *viscosity (asphalt)*.

viscosity-gravity constant (VGC) — indicator of the approximate hydrocarbon composition of a petroleum oil. As described in test method ASTM D 2501, VGC is calculated from one of the following equations, depending on the temperature at which viscosity is determined (VGC at 37.8°C [100°F] is the preferred equation):

$$\text{VGC @ 37.8°C (100°F)} = \frac{10G - 1.0752 \log (V-38)}{10 - \log (V-38)}$$

where: G = specific gravity @ 15.6°/15.6°C (60/60°F),
and
V = Saybolt Universal viscosity @ 37.8°C (100°F).

or

$$\text{VGC @ 98.9°C (210°F)} = \frac{G - 0.1244 \log (V_1 - 31)}{0.9255 - 0.0979 \log (V_1 - 31)} - 0.0839$$

where: G = specific gravity at 15.6°/15.6°C (60/60°F), and
V_1 = Saybolt Universal viscosity at 98.9° (210°F).

Values of VGC near 0.800 indicate an oil of paraffinic character (see *paraffin*); values close to 1.00 indicate a preponderance of *aromatic* structures. Like other indicators of hydrocarbon composition (as opposed to a specific laboratory analysis), VGC should not be indiscriminately applied to residual oils (see *bottoms*), asphaltic materials, or samples containing appreciable quantities of non-hydrocarbons. See *Saybolt Universal Viscosity*, *specific gravity*.

viscosity index (V.I.) — empirical, unitless number indicating the effect of temperature change on the *kinematic viscosity* of an oil. Liquids change *viscosity* with temperature, becoming less viscous when heated; the higher the V.I. of an oil, the lower its tendency to change viscosity with temperature. The V.I. of an oil — with known viscosity at 40°C and at 100°C — is determined by comparing the oil with two standard oils having an arbitrary V.I. of 0 and 100, respectively, and both having the same viscosity at 100°C as the test oil. The following formula is used, in accordance with test method ASTM D 2270:

$$\text{V.I.} = \frac{L-U}{L-H} \times 100$$

where L is the viscosity at 40°C of the 0-V.I. oil, H is the viscosity at 40°C of the 100-V.I. oil, and U is the viscosity at 40°C of the test oil. There is an alternative calculation, also in ASTM D 2270, for oils with V.I's above 100. The V.I. of paraffinic oils (see *paraffin*) is inherently high, but is low in naphthenic oils (see *naphthene*), and even lower in *aromatic* oils (often below 0). The V.I. of any petroleum oil can be increased by adding a *viscosity index improver*. High-V.I. lubricants are needed wherever relatively constant viscosity is required at widely varying temperatures. In an automobile, for example, an engine oil must flow freely enough to permit cold starting, but must be viscous enough after warm-up to provide full lubrication. Similarly, in an aircraft hydraulic system, which may be exposed to temperatures above 38°C at ground level and temperatures below −54°C at high altitudes, consistent hydraulic fluid performance requires a high viscosity index.

viscosity index (V.I.) improver — lubricant *additive,* usually a high-molecular-weight *polymer,* that reduces the tendency of an oil to change *viscosity* with temperature. *Multi-grade oils,* which provide effective lubrication over a broad temperature range, usually contain V.I. improvers. See *viscosity index.*

viscosity-temperature relationship — the manner in which the *viscosity* of a given fluid varies inversely with temperature. Because of the mathematical relationship that exists between these two variables, it is possible to predict graphically the viscosity of a petroleum fluid at any temperature within a limited range if the viscosities at two other temperatures are known. The charts used for this purpose are the *ASTM* Standard Viscosity-Temperature Charts for Liquid Petroleum Products, available in 6 ranges. If two known viscosity-temperature points of a fluid are located on the chart and a straight line drawn through them, other viscosity-temperature values of the fluid will fall on this line; however, values near or below the *cloud point* of the oil may deviate from the straight-line relationship.

wash oil — see *absorber oil.*

wax (petroleum) — any of a range of relatively high-molecular-weight hydrocarbons (approximately C_{16} to C_{50}), solid at room temperature, derived from the higher-boiling petroleum fractions. There are three basic categories of petroleum-derived wax: **paraffin** (crystalline), **microcrystalline** and **petrolatum.** ● Paraffin waxes are produced from the lighter lube oil *distillates,* generally by chilling the oil and filtering the crystallized wax; they have a distinctive crystalline structure, are pale yellow to white (or colorless), and have a melting point range between 48°C (118°F) and 71°C (160°F). Fully refined paraffin waxes are dry, hard, and capable of imparting good gloss. ● Microcrystalline waxes are produced from heavier lube distillates and residua (see *bottoms*) usually by a combination of solvent dilution and chilling. They differ from paraffin waxes in having poorly defined crystalline structure, darker color, higher *viscosity*, and higher melting points —ranging from 63°C (145°F) to 93°C (200°F). The microcrystalline grades also vary much more widely than paraffins in their physical characteristics: some are ductile and others are brittle or crumble easily. Both paraffin and microcrystalline waxes have wide uses in food packaging, paper coating, textile moisture proofing, candle-making, and cosmetics. ● Petrolatum is derived from heavy residual lube stock by propane dilution and filtering or centrifuging. It is microcrystalline in character and semi-solid at room temperature. The best known type of petrolatum is the "petroleum jelly" used in ointments. There are also heavier grades for industrial applications, such as corrosion preventives, carbon paper, and butcher's wrap. Traditionally, the terms *slack wax, scale wax,* and *refined wax* were used to indicate limitations on oil content. Today,

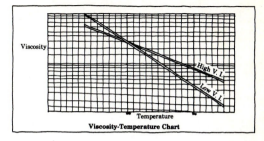

Viscosity-Temperature Chart

VM&P Naphtha — Varnish Makers and Painters Naphtha; term for a *naphtha* commonly used as a solvent in paints and varnishes.

volatility — expression of evaporation tendency. The more volatile a petroleum liquid, the lower its boiling point and the greater its flammability. The volatility of a petroleum product can be precisely determined by tests for evaporation rate; also, it can be estimated by tests for *flash point* and *vapor pressure,* and by *distillation tests.*

volumetric efficiency — ratio of the weight of air drawn into the cylinder of an operating *internal combustion engine* to the weight of air the cylinder could hold at rest when the piston is at the bottom of the stroke and the valves are fully closed. Any restriction of air flow into the cylinder reduces volumetric efficiency, which, in turn, reduces power output. The volumetric efficiency of an automotive engine is usually slightly more than 80% at about half the rated speed of the engine, then decreases considerably at higher speed, thus limiting the power output of the engine. The air charge to the cylinder can be increased at high speeds by means of supercharging. See *supercharger.*

these classifications are less exact in their meanings, especially in the distinction between slack wax and scale wax. For further information relating to wax, see *blocking point, gloss, laminating strength, melting point of wax, oil content of petroleum wax, paraffin wax, petrolatum, refined wax, scale wax, scuff resistance, sealing strength, slack wax, strike-through.*

wax appearance point (WAP) — temperature at which wax begins to precipitate out of a *distillate* fuel, when the fuel is cooled under conditions prescribed by test method ASTM D 3117. WAP is an indicator of the ability of a distillate fuel, such as *diesel fuel*, to flow at cold operating temperatures. It is very similar to *cloud point.*

way — longitudinal surface that guides the reciprocal movement of a machine part. See *stick-slip motion.*

weed killer — *pesticide* product, often derived from petroleum, used for destroying weeds by direct application to the plants. "Selective" weed killers are those that, when applied in accordance with instructions, are not destructive to specified crops. Nonselective weed killers kill all vegetation.

weld point — the lowest applied load in kilograms at which the rotating ball in the Four Ball EP test either seizes and welds to the three stationary balls, or at which extreme *scoring* of the three balls results. See *four-ball method.*

wellbore — see *borehole.*

wellhead — the point at which oil and gas emerge from below ground to the surface; also, the equipment used to maintain surface control of a well.

wet gas — natural gas containing a high proportion of hydrocarbons that are readily recoverable from the gas as liquids. See *natural gas liquids*.

white oil — highly refined *straight mineral oil*, essentially colorless, odorless, and tasteless. White oils have a high degree of chemical stability. The highest purity white oils are free of unsaturated components (see *unsaturated hydrocarbon*) and meet the standards of the United States Pharmacopeia *(USP)* for food, medicinal, and cosmetic applications. White oils not intended for medicinal use are known as **technical white oils** and have many industrial applications — including textile, chemical, and plastics

manufacture — where their good color, non-staining properties, and chemical inertness are highly desirable.

wide cut — see *distillation test*.

wildcat — exploratory oil or gas well drilled in an area not previously known to be productive.

wood alcohol — see *methanol*.

worked penetration — see *penetration (grease)*.

workover — well that has been put back into production following remedial actions (such as deepening, clearing, *acidizing*) to restore or increase production.

worm gear — see *gear*.

xylene — aromatic hydrocarbon, C_8H_{10}, with three *isomers* plus ethylbenzene. It is used as a *solvent* in the manufacture of *synthetic rubber* products, printing inks for textiles, coatings for paper, and adhesives, and serves as a raw material in the chemical industry.

yield point — the minimum force required to produce flow of a plastic material.

ZDDP (zinc dialkyl dithiophosphate or zinc diaryl dithiophosphate) — widely used as an anti-wear agent in motor oils to protect heavily loaded parts, particularly the valve train mechanisms (such as the camshaft and cam followers) from excessive wear. It is also used as an anti-wear agent in *hydraulic fluids* and certain other products. ZDDP is also an effective *oxidation inhibitor*. Oils containing ZDDP should not be used in engines that employ silver alloy bearings. All car manufacturers now recommend the use of dialkyl ZDDP in motor oils for passenger car service.

ZN/P curve — general graphic representation of the equation: μ =(f) ZN/P, where μ (the *coefficient of friction* in a *journal* bearing) is a function (f) of the dimensionless parameter ZN/P, (viscosity × speed)/pressure. This is the fundamental lubrication equation, in

which the coefficient of friction is the friction per unit load, Z the viscosity of the lubricating oil, N the rpm of the journal, and P the pressure (load per unit area) on the bearing. The ZN/P curve illustrates the effects of the three variables (viscosity, speed, and load) on friction and, hence, on lubrication. See *boundary lubrication, full-fluid-film lubrication*.

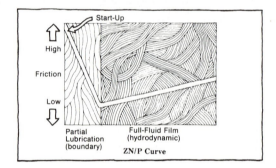

ZN/P Curve

D

Summary of Pump Engineering Data

This appendix has been prepared by the ITT Fluid Technology Corporation (PO Box 200, Midland Park, NJ 07432). It contains an extensive summary of key definitions and engineering data useful to the selection and sizing of all types of pumps. It is important to note that the information compiled in this appendix was gathered from numerous sources, i.e., various manufacturers' data, engineering handbooks, practical experience, and laboratory tests. Although the information presented is accurate to the best of ITT Fluid Technology Corporation's and the authors' knowledge, no responsibility is claimed for any information reported herein.

COMMON PUMP HEAD TERMS

The term "head" by itself is rather misleading. It is commonly taken to mean the difference in elevation between the suction level and the discharge level of the liquid being pumped. Although this is partially correct, it does not include all of the conditions that should be included to give an accurate description.

FRICTION HEAD

—is the pressure expressed in lbs./sq. in. or feet of liquid needed to overcome the resistance to the flow in the pipe and fittings.

SUCTION LIFT

—exists when the source of supply is below the center line of the pump.

SUCTION HEAD

—exists when the source of supply is above the center line of the pump.

STATIC SUCTION LIFT

—is the vertical distance from the center line of the pump down to the free level of the liquid source.

STATIC SUCTION HEAD

—is the vertical distance from the center line of the pump up to the free level of the liquid source.

STATIC DISCHARGE HEAD

—is the vertical elevation from the center line of the pump to the point of free discharge.

DYNAMIC SUCTION LIFT

—includes static suction lift, friction head loss, and velocity head.

DYNAMIC SUCTION HEAD

—includes static suction head minus friction head minus velocity head.

DYNAMIC DISCHARGE HEAD

—includes static discharge head plus friction head plus velocity head.

TOTAL DYNAMIC HEAD

—includes the dynamic discharge head plus dynamic suction lift or minus dynamic suction head.

VELOCITY HEAD

—is the head needed to accelerate the liquid. Knowing the velocity of the liquid, the velocity head loss can be calculated by a simple formula Head $= V^2/2g$ in which g is acceleration due to gravity or 32.16 ft./sec. Although the velocity head loss is a factor in figuring the dynamic heads, the value is usually small and in most cases negligible. See table

VELOCITY — VELOCITY HEAD

Velocity Ft./Sec.	4	5	6	7	8	9	10	11	12	13	14	15
Velocity Head-Feet	.25	.39	.56	.76	1.0	1.25	1.55	1.87	2.24	2.62	3.05	3.25

The term "head" is usually expressed in feet whereas pressure is usually expressed in pounds per square inch. Quite often the suction lift is expressed in inches of vacuum (mercury). The formulii for converting these factors follows:

$$\frac{\text{lbs./sq. in.}}{\text{Pressure}} = \frac{\text{Head (Feet)} \times \text{Specific Gravity}}{2.31}$$

$$\text{Head (Feet)} = \frac{\text{Pressure (PSI)} \times 2.31}{\text{Specific Gravity}}$$

Vacuum
inches of mercury = Dynamic suction lift (feet) × 0.883 × Specific Gravity

DEFINITIONS AND OTHER INFORMATION

SPECIFIC GRAVITY

Direct ratio of any liquid's weight to the weight of water at 62°F. Water at 62°F weighs 8.33# per gallon and is designated 1.0 sp. gr.

NOTE: A centrifugal pump develops head, not pressure. All pressure figures should be converted to feet of head taking into consideration the sp. gr. (Ft. HD = PSI × 2.31 ÷ Sp. Gr.)

VISCOSITY

Property of a liquid that resists any force tending to produce flow. It is the evidence of cohesion between the particles of a fluid which causes a liquid to offer resistance analogous to friction. An increase in the temperature reduces the viscosity; conversely, a temperature reduction increases the viscosity. Pipe friction loss increases as viscosity increases.

EFFECTS OF VISCOSITY

Viscous liquids tend to increase pump HP, reduce efficiency, reduce capacity and head and increase pipe friction. (See pages 17 through 20.)

VISCOSITY-CENTRIFUGAL PUMP PERCENTAGE OF WATER CHARACTERISTICS— APPROXIMATE GUIDE

Viscosity S.S.U.	Capacity	Head	Efficiency
31.5	100%	100%	100%
500	95%	98%	80%
1000	92%	97%	70%
2000	89%	94%	60%

NOTE: Pump applications on viscosities higher than 500 SSU should be referred to factory.

ELECTROLYSIS

Any combination of dissimilar metals should be avoided whenever practicable when the liquid being pumped is an electrolyte. Such metals tend to promote galvanic action and may deteriorate rapidly. An electrolyte solution is one similar to brackish sea water, certain acids, etc.

pH is a measure of hydrogen ion concentration. pH of 7 is neutral—below 7 acid—above 7 alkaline

General Guide—Pump Construction

pH VALUE OF SOLUTIONS—GENERAL GUIDE FOR PUMP CONSTRUCTION

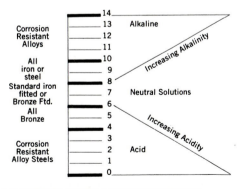

NOTE: There is no substitute for experience in selecting materials for pump construction. Previous experience in handling a particular solution should be the main criterion in determining the optimal construction.

POWER REQUIRED FOR PUMPING

$$\text{Water Horsepower} = \frac{\text{GPM} \times \text{Feet Head} \times \text{Sp. Gr.}}{3960}$$
(Theoretical H.P.)

$$\text{Brake Horsepower} = \frac{\text{GPM} \times \text{Feet Head} \times \text{Sp. Gr.}}{3960 \times \text{Pump Efficiency}}$$

$$\text{Over-All-Horsepower} = \frac{\text{BHP}}{\text{Motor Efficiency}}$$
(wire to water HP)

MOTOR DATA

The syncronous speed of any alternating current (AC) motor is set by the frequency of the line voltage and the number of poles.

$$\text{Current Consumption in Kilowatt Hours} = \frac{\text{Feet Head} \times 0.00315}{\text{Pump Eff.} \times \text{Motor Eff.}}$$
(per 1000 gallons of water)

SYNCHRONOUS SPEED		
POLES	60 HERTZ	50 HERTZ
2	3600	3000
4	1800	1500
6	1200	1000
8	900	750
10	720	600
12	600	500

Squirrel-cage or wound-rotor induction motors lose speed as the load increases. Decrease the syncronous speed about 3% to 5% to obtain the actual shaft speed.

MOTOR TERMINAL AMPERES AT FULL LOAD†
Average Values for All Speeds and Frequencies

| | SINGLE-PHASE A-C | | POLYPHASE A-C (INDUCTION TYPE) SQUIRREL-CAGE AND WOUND ROTOR | | | | | | | | | | DIRECT CURRENT | | |
| | | | 115 Volts | | 230 Volts | | 460 Volts | | 575 Volts | | | | | |
HP	115 Volts	230 Volts*	3-Ph.	2-Ph. 4-Wire†	3-Ph.	2-Ph. 4-Wire‡	3-Ph.	2-Ph. 4-Wire‡	3-Ph.	2-Ph. 4-Wire‡	180 Volts	240 Volts	500 Volts
¼	5.8	2.9	....	....	....	....	....	....	....	....	2.0	....	....
⅓	7.2	3.6	....	....	....	....	....	....	....	....	2.6	....	....
½	9.8	4.9	4.0	4.0	2.0	2.0	1.0	1.0	.8	.8	3.4	2.7	....
¾	13.8	6.9	5.6	4.8	2.8	2.4	1.4	1.2	1.1	1.0	4.8	3.8	....
1	16	8	7.2	6.4	3.6	3.2	1.8	1.6	1.4	1.3	6.1	4.7	....
1½	20	10	10.4	9.0	5.2	4.5	2.6	2.3	1.8	1.8	8.3	6.6	....
2	24	12	13.6	11.8	6.8	5.9	3.4	3.0	2.7	2.4	10.8	8.5	....
3	34	17	....	....	9.6	8.3	4.8	4.2	3.9	3.3	16	12.2	....
5	56	28	....	....	15.2	13.2	7.6	6.6	6.1	5.3	27	20	13.6
7½	80	40	....	....	22.0	19.0	11.0	9	9	8	....	29	18
10	100	50	....	....	28.0	24.0	14.0	12	11	10	....	38	27
15	....	....	....	....	42.0	36.0	21	18	17	14	....	55	34
20	....	....	....	....	54	47	27	23	22	19	....	72	43
25	....	....	....	....	68	59	34	29	27	24	....	89	51
30	....	....	....	....	80	69	40	35	32	28	....	106	67
40	....	....	....	....	104	90	52	45	41	36	....	140	83
50	....	....	....	....	130	113	65	56	52	45	....	173	99
60	....	....	....	....	154	133	77	67	62	53	....	206	123
75	....	....	....	....	192	166	96	83	77	66	....	255	

‡ Current in common conductor of 2-phase, 3 wire system will be 1.41 times value given.
* For full-load currents of 208 and 200-volt motors increase the corresponding 230-volt motor full-load current by 10 and 15 per cent respectively.
† These values of full-load current are for motors running, at speeds usual for belted motors and motors with normal torque characteristics. Motors built for especially low speeds or high torques may require more running current, in which case the nameplate current rating should be used.

AFFINITY LAWS

Effect on centrifugal pumps of change of speed or impeller diameter.

Capacity varies directly as the speed or impeller diameter. (GPM x rpm x D)
Head varies as the square of the speed or impeller diameter. (GPM x rpm² x D²)
BHP varies as the cube of the speed or impeller diameter. (BHP x rpm³ x D³)

CENTRIFUGAL PUMPS

A centrifugal pump is one that employs centrifugal force for pumping liquids. Liquid entering the eye (center) of the impeller is accelerated by the impeller vanes to a high velocity and is thrown out from the rotating vanes by centrifugal force into an annular channel or volute to the discharge. It differs from the Rotary positive displacement pump in that there is no close internal fit of rubbing or sliding parts.

The self-priming centrifugal pump combines the smooth, quiet and efficient operation of the straight centrifugal pump with the self-priming characteristics of a rotary pump. It will operate on suction lifts where straight centrifugals cannot be used and yet work equally as efficient on flooded suctions as well. For handling volatile liquids or "dry" liquids it is especially desirable. No other pump is as versatile.

STRAIGHT CENTRIFUGAL

 This single stage, end suction straight centrifugal pump has an impeller rotating within the volute casing.

SELF PRIMING CENTRIFUGAL

This patented Marlow diffuser prime pump has an impeller rotating within a stationary diffuser. Air is mixed with liquid at the vane tips during the priming action. When primed, diffuser acts as several volutes.

The diffuser prime action picks up air from the impeller. The air is expelled through the diffuser passageways and escapes by bubbling to the surface of the liquid. Here it is eliminated from the pump system while the reservoir liquid is returned to the impeller tips to capture more air. This action continues very rapidly until all of the air is eliminated at which time the suction line is fully primed and pumping begins. This type of self-priming centrifugal pump retains priming efficiency since there are no recirculating priming valves or pealers to wear.

CENTRIFUGAL PUMP ADVANTAGES

1. Low initial cost
2. Low maintenance cost
3. Extremely long life
4. Reduces weight and size
5. Quietness
6. Greater performance

GENERAL INFORMATION ON PUMP INSTALLATION

Care must be exercised in the layout and installation of pumping systems for handling industrial and petroleum liquids, especially systems handling volatile liquids.

The suction piping should be as short and direct as possible. It should be the full size called for by the pump, and if the line is long, the size should be increased. All horizontal runs should slope up to the pump so as to avoid air pockets in the suction line.

Many pumps are installed as replacements where it is necessary to utilize existing suction lines. In such places, the lift and friction of the piping should be carefully analyzed so that the pump selected will be one that does not exceed the pipe capacity.

On installations involving suction lifts a good foot valve or line check, located at the beginning of the suction lift or an angle check valve at ground level will help insure flow as soon as the pump is started. Careful consideration should be given to the friction loss through the valve under consideration.

Fittings should be provided to permit the installation of vacuum and pressure gauges on each side of the pump if provision has not already been made in the pump for these gauges.

Quick-closing valves or nozzles should not be used on the discharge lines.

A check-valve should be installed in the discharge line as close as possible to the pump when the static discharge head exceeds 25 feet.

INFORMATION NEEDED FOR SELECTING PUMP

GPM — (Flow)

DISCHARGE HEAD — (Push) PSI or Ft.

SUCTION LIFT — (Pull)

FRICTION — (Resistance)

TEMPERATURE — (How hot)

VISCOSITY — (How thick)

LIQUID — (What liquid)

NATURE OF LIQUID — (Clear—with solids)

SERVICE FACTOR — (Continuous or intermittent)

SP. GR. — (Weight)

TYPE OF DRIVE — (Motor—Engine, Etc.)

NOTE: *If the Total Dynamic Suction Lift (or head) and Total Dynamic Discharge Head have not been figured— list separately for suction and discharge the length and size of piping including list of all fittings, static elevation, and pressure required at discharge nozzle, if any.*

COMPUTING SUCTION AND DISCHARGE CONDITIONS

PROBLEM An oil company wishes to pump 200 GPM of kerosene from a 12 ft. tank buried three feet under the ground to a loading rack assembly. The spout is sixteen feet above the same ground level and approximately 200 ft. from the tank. The suction line consists of 25' of 3" pipe and a 3" angle check valve. The discharge line consists of 200' of 3" pipe with three 90° elbows; 3" strainer, meter and air eliminator; and 3" loading arm assembly.

DYNAMIC SUCTION LIFT

1. Static Suction Lift (Maximum)
   (12' Tank + 3' below ground level)   15   Ft.

   Suction Pipe Length .... 25'
   Equivalent Length of Angle Check Valve
   (See Page 26) .... 18'
                                                          ————
        Total Equivalent Pipe .... 43'

2. Friction Loss 200 GPM — 3" pipe
   Ft. loss per 100' Pipe (See Page 23) .... 12.2' × .43 =   5.2   Ft.

3. Velocity Head — 9.08 Ft/sec Velocity ....   =   1.25 Ft.
                                                              ————
                            Dynamic Suction Lift ....   21.45 Ft.

DYNAMIC DISCHARGE HEAD

1. Static Discharge Head ....   16.0 Ft.

	Pipe Length	Head Loss
Discharge Pipe Length	200'	
Equivalent Pipe Length:		
Meter (See Page 28)		7 Ft. Loss
Air Eliminator (See Page 27)....		3 Ft. Loss
Strainer (See Page 27)		4 Ft. Loss
Loading Arm (See Page 28)		9.2 Ft. Loss
Three 90° Ells (See Page 26)....	33'	
Total Equivalent Pipe Lgth. ..	233	

2. Head Loss of Accessories ....   23.2 Ft. Loss = 23.2 Ft.

3. Friction Loss 200 GPM—3' Pipe based on 233' pipe .... = 28.4 Ft.
   Ft. Loss per 100' Pipe (See Page 23) 12.2 × 2.33
                                                                       ————
   Dynamic Discharge Head ....   67.6 Ft.

   Total Dynamic Head 21.45 + 67.6 = 89 Ft.

A pump to handle 200 GPM against a Total Dynamic Head of 89 Ft. (including 21' suction) can now be selected from the Head-capacity charts or curves on the individual pump models. In the selection, the dynamic suction lift must be considered to make sure that the pump is capable of this lift at the flow rate of 200 GPM. If not, the suction must be reduced or a larger pump selected.

NOTE

The Dynamic Suction Lift calculated in this problem represents the maximum condition (empty tank). The average condition would be a tank half full in which case the static lift would be 9' and the Dynamic Suction Lift 15.45 ft. The Total Dynamic Head would then be 83 ft. This average condition is often used in pump selection.

SUCTION LIMITATIONS

(ANY PUMP)

The importance of keeping within the suction limitations of any pump (centrifugal, rotary, piston) cannot be emphasized too greatly. A pump, by creating a vacuum at the suction (impeller eye on a centrifugal) utilizes atmospheric pressure (14.7# at sea level) to push the liquid into the pump. Because of this, the suction lift is limited theoretically to 33.9 ft. of water maximum (14.7# × 2.31 ÷ SG (1.0) = 33.9′ water). Internal pump losses reduce this limitation even more. The dynamic suction lift should be calculated carefully at the required capacity to make sure that it is within the pump's capabilities. Even systems taking suction from a source above the pump can cause trouble when friction losses are too great. *Always keep the pump as close to the liquid source as possible.* Many pump performance curves will show the maximum practical dynamic suction lifts for a given pump or for given capacities from that same pump. Since the limitation is based on internal pump losses also, it can be seen that in any given pump the recommended suction lift is reduced as flow increases.

CAVITATION

When pressure in suction line falls below vapor pressure, vapor is formed and moves along with the stream. These vapor bubbles or "cavities" collapse when they reach regions or higher pressure on their way through the pump.

The most obvious effects of cavitation are noise and vibration. This is caused by the collapse of the vapor bubbles as they reach the high pressure side of the pump. The bigger the pump, the greater the noise and vibration. If operated under cavitating conditions for a sufficient length of time, especially on water service, impeller vane pitting will take place. The violent collapse of vapor bubbles forces liquid at high velocity into vapor filled pores of the metal, producing surge pressures of high intensity on small areas. These pressures can exceed the tensile strength of the metal, and actually blast out particles, giving the metal a spongy appearance. This noise and vibration also can cause bearing failure, shaft breakage and other fatigue failures in the pump.

The other major effect of cavitation is a drop in efficiency of the pump, apparent as a dropoff in capacity.

The drop in efficiency and head capacity curve may occur before the vapor pressure is reached, particularly in petroleum oils, because of the liberation of light fractions and dissolved and entrained air.

The harmful pitting is not as serious when handling oils due to the cushioning effect of the more viscous liquid.

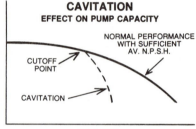

In general, cavitation indicates insufficient available N.P.S.H. Excessive suction pipe friction, combined with low static suction head and high temperatures contribute to this condition. If the system cannot be changed, it may be necessary to change conditions so that a different pump with lower N.P.S.H. requirements can be used. Larger pumps might require the use of a booster pump to add pressure head to the available N.P.S.H.

VAPOR PRESSURE

Another factor that can limit the suction lift is the vapor pressure of the liquid being handled. Vapor pressure denotes the lowest absolute pressure witnessed with a given liquid at a given temperature. If the pressure in a pump system is not equal to or greater than the vapor pressure of the liquid, the liquid will flash into a gas. It is for this same reason that we must have pressure available on the suction side of a pump when handling hot water or volatile liquids such as gasoline. Without sufficient pressure, the liquid will flash into a gas and become, of course, unpumpable.

Many process applications use pressurized vessels on the suction side to overcome vapor pressure of some liquids. The amount of pressure needed depends on the liquid and liquid temperature. The higher the temperature, the higher the vapor pressure. On applications involving an above ground or underground vented tank or a sump, care must be taken when handling volatile liquids to keep within the atmospheric pressure limitations.

Consider, for example a ball of liquid that has a VP of 6# absolute. This means that at least 6# pressure is needed to maintain the liquid state. Since atmospheric pressure is only 14.7# pressure (sea level) we have only 8.7# left to cover suction static lift and friction besides internal pump losses.

NOTE: *Water boils at 212° F. at sea level because its vapor pressure is 14.7# at that temperature. Since atmospheric pressure does not exceed 14.7#, there is no extra pressure to maintain a liquid state.*

NOTE: *V.P. is measured in pounds absolute. Absolute pressure is pressure above a perfect vacuum.*

NET POSITIVE SUCTION HEAD

(NPSH)

NPSH combines all of the factors limiting the suction side of a pump: internal pump losses, static suction lift, friction losses, vapor pressure and atmospheric conditions. It is important to differentiate between *Required NPSH* and *Available NPSH*.

Required NPSH — this refers to internal pump losses and is determined by laboratory test. It varies with each pump and with each pump capacity and speed change. The greater the capacity, the greater the required NPSH. *Required NPSH must always be given by the manufacturer.*

Available NPSH —this is a characteristic of the suction system. It can be calculated, or on an existing installation, it can be determined by field vacuum gauge readings. By definition, it is the net positive suction head above the vapor pressure available at the suction flange of the pump to maintain a liquid state. Since there are also internal pump losses (required NPSH) the available NPSH in a system must exceed the pump required NPSH — otherwise, reduction in capacity, loss of efficiency, noise, vibration and cavitation will result.

AVAILABLE NPSH IN A SYSTEM

Available NPSH = Positive factors (less) negative factors

Positive Factors	Negative Factors
Static Suction Head (if any)	Vapor Pressure—PSIA
Atmospheric Pressure (if open or vented tank or sump)	*All Friction Losses (including velocity head) Total Dynamic Suction Lift
Positive Pressure (if closed pressurized tank)	*Static Suction Lift (if any)
This side (less)	This Side = Balance or Available NPSH

Note: Always convert all terms to feet taking into consideration the Sp. Gr. of the liquid being handled.

An ideal suction condition pumping from a vented tank would be:

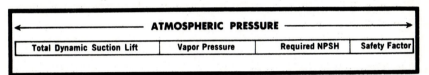

ATMOSPHERIC PRESSURE			
Total Dynamic Suction Lift	Vapor Pressure	Required NPSH	Safety Factor

NPSH FORMULAS

PROPOSED INSTALLATION — EXISTING INSTALLATION

To determine the N.P.S.H. available in a proposed application, the following formula is recommended:

$$Hsv = Hp \pm Hz - Hf - Hvp$$

Hsv = Available N.P.S.H. expressed in feet of fluid.

Hp = Absolute pressure on the surface of the liquid where the pump takes suction, expressed in "feet". This could be atmospheric pressure or vessel pressure (pressurized tank).

Hz = Static elevation of the liquid above or below the centerline of the impeller, expressed in feet.

Hf = Friction and velocity head loss in the piping, also expressed in feet.

Hvp = Absolute vapor pressure of the fluid at the pumping temperature, expressed in feet of fluid.

To determine the N.P.S.H. available in an *existing installation*, the preceding formula can be used or the following can be employed in which case it is not necessary to figure elevations and friction losses because the suction gauge reading accounts for these factors.

$$Hsv = Pa \pm Ps - \frac{Vs2}{2g} - Hvp$$

Hsv = N.P.S.H. expressed in feet of fluid.

Pa = Atmospheric pressure for the elevation of the installation, expressed in feet.

Ps = Gauge pressure or vacuum at the suction flange of the pump corrected to the pump centerline and expressed in feet. (+ if pressure or— if vacuum).

$\dfrac{Vs2}{2g}$ = Velocity head at the point of measurement of Ps.

Hvp = Absolute vapor pressure, expressed in feet.

MISCELLANEOUS DATA ON STANDARD PIPE

Nominal Diameter, Inches	Actual Outside Diameter, Inches	Actual Inside Diameter, Inches	Inside Area, Square Inches	Weight per Foot, Pounds	Length in Feet Containing One U. S. Gallon	Length in Feet Containing One Cubic Foot	U. S. Gallons in One Lineal Foot	Pounds of Water in One Lineal Foot
1/2	0.840	0.622	0.304	0.846	63.322	473.910	0.0158	0.1316
3/4	1.050	0.824	0.533	1.119	36.116	270.030	0.0277	0.2309
1	1.315	1.049	0.864	1.660	22.280	166.620	0.0449	0.3742
1 1/4	1.660	1.380	1.496	2.244	12.867	96.275	0.0777	0.6477
1 1/2	1.900	1.610	2.036	2.684	9.454	70.733	0.1058	0.8816
2	2.375	2.067	3.356	3.609	5.736	42.913	0.1743	1.4530
2 1/2	2.875	2.469	4.788	5.725	4.020	30.077	0.2487	2.0732
3	3.500	3.068	7.393	7.486	2.593	19.479	0.3840	3.2012
3 1/2	4.000	3.548	9.887	9.001	1.947	14.565	0.5136	4.2812
4	4.500	4.026	12.730	10.665	1.512	11.312	0.6613	5.5125
4 1/2	5.000	4.506	15.947	12.392	1.207	9.030	0.8284	6.9053
5	5.563	5.047	20.006	14.448	0.962	7.198	1.0398	8.6629
6	6.625	6.065	28.890	18.755	0.666	4.984	1.5008	12.5101
7	7.625	7.023	38.738	23.271	0.497	3.717	2.0124	16.7743
8	8.625	7.981	50.027	28.221	0.384	2.878	2.5988	21.6627

PRESSURE EQUIVALENTS

PSI	Feet of Water	Inches of Water	Meters of Water	Inches of Mercury	MM of Mercury	Kilograms per Sq. CM
1.0	2.306	27.68	.704	2.036	51.712	.0703
.433	1.0	12.0	.305	.8826	22.418	.0305
.0361	.0833	1.0	.0254	.0736	1.868	.00254
1.421	3.28	39.37	1.0	2.89	—	.1
.4912	1.133	13.596	.0346	1.0	25.4	.0345
.01934	.0446	.5354	.0136	.03937	1.0	.001360
14.223	32.81	393.7	10.0	28.96	735.51	1.0

DECIMAL EQUIVALENTS

1/64	.015625	17/64	.265625	33/64	.515625	49/64	.765625
1/32	.03125	9/32	.28125	17/32	.53125	25/32	.78125
3/64	.046875	19/64	.296875	35/64	.546875	51/64	.796875
1/16	.0625	5/16	.3125	9/16	.5625	13/16	.8125
5/64	.078125	21/64	.328125	37/64	.578125	53/64	.828125
3/32	.09375	11/32	.34375	19/32	.59375	27/32	.84375
7/64	.109375	23/64	.359375	39/64	.609375	55/64	.859375
1/8	.125	3/8	.375	5/8	.625	7/8	.875
9/64	.140625	25/64	.390625	41/64	.640625	57/64	.890625
5/32	.15625	13/32	.40625	21/32	.65625	29/32	.90625
11/64	.171875	27/64	.421875	43/64	.671875	59/64	.921875
3/16	.1875	7/16	.4375	11/16	.6875	15/16	.9375
13/64	.203125	29/64	.453125	45/64	.703125	61/64	.953125
7/32	.21875	15/32	.46875	23/32	.71875	31/64	.96875
15/64	.234375	31/64	.484375	47/64	.734375	63/64	.984375
1/4	.25	1/2	.50	3/4	.75	1	1.

SPECIFIC GRAVITY CONVERSION TABLES

NOTE: *To convert degrees API to specific gravity (liquids lighter than water)*

$$\text{Sp. Gr.} = \frac{141.5}{131.5 + \text{Degrees API}}$$

To convert degrees Baumé to specific gravity (liquids heavier than water)

$$\text{Sp. Gr.} = \frac{145}{145 - \text{Degrees Baumé}}$$

CONVERSION TABLE BAUME

Specific Gravity — Weight per Gallon for liquids HEAVIER than water

Baumé	Specific Gravity	Wght. per Gal.	Baumé	Specific Gravity	Wght. per Gal.	Baumé	Specific Gravity	Wght. per Gal.	Baumé	Specific Gravity	Wght. per Gal.	Baumé	Specific Gravity	Wght. per Gal.
0	1.000	8.33	10	1.074	8.95	20	1.160	9.67	30	1.260	10.50	40	1.381	11.51
1	1.006	8.38	11	1.082	9.02	21	1.169	9.74	31	1.271	10.59	45	1.450	12.08
2	1.014	8.45	12	1.090	9.08	22	1.178	9.82	32	1.283	10.69	50	1.526	12.72
3	1.021	8.51	13	1.098	9.15	23	1.188	9.90	33	1.294	10.78	55	1.611	13.42
4	1.028	8.57	14	1.106	9.22	24	1.198	9.98	34	1.306	10.88	60	1.705	14.21
5	1.035	8.62	15	1.115	9.29	25	1.208	10.07	35	1.318	10.98	65	1.812	15.10
6	1.043	8.69	16	1.125	9.37	26	1.218	10.15	36	1.330	11.08	70	1.933	16.11
7	1.050	8.75	17	1.132	9.43	27	1.228	10.23	37	1.342	11.18	....	....	....
8	1.058	8.82	18	1.141	9.51	28	1.239	10.32	38	1.355	11.29	....	....	....
9	1.066	8.88	19	1.150	9.58	29	1.250	10.42	39	1.367	11.39	....	....	....

CONVERSION TABLE API

Specific Gravity — Weight per Gallon for liquids LIGHTER than water

A.P.I.	Specific Gravity	Wght. per Gal.	A.P.I.	Specific Gravity	Wght. per Gal.	A.P.I.	Specific Gravity	Wght. per Gal.	A.P.I.	Specific Gravity	Wght. per Gal.	A.P.I.	Specific Gravity	Wght. per Gal.
10	1.000	8.33	31	0.871	7.25	52	0.7712	6.42	73	0.6926	5.76	91	.636	5.29
11	0.993	8.27	32	0.865	7.21	53	0.7670	6.39	74	0.6893	5.73	92	.633	5.27
12	0.986	8.21	33	0.860	7.16	54	0.7637	6.35	75	0.6859	5.70	93	.630	5.25
13	0.979	8.16	34	0.855	7.12	55	0.7597	6.32	76	0.6826	5.68	94	.628	5.22
14	0.973	8.10	35	0.850	7.08	56	0.7556	6.28	77	0.6793	5.65	95	.625	5.20
15	0.966	8.04	36	0.845	7.03	57	0.7516	6.28	78	0.6750	5.62	96	.622	5.18
16	0.959	7.99	37	0.840	6.99	58	0.7476	6.22	79	0.6728	5.60	97	.619	5.15
17	0.953	7.94	38	0.835	6.95	59	0.7437	6.18	80	0.6696	5.57	98	.617	5.13
18	0.946	7.88	39	0.830	6.91	60	0.7398	6.15	81	0.6665	5.54	99	.614	5.11
19	0.940	7.83	40	0.825	6.87	61	0.7359	6.12	82	0.6634	5.52	100	.611	5.09
20	0.934	7.78	41	0.820	6.83	62	0.7310	6.09	83	0.6603	5.49	.....	....	....
21	0.928	7.73	42	0.816	6.79	63	0.7283	6.06	84	0.6572	5.47			
22	0.921	7.68	43	0.811	6.75	64	0.7246	6.03	85	0.6541	5.44			
23	0.916	7.63	44	0.806	6.71	65	0.7209	5.99	86	0.6511	5.42			
24	0.910	7.58	45	0.802	6.68	66	0.7172	5.96	87	0.6481	5.39			
25	0.904	7.53	46	0.797	6.64	67	0.7136	5.93	88	0.6452	5.37			
26	0.898	7.48	47	0.793	6.60	68	0.7090	5.90	89	0.6422	5.34			
27	0.893	7.43	48	0.788	6.56	69	0.7065	5.87	90	0.6393	5.32			
28	0.887	7.39	49	0.784	6.53	70	0.7020	5.85	....					
29	0.882	7.34	50	0.780	6.49	71	0.6995	5.82	....					
30	0.876	7.30	51	0.775	6.46	72	0.6950	5.79	....					

ATMOSPHERIC PRESSURE CONDITIONS—

ELEVATIONS ABOVE SEA LEVEL

Altitude Above Sea Level	Atmospheric Pressure Pounds/sq. in.	Barometer Reading Inches of Mercury	Equivalent Head or Water Feet	Reduction to Max. Practical Dyn. Suction Lift
0	14.7	29.929	33.95	0 Ft.
1000	14.2	28.8	32.7	1.2 "
2000	13.6	27.7	31.6	2.3 "
3000	13.1	26.7	30.2	3.7 "
4000	12.6	25.7	29.1	4.8 "
5000	12.1	24.7	27.9	6 "
6000	11.7	23.8	27.	6.9 "
7000	11.2	22.9	25.9	8 "
8000	10.8	22.1	24.9	9 "

MAXIMUM PRACTICAL DYNAMIC SUCTION LIFT AND VAPOR PRESSURE

WATER CHARACTERISTICS

Temp. of °F	Vapor Pressure PSI Abs	Vapor Pressure Feet	Specific Gravity	Approx. Maximum Theoretical Suction Lift — Feet	Maximum Practical Dyn. Suction Lift — Feet
40	.1217	0.281	1.0000	33.7	25
50	.1781	0.4115	.9997	33.5	25
60	.2563	0.592	.9990	33.4	25
70	.3631	0.815	.9980	33.1	25
80	.5069	1.17	.9966	32.7	24
90	.6982	1.612	.9950	32.3	24
100	.9492	2.191	.9931	31.4	23
110	1.275	2.942	.9906	31	22
120	1.692	3.91	.9888	30	21
130	2.223	5.145	.9857	288	20
140	2.889	6.675	.9833	27.2	18
150	3.718	8.56	.9803	25.3	16
160	4.741	10.945	.9773	23	14
170	5.992	13.84	.9738	20	11
180	7.510	17.35	.9702	16.5	7
190	9.339	21.55	.9667	12.4	3
200	11.53	26.65	.9632	7.2	2' Positive
210	14.12	32.6	.9592	1.3	8' Positive
220	17.19	39.7	.9552	0	15' Positive

MAXIMUM PRACTICAL DYNAMIC SUCTION LIFTS

GASOLINES • JET FUEL • KEROSENE • SEA LEVEL ATMOSPHERE CONDITIONS

TEMPERATURE	MOTOR GASOLINE Winter Gas Reid Vapor Pressure 14#	Summer Gas 12#	10#	8#	AVIATION GAS	JET FUEL AND KEROSENE
50° F.	15	18	21	23	25	25'
60° F.	11	15	18'	21	23	25'
70° F.	6'	11	15'	18'	20	25'
80° F.	—	5	11	15	18	24'
90° F.	—	—	6	11	14	23'
110° F.	—	—	—	6	10	22'

NOTE: Average temperature of gasoline in underground storage tank is 55° F., not ambient.

APPROX. VAPOR PRESSURE — *PSI Absolute*

FUEL CHARACTERISTICS

TEMPERATURE	MOTOR GASOLINE Winter Gas 14 Reid	12 Reid	Summer Gas 10 Reid	8 Reid	AVIATION GAS	JET FUEL AND KEROSENE
40	4.8	4.0	3.4	2.8	2.2	.9
50	5.9	4.9	4.1	3.4	2.8	1.0
60	7.4	6.0	5.0	4.1	3.5	1.0
70	8.9	7.4	6.0	5.0	4.2	1.2
80	10.7	9.0	7.1	5.9	5.1	1.4
90	12.8	10.6	8.6	7.0	6.2	1.7
100	14	12	10	8	7.4	2.0

EFFECTS OF ALTITUDE ON INTERNAL COMBUSTION ENGINES

There is a power loss of approximately 3 per cent for every 1000 of elevation above sea level; also a power loss of 1 per cent for every 10° F. over 60° F. This will result in a loss of speed, flow and pressure on the pump. The following table gives the percentage of sea level performance that can be expected at various elevations.

Altitude	Flow	Head	Altitude	Flow	Head
Sea Level	100%	100%	6000 Ft.	93%	87%
2000 Ft.	97%	95%	8000 Ft.	91%	83%
4000 Ft.	95%	91%	10000 Ft.	88%	78%

Approx. Characteristics

NATURAL GAS OR BUTANE

a. Air cooled engines — (Wisconsin).
Engines built at the factory with high compression heads, special carburetor, manifold, and gas valve and regulator will develop the same power as the engines when using gasoline. The BTU content of the gas should be above 1000. Reductions in BTU content will reduce the power.

b. Water Cooled Engines.
Converted at the factory with higher compression ratios and other equipment, there will be no power loss from the gasoline fuel performance. Field conversions without change of head (compression ratio) or manifolds will result in a 10 - 15% power loss.

CONVERSION CHART LIFT IN FEET VS. VACUUM IN INCHES OF MERCURY

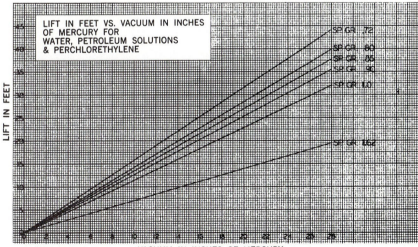

CONVERSION CHART HEAD IN FEET VS. HEAD IN P.S.I.

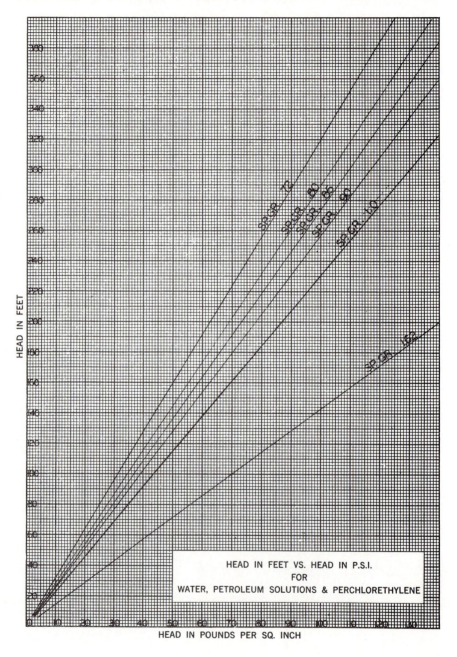

HEAD IN FEET VS. HEAD IN P.S.I.
FOR
WATER, PETROLEUM SOLUTIONS & PERCHLORETHYLENE

HEAD IN POUNDS PER SQ. INCH

THEORETICAL DISCHARGE OF NOZZLES IN U. S. GALLONS PER MINUTE

Pounds	Feet	Velocity of Discharge Feet Per Second	1/16	1/8	3/16	1/4	3/8	1/2	5/8	3/4	7/8
	HEAD		DIAMETER OF NOZZLE IN INCHES								
10	23.1	38.6	0.37	1.48	3.32	5.91	13.3	23.6	36.9	53.1	72.4
15	34.6	47.25	0.45	1.81	4.06	7.24	16.3	28.9	45.2	65.0	88.5
20	46.2	54.55	0.52	2.09	4.69	8.35	18.8	33.4	52.2	75.1	102
25	57.7	61.0	0.58	2.34	5.25	9.34	21.0	37.3	58.3	84.0	114
30	69.3	66.85	0.64	2.56	5.75	10.2	23.0	40.9	63.9	92.0	125
35	80.8	72.2	0.69	2.77	6.21	11.1	24.8	44.2	69.0	99.5	135
40	92.4	77.2	0.74	2.96	6.64	11.8	26.6	47.3	73.8	106	145
45	103.9	81.8	0.78	3.13	7.03	12.5	28.2	50.1	78.2	113	153
50	115.5	86.25	0.83	3.30	7.41	13.2	29.7	52.8	82.5	119	162
55	127.0	90.4	0.87	3.46	7.77	13.8	31.1	55.3	86.4	125	169
60	138.6	94.5	0.90	3.62	8.12	14.5	32.5	57.8	90.4	130	177
65	150.1	98.3	0.94	3.77	8.45	15.1	33.8	60.2	94.0	136	184
70	161.7	102.1	0.98	3.91	8.78	15.7	35.2	62.5	97.7	141	191
75	173.2	105.7	1.01	4.05	9.08	16.2	36.4	64.7	101	146	198
80	184.8	109.1	1.05	4.18	9.39	16.7	37.6	66.8	104	150	205
85	196.3	112.5	1.08	4.31	9.67	17.3	38.8	68.9	108	155	211
90	207.9	115.8	1.11	4.43	9.95	17.7	39.9	70.8	111	160	217
95	219.4	119.0	1.14	4.56	10.2	18.2	41.0	72.8	114	164	223
100	230.9	122.0	1.17	4.67	10.5	18.7	42.1	74.7	117	168	229
105	242.4	125.0	1.20	4.79	10.8	19.2	43.1	76.5	120	172	234
110	254.0	128.0	1.23	4.90	11.0	19.6	44.1	78.4	122	176	240
115	265.5	130.9	1.25	5.01	11.2	20.0	45.1	80.1	125	180	245
120	277.1	133.7	1.28	5.12	11.5	20.5	46.0	81.8	128	184	251
125	288.6	136.4	1.31	5.22	11.7	20.9	47.0	83.5	130	188	256
130	300.2	139.1	1.33	5.33	12.0	21.3	48.0	85.2	133	192	261
135	311.7	141.8	1.36	5.43	12.2	21.7	48.9	86.7	136	195	266
140	323.3	144.3	1.38	5.53	12.4	22.1	49.8	88.4	138	199	271
145	334.8	146.9	1.41	5.62	12.6	22.5	50.6	89.9	140	202	275
150	346.4	149.5	1.43	5.72	12.9	22.9	51.5	91.5	143	206	280
175	404.1	161.4	1.55	6.18	13.9	24.7	55.6	98.8	154	222	302
200	461.9	172.6	1.65	6.61	14.8	26.4	59.5	106	165	238	323

Pounds	Feet	Velocity of Discharge Feet Per Second	1	1⅛	1¼	1⅜	1½	1¾	2	2¼	2½
	HEAD		DIAMETER OF NOZZLE IN INCHES								
10	23.1	38.6	94.5	120	148	179	213	289	378	479	591
15	34.6	47.25	116	147	181	219	260	354	463	585	723
20	46.2	54.55	134	169	209	253	301	409	535	676	835
25	57.7	61.0	149	189	234	283	336	458	598	756	934
30	69.3	66.85	164	207	256	309	368	501	655	828	1023
35	80.8	72.2	177	224	277	334	398	541	708	895	1106
40	92.4	77.2	188	239	296	357	425	578	756	957	1182
45	103.9	81.8	200	253	313	379	451	613	801	1015	1252
50	115.5	86.25	211	267	330	399	475	647	845	1070	1320
55	127.0	90.4	221	280	346	418	498	678	886	1121	1385
60	138.6	94.5	231	293	362	438	521	708	926	1172	1447
65	150.1	98.3	241	305	376	455	542	737	964	1220	1506
70	161.7	102.1	250	317	391	473	563	765	1001	1267	1565
75	173.2	105.7	259	327	404	489	582	792	1037	1310	1619
80	184.8	109.1	267	338	418	505	602	818	1070	1354	1672
85	196.3	112.5	276	349	431	521	620	844	1103	1395	1723
90	207.9	115.8	284	359	443	536	638	868	1136	1436	1773
95	219.4	119.0	292	369	456	551	656	892	1168	1476	1824
100	230.9	122.0	299	378	467	565	672	915	1196	1512	1870
105	242.4	125.0	306	388	479	579	689	937	1226	1550	1916
110	254.0	128.0	314	397	490	593	705	960	1255	1588	1961
115	265.5	130.9	320	406	501	606	720	980	1282	1621	2005
120	277.1	133.7	327	414	512	619	736	1002	1310	1659	2050
125	288.6	136.4	334	423	522	632	751	1022	1338	1690	2090
130	300.2	139.1	341	432	533	645	767	1043	1365	1726	2132
135	311.7	141.8	347	439	543	656	780	1063	1390	1759	2173
140	323.3	144.3	354	448	553	668	795	1082	1415	1790	2212
145	334.8	146.9	360	455	562	680	809	1100	1440	1820	2250
150	346.4	149.5	366	463	572	692	824	1120	1466	1853	2290
175	404.1	161.4	395	500	618	747	890	1210	1582	2000	2473
200	461.9	172.6	423		660	790	950	1294		2140	2645

NOTE: The actual quantities will vary from these figures, the amount of variation depending upon the shape of nozzle and size of pipe at the point where the pressure is determined. With smooth taper nozzles the actual discharge is about 94 per cent. of the figures given in the tables.

VOLUME AND CAPACITY EQUIVALENTS

U. S. Gallons	Imperial Gallons	Cubic Inches	Cubic Feet	Cubic Meters	Liters
1.	0.8327	231.0	0.1337	0.003785	3.785
1.201	1.	277.4	0.1605	0.004545	4.546
0.004329	0.003604	1.	0.0005787	0.00001639	0.01639
7.481	6.229	1728.	1.	0.02832	28.32
264.2	220.0	61023.	35.31	1.	1000.
0.2642	0.2200	61.02	0.03531	0.001000	1.

VISCOSITY CONVERSION TABLE

SSU Seconds Saybolt Universal	SSF Seconds Saybolt Furol	Absolute Viscosity Centipoises	Seconds Redwood (Standard)
31	....	1.00	29
35	....	2.56	32.1
40	....	4.30	36.2
50	....	7.40	44.3
60	....	10.20	52.3
70	12.95	12.83	60.9
80	13.70	15.35	69.2
90	14.44	17.80	77.6
100	15.24	20.20	85.6
150	19.30	31.80	128
200	23.5	43.10	170
250	28.0	54.30	212
300	32.5	65.40	254
400	41.9	87.60	338
500	51.6	110.0	423
600	61.4	132	508
700	71.1	154	592
800	81.0	176	677
900	91.0	198	762
1000	100.7	220	896
1500	150	330	1270
2000	200	440	1690
2500	250	550	2120
3000	300	660	2540
4000	400	880	3380
5000	500	1100	4230
6000	600	1320	5080
7000	700	1540	5920
8000	800	1760	6770
9000	900	1980	7620
10000	1000	2200	8460
15000	1500	3300	13700
20000	2000	4400	18400

$$\text{Kinematic Viscosity (Centistokes)} = \frac{\text{Absolute Visc. (Centipoise)}}{\text{Specific Gravity}}$$

$$1 \text{ Centistoke} = \frac{\text{Stoke}}{100}$$

$$1 \text{ Centipoise} = \frac{\text{Poise}}{100}$$

1 Stoke   = 100 Centistokes

1 Poise   = 100 Centipoises

*Centipoises

The term "Centipoises" is referred to commonly as a measure of **Absolute Viscosity.** Convert centipoises to centistokes by dividing by the Specific Gravity of the solution at the operating temperature.

NOTE: Plotting Viscosity = If viscosity is known at any two temperatures, the viscosity at other temperatures can be obtained by plotting the viscosity against temperature in degrees fahrenheit on log paper. The points lie in a straight line.

VISCOSITY

IN

SECONDS

SAYBOLT

UNIVERSAL

PERCENTAGE OF WATER CHARACTERISTICS

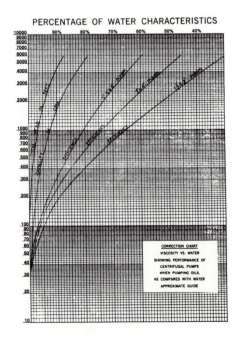

CORRECTION CHART
VISCOSITY VS. WATER
SHOWING PERFORMANCE OF
CENTRIFUGAL PUMPS
WHEN PUMPING OILS,
AS COMPARED WITH WATER
APPROXIMATE GUIDE

VISCOSITY AND SPECIFIC GRAVITY OF COMMON LIQUIDS

LIQUID	SPECIFIC GRAVITY	VISCOSITY S.S.U.						
		40° F.	60° F.	80° F.	100° F.	120° F.	140° F.	160° F.
Miscellaneous Liquids								
Water	1.0	31.5	31.5	31.5	31.5	31.5	31.5	31.5
Gasoline	.68 -.74	30	30	30	30	30	30	30
Jet Fuel	.74 -.85	35	35	35	35	35	35	35
Kerosene	.78 -.82	42	38	34	33	31	30	30
Turpentine	.86 -.87	34	33	32.8	32.6	32.4	32	32
Varnish Spar	.9	3500	1600	1000	650	530	250	230
Fuel Oil and Diesel Oil								
No. 1 Fuel Oil	.82 -.95	40	38	35	33	31	30	30
No. 2 Fuel Oil	.82 -.95	70	50	45	40	—	—	—
No. 3 Fuel Oil	.82 -.95	90	68	53	45	40	—	—
No. 5A Fuel Oil	.82 -.95	1000	400	200	100	75	60	40
No. 5B Fuel Oil	.82 -.95	1300	600	490	400	330	290	240
No. 6 Fuel Oil	.82 -.95	—	70000	20000	9000	1900	900	500
No. 2D Diesel Fuel Oil	.82 -.95	100	68	53	45	40	36	35
No. 3D Diesel Fuel Oil	.82 -.95	200	120	80	60	50	44	40
No. 4D Diesel Fuel Oil	.82 -.95	1600	600	280	140	90	68	54
No. 5D Diesel Fuel Oil	.82 -.95	15000	5000	2000	900	400	260	160
Crankcase Oils — Automobile Lubricating Oils								
SAE 10	.88 -.935	1500-2400	600-900	300-400	170-220	110-130	75-90	60-65
SAE 20	.88 -.935	2400-9000	900-3000	400-1100	220-550	130-280	90-170	65-110
SAE 30	.88 -.935	9000-14000	3000-4400	1100-1800	550-800	280-400	170-240	110-150
SAE 40	.88 -.935	14000-19000	4400-6000	1800-2400	800-1100	400-550	240-320	150-200
SAE 50	.88 -.935	19000-45000	6000-10000	2400-4000	1100-1800	550-850	320-480	200-280
SAE 60	.88 -.935	45000-60000	10000-17000	4000-6000	1800-2500	850-1200	480-580	280-380
SAE 70	.88 -.935	60000-120000	17000-45000	6000-10000	2500-4000	1200-1800	580-900	380-500

LIQUID	SPECIFIC GRAVITY	VISCOSITY S.S.U.						
		40° F.	60° F.	80° F.	100° F.	120° F.	140° F.	160° F.
Transmission Oils — Automobile Transmission Gear Lubricants								
SAE 90	.88 -.935	14000	5500	2200	1100	650	380	240
SAE 140	.88 -.935	35000	12000	5000	2200	1200	650	400
SAE 250	.88 -.935	160000	50000	18000	7000	3300	1700	1000
Other Oils								
Castor Oil	.96	36000	9000	3000	1400	900	400	300
Chinawood	.943	4000	1800	1000	580	400	300	200
Cocoanut	.925	1500	500	250	140	100	70	60
Cod	.928	1800	600	300	175	110	80	70
Corn	.924	1600	700	400	250	175	100	80
Cotton Seed	.88 -.925	1500	600	300	176	125	80	70
Cylinder	.82 -.95	60000	14000	6000	2700	1400	1000	400
Navy No. 1 Fuel Oil	.989	4000	1100	600	380	200	170	90
Navy No. 2 Fuel Oil	1.0	—	24000	8700	3500	1500	900	480
Gas	.887	180	90	60	50	45	—	—
Insulating		350	150	90	65	50	45	40
Lard	.912-.925	1100	600	380	287	180	140	90
Linseed	.925-.939	1500	500	250	143	110	85	70
Raw Menhadden	.933	1500	500	250	140	110	80	70
Neats Foot	.917	—	1000	430	230	160	100	80
Olive	.912-.918	1500	550	320	200	150	100	80
Palm	.924	1700	700	380	221	160	120	90
Peanut	.920	1200	500	300	195	150	100	80
Quenching	—	2400	900	450	250	180	130	90
Rape Seed	.919	2400	900	450	250	180	130	90
Rosin	.980	28000	7800	3200	1500	900	500	300
Rosin (Wood)	1.09	Extremely Viscose						
Sesame	.923	1100	500	290	184	130	90	60
Soya Bean	.927-.98	1200	475	270	165	120	80	70
Sperm	.883	360	250	170	110	90	70	60
Turbine (Light)	.91	500	350	230	150			
Turbine (Heavy)	.91	3000	1400	700	330	200	150	100
Whale	.925	900	450	275	170	140	100	80

LIQUID	SPECIFIC GRAVITY	VISCOSITY S.S.U.		
		70° F.	100° F.	130° F.
Sugar, Syrups, Molasses, etc.				
Corn Syrups	1.4 -1.47	—	5,000-500,000	1,500-60,000
Glucose	1.35-1.44	—	35,000-100,000	10,000-13,000
Honey (Raw)	—	—	340	—
Molasses	1.40-1.49	—	1,300-250,000	700-75,000
Sugar Syrups 60 Brix	1.29	230	92	—
Sugar Syrups 62 Brix	1.30	310	111	—
Sugar Syrups 64 Brix	1.31	440	148	—
Sugar Syrups 66 Brix	1.33	650	195	—
Sugar Syrups 68 Brix	1.34	1000	275	—
Sugar Syrups 70 Brix	1.35	1650	400	—
Sugar Syrups 72 Brix	1.36	2700	640	—
Sugar Syrups 74 Brix	1.38	5500	1100	—
Sugar Syrups 76 Brix	1.39	10000	2000	—
Corn Starch 22 Baume	1.18	150	130	—
Corn Starch 24 Baume	1.20	600	440	—
Corn Starch 25 Baume	1.21	1400	800	—
Ink—Printers	1.0 -1.38	—	2,500-10,000	1,100- 3,000
Ink—Newspaper	—	—	5,500- 8,000	2400
Tallow	.918	56 SSU at 212° F.		—
Tars				
Coke Oven—Tar	1.12+	3000- 8000	650- 1,400	—
Gas House—Tar	1.16-1.3	15,000-300,000	2,000-20,000	—
Crude Oils				
Texas, Oklahoma	.81-.916	100- 700	34-210	—
Wyoming, Montana	.86-.88	100-1100	46-320	—
California	.78-.92	100-4500	34-700	—
Pennsylvania	.8 - 85	100- 200	38-86	—
Glycol				
Propylene	1.038	240.6	—	—
Triethylene	1.125	185.7	—	—
Diethylene	1.12	149.7	—	—
Ethylene	1.125	88.4	—	—
Glycerine (100%)	1.26	2900	813	—
Phenol (Carbolic Acid)	.95-1.00	60	—	—
Silicate of Soda	—	—	365-640	—
Sulfuric Acid (100%)	1.83	75	—	—

FRICTION SECTION

The size of pipe and pipe fittings for any installation should be large enough to keep friction losses reasonably low. The velocity should be kept within 10 ft./sec. for good practical results. There are, however, many other factors to consider such as the length and cost of pipe vs cost of pump power. The cost factor can be especially important on installations involving long pipe runs and numerous valves and fittings.

PIPE & FITTINGS

All charts are based on friction losses for clean steel pipe on Schedule 40 and show average values for new pipe including adjustment of 15% for commercial installation.

To obtain approximate values for other types of pipe use multiplier correction factor of 0.9 for smooth pipe, — for 15 year old pipe use 1.43. At best, these are rough estimates.

FRICTION LOSS FOR WATER IN FEET PER 100 FT. PIPE
Schedule 40

SIZES ½" THRU 1¼"　　　　　　　　　　　　　　CLEAN STEEL PIPE

Flow US GPM	Size ½" Velocity Ft./Sec.	Friction Head in Ft.	Size ¾" Vel.	Frict.	Size 1" Vel.	Frict.	Size 1¼" Vel.	Frict.
2	2.11	5.50	—	—	—	—	—	—
3	3.17	11.50	—	—	—	—	—	—
4	4.22	19.67	2.41	4.84	—	—	—	—
5	5.28	29.67	3.01	7.27	—	—	—	—
6	6.34	41.98	3.61	10.20	2.23	3.08	—	—
8	8.45	72.11	4.81	17.25	2.97	5.22	—	—
10	10.56	110.29	6.02	26.45	3.71	7.98	—	—
12	12.7	156.40	7.22	37.99	4.45	11.06	2.57	2.85
14	14.8	210.45	8.42	50.02	5.20	14.72	3.00	3.77
16	16.9	270.35	9.63	64.75	5.94	18.98	3.43	4.83
18	—	—	10.8	80.85	6.68	23.69	3.86	6.00
20	—	—	12.0	99.01	7.42	28.87	4.29	7.29
22	—	—	13.2	119.60	8.17	34.73	4.72	8.72
24	—	—	14.4	140.30	8.91	40.94	5.15	10.26
26	—	—	15.6	164.45	9.65	47.84	5.58	11.93
28	—	—	16.8	188.60	10.39	55.09	6.01	13.69
30	—	—	—	—	11.1	62.79	6.44	15.64
35	—	—	—	—	13.0	89.30	7.50	20.93
40	—	—	—	—	14.8	109.25	8.58	27.02
45	—	—	—	—	16.7	136.85	9.65	33.70
50	—	—	—	—	18.6	167.90	10.7	41.40
55	—	—	—	—	—	—	11.8	49.68
60	—	—	—	—	—	—	12.9	58.65
65	—	—	—	—	—	—	13.9	68.54
70	—	—	—	—	—	—	15.0	79.12
75	—	—	—	—	—	—	16.1	90.51
80	—	—	—	—	—	—	17.2	102.58

FRICTION LOSS IN FEET OF LIQUID PER 100' STEEL PIPE

SIZES 1½" – 2½"

PIPE SIZE	VELOCITY FT/SEC	U.S. GPM	31.5	35	40	50	60	80	100	150	200	300	500
								VISCOSITY S.S.U.					
1½"	.94	6	.3	.5	.6	.5	.7	1.0	1.4	2.1	2.9	4.3	7.0
	1.26	8	.7	.8	.9	1.0	.9	1.4	1.8	2.8	3.8	5.6	9.3
	1.57	10	.9	1.2	1.4	1.5	1.2	1.7	2.2	3.5	4.8	7.0	11.7
	1.89	12	1.4	1.6	1.8	2.1	2.3	2.1	2.6	4.1	5.8	8.4	14.0
	2.36	15	2.1	2.4	2.6	3.1	3.3	2.4	3.3	5.2	7.1	10.6	17.5
	3.15	20	3.3	4.0	4.4	5.1	5.5	6.2	4.5	6.9	9.5	14.0	23.3
	3.80	25	5.2	6.0	6.7	7.5	8.1	9.1	9.8	8.6	11.8	17.5	29.2
	4.72	30	7.2	8.3	9.1	10.2	11.2	12.5	13.5	10.4	14.3	21.0	35.0
	6.30	40	12.4	14.0	15.2	17.3	18.4	20.5	22.0	24.8	19.1	28.1	46.7
	7.87	50	18.9	21.0	22.8	25.5	27.3	30.2	32.3	36.5	39.3	35.0	58.3
	9.44	60	26.7	29.3	32.1	35.2	38.0	41.6	44.6	45.39	48.06	56.07	74.76
	11.02	70	35.9	39.4	42.4	46.2	50.1	54.9	58.4	61.03	64.62	75.39	100.52
	12.59	80	46.3	50.6	54.1	59.3	63.4	69.2	73.9	78.71	83.34	97.23	
	14.2	90	58.5	63.3	67.3	72.9	78.3	85.8	91.1	99.45	105.3		
	15.74	100	71.5	77.1	82.1	89.0	94.3	103.3	109.1				
2"	2.39	25	1.5	1.8	2.0	2.3	2.5	2.8	2.1	3.2	4.3	6.4	10.8
	2.87	30	2.1	2.5	2.8	3.1	3.5	3.8	4.1	3.8	5.2	7.7	12.9
	3.35	35	2.8	3.2	3.6	4.1	4.5	5.1	5.4	4.5	6.0	9.1	15.1
	3.82	40	3.6	4.1	4.6	5.2	5.6	6.2	7.0	5.1	6.9	10.4	17.3
	4.78	50	5.4	6.2	6.8	7.7	8.3	9.1	9.9	11.2	8.6	12.9	21.5
	5.74	60	7.6	8.5	9.4	10.6	11.4	12.8	13.7	15.4	16.8	15.5	25.9
	6.69	70	10.2	11.5	12.4	13.8	15.0	16.7	17.9	20.2	21.7	18.1	30.1
	7.65	80	13.1	14.6	15.9	17.7	19.1	20.9	22.4	25.2	27.4	20.7	34.5
	8.60	90	16.3	18.2	19.7	22.1	23.5	25.9	27.5	31.1	33.7	37.6	38.8
	9.56	100	20.0	22.2	23.9	26.5	28.1	30.9	33.2	37.0	40.5	44.6	43.0
	10.52	110	24.3	26.6	28.3	31.6	33.4	36.7	38.5	43.6	47.4	53.5	47.4
	11.5	120	28.6	31.4	33.4	36.9	39.2	43.1	45.5	51.2	53.2	61.8	51.6
	12.4	130	33.4	36.3	38.5	42.8	45.1	49.7	52.8	58.7	63.1	71.1	56.0
	13.4	140	39.9	41.9	44.0	48.8	51.6	56.5	59.9	66.7	72.1	80.8	60.3
	14.3	150	43.9	47.7	50.3	55.1	58.7	63.6	67.4	75.3	81.1	91.1	104.5
	15.3	160	49.5	53.9	56.8	61.6	65.8	71.4	75.8	84.8	91.1	102.0	
	16.3	170	55.8	60.0	63.6	69.1	73.5	79.8	83.8	92.9	101.4		
	17.2	180	62.4	67.0	71.2	76.5	82.1	88.2	92.5	100.2	110.6		
	18.2	190	69.2	74.6	79.1	84.3	91.1	97.3	101.7	113.4			
	19.1	200	76.6	82.3	86.9	92.9	97.9	106.6	112.4				
2½"	2.01	30	.9	1.0	1.2	1.4	1.5	1.6	1.3	1.8	2.5	3.8	6.3
	2.35	35	1.2	1.4	1.5	1.7	2.0	2.2	2.3	2.2	3.0	4.5	7.5
	2.68	40	1.5	1.7	2.0	2.2	2.4	2.6	2.9	2.5	3.3	5.1	8.5
	3.02	45	1.8	2.2	2.4	2.8	3.0	3.3	3.6	2.9	3.8	5.8	9.5
	3.35	50	2.2	2.6	2.9	3.2	3.6	3.9	4.3	3.1	4.3	6.3	10.6
	4.02	60	3.1	3.6	3.9	4.5	4.8	5.4	5.9	6.6	5.1	7.6	12.7
	4.69	70	4.1	4.8	5.2	6.0	6.4	7.1	7.6	8.5	5.9	8.9	14.8
	5.36	80	5.4	6.1	6.6	7.5	8.1	9.0	9.5	10.8	11.6	10.1	16.9
	6.03	90	6.8	7.6	8.2	9.2	10.0	10.9	11.8	13.2	14.4	11.4	19.1
	6.70	100	8.3	9.2	9.9	11.0	12.0	13.2	14.1	15.8	17.0	12.7	21.2
	7.37	110	9.9	10.9	11.8	13.0	13.9	15.5	16.8	18.9	20.2	22.3	23.2
	8.04	120	11.7	12.9	13.9	15.3	16.4	18.2	19.4	21.6	23.6	26.0	25.4
	8.71	130	13.6	15.0	16.0	17.8	19.1	20.8	22.3	24.8	26.7	30.2	27.5
	9.38	140	15.6	17.3	18.4	20.2	21.6	24.0	25.4	28.4	30.7	34.4	29.6
	10.05	150	17.8	19.6	20.8	22.8	24.4	26.9	28.8	32.1	35.1	38.3	31.7
	10.7	160	20.1	22.0	23.3	25.8	27.4	30.1	32.3	35.9	38.9	43.0	33.8
	11.4	170	22.5	24.6	26.0	28.5	30.7	33.4	35.9	39.7	42.9	47.7	35.9
	12.1	180	25.2	27.5	29.1	31.9	34.2	37.0	39.7	44.2	47.4	53.0	61.0
	12.7	190	27.9	30.5	32.2	35.2	37.7	40.9	43.5	48.3	52.2	59.1	67.5
	13.4	200	30.0	33.6	35.4	38.4	41.1	45.0	48.0	52.6	56.7	63.3	72.2
	14.7	220	37.0	40.4	42.4	46.0	49.1	53.0	55.9	62.1	67.4	75.7	86.5
	16.1	240	43.8	47.8	50.1	53.9	56.8	62.0	65.9	72.8	78.4	87.4	99.6
	17.4	260	51.2	55.4	60.4	62.4	6.61	71.4	75.9	84.2	89.9	99.7	
	18.8	280	59.0	63.9	66.7	72.5	75.6	81.7	87.2				
	20.1	300	67.4	72.7	75.8	81.1	86.0	92.3	97.6				

Values to the right of the solid line are for any pipe (laminar flow). Values to the left apply only to clean steel pipe (turbulent flow).

Chart friction loss shows average value for clean steel pipe

SIZES 3" – 6" ——— VISCOSITY S.S.U.

VELOCITY FT/SEC	U.S. GPM	31.5	35	40	50	60	80	100	150	200	300	500	SIZE PIPE
2.17	50	.8	.9	1.0	1.2	1.3	1.4	1.5	1.4	1.7	2.6	4.5	
2.60	60	1.0	1.3	1.4	1.6	1.7	2.0	2.1	1.6	2.2	3.2	5.3	
3.04	70	1.5	1.7	1.8	2.1	2.3	2.5	2.8	3.1	2.5	3.7	6.2	
3.47	80	1.8	2.2	2.3	2.6	2.9	3.2	3.5	3.9	2.9	4.3	7.1	
3.91	90	2.4	2.6	2.9	3.2	3.6	3.9	4.1	4.7	5.2	4.8	7.9	
4.34	100	2.8	3.2	3.5	3.9	4.3	4.7	5.1	5.8	6.1	5.3	8.9	
5.21	120	3.9	4.5	4.8	5.4	5.9	6.6	6.9	7.8	8.4	6.4	10.7	
6.08	140	5.3	5.9	6.4	7.6	7.8	8.4	9.1	10.2	10.9	12.4	12.4	
6.94	160	6.7	7.6	8.2	9.1	9.7	10.6	11.3	12.8	13.8	15.5	14.1	
7.81	180	8.4	9.4	10.0	11.2	12.0	13.1	14.0	15.8	17.0	18.9	16.0	
8.68	200	10.2	11.5	12.2	13.5	14.4	15.8	16.8	18.7	20.6	22.8	17.7	
9.76	225	12.9	14.3	15.2	16.7	17.8	19.4	20.7	23.5	25.6	27.8	32.1	3"
10.85	250	15.8	17.4	18.4	20.1	21.6	23.5	25.1	27.8	30.0	33.5	38.2	
11.93	275	18.9	20.7	22.0	23.9	26.0	27.9	29.7	32.8	35.8	39.4	45.5	
13.0	300	22.2	24.4	25.9	28.1	30.0	32.3	34.3	38.1	41.5	45.8	52.4	
14.10	325	25.9	28.4	30.0	32.5	34.5	37.4	39.7	43.8	47.8	52.7	60.6	
15.20	350	29.9	32.5	34.5	37.3	39.6	42.9	45.5	50.1	53.7	60.1	68.2	
16.27	375	34.3	37.0	39.2	42.2	44.6	48.8	51.5	56.7	60.7	67.4	77.3	
17.40	400	39.0	41.7	44.3	47.6	50.1	54.7	57.3	63.3	68.1	75.4	86.4	
18.44	425	43.8	46.8	49.7	53.2	56.1	60.8	63.3	70.4	75.4	83.7	95.9	
19.53	450	49.1	52.3	55.2	59.2	62.4	67.5	71.2	77.6	83.7	92.2	106.7	
20.61	475	54.4	58.0	61.2	65.7	68.9	74.3	78.4	85.6	92.1	102.2		
21.7	500	60.3	63.8	67.5	72.1	75.7	81.3	86.3					
22.78	525	66.0	70.0	74.1	79.1	82.8	88.6	94.4					
23.9	550	72.1	76.5	80.8	86.0	90.0	96.4	102.4					
1.51	60	.3	.3	.3	.5	.5	6	.6	.6	.7	1.0	1.8	
1.76	70	.3	.5	.5	.6	.6	.7	.8	.6	.8	1.3	2.1	
2.02	80	.5	.6	.7	.7	.8	.9	.9	.7	.9	1.5	2.4	
2.27	90	.6	.7	.8	.9	.9	1.0	1.2	1.3	1.0	1.6	2.8	
2.52	100	.7	.8	.9	1.0	1.2	1.3	1.4	1.6	1.3	1.8	3.0	
3.02	120	1.0	1.2	1.3	1.5	1.6	1.7	2.0	2.2	2.3	2.2	3.6	
3.53	140	1.4	1.6	1.7	2.0	2.1	2.3	2.5	2.8	3.1	2.5	4.3	
4.03	160	1.7	2.0	2.2	2.4	2.6	2.9	3.1	3.5	3.8	2.9	4.8	
4.54	180	2.2	2.5	2.6	3.0	3.2	3.6	3.8	4.4	4.7	5.3	5.4	
5.04	200	2.6	3.0	3.2	3.7	3.9	4.3	4.6	5.2	5.6	6.3	6.0	
5.54	220	3.1	3.6	3.8	4.3	4.6	5.1	5.4	6.1	6.6	7.5	6.6	
6.05	240	3.7	4.1	4.5	5.1	5.4	6.0	6.3	7.1	7.5	8.6	7.2	4"
6.55	260	4.3	4.8	5.2	5.8	6.2	6.8	7.2	8.2	8.9	9.9	7.8	
7.06	280	4.9	5.5	6.0	6.7	7.1	7.8	8.2	9.3	10.0	11.2	8.4	
7.56	300	5.6	6.3	6.8	7.5	8.1	8.7	9.3	10.5	11.4	12.7	14.6	
8.82	350	7.6	8.4	9.0	9.9	10.5	11.6	12.3	13.6	14.7	16.4	19.1	
10.10	400	9.8	10.8	11.5	12.5	13.3	14.6	15.5	17.1	18.6	20.7	23.9	
11.34	450	12.2	13.5	14.3	15.5	16.6	18.1	19.2	21.2	22.8	25.5	29.2	
12.6	500	15.0	16.3	17.4	18.9	20.0	21.7	23.1	25.4	27.3	30.5	35.1	
13.9	550	18.1	19.6	20.7	22.4	23.7	25.8	27.5	30.2	32.2	36.0	41.4	
15.1	600	21.3	23.0	24.4	26.3	27.8	30.2	32.0	35.1	37.7	41.9	48.0	
16.4	650	24.8	26.6	28.3	30.6	32.2	34.8	36.7	40.6	43.5	48.2	55.2	
17.6	700	28.6	30.7	32.5	35.2	37.0	39.8	42.0	46.2	49.8	54.5	62.8	
18.9	750	32.8	35.1	36.9	40.0	42.1	45.0	47.6	52.3	56.1	61.3	70.8	
20.2	800	37.3	39.7	41.6	45.1	47.3	50.5	53.5	58.7	62.8	65.6	79.0	
3.05	275	.7	.7	.8	.9	.9	1.0	1.2	1.3	1.4	1.6	1.6	
3.33	300	.7	.8	.9	1.0	1.2	1.3	1.4	1.5	1.6	1.8	1.7	
3.90	350	1.0	1.2	1.3	1.4	1.5	1.6	1.7	2.0	2.1	2.4	2.1	
4.44	400	1.3	1.5	1.6	1.7	1.8	2.1	2.2	2.4	2.6	3.0	2.3	
5.00	450	1.6	1.8	2.0	2.2	2.3	2.5	2.8	3.0	3.3	3.7	4.3	
5.55	500	2.0	2.2	2.4	2.6	2.8	3.1	3.3	3.6	3.9	4.4	5.1	
6.11	550	2.3	2.5	2.8	3.1	3.3	3.6	3.9	4.3	4.6	5.2	6.0	
6.66	600	2.6	3.0	3.3	3.6	3.8	4.1	4.5	4.9	5.4	6.0	6.8	
7.22	650	3.1	3.6	3.8	4.1	4.5	4.8	5.2	5.8	6.2	6.9	8.1	
7.77	700	3.6	4.0	4.4	4.7	5.1	5.5	5.9	6.6	7.0	7.9	9.1	6"
8.33	750	4.1	4.6	4.9	5.4	5.8	6.2	6.6	7.5	7.9	8.9	10.2	
8.88	800	4.7	5.2	5.5	6.1	6.4	7.0	7.4	8.3	8.9	9.9	11.3	
9.99	900	5.9	6.4	6.8	7.6	7.9	8.6	9.2	10.2	10.9	12.1	13.8	
11.1	1000	7.1	7.8	8.4	9.1	9.7	10.5	11.0	12.2	13.2	14.5	16.6	
12.2	1100	8.6	9.3	10.0	10.9	11.5	12.5	13.1	14.4	15.6	17.0	19.7	
13.3	1200	10.1	11.0	11.7	12.8	13.5	14.5	15.3	16.8	18.2	19.9	23.0	
15.5	1400	13.6	14.7	15.5	16.8	17.8	19.2	20.4	22.2	23.7	26.5	29.7	
17.8	1600	17.6	18.9	19.9	21.5	22.9	24.5	25.8	28.1	29.9	33.1	37.1	
20.0	1800	22.2	23.7	24.8	26.7	28.3	30.4	32.0	34.4	37.0	40.3	45.9	
22.2	2000	27.3	29.1	30.4	32.7	34.0	36.8	38.4	41.9	44.5	48.9	55.3	

SIZE 8" – 12"

VISCOSITY S.S.U.

PIPE SIZE	VELOCITY FT/SEC	U.S. GPM	31.5	35	40	50	60	80	100	150	200	300	500
8"	3.21	500	.5	.6	.6	.7	.8	.8	.9	1.0	1.0	1.3	1.0
	3.85	600	.7	.8	.9	.9	1.0	1.2	1.3	1.4	1.5	1.6	2.0
	4.49	700	.9	1.0	1.2	1.3	1.4	1.5	1.6	1.8	2.0	2.2	2.5
	5.13	800	1.2	1.4	1.5	1.6	1.7	2.0	2.1	2.3	2.4	2.8	3.1
	5.77	900	1.5	1.7	1.8	2.0	2.2	2.3	2.5	2.8	3.0	3.3	3.8
	6.41	1000	1.8	2.1	2.2	2.4	2.5	2.9	3.0	3.3	3.6	4.0	4.6
	7.70	1200	2.5	2.9	3.1	3.3	3.6	3.9	4.1	4.6	4.9	5.4	6.2
	8.98	1400	3.5	3.8	4.1	4.4	4.7	5.1	5.5	6.0	6.4	7.1	8.2
	10.3	1600	4.5	4.8	5.2	5.6	6.0	6.4	6.8	7.6	8.2	9.0	10.2
	11.50	1800	5.5	6.0	6.4	7.0	7.5	8.1	8.5	9.3	10.0	11.0	12.7
	12.8	2000	6.8	7.4	7.7	8.5	9.0	9.7	10.2	11.3	12.1	13.3	15.2
	14.1	2200	8.1	8.7	9.3	10.1	10.6	11.5	12.1	13.3	14.3	15.8	17.8
	15.4	2400	9.5	10.4	10.9	11.8	12.4	13.5	14.1	15.5	16.7	18.3	20.8
	16.7	2600	11.3	12.1	12.7	13.7	14.4	15.5	16.3	17.8	19.3	21.0	23.9
	18.0	2800	12.9	13.9	14.6	15.6	16.6	17.8	18.7	20.5	22.0	24.2	27.1
10"	3.66	900	.5	.6	.6	.7	.7	.8	.8	.9	1.0	1.2	1.4
	4.07	1000	.6	.7	.7	.8	.8	.9	1.0	1.2	1.3	1.4	1.6
	4.48	1100	.7	.8	.8	.9	1.0	1.2	1.2	1.4	1.4	1.6	1.8
	4.88	1200	.8	.9	1.0	1.2	1.2	1.3	1.4	1.6	1.7	1.8	2.2
	5.29	1300	.9	1.0	1.2	1.3	1.4	1.5	1.6	1.8	2.0	2.2	2.4
	5.70	1400	1.2	1.3	1.4	1.5	1.6	1.7	1.8	2.1	2.2	2.4	2.8
	6.10	1500	1.3	1.4	1.5	1.6	1.7	2.0	2.1	2.3	2.4	2.8	3.1
	6.51	1600	1.4	1.6	1.7	1.8	2.0	2.2	2.3	2.5	2.8	3.0	3.6
	7.32	1800	1.7	2.0	2.1	2.3	2.4	2.6	2.8	3.1	3.3	3.7	4.3
	8.14	2000	2.2	2.4	2.5	2.8	3.0	3.2	3.5	3.8	4.0	4.5	5.2
	8.95	2200	2.5	2.9	3.0	3.3	3.5	3.8	4.0	4.5	4.7	5.3	6.0
	9.76	2400	3.0	3.3	3.6	3.9	4.1	4.5	4.7	5.2	5.5	6.2	7.0
	10.60	2600	3.6	3.9	4.1	4.5	4.7	5.2	5.4	6.0	6.4	7.1	8.1
	11.40	2800	4.0	4.5	4.7	5.2	5.4	5.9	6.2	6.8	7.4	8.1	9.2
	12.20	3000	4.6	5.1	5.4	5.9	6.1	6.7	7.0	7.7	8.3	9.1	10.4
12"	2.29	800	.1	.2	.2	.2	.2	.2	.3	.3	.3	.5	.3
	2.58	900	.2	.2	.2	.3	.3	.3	.3	.5	.5	.5	.6
	2.87	1000	.2	.2	.3	.3	.3	.5	.5	.5	.5	.6	.7
	3.44	1200	.3	.3	.5	.5	.5	.6	.6	.7	.7	.8	.9
	4.01	1400	.5	.5	.6	.6	.7	.7	.8	.8	.9	1.0	1.2
	4.59	1600	.6	.7	.7	.8	.8	.9	.9	1.0	1.2	1.3	1.5
	5.16	1800	.7	.8	.9	.9	1.0	1.2	1.3	1.4	1.5	1.6	1.8
	5.73	2000	.9	1.0	1.0	1.2	1.3	1.4	1.5	1.6	1.7	2.0	2.2
	7.17	2500	1.4	1.5	1.6	1.7	1.8	2.1	2.2	2.4	2.5	2.8	3.2
	8.60	3000	2.0	2.1	2.2	2.3	2.5	2.9	3.0	3.3	3.5	3.9	4.4

FRICTION LOSS IN FEET HEAD
PER 100 FEET OF SMOOTH BORE RUBBER HOSE

U. S. Gal. Per Min.	ACTUAL INSIDE DIAMETER IN INCHES							
	¾	1	1¼	1½	2	2½	3	4
15	70	23	5.8	2.5	.9	.2		
20	122	32	10	4.2	1.6	.5		
25	182	51	15	6.7	2.3	.7		
30	259	72	21.2	9.3	3.2	.9	.2	
40		122	35	15.5	5.5	1.4	.7	
50		185	55	23	8.3	2.3	1.2	
60		233	81	32	11.8	3.2	1.4	
70			104	44	15.2	4.2	1.8	
80			134	55	19.8	5.3	2.5	
90			164	70	25	7	3.5	.7
100			203	85	29	8.1	4	.9
125			305	127	46	12.2	5.8	1.4
150			422	180	62	17.3	8.1	1.6
175				230	85	23.1	10.6	2.5
200				308	106	30	13.6	3.2
250					162	44	21	4.9
300					219	62	28	6.7
350					292	83	39	9.3
400						106	49	11.8
500						163	74	17.1
600						242	106	23
700						344	143	30
800						440	182	40
900							224	51
1000							270	63
1250							394	100
1500							525	141
1750								185
2000								230

Chart losses based on 31.5 SSU Liquid Viscosity. For other liquids, multiply these losses by friction factor for other viscosities shown in chart on page 26 of this manual.

LOSS OF HEAD IN FEET PER 100 FEET OF ALUMINUM PIPE

Table based on Scobey's Formula

KS = .34 for 2" pipe,

KS = .33 for 3" pipe,

KS = .32 for other sizes

1 cubic ft. per sec. = 7.48 gal. per second

= 448.8 gal. per minute (commonly used as 450 gpm)

= 646,272 gal. per day (24 hours)

G.P.M.	C.F.S.	2" O.D. .05" Wall	3" O.D. .05" Wall	4" O.D. .063" Wall	5" O.D. .063" Wall	6" O.D. .063" Wall	7" O.D. .078" Wall	8" O.D. .094" Wall
5	.01	.07						
10	.02	.32	.04					
20	.04	1.20	.15	.04				
30	.07	2.58	.32	.08				
40	.09	4.49	.56	.13	.04			
50	.11	6.85	.85	.20	.07	.04		
60	.13	9.67	1.21	.28	.09	.04		
70	.16	12.95	1.61	.38	.12	.05		
80	.18	16.70	2.06	.49	.16	.06	.03	
90	.20	20.80	2.58	.60	.20	.08	.04	
100	.22	25.40	3.18	.74	.24	.10	.05	.03
120	.27		4.51	1.06	.34	.14	.07	.04
140	.31		6.00	1.41	.46	.19	.09	.05
160	.36		7.76	1.82	.59	.24	.11	.06
180	.40		9.67	2.27	.73	.30	.14	.07
200	.45		11.83	2.78	.89	.36	.17	.09
220	.49		14.12	3.31	1.07	.44	.20	.11
240	.54		16.72	3.91	1.27	.52	.24	.13
260	.58		19.42	4.56	1.47	.60	.28	.15
280	.62		22.40	5.26	1.71	.69	.33	.17
300	.67		25.45	5.98	1.93	.79	.37	.19
350	.78			8.03	2.59	1.05	.50	.26
400	.89			10.36	3.33	1.35	.64	.33

G.P.M.	C.F.S.	2" O.D. .05" Wall	3" O.D. .05" Wall	4" O.D. .063" Wall	5" O.D. .063" Wall	6" O.D. .063" Wall	7" O.D. .078" Wall	8" O.D. .094" Wall
450	1.00			12.90	4.15	1.69	.80	.41
500	1.12			15.73	5.07	2.06	.97	.50
550	1.23			19.12	6.16	2.50	1.18	.62
600	1.34			22.46	7.24	2.94	1.38	.72
650	1.45			26.10	8.42	3.41	1.62	.84
700	1.56				9.68	3.92	1.86	.97
750	1.67				11.05	4.46	2.11	1.10
800	1.78				12.48	5.03	2.38	1.24
850	1.90				13.95	5.64	2.67	1.39
900	2.01				15.65	6.35	2.98	1.56
950	2.12				17.35	7.02	3.32	1.73
1000	2.23				19.10	7.72	3.64	1.90
1100	2.46				22.85	9.22	4.37	2.27
1200	2.68				26.95	10.88	5.16	2.68
1300	2.90					12.62	5.96	3.10
1400	3.12					14.65	6.90	3.60
1500	3.34					16.67	7.87	4.07
1600	3.57					18.80	8.89	4.62
1700	3.79					20.95	9.95	5.16
1800	4.01					23.60	11.15	5.79
1900	4.24						12.35	6.42
2000	4.46						13.65	7.10

(Above table computed for Aluminum Pipe with Coupler)

CARRYING CAPACITY AND FRICTION LOSS FOR SCHEDULE 40 THERMOPLASTIC PIPE

(Independent variables: Gallons per minute and nominal pipe size O.D. — Dependent variables: Velocity, friction head and pressure drop per 100 feet of pipe, interior smooth.)

In each column pair below, the first figure is **Velocity (Ft. Per Second)** and the second is **Friction Head (Feet)**. Pipe sizes follow a staircase across the columns; the size labels ("½ in.", "¾ in.", etc.) mark where each size's data begins.

Gallons Per Minute	Vel.	Fric.	Vel.	Fric.	Vel.	Fric.	Vel.	Fric.	Vel.	Fric.	Vel.	Fric.	Vel.	Fric.	Vel.	Fric.
	½ in.		¾ in.		1 in.		1¼ in.		1½ in.		2 in.		2½ in.		3 in.	
1	1.13	2.08	0.63	0.51												
2	2.26	4.16	1.26	1.02	0.77	0.55	0.44	0.14	0.33	0.07						
5	5.64	23.44	3.16	5.73	1.93	1.72	1.11	0.44	0.81	0.22	0.49	0.066	0.30	0.038	0.22	0.015
7	7.90	43.06	4.43	10.52	2.72	3.17	1.55	0.81	1.13	0.38	0.69	0.11	0.49	0.051	0.31	0.021
10	11.28	82.02	6.32	20.04	3.86	6.02	2.21	1.55	1.62	0.72	0.98	0.21	0.68	0.09	0.44	0.03
15	4 in. 0.51	0.03	9.48	42.46	5.79	12.77	3.31	3.28	2.42	1.53	1.46	0.45	1.03	0.19	0.66	0.07
20	0.64	0.04	12.65	72.34	7.72	21.75	4.42	5.59	3.23	2.61	1.95	0.76	1.37	0.32	0.88	0.11
25	0.77	0.06	5 in. 0.49	0.02	9.65	32.88	5.52	8.45	4.04	3.95	2.44	1.15	1.71	0.49	1.10	0.17
30	0.89	0.08	0.57	0.03	11.58	46.08	6.63	11.85	4.85	5.53	2.93	1.62	2.05	0.68	1.33	0.23
35							7.73	15.76	5.66	7.36	3.41	2.15	2.39	0.91	1.55	0.31
40	1.02	0.11	0.65	0.03			8.84	20.18	6.47	9.43	3.90	2.75	2.73	1.16	1.77	0.40
45	1.15	0.13	0.73	0.04			9.94	25.10	7.27	11.73	4.39	3.43	3.08	1.44	1.99	0.50
50	1.28	0.16	0.81	0.05			11.05	30.51	8.08	14.25	4.88	4.16	3.42	1.75	2.21	0.60
60	1.53	0.22	0.97	0.07	6 in. 0.56	0.02			9.70	19.98	5.85	5.84	4.10	2.46	2.65	0.85
70	1.79	0.30	1.14	0.10	0.67	0.03	0.79	0.04	6.83	7.76	4.79	3.27	3.09	1.13		
75	1.92	0.34	1.22	0.11	0.84	0.05			7.32	8.82	5.13	3.71	3.31	1.28		
80	2.05	0.38	1.30	0.13	0.90	0.05			7.80	9.94	5.47	4.19	3.53	1.44		
90	2.30	0.47	1.46	0.16	1.01	0.06			8.78	12.37	6.15	5.21	3.96	1.80		
100	2.56	0.58	1.62	0.19	1.12	0.08	8 in. 0.65	0.03	9.75	15.03	6.84	6.33	4.42	2.18		
125	3.20	0.88	2.03	0.29	1.41	0.12	0.81	0.035			8.55	9.58	5.52	3.31		
150	3.84	1.22	2.44	0.40	1.69	0.16	0.97	0.04			10.26	13.41	6.63	4.63		
175	4.48	1.63	2.84	0.54	1.97	0.22	1.14	0.055					7.73	6.16		
200	5.11	2.08	3.25	0.69	2.25	0.28	1.30	0.07	10 in. 0.82	0.027			8.83	7.88		
250	6.40	3.15	4.06	1.05	2.81	0.43	1.63	0.11	1.03	0.035			11.04	11.93		
300	7.67	4.41	4.87	1.46	3.37	0.60	1.94	0.16	1.23	0.05						
350	8.95	5.87	5.69	1.95	3.94	0.79	2.27	0.21	1.44	0.065	12 in. 1.01	0.027				
400	10.23	7.52	6.50	2.49	4.49	1.01	2.59	0.27	1.64	0.09	1.16	0.04				
450			7.31	3.09	5.06	1.26	2.92	0.33	1.85	0.11	1.30	0.05				
500			8.12	3.76	5.62	1.53	3.24	0.40	2.05	0.13	1.45	0.06				
750					8.43	3.25	4.86	0.85	3.08	0.28	2.17	0.12				
1000					11.24	5.54	6.48	1.45	4.11	0.48	2.89	0.20				
1250							8.11	2.20	5.14	0.73	3.62	0.31				
1500							9.72	3.07	6.16	1.01	4.34	0.43				
2000									8.21	1.72	5.78	0.73				
2500									10.27	2.61	7.23	1.11				
3000											8.68	1.55				
3500											10.12	2.07				
4000											11.07	2.66				

CARRYING CAPACITY AND FRICTION LOSS FOR SCHEDULE 80 THERMOPLASTIC PIPE

(Independent variables: Gallons per minute and nominal pipe size O.D. — Dependent variables: Velocity, friction head and pressure drop per 100 feet of pipe, interior smooth.)

For each pipe size: Vel = Velocity Ft. Per Second, Fric = Friction Head Feet.

GPM	½ Vel	½ Fric	¾ Vel	¾ Fric	1 Vel	1 Fric	1¼ Vel	1¼ Fric	1½ Vel	1½ Fric	2 Vel	2 Fric	2½ Vel	2½ Fric	3 Vel	3 Fric	4 Vel	4 Fric	5 Vel	5 Fric	6 Vel	6 Fric	8 Vel	8 Fric	10 Vel	10 Fric	12 Vel	12 Fric
1	1.48	4.02	0.74	0.86																								
2	2.95	8.03	1.57	1.72	0.94	0.88	0.52	0.21	0.38	0.10																		
5	7.39	45.23	3.92	9.67	2.34	2.75	1.30	0.66	0.94	0.30	0.56	0.10	0.39	0.05	0.25	0.02												
7	10.34	83.07	5.49	17.76	3.28	5.04	1.82	1.21	1.32	0.55	0.78	0.15	0.54	0.07	0.35	0.028												
10			7.84	33.84	4.68	9.61	2.60	2.30	1.88	1.04	1.12	0.29	0.78	0.12	0.50	0.04												
15					7.01	20.36	3.90	4.87	2.81	2.20	1.68	0.62	1.17	0.26	0.75	0.09												
20					9.35	34.68	5.20	8.30	3.75	3.75	2.23	1.06	1.56	0.44	1.00	0.15	0.57	0.04										
25					11.69	52.43	6.50	12.55	4.69	5.67	2.79	1.60	1.95	0.67	1.25	0.22	0.72	0.06										
30					14.03	73.48	7.80	17.59	5.63	7.95	3.35	2.25	2.34	0.94	1.49	0.31	0.86	0.08	0.54	0.03								
35							9.10	23.40	6.57	10.58	3.91	2.99	2.73	1.25	1.74	0.42	1.00	0.11	0.63	0.04								
40							10.40	29.97	7.50	13.55	4.47	3.83	3.12	1.60	1.99	0.54	1.15	0.14	0.72	0.04								
45							11.70	37.27	8.44	16.85	5.03	4.76	3.51	1.99	2.24	0.67	1.29	0.17	0.81	0.06								
50							13.00	45.30	9.38	20.48	5.58	5.79	3.90	2.42	2.49	0.81	1.43	0.21	0.90	0.07	0.63	0.03						
60									11.26	28.70	6.70	8.12	4.68	3.39	2.99	1.14	1.72	0.30	1.08	0.10	0.75	0.04						
70											7.82	10.80	5.46	4.51	3.49	1.51	2.01	0.39	1.26	0.13	0.88	0.05						
75											8.38	12.27	5.85	5.12	3.74	1.72	2.15	0.45	1.35	0.14	0.94	0.06						
80											8.93	13.83	6.24	5.77	3.99	1.94	2.29	0.50	1.44	0.16	1.00	0.07						
90											10.05	17.20	7.02	7.18	4.48	2.41	2.58	0.63	1.62	0.20	1.13	0.08						
100											11.17	20.90	7.80	8.72	4.98	2.93	2.87	0.76	1.80	0.24	1.25	0.10						
125													9.75	13.21	6.23	4.43	3.59	1.16	2.25	0.37	1.57	0.16	0.90	0.045				
150													11.70	18.48	7.47	6.20	4.30	1.61	2.70	0.52	1.88	0.22	1.07	0.05				
175															8.72	8.26	5.02	2.15	3.15	0.69	2.20	0.29	1.25	0.075				
200															9.97	10.57	5.73	2.75	3.60	0.88	2.51	0.37	1.43	0.09	0.90	0.036		
250															12.46	16.00	7.16	4.16	4.50	1.34	3.14	0.56	1.79	0.14	1.14	0.045		
300																	8.60	5.83	5.40	1.87	3.76	0.78	2.14	0.20	1.36	0.07		
350																	10.03	7.76	6.30	2.49	4.39	1.04	2.50	0.27	1.59	0.085	1.12	0.037
400																	11.47	9.93	7.19	3.19	5.02	1.33	2.86	0.34	1.81	0.11	1.28	0.05
450																			8.09	3.97	5.64	1.65	3.21	0.42	2.04	0.14	1.44	0.06
500																			8.99	4.82	6.27	2.00	3.57	0.51	2.27	0.17	1.60	0.07
750																					9.40	4.25	5.36	1.08	3.40	0.36	2.40	0.15
1000																					12.54	7.23	7.14	1.84	4.54	0.61	3.20	0.26
1250																							8.93	2.78	5.67	0.92	4.01	0.40
1500																							10.71	3.89	6.80	1.29	4.81	0.55
2000																									9.07	2.19	6.41	0.94
2500																									11.34	3.33	8.01	1.42
3000																											9.61	1.99
3500																											11.21	2.65
4000																											12.82	3.41

FRICTION LOSS IN STEEL VALVES AND FITTINGS
EQUIVALENT LENGTH OF STRAIGHT PIPE
TURBULENT FLOW

Size in Inches		1"	1½"	2"	2½"	3"	4"	6"	8"	10"	12"
Elbow 45°	Screwed	1.3	2.1	2.7	3.2	4.0	5.5	—	—	—	—
	Flanged	0.81	1.3	1.7	2.0	2.6	3.5	5.6	7.7	9.0	11
Elbow 90°	Screwed	5.2	7.4	8.5	9.3	11	13	—	—	—	—
	Flanged	1.6	2.4	3.1	3.6	4.4	5.9	8.9	12	14	17
Elbow 90° Long Radius	Screwed	2.7	3.4	3.6	3.6	4.0	4.6	—	—	—	—
	Flanged	1.6	2.3	2.7	2.9	3.4	4.2	5.7	7	8	9
Tee — Run Thru	Screwed	3.2	5.6	7.7	9.3	12	17	—	—	—	—
	Flanged	1.0	1.5	1.8	1.9	2.2	2.8	3.8	4.7	5.2	6.0
Tee — Thru Side	Screwed	6.6	9.9	12	13	17	21	—	—	—	—
	Flanged	3.3	5.2	6.6	7.5	9.4	12	18	24	30	34
180° Return Bend	Screwed	5.2	7.4	8.5	9.3	11	13	—	—	—	—
	Flanged	1.6	2.4	3.1	3.6	4.4	5.9	8.9	12	14	17
Gate Valve	Screwed	.84	1.2	1.5	1.7	1.9	2.5	—	—	—	—
	Flanged	—	—	2.6	2.7	2.8	2.9	3.2	3.2	3.2	3.2
Globe Valve	Screwed	29	42	54	62	79	110	—	—	—	—
	Flanged	45	59	70	77	94	120	190	260	310	390
Swing Check Valve	Screwed	11	15	19	22	27	38	—	—	—	—
	Flanged	7.2	12	17	21	27	38	63	90	120	140
Angle Valve	Screwed	17	18	18	18	18	18	—	—	—	—
	Flanged	17	18	21	22	28	38	63	90	120	140
Plug Valve		—	—	6	7	8	17	65	110	150	—
Foot Valve		—	38	46	55	64	71	77	79	81	83
Enlargement	½	—	2.6	3.2	3.8	4.7	6.2	9.5	13	16	19
	¾	—	1.0	1.2	1.3	1.7	2.3	3.4	4.5	5.6	6.8
Contraction	½	—	1.5	1.8	2.2	2.8	3.6	5.6	24	9.5	11
	¾	—	1.0	1.2	1.3	1.7	2.3	3.4	4.5	5.6	6.8

VISCOSITY CORRECTION

FOR EQUIPMENT ON PAGES 27, 28, 29
MULTIPLY FRICTION LOSS BY:

31.5 SSU	1.0	150 SSU	1.7
40 SSU	1.15	200 SSU	1.8
60 SSU	1.3	300 SSU	2.1
80 SSU	1.4	400 SSU	2.5
100 SSU	1.5	500 SSU	2.8

NOZZLE LOSS (HEAD FEET)

GPM	FUEL OIL TYPE		AVIATION TYPE		
	1¼"	1½"	1½"	2"	2½"
50	19'	9'	—	—	—
75	33'	16'	—	—	—
100	—	28'	8'	5'	—
150	—	—	16'	7'	—
200	—	—	30'	12'	7'
300	—	—	—	25'	12'
400	—	—	—	—	18'
500	—	—	—	—	29'

PRESSURE DROPS (HEAD FEET)   AIR ELIMINATORS

GPM	1½"	2"	2½"	3"	4"	6"
50	3'	—	—	—	—	—
100	5'	3'	—	—	—	—
150	—	5'	3'	—	—	—
200	—	7'	5'	3'	—	—
250	—	—	10'	5'	—	—
300	—	—	—	6'	3'	—
350	—	—	—	7'	4'	—
400	—	—	—	10'	5'	3'
450	—	—	—	—	6'	3'
500	—	—	—	—	7'	4'
550	—	—	—	—	8'	4'
600	—	—	—	—	—	5'

LINE STRAINERS

	2"	2½"	3"	4"	6"
50	1'	—	—	—	—
100	4'	—	—	—	—
150	6'	3'	3'	—	—
200	—	5'	4'	—	—
250	—	7'	5'	—	—
300	—	—	6'	3'	—
350	—	—	7'	4'	—
400	—	—	10'	5'	1'
450	—	—	—	6'	1'
500	—	—	—	7'	3'
550	—	—	—	—	·3'
600	—	—	—	—	4'

PRESSURE DROP IN FEET LOADING RACK EQUIPMENT

CAPACITY GPM	COMPLETE LOADING ASSEMBLY				SWING JOINT				LOADING VALVE			
	2"	2½"	3"	4"	2"	2½"	3"	4"	2"	2½"	3"	4"
100	13.8	6.9	—	—	1.1	—	—	—	6.9	3.4	1.1	—
150	27.7	12.7	4.6	—	2.3	1.1	—	—	13.8	5.7	3.4	1.1
200	41.5	20.7	9.2	—	3.4	2.3	1.1	—	23.1	9.2	4.6	2.3
250	64.6	30.0	13.8	4.6	4.6	3.4	2.3	1.1	—	16.1	6.9	3.4
300	87.7	43.8	18.4	5.7	6.9	4.6	3.4	2.3	—	—	9.2	3.4
350	—	57.7	25.4	8.0	11.5	6.9	5.7	2.3	—	—	11.5	4.6
400	—	—	33.4	10.3	13.8	9.2	6.9	3.4	—	—	13.8	4.6
450	—	—	46.2	13.8	16.1	11.5	9.2	3.4	—	—	20.7	6.9
500	—	—	57.7	16.1	20.7	16.1	11.5	4.6	—	—	25.4	9.2
550	—	—	—	18.4	—	—	16.1	4.6	—	—	—	11.5
600	—	—	—	21.9	—	—	—	6.9	—	—	—	13.8
650	—	—	—	27.7	—	—	—	8.0	—	—	—	16.1
700	—	—	—	32.3	—	—	—	8.0	—	—	—	18.4

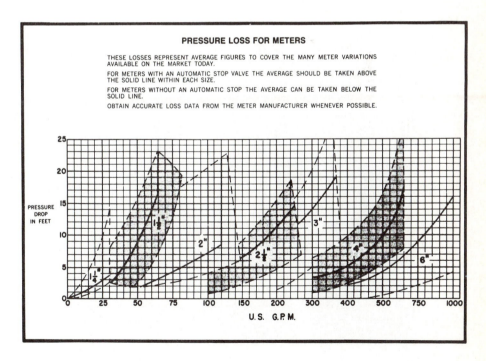

PRESSURE LOSS FOR METERS

THESE LOSSES REPRESENT AVERAGE FIGURES TO COVER THE MANY METER VARIATIONS AVAILABLE ON THE MARKET TODAY.

FOR METERS WITH AN AUTOMATIC STOP VALVE THE AVERAGE SHOULD BE TAKEN ABOVE THE SOLID LINE WITHIN EACH SIZE.

FOR METERS WITHOUT AN AUTOMATIC STOP THE AVERAGE CAN BE TAKEN BELOW THE SOLID LINE.

OBTAIN ACCURATE LOSS DATA FROM THE METER MANUFACTURER WHENEVER POSSIBLE.

TRANSPORT UNLOADING
SUCTION HEAD LOSS IN FEET (31.5 SSU)

FLOW G.P.M.	MANIFOLD LOSSES				HOSE LOSSES (15" LENGTH)		
	2"	2½"	3"	4"	2"	2½"	3"
50	2.2'	1.1'	.5'	—	1.3'	.4'	—
100	8.0'	4.2'	1.4'	.4'	4.4'	1.2'	—
150	17.5'	8.9'	3.0'	1.1'	9.3'	2.7'	1.2'
200	30.0'	15.5'	.5.1'	1.8'	16.0'	4.5'	2.0'
250	—	24.0'	7.9'	2.8'	—	6.6'	3.2'
300	—	—	11.1'	3.9'	—	9.3'	4.3'
400	—	—	19.5'	6.8'	—	16.0'	7.4'
500	—	—	30.2'	10.5'	—	—	11.3'
600	—	—	—	15.0'	—	—	16.0'

(CHART 1)

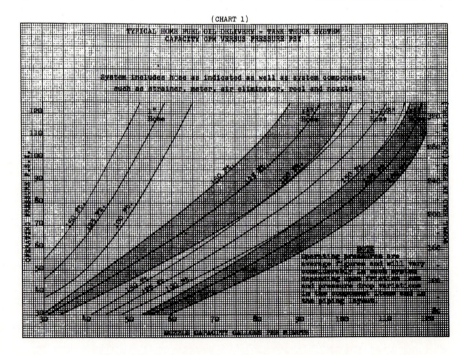

MATERIALS RECOMMENDED FOR PUMPING VARIOUS LIQUIDS

Liquid	Condition	Specific Gravity	Pump Material
Acetaldehyde	Moisture Free	0.98	All Iron
Acetaldehyde	Presence of Moisture	0.78	Bronze Liquid End, 304
Acetone		0.79	All Iron, Standard Fitted
Acetate Solvents			Std. Ftd. All Bronze, All Iron
Acid, Acetic	5% Room Temp.	1.05	Bronze Liquid End, 304, 316
Acetic	20% Room Temp.		304, 316
Acetic	50% Room Temp.		304, 316
Acetic	50% Boiling		Hastelloy, 316 with Caution
Acetic	100% Room Temp.		Aluminum, 304, 316, Alum Bronze
Acetic	100% Boiling		Hastelloy, 316 with Caution
Arsenic	Room Temp.	2.0-2.5	All Iron, 304, 316
Arsenic	90% at 225°F		304 with Caution
Benzoic		1.27	304, 316, Aluminum
Boric			Bronze Liquid End, 304
Butyric	5%-100% Room Temp.	0.96	Aluminum, 304, 316
Carbolic	Concentrated	1.07	All Iron, 304, 316
Carbolic in H₂O			All Iron, Standard Fitted
Carbonic	Aqueous Solution		Bronze Liquid End, Alum., 304, 316
Chromic			All Iron, 304, 316
Chromic with H₂SO₄			304, 316
Chromic	50% Boiling		Hastelloy C
Citric			Bronze Liquid End, 304, 316
Citric	Concentrated, Boiling		316, Hastelloy (All)
Fatty (Oleic, Palmitic & Stearic)			Bronze Liquid End, 304, 316 Alum Bronze, Monel
Formic		1.22	Bronze Liquid End, 304, 316
Fruit			Bronze Liquid End, 304, 316, Alum Bronze
Gallic	5% Room to Boiling		Aluminum, Std. Ftd. 304, 316
Hydrobromic	Boiling		Hastelloy B
Hydrochloric	5% Unaerated-Room Temp.	1.19(38%)	Hastelloy All, High Silicon
Hydrochloric	10% Unaerated-Room Temp.		Hastelloy All, Monel, Nickel High Silicon
Hydrochloric	All 100°F		Hastelloy All, High Silicon Stoneware
Hydrochloric	All 160°F		Hastelloy A & B, High Silicon
Hydrochloric	Fumes		Illium, Cupro-Nickel with Caution
Hydrocyanic		.70	All Iron, Aluminum, 304, 316
Hydrofluoric			Bronze or Monel with Caution
Hydrofluosilicic		1.30	Bronze or Monel with Caution
Lactic	Room Temp.	1.25	Bronze Liquid End, 304, Aluminum
Minewater			Bronze Liquid End, 304
Mixed			Full Range from All Iron, 304, 316, Lead, Steel Depending on Acid Conc. & Percent Water
Naphthenic			Aluminum, 304, 316
Nitric	Concentrated-Room Temp.	1.50	Aluminum, 304, 316
Nitric	95% Room Temp.		Aluminum
Nitric	65% Boiling		304, 316
Nitric	Dilute Room Temp.		304, 316
Oxalic	5%-10% Room or Hot	1.65	Bronze Liquid End, 304, 316
Oxalic	10% Boiling		Illium, Silicon, Bronze, Hastelloy
Ortho-Phosphoric	Crude	1.87	316
Phosphoric	Dilute Room Temp.		304, 316
Pickling			Depending on Conditions
Picric	Concentrated-Room Temp.	1.76	304, 316, High Silicon Iron
Pyrogallic		1.45	304, 316
Pyroligneous			Bronze Liquid End, 304, 316
Sulfuric	77%-Room Temp.	1.69-1.84	All Iron, 304, 316
Sulfuric	77%-Hot		304, 316
Sulfuric	Very Dilute-Room Temp.		304, 316
Sulfuric	Very Dilute-Boiling		Hastelloy B
Sulfuric Fuming (Oleum)			Steel
Sulfurous		1.92-1.94	Bronze, Liquid End, 316, Alum Bronze, Lead
Tannic			Bronze Liquid End, 304, 316, Monel
Tartaric	Solution		Bronze Liquid End, 304
Alcohol, Grain Ethyl			Bronze Liquid End, Std. Ftd.
Alcohol, Wood Methyl			Bronze Liquid End, Std. Ftd.
Aluminum Sulfate	Solution		High Silicon, Lead, 304, 316
Ammonium Bicarbonate	H₂O Solution		All Iron
Ammonium Chloride	H₂O Solution		All Iron, 316
Ammonium Hydroxide	H₂O Solution		All Iron
Ammonium Nitrate	H₂O Solution		All Iron, 304, 316
Ammonium Phosphate	H₂O Solution		All Iron, 304, 316
Ammonium Sulphate with H₂SO₄			Bronze Liquid End, Lead
Ammonium Sulphate	H₂O Solution		All Iron, 304, 316
Aniline		1.02	All Iron
Aniline Hydrochloride	H₂O Solution		High Silicon Iron
Asphalt	Hot	.98-1.4	All Iron
Barium Chloride	H₂O Solution	All Iron, 316	
Barium Nitrate	H₂O Solution	All Iron, 304, 316	
Beer			Bronze Liquid End, 304, 316
Beer Wort			Bronze Liquid End, 304, 316
Beet Juice			Bronze Liquid End, 304, 316
Benzene			See Benzol
Benzol		.88	All Iron, Standard Fitted

Liquid	Condition	Specific Gravity	Pump Material
Benzine			See Petroleum Ether
Bichloride of Mercury			See Mercuric Chloride
Bittern			Ni Resist
Bitterwasser			Bronze Liquid End, 304
Black Liquor			All Iron, 304, 316, Ni Resist
Bleach Solutions			See Respective Hypochlorites
Blue Vitriol			See Copper Sulfate
Boiler Feed Water			All Iron, Bronze Liquid End, Standard Fitted Depending on pH
Borax			All Iron
Brine, Calcium Chloride			All Iron if pH is 8.5, otherwise Bronze Liquid End
Brine, Calcium & Magnesium			Bronze Liquid End, Ni Resist
Brine, Calcium & Sodium Chloride			Bronze Liquid End
Brine, Sodium Chloride	3% Salt	1.02-1.20	All Iron, Bronze Liquid End
Brine, Sodium Chloride	Over 3%		Bronze Liquid End, 304, 316
Brine, Sea Water		1.03	All Iron, Bronze Liquid End
Butane		.60 (32°F.)	Standard Fitted Iron Fitted
Cachaza			Standard Fitted
Cadium Plating Solution			Chrome-Nickel Iron, High Silicon,
Calcium Bisulfite		1.06	Rubber, Stone
Calcium Chlorate			316
Calcium Chlorite			304, 316
Calcium Hypochlorite			See Brine All Iron, High Silicon Iron
Cane Juice			Standard Fitted Bronze Liquid End
Carbonated Water			See Acid, Carbonic
Carbon Bisulfide		1.26	All Iron
Carbonate of Soda			See Soda Ash
Carbon Dioxide			See Acid, Carbonic
Carbon Tetrachloride	Moisture Free	1.50	All Iron
Carbon Tetrachloride	In Presence of Moisture		Bronze Liquid End
Caustic Potash			See Potassium Hydroxide
Caustic Soda			See Sodium Hydroxide
Cellulose Acetate			316, High Silicon Iron
Chlorate of Lime			See Calcium Chlorate
Chloride of Lime			See Calcium Hypochlorite
Chlorinated Solvents	Moisture Free		All Iron
Chlorinated Solvents	Presence of Moisture		Bronze Liquid End
Chlorine			Hastelloy, High Silicon
Chlorobenzene		1.1	Stoneware, Rubber
Chloroform		1.5	Standard Fitted 304 Bronze Liquid End, 304, 316, Lead
Chrome Alum	H₂O Solution		High Silicon Iron, Chrome-Nickel Iron
Copperas Green			See Ferrous Sulfate
Copper Acetate			316
Copper Chloride	H₂O Solution		Hastelloy, High Silicon, Rubber Stoneware
Copper Nitrate			304, 316
Copper Sulfate, Blue Vitriol			304, 316, High Silicon, Lead
Clay Slip, (Paper Mill)			All Iron, (Hardened Fittings)
Condensate			Standard Fitted
Creosote		1.04-1.10	All Iron, Std. Ftd.
Cresol, Meta		1.03	All Iron
Cyanide			All Iron
Cyanogen	In H₂O		All Iron
Developing Solutions			304
Diethanolamine			All Iron
Diethylene Glycol			All Iron, Bronze Liquid End,
Diphenyl			Standard Fitted
Diphenyl Oxide,	Moisture Free	.99	All Iron, Steel
Diphenyl	Moisture Free		All Iron, Steel
Distillery Wort	In Alcohol		All Iron Bronze Liquid End
Enamel			All Iron
Epsom Salts			See Magnesium Sulfate
Ethyl Acetate			All Iron 316
Ethyl Alcohol (Ethanol)			See Alcohol, Ethyl
Ethylene Chloride	Cold	1.28	Lead, High Silicon, High Chrome,
Ferric Chloride			Nickel Iron Hastelloy, High Silicon, Rubber, Stoneware
Ferric Sulfate			304, High Silicon
Ferrous Chloride			Bronze Liquid End, 304
Ferrous Sulfate			Bronze Liquid End, 304, Rubber, Stoneware, Lead
Formaldehyde		1.08	Bronze Liquid End, 304
Fruit Juices			Bronze Liquid End, 304
Fuel Oil			See Oil, Fuel
Furfural		1.16	All Iron, Bronze Liquid End, Standard Fitted
Fusel Oil			Bronze Liquid End
Gasoline, Refined		.68-.75	Standard Fitted Iron Ftd.
Glaubers Salt			See Sodium Sulfate
Glue			Standard Fitted
Glue Sizing			Bronze Liquid End

Liquid	Condition	Specific Gravity	Pump Material
Glycerin			See Glycerol
Glycerol		1.26	Std. Fitted, Bronze Liquid End, All Iron. 304, 316
Green Liquor			Standard Fitted
Heptane		0.69	Aluminum, 304, 316
Hydrogen Peroxide			
Hydrogen Sulfide			Aluminum, 304
Hydrosulfite of Soda			See Sodium Hydrosulfite
Hyposulfite of Soda			See Sodium Thiosulfate
Kaolin Slip			Bronze Liquid End
Kerosene			Std. Fitted, Bronze Liquid End
Lacquer			Std. Fitted, All Iron, Bronze Liquid End
Lacquer Solvents			Std. Fitted, All Iron, Bronze Liquid End
Lard			Std. Fitted, All Iron
Latex			All Iron
Lead (Molten)			All Iron, Steel
Lead Acetate	H_2O Solution		304, 316, Rubber, Stoneware
Lime Water			All Iron
Lithium Chloride	H_2O Solution		All Iron, Bronze Liquid End
Lye			See Sodium Hydroxide
Magnesium Chloride			Bronze Liquid End, Lead, High Silicon
Magnesium Sulfate			Bronze Liquid End, All Iron
Magma (Thick Residue)			Bronze Liquid End, All Iron, 304
Manganese Chloride	H_2O Solution		Ni Resist Bronze Liquid End, 304, 316
Manganous Sulfate	H_2O Solution		All Iron, Bronze Liquid End, 304, 316
Mash			Standard Fitted, Bronze Liquid
Mercuric Chloride	Very Dilute H_2O Solution		End, 304 304, High Silicon Iron
Mercuric Chloride	Comm'l Conc. H_2O Solution		Hastelloy, High Silicon, Stoneware
Mercuric Sulfate	In H_2SO_4		High Silicon Iron, Stoneware
Mercurous Sulfate	In H_2SO_4		High Silicon Iron, Stoneware
Mercury			All Iron
Methyl Acetate			316
Methyl Alcohol			See Alcohol, Wood
Methyl Chloride		0.52	All Iron
Methylene Chloride		1.34	All Iron, 304
Milk		1.03-1.04	Tinned Bronze, 304
Milk of Lime			See Lime Water
Mine Water			See Acid, Mine Water
Miscella		0.75	All Iron
Molasses			Std. Fitted, Bronze Liquid End
Mustard			Bronze Liquid End, 304
Naphtha		.78-.88	Standard Fitted
Naphtha, Crude		.92-.95	Standard Fitted
Nickel Chloride	Low pH Plating Solutions		Chrome-Nickel Iron, High Silicon
Nickel Sulfate	Low pH Plating Solutions		Chrome-Nickel Iron, High Silicon
Nicotine Sulfate			Chrome-Nickel Iron, High Silicon
Nitre			See Potassium Nitrate
Nitre Cake			See Sodium Bisulphate
Nitro-Ethane		1.04	304, 316
Nitro-Methane		1.14	304, 316
Oil, Cocoanut		.91	All Iron, Br. Ftd. All Bronze
Oil Crude (Asphalt Base)			Standard Fitted
Crude (Paraffin Base)			Standard Fitted
Fuel (Furnace)			All Iron, Standard Fitted
Fuel (Kerosene)			All Iron, Standard Fitted
Lubricating			Standard Fitted
Mineral			Standard Fitted
Mineral	U. S. P.		304
Palm		.90	All Iron, Bronze Liquid End, Monel
Quenching		.91	Standard Fitted, All Iron
Soya Bean			All Iron, Bronze Liquid End, Monel
Vegetable			All Iron, Standard Fitted
Coal Tar			All Iron, Standard Fitted
Creosote			All Iron, Standard Fitted
Turpentine		.87	All Iron, Standard Fitted
Linseed		.94	All Iron, Std. Ftd.
Rapeseed		.92	Bronze Liquid End, 304, Monel
Olive Oil		.90	All Iron, Std. Ftd.
Paraffin			Standard Fitted, All Iron
Perchlorethlene		1.62	Std. Ftd. Iron Ftd.
Petroleum			See Oil, Crude
Petroleum Solvent		0.80	Std. Ftd. Iron Ftd.
Perhydrol			See Hydrogen Peroxide
Peroxide of Hydrogen			See Hydrogen Peroxide
Petroleum Ether			All Iron, Standard Fitted
Phenol		1.07	See Acid, Carbonic
Photographic Developers			See Developing Solutions
Pickling Acids			See Acid, Pickling
Potash			See Potassium Carbonate
Potash Alum			Bronze Liquid End
Potassium Bichromate			All Iron
Potassium Carbonate			All Iron
Potassium Chloride			Bronze Liquid End
Potassium Cyanide			All Iron
Potassium Hydroxide			All Iron, 304
Potassium Nitrate			All Iron
Potassium Sulfate			All Iron
Propane		.59 (48°F.)	Std. Ftd. All Iron

Liquid	Condition	Specific Gravity	Pump Material
Pyridine Sulfate			High Chrome-Nickel Iron, Lead
Rectifying Pump (Distillery)			Bronze Liquid End
Rhigolene (Oil Distillery)			Standard Fitted
Rosin (Colophony)			All Iron
Sal Ammoniac			See Ammonium Chloride
Salt			See Brines
Salt Cake			Bronze Liquid End
Sea Water			See Water, Sea
Sewage			Standard Fitted All Iron
Silicate of Soda			All Iron, Standard Fitted, Bronze
Silver Nitrate			Liquid End
Slop, Brewery			304, 316, High Silicon Iron
Slop, Distillery			Bronze Liquid End
Soap Liquor			Bronze Liquid End
			All Iron
Soda Ash			See Sodium Carbonate
Sodium Bicarbonate			All Iron, Bronze Liquid End
Sodium Bisulfate			High Silicon, Lead
Sodium Carbonate			All Iron
Sodium Chloride			See Brine, Sodium
Sodium Cyanide			All Iron
Sodium Hydroxide			All Iron
Sodium Hydrosulfite			304, 316, Lead
Sodium Hypochlorite			High Silicon, Stoneware, Lead
Sodium Hyposulfite			See Sodium Thiosulfate
Sodium Nitrate			All Iron, 304
Sodium Phosphate (mono)			Bronze Liquid End, 304
Sodium Phosphate (di)			Bronze Liquid End
Sodium Phosphate (tri)			All Iron
Sodium Phosphate (meto)			304
Sodium Silicate (water glass)			All Iron
Sodium Sulfate			Bronze Liquid End, All Iron
Sodium Sulfide			All Iron, 304
Sodium Sulfite			Bronze Liquid End
Sodium Tetraborate			All Iron
Sodium Thiosulfate			Bronze Liquid End, 304
Stannic Chloride			Hastelloy, High Silicon, Rubber, Stoneware
Stannous Chloride			Hastelloy, High Silicon, Rubber, Stoneware
Starch			Standard Fitted
Stock, Paper			Bronze Liquid End or Std. Fitted, Depending on pH value
Strontium Nitrate			All Iron, 304
Sugar			Bronze Liquid End
Sulfite Liquor (paper mill)			316, Bronze Liquid End
Sulfur	In Water		All Iron, Bronze Liquid End, Ni Resist
Sulfur	Molten		All Iron
Sulfur Chloride	Cold		All Iron, Lead
Syrup			Bronze Liquid End
Tallow		0.90	All Iron
Tanning Liquor			Bronze Liquid End, 304
Tar			All Iron
Tar & Ammonim	In Water		All Iron
Tetraethyl Lead		1.66	Std. Fitted, All Iron
Toluene (Toluol)		0.87	Std. Fitted, All Iron
Trichloroethylene	Moisture Free	1.47	All Iron
Trichloroethylene	Moisture Present	1.47	Bronze Liquid End
Trisodium Phosphate			See Sodium Phosphate (tri)
Turpentine			All Iron, Standard Fitted
Urine			Bronze Liquid End
Varnish			Standard Fitted
Vinegar			Bronze Liquid End, 304
Vitriol, Blue			See Copper Sulfate
Vitriol, Green			See Ferrous Sulfate
Vitriol, Oil of			See Acid, Sulfuric
Vitriol, White			See Zinc Sulfate
Water, Acid Mine			See Acid, Mine Water
Water, Distilled			Bronze Liquid End, Std. Fitted
Water, Fresh			Std. Fitted, Bronze Liquid End
Water, Salt			Bronze Liquid End, All Iron
Water, Sea			Bronze Liquid End, All Iron
Whiskey			Bronze Liquid End, 304
White Liquor			Bronze Liquid End or Std. Fitted
Wine			Depending On pH
Wood Vinegar			Bronze Liquid End, 304
Wort			See Pyroligneous Acid
Xylol (Xylene)		0.87	Standard Fitted, All Iron
Yeast			Bronze Liquid End, Std. Fitted
Zinc Chloride			316, High Silicon Iron
Zinc, Plating Solution			High Silicon Iron, Lead
Zinc Sulfate			Bronze Liquid End, 304

NOTE: **ALL IRON PUMP**—All parts of the pump coming in direct contact with the liquid pumped are to be made of iron or ferrous metal.

STANDARD FITTED PUMP—Iron Fitted or Bronze Fitted includes cast iron casing, steel shaft, either iron or bronze impeller and usually bronze wearing rings and shaft sleeves (when used).

BRONZE LIQUID END—All parts of the pump coming in direct contact with the liquid pumped are to be made of bronze, with stainless steel fastenings.

PUMP CHARACTERISTICS—COMPARISON

CHARACTERISTICS	ROTARY PUMPS	STRAIGHT CENTRIFUGALS	SELF-PRIMING CENTRIFUGALS
Rotary Pump Advantages Straight Centrifugal Advantages Self-Priming Centrifugal Pump Advantages			
Use on Sucton Lifts (Self-Priming)	Excellent (when new)	Auxiliary Equipment Needed	Excellent
Pump Life	Poor on Dry Liquids	Excellent	Excellent
Quiet Operation	No	Yes	Yes
Pressure Relief	Bypass Valve Required	Unnecessary	Unnecessary
Resists Vapor Lock	Yes	No	Yes
Direct Drive Possible	No	Yes	Yes
High Capacity	No	Yes	Yes
H. P. Required	High	Low	Low
High Viscosity	Yes	No	No

ADDED SELF PRIMING CENTRIFUGAL ADVANTAGES

1. Ability to prime with suction lift.
2. Ability to handle air.
3. Ability to handle volatile and dry liquids such as gasoline.

SELF PRIMING CENTRIFUGAL PUMP APPLICATION

Because of its vapor handling and self priming abilities, the self priming centrifugal is **ideally** suited for applications involving suction lifts, volatile liquids or stripping requirements. These include:

Sump applications
Pumping from Underground Tanks
Stripping a compartment
De-watering applications
Tank Truck Delivery

Unloading Tank Cars
Unloading tank trucks or trailers
Dual operations — loading and unloading
Filter Application
Aviation Refueling

NOTE: *On applications such as unloading tank cars or tank trucks (trailers) a self priming pump is needed even though the source of supply is above the pump. Introduction of air into the suction system occurs frequently from vortex conditions especially when the liquid level is low. This air must be manually relieved before pumping is resumed. In addition, it is next to impossible to strip the tank no matter what air release devices are used (manual or mechanical). In explanation of this, the static head is not high enough to develop the pressure needed to push the air through the suction pipe and pump to the discharge side. The "self primer" is inherently capable of handling some air during pumping and when loss of prime occurs, it is capable to developing the vacuum needed to transfer air to the discharge side to reprime the system.*

CONVERSION UNITS

Multiply	By	To Obtain
ACRES	160	Square rods
Acres	4840	Square yards
Acres	43.560	Square feet
ACRES INCHES	27,154	Gallons
ACRES INCH/HR.	452	GPM
ATMOSPHERES (STD.) 760 MM of Mercury at 32°F.	14.696	Lbs./sq. inch
ATMOSPHERES	76.0	Cms. of mercury
Atmospheres	29.92	Inches of mercury
Atmospheres	33.90	Feet of water
Atmospheres	1.0333	Kgs./sq. cm.
Atmospheres	14.70	Lbs./sq inch
Atmospheres	1.058	Tons/sq. ft.
BARRELS-OIL	42	Gallons-Oil
BARRELS (Beer)	31.5	Gallons
(Wine)	31.0	Gallons
BRIT. THERM. UNITS	0.2520	Kilogram-calories
Brit. Therm. Units	777.5	Foot-lbs.
Brit. Therm. Units	3.927×10^{-4}	Horse-power-hrs.
Brit. Therm. Units	107.5	Kilogram-meters
Brit. Therm. Units	2.928×10^{-4}	Kilowatt-hrs.
B.T.U./MIN	12.96	Foot-lbs./sec.
B.T.U./min	0.02356	Horse-power
B.T.U./min	0.01757	Kilowatts
B.T.U./min	17.57	Watts
CENTARES (CENTIARES)	1	Square meters
CENTIGRAMS	0.01	Grams
CENTILETERS	0.01	Liters
CENTIMETERS	0.3937	Inches
Centimeters	0.03280	Feet
Centimeters	0.01	Meters
Centimeters	10	Millimeters
CENTIMTRS. OF MERC.	0.01316	Atmospheres
Centimtrs. of merc.	0.4461	Feet of water
Centimtrs. of merc.	136.0	Kgs./sq. meter
Centimtrs. of merc.	27.85	Lbs./sq. ft.
Centimtrs. of merc.	0.1934	Lbs./sq. inch
CENTIMTRS./SECOND	1.969	Feet/min.
Centimtrs./second	0.03281	Feet/sec.
Centimtrs./second	0.036	Kilometers/hr.
Centimtrs./second	0.6	Meters/min.
Centimtrs./second	0.02237	Miles/hr.
Centimtrs./second	3.728×10^{-4}	Miles/min.
CMS./SEC./SEC.	0.03281	Feet/sec./sec.
CUBIC CENTIMETERS	3.531×10^{-5}	Cubic feet
Cubic centimeters	6.102×10^{-2}	Cubic inches
Cubic centimeters	10^{-6}	Cubic meters
Cubic centimeters	1.308×10^{-6}	Cubic yards
Cubic centimeters	2.642×10^{-4}	Gallons
Cubic centimeters	10^{-3}	Liters
Cubic centimeters	2.113×10^{-3}	Pints (liq.)
Cubic centimeters	1.057×10^{-3}	Quarts. (liq.)

Multiply	By	To Obtain
CUBIC FEET	2.832×10^{4}	Cubic cms.
Cubic feet	1728	Cubic inches
Cubic feet	0.02832	Cubic meters
Cubic feet	0.03704	Cubic yards
Cubic feet	7.48052	Gallons U.S.
Cubic feet	6.23	Imper. Gallons
Cubic feet	28.32	Liters
Cubic feet	59.84	Pints (liq.)
Cubic feet	29.92	Quarts (liq.)
CUBIC FEET/MINUTE	472.0	Cubic cms./sec.
Cubic feet/minute	0.1247	Gallons/sec.
Cubic feet/minute	0.4720	Liters/sec.
Cubic feet/minute	62.43	Lbs. of water/min.
CUBIC FEET/SECOND	0.646317	Million gals./day
Cubic feet/second	448.831	Gallons/min.
CUBIC FOOT WATER	62.4	Pounds
Cubic foot water	998.8	Ounces
Cubic foot water	28.315	Kilograms
CUBIC INCHES	16.39	Cubic centimeters
Cubic inches	5.787×10^{-4}	Cubic feet
Cubic inches	1.639×10^{-5}	Cubic meters
Cubic inches	2.143×10^{-5}	Cubic yards
Cubic inches	4.329×10^{-3}	Gallons
Cubic inches	1.639×10^{-2}	Liters
Cubic inches	0.03463	Pints (liq.)
Cubic inches	0.01732	Quarts (liq.)
CUBIC METERS	10^{6}	Cubic centimeters
Cubic meters	35.31	Cubic feet
Cubic meters	61,023	Cubic inches
Cubic meters	1.308	Cubic yards
Cubic meters	264.2	Gallons U.S.
Cubic meters	220	Imperial Gallons
Cubic meters	10^{3}	Liters
Cubic meters	2113	Pints (liq.)
Cubic meters	1057	Quarts (liq.)
CUBIC YARDS	7.646×10^{5}	Cubic centimeters
Cubic yards	27	Cubic feet
Cubic yards	46,656	Cubic inches
Cubic yards	0.7646	Cubic meters
Cubic yards	202.0	Gallons
Cubic yards	764.6	Liters
Cubic yards	1616	Pints (liq.)
Cubic yards	807.9	Quarts (liq.)
CUBIC YARDS/MIN.	0.45	Cubic feet/sec.
Cubic yards/min.	3.367	Gallons/sec.
Cubic yards/min.	12.74	Liters/sec.
DECIGRAMS	0.1	Grams
DECILITERS	0.1	Liters
DECIMETERS	0.1	Meters
DEGREES (ANGLE)	60	Minutes
Degrees (angle)	0.01745	Radians
Degrees (angle)	3600	Seconds
DEGREES/SEC.	0.01745	Radians/sec.
Degrees/sec.	0.1667	Revolutions/min.
Degrees/sec.	0.002778	Revolutions/sec.
DEKAGRAMS	10	Grams

CONVERSION UNITS

Multiply	By	To Obtain
DEKALITERS	10	Liters
DEKAMETERS	10	Meters
DRAMS	27.34375	Grains
Drams	0.0625	Ounces
Drams	1.771845	Grams
FATHOMS	6	Feet
FEET	30.48	Centimeters
Feet	12	Inches
Feet	0.3048	Meters
Feet	1/3	Yards
FEET OF WATER	0.02950	Atmospheres
Feet of water	0.8826	Inches of mercury
Feet of water	0.03048	Kgs./sq. cm.
Feet of water	62.43	Lbs./sq. ft.
Feet of water	0.4335	Lbs./sq. inch
FEET/MIN.	0.5080	Centimeters/sec.
Feet/min.	0.01667	Feet/sec.
Feet/min.	0.01829	Kilometers/hr.
Feet/min.	0.3048	Meters/min.
Feet/min.	0.01136	Miles/hr.
FEET/SEC./SEC.	30.48	Cms./sec./sec.
Feet/sec./sec.	0.3048	Meters/sec./sec.
FOOT-POUNDS	1.286×10^{-3}	Br. Thermal Units
Foot-pounds	5.050×10^{-7}	Horse-power-hrs.
Foot-pounds	3.241×10^{-4}	Kilogram-calories
Foot-pounds	0.1383	Kilogram-meters
Foot-pounds	3.766×10^{-7}	Kilowatt-hrs.
FOOT-POUNDS/MIN.	1.286×10^{-3}	B. T. Units/min.
Foot-pounds/min.	0.01667	Foot-pounds/sec.
Foot-pounds/min.	3.030×10^{-5}	Horse-power
Foot-pounds/min.	3.241×10^{-4}	Kg.-calories/min.
Foot-pounds/min.	2.260×10^{-5}	Kilowatts
FOOT-POUNDS/SEC.	7.717×10^{-2}	B. T. Units/min.
Foot-pounds/sec.	1.818×10^{-3}	Horse-power
Foot-pounds/sec.	1.945×10^{-2}	Kg.-calories/min.
Foot-pounds/sec.	1.356×10^{-3}	Kilowatts
GALLONS	3785	Cubic centimeters
Gallons	0.1337	Cubic feet
Gallons	231	Cubic inches
Gallons	3.785×10^{-3}	Cubic meters
Gallons	4.951×10^{-3}	Cubic yards
Gallons	128	Fluid ounces
Gallons	3.785	Liters
Gallons	8	Pints (liq.)
Gallons	4	Quarts (liq.)
GALLONS, IMPERIAL	1.20095	U.S. Gallons
Gallons, U.S.	0.83267	Imperial gallons
Gallons Imperial	277.3	Cubic inches
Gallons Imperial	0.16	Cubic foot
Gallons Imperial	4.546	Liters
Gallons Imperial	0.00454	Cubic meter
GALLONS WATER	8.3453	Pounds of water
GALS. WATER (U.S.)	3.785	Kilograms

Multiply	By	To Obtain
GALS. WATER (IMP.)	10.02	Pounds
Gals. water (Imp.)	4.54	Kilograms
GALLONS/MIN.	2.228×10^{-3}	Cubic feet/sec.
Gallons/min.	0.06308	Liters/sec.
Gallons/min.	8.0208	Cu. ft./hr.
GALLONS WATER/MIN.	6.0086	Tons water/24 hrs.
GRAINS (TROY)	1	Grains (avoir.)
Grains (troy)	0.06480	Grams
Grains (troy)	0.04167	Pennyweights (troy)
Grains (troy)	2.0833×10^{-3}	Ounces (troy)
GRAINS/U.S. GAL.	17.118	Parts/million
Grains/U.S. gal.	142.86	Lbs./million gal.
GRAINS/IMP. GAL.	14.286	Parts/million
GRAMS	980.7	Dynes
Grams	15.43	Grains
Grams	10^{-3}	Kilograms
Grams	10^{3}	Milligrams
Grams	0.03527	Ounces
Grams	0.03215	Ounces (troy)
Grams	2.205×10^{-3}	Pounds
GRAMS/CM.	5.600×10^{-3}	Pounds/inch
GRAMS/CU. CM.	62.43	Pounds/cubic foot
Grams/cu. cm.	0.03613	Pounds/cubic inch
GRAMS/LITER	58.417	Grains/gal.
Grams/liter	8.345	Pounds/1000 gals.
Grams/liter	0.062427	Pounds/cubic foot
Grams/liter	1000	Parts/million
HECTOGRAMS	100	Grams
HECTOLITERS	100	Liters
HECTOMETERS	100	Meters
HECTOWATTS	100	Watts
HORSE-POWER	42.44	B. T. Units/min.
Horse-power	33,000	Foot-lbs./min.
Horse-power	550	Foot-lbs./sec.
Horse-power	1.014	H-power (Metric)
Horse-power	10.70	Kg.-calories/min.
Horse-power	0.7457	Kilowatts
Horse-power	745.7	Watts
HORSE-POWER (BOILER)	33,479	B. T. U./hr.
Horse-power (boiler)	9.803	Kilowatts
HORSE-POWER-HOURS	2547	Br. Thermal Units
Horse-power-hours	1.98×10^{6}	Foot-lbs.
Horse-power-hours	641.7	Kilogram-calories
Horse-power-hours	2.737×10^{5}	Kilogram-meters
Horse-power-hours	0.7457	Kilowatt-hours
INCHES	2.540	Centimeters
Inches	25.4	Millimeters
Inches	.0254	Meters
Inches	.0833	Foot
INCHES OF MERCURY	0.03342	Atmospheres
Inches of mercury	1.133	Feet of water

CONVERSION UNITS

Multiply	By	To Obtain
Inches of mercury	0.03453	Kgs./sq. cm.
Inches of mercury	70.73	Lbs./sq. ft.
Inches of mercury	0.4912	Lbs./sq. inch
INCHES OF WATER	0.002458	Atmospheres
Inches of water	0.07355	Inches of mercury
Inches of water	0.002540	Kgs./sq. cm.
Inches of water	0.5781	Ounces/sq. inch
Inches of water	5.202	Lbs./sq. foot
Inches of water	0.03613	Lbs./sq. inch
KILOGRAMS	980,665	Dynes
Kilograms	2.205	Lbs.
Kilograms	1.102×10^{-3}	Tons (short)
Kilograms	10^3	Grams
KGS./METER	0.6720	Lbs./foot
KGS./SQ. CM.	0.9678	Atmospheres
Kgs./sq. cm.	32.81	Feet of water
Kgs./sq. cm.	28.96	Inches of mercury
Kgs./sq. cm.	2048	Lbs./sq. foot
Kgs./sq. cm.	14.22	Lbs./sq. inch
KGS./SQ. MILLIMETER	10^6	Kgs./sq. meter
KILOLITERS	10^3	Liters
KILOMETERS	10^5	Centimeters
Kilometers	3281	Feet
Kilometers	10^3	Meters
Kilometers	0.6214	Miles
Kilometers	1094	Yards
KILOMETERS/HR.	27.78	Centimeters/sec.
Kilometers/hr.	54.68	Feet/min.
Kilometers/hr.	0.9113	Feet/sec.
Kilometers/hr.	0.5396	Knots
Kilometers/hr.	16.67	Meters/min.
Kilometers/hr.	0.6214	Miles/hr.
KMS./HR./SEC.	27.78	Cms./sec./sec.
Kms./hr./sec.	0.9113	Ft./sec./sec.
Kms./hr./sec.	0.2778	Meters/sec./sec.
KILOWATTS	56.92	B. T. Units/min.
Kilowatts	4.425×10^4	Foot-lbs./min.
Kilowatts	737.6	Foot-lbs./sec.
Kilowatts	1.341	Horse-power
Kilowatts	14.34	Kg.-calories/min.
Kilowatts	10^3	Watts
KILOWATT-HOURS	3415	Br. Thermal Units
Kilowatt-hours	2.655×10^6	Foot-lbs.
Kilowatt-hours	1.341	Horse-power-hrs.
Kilowatt-hours	860.5	Kilogram-calories
Kilowatt-hours	3.671×10^5	Kilogram-meters
LITERS	10^3	Cubic centimeters
Liters	0.03531	Cubic feet
Liters	61.02	Cubic inches
Liters	10^{-2}	Cubic meters
Liters	1.308×10^{-3}	Cubic yards
Liters	0.2642	Gallons
Liters	2.113	Pints (liq.)
Liters	1.057	Quarts (liq.)
LITERS/MIN.	5.886×10^{-4}	Cubic/ft./sec.
Liters/min.	4.403×10^{-3}	Gals./sec.

Multiply	By	To Obtain
LUMBER WIDTH (IN.) X THICKNESS (IN.)	Length (ft.)	Board Feet
	12	
METERS	100	Centimeters
Meters	3.281	Feet
Meters	39.37	Inches
Meters	10^{-3}	Kilometers
Meters	10^3	Millimeters
Meters	1.094	Yards
METERS/MIN.	1.667	Centimeters/sec.
Meters/min.	3.281	Feet/min.
Meters/min.	0.05468	Feet/sec.
Meters/min.	0.06	Kilometers/hr.
Meters/min.	0.03728	Miles/hr.
METERS/SEC.	196.8	Feet/min.
Meters/sec.	3.281	Feet/sec.
Meters/sec.	3.6	Kilometers/hr.
Meters/sec.	0.06	Kilometers/min
Meters/sec.	2.237	Miles/hr.
Meters/sec.	0.03728	Miles/min.
METRIC TONS	2204.6	Pounds
Metric tons	1.1023	Short tons
MICRONS	10^{-6}	Meters
MILES	1.609×10^5	Centimeters
Miles	5280	Feet
Miles	1.609	Kilometers
Miles	1760	Yards
MILES/HR.	44.70	Centimeters/sec.
Miles/hr.	88	Feet/min.
Miles/hr.	1.467	Feet/sec.
Miles/hr.	1.609	Kilometers/hr.
Miles/hr.	0.8684	Knots
Miles/hr.	26.82	Meters/min.
MILES/MIN.	2682	Centimeters/sec.
Miles/min.	88	Feet/sec.
Miles/min.	1.609	Kilometers/min.
Miles/min.	60	Miles/hr.
MILLIERS	10^3	Kilograms
MILLIGRAMS	10^{-3}	Grams
MILLILITERS	10^{-3}	Liters
MILLIMETERS	0.1	Centimeters
Millimeters	0.03937	Inches
MILLIGRAMS/LITER	1	Parts/million
MILLION GALS./DAY	1.54723	Cubic ft./sec.
MINER'S INCHES	1.5	Cubic ft./min.
Miner's inches	11.25	G.P.M.
(Arizona, Cal., Mont., Nevada, Oregon)		

CONVERSION UNITS

Multiply	By	To Obtain
(Idaho, Kansas, Neb., N.M., N.D., S.D., Utah)	9	G.P.M.
MINUTES (ANGLE)	2.909x10⁻⁴	Radians
OUNCES	16	Drams
Ounces	137.5	Grains
Ounces	0.0625	Pounds
Ounces	28.349527	Grams
Ounces	0.9115	Ounces (troy)
Ounces	2.790x10⁻⁵	Tons (long)
Ounces	2.835x10⁻⁵	Tons (metric)
OUNCES, TROY	480	Grains
Ounces, troy	20	Pennywghts. (troy)
Ounces, troy	0.08333	Pounds (troy)
Ounces, troy	31.103481	Grams
Ounces, troy	1.09714	Ounces, avoir.
OUNCES (FLUID)	1.805	Cubic inches
Ounces (fluid)	0.02957	Liters
OUNCES/SQ. INCH	0.0625	Lbs./sq. inch
PARTS/MILLION	0.0584	Grains/U.S. gal.
Parts/million	0.07016	Grains/Imp. gal.
Parts/million	8.345	Lbs./million gal.
PENNYWGHTS. (TROY)	24	Grains
Pennywghts. (troy)	1.55517	Grams
Pennywghts. (troy)	0.05	Ounces (troy)
Pennywghts. (troy)	4.1667x10⁻³	Pounds (troy)
PINTS	0.4732	Liter
POUNDS (AVOIR.)	16	Ounces
Pounds (avoir.)	256	Drams
Pounds (avoir.)	7000	Grains
Pounds (avoir.)	0.0005	Tons (short)
Pounds (avoir.)	453.5924	Grams
Pounds (avoir.)	1.21528	Pounds (troy)
Pounds (avoir.)	14.5833	Ounces (troy)
Pounds (avoir.)	0.454	Kilograms
POUNDS (TROY)	5760	Grains
Pounds (troy)	240	Pennywghts. (troy)
Pounds (troy)	12	Ounces (troy)
Pounds (troy)	373.24177	Grams
Pounds (troy)	0.822857	Pounds (avoir.)
Pounds (troy)	13.1657	Ounces (avoir.)
Pounds (troy)	3.6735x10⁻⁴	Tons (long)
Pounds (troy)	4.1143x10⁻⁴	Tons (short)
Pounds (troy)	3.7324x10⁻⁴	Tons (metric)
POUNDS OF WATER	0.01602	Cubic feet
Pounds of water	27.68	Cubic inches
Pounds of water	0.1198	Gallons
Pounds of water	0.10	Imp. gallon
LBS. OF WATER/MIN.	2.670x10⁻⁴	Cubic ft./sec.
POUNDS/CUBIC FOOT	0.01602	Grams/cubic cm.
Pounds/cubic foot	16.02	Kgs./cubic meter
Pounds/cubic foot	5.787x10⁻⁴	Lbs./cubic inch

Multiply	By	To Obtain
POUNDS/CUBIC INCH	27.68	Grams/cubic cm.
Pounds/cubic inch	2.768x10⁴	Kgs./cubic meter
Pounds/cubic inch	1728	Lbs./cubic foot
POUNDS/FOOT	1.488	Kgs./meter
Pounds/inch	178.6	Grams/cm.
POUNDS/SQ. FOOT	0.01602	Feet of water
Pounds/sq. foot	4.883x10⁻⁴	Kgs./sq. cm.
Pounds/sq. foot	6.945x10⁻³	Pounds/sq. inch
POUNDS/SQ. INCH	0.06804	Atmospheres
Pounds/sq. inch	2.307	Feet of water
Pounds/sq. inch	2.036	Inches of mercury
Pounds/sq. inch	0.07031	Kgs./sq. cm.
QUARTS (DRY)	67.20	Cubic inches
QUARTS (LIQ.)	57.75	Cubic inches
QUINTAL, ARGENTINE	101.28	Pounds
Quintal, Brazil	129.54	Pounds
Quint., Castile, Péru	101.43	Pounds
Quintal,Chile	101.41	Pounds
Quintal, Mexico	101.47	Pounds
Quintal, Metric	220.46	Pounds
$\dfrac{1}{\text{SQ. FT./GAL./MIN.}}$	8.0208	Overflow rate (ft./hr.)
TEMP. (°C.)+273	1	Abs. temp. (°C.)
Temp. (°C.)+17.78	1.8	Temp. (°F.)
Temp. (°F.)+460	1	Abs. temp. (°F.)
Temp. (°F.)−32	5/9	Temp. (°C.)
TONS (LONG)	1016	Kilograms
Tons (long)	2240	Pounds
Tons (long)	1.12000	Tons (short)
TONS (METRIC)	10³	Kilograms
Tons (metric)	2205	Pounds
TONS (SHORT)	2000	Pounds
Tons (short)	32000	Ounces
Tons (short)	907.18486	Kilograms
Tons (short)	2430.56	Pounds (troy)
Tons (short)	0.89287	Tons (long)
Tons (short)	29166.66	Ounces (troy)
Tons (short)	0.90718	Tons (metric)
TONS OF WATER/24 HRS.	83.333	Pounds water/hr.
Tons of water/24 hrs.	0.16643	Gallons/min.
Tons of water/24 hrs.	1.3349	Cu. ft./hr.
WATTS	0.05692	B. T. Units/min.
Watts	44.26	Foot-pounds/min.
Watts	0.7376	Foot-pounds/sec.
Watts	1.341x10⁻³	Horse-power
Watts	0.01434	Kg.-calories/min.
Watts	10⁻³	Kilowatts
WATT-HOURS	3.415	Br. Thermal Units
Watt-hours	2655	Foot-pounds
Watt-hours	1.341x10⁻³	Horse-power hrs.
Watt-hours	0.8605	Kilogram-calories
Watt-hours	367.1	Kilogram-meters
Watt-hours	10⁻³	Kilowatt-hours

Index

D

E